Advances in Oil and Gas Exploration & Production

Series Editor

Rudy Swennen, Department of Earth and Environmental Sciences, K.U. Leuven, Heverlee, Belgium

W0111833

The book series Advances in Oil and Gas Exploration & Production publishes scientific monographs on a broad range of topics concerning geophysical and geological research on conventional and unconventional oil and gas systems, and approaching those topics from both an exploration and a production standpoint. The series is intended to form a diverse library of reference works by describing the current state of research on selected themes, such as certain techniques used in the petroleum geoscience business or regional aspects. All books in the series are written and edited by leading experts actively engaged in the respective field.

The Advances in Oil and Gas Exploration & Production series includes both single and multi-authored books, as well as edited volumes. The Series Editor, Dr. Rudy Swennen (KU Leuven, Belgium), is currently accepting proposals and a proposal form can be obtained from our representative at Springer, Dr. Alexis Vizcaino (Alexis.Vizcaino@springer.com).

More information about this series at http://www.springer.com/series/15228

Shell International B.V. ·
The Development Research Center
(DRC) of the State Council of the
People's Republic of China
Editors

China's Energy Revolution in the Context of the Global Energy Transition

Editors
Shell International B.V.
The Hague, the Netherlands

The Development Research Center
(DRC) of the State Council of the
People's Republic of China
Beijing, China

ISSN 2509-372X   ISSN 2509-3738   (electronic)
Advances in Oil and Gas Exploration & Production
ISBN 978-3-030-40156-6   ISBN 978-3-030-40154-2   (eBook)
https://doi.org/10.1007/978-3-030-40154-2

The data and analysis contained in this publication was produced in 2019 when a Mandarin version of this book was published. The English translation of this book, published in 2020, does not contain updated data or analysis.

This Springer imprint is published by the registered company Springer Nature Switzerland AG
The registered company address is: Gewerbestrasse 11, 6330 Cham, Switzerland

Foreword 1

In June 2014, at a conference of China's Central Leading Group for Financial and Economic Affairs Commission, General Secretary Xi Jinping launched the idea that China should initiate an energy revolution. The revolution would be comprehensive in scope. It would encompass demand, supply, technology and the energy system itself, and it would strengthen international cooperation and guide China's energy reforms. As part of this reform process, the Development Research Center (DRC) of the State Council of China and Shell International, building on their long-term collaboration, started a joint research programme on China's Energy Revolution in the Context of the Global Energy Transition in late 2015.

The research focuses on how to promote China's energy revolution by reforming the energy system, bolstering innovation policy and motivating all stakeholders, including government, industry, companies and citizens. Our findings show that China will learn from international energy transition experience and strive to improve and build a modern, high-quality energy system. This will improve people's living standards, help make China a high-value manufacturer, protect the environment, and drive China's economic development. In short, China will provide high-quality energy for high-quality growth.

A high-quality energy system should have the following three features. First, energy should be clean and low carbon. The entire energy life cycle—from production and conversion to transmission and consumption—should be low pollution with minimal emissions of harmful local pollutants and with CO_2 from energy production and consumption minimised. Second, energy should be efficiently priced and affordable. China has not yet completed its industrialisation process and is still in a critical period of upgrading the manufacturing value chain. As energy is a key component of production and circulation, the price of energy should be competitive internationally and bolster Chinese manufacturing. Third, energy should be secure and reliable. The energy system should guarantee basic and stable supply, even during abnormal conditions like natural disasters or geopolitical tensions. It should also be sufficiently flexible to integrate ever-increasing volumes of renewables in the energy mix.

To build a modern and high-quality energy system, China needs to promote energy revolution in a comprehensive way through supply-side structural reform and improvement of the system itself. In terms of energy supply, China should take effective action in the following four areas:

First, by systematically cutting coal overcapacity and through efficient and clean coal utilisation. To address air pollution and reduce greenhouse gas emissions, China should continue to cut coal overcapacity and reduce coal's share of the energy mix. However, as coal will remain a main energy source in the longer term, China should continue to promote efficient and clean coal utilisation.

Second, by coordinated development of fossil fuels and renewable energy. Thanks to technological progress and business model innovation, the cost of renewable energy is decreasing rapidly, and its share of the energy mix has risen significantly. Renewables have great growth potential. As energy transition is a progressive process, fossil fuels will remain important in the near future, and renewable energy must rely on them for support.

Third, by coordinated development of centralised and distributed energy systems. Centralised energy supply is the long-standing mainstream model. But as demand changes, China should focus more on distributed energy and gradually shift to a modern energy supply system that adapts centralised and decentralised energy supply to local needs and conditions.

Fourth, by increasing the share of natural gas in the energy mix. In the short term, there are still many constraints on the development of renewable energy, and natural gas has great potential to offset these constraints. China will strive to ensure that the share of natural gas in primary energy consumption will reach 10% by the end of the 13th Five-Year Plan in 2020 and increase to 15% by 2030, turning natural gas into the third largest energy source after coal and oil.

Innovation is the foundation of the energy revolution. China should firmly implement energy system reform, consider energy a commodity, and build a market structure and system that feature effective competition. Moreover, China should strengthen regulatory and policy incentives, and vigorously promote the transition to clean and low-carbon energy that is efficiently priced and affordable, secure and reliable. In addition, China should systematically consider how to open itself to the outside world and make greater efforts to improve its participation in international energy cooperation and governance.

As China enters a new era of economic development, and in the face of advancing energy technologies and changing business models, the question of how to promote energy revolution should be continuously explored. As China can achieve its energy revolution only in phases, the research needs to be continuously deepened over time. Dozens of experts at home and abroad have contributed to this book, which means it will inevitably contain errors or oversights. We kindly invite readers to provide corrections and suggestions.

Li Wei
Director of the Development Research
Center of the State Council of China

Foreword 2

It is an honour for Royal Dutch Shell to have worked with the Development Research Center (DRC) of the State Council on this book. The fact that this is the third book we have produced together on China's energy system is a matter of great pride to me and to the company. It is a partnership of mutual respect and understanding.

The book you are holding is the greatest achievement so far in a collaboration which started in 2011. This collaboration brings together the DRC's deep understanding of China's energy system and the development challenges which need to be addressed, with Shell's international experience and knowledge of energy markets, regulatory mechanisms and the drivers of energy demand.

In the first publication, the DRC and Shell took a broad look across China's energy system. The second focused on the role of natural gas in diversifying China's energy mix. This book explores China's energy revolution in the context of a changing world energy system.

It is hard to miss the changes taking place in China's energy landscape. In 2019 China produced 40% of all the wind turbines in the world. It made three-quarters of the world's solar panels. Nearly half of the electric vehicles on the planet today, and half the hydrogen-fuelled vehicles, are owned by Chinese people.

The role of renewables in China's energy system is also gaining ground fast. The amount of wind energy could double and the amount of solar quadruple between 2015 and 2020. And the Chinese government's move towards establishing a national carbon pricing mechanism is yet another sign of the country making progress towards a new era of cleaner energy.

Of course, China's challenge is the world's challenge: to meet growing energy demand while causing as little harm as possible to the environment.

The world population is growing, from 7.5 billion today to 9.8 billion in 2050, according to the United Nations. That population, which currently includes around 1 billion people without access to basic electricity, is seeking a better standard of living. Many of the people who improve their lives will do so by consuming more energy. So, we can expect global energy demand to rise.

Yet, if the world is also to achieve the goals of the Paris Agreement, credible scenarios suggest it must collectively stop adding greenhouse gases to the stock in the atmosphere by 2070. It needs both more energy and cleaner energy. There is a lot of work for the world to do if it is to succeed.

China's goal, as President Xi told the 19th Party Congress, is a "New Era" with a better quality of life. This is in tune with Paris. For China's energy system it means cleaner energy with better air quality and lower greenhouse gas emissions. It also means affordable energy, secure energy supplies and reliable delivery to consumers.

The main message of this book is that this is a goal that is well within China's grasp—even if reaching that goal does require new thinking of the sort that has been laid out in these pages. The recommendations made in this book are intended to be a route to reaching this goal. It is important to note that they are recommendations that are only possible because of the transformational moves China has already made in its energy system.

They draw on the lessons from other countries going through energy transition, taking on-board what has worked and avoiding the less successful paths.

They also point out opportunities to move towards a system which harnesses the best of both government and the free market. The direction and policy interventions that only government can provide, with the market system's inbuilt drive towards efficiency.

One of the findings is that China has the chance to build a unified, efficient and flexible electricity market. Electrification enables an increasing amount of energy use to be powered by renewables. So, electrification can be a critical part of any shift to a low-carbon energy system, as long as electricity is increasingly generated by zero-carbon sources.

Of course, coal is likely to remain in China's energy mix for some time to come, and the report suggests reforms aimed at driving out inefficiencies and ensuring high quality. Ultimately, however, coal is expected to decline in importance within China's energy system and other cleaner sources of energy such as renewables and gas will rise. The book goes on to look at ways to deepen the reform of the oil and gas landscape to encourage investment and development. It looks at reforms to the mining rights system, the natural gas pipeline network, pricing and industry regulation.

It recommends establishing new regulatory authorities and suggests an enhanced system of energy laws to back this up. These suggestions build on, and deepen, the recommendations made in the second book that came from the DRC-Shell collaboration.

The study also explores how to make the most of the plans for a national carbon market. It is here that it becomes most clear how interconnected China's energy system is, and also how a harmonious approach can have a greater impact than more dramatic, but less coordinated, action. Finally, the book looks at the opportunities China can take by playing an ever-more active role in shaping global energy governance. It suggests ways that China can help address global energy concerns as part of the international community: making China's voice heard clearly and influencing the outcomes.

It is right that China's voice is heard. The change it has already gone through is impressive, and the change proposed in this book can bring the country's energy goals within reach. The world should take note. It should also take note of what can be achieved when governments and companies collaborate as effectively as the DRC and Shell have done since 2011. I look forward to what comes next.

Ben Van Beurden
Chief Executive Officer of Royal
Dutch Shell plc

Acknowledgements

Project Chairs

Li Wei	Director and Researcher of the Development Research Center (DRC) of the State Council of the People's Republic of China.
Ben van Beurden	Chief Executive Officer of Royal Dutch Shell plc.

Project Executives

Long Guoqiang	Deputy Director and Researcher of DRC, the State Council of China.
Maarten Wetselaar	Member of the Executive Committee and Director of Integrated Gas and New Energies, Royal Dutch Shell.
Zhang Xinsheng	Executive Chairman, Shell Companies in China.
Simon Henry	Former Chief Financial Officer of Royal Dutch Shell.

Project Core Advisors

Xu Kuangdi	Vice Chairman of the 10th Chinese People's Political Consultative Conference and Former President and Academician of the Chinese Academy of Engineering.
Chen Qingtai	Former Secretary of the Party Leadership Group and Deputy Director of DRC, the State Council of China.
Fu Chengyu	Former Chairman of China Petrochemical Corporation (Sinopec Group).
Wang Jiaxiang	Former General Manager and Member of the Party Leadership Group of China National Offshore Oil Corporation (CNOOC).

Huang Weihe	Vice President of China National Petroleum Corporation (CNPC) and Academician of the Chinese Academy of Engineering.
Zhai Guangming	Former Director of CNPC Consulting Center and Academician of the Chinese Academy of Engineering.
Xu Lin	Chairman of the U.S.-China Green Fund and Former Director of the City and Small-Town Reform and Development Center of the National Development and Reform Commission.

Project Review Expert Panel

Shi Dan	Secretary of the Institute of Industrial Economics of the Chinese Academy of Social Sciences.
An Fengquan	Deputy Director of the International Department of the National Energy Administration.
Li Junfeng	Former Director of the National Center for Climate Change Strategy and International Cooperation.
Sun Jinhua	Vice President of the Central Research Institute of State Power Investment Corporation.
Zhao Lianzeng	Vice President of China Petroleum Planning and Engineering Institute.
Jiang Liping	Vice President of State Grid Energy Research Institute (SGERI).
Wu Guogan	Deputy Director of CNPC Consulting Center.

Project Sponsors

| Zhao Changwen | Director and Researcher of the Research Department of Industrial Economy, DRC of the State Council of China. |
| Jeremy Bentham | Vice President Global Business Environment, Shell International B.V. |

Project Team Leads

| Yang Jianlong | Deputy Director and Researcher of the Research Department of Industrial Economy, DRC of the State Council of China. |
| Mallika Ishwaran | Senior Economist and Policy Advisor, Shell International B.V. |

Shi Yaodong	Deputy Director and Researcher of the Research Department of Industrial Economy, DRC of the State Council of China.
Wang Ling	General Manager Government Affairs, Shell China Ltd.

DRC Project Team Members

Xu Zhaoyuan	Director and Researcher of the Research Office, the Research Department of Industrial Economy, DRC of the State Council of China.
Wang Xiaoming	Former Director and Researcher of the Research Office, the Research Department of Industrial Economy, DRC of the State Council of China.
Wei Jigang	Director and Researcher of the Research Office, the Research Department of Industrial Economy, DRC of the State Council of China.
Song Zifeng	Director and Researcher of the Research Office, the Research Department of Industrial Economy, DRC of the State Council of China.
Guo Jiaofeng	Assistant to the Director of the Research Department of Resource and Environment Policies, DRC of the State Council of China.
Hong Tao	Director and Researcher of the Research Department of Resource and Environment Policies, DRC of the State Council of China.
Chen Jianpeng	Director and Researcher of the Research Department of Resource and Environment Policies, DRC of the State Council of China.
Zhou Jianqi	Director and Associate Researcher of the Research Department of Business, DRC of the State Council of China.
Li Weiming	Deputy Director and Associate Researcher of the Research Department of Resource and Environment Policies, DRC of the State Council of China.
Zhou Yi	Associate Researcher of the Research Department of Industrial Economy, DRC of the State Council of China.
Li Jifeng	Deputy Director and Associate Researcher of the Policy Simulation Laboratory, the Department of Economic Forecast, State Information Center.
Zeng Ming	Professor of North China Electric Power University.
Liu Xiaoli	Researcher of the Energy Research Institute of the National Development and Reform Commission.

Shi Shude	Deputy Director of the Research Department of Management and Consulting, State Grid Energy Research Institute Co., Ltd.
Yang Guang	Associate Researcher of the Energy Research Institute of the National Development and Reform Commission.
Liu Ying	Associate Professor of the School of Economics and Management, University of Chinese Academy of Sciences.
Duan Hongbo	Assistant Professor of the School of Economics and Management, University of Chinese Academy of Sciences.
Ji Qiang	Associate Researcher of the Institutes of Science and Development, Chinese Academy of Sciences.
Mo Jianlei	Associate Researcher of the Institutes of Science and Development, Chinese Academy of Sciences.
Chen Jinxiao	Associate Researcher of the Institute of Quantitative & Technical Economics, Chinese Academy of Social Sciences.
Liu Bing	President of Shanghai AILNG Energy Technology Co., Ltd.
Zhang Wenqiang	Manager of IoT, Shanghai AILNG Energy Technology Co., Ltd.
Li Zhenyu	Senior Engineer of the Petrochemical Research Institute, CNPC.
Xing Lu	China National Petroleum Corporation (CNPC).
Zhang Jun	State Grid Energy Research Institute Co., Ltd.
Huang Bibin	State Grid Energy Research Institute Co., Ltd.
Tu Junming	Strategic Investment and Business Management, China National Travel Service Group Corporation Limited [China Travel Service (Holdings) Hong Kong Limited].
Kang Xiaowan	Energy Research Institute of the National Development and Reform Commission.
Huang Yanghua	Associate Researcher of the Institute of Industrial Economics of the Chinese Academy of Social Sciences.
Feng Yujia	Ph.D. candidate of the School of Economics and Management, Tsinghua University.
Fan Jingli	Ph.D. candidate of China University of Mining and Technology.
Wu Lin	Ph.D. candidate of China University of Mining and Technology.
Wang Yuqing	Ph.D. candidate of North China Electric Power University.
Long Zhuhan	Master's degree candidate of North China Electric Power University.
Gu Lin	Researcher of the Liaowang Institute.
Meng Yiming	Ph.D. candidate of the Graduate School of Chinese Academy of Social Sciences.

Shell Project Team Members

Wang Wei	Vice President Government Affairs and Business Support, Shell Companies in China.
Angus Gillespie	Former Vice President Group CO2, Shell Global Solutions.
Nie Shangyou	Business Development Manager, Shell Exploration and Production Company.
Martin Haigh	Senior Energy and Climate Change Advisor, Shell International B.V.
Nigel Dickens	Economics and Analysis Manager—New Fuels, Shell International Petroleum Company Ltd.
Peter Webb	Government Relations Advisor, Shell International B.V.
Gu Jing	Former General Manager Coal-to-Gas Business Development Technology, Shell China Ltd.
Ren Xianfang	Gas Strategy and Portfolio Manager, Shell China Ltd.
Su Wu	General Manager New Ventures Development, Shell China Exploration and Production Company Ltd.
Yuan Yuan	Shell China Downstream LNG Business Lead, Shell China Ltd.
Fu Xiao	Chemical Research Engineer of Shell Projects & Technology.
Tobias Chen	Energy Transition Manager, Shell China Ltd.
Cameron Hepburn	Project Director, Vivid Economics Ltd.
Philip Gradwell	Project Manager of Vivid Economics Ltd.
Thomas Nielsen	Project Manager, Vivid Economics Ltd.
Rob Bailey	Head of the Department of Energy, Resources and Environment, Chatham House.
Felix Preston	Project Officer, Chatham House.
Daniel Quiggin	Researcher, Chatham House.
Wang Yunshi	Professor, University of California Davis.

Contents

List of Figures

Special Report 1: A Study of China's Energy Supply Revolution

Special Report 2: Research on China's Energy Demand Revolution

Special Report 3: A Study of China's Technology Revolution

Special Report 4: China's Energy System Revolution

Special Report 5: International Energy Cooperation and Governance

List of Tables

Special Report 2: Research on China's Energy Demand Revolution

Special Report 3: A Study of China's Technology Revolution

Special Report 4: China's Energy System Revolution

Special Report 5: International Energy Cooperation and Governance

Overview: High-Quality Energy for High-Quality Growth: China's Energy Revolution in the New Era

Xu Zhaoyuan and Mallika Ishwaran

In June 2014, General Secretary Xi Jinping remarked at a conference of the Communist Party of China's (CPC) Central Leading Group for Financial and Economic Affairs Commission that a revolution in energy production and consumption was needed to safeguard national energy security. New patterns in supply and demand, compounded by changing trends in international energy development, were presenting China with opportunities to develop and drive a new energy era.

In December 2016, the Chinese government published its Energy Production and Consumption Revolution Strategy (2016–30), which set out the specific actions needed to promote the energy revolution that President Xi had referred to. In October 2017, the report of the 19th CPC National Congress noted that:

> Socialism with Chinese characteristics has crossed the threshold into a new era. China's economy is transitioning from a phase of rapid growth to a stage of high-quality development.
>
> China must put quality first and prioritise performance; and China should make supply-side structural reform its main task and work hard to achieve better quality, higher efficiency, and more robust drivers of economic growth.

The new era provides a new background and poses new requirements for China to deepen its study of the energy revolution. Global practices show that all countries have undergone an energy transition in the course of their development, albeit along different paths. From the perspective of the global energy system as a whole, the current energy transition towards a greener and more sustainable energy system is significantly different from the energy transitions of the past.

China needs to learn from other countries' practices in promoting energy transition. It must keep up with the latest trends in global energy transitions, take account of the needs of economic, social and environmental development in China, vigorously develop energy technologies, and conduct an in-depth study on how to achieve China's energy revolution.

DRC Team Lead for the Overview:
Xu Zhaoyuan, Research Department of Industrial Economy, DRC of the State Council of China.

Shell Team Lead for the Overview:
Mallika Ishwaran, Global Business Environment, Shell International B.V.

Contributors:
Zhou Yi, Research Department of Industrial Economy, DRC of the State Council of China.

X. Zhaoyuan (✉)
Research Department of Industrial Economy, DRC of the State Council of China, Beijing, China
e-mail: xuzhaoyuan@gsm.pku.edu.cn

M. Ishwaran
Global Business Environment, Shell International B.V., The Hague, the Netherlands

1 Global Energy Transitions: Historical Experience and the Latest Trends

The energy system is comprehensive in scope, integrating energy production, conversion, transmission, consumption and management into a single system.

© The Author(s) 2020
Shell International B.V. and the Development Research Center (DRC) of the State Council of the People's Republic of China (eds.), *China's Energy Revolution in the Context of the Global Energy Transition*, Advances in Oil and Gas Exploration & Production, https://doi.org/10.1007/978-3-030-40154-2_1

Energy transition is a long-term structural change to the energy system, where entirely new components arise or old patterns fundamentally change. The energy system is in a constant state of evolution. For example, China's and India's energy systems have been organised to date around low-cost energy; France's transition to nuclear power in the 1970s was driven by the desire for security after the oil price shocks of that decade; and Germany's energy transition in the 2010s was propelled by the desire for clean energy. Transition also occurs when new end uses require new forms of energy. For example, the development of electric vehicles changes the transport industry's demand for oil into the need for electric power.

1.1 Energy Demand Changes with Economic Development

Countries generally go through an energy transition as they develop. Energy demand can be divided into three broad categories (Fig. 1). In the first category, countries with an income per person of less than $5,000 in purchase power parity (PPP) have less economic development and therefore low energy consumption. Second, countries with an income per person of between $5,000 and $15,000: as these countries industrialise, energy demand growth accelerates due to

the high energy intensity of industrialisation, urbanisation and large-scale infrastructure construction. And third, countries with an income per person of more than $15,000: once these countries have industrialised, the growth rate of energy demand starts to slow down.

While there is a general trend of energy transition that countries go through as they develop, national experience shows that the energy demand path that a country follows can vary significantly.

The USA and Canada are typical examples of high-income countries with high energy consumption. As Fig. 2 shows, the USA and Canada rapidly increased their energy consumption per person between 1960 and the late 1980s, during which time income per capita doubled. This rapid increase in energy consumption was the outcome of fast growth in energy-intensive industries and high domestic energy consumption. Energy consumption in the transport sector was high due to both countries' low population densities. Since the late 1980s, incomes continued to grow, albeit more slowly, while energy consumption was relatively flat.

Japan and major European countries experienced a slower increase in energy consumption than the USA and Canada. These countries have reached similar levels of income per capita as the USA and Canada, but with around half the level of energy consumption per person. In addition to having more light industry and less transport

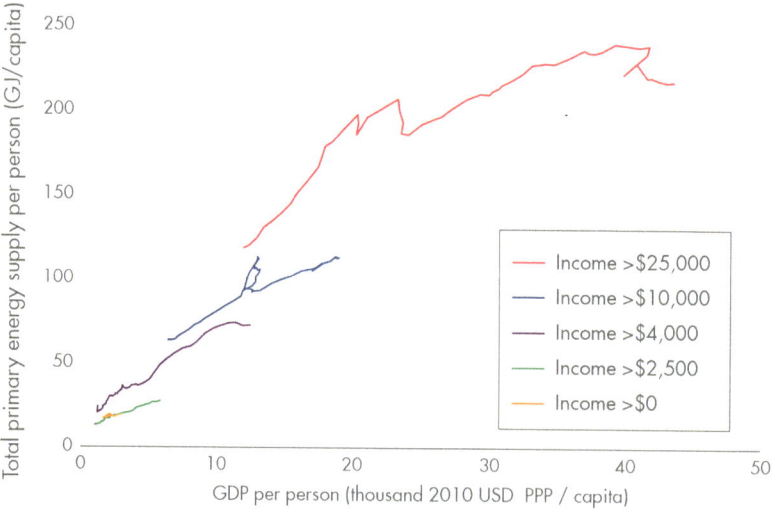

Fig. 1 Countries go through energy transition as they develop. *Note* Average energy supply per capita since 1960 of those countries that had GDP per capita in one of five income classes in 2015. *Source* Shell International

demands, these economies generally focus more on energy efficiency, often driven by policy. Spain and Italy have lower energy use per person than other developed economies. This is because they are more service-oriented, with fewer energy-intensive industries and a Mediterranean climate that reduces the need for heating and cooling relative to other countries.

The data for Australia, South Korea and Sweden show how countries with energy-intensive industries can have higher energy consumption. These three countries form a cluster between the USA and Canada, which have high energy consumption, and Japan and major European countries, which have lower energy consumption, as shown in Fig. 2. This is because although Australia, South Korea and Sweden have large energy-intensive industries, they are more energy efficient than the USA and Canada. Australia has built an energy-intensive industrial system on its extensive natural resources, primarily non-ferrous metals, iron and steel, mining and chemicals. It also has a high transport energy use per person due to its low population density and large size. South Korea is an example of a country that has increased GDP per capita through high-value industrialisation, despite not having domestic energy resources. Sweden is somewhat unique as its energy use is increased by a large pulp and paper industry, which is very energy intensive, and a very high level of energy use in buildings for heating.

1.2 Previous Global Transitions in Energy Supply

The global energy system undergoes transitions. There have been three global energy transitions since 1800 (Fig. 3). The first was the rise of coal to fuel the Industrial Revolution; the second was the use of oil for mass transport; and the third was the rise of gas, hydropower and nuclear power in electrification. In recent years, a fourth energy transition has begun with the use of renewables to provide clean and sustainable energy.

History shows that rising societal demand for energy and advances in technology can lead to global energy transitions. For example, energy supply shifted from traditional biofuels to coal in the 19th century to fuel industrialisation in the UK and other countries in Europe. After the oil price shocks in the 1970s, many countries needed to secure energy supply, stimulating the transition from oil to gas and nuclear power in electricity. Since the beginning of this century, demand for clean energy is also driving an energy transition. Transitions also occur when significant improvements are made in energy technologies or when new energy carriers offer the flexibility of providing a range of increasingly sophisticated services and end uses, as electricity does for buildings and industry.

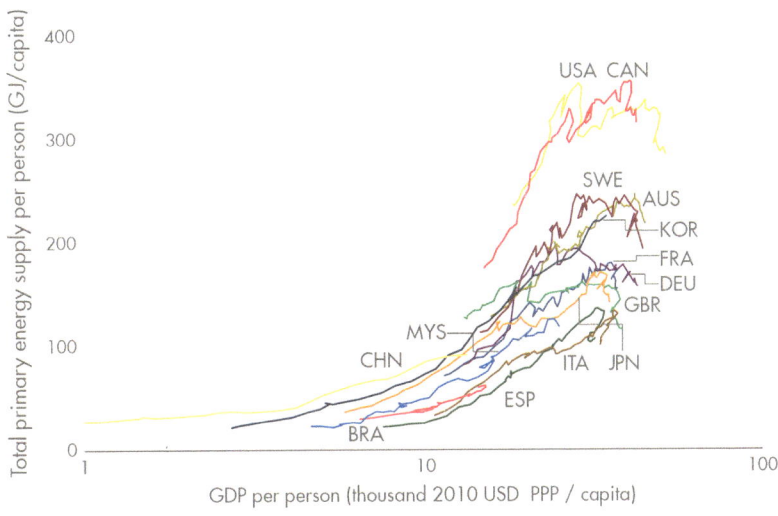

Fig. 2 Energy transition paths of various countries. *Note* Data are from 1960 to 2015; note the log scale on the X axis. *Source* Vivid Economics based on IEA data

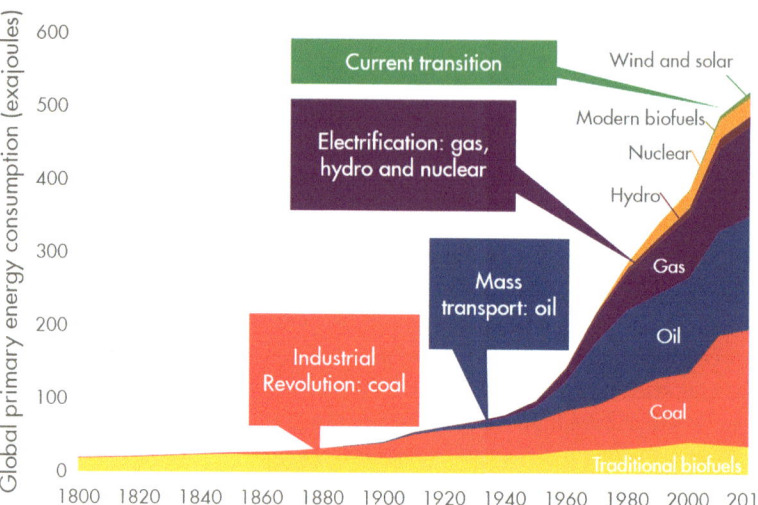

Fig. 3 There have been four global transitions in energy supply since 1800. *Source* Vivid Economics based on data in energy transitions: Global and National Perspectives by Vaclav Smil (2017)

1.3 Energy Technologies are Undergoing Significant Change

New energy and information technologies are developing rapidly and having a significant impact on energy production and consumption.

1.3.1 The Cost of Clean Energy Technologies is Declining Rapidly

Since the early 2010s, the cost of renewable energy technologies such as wind and solar have fallen by more than half. Lithium-ion batteries, which have the potential for use in the transport and power sectors, have also seen similar cost reductions (Fig. 4).

1.3.2 New Information and Communications Technologies (Digitalisation) are Increasingly Being Used in the Energy System, with Several Important Implications

To begin with, digitalisation increases the demand-response potential for electricity. According to International Energy Agency

(IEA) forecasts, digitalisation could increase global electricity demand-response potential from 3,900 terrawatt-hours (TWh) in 2015 to 6,900 TWh in 2040, up 77%. This level of demand-response can free up some 185 GW of generating capacity and reduce the need for investment in new generation, transmission and distribution on a cumulative basis by $270 billion (calculated in 2016 $).

Second, digitalisation has improved the flexibility of the power system to integrate renewable energy. For example, the IEA estimates that digitalisation and demand-side response technologies can limit total wind and solar power curtailment to less than 1.6%. By 2040, total wind and solar curtailment will be 79% lower than in 2015. This will enable the global power system to accommodate 67 TWh of new renewable energy annually by around 2040 and avoid about 30 Mt of CO_2 emissions per year.

Third, digitalisation can help improve electrification in the transport sector. Smart charging of electric vehicles can greatly reduce the demand for power generation. With the added flexibility that smart charging provides to power grids, investment in grids can be reduced by $100 to $280 billion by 2040 (in 2016 $).

Fourth, digitalisation may affect the energy use and consumption patterns of manufacturing.

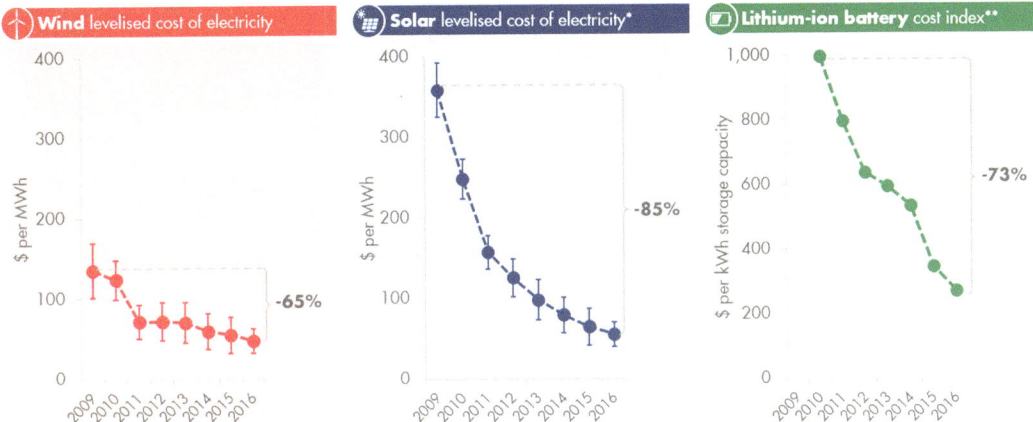

Fig. 4 The cost of many clean energy technologies is declining rapidly. *Note* Cost does not include subsidies. *The calculated cost of solar energy is based on utility-scale monocrystalline silicon solar panels. *Source* The cost data of wind and solar energy are from Lazard (2016), while the **cost data of batteries are from Bloomberg New Energy Finance (2017)

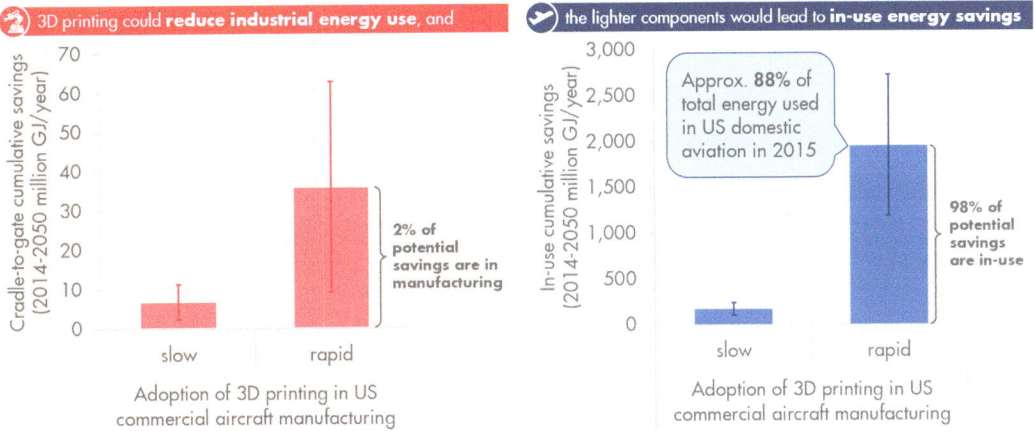

Fig. 5 Industry 4.0 will have a significant impact on energy demand. *Note* The data above are based solely on the adoption of 3D printing in US commercial aircraft manufacturing. *Source* Huang et al. (2016)

Industry 4.0—the trend towards greater use of automation and data exchange—will have a significant impact on energy demand. As shown in Fig. 5, the adoption of 3D printing in US commercial aircraft manufacturing can reduce energy use in production and in flight, thanks to the use of more lightweight components.

1.4 Main Characteristics of the New Global Energy Transition

The global energy system is undergoing a major transition, after a period of relative stability. Before 1985, the energy systems of the G7 countries experienced significant changes as a

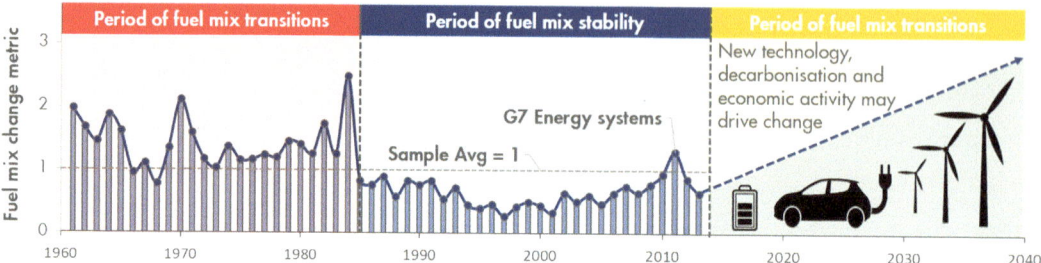

Fig. 6 Changes in the energy systems of G7 countries. *Note* Information on the energy mix change metric can be found in Special report 3 on the technology revolution. *Source* Vivid Economics

result of oil price fluctuations, the discovery of new oil and gas reserves, the emergence of new energy carrier networks and the development of new technologies like nuclear power. However, the global energy system has been relatively stable over the past 30 years, with little change in the energy mix of the G7 countries (Fig. 6). Over the next 30 years, digitalisation, new technologies, decarbonisation, more stringent environmental requirements and new forms of economic activity are expected to transform the energy system.

1.4.1 Clean and Low-Carbon Energy are Driving the New Global Energy Transition

The unprecedented attention paid by countries to climate change and environmental protection, along with increasing consumer demand for cleaner energy services, is driving the global transition to clean and low-carbon energy.

Compared with previous drivers of energy transition, climate change is a global concern, the solution to which requires the engagement of all countries. Both energy consumption and CO_2 emissions per unit of GDP are now declining. There is a sign that energy consumption and CO_2 emissions have been decoupling from economic growth increasingly rapidly since 2010. The Paris Agreement of 2015 shows that the world is paying more attention to climate change and strengthening its efforts to disconnect CO_2 emissions from economic growth. In this respect, the European Union is in a leading position, with GDP increasing by 50% between 1990 and 2015 and CO_2 emissions dropping by more than 20%

over the same period. The USA and Japan have also remained generally stable in CO_2 emissions (Fig. 7), but the USA's withdrawal from the Paris Agreement and President Trump's new energy policies bring some uncertainties.

1.4.2 Significantly More Electrification Characterises the New Global Energy Transition

To achieve the ambitions of the Paris Agreement, more electrification is required to decarbonise the global economy. This process is accelerating in the major economies, especially in transport. Figure 8 shows growth in electric car registrations in several countries over recent years. Electric vehicles remain an important area of growth, despite their current small market share (except in some Nordic countries).

Economic restructuring is also driving electrification. As developing countries like China move towards a service-oriented economy, and consumers are keen to get cleaner and more flexible types of energy, such as gas and electricity, electrification continues to increase. During the current energy transition, this trend has become more and more obvious—emerging economies such as China and Brazil have achieved higher levels of electrification at a lower level of per capita income—and is expected to continue (Fig. 9).

1.4.3 Policy Plays a More Important Role in This Energy Transition

While policy has been a key driver of previous energy transitions, the scope of the policy

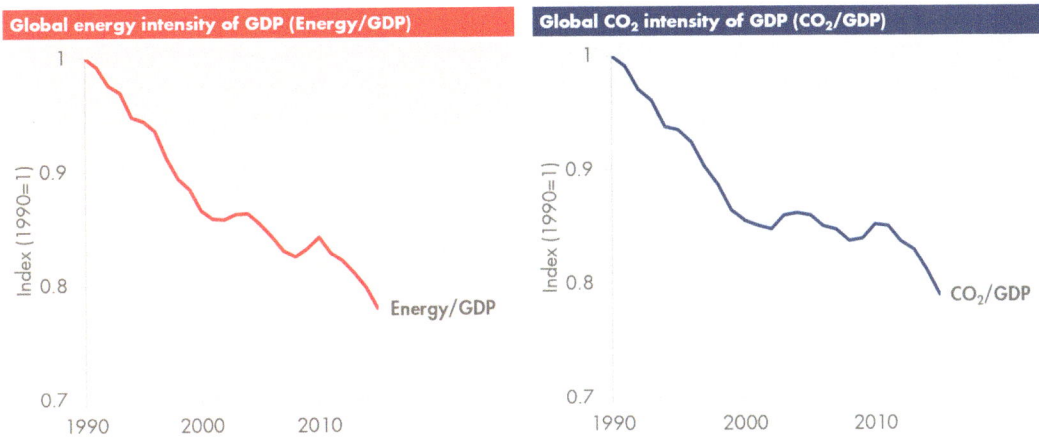

Fig. 7 Energy consumption and CO_2 emissions per unit of GDP are declining globally. *Note* GDP data are calculated in 2010 \$; CO_2 emissions refer only to energy-related CO_2 emissions. *Source* IEA (2017) and World Bank (2017)

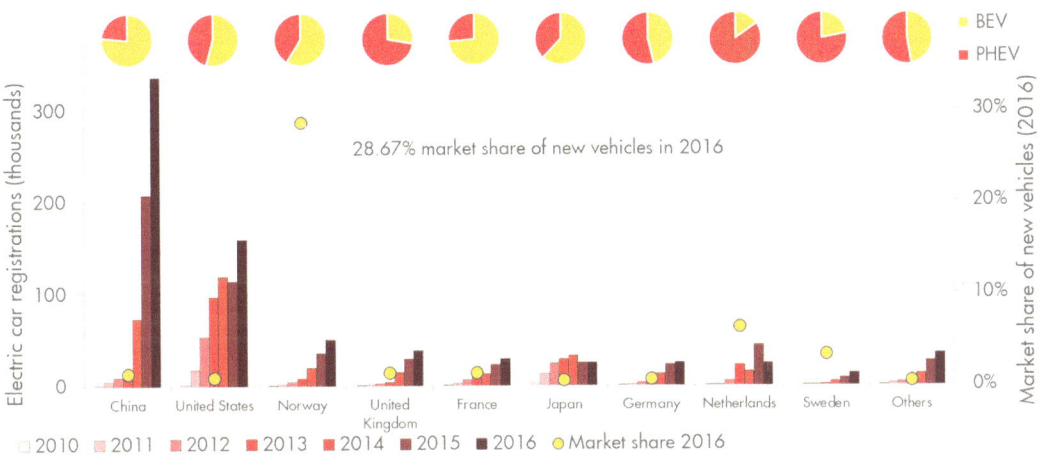

Fig. 8 Electrification in the transport sector is accelerating. *Note* BEV = battery electric vehicles; PHEV = plug-in hybrid electric vehicles. *Source* IEA, Global EV Outlook (2017)

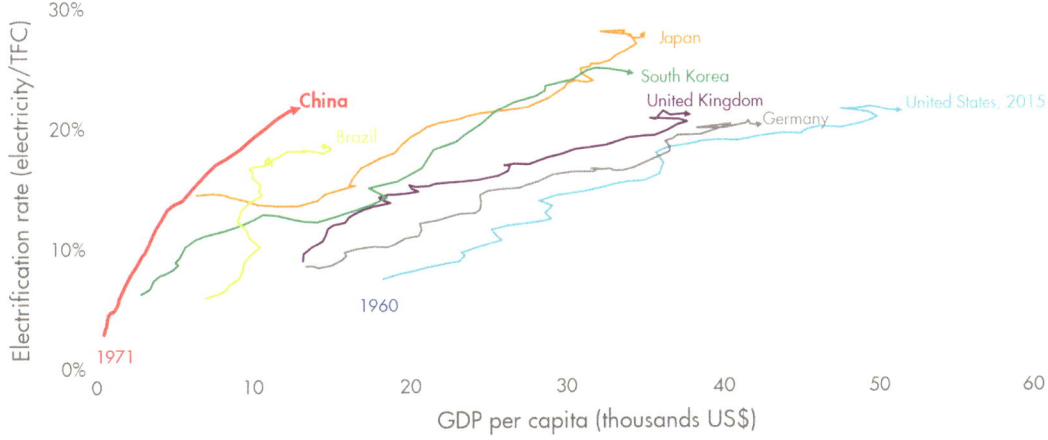

Fig. 9 Electrification increases with income. *Note* Data for all countries except China are from 1960–2015; data for China are from 1971–2015. *Source* Vivid Economics, based on IEA data

challenge this time is vast. This is because the goal of the new transition is to promote the development of clean and renewable energy. To level the playing field, the negative environmental effects of fossil fuels, such as air pollution and carbon emissions, compared to the positive environmental effects of clean energy, must be reflected in economic decisions. Policy interventions are

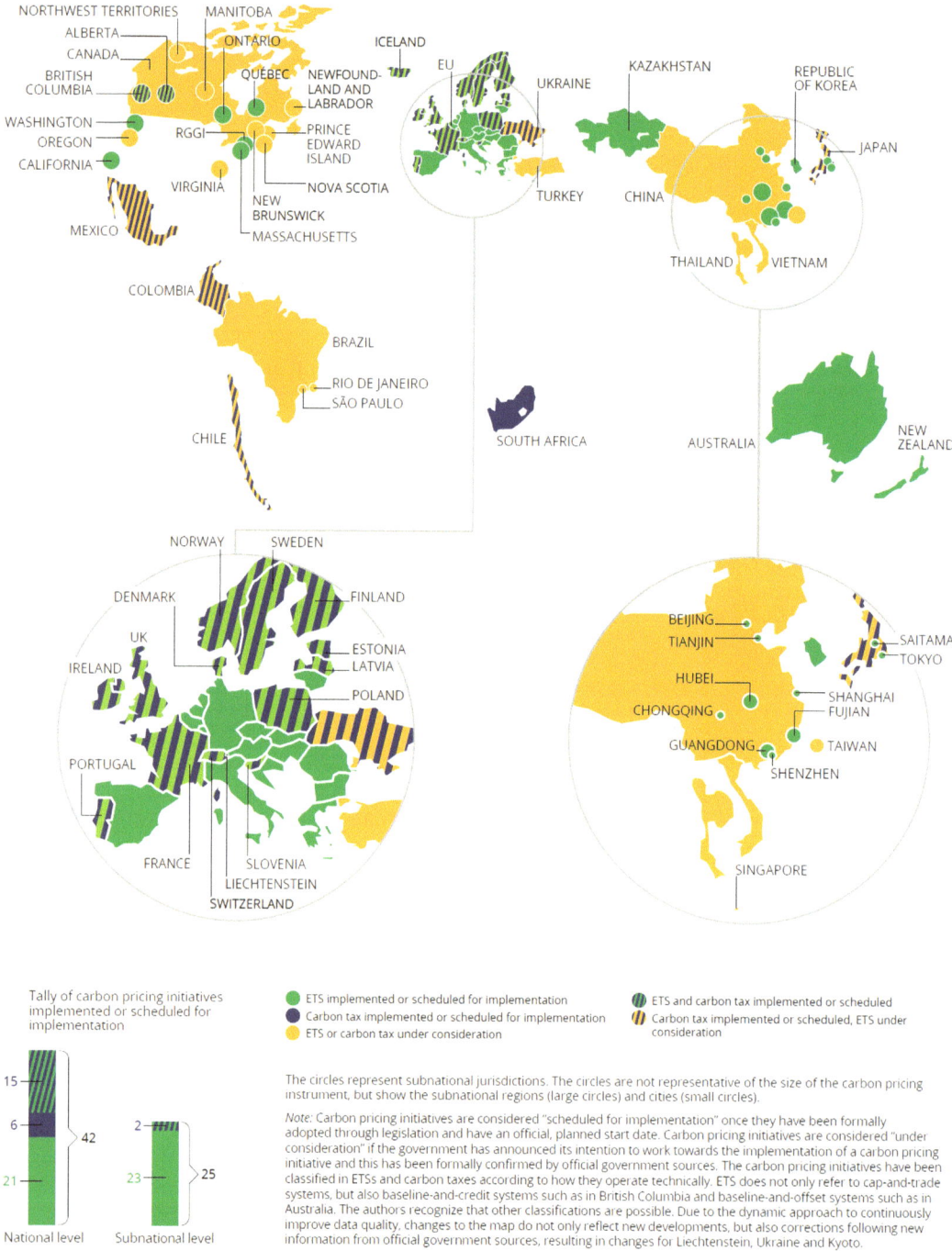

Fig. 10 Carbon pricing is accelerating globally. *Source* World Bank, State and Trends of Carbon Pricing (2017)

needed to rectify the economic distortions caused by such market failures. In addition, the introductory and growth periods of new energy need support policies. Due to the complexity of the energy system, the energy transition must use market forces to promote innovation and find effective and flexible ways to deliver desired goals. Many governments have realised this and are taking measures to ensure market failures are corrected. Figure 10 shows the countries that have implemented, are about to implement, or are considering implementing carbon pricing (in the form of a carbon tax or emissions trading system).

1.5 Developing and Emerging Economies Can Leapfrog Ahead

International experience shows that while there is a general route for countries to follow in energy transition as they develop, the path that a country takes can vary significantly, depending on its economic structure, improvements in energy efficiency, consumption patterns, population density and climate, among other things.

Countries such as China, Brazil and Malaysia are at a crossroads. Their path to energy transition may diverge from international experience. These countries need to choose between the energy development paths available to them: whether to have a level of energy consumption per person similar to that of the USA (high), South Korea (medium) or Spain (low) as income per capita increases. Alternatively, they can use policy to leapfrog historical patterns of energy system development to achieve high income per capita and lower and cleaner energy consumption.

By leveraging advanced technologies and learning from, then adapting, the policies and institutional frameworks of other countries, developing and emerging economies can achieve lower energy consumption and greenhouse gas emissions per capita earlier than high-income countries (as shown in Fig. 11). Developing and emerging economies have the potential to provide advanced energy services without the negative environmental impact that advanced economies have had, and they can exploit their late-mover advantage in the growing global market for energy services.

2 From Quantity to Quality: The Goal and Approach of China's Energy Revolution

2.1 The Goal of China's Energy Revolution

In 2015, China's 13th Five-Year Plan (2016–20) proposed to empower low-carbon cyclic development by promoting an energy revolution, accelerating energy technology innovation, and setting up a clean, low-carbon, secure and efficient energy system. In 2017, the report of the 19th National Congress of the Communist Party of China noted that China's economy has moved from a phase of rapid growth to one of high-quality development. China, it said, must focus on improving the supply system and strengthening the economy in terms of quality. Quality energy is an important part of the supply system and a key objective of the energy revolution.

2.1.1 What is a High-Quality Energy System?

Based on the 13th Five-Year Plan, the report of the 19th National Congress of the CPC and development strategies such as Made in China 2025 state that China's high-quality energy system should feature the following three characteristics:

First, it should be clean and low carbon. A clean energy system is one in which the entire energy life cycle—from production and conversion to transmission and consumption—achieves the lowest possible levels of pollution and emissions. Low carbon is a key part of this. CO_2 itself is not a polluting gas, but it has significant impacts on the environment. Moreover, the Chinese government has made a solemn commitment to the international community, as part of its nationally determined contributions, to reduce global carbon emissions. The energy

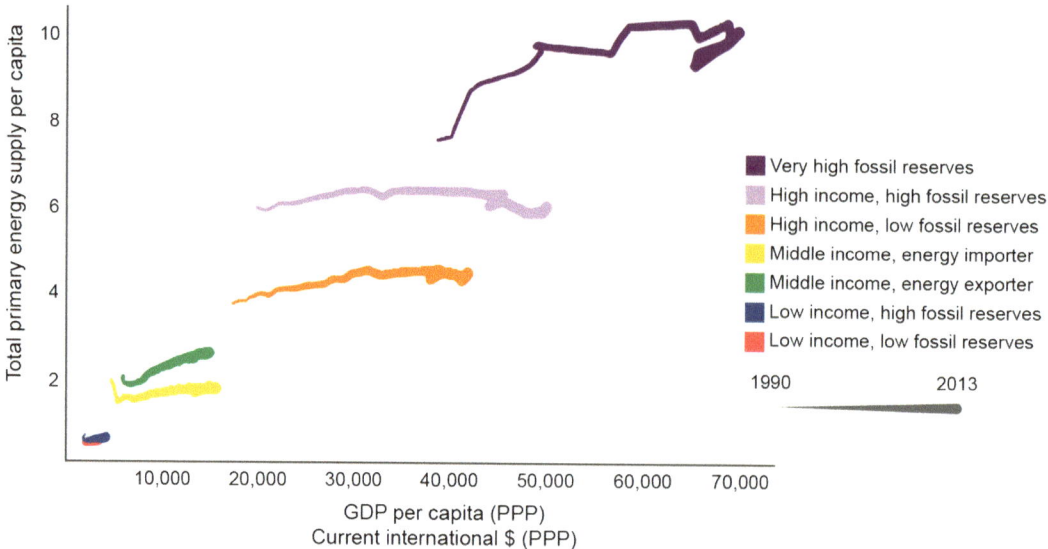

Fig. 11 Low- and middle-income countries can leapfrog the historical patterns of energy consumption of high-income countries. *Note* PPP = purchasing power parity. *Source* Vivid Economics, based on World Bank and WRI data

system is the largest carbon emitter, so low carbon is undoubtedly an important characteristic of a high-quality energy system.

Second, it should be efficiently priced and affordable. Affordability means that the price of energy should be competitive internationally to allow manufacturing to compete with global manufacturing powers like the USA, Japan and Germany. China is now in an important period of making itself a manufacturing power. Energy cost is a key element of the real economic cost, so a high-quality energy system should be able to supply energy at a competitive price. Efficiency means that existing technologies should be leveraged in energy production, conversion, transmission and consumption to save energy and improve efficiency.

Third, it should be secure and reliable. Secure means that energy sources are diverse and that they ensure a stable supply for economic development, even during natural disasters or times of geopolitical change. Reliable means that as the amount of renewable energy increases, the energy system has the flexibility to adapt and maintain a sustainable and stable energy supply for the national economy.

2.1.2 Three Characteristics of the Energy Revolution

(1) Clean and low carbon

According to research by the Institute of Resources and Environmental Policy of the Development Research Center of the State Council, total emissions of major air pollutants in China are likely to peak during the 13th Five-Year Plan (2016–20), after which the country's energy system will be much cleaner. Total inhalable particulate matter (PM10) emissions have been declining since the 1990s. Sulphur dioxide emissions reached their highest point in 2006, and have since declined steadily. Nitrogen oxide (NO_x) emissions dropped for the first time (in 2012) since reliable databases were established and are expected to flatten, then fall. The research initially concludes, therefore, that air pollutant emissions have reached a turning point. Combined emissions of major water pollutants are estimated to peak in 2016–20, then flatten, flowed by a gradual decline.

In terms of CO_2 emissions, more effort is needed to achieve a turning point by 2030. According to the Energy Production and Consumption Revolution Strategy (2016–30), by

2020, China's CO_2 emissions per unit of GDP will decrease by 18% compared to 2015, and by 2030 they will drop by 60–65% compared to 2005. The findings of this report show that to achieve the goal of carbon intensity reduction, additional policies and measures—including carbon pricing and subsidies for non-fossil energy sources—need to be gradually introduced to achieve tangible results.

In terms of the energy mix, the share of clean energy should increase significantly. According to the Energy Production and Consumption Revolution Strategy (2016–30), the share of non-fossil energy in China will be 15% by 2020, increasing to about 20% by 2030. In the Recommended scenario of this report, in which the energy revolution is strongly promoted, the share of non-fossil energy is expected to reach 15.7% by 2020, 22.5% by 2030, and more than 40% by 2050.

(2) Affordable and efficient

According to the above-mentioned strategy, energy consumption per unit of GDP is expected to drop by 15% by 2020 (compared to 2015), reach the current world average by 2030 (based on current prices), and achieve stability by 2050.

In the Recommended scenario of this report, energy consumption per unit of GDP will be 19.7% lower in 2020 than in 2015. This exceeds the 15% reduction target of the Energy Production and Consumption Revolution Strategy (2016–30). Energy intensity in 2030 will be 35.1% lower than in 2020, and 54.1% lower in 2050 than in 2030 (Fig. 12).

Energy cost per unit of GDP, which reflects the quantity and price of energy consumed to produce a unit of GDP, needs to be reduced significantly. Energy consumption per unit of GDP is an indicator of energy consumption efficiency but does not take price into consideration. In fact, energy price has an obvious impact on economic and social development. On the one hand, the price of energy can incentivise the whole economy and society to save energy. On the other hand, it constitutes a major cost to the real economy. To sharpen China's international competitive edge, energy cost per unit of GDP should be significantly lower.

Historical statistics show that China's energy cost per unit of GDP has followed an inverse U-shape path. It increased rapidly from 0.1 in 1990 to 0.18 in 2005, up by 80%. It then peaked around 2005 and decreased to 0.12 in 2012.

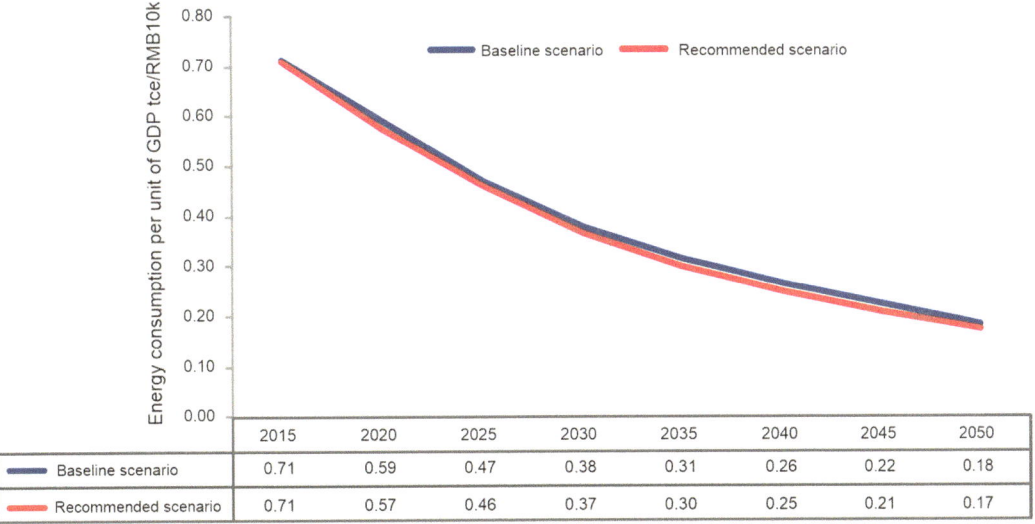

Fig. 12 Evolution pathways of China's energy consumption per unit of GDP in two scenarios. *Source* Findings of the DRC research team

The current energy cost per unit of GDP in China is comparable to that of South Korea, but notably higher than the USA and Japan. South Korea's energy cost per unit of GDP in 2011 was 0.18, slightly higher than China's. But the figure in Japan is only 0.09 and 0.08 in the USA, about 66% lower than in China. Since 2011, energy cost per unit of GDP in the USA declined rapidly, down to 0.04 in 2016 (Fig. 13).

(3) Secure and reliable

Energy supply sources should be diverse. To achieve energy supply diversity, China should develop new and unconventional energy sources domestically by leveraging its own energy resources and reducing the use of coal, while increasing the share of clean coal facilities in its fleet of coal-fired power plants. Externally, China should seize the opportunities of economic globalisation and diversify its international energy supply sources, by increasing imports of oil and gas from regions other than the Middle East to spread risk and secure supply.

The reliability of the energy system should be improved. Efforts should be made to:

(i) promote Internet+ smart energy development and establish an Energy Internet by 2025;

(ii) build smart wind farms and smart solar photovoltaic power plants; a smart system for the recovery, processing and use of coal, oil and gas; and a cloud-based platform to enable intelligent energy production;

(iii) deploy grid-scale energy storage of appropriate size at large-scale power generation sites to coordinate and optimise the operation of energy storage systems, renewable energy sources and power grids;

(iv) build smart homes, smart buildings, smart communities and smart factories, featuring smart end-use technologies and flexible energy trading to help create smart cities; and

(v) strengthen demand-side management, popularise intelligent energy consumption metering and diagnostic technologies, accelerate the deployment of energy management centres for industrial companies, and build an Internet-based information service platform.

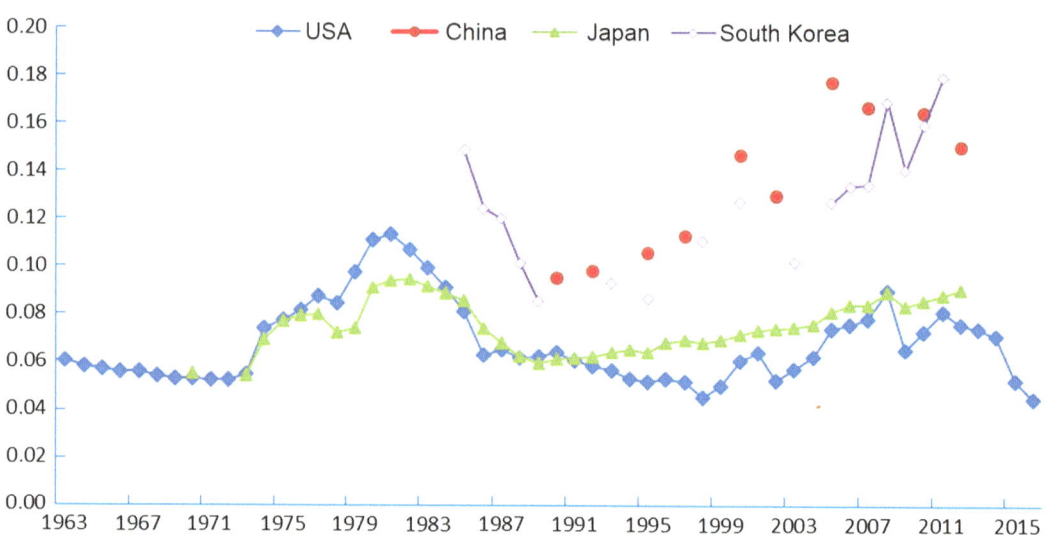

Fig. 13 Comparison of energy cost per unit of GDP between China and several major countries. *Source* Results calculated by the DRC research team based on the input-output table by country

2.2 To Achieve the Energy Revolution, China Needs to Get Five Driving Forces into Play: Four Pillars and International Cooperation

To achieve an energy revolution requires identifying, guiding and strengthening the driving forces behind it. Analysis of energy transitions in G20 countries since the 1970s shows that the drivers of energy transition include four pillars and international cooperation. Often several drivers must be in place before transition can develop momentum. Individually, these drivers are insufficient to start the process of systemic change. The G20 energy transitions were driven mainly by economic growth, energy security concerns, new market incentives or price shocks, with new technology playing a supporting role. They occurred primarily in the upstream and midstream sectors, with the energy mix remaining relatively constant (Fig. 14).

Supply

Historically, the abundance or scarcity of local energy resources is a fundamental driver of transition. The greatest transitions occur at the extremes, either when resources are plentiful or extremely scarce. In the current transition, technology is making unconventional and renewable energy resources increasingly available and competitive, and doing so in geographies that do not have access to conventional resources.

Demand

Rapid economic growth is usually accompanied by a change in industrial structure, often to higher added-value economic activities that favour a new energy mix based on natural gas and renewable electricity. Furthermore, consumers tend to choose cleaner and more flexible fuels as their income rises. This is a major trigger of energy transition, especially when accompanied by new low-cost supplies of energy. Energy security is also an important driver. For example, the oil crisis of the 1970s and the shutting down of Japan's fleet of nuclear power plants following the Fukushima tsunami and nuclear accident in 2011 increased people's awareness of the need for energy security, resulting in significant energy transition events.

Technology

Many energy technologies like nuclear power require vast investment in research and development. Once deployed, such technologies can transform the energy system through scale effects. International experience shows that government deployment of capital-intensive technologies is a common way to deliver energy system requirements such as supply security. Technologies that can plug and play into existing networks tend to be more successful than those requiring new networks to be built, thereby contributing more to the energy transition.

Markets

Markets have an important role to play. They can accelerate the process of innovation and adoption of new technologies through liberalisation and reform. Governments also have an important role to play. They can ensure markets operate efficiently through pricing externalities and regulatory oversight. Policies and institutions that increase the efficiency of markets like market liberalisation and reform can help the energy transition progress faster.

Fig. 14 The drivers behind energy revolution: four pillars and international cooperation. *Source* Vivid Economics

International cooperation

An increase in a country's participation in global energy trade can also trigger an energy revolution. For example, new gas pipelines or liquefied natural gas (LNG) infrastructure could increase trade in natural gas. New technologies and infrastructure can deepen international cooperation and have a transformational impact on future development. For example, extra high voltage power transmission enables long-distance power trade between countries.

2.3 Accelerating the Energy Transition Requires Four Intensifiers

Once the drivers for transition are in place, positive feedback loops (intensifiers) can accelerate transition or increase its scale. Our review of historical transitions concludes that while the four pillars and one cooperation can provide the necessary conditions for a transition and the momentum for change, these drivers only cause real transition when reinforced by positive feedback (intensifiers). Policymakers can often directly influence these intensifiers and should use them to control the speed and scale of transition (Fig. 15).

Preferences cause society to act

Transition can be accelerated and intensified if people's preferences are influenced by the belief that energy transition is good. The reason is simple: when some people think something is good or bad, they can influence or persuade others, which can cause their preferences to spread. This creates a positive feedback loop, which makes the next action easier. Value preferences about what society should or should not be doing have been powerful forces in past energy transitions. For example, Japan and France's embrace of nuclear power was a signifier of a technologically advanced society, whereas Germany's anti-nuclear movement saw the same technology as an environmental threat.

Expectations cause consumers to act

Expectations are also an intensifier: as more and more people expect transition to happen, they will act as if it is already happening. As with preferences, when people expect the world to change they will act accordingly and convince others that it will change. This creates a positive feedback loop. An expectation describes what is likely to happen, while a preference is a view about what ought to happen. For example, a person can expect something to happen even if he or she has no preference for it. Consumer expectations were important in past energy transitions, as it is consumers who select and purchase the products they expect to serve them best in the future. For example, consumers in the present transition may purchase an electric vehicle in expectation of future climate policies and legislation.

Fig. 15 The four intensifiers for transition act together. *Source* Vivid Economics

Private economics cause businesses to act

Energy transition may improve the private economics of businesses through positive spillover effects, which encourage further action. Businesses will act if it creates value for them, or if the benefits from energy transition are greater than the cost. When businesses take action this can change the private value of future action via externalities, positive spillovers or complementary goods. For example, the cost of wind power decreased as deployment increased due to the positive spillover of learning effects. Because action now increases the net private value of future action, this encourages future action, which then further increases the net private value of future action, creating a positive feedback loop.

Public good causes government to act

Government tends to act if the economics or the politics of the transition improve.

If energy transition generates value for society, it will win the public's backing, and in turn it can generate national political support or directly drive policy action. International political support can also modify national political support. For example, the Paris Agreement and the International Energy Agency have affected national political support for climate action and energy security respectively.

During energy transition, there is a strong positive feedback loop between private economics and the public good. As Fig. 16 shows, private economics and the public good are closely interrelated. If private actions by businesses can have positive externalities, such as reducing carbon emissions, or negative externalities, such as pollution, then government action will increase net private value, thereby creating an additional feedback loop between public and private action in a virtuous and reinforcing cycle.

The four intensifiers of the transition come together to accelerate and scale change. As Fig. 15 shows, social preferences, consumer expectations, the private economics of businesses and the public good all contain positive feedback loops that can intensify the motivation to act. Furthermore, these all work together so that if, for example, social preferences are causing people to act in a way (such as demanding low-carbon energy and related products and services) that makes a particular investment more valuable, then businesses will act. This, in turn, may create political conditions and support for government actions intended to promote energy transition.

2.4 Policy Plays a Crucial Role in Effectively Leveraging the Drivers and Intensifiers of Energy Transition

Compared with previous transitions, the current energy transition is more complex and has stronger externalities. For example, in the current transition the increasing complexity of a diverse and decentralised power sector requires system-wide change to drive down the cost of renewables and integrate them effectively into the power grid. The need for change affects the entire power value chain—flexible low-carbon generation, dispatch, balancing and ancillary

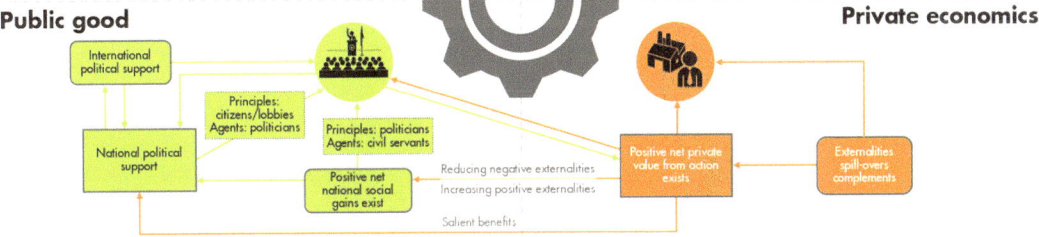

Fig. 16 Private economics and the public good have strong positive feedback loops between them. *Source* Vivid Economics

services, transmission and distribution networks, and consumer participation in retail markets through mechanisms such as demand-response to prices—and the incentives to support a more flexible and complex system.

In this case, policy plays a crucial role in supporting the drivers of transition. For example, policies help shift energy demand to cleaner and more efficient end uses, support a cleaner energy mix, develop efficient market mechanisms, stimulate innovation in clean and efficient technologies, and support national energy transition by leveraging international energy cooperation and governance.

Policies accelerate transition by giving the intensifiers momentum. As shown in Fig. 17, given the policy-directed nature of the current transition, government must take action to address environmental market failures, encourage change in consumer behaviour (and in broader society) and prompt businesses to invest in new low-carbon technologies. For example, subsidies or other policy tools could be offered to improve businesses' risk-return structure.

In short, policy plays a crucial role in energy transition. Perhaps a more important role for policy than giving the intensifiers momentum is keeping the rate and scale of change in social preferences, consumer expectations, private economics and the public good in sync as energy transition happens.

3 Adopt Multiple Measures: A Roadmap for China's Energy Revolution

3.1 Continuously Improve Energy Consumption Efficiency by Saving First

3.1.1 Optimise China's Industrial Structure by Reducing the Proportion of Energy-Intensive Industries

As China's economic development enters the later stages of industrialisation, there is a drive to upgrade the country's industrial structure, which, together with guiding policies, may lead to cleaner and more environmentally friendly industries.

It can be seen from the demand structure of investment, consumption and export that the share of investment in economic growth gradually declines as the share of consumption steadily grows. In this way, China is undergoing a shift from production-driven to consumption-driven economic growth, leading to accelerated industrial restructuring.

Moreover, there is great potential for adjusting the energy consumption mix.

First, with mechanisation at saturation point, agricultural modernisation will shift focus to biotechnology and digitalisation, which is

Enabling **PREFERENCES** for clean, efficient energy to spread, e.g., through smart, responsive infrastructure, innovative products at lower cost

Setting **EXPECTATIONS** for the transition, e.g., through long-term policy goals, roadmap for achieving those goals, maintaining policy certainty and credibility

Investing in **PUBLIC GOODS** necessary for the transition, e.g., coordination/investment in strategic infrastructure, urban planning, addressing environmental market failures

Developing policy frameworks supporting **PRIVATE ECONOMICS**, e.g., through pricing environmental externalities, innovation support, efficient regulatory frameworks

Fig. 17 Policy plays a crucial role in driving energy transition. *Source* Shell International

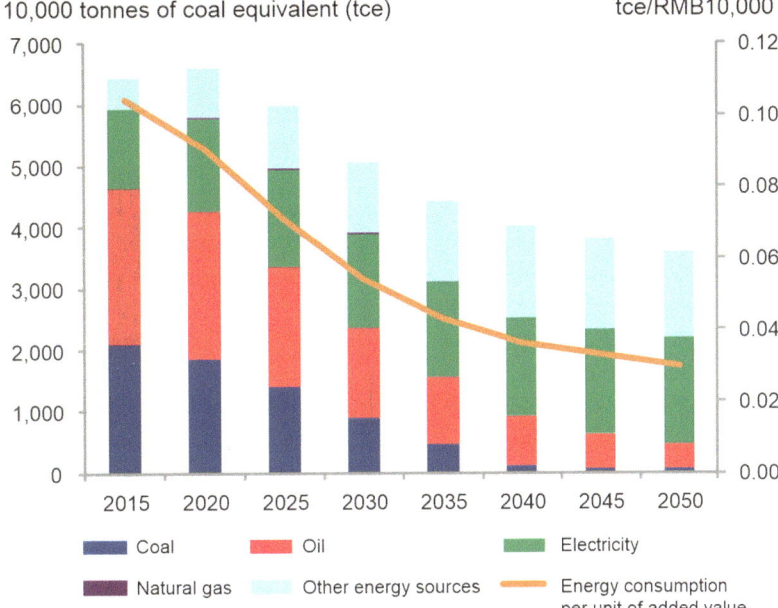

Fig. 18 Forecast of China's end-use energy consumption in agriculture. *Source* Findings of the DRC research team

expected to further reduce the energy consumption per unit of added value (Fig. 18).

Second, optimisation of the country's industrial structure and changes to production processes will significantly improve the efficiency of industrial energy consumption. Starting with the 13th Five-Year Plan (2016–20), the amount of energy-intensive products—such as steel, cement, glass, aluminium and synthetic ammonia—has begun to peak and may drop.

Third, energy consumption in buildings is expected to grow steadily. Based on the experience of developed countries, China's urban residential floor space per person is likely to peak at about 40 square metres and public floor space per person at about 20 square metres. Estimates based on future population trends show that the floor space peak in China should remain at about 90 billion square metres.[1] To ensure steady and sustainable development of the real estate industry, the peak should occur around 2040. However, with the development of energy-efficient

buildings, energy consumption per unit of building area will be significantly reduced, even though the total energy consumption of buildings in China is expected to increase steadily.

Fourth, energy consumption in the service sector is expected to continue to grow as it scales up, but will be constrained by the total stock of commercial floor space. Based on forecasts of added value in the service sector and commercial floor area, end-use energy consumption will increase to 350 Mt of coal equivalent (Mtce) by 2035 and 470 Mtce by 2050.

Fifth, residential end-use energy consumption will continue to grow, but could remain significantly lower than that in developed countries. With people's living standards improving, household energy consumption is expected to reach 520 Mtce by 2035 and 660 Mtce by 2050.

3.1.2 Use New Technologies, Processes and Products to Save Energy

Efforts should be made to promote the development of green and low-carbon buildings. In China, energy consumption in buildings accounts for nearly half of the country's total energy consumption, far higher than that of developed economies.

---

[1]China's floor space stock in 2015 was close to 60 billion square metres, including 17.6 billion square metres of urban residential floor space, 27.6 billion square metres of rural residential floor space, and 14.0 billion square metres of urban commercial floor space.

Government ministries and agencies introduced a series of plans, including the 12th Five-Year Plan for Developing Green Buildings and Eco-Cities (2012–17). However, further improvements in green building systems and standards and increased R&D of green building technologies and life cycle management of green buildings are required. Technology innovation in green buildings is currently focused on lighting and heating. Energy- efficient lighting is one of the most effective ways to reduce greenhouse gas emissions from buildings in almost all countries. The Roadmap for the Phase-out of Incandescent Lamps in China, issued by the National Development and Reform Commission (NDRC), banned the import and sale of incandescent lamps of 15 W or more from October 2016. If all existing incandescent lamps are replaced with energy-efficient lamps, 48,000 GWh of power would be saved annually, equivalent to a reduction in CO_2 emissions of 48 Mt. It is estimated that China's cumulative newly built urban residential areas will exceed 5 billion square metres by 2020, and that the newly added energy consumption from heating in north China (the coldest part of the country) will be about 125 Mtce. If heating from renewable sources is deployed in all these new residential areas, the resulting reductions in CO_2 emissions would be 375 Mt.

Centralised coal-fired power generation and coal-fired combined heat and power (CHP) should be increased to save energy and reduce emissions. Currently, centralised coal-fired power generation at large power plants accounts for only 48% of total coal consumption in China, compared to 99% in the USA. The extremely large number of distributed, small-scale coal-fired facilities in China, which do not have the capability to treat pollutants, offers great potential for energy saving and emissions reduction. China needs to take several measures to significantly shift coal use from small-scale to large-scale centralised generation. This will reduce pollutant emissions from coal combustion and improve the heat to electricity conversion efficiency of coal.

3.1.3 Introduce Carbon Pricing to Improve Energy Consumption Efficiency

Carbon pricing can have a significant energy saving effect as it increases the cost of fossil fuels and causes a shift in the energy mix towards lower-carbon fuels. The increase in the price of energy will also drive energy efficiency and reduce total energy consumption—as the carbon price goes up, total energy consumption will decrease. In the policy scenarios of $30 per tonne CO_2 equivalent (tCO_2e), $60/$tCO_2$e and $90/$tCO_2$e, total energy consumption will be 7.4%, 14.0% and 19.4% lower respectively than in the zero-carbon-tax scenario. Furthermore, the effects of carbon pricing policy will become significant over time.

3.2 Enable Cleaner Energy Consumption by Using Less Scattered Coal and by Increasing Electrification

3.2.1 Substitute Electricity and Gas for Scattered Coal

In 2015, China's scattered coal consumption reached 617 million tonnes (Mt), mainly used in coal mining (120 Mt), household heating (93 Mt) and chemical production (90 Mt). A further 260 Mt of scattered coal was used by light industries—food, textiles, equipment manufacturing and services.

To replace scattered coal in residential heating, China will encourage central heating (gas-fired boilers, geothermal heating and waste heat recovery) and increasingly substitute electricity and gas for scattered coal (such as wall-mounted gas-fired heaters) in areas where central heating is not possible. The measures for substituting electricity and gas for scattered coal in the industrial and commercial sectors include replacing small coal-fired boilers with gas-fired boilers or installing waste heat recovery or other intensive heating methods. According to the Energy Production and Consumption Revolution Strategy (2016 30), more than 35% of scattered coal will be replaced by 2020 and about 70% by 2030. In this way, scattered coal consumption will decrease to 400 Mt in 2020, 180 Mt in 2030 and 60 Mt in 2050.

3.2.2 Speed Up Electric Vehicle Development to Promote Clean Energy Consumption

In the Recommended scenario, the number of registered electric vehicles (EV) in China will reach 3 million in 2020, 80 million in 2030 and 270 million in 2050. However, the industry's current strong momentum indicates the possibility of faster development.

First, there is a global boom in EV R&D, mainly around such critical technologies as batteries, automated driving systems and charging facilities. New concepts in leasing, business models and financial services are expected to accelerate developments in the EV market.

Second, as artificial intelligence technologies evolve, automated driving +EV is expected to become a standard travel model in the future, extending vehicles from a travel tool to a home on wheels. This new way of travelling can speed up the transition from conventional vehicles to EVs in the medium and long terms.

Third, EVs can serve as both a means of transport and a distributed energy storage facility. With expanding battery (energy storage) size and support from smart grid technologies, EVs can

Table 1 Two EV development scenarios *Source* Findings of the DRC research team

Scenario		Year		
		2020	2030	2050
Recommended scenario	Ownership (million)	5.24	83	270
	Proportion (%)	1.9	18.4	50.0
Accelerated scenario	Ownership (million)	5.24	200	500
	Proportion (%)	2.0	44	93
Vehicle ownership (million, including fuel vehicles and EVs)		272.23	450	540

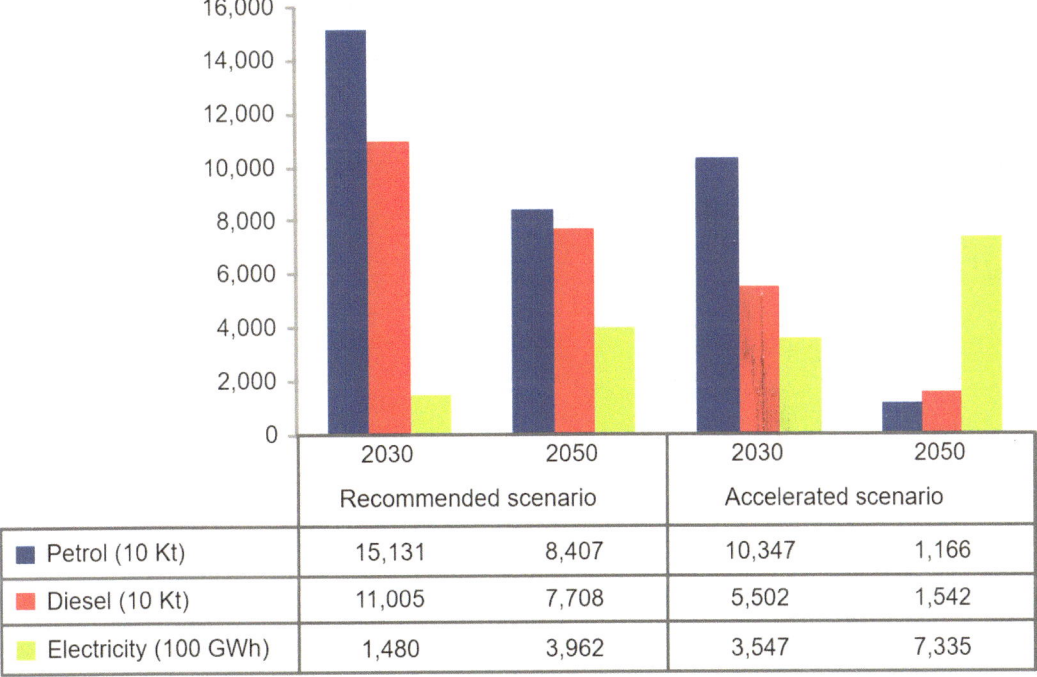

	2030 Recommended scenario	2050 Recommended scenario	2030 Accelerated scenario	2050 Accelerated scenario
■ Petrol (10 Kt)	15,131	8,407	10,347	1,166
■ Diesel (10 Kt)	11,005	7,708	5,502	1,542
■ Electricity (100 GWh)	1,480	3,962	3,547	7,335

Fig. 19 China's total vehicle energy demand in two EV development scenarios

make the most of renewable energy and support its long-term development.

To this end, and based on its findings in the recommended scenario, this report proposes an Accelerated development scenario in which we investigate energy demand in transport and its impact on China's energy supply mix. EV development scenarios are set up as follows (Table 1).

Estimates for vehicle energy consumption in the Accelerated scenario are shown in Fig. 19. They assume vehicle ownership remains unchanged and newly added EVs replace mainly diesel and petrol vehicles and do not affect the number of natural gas-fuelled vehicles. Petrol and diesel consumption are expected to decrease by 100 Mt by 2030 and by 130 Mt by 2050. Meanwhile, electricity demand is projected to increase by 200,000 GWh by 2030 and by 340,000 GWh by 2050, compared to the recommended scenario.

3.2.3 Accelerate Electrification by Decarbonisation

In 2014, electricity accounted for nearly a quarter of China's end-use energy demand. The country's electrification rate is comparable to that of many OECD economies. As China's society and economy develop, consumer demand for higher quality energy will grow, and technological and structural shifts will drive further change. Historical experience suggests that the electrification rate in 2030 will grow by 5.5% compared to 2015, thanks to higher incomes; and by 6.4% due

to technological progress; but shrink by −2.6% as a result of economic restructuring. Together, these factors may lead the electrification rate to rise from 23% today to 32% by 2050.

Decarbonisation will further drive the process of electrification. The electrification rate is expected to increase by 5–10% due to accelerated EV development in the transport sector, by 2.5–5% from decarbonisation in construction projects, and by 0.5% from decarbonisation in industry, reaching more than 40% by 2050 (Fig. 20).

Through the rapid development of clean energy and the use of clean coal, the share of electricity in the end-use energy mix will gradually increase from 22% in 2015 to 30% by 2030 and to around 40% by 2050, significantly improving the electrification rate.

3.3 Develop a Clean Energy Production Mode Featuring the Efficient Development of Conventional Energy and a Combination of Centralised and Distributed Energy Systems

3.3.1 Increase the Proportion of Scientific Coal Capacity

Coal-dominated conventional energy will remain the main energy source in China, both now and

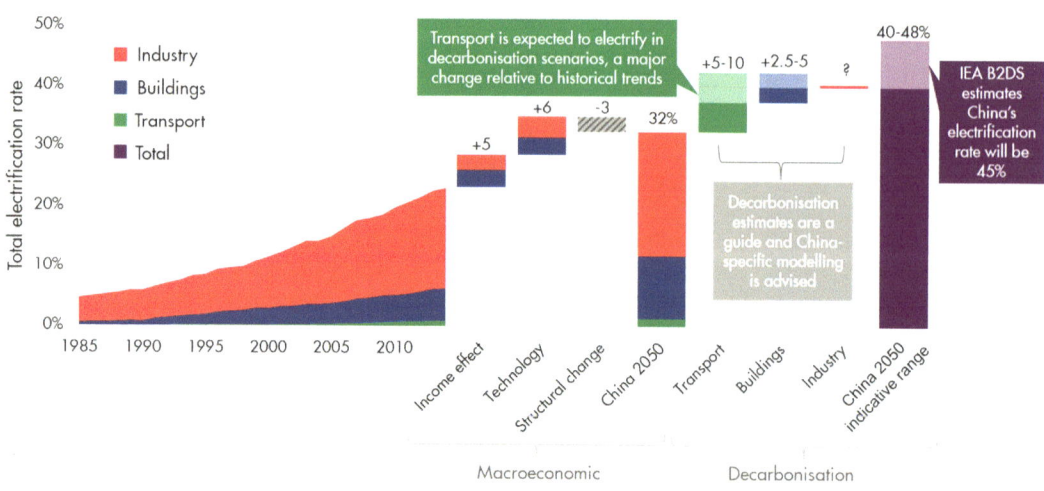

Fig. 20 China's electrification rate growth potential

in the longer term. To address the requirements of the energy revolution, production needs to be safe, efficient and sustainable. In mining, coal production will shift from extensive development to intensive, green production. In conversion, priority will be given to upgrading coal-fired generation from high to ultra-low emissions.

Taking into account the geology, coal reserves, water resources and ecology of China's coal-producing regions, the one-third of coal mines that meet the criteria for scientific capacity will be retained, the one-third that fail to meet the criteria will be upgraded, and the one-third that are backward or not possible to be upgraded will be gradually shut down.

3.3.2 Maintain Steady Development of Oil Supply Capacity

Although EV development can significantly reduce future oil demand, in the Recommended scenario China's oil demand and oil processing capacity peaks will exceed 650 Mt and 720 Mt respectively in 2030. Even in 2050, China's oil demand will still be around 650 Mt. Currently, China's oil processing capacity has been close to its peak, so it is necessary to continue to constrain total demand and accelerate the shift from oil refining to chemical production.

3.3.3 Significantly Increase Gas Supply Capacity

In the future, China's natural gas demand is expected to see fast growth. To ensure that demand can be met, which will increase as gas is substituted for scattered coal, supply capacity must be significantly improved. It is forecast that 260 billion cubic metres (bcm), 450 bcm and 490 bcm of natural gas will be needed by 2020, 2030 and 2050 respectively.[2] Therefore, China needs to vigorously explore and exploit conventional gas, tight gas, shale gas and coalbed methane, and drive research forward into natural gas hydrate exploitation technologies.

3.3.4 Develop Clean Energy (Mostly Renewable Energy) in a Well-Planned Manner

China's clean energy sources and load centres are located in west and east China respectively. As the energy supply revolution progresses, the development of clean energy in these two regions should be coordinated.

First, there needs to be coordination between renewable energy generation and transmission and distribution networks to connect regions rich in renewables to demand centres, provide energy storage to manage intermittency, and optimise demand (including the use of demand-side response to manage peaks and intermittency).

Second, a mix of multiple sources of energy and technologies is needed to ensure a stable supply of electricity. These would include small wind farms; solar photovoltaic; combined cooling, heat and power plants; and others.

Third, collaboration across the entire electricity supply chain (generators, grid operators and consumers) is essential to optimise supply, distribution and consumption.

Fourth, in west China, energy should be generated at large utility-scale renewables plants and connected to national and regional transmission and distribution networks. In east and central China, generation should be smaller scale —distributed renewables and low carbon in microgrids or community power grids.

3.4 Gradually Establish an Energy Mix Centred on Conversion to Electricity

In the end-use energy mix, electricity's share will progressively increase from 22% in 2015 to 30% by 2030, and then to around 38% by 2050. This electricity-centred energy mix is made possible mainly by the rapid development of clean power technologies and clean coal utilisation (Fig. 21).

---

[2]Research on China's Gas Development Strategies, published by China Development Press in 2016 and by Springer in 2017, forecasts China's gas output to reach 270 bcm in 2020 and 470 bcm in 2030.

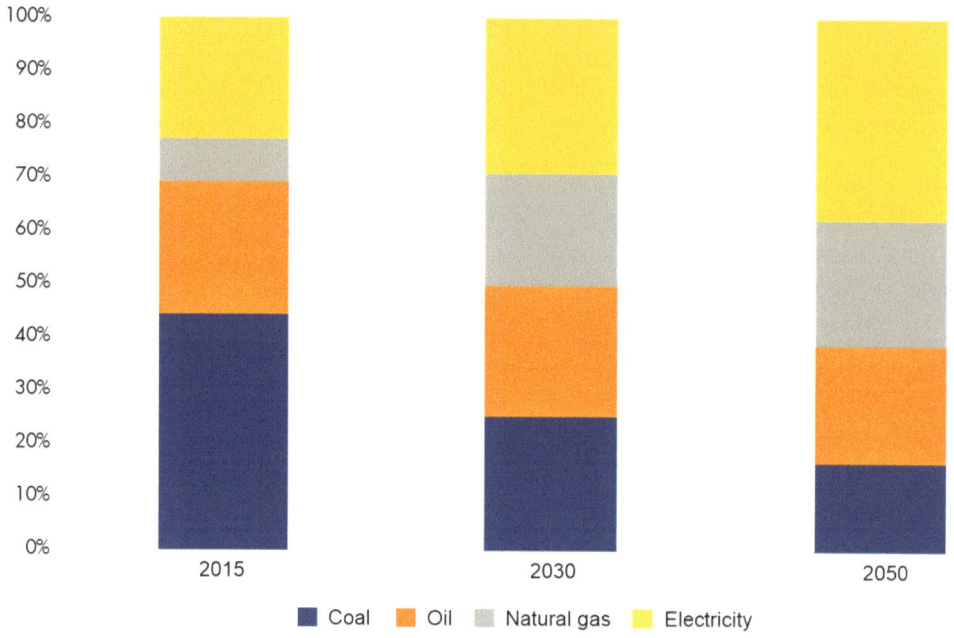

Fig. 21 Changes in China's end-use energy mix in the Recommended scenario. *Source* Findings of the DRC research team

3.4.1 Increase the Proportion of Renewable Energy (Mostly Wind, Solar and Biomass) and Nuclear Power

Based on an optimal design of the future power supply mix, the research team recommends the following roadmap for clean energy development:

- the proportion of wind power and solar photovoltaic will increase from 4.9% in 2015 to about 15% in 2030 and more than 25% by 2050;
- the share of gas-fired power will increase steadily from 3% in 2015 to 9% in 2030 and around 10% by 2050; and
- although hydropower capacity will continue to grow, the rapid development of other power sources will lower its share of the energy mix from 17% in 2015 to 14% in 2030 and around 13.5% by 2050.

Overall, the proportion of clean power, not fuelled by coal or oil, will increase from 30% in

2015 to 38% in 2020 and 52% by 2030. By 2050, non-fossil fuel generation will increase to more than 70% of installed capacity and 66% of power output.

In the medium term, total consumption of coal as a primary energy source will not change greatly, but its share of energy supply will decrease sharply. The energy supply revolution will drive the shift from direct coal use to its conversion into an electricity-oriented secondary energy source of efficient and clean coal utilisation.

3.4.2 Encourage the Substitution of Non-fossil Fuel Energy for Oil and Coal

Coal consumption will fall sharply from about 44% in 2015 to 27% by 2050. Oil consumption will first increase and then decrease, returning to its current share of around 16% at mid-century. Natural gas, which is a cleaner energy source than coal and oil, will enjoy rapid growth in the medium term, rising to 15% in 2030 and staying stable thereafter. Non-fossil fuel energy will increase substantially, rising to more than 20% in

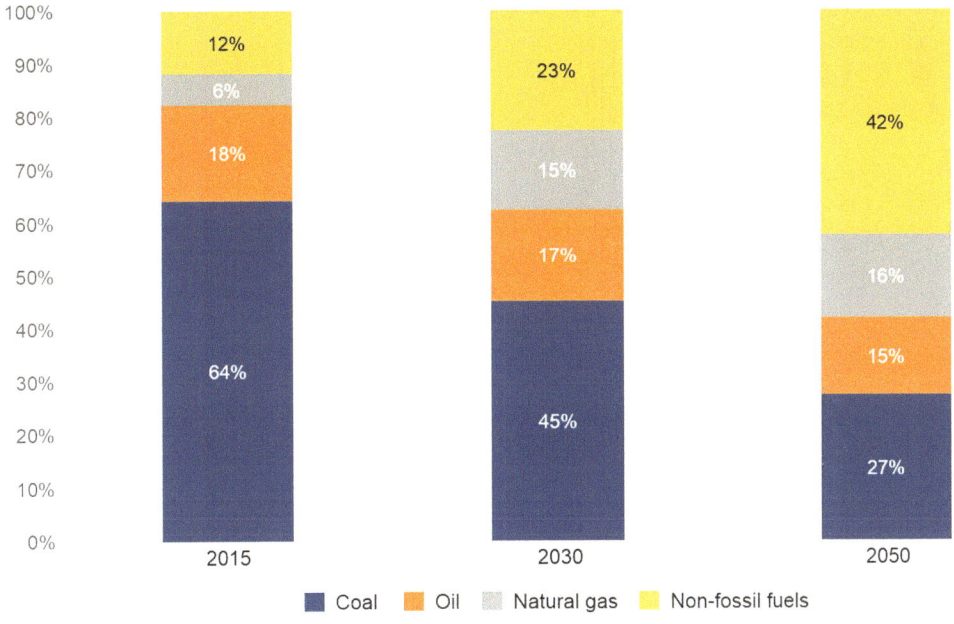

Fig. 22 Changes in China's primary energy mix in the Recommended scenario. *Source* Findings of the DRC research team

2030 to replace oil as the second-largest energy source. By 2050, non-fossil fuels' share of the energy mix will be more than 40%, replacing coal as the number-one energy source (Fig. 22).

3.5 Build an Internet+ Intelligent Energy System

Internet+ is an intelligent energy system of the future that uses digitalisation to integrate a greater share of distributed renewables into the power system. It will need to incorporate micro-, community and regional grids and coordinate their activities into a greater, harmonised whole. It will use energy storage and demand-side response to manage solar and wind intermittency. And it will encompass a range of low-carbon energy sources and technologies (horizontal integration), as well as the various parts of the electricity value chain—generation, transmission, distribution and consumers (vertical integration). This will help manage the impacts of intermittency that distributed renewable energy systems can have on local grids and

provide a feasible pathway for connection of distributed renewable energy at large scale.[3]

3.5.1 Promote Intelligent Energy Consumption

The development of smart homes, buildings, communities and factories featuring smart meters and flexible energy trading should be encouraged to help create smart cities and leverage intelligent technologies to reduce residential energy consumption.

A smart energy system, through the use of smart grids and smart meters, should be developed to enable real-time measurement, information exchange and active control of energy consumption for electricity, heating and cooling. The implementation of such an intelligent and advanced measurement system will be enriched to enable remote, automatic and centralised acquisition and collection of water, gas, heating and electricity consumption data, thereby realising multi-meter integration. The networking structure

---

[3]Gao Shiji and Guo Jiaofeng et al., Energy Internet Boosts Energy Transformation and Institutional Innovation in China, China Development Press, 2017.

and information interface of the advanced measurement system will be standardised throughout the whole system to achieve secure, reliable and fast two-way communication with all users.

Efforts should also be made to strengthen demand-side management and increase the use of intelligent energy consumption monitoring and diagnosis technologies. Additionally, the creation of energy management centres for industrial companies needs to accelerate and an Internet-based information service platform should be built, so that businesses can monitor and analyse energy consumption in each production process and intelligently dispatch water, electricity, gas and fuel whenever changes in production parameters are made. This will enable real-time monitoring, timely adjustment, automatic alarm and other functions throughout the production process (from procurement to use), thereby realising intelligent energy management and tapping energy saving potential.

3.5.2 Establish Micro-Balancing Systems that Allow Energy End Users to Participate in Energy Markets

Equipment, facilities and platforms need to be established to allow energy users—homes, businesses, industrial parks and communities—to participate in the energy market and provide balancing services to the microgrid or community grid to which they are connected. This would be in the form of, for example, behind-the-meter storage, electric vehicle batteries as storage, or demand-side response. It would promote flexible and interactive energy use, support distributed energy trading and feature multi-energy source integration, openness and sharing, real-time two-way communication and intelligent control.

3.5.3 Accelerate the Construction of Integrated Energy Network Infrastructure

An integrated energy network, based on the smart grid concept, should interconnect with other networks such as those for district heating or cooling, natural gas distribution and various transport networks. It would enable efficient conversion from one energy form to another, such as natural gas

into heating or cooling, and allow centralised and distributed energy operations to be coordinated in a single smart system. Deployment would initially be made in new urban areas, new industrial parks, or in districts affected by air pollution. The objective would be to create a highly integrated energy system that provides flexible, controllable, safe and stable energy transmission.

3.5.4 Set up Internet+ Intelligent Energy Development

In 2017–20: (i) distributed power generation and storage technologies will be promoted and deployed at scale; (ii) the digital multi-energy trading system will go live; and (iii) pilot and demonstration projects featuring interconnectivity among various energy networks, energy sources and technologies will be launched.

In 2021–25: (i) optimisation across diversified energy carriers will be gradually made possible, and distributed power generation and storage systems widely deployed; and (ii) urban smart and diversified energy networks will be established to optimise energy from different sources and address various energy requirements.

In 2026–30: (i) new electricity microgrids and an interconnected non-fossil energy network that features multiple and complementary energy sources and technologies will be promoted and constructed across China; and (ii) an open and sharing smart energy ecosystem will take shape to significantly improve overall energy efficiency.

After 2030: (i) renewable energy will be widely used in such sectors as agriculture, industry, transport, commercial and residential; and (ii) the industry ecosystem supporting rapid and sound development of renewable energy will continue to fast-track renewable energy development.

3.6 Develop New Energy Technologies that Fully Support the Energy Revolution

3.6.1 Continuously Promote the Smart Power Grid

The smart power grid is an important means to integrate energy production and consumption, as well as new technologies and system revolutions. It is also the enabler of the Energy Internet. In April 2016, the National Development and Reform Commission (NDRC) and the National Energy Administration (NEA) published their Action Plan for Innovation in the Energy Technology Revolution (2016–30)[4] and their Roadmap for Major Innovation Actions in the Energy Technology Revolution,[5] which launched a plan to develop smart grid power transmission and smart end-user devices.

To develop a smart grid, China will: (i) strive to make breakthroughs in key technologies and core equipment that it currently imports from other countries, especially in the fields of direct current, power electronics and renewable energy; and (ii) develop a national and industry standards system for smart grids as soon as possible to support the building of smart grids.

3.6.2 Develop New Energy Technologies

In wind power, China will work to achieve breakthroughs in fields such as aerodynamics, flow field analysis, load calculation, as well as high-end technologies like large wind turbine design, wind turbine bearings, and wind farm control and pitch control systems. This will address the country's current weakness in conducting basic research on wind power and critical wind power equipment.

In solar photovoltaic, China needs to significantly improve photovoltaic conversion efficiency and make breakthroughs in areas like interdigitated back contact, heterojunction with intrinsic thin layer cells, passivated emitter cells and rear cell technology, metallisation wrap through, bifacial modules and boost the efficiency of battery technologies.

In new energy technologies that have great potential in the short and medium terms—such as solar thermal power and ocean power—China

will continue to develop demonstration projects to rapidly accumulate experience in planning, design, construction, operation and management. Such an approach will lay a solid foundation for policy studies and the development of industrial-scale technologies. It will also improve international competitiveness and reduce costs.

3.6.3 Increase Support for the Development of Energy Storage Technologies

Energy storage technologies are the key to the electricity revolution. They are also at the cutting edge and the site of fierce competition. Currently, China sees severe wind, solar and hydro curtailment and nuclear power restriction, which result in more than 100,000 GWh of wasted power output. Energy storage is an important means to use currently wasted output and integrate unstable energy supply and consumption. Technologies such as physical and chemical energy storage, hydrogen fuel cells and heat storage are at the forefront. Ultimately, one or two of these technologies will survive the competition and grow.

3.6.4 Prioritise Nuclear Power Development

Nuclear power is indispensable to China's development and one of the energy pillars China must secure strategically. To this end, China will: (i) invest more in scientific research to enhance basic capabilities in nuclear technologies; and (ii) establish scientifically based decision-making and interaction mechanisms to win the public's recognition of the importance of nuclear power and enable the safe development of nuclear energy in China.

3.6.5 Make Unconventional Gas a Major Component of New Gas Capacity

China will increase its collaborative R&D efforts with overseas organisations to innovate advanced technologies, especially exploration and production technologies suitable for China's unconventional oil and gas resources. This will speed up the development and scale-deployment of

---

[4]NDRC and NEA, Action Plan for Innovation in the Energy Technology Revolution (2016–30), 2016, p.6.

[5]NDRC and NEA, Roadmap for Major Innovation Actions in the Energy Technology Revolution, 2016, pp.67–69.

critical technologies with Chinese characteristics. Moreover, China will focus on the development of unconventional natural gas resources like tight gas, shale gas and natural gas hydrates.

3.7 Strengthen China's Energy Security by Improving Global Energy Governance

3.7.1 Cooperate with and Reform Existing International Energy Governance Organisations

Make sure energy security is in line with countries' interests. With the strengthening of its global influence, China should participate deeply in global energy governance systems as soon as possible. This, to make them more widely representative and better reflect the interests of developing and emerging countries, thereby promoting a global energy community with a common future. China should implement international energy cooperation strategies that integrate multi-level international energy cooperation partners, diversified international energy cooperation forms, multi-channel international energy cooperation methods, multi-area international energy cooperation content and multi-task international energy cooperation processes to adapt to the changes in the global energy landscape and meet its own development demand. High-quality development of China's energy system should be promoted while accelerating the global transition to clean, low-carbon, affordable, efficient, secure and reliable energy.

3.7.2 Seek G20 Support to Facilitate the Energy Transition by Aligning Global Energy and Climate Governance

Attempts should be made to get the G20 to attach more importance to the energy transition and reach agreement on the transition to a low-carbon and secure energy future. This could be achieved by the G20 issuing statements and commitments on energy security and long-term decarbonisation and by making collective efforts to raise the ambitions of nationally determined contributions

(NDCs) to reduce greenhouse gas emissions. Long-term G20 ministerial conferences on energy should be set up and institutional capacity built by establishing energy secretariats in relevant international organisations.

3.7.3 Reduce the Risk of Investing in Partner Countries to Improve China's Energy Security

Over the past decade, China's Go Out strategy to encourage its enterprises to invest abroad has rapidly increased foreign direct investment (FDI) in overseas energy sectors. In the coming decade, the Belt and Road Initiative (BRI) will further channel Chinese capital into energy resources and infrastructure. It is necessary for China to ensure good energy sector governance in those partner countries where it invests. This will mitigate the risk of instability and underinvestment in those countries, enhancing China's energy security as a result. China can also cooperate with partner countries to support them through the energy transition, especially by providing access to China's low-carbon technologies.

In-depth cooperation on energy between China and the BRI countries should be strengthened. Through its cooperation in the BRI, China plays a unique role in low-carbon energy, energy security and the energy transition. A framework for FDI and energy infrastructure development in clean energy technologies should be created. At the same time, China should work with others to enhance the governance and transparency of energy sector investment.

The way that China invests in countries in the BRI will be a measure of its commitment to green and sustainable growth, both at home and abroad. The risk of investment disputes with host countries would be minimised by learning from historical experience, adhering to high standards of social and environmental governance, and developing appropriate risk management tools.

3.7.4 Strengthen Global Electricity Cooperation

One of the principal solutions to achieve power supply security, reduce curtailment and ensure balance in power systems with a high penetration of renewables is to increase electricity trading through interconnected grids. As the largest battery manufacturer and generator of renewable power in the region, China could take the lead in cooperation and governance reforms to facilitate cross-border grid interconnections with neighbouring countries.

China and its regional partners should harmonise national electricity markets to enable power trading through such interconnections. Connecting high renewable supply areas with load centres requires regional planning to avoid lock-in of fossil-based assets and infrastructure and lock-out of renewables. Further, as electricity markets become increasingly interconnected the risk of cyberattack is best mitigated by strengthening and harmonising regulations across jurisdictions.

China can benefit from the opportunities to export surplus electricity and reduce electricity costs through grid interconnections and balancing: regional electricity trading saves consumers money and decreases the capacity margin requirements for China and its partners. Cooperation on energy interdependence can build trust among partners, creating wider benefits.

Regarding electricity market reform (EMR), regional differences in capacity, dispatch and balancing and political willingness to align reforms may be difficult to overcome. A stepped approach to EMR could be adopted, so that the benefits of progressive reform are tangible to each partner country.

4 Systematically Build a High-Quality Energy System: Policy Suggestions for Promoting the Energy Revolution

4.1 Structural Change Is Necessary for China's Energy Revolution

China's strategic goal of supporting high-quality economic development with high-quality energy requires China to create a global and modern energy system with effective market mechanisms, moderate macro-control, vibrant enterprises and clean, low-carbon, economic, efficient, secure and reliable energy. To achieve this, the government has a key role to play in setting the policy framework that levels the playing field between low- and high-carbon sources of energy. Once the government does this, markets will seek out the cheapest energy sources and drive innovation in low-carbon technologies and investment in enabling infrastructure.

4.1.1 Strategic Goals

- *Well-designed market system.* The objective is to create a modern energy market system based on harmonisation, openness and orderly competition. Market monopolies will be eliminated. Instead, a market will be shaped comprising large energy companies cooperating, coexisting and competing fairly with a range of companies of different scale and ownership across the energy value chain—in energy production, transmission and sales. This will address issues such as the misallocation of resources caused by the unequal status of market players and disorderly competition.

- *Sound price mechanism.* In competitive segments, prices will be decided by the market. In natural monopoly segments such as pipeline transmission, prices will be controlled by the government. A price mechanism and a fiscal and taxation system that truly reflect market supply and demand, resource scarcity and environmental impact will be created to address current unreasonable price formation mechanisms.

- *Well-regulated government management.* To clearly define the boundary between government and market and the new round of government reforms, several high-level, consolidated and independent energy management authorities will be created, integrating functions such as industrial development, Five-Year Plans, and energy policies and regulations. They will operate according to the principles of "what is not mandated by the law shall not be done, what is not prohibited can be done, and what is defined as legal responsibility must be done".

- *Effective market regulation.* Unified, independent and specialised regulatory authorities and a modern energy regulation system will be established. The latter will demonstrate clear rights and responsibilities, fairness and impartiality, transparency and efficiency, and effective regulation. It will address issues such as policymaker-regulator overlaps, decentralisation, the current lack of regulatory functions and the shortage of regulatory personnel and powers.
- *Well-developed legal system.* A system of regulations and standards will be developed—based on the Energy Law and supported by legislation in the power, coal, oil, and gas sectors—to ensure national energy security and sustainable development. It will address issues across the energy system such as the lack of consistent guidelines and principles for legislation, and the present inconsistency and lack of alignment of the current legal system.

4.1.2 Strategic Priorities

A modern energy market system should be created by:

(i) separating natural monopoly enterprises from competitive businesses, improving market access and encouraging investors to enter the energy sector in an orderly manner;

(ii) establishing and enhancing an energy trading system that integrates the national market with multiple regional markets and applies consistent rules, complementary functions and multi-level coordination;

(iii) establishing power system operators with independent scheduling and trading to effectively separate ownership and operation of power transmission and distribution networks;

(iv) encouraging oil and gas pipeline operators to gain exclusive rights, separating pipeline transport services from downstream retail sales, and providing fair access for third parties to pipeline networks; and

(v) accelerating the development of smart energy systems and an integrated energy services market, and building an energy system where centralised generation, distributed energy, energy storage and demand-side load management enjoy equal access.

The energy market's price mechanism should be reshaped by:

(i) applying the principle of "allowable costs plus a reasonable profit" to natural monopolies like pipelines, as this incentivises monopoly industries to reduce prices and other companies to innovate;

(ii) deregulating market prices in competing segments to achieve a market-determined price mechanism;

(iii) implementing an electricity price mechanism—that controls power transmission and distribution and deregulates power generation, sales and use—by establishing a well-designed, independent and performance-based power pricing system;

(iv) determining the methods needed to disclose power transmission and distribution costs and pricing;

(v) deregulating the price of oil and natural gas (which will be based on market competition) and taking steps to deregulate pricing in infrastructure like oil and gas pipelines (but not gas distribution networks);

(vi) establishing and improving directional subsidy and relief mechanisms for people in need and some non-profit industries; and

(vii) eliminating cross-subsidies in energy prices.

The energy management system should be improved by:

(i) establishing and strengthening regulatory authorities to better manage state-owned natural resources and monitor natural ecosystems. This should be in line with the new round of institutional reforms to implement responsibilities relating to publicly owned natural resources;

(ii) defining the boundary between government and the market, standardising and simplifying approval procedures, and reducing the administrative burden on companies subject to these approvals;

(iii) addressing grid integration of renewable power and the need for inter-provincial and interregional power transmission and distribution under the national energy strategy; as well as fully deregulating the power generation sector and ensuring all regions purchase and use the legally binding minimum number of hours of wind and solar power generation annually; and

(iv) improving peak shaving and the backup auxiliary service market mechanisms, upgrading thermal power flexibility and increasing energy storage power sources.

The energy regulatory system should be improved by:

(i) promoting policymaker-regulator separation, setting up independent, unified and specialised regulatory authorities, and improving the regulatory system at the national and provincial levels;

(ii) defining regulatory responsibilities (mainly for economic regulation), and strengthening social regulation to ensure competitive outcomes in natural monopoly segments such as network infrastructure (mostly pipelines); and

(iii) improving regulatory capabilities, innovating regulatory approaches, improving regulatory effectiveness and maintaining a fair and competitive market.

A modern energy legal system should be accelerated by:

(i) developing the Energy Law to provide a basis for the creation and revision of other laws and regulations in the energy sector;

(ii) revising the Electric Power Law, developing the Oil and Gas Law, improving the Coal Law as soon as possible, and clarifying the basis for the creation, implementation, assessment, supervision and adjustment of plans and strategies for electricity, coal, oil and gas;

(iii) implementing the Energy Conservation Law and the Renewable Energy Law, and establishing consistent regulatory, coordination, decision-making and social participation mechanisms; and

(iv) developing energy regulations and establishing and improving existing energy rules, requirements, approaches and procedures.

4.2 Create a Nationally Unified and Dynamic Carbon Trading Market

China has announced the launch of a nationally unified carbon trading market, covering key carbon-emitting industries such as steel, electricity, chemicals, building materials, paper and non-ferrous metals. From the progress of ongoing pilot programmes, the following challenges were observed.

First, the failure to achieve market-based electricity pricing is preventing the carbon trading market from working effectively. China hasn't yet established a new mechanism that enables the market to determine electricity prices. Electricity price control limits the power sector's potential to tap low-cost emissions reduction.

Second, current regulations on carbon emissions are not sufficiently advanced to support the creation of an effective carbon trading market, as there are no uniform standards yet on assessment and approval, trading and settlement, and their effective supervision is not possible.

Third, the regional carbon trading mechanisms are immature. Domestic products are isolated from the international carbon trading market. It is still under discussion whether China's future national carbon trading market should choose centralised trading at a single exchange or decentralised trading at several exchanges.

Fourth, the liquidity of the carbon trading market needs further improvement. Pilot carbon trading markets experience the same problem of poor liquidity, with both low volume and low turnover.

To overcome these challenges, efforts need to be made to:

4.2.1 Improve the System of Laws and Regulations to Increase Regulatory Capacity

Relevant legislative procedures should be strictly implemented and the legal system for carbon trading improved. In particular, the system design for the national carbon trading market should be based on experience from pilot schemes. The rights and obligations of carbon emission allowance trading participants, trading methods and rules, dispute resolution mechanisms, type and extent of penalties for violations, and legal authorisation from jurisdictions should be clarified—based on the Measures for the Administration of National Carbon Emission Trading issued by the NDRC.

Moreover, a regulatory mechanism for a carbon emission allowance trading market should be established and a dedicated regulatory authority set up to supervise market operations and participants. The building of a comprehensive risk control system for the carbon trading market should be explored to prevent illegal operations and maintain a well-functioning carbon trading market. A legal supervisory system covering various parties—including the government, service providers, trading platforms and businesses—should be developed. While fulfilling its regulatory role, the government should avoid controlling and commanding the market directly. Rather, it should focus on macro-policy planning and supervision and leave the market to play its intended leading role.[6]

4.2.2 Coordinating the Cap and Quota Structure Correctly

The initial allocation of carbon trading quotas is a core function of a carbon trading market. The regional quota cap should be determined by a combination of factors, including greenhouse gas emissions, economic growth, industrial structure, energy structure, and inclusion of the businesses that fall within the emissions trading scheme. Some quotas should be reserved for auction, held in reserve to regulate the market if needed, and for major projects.

First, the cap should take into account economic growth, technological progress and emissions reduction targets, and follow the principles of "rigid cap, flexible structure, tight stock and optimal increment" in the functioning and evolution of the carbon trading market. Given economic volatility and the uncertainty of technological progress, an adjustment mechanism should be designed to deal with these uncertainties.

Second, industries vary in their emissions reduction costs, emissions reduction potential, competitiveness and carbon leakage effects, which should be taken into account when designing industry-specific emissions control coefficients.

Third, the design of a 3–5-year trading cycle, early determination of cap and adjustment measures, and a quota-relevant reserve system, are necessary to meet long-term market expectations and businesses' inter-temporal quota management needs and hence reduce compliance costs.

4.2.3 Establish a Unified Trading Platform and Pricing Mechanism

A unified carbon trading platform should include regulatory, trading and support systems. Given the inconsistencies in trading rules, trading procedures and price control in different carbon trading markets, it is necessary to improve the pricing mechanism (Fig. 23).

Auctioning initial carbon emission rights enables the fair and effective allocation of those rights. Government can set the minimum price and businesses can acquire emission allowances

---

[6]Liu Huiping, Song Yan, Challenges for Launching the National Carbon Emission Rights Trading Market and Solutions, in Economic Review, 2017, (01):pp.40–45.

Fig. 23 China's carbon trading market system

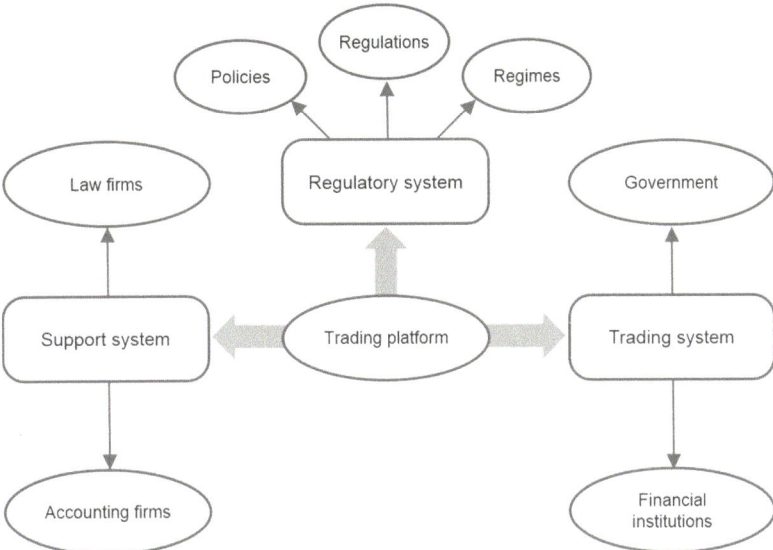

through bidding. Given the status quo of Chinese businesses, the creation of a pricing mechanism for auctioning initial emission rights will need to evolve from mostly free allowances to a growing number of auctioned allowances.

4.3 Create a Unified, Efficient and Flexible Electricity Market

4.3.1 The Goal of Electricity Market Reform

Electricity market reform is intended to:

(i) design a unified nationwide electricity market to allow the free flow of resources across the country, integrate provincial markets in a phased manner and eliminate the barriers to energy trading between them;

(ii) increase the types of energy traded between provinces and regions, and enlarge the number of participants in the market;

(iii) create a market mechanism that promotes inter-provincial and interregional grid

connections for the exchange of clean energy; and gradually introduce auxiliary service market mechanisms, like inter-provincial and interregional peak shaving, to increase clean power trading volumes across domestic borders;

(iv) expand the size of the electricity market through mergers or fusions to reduce the concentration ratio of power generators; and

(v) develop a variety of power products such as contracts for difference and medium- and long-term contracts and futures to mitigate fluctuations in electricity prices and ensure stable revenues for all market participants.

4.3.2 Establish an Efficient Pricing Mechanism

- *Deregulate prices.* Price deregulation should proceed step by step. It could start upstream and develop into downstream, beginning with input fuels and progressing through to generation and network access and eventually to retail.

- *Harmonise trading arrangements between different transmission systems.* Use price to determine flows between interconnected provincial and regional transmission systems, signalling which provinces or regions could benefit from new investment. Price can stimulate lower-cost power plants to respond more efficiently to demand.
- *Implement time-of-use pricing.* Time-of-use pricing gives consumers the flexibility to respond to variations in electricity prices and demand. Again, it can proceed gradually, starting with larger consumers, such as industrial facilities with flexible production schedules, and ending with smart household appliances.

4.3.3 Launch Market Trials Progressively

- *Carry out small-scale trials.* Carry out small-scale trials to competitively procure new transmission investments in non-network alternatives like microgrids or distributed energy resources, ancillary services and so on. Use competitive bidding or auctions for the trials. These could provide valuable proof-of-concept experience, as well as innovative and cost-effective solutions. To be successful, procurement must be open and transparent.
- *Progressively introduce market-oriented procurement.* If the competitive procurement trials are successful, they can be scaled up and wider market procurement progressively introduced, where appropriate, across each electricity system. Market procurement can reveal information about the relative cost and benefits of a range of power generation technologies.

4.3.4 Optimise the Power Management Structure

The power management structure includes an enhanced role for market procurement and a regulated transmission system operator (TSO) or independent system operator (ISO). International experience of the regulated TSO or ISO model is yet to reveal the best performer of the two, so it is more important to adopt a good quality model early than to choose between the options.

4.4 Reform and Improve New Energy Subsidy Policies

4.4.1 The Combination of Carbon Pricing and New Energy Subsidy Policies Can Deliver Better Results

The present study shows that the implementation of carbon pricing policies will have a significant impact on total energy consumption. As the initial carbon pricing level increases, total energy consumption will decrease. The effects of carbon pricing policies will be evident in the short term and become significant over time.

In contrast to carbon pricing policies, the introduction of renewable energy subsidies will increase total energy consumption. Subsidy policies will have a less significant impact on total energy consumption than carbon pricing policies. In the scenario that incorporates both carbon pricing and renewable energy subsidies, total energy consumption will be higher than in the pure carbon tax scenario but lower than in the subsidy policy scenario.

The scenarios reveal an important insight. If governments are concerned that the introduction of carbon pricing will have a significant negative impact on energy consumption, they can introduce non-fossil energy subsidy policies at the same time. This will increase non-fossil energy consumption and reduce fossil energy use, thereby ensuring a smooth transition in the overall level and patterns of energy consumption (Fig. 24).

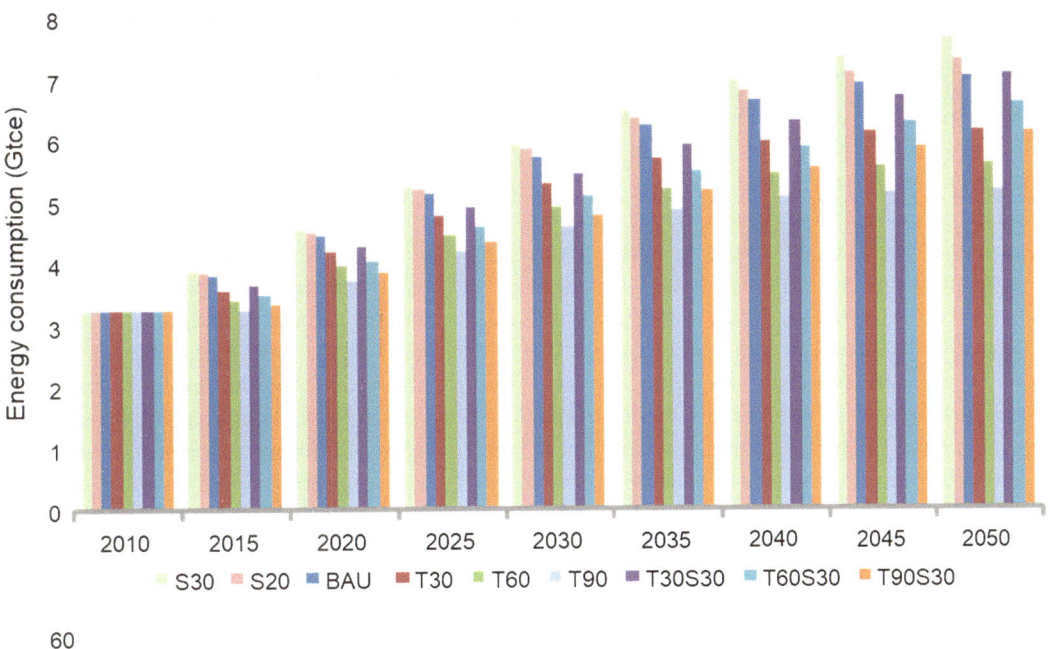

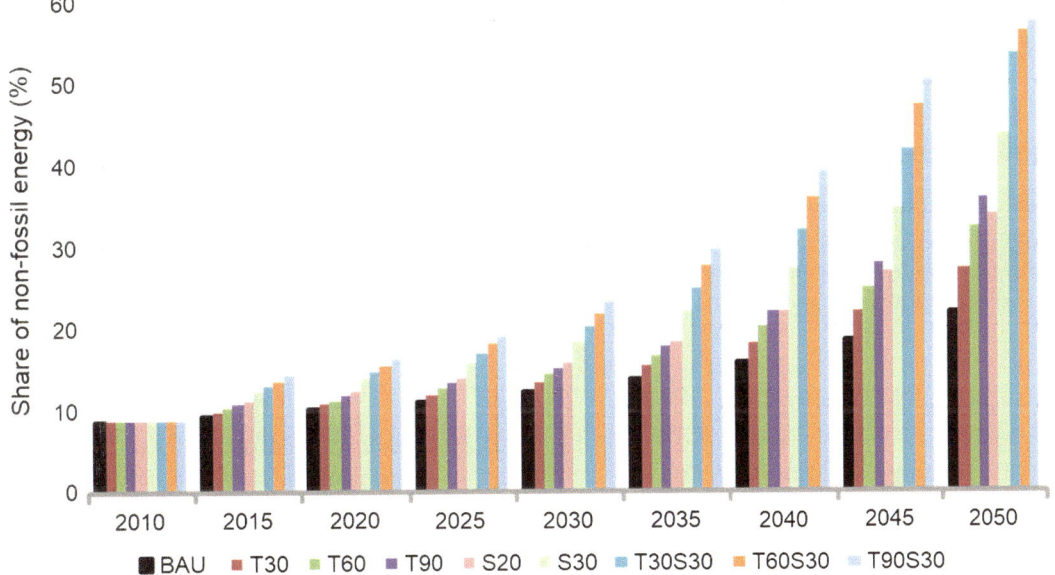

Fig. 24 Influence of carbon pricing and non-fossil energy subsidies on total energy consumption and the energy mix. *Note* BAU represents a carbon- and subsidy-free baseline scenario. T90, T60 and T30 represent three carbon price scenarios ($90/tCO$_2$e, $60/tCO$_2$e and $30/tCO$_2$e respectively). S30 and S20 represent two scenarios (non-fossil energy price subsidy of 20% and 30% respectively)

4.4.2 Continuously Improve and Implement Non-fossil Energy Subsidy Policies

With the introduction of carbon pricing and non-fossil energy subsidy policies, the proportion of non-fossil energy in the energy mix will increase. Non-fossil energy subsidies will have a greater impact on energy mix adjustment than carbon pricing. This may be due to the slower reduction in cost of non-fossil energy technologies in the absence of policy support. Even though carbon pricing policies reduce fossil energy consumption, if non-fossil energy subsidies are not introduced the widespread use of non-fossil energy would be difficult to achieve and the adjustment of the energy mix would be slow.

In the combined policy scenario, fossil energy consumption can be reduced and subsidy support given to non-fossil energy. Under the joint action of the two policy types, the effects of energy mix adjustments can be very significant. Model analysis results show that by combining carbon pricing at $30/tCO_2e$ with subsidies of 30%, the proportion of non-fossil energy in the energy mix can be increased to 50% by 2050. This indicates that a combined policy mechanism is both effective and necessary to promote future energy transformation and an energy revolution.

4.5 Build a New System for Oil and Gas Management and Operation

In line with the overall requirements of Several Opinions on Deepening Oil and Gas Sector Reform published by the Central Committee of the CPC and the State Council, China will promote market-oriented reform throughout the entire oil and gas value chain. The objective is to build a modern oil and gas market system that features fair competition, openness and orderliness, a strong legal foundation and effective regulation by around 2030.

4.5.1 Reform the Mining Rights Management System for Oil and Gas to Help Create a Mining Rights Market

Reform the franchise system for oil and gas exploration and production and deregulate the mining rights market in an orderly manner. While single-level mining rights management continues, a shift from allocation to tendering should be made. In high-risk areas, exploration rights may be freely transferred by means of invitation for bids, open bidding, exploration plans or investment, or joint venture/cooperation.

In proven or low-risk areas, mining rights should be transferred (sold) according to their valuation.

Set a reasonable cost of ownership for mining rights and resolve the relationship between resources and income. Different minimum exploration investment standards should be set for different regions, different minerals and different oil and gas resource areas at different stages of exploration. In particular, the minimum exploration investment thresholds for the first and second years should be raised, while those for the third year should be raised if needed.

Establish an exit compensation mechanism and speed up the transfer of mining rights and reserves to improve resource utilisation. To acquire the exploration rights after a licence holder exits, the prospective owner (the relevant national government authority or other applicant) should compensate the licence holder accordingly.

Reform the franchise for joint operation with overseas companies and improve domestic oil and gas capacity. China will deregulate the franchise of the three major state-owned oil and gas companies (CNPC, Sinopec and CNOOC) for cooperation with foreign companies in onshore exploration and allow other companies to make independent decisions consistent with government regulations. The franchise for offshore cooperation should also be steadily deregulated.

4.5.2 Accelerate Reform of the Natural Gas Pipeline Network and Build an Independent and Diversified Oil and Gas Infrastructure Market

Take steps to promote orderly pipeline independence. In the medium and long terms, pipeline transmission companies will take steps towards financial, legal and ownership independence to accelerate the separation of pipeline management, sales and storage. This will ensure open and fair access to third parties and improve pipeline utilisation.

Improve laws and regulations and strengthen supervision to ensure fair access. The Measures for the Regulation of Fair and Open Access to Oil and Gas Pipeline Networks and the Measures for the Administration of Natural Gas Infrastructure Deployment and Operation will be improved, and the Rules for the Implementation of Fair, Open Access to Natural Gas Infrastructure will be developed. These rules and regulations will define how infrastructure operators will maintain independent operation and provide various services, such as pipeline transmission, to all users in a fair and just manner. They will also define rules for pipeline access, information disclosure requirements and legal liability for breach of fairness and openness.

Progressively eliminate restrictions to diversify investment. Restrictions will be progressively eliminated to allow and encourage the investment of private capital in pipeline deployment and operation.

Strengthen the supervision and approval of transmission and distribution pricing to reduce transmission and distribution costs. Following the principle of "controlling power transmission and distribution, and deregulating power generation, sales and use" in the power industry, a pricing system should be established to verify independent gas transmission and distribution costs. It should be based on allowable cost, plus reasonable profit, and include constraints and incentives,

Create standards and plans to interconnect pipeline networks. The government should make plans regarding investment, construction, operation and pricing criteria for crude oil, natural gas and oil product pipelines.

4.5.3 Improve Oil and Gas Pricing Mechanisms and Progressively Deregulate Oil and Gas Pricing

Improve the oil product pricing mechanism. It is advisable to deregulate domestic oil product pricing step by step, allowing it to be determined by supply and demand.

Progressively deregulate natural gas pricing. In the near term, China will:

(i) carry out pilot programmes to deregulate the end-use retail price of natural gas;

(ii) introduce price incentive policies, such as interruptible supply pricing and peak and off-peak gas prices;

(iii) determine the relationship between residential and non-residential gas prices and gradually eliminate cross-subsidies between residential and industrial/commercial gas consumption; and

(iv) improve directional subsidy and relief mechanisms for the most vulnerable households and some non-profit industries.

In the medium to long terms, once upstream operators are diversified, third parties have fair access to infrastructure, and prices at the natural gas exchange reflect supply and demand, China will fully deregulate gas supply and sales pricing, leaving it to be determined by the market.

4.5.4 Standardise Government Administration and Create an Effective Regulatory System for Oil and Gas

China will:

(i) promote policymaker-regulator separation; set up independent, unified and specialised regulatory authorities; and improve the vertical regulatory system at national and provincial levels;

(ii) consider setting up an independent centralised energy supervision system within the National People's Congress, define regulatory responsibilities (mainly economic) and strengthen social regulation to ensure fair competition in natural monopoly segments such as network infrastructure (mostly pipelines); and

(iii) improve the efficiency of government regulation; give regulatory priority to processes and activities such as market access, trading behaviour, monopoly segments, tax payments, pricing, security and environmental protection; and use a combination of approaches including laws, administration, regulation and public engagement to enable collaborative supervision.

4.5.5 Improve China's Energy Emergency Response System and Increase Its Strategic Oil Reserves

China should accelerate its introduction of the laws and regulations on energy emergency response and strategic reserve management to improve coordination and command in response to energy emergencies and the ensuing risk to energy security. Energy emergency response tools should be strengthened. The use and storage of strategic oil reserves should be improved, as should supervision, post-event evaluation, and information, technology and capital management systems. Private capital's desire to enter the commercial reserve sector should be stimulated with a reasonable reserve price mechanism.

4.6 Deepen Reform of the Coal Industry

4.6.1 Restructure National Coal Authorities

Government should guide and serve the coal industry. By grasping the opportunities provided by "deepening CPC and national institutional restructuring", national coal authorities should be restructured to better implement coal industry and market strategies and reduce administrative intervention in individual entities.

4.6.2 Improve the Regulatory System

First, the coal industry regulatory system should be strengthened and include input from regulatory and government authorities, central and local governments, trade associations and social organisations.

Second, the regulatory regime should cover the various parts of the coal market. In addition to coal quality, the regulation of commercial transactions, logistics, capital and information flow should be considered.

Third, regulation should efficiently target key processes, such as quality standards, environmental protection, performance of contract, credit supervision, risk supervision, flow rate and flow direction control, and coal mine safety.

Fourth, supervision of coal mine safety needs to be aligned with reform of the state-owned capital management system. Information and smart technologies should be used to do this. The limits of corporate governance should be defined to reduce the burden of management on shareholders and investors.

4.6.3 Build a Market Information Network

First, coal circulation (from mine to end user) should become increasingly smart to provide insights into logistics and processing, deliver important information on the coal market, and enable key performance indicators to be measured.

Second, smart technologies should be used to create integrated coal distribution parks, and centralised real-time data acquisition points should be set up at different distribution nodes.

Third, administrative silos should be dismantled and the regulatory IT platform should be able to aggregate data from different authorities and local government departments.

Fourth, cooperation between coal industry associations and big data companies should be enabled to create an intelligent coal market information network and enhance coal distribution regulations.

4.6.4 Deepen the Reform of State-Owned Coal Companies

China should seize the policy opportunities provided by the structural reform of the coal industry and the state-owned capital management system to speed up the optimisation of capital stock and eliminate inefficiencies.

At the same time, efforts should be made to promote the two-way flow of capital, coordinate the exit of inefficient assets, facilitate the entry of new companies, optimise the industry's structure, improve the operation of coal companies, and accelerate innovation and transition through reform. A disposal platform for inefficient assets should be created to promote their repackaging and restructuring. Financial policies to reduce capacity should be used to their full extent. The ability to exit capital should be made easier and in conformity with market principles by leveraging the reforms to the state-owned capital investment management system and through mixed ownership.

4.7 Speed Up Reform of State-Owned Energy Companies

4.7.1 Accelerate Reform of State-Owned Energy Companies by Improving Investment Efficiency

The government's responsibilities as the investor of state-owned capital should be defined to achieve a shift from corporate management to capital management. Within the framework of reforming the state-owned capital management system, state-owned energy companies should set up state-owned capital investment and operations companies as soon as possible. They should also establish modern corporate systems and internal mechanisms adapted to the new management system and carry out mixed state-private ownership reforms.

4.7.2 Help State-Owned Energy Companies Become Stronger and More Competitive

To ensure national energy security and social stability, state-owned energy companies have a social responsibility to become stronger and better. They should make the most of the country's strong industrial base. Issues such as high interest on debt, weak operational ability and low profit margins should be resolved to strengthen their businesses and improve their global competitiveness.

4.7.3 Make Reform Breakthroughs by Tackling Key Issues

First, a mechanism for state-owned capital's exit from inefficient and ineffective energy assets should be created. State-owned energy companies could limit losses by exiting state-owned capital to become leaner and fitter. The central government, local government, businesses and the general public should work together to resolve the issues of "where will the laid-off workers go, how will the debt be paid, and how will the value of assets be managed".

Second, the pace of reform in the power and oil and gas industries should be accelerated. A dynamic and unified market system should be created by eliminating the impact of existing interests on conventional businesses. Energy market supervision and regulation should be rapidly introduced to improve energy supply capability.

Third, the investment management capability of state-owned energy companies should be improved, and the investment function of state-owned capital given full play. More people should be employed by state-owned energy companies, especially those with capital investment experience. To help state-owned energy companies become dynamic, their internal management systems should be optimised and incentives increased.

Fourth, mixed ownership reform should be deepened. Given the strategic significance of the energy sector, state-owned holdings will continue to exist in the long term. But investment openness and collaboration between state-owned and private capital should be realised at different levels and in different ways. Most importantly, the different capital governance and management mechanisms that apply to public and private companies respectively should be aligned.

4.8 Enhance China's Engagement in Global Energy Governance

4.8.1 Develop Strategies for, and Engage Deeply in, International Energy Governance Organisations

It is in China's interests to firmly adhere to multilateral norms and international rules, seize opportunities and address challenges together with the international community, and actively participate in global energy governance.

China should:

- establish comprehensive strategies and a roadmap for participation in global energy governance; assess the cost, benefits and risks of its participation in global energy governance in different scenarios; and define the different methods for participation and the support systems required;
- respond actively to the expectations of the international community on China's participation in global energy governance by adopting the International Energy Agency's oil emergency response mechanism, improving data quality and integrating international energy strategies;
- establish an internal conference and consultation mechanism to reach agreement on major international energy issues and be represented at international conferences to express China's point of view in a more powerful way;
- decide which international energy conferences and activities to participate in and identify the departments and officials to attend and the goals to be achieved;
- release transparent national energy policies on a regular basis. This means publishing a national energy white paper at regular intervals and publicising China's domestic energy policies and external energy relations; and
- set up a minimum supply and emergency system (like strategic oil reserves) for potential energy security problems (like wars).

4.8.2 Strengthen China's Capacity to Participate in Global Energy Governance

China's ability to modernise its energy governance is a prerequisite for its participation in global energy governance. China should:

- work actively on the international energy agenda, especially those issues that affect emerging economies and developing countries (soft power);
- develop the ability to apply international energy rules expertly and be familiar with the laws and regulations of international energy trade and financial investment;
- reform the domestic energy governance system and international energy cooperation mechanism and introduce a modern governance structure for energy diplomacy;
- develop a pool of expertise on international energy governance;
- improve the ability of energy companies to participate in international energy market activities (hard power); and
- form discussion platforms with the help of non-official and international organisations and enhance research in—and build capacity to participate in—global energy governance.

4.8.3 Create a Spirit of Openness and Accept that the International Energy Market Can Ensure Energy Security Under Normal Conditions

In the future, China and the world will be increasingly open towards each other in global energy governance. China should be more aware that the international energy market can ensure energy security under normal conditions and support efficient and competitive global markets.

There is global consensus that the international energy market can ensure energy security under normal conditions. Dependence on imported energy (mainly oil and gas) no longer means weak energy security. In the future of globalisation, thanks to mutual opening up, the global energy transition and China's energy revolution, the effects of an energy shortage may not be as critical as before and energy commodities procured from the international market may be cheaper than those produced in China.

Win-win energy development can be achieved by identifying areas for international cooperation, learning from international energy development experience, applying advanced technologies, and using foreign investment to make the domestic energy market more competitive. China's further integration into the global energy market will contribute to the energy security of the whole world, not only to China and East Asia.

Special Report 1: A Study of China's Energy Supply Revolution

Wang Xiaoming, Nie Shangyou and Wang Yunshi

1 Definition Implications of the Energy Revolution

1.1 Elements of the Energy Revolution

1.1.1 Defining the Energy Revolution

Energy supply goes through significant change as production and productivity evolve. When an energy revolution occurs it can drive social development and move society forward.

Scientists and research institutes define energy revolution differently. Here are four examples:

- "The purpose of a revolution in energy production is to accelerate the transition away from fossil fuels to clean and low-carbon energy, and to accelerate the growth of nuclear energy, renewables and natural gas in the primary energy consumption mix in China."[1]

- "The heart of the energy revolution is to improve efficiency and societal benefits. Restrictions limiting environmental impacts should be the red lines for energy development and investment to enable a clean and low-carbon energy economy to thrive."[2]

- "The energy revolution makes fundamental changes to the energy mix and energy industry. The excessively high cost of developing and using the predominant types of primary energy, and the mismatch between the benefits they provide and the negative impacts they make, drive innovation and the substitution of new energy carriers for old ones."[3]

- "Energy revolution is the evolution of the energy system by society. Unlike social and political revolutions, energy revolution is

DRC Team Lead of Special Report 1:
Wang Xiaoming from the Research Department of Industrial Economy, DRC of the State Council of China.

Shell Team Lead of Special Report 1:
Nie Shangyou and Professor Wang Yunshi from University of California, Davis, USA.

Contributors:
Guo Jiaofeng from DRC of the State Council of China; Liu Bing and Zhang Wenqiang from Shanghai AILNG Energy Technology Co., Ltd.; Shi Shude from the Research Department of Management and Consulting, State Grid Energy Research Institute Co., Ltd.; Yang Guang from the Energy Research Institute, National Development and Reform Commission; Li Jifeng from the State Information Center, Government of China; Zeng Ming, Wang Yuqing and Long Zhuhan from the North China Electric Power University, Beijing; and Feng Yujia from Tsinghua University, Beijing.

W. Xiaoming (✉)
Research Department of Industrial Economy, DRC of the State Council of China, Beijing, China

N. Shangyou · W. Yunshi
University of California, Davis, USA
e-mail: Shangyou.Nie@shell.com

W. Yunshi
e-mail: yunwang@ucdavis.edu

[1]Hua Ben, Urbanization—the Main Battlefield for Energy Revolution, in China and Foreign Energy, Issue 4, 2013.

[2]Zhou Dadi, Improvement in Efficiency and Benefits—the Core of Energy Revolution.

[3]Han Xingwang, A Discussion on the Elements of Energy Revolution, in Citizen and Law, Issue 12, 2014.

© The Author(s) 2020
Shell International B.V. and the Development Research Center (DRC) of the State Council of the People's Republic of China (eds.), *China's Energy Revolution in the Context of the Global Energy Transition*, Advances in Oil and Gas Exploration & Production, https://doi.org/10.1007/978-3-030-40154-2_2

evolutionary. When the existing energy system can no longer adapt to social and economic change, a new energy system is required to replace it. This triggers transformation in supply and demand, including significant change in energy resources, technologies, management and public perception. The current energy revolution is adopting a scientific approach to energy use and replacing the conventional energy system with one that is efficient, clean, low-carbon and smart to help drive society forward."[4]

At the 6th meeting of the Central Leading Group for Financial and Economic Affairs in June 2014, President Xi launched his Four Types of Energy Revolution and One Cooperation as the new energy strategy for China's energy security and development. This strategy requires a revolution in energy demand to curb unreasonable consumption; a revolution in energy supply to develop a diversified supply system of multiple sources; a revolution in energy technology to stimulate innovation and modernisation; a revolution in the energy system to fast-track development; and strengthening international cooperation to guarantee energy security. President Xi made clear that whichever definition is adopted, energy revolution is a combination of energy supply diversification, energy demand changes, energy technology advances, energy system innovations and far-reaching changes in the global energy landscape.

1.1.2 Definition and Elements of the Energy Supply Revolution

China's energy revolution is a significant attempt to accelerate the energy transition, guarantee national energy security, meet the nation's demand for energy, improve energy efficiency and combat global climate change. Some researchers have proposed[5] that this involves

optimising the energy mix, reducing the use of coal and coal's share in primary energy consumption, and developing a modern energy supply system that is efficient, clean, low-carbon and diversified. To understand the elements of the energy revolution, the following three factors need to be taken into consideration:

(1) The energy supply revolution will be a revolution in the true sense of the word
Revolution implies the transition from something old to something new, triggered by fundamental change. There are many factors that may cause an energy revolution, such as technological advances, changes in resource availability, environmental constraints or geopolitical disruptions. In the energy revolution, the old refers to the high-carbon and high-pollution energy supply structure dominated by coal and oil, while the new refers to a low-carbon, clean, diversified, stable and smart energy system. With regard to the objectives of energy revolutions, there are sharp differences between countries. China, for example, has an abundance of coal and a shortage of oil and gas. Despite recent technical advances, the economic feasibility of renewable energy still falls far behind fossil fuels. Therefore, in addition to focusing on renewables in its energy revolution, China also needs to attach great importance to the clean and efficient use of coal.

Energy revolution also implies fundamental change in energy supply. This does not happen overnight, but systematically and gradually in politics, technology, the economy and the environment. It is, therefore, important to weigh the pros and cons of each factor to understand its role in the energy revolution.

(2) The energy supply revolution will lead to major changes in the supply chain
Revolution can optimise the effectiveness and flexibility of supply and increase the share of high-quality energy in the mix to better meet demand and change the way energy is used. Optimisation helps remove obstacles to technological advances and encourages system innovations, driving the energy revolution forward. A revolution in supply is a cornerstone of the broader energy revolution. But it needs to be

[4]Ren Dongming, An Analysis of Energy Revolution, in China Energy News.
[5]Ying Guangwei, Guo Jiaofeng and Wu Xun, China's Energy Supply Revolution Urgently Requires Acceleration To Develop Gas as a Clear Energy, in China Economic Times, August 17, 2015.

supported by similar revolutions in consumption, technology, demand and the energy system itself.

(3) The aim of the energy supply revolution is to combat climate change, meet demand and guarantee energy security

The most important role of the energy revolution is to meet the demand for energy and improve people's quality of life, while combating global climate change. In the transition from high-carbon to low-carbon energy, the energy revolution focuses on the supply of cleaner conventional energy and the development of renewable energy. This will enable energy supply to match demand more effectively and adapt to change more flexibly.

The energy revolution aims for energy security. Close to 70% of China's energy comes from coal, which makes China the country suffering the most from coal-smoke pollution and the world's largest greenhouse gas emitter. As a result, China shoulders increasing pressure from the international community to reduce its emissions. As the scramble for oil and gas intensifies, the difficulty of accessing resources in international markets is mounting. China's oil imports are surging, around 80% of which come from the Middle East and North Africa. Such a high dependency on one region, in combination with a lack of strategic oil reserve capacity, weakens China's ability to respond to emergencies. A weak emergency response system increases the risk of long-term dependency on international oil resources. The energy revolution must therefore focus on China's specific conditions and aim to ensure energy security.

1.1.3 Implications of the Energy Revolution for China

At the 6th meeting of the Central Leading Group for Financial and Economic Affairs in 2014, President Xi defined the specific requirements of the energy revolution: to develop a diversified energy supply system; create a diversified domestic supply system to ensure energy security; vigorously promote clean and efficient coal use; focus on more non-coal energy supply sources to develop an energy supply system based on coal, oil and gas, nuclear power and renewable energy; strengthen the power transmission and distribution systems; and build energy storage facilities.

These requirements reflect China's actual conditions. China has many challenges to address, including excess capacity in fossil energy (mainly coal), insufficient supplies of renewable energy, intense environmental constraints, and severe defects in the energy supply system. Alleviation of excess capacity in the coal sector, optimisation of the energy supply system, and robust development of clean energy are, therefore, high on the energy revolution agenda.

1.2 Characteristics of the Energy Revolution

1.2.1 Key Characteristics

(1) Changes in primary energy

(1) Fossil fuels

Fossil fuels include coal, oil and gas. Coal mining and coal utilisation technologies are relatively mature, but due to high levels of pollution coal's share of the energy mix will decrease dramatically as the energy revolution moves forward. In the future, clean coal mining technologies will be widely deployed to enable scientific and efficient exploitation, and investments in safety will also increase. This will enable the cost-effective and safe exploitation of coal resources.

China has limited proven resources of conventional oil and gas. There is great potential for unconventional oil and gas in China. However, due to the limitations of technology and other factors, the utilisation rate of unconventional oil and gas remains low. The energy revolution, therefore, needs to shift to more unconventional oil and gas, deep-sea and deep-shale oil and gas resources. It needs to accelerate the exploration and production of shale gas and coalbed methane through technology intensification and innovation, and it needs to promote a low-cost, clean and green energy strategy.

(2) Non-fossil energy

Non-fossil energy includes renewables and nuclear power. The strategy of the energy revolution for renewables is as follows:

(1) hydropower—construct a fleet of hydropower plants, as planned, that comply with stringent environmental protection standards, as the ecologic protection redline should never be crossed;

(2) wind power—develop a wind power control and management system that fulfils the requirements of large-scale grid connection, improve equipment quality, establish a sector-specific mechanism to remove inefficiency, and construct support facilities like energy storage;

(3) solar power—build solar power plants in the north-west desert region of China and grid-connected rooftop solar power in east and central China to ensure the development of complementary solar energy technologies;

(4) bioenergy—increase R&D investment in non-grain biofuels to replace petroleum fuel with liquid biofuels at scale in several industries; and

(5) heating—deploy proven solar, biomass and geothermal energy technologies in heating applications. The pace of solar and wind energy development should match the speed of the energy transition to prevent excessive capacity build-up.

In nuclear, China should increase investment in nuclear power stations in the eastern coastal region, create Chinese nuclear power technology brands and become a global centre for nuclear power plant development and manufacturing.

(2) Changes in secondary energy

(1) Transition from primary to secondary energy

The energy revolution increases the share of electricity (from fossil fuels and renewables) in the energy mix. As electricity's share increases, and the proportion of non-fossil energy use rises, the power supply system becomes cleaner and more efficient. Power consumption in the service

industry and households will increase, while that in manufacturing gradually decreases.

(2) Combination of centralised and distributed energy systems

China has adopted a centralised approach to energy supply and demand, which will remain dominant in China's energy system for some time. Distributed energy systems provide less power per generating unit and much lower generating efficiency than centralised energy systems. However, their flexibility and environment-friendliness make them an important means of addressing pollution and energy shortage. As such, they are indispensable to the energy revolution and integral to China's future energy system. An optimal combination of centralised and distributed energy systems would guarantee a stable and efficient supply of energy and optimise China's energy mix.

1.2.2 Drivers

(1) Less pressure on global energy supply

Growth in global energy demand is slowing down. Influenced by the shale gas revolution in the USA, some Latin American countries are developing their domestic oil and gas resources and have made clear progress to reduce their dependency on imports from other states. This eases the pressure on global supply and demand and encourages countries to diversify their energy supply. As the share of renewable energy grows globally, diversified energy supply will become an irreversible trend.

(2) Opportunities in the energy revolution

At the opening ceremony of the 2018 Summer Davos Forum, Li Keqiang, Premier of China's State Council, said the Chinese economy is at a crucial stage of transition, from old to new growth drivers. It is faced with structural contradictions and downward pressure in several regions and industries. However, opportunities coexist with challenges. In the energy industry, China continues to implement its Internet + strategy and roll out the Energy Internet which, by integrating the energy industry with digitalisation, boosts innovation and new energy forms and revitalises energy supply. Natural gas, a clean energy carrier, is another of China's energy priorities. China's proven gas reserves are increasing rapidly, and exploration of natural gas hydrates has achieved major breakthroughs. As technology and innovation advance, natural gas has great potential to partly replace coal and oil.

(3) Environmental constraints are driving the energy revolution

As Wang Jinnan, Vice Director and Chief Engineer of the Chinese Academy for Environmental Planning points out, China tops the world in terms of emissions of almost all air pollutants, including carbon dioxide. Despite measures to control and prevent pollution, thick haze often hangs over large parts of China. Along with almost 200 other signatories, China ratified the United Nations Framework Convention on Climate Change (UNFCCC) at the 2015 Paris Climate Conference, which took effect in December 2016. Air pollution is inextricably linked to energy supply. In the face of unprecedented pressure, the need to protect the environment forces the energy revolution to accelerate the transition to clean, low-carbon energy.

1.3 Evaluating the Energy Revolution

As energy production transits from coal to low carbon, the latter's share of the energy mix becomes a key criterion for judging the success of the energy revolution. However, a revolution cannot be achieved overnight. Rather, it is an evolving process that takes place under certain conditions. It is influenced by specific factors and has significant impacts on energy supply in certain geographies. And it can create profound change over time. To decide whether the changes China is experiencing can be considered an energy supply revolution, we look at four countries that have undergone energy revolutions: the UK, France, Japan and Germany.

1.3.1 Energy Revolutions in Four Countries

The UK's energy revolution started with the discovery of North Sea oil in the 1960s. In 1973, coal made up 84% of the country's energy production and oil accounted for 44% of total energy consumption. As the production capacity of UK North Sea hydrocarbons expanded, the share of oil and gas in energy production and consumption rose rapidly, while that of coal decreased. By 1983, the UK was one of the world's top 10 oil producing countries and an important oil and gas exporter across the globe. At a time when overall energy demand was stable, the share of gas in the energy mix increased rapidly in the UK, becoming a major alternative to coal.

Japan's energy revolution can be clearly divided into three stages based on energy mix share. Stage 1 is changes in the supply of primary energy to the power industry, reflected by a higher share for oil and a smaller share for coal. In 1973, the first oil crisis spurred a transition from oil to gas and nuclear power, which started stage 2 of Japan's energy revolution. Use of gas in the energy mix grew from 1.6 to 19.2%, and that of nuclear power from 0.6 to 11.8%. Thanks to reduced costs and improvements in equipment efficiency, the use of coal rose from 16.9% to 22.6%. Following the Fukushima nuclear accident in 2011, Japan initiated stage 3 of its energy revolution by phasing out nuclear power.

In France, oil and coal made up a high share of the energy mix before the energy transition. As a result of the first oil shock, Prime Minister Pierre Messmer announced a large nuclear power development programme (the Messmer Plan) in 1974, which included 13 1,000-megawatt nuclear power plants. This not only completely changed France's electricity sector, it made France a major force in nuclear power and related technologies. Currently, nuclear makes up 75% of total power output in France, which is a substantial contribution to France's energy self-sufficiency.

Germany passed the Renewable Energy Sources Act in 2000, pioneering the transition from fossil fuels to renewable energy. Coal and nuclear power were the major sources of energy in Germany at that time. According to Stephan Kohler, former head of the German Energy Agency and director of the Center of Sino-German Renewable Energy Partnership, the goal of Germany's energy transition is mainly to phase out nuclear power and fossil fuels by 2050. After the Fukushima disaster in 2011, the German government decided to shut down nuclear reactors permanently and make wind and solar power the basis of Germany's future low-carbon energy system. As Stephan Kohler pointed out in 2000, 5% of Germany's then energy consumption came from renewables, especially hydropower. By the end of 2016, this had risen to 33%, with solar power especially registering explosive growth.

1.3.2 Timelines of the Energy Revolutions

Starting with the discovery of North Sea oil in 1968, the UK's natural gas revolution can be divided into two stages. In stage 1 from 1968 to 1990, the supply of natural gas grew dramatically. It was used mainly by households and industry, with a very small amount used for power generation. In 1991, a large gas field with sufficient reserves for 15 years was discovered along the North Sea coast, marking the outset of stage 2 of the UK's energy revolution. The UK government increased the use of natural gas in power generation, which far exceeded that used by industry.

Japan's energy revolution began in 1960. It is now going through its third energy transition. The first transition, from 1960–73, substituted oil for coal. As a result of the first oil crisis, Japan underwent a second transition, from oil to gas and nuclear power, from 1973–2011, which is the focus of this report. The third transition took place after the Fukushima disaster in 2011, when Japan started to phase out nuclear power.

The first oil crisis also triggered France's transition to nuclear power from 1974. In June 2017, Nicolas Hulot, French energy and environment minister, announced that the government planned to shut down several nuclear reactors. The move was intended to decrease the share of nuclear power in total power output from 75% to 50%, but the timeline remained unclear.

In 2000, Germany embarked on a transition to renewable energy. In September 2010, the minister responsible for the Federal Ministry for Economic Affairs and Energy announced Germany's medium- and long-term energy policies and identified the energy transition targets for 2050.

1.3.3 Pathway Options for Energy Revolution

The UK discovered North Sea oil when its domestic energy resources were scarce. The first oil crisis accelerated drilling and exploitation in the North Sea and escalated the use of natural gas. The UK's energy revolution was therefore driven by resource shock.

During the first oil crisis, oil prices soared and the cost of oil-fired power generation in France and Japan spiralled. High oil consumption and a lack of domestic resources forced the two countries to carry out an energy revolution. France has few oil, coal and gas resources, but relatively abundant uranium ore. France decided to transition to nuclear power, while Japan chose to replace oil with gas and nuclear power.

Germany spearheaded the transition from high-carbon to low-carbon energy to reduce air pollution from fossil fuels and to strengthen energy security. The transition to renewable energy is now a common trend in energy revolutions.

1.4 Pathways to Energy Revolution

1.4.1 Economic Development, Energy Security and Environmental Protection

According to the above analysis of energy revolutions in four countries, the UK, France and Japan sought an energy transition before and after the first oil crisis due to a lack of domestic energy resources. They needed to limit the potential impacts of an energy supply shortage on their economies and energy security. Germany's energy supply revolution occurred relatively late. In addition to economic development and energy security, Germany also took environmental pollution from fossil fuels into consideration, and eventually decided to point its energy revolution towards renewable energy.

All four energy revolutions targeted economic development, energy security and environmental protection. These same three factors apply to China's energy revolution as well.

First, economic development is a permanent goal. China has many challenges that need to be urgently addressed, such as overpopulation, large income inequality and regional imbalances in economic development. The energy revolution can satisfy people's energy demands and create high-tech and low-energy-consumption industries, bringing new employment opportunities. More jobs will reduce income inequality and regional development imbalances, helping to deliver the government's goal of economic prosperity.

Second, the energy revolution can reduce China's dependency on other countries for energy and lower potential supply risks, thus improving energy security.

Finally, the severe haze and air pollution challenges facing China are closely related to its underdeveloped energy system. In addition, pressure from the international community on greenhouse gas emissions makes environmental protection another goal of China's energy revolution.

In recent years, as China's economic development entered the new normal of lower growth rates, the energy industry also entered a new phase. China's energy system is shifting from growth in quantity to a phase consistent with the new era of socialism with Chinese characteristics and China's new normal economy. As new businesses start up, and energy technology innovation and deployment accelerate, the three factors of economic development, energy security and environmental protection become increasingly interdependent.

1.4.2 Scenarios for China's Energy Revolution

The scenarios for China's energy revolution are based on the following four factors.

First, in alignment with the government's Chinese Dream strategy, China's socioeconomic development scenario forecasts socioeconomic development and energy service demand for 2020, 2030 and 2050.

Second, in accordance with the Energy Supply and Consumption Revolution Strategy (2016–30) to forecast end-use energy demand and to analyse the possible and recommended pathway.

Third, we have conducted research on the impact on the energy supply system in three scenarios (High, Medium and Low) in which electricity and gas replace scattered coal and the Recommended and Limit scenarios in which electric vehicles replace internal combustion engines.

Fourth, we forecast China's future primary energy supply in the Recommended scenario for its energy revolution pathway.

1.4.3 Outlook for China's Energy Revolution and Scenario Analysis

China's future population, floor space growth and trends in major industrial products are analysed to forecast China's socioeconomic development and energy service demand in 2050. Energy consumption in agriculture, industry and buildings, transport, the service industry and households, is predicted in order to estimate future total end-use energy demand. The scenarios for replacing scattered coal and internal combustion engines are analysed to estimate electricity and natural gas consumption and total energy demand for vehicles in 2050.

A review of China's socioeconomic development and energy service demand reveals that China's population will peak by 2030. As the population ages, China's future workforce will decline in absolute terms. China's floor space is forecast to peak at about 92 billion cubic metres by 2040. The rapid expansion of the real-estate industry will come to an end, and its ability to stimulate economic growth will weaken. China's automobile industry maintains strong momentum, but growth is expected to slow after 2030 due to gradual market saturation. Iron and steel output will fall in line with declining annual average floor space, and the steel recycling rate will gradually improve. The nonferrous metal industry will see continuous growth in energy consumption to 2030, and new building materials will gradually replace old ones. By 2030, China will achieve the government's goal of "common prosperity" and avoid the middle-income trap of developing countries. In 2030, China's GDP will exceed that of the USA, making it the world's largest economy. In 2050, China's economic growth will shift from production-driven to consumption-driven. As industrial restructuring accelerates, China's GDP will reach and exceed the level of moderately developed countries.

In agriculture, machinery will be powered by electricity, especially by renewables and biofuels, instead of by fuel oil. Use of scattered coal will start to disappear. Energy use in industry and buildings is expected to peak in 2025–30, and then steadily decline. In iron and steel, as the mainstream technologies gradually shift to a more balanced use of long-term and short-term objectives, the use of coal will fall sharply and that of natural gas, electricity and heat increase significantly. In the transport sector, petrol, diesel and kerosene remain dominant, but the rate of replacing them with electricity, natural gas and biofuels will increase significantly. Energy consumption in transport will continue to grow to 2030, then slowly decline. Energy demand in the service industry will increase continuously throughout the period to 2050 and use of coal will decrease rapidly as the service industry modernises and replaces scattered coal with electricity and gas. Household electricity demand will remain steady.

In summary, China's end-use energy demand will stay at peak level between 2030–50. Specifically, energy demand in agriculture, industry, building construction and transport will peak in 2020, 2025–30 and 2030–40 respectively. In the service and household sectors, peak energy demand will not be reached until 2050.

Three rate of change scenarios (Fast, Medium and Slow) are also analysed for substituting electricity and gas for scattered coal (SEGFSC). For the pathway for the electric vehicle sector, a recommended scenario and an extreme scenario (featuring accelerated development) are analysed,

leading to energy demand forecasts for petrol, diesel and electricity.

In the Recommended scenario, the share of electricity in end-use energy demand rises steadily. In the SEGFSC and electric vehicle scenarios total energy consumption will decline, although not significantly, as electricity and gas are more efficient than coal and oil. This will increase demand for electricity and gas, but to a manageable extent. However, a rapid increase in the number of electric vehicles will boost demand for electricity. The power system will therefore require sufficient backup capacity or better demand-side management to meet this uncertainty.

In the Recommended scenario, electricity demand is divided into four segments—peak load and ancillary services, intermediate load, base load and distributed load. Clean power's share of energy supply will gradually increase, wind and solar will rise rapidly, and nuclear power steadily. Gas-fired power generation's share will remain around 10% in 2050, while hydropower's will gradually decrease despite its growing generating capacity.

In the Recommended scenario, China's total primary energy demand will continue to grow. Coal and oil will peak then decline. The share of non-coal in energy supply is expected to increase to 73% by 2050, whereas oil will rise then gradually fall as the scale of replacing the internal combustion engine with electric vehicles rapidly expands. Meanwhile, clean energy will gradually play a critical role in meeting energy demand. In addition, the rising rate of electrification will increase the share of power generation in energy supply.

1.5 Impacts of the Energy Revolution

The Recommended scenario analyses the impacts of the energy revolution on the following four aspects: energy supply, industry and capacity conversion, investment, and employment.

(1) The energy revolution changes the energy supply system by transforming production and innovating new energy supply technologies to

safeguard national energy security. The shift in production from centralised generation to an optimal combination of centralised and distributed energy improves efficiency and encourages the development and use of clean energy. Technology advances are reflected in the development and deployment of the Energy Internet (Internet +). Internet + is a combination of complementary energy sources in an optimised mix of centralised grids and distributed energy networks that is low carbon and intelligent. With support from Energy Internet technologies, demand response becomes a crucial forcing mechanism to accelerate supply-side change and improve energy management. In the Recommended scenario, China's energy imports will continue to rise in the short term, but as renewable energy and nuclear power advance, China's dependency on energy imports will fall.

(2) The energy revolution drives change in the energy system and guides development of the energy industry. As coal will continue to dominate China's short-term energy mix, low-carbon energy and clean coal should be the focus in the near future. The oil and gas sector will concentrate on innovating greener technologies and accelerating the exploitation of unconventional resources. China will make vigorous efforts to drive the development of clean energy. It will invest more in technology innovation to lower costs and improve energy efficiency. Power conversion will be cost-efficient, clean and green. It will replace fossil fuels with clean energy and increase the amount of electricity in end-use consumption.

(3) The energy revolution drives the energy industry to invest in clean energy and the cleaner use of fossil fuels. To deliver the energy strategy of a diversified supply system, the participation of various types of investor is needed.

(4) The energy revolution will make the structural imbalances in the labour market more severe. There will be more job opportunities for highly skilled people, but rising unemployment for those in the conventional energy sector. This applies to those regions and provinces that rely on conventional energy, less so on those developing new energy sectors. It may force some

provinces to carry out an energy transition and reduce their dependence on conventional energy. So doing would spur their economy, ease unemployment and make for healthier and more stable development.

2 Precedents and Prospects of International Energy Revolutions

2.1 Energy Companies in Transition —Responses to Future Trends

2.1.1 Introduction

Strategies and organisational structures within the oil and gas industry are the result of historical market conditions that are likely to change over the coming decade. Oil and gas markets have traditionally been characterised by rising demand and prices. This has led companies to focus on megaprojects that require intense coordination. As a result, pyramid organisational structures with multiple levels of oversight have become the industry norm. However, the oil and gas industry faces large structural shifts that can affect its organisational and strategic structures in the future.

Specifically, two trends are likely to trigger change. First, structurally lower oil and gas prices are becoming increasingly likely. This is due to a combination of technology and policy that increases supply and reduces demand. Supply is increasing as horizontal drilling and hydraulic fracturing techniques are diffused and climate policies—together with energy efficiency improvements—curb demand. Second, technology disruptions in energy markets are becoming more likely. Digitalisation and innovations in decarbonisation will increasingly change how energy is produced and consumed, which will ultimately lead to changes in existing markets.

Oil and gas companies have already begun to adjust their strategies and organisational structures. The industry is currently characterised by two strategic themes. First, oil and gas companies, including Equinor, Total and Shell, have begun to invest in renewable energy technologies, such as solar and wind, to diversify their portfolios. Second, companies are shifting their hydrocarbon investments away from conventional mega-ventures towards smaller more flexible projects, such as shale wells, which can be scaled up or down in response to market changes. These strategic changes have been accompanied by organisational adjustments. New, smaller investments are more reliant on trial and error, agile local teams and rapid coordination. Recognising this, oil and gas companies like BP and ExxonMobil, are adopting flatter organisational structures that allow for greater local autonomy.

However, oil and gas companies are at the start of a long journey. It is, therefore, useful to study responses in comparable sectors that have faced similar structural shifts for longer. The postal and power sectors are good examples as they, like oil and gas, are heavily regulated, capital-intensive industries dominated by large state-owned enterprises. Furthermore, they have faced structural shifts similar to those expected in oil and gas. The postal industry has faced falling letter volumes in the past decade and companies in the sector have already adapted their organisational structures and strategies. Likewise, the electric utility sector can be seen as a frontrunner for oil and gas; it has long been challenged by large technology disruptions in the form of increasingly price competitive renewables. These parallels make postal and power companies' useful cases for exploring future strategic and organisational changes in the oil and gas industry.

The postal and power cases show that company-specific motives and context can lead to different but rational strategic responses to the same structural shift. Some companies, such as United States Postal Service (USPS) and RWE (circa 2004), changed very little in response to structural shifts, while others, such as Deutsche Post and DONG Energy (now Ørsted), transformed or diversified their companies completely. Little strategic change was observed in situations where owners valued security of the existing service and nearer term profits. In the case of USPS and RWE in the early 2000s, their strategy was to harvest the short-term value of legacy assets. In companies, such as the UK's Royal Mail—where stakeholders valued longer-term profitability but were constrained by regulation, conflicting interests or limited capital—business strategy focused on streamlining and redirecting resources to new capabilities. In cases where owners sought to expand the business, had relevant capabilities and sufficient capital, the strategic shift was more aggressive than for those acting later or facing resistance. As an example, both Deutsche Post and DONG Energy moved early and used rents from legacy activities to finance their strategic transformation to international logistics and offshore wind respectively. In contrast, late movers such as RWE and Innogy SE suffered from capital constraints, which slowed their strategic transition.

Organisational change needs to align with the adopted strategies and business models. In all our cases, companies started out with relatively hierarchical, top-down organisational structures. Where the strategic focus remained on the traditional business model, as with USPS and RWE, organisational change was limited to making the prevailing structure more effective. Where new business growth was the main priority, all companies first ensured some separation between new and old—this gave the new business sufficient autonomy to grow. Where new businesses had similar underlying characteristics to the traditional business, as with Royal Mail and DONG Energy, organisational change focused on reorienting the existing structure and flattening the hierarchy, but not on a radical shift to a horizontal structure. In contrast, radically new product or service models involved far-reaching organisational change. As Deutsche Post transformed from a German mail delivery firm to an international logistics company it also changed its organisational model. Notably, it shifted focus from production to consumers and organised itself around a series of relatively autonomous divisions supported by cross-cutting service divisions. Innogy's split from RWE in 2016 resulted in a more radical turn towards business unit autonomy focused on customers, allowing a more agile shift in investment across units based on results.

Motive and context will likely determine how Chinese oil and gas companies respond to lower prices and new technology disruptions. Drawing on the postal and power cases, four archetypes of potential strategy response arise:

- *Supply security focused harvesters*: These companies ensure supply security or near-term employment, and harvest value from legacy assets during the transition period. However, as was the case with USPS, they face the risk of shrinking considerably as the market changes around them.
- *Constrained niche growers*: These companies have aspirations for growth, but regulation or capital constraints prevent them from moving beyond conventional oil and gas business areas. They focus on streamlining their core functions and growing within niche markets such as chemicals.
- *Dividing conquerors*: Acknowledging the potential for misaligned internal incentives, these companies split into two with one focused on harvesting value from legacy hydrocarbon assets to deliver near-term supply security, dividends or employment, and the other focused on delivering long-term growth by developing new technologies or wider energy sector diversification.
- *Transformative diversifiers*: In contrast to the above, these companies move quickly to a new energy service model, catalysing change and gradually selling off their least profitable legacy assets to finance the transition. They have the potential to deliver long-term growth but are riskier; supply security, employment and profitability outcomes are uncertain.

The different archetypes for oil and gas company strategy will be accompanied by several organisational considerations. Supply security focused harvesters will likely maintain a top-down approach with central control of big decisions. In contrast, constrained growers will have to enable agile decentralised units to efficiently capture niche markets. Their central functions will be focused on portfolio optimisation, as well as asset management and organisational efficiency. Companies that face significant internal cannibalisation between old and new business units will likely become dividing conquerors. Like Royal Mail and DONG Energy, these companies will strategically and operationally split, but could be managed as a conglomerate that carries over existing organisational strengths. Finally, transformative diversifiers will likely adopt more horizontal organisations structured around energy services rather than production. Like Deutsche Post, their organisations will likely be flatter with a series of relatively autonomous units supported by service divisions that leverage synergies across the group.

Motive and context are not set in stone but controlled by stakeholders, and the Chinese government can therefore shape strategic and organisational outcomes in its oil and gas industry. As the owner of national oil companies, the Chinese government can shift the motive driving strategic and organisational change. Supply security requirements could be lifted by reducing fixed reserve targets; incentives to maintain unproductive workers could be minimised by reducing local political oversight; and near-term profitability motives could be eased by adopting longer-term key performance indicators for managers. Furthermore, as a regulator and key stakeholder, the Chinese government influences the context in which oil and gas companies will respond to structurally lower prices and technology disruptions. For example, capacity development could be supported through public-private R&D schemes, regulatory constraints could be reduced by lifting natural monopolies, and capital limitations could be minimised by providing policy clarity (this would allow rents from legacy assets to be channelled sooner towards financing new investments). An important lesson from both the postal and power sectors is that the government is key in setting the conditions that determine strategic and organisational change. Policymakers who recognise this will be able to directly or indirectly shape the future of the Chinese oil and gas industry.

2.1.2 Future Trends in the Oil and Gas Industry

The oil and gas sector faces new trends that could induce change in business strategy and companies' organisational structure. Trends have always arisen to create new challenges and opportunities in the oil and gas sector, as the discovery of new resources, technologies and political shocks have changed the fundamental characteristics of the market. For example, the emergence of powerful state-owned oil companies in the 1960s and 1970s forced many international oil companies to transform their business models to focus on megaprojects in hard-to-reach places. The future appears to hold even more substantial changes due to advances in technology and decarbonisation, which have already begun to affect energy markets and will become more prevalent over time. Oil and gas companies are already considering changing their strategies and organisational models in response to these trends and will need to continue to do so in the future.

This report focuses on the two main trends facing the oil and gas sector: (1) structurally lower oil and gas prices, and (2) an increasing set of disruptive technologies. The trend of structurally lower prices reduces margins and makes high-cost supply uncompetitive. It is already occurring, as new shale resources continue to increase supply, and is likely to strengthen as decarbonisation policies reduce fossil fuel use. Several technology disruptions are also impacting the sector, including those driven by digitalisation and increasing low-carbon R&D flows. These can rapidly shift value within the energy sector, which will require oil and gas companies to be more versatile if they are to capitalise on opportunities in areas of new value and avoid risks in areas of declining value. This report examines the broad impact of these shifts in business strategies, and the implications for future organisational change within oil and gas companies (Fig. 1).

Companies will need to understand the implications of these trends and how to respond best to them in order to thrive. It is possible to draw parallels between the trends facing the oil and gas industry today and the trends that other sectors have already experienced and responded to. This report aims to use the learnings from cross-sectoral responses to better understand the possible set of strategic and organisational

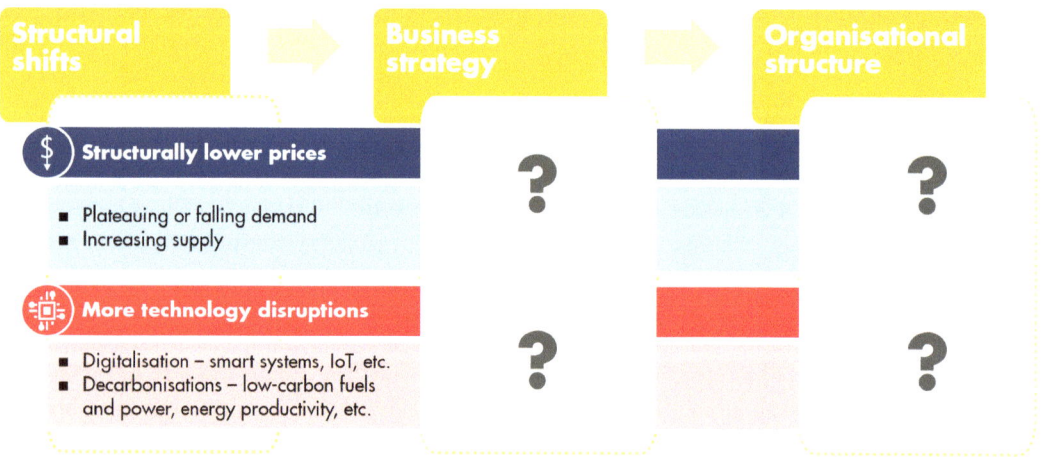

Fig. 1 Structural shifts will alter the business strategies and organisational structure of oil and gas companies

responses of oil and gas companies to structurally lower prices and increased technology disruption, as well as the factors that will affect those responses.

(1) Current strategies

Current strategies and organisational structures within the oil and gas industry have been shaped by historical market conditions. The main strategy employed in recent years has been the development and management of megaprojects with large volumes of recoverable resources: in 2005, 60% of oil production was from giant fields of more than 0.5 billion barrels of recoverable resources. This focus on quantity has been driven by the historical trends of growing demand and generally increasing, if fluctuating, prices, as seen in Fig. 2. Given these trends, replacement ratios have been of key importance for oil and gas companies, which have focused on accumulating reserves and been less concerned about the cost of extracting those reserves.

Effectively deploying large, complex projects requires extensive coordination that has encouraged pyramid organisational structures with multiple layers of oversight. These megaprojects are technologically demanding and usually involve complex multiparty relationships, as well as high commercial and environmental risk. Given these features, a pyramid structure that allows for strong oversight is required to ensure all aspects of the project are under control and that coordination between them is achieved.

(2) Structurally lower prices

Rising supply from unconventional sources and plateauing demand make it likely that prices will remain lower for longer. Recent advances in drilling technology, such as hydraulic fracturing and horizontal drilling, have made new oil and gas resources available. This has shifted the global supply curve outwards, putting downward pressure on prices, and is likely to continue as these technologies are deployed globally.

The economic logic for this is described in Fig. 3, where a shift out of the supply curve and a contraction of the demand curve can lead to significantly reduced producer surplus, illustrated by the difference in size between the original red area of producer surplus and the new blue area. This is accentuated by the steepness of the global oil and gas supply curve at the higher levels of quantity—this means that a small fall in demand can cause the price and surplus to fall rapidly.

There have already been sharp falls in oil and gas prices in recent years, demonstrating the potential for structurally lower prices. The trend is already in place. Oil prices have been consistently around $50 per barrel since 2015, as US oil shale resources have contributed to a supply glut

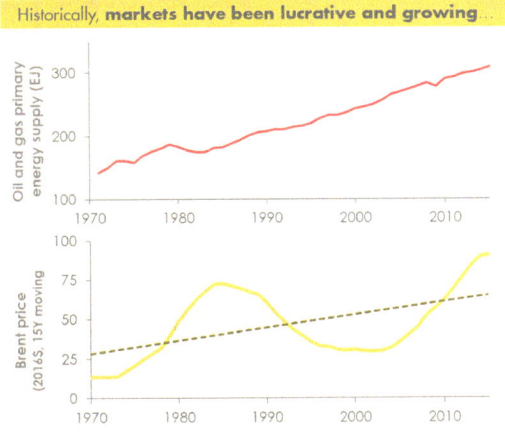

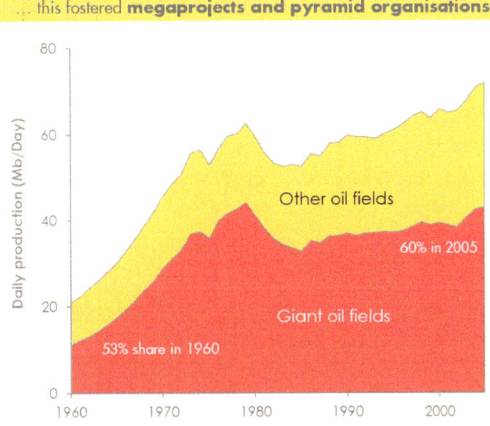

Fig. 2 Megaprojects, and the pyramid organisational structure that facilitates them, have been motivated by historical trends. *Note* Giant fields are defined as having

more than 0.5 billion barrels of recoverable resources. *Source* International Energy Agency, BP, Robelius (2007)

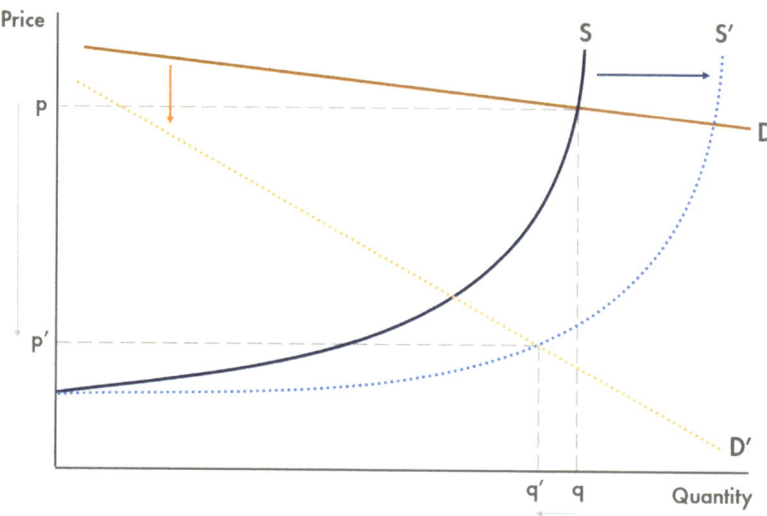

Fig. 3 The trends of the oil and gas industry will likely result in significantly tighter margins for producers. *Source* Vivid Economics

Fig. 4 The fall of oil prices in late 2014 coincided with sharp drops in the share price of major oil companies. *Source* BP, Google Finance

that has driven down prices to pre-2005 levels. The overall impact of this on oil and gas companies can be seen in their share price during and after the final quarter of 2014 in Fig. 4. The average share price of five major oil and gas companies fell by more than 30% from 2014 to the trough in 2016 and has yet to recover.

New, relatively low-cost supply sources are becoming increasingly available as technology improves, driving production above predicted levels. New technologies have reversed persistent declines in production and have exceeded historical expectations, as illustrated in Fig. 5. The U.S. Energy Information Administration (EIA) forecasts for oil production in both the 2002 and 2012 Annual Energy Outlook did not fully anticipate the impact that new fracking methods would have on increasing tight oil supply, as conventional oil production has continued to diminish. The result has been a near doubling of oil production from its trough in 2008 to its peak in 2015.

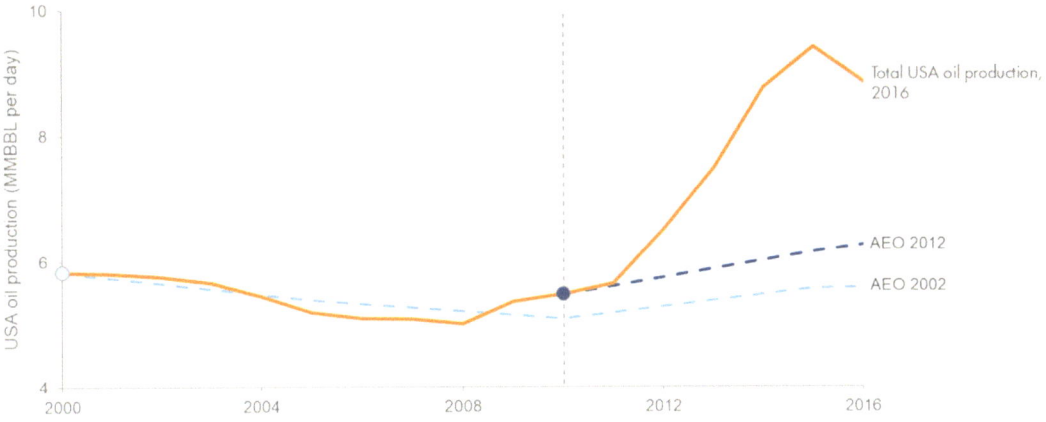

Fig. 5 New tight oil sources have pushed total US oil production to heights that exceeded forecasts. *Note* Total US oil production includes conventional and tight oil. *Source* EIA (2017)

The low-carbon transition will reduce demand for fossil fuels in favour of low-carbon alternatives over time. Decarbonisation has been an increasingly important priority in energy in recent years and now has widespread support, as demonstrated by the signing of the Paris Agreement in 2015. As a result, policies are in place globally to phase out fossil fuels and promote low-carbon energy sources. Figure 6 shows potential pathways for primary oil and gas demand in two EIA scenarios: a reference scenario that reflects the world's announced decarbonisation policies and a 2-degree scenario (2DS), which assumes additional policies to limit global warming to 2°C. In the reference scenario, demand for oil and gas rises, but if additional policies are put in place to limit global warming to 2°C as intended, then demand for both oil and gas is expected to fall sharply in the future. This reduction in oil and gas demand due to decarbonisation will likely reduce prices, and so contribute to structurally lower prices.

In a future of structurally lower prices, the decision of whether to divest or diversify into new areas will gain greater importance. Oil and gas companies face the risk of lower margins in their main areas of operation. As a result, there may be value in pursuing strategies such as divesting or diversifying to attempt to maintain profits in the future. The potential strategy options and the factors that induce different types of responses are discussed in Sect. 2.1.3.

(3) Disruptive technologies
Disruptive technologies are shifting the sources of value in energy markets, driven by digitalisation and decarbonisation. Digitalisation is the automatic collection of large quantities of data and the application of computing power to the data to enable better decision-making. Its use in other (non-energy) sectors has driven rapid advances. Collecting big data with remote sensors is already prevalent in the oil and gas industry, and there is enormous potential for digitalisation in other areas of the energy system through smart technology that enables dynamic, autonomous energy systems. Decarbonisation policies have increased R&D spending (01.2.7) and raised market expectations about the long-term value of new innovation across various parts of the energy system. This has resulted in major advances, including those related to power generation (like wind and solar), oil and gas exploitation (floating liquefied natural gas and advanced seismic analysis for oil and gas exploration), and energy demand (electric vehicles and smart homes with demand-side response).

This has created opportunities and risks for the oil and gas sector. The impact of these technologies, whether marginal or revolutionary, will have repercussions on oil and gas

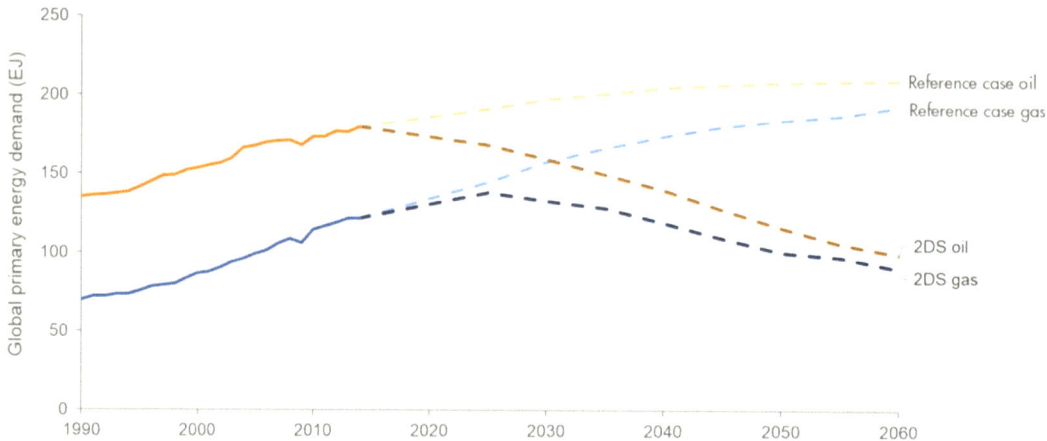

Fig. 6 Primary demand for oil and gas falls sharply under the IEA's 2-degree scenario (2DS), presenting a challenge to the oil and gas sector. *Source* IEA Energy Technology Perspectives 2017

Fig. 7 R&D is shifting towards new areas, increasing the potential for disruptive technologies to emerge. *Source* OECD

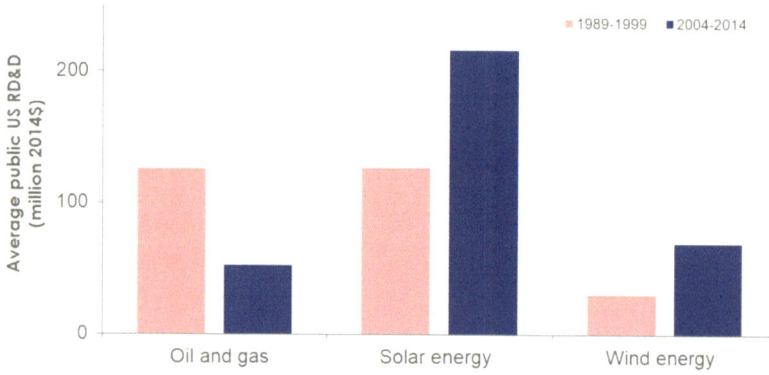

companies. For example, sensors on drills are collecting masses of high-quality real-time geological data that allow for more cost-effective extraction. Smart grids can better coordinate decentralised variable generation and minimise curtailment rates—this makes renewables more productive and reduces the need for fossil fuel peaking plants to balance the energy system. And reductions in carbon capture and storage (CCS) costs could make the large-scale use of gas in power and industry low carbon (Fig. 7).

(4) Initial responses

Oil and gas companies have started to adjust their strategies, shifting towards smaller-sized projects and investing in renewables. Structurally lower prices and technology disruption have already begun to appear and affect the oil and gas

industry, driving initial responses. Some major companies have begun to integrate renewables into their portfolio to gain presence in a market that is rapidly growing, thanks to widespread policy incentives and improving renewable technologies—both Equinor and Total have made recent renewable energy investments. There has also been a shift away from the historical focus on megaprojects, as shown in Fig. 8. Investment in smaller fields has been consistently higher since 2009 and is expected to continue into the near future as flexible shale projects and lower extraction costs become increasingly prominent in a future of lower prices.

These changes in strategy have also led to organisational adjustments—this has been prominent as companies have implemented more

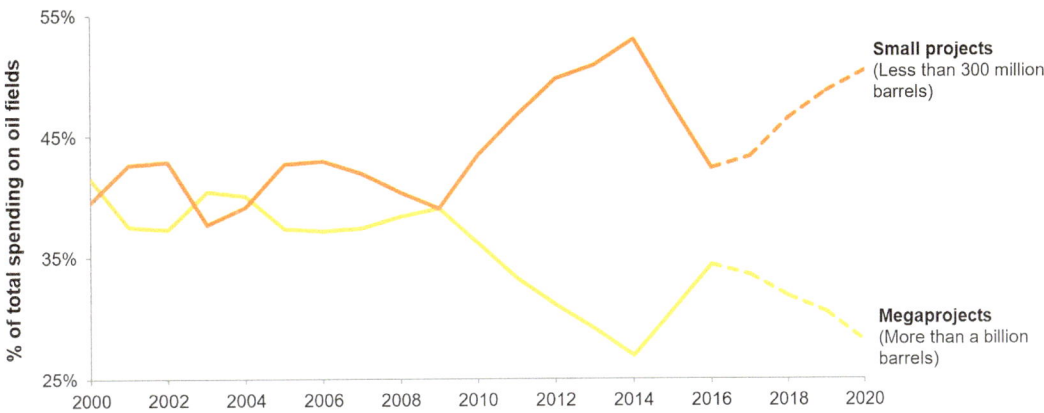

Fig. 8 Investment in smaller projects has been greater than investment in megaprojects since 2009. *Source* Rystad Energy, Bloomberg (2017)

shale projects. Successful shale projects rely on a very different approach to that of megaprojects. Shale projects are smaller and far less complex but require trial and error and multiple iterations to be successful. This is more suited to a flatter organisational structure with agile, local teams and rapid coordination between business development and exploration divisions. Given this, large companies have tended to separate their shale operations into autonomous subsidiaries to preserve the flatter structure that makes them successful, rather than integrate them. This is just one example, but the expectation is that oil and gas companies will be faced with scenarios of a similar nature and possibly on a larger scale.

2.1.3 Case Studies of Responses in Other Sectors

(1) Introduction

The prevailing structural shifts that are facing oil and gas are not unique—other sectors have already experienced and responded to similar transitions. The postal sector has faced structurally lower demand since the early 2000s due to steadily declining letter volumes. Letters have historically been a major source of value, and their decline has had a severe impact on the financial performance of postal companies. Power generation utilities have faced disruptive

technologies in the form of ever-cheaper renewable generation backed by favourable policies, impacting the profitability and longevity of their traditional assets.

Figure 9 shows the perspective we are taking for the case study analysis: structural shifts can lead to changes in business strategy, which then motivate change in organisational structure. The trends of lower prices and increasing technology disruptions have caused fundamental structural shifts in the postal and power utility sectors. Companies have employed a range of business strategies to respond to these trends and subsequently have often had to adjust their organisational structure to facilitate their new strategies. We analyse the response of each company at specific points in time to better understand the factors influencing their adoption of different strategies in response to similar sector-wide trends.

These cases provide examples of how the oil and gas industry can respond to trends, and what the outcomes of different responses may be. Studying other sectors that have experienced similar price and technology trends can help to guide the future strategy and organisational decisions of oil and gas companies. It provides an understanding of the potential consequences from different types of responses and the factors

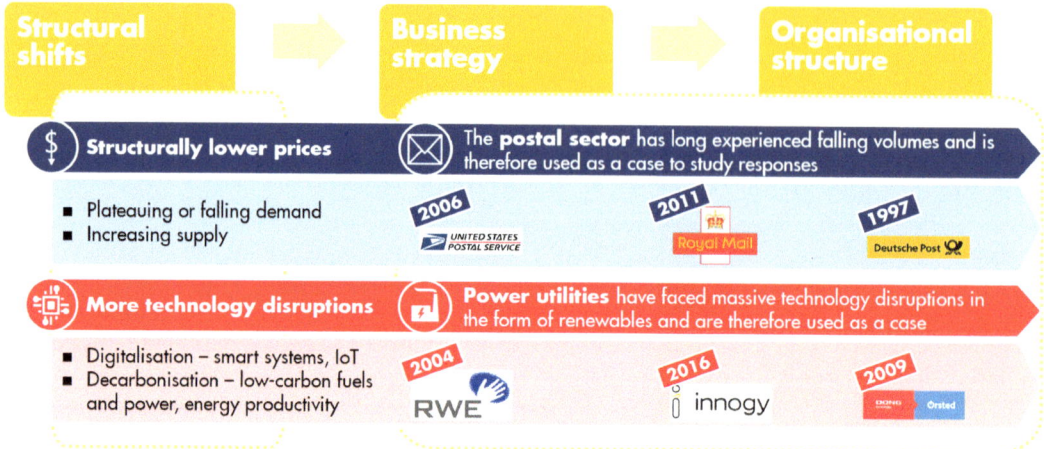

Fig. 9 We study archetypical responses in comparable sectors to understand how oil and gas might change. *Source* Vivid Economics

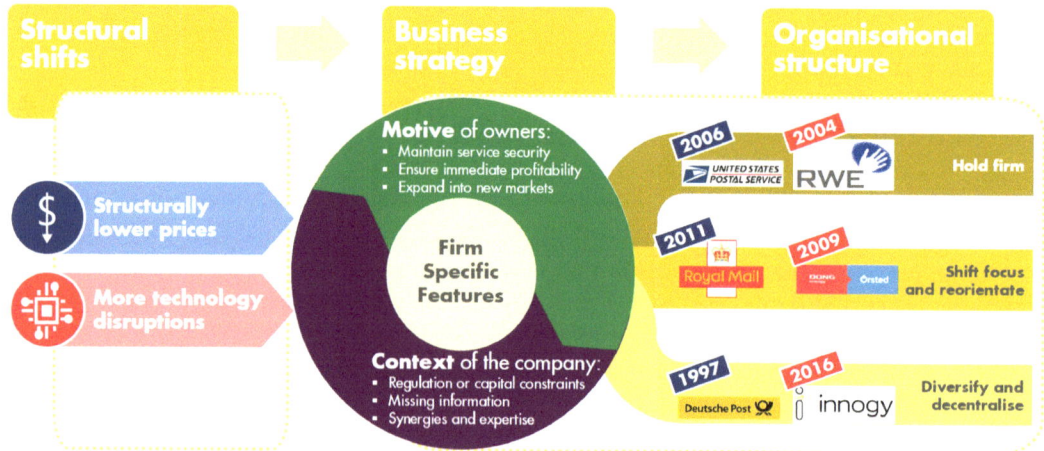

Fig. 10 Structural shifts lead to changes in business strategy, which in turn lead to changes in organisational structure —but the exact changes depend on motive and context. *Source* Vivid Economics

that motivate these responses, particularly when there is no clear dominant strategy for adapting to these structural shifts (Fig. 10).

(2) Motive and context framing

The cases show that company-specific motives and context can lead to a range of different but rational responses to the same structural shift. Companies can rationally have different strategy responses to the same trend if they face different decision factors, namely the motive of the owners and context of the company. For example,

even if two companies have the same motives and overarching goals, the different regulatory environments each faces may force them to choose different strategies in response to a trend. The motive of a company is derived from the owners and broadly encompasses the desire to maintain service and employment security, the time preferences of cash flows (the weight placed on profit today versus profit in the future) and whether the risk preferences of the company are suitable to encourage expansion into new areas of value. The context of a company is the operational conditions that affect its ability to implement different strategies—this includes the regulation it faces, internal or external capital constraints, and the potential synergies between current business areas and future ventures.

We have analysed the different responses of three companies in each sector. Understanding the motive and context allows for the outcomes of each company's response to be understood holistically and deeper parallels to be drawn to the potential future responses of the oil and gas industry.

Two companies demonstrate a relatively minimal response, with a continued focus on improving their current business through the existing organisational model. USPS, protected by a monopoly on letters but banned from entering new non-postal markets, represents the minimal response or "hold firm" approach to industry-wide trends, both in terms of business strategy and organisational structure. RWE is the parallel example in the utilities sector, preserving a focus on its traditional assets rather than on integrating renewables.

Our four other cases all chose to split their old from their new businesses organisationally, although in different ways. Both the UK's Royal Mail and Innogy were legally split from larger companies to help them reorient their business; Deutsche Post split its businesses organisationally but did not divest; while DONG Energy separated its businesses and then divested old business units.

Two companies pursued business strategies that led to a deep shift towards flatter, less hierarchical organisations. Innogy was separated from RWE in 2016 to focus on new energy markets and has taken measures to increase the autonomy of its business units (based on target customers) and enable more agile investment and divestment in rapidly changing markets. Deutsche Post was privatised in 2000; it then diversified from its former core business, both geographically and in terms of products offered. To support this, it took an organisational shift towards decentralised and autonomous divisions.

Two other companies pursued quite different business strategies, with both seeking to change their organisations, yet without radically changing their organisational structure. Royal Mail was unbundled and privatised in 2013 to increase focus on competitive mail and parcel services; and it took a range of measures to de-layer the organisation, reduce operational costs and reorient its corporate culture and processes towards customer service. DONG Energy is an even more extreme example of business strategy change, completely divesting its oil and gas exploration assets to focus on offshore wind generation. While this was accompanied by various changes in organisational culture and responsibilities, it has not (yet) resulted in the degree of diversification and decentralisation seen in Innogy and Deutsche Post.

Table 1 illustrates the role that motive and context plays and the general response archetypes that each of our cases falls into. In response to structural shifts, companies within the same sector and facing the same shifts have chosen alternative strategies, due in part to their differing motives and context. This has led to a range of outcomes and organisational shifts—oil and gas companies facing their own unique motives and context can learn from these cases to better guide their future responses.

(3) Lessons from the case studies—the postal sector

(1) United States Postal Service (USPS)

The US government's priority was for USPS to provide service security. This motive, combined with legislation that constrained USPS to its existing core business areas, meant the business strategy and hierarchical structure remained unchanged in response to declining letter volumes.

- *Context:* The Postal Accountability and Enhancement Act of 2006 was passed after USPS posted several years of strong earnings. This act prevented diversification into non-postal areas due to potential cross-subsidisation, with monopoly earnings leading to unfair competition and market distortion. However, the timing of this act was unfortunate, as mail volumes began to decline from 2006 onwards.

Table 1 Differences in motive and context can lead to a range of organisational and strategic responses

FIRM	MOTIVE	CONTEXT	STRATEGIC RESPONSE	ORGANISATIONAL STRUCTURE
2006 UNITED STATES POSTAL SERVICE	**Keep service security** The owner (government) put strong emphasis on maintaining a cheap universal service	**Tightly restricted to current model** Strong profits in the early 2000s, and the 2006 Postal Act, prevented diversification	**Hold firm** Did not expand into new business areas, but focused on the efficiency of the existing model during a steady decline	**Hierarchical and lean** Still under control of Congress, with clear top-down decision-making but lean management structure
2011 Royal Mail	**Return to profits** Facing declining letter volumes, the owner (government) sought to reduce its liability	**Capital constraint** Losses from 2008-2011, large pension liabilities and service obligation limited ability to diversify	**Reorient core business** Profitability restored through fewer post offices and mail centres, intense cost-cutting, and a focus on customer service	**Division and delayering** The formation of separate companies for distinct business models and a de-layering of mail services to focus on customers
1997 Deutsche Post	**Expand market opportunities** Regulation increased competition in domestic post, motivating a search for new value areas	**Ability to invest** Large cash flows from conventional business allowed internal financing of new investments	**Diversify** Aggressive programme of acquisitions, geographically and into new logistic services such as freight and parcels	**Adaptive, Cross-cutting** Service-focused divisions, with decentralised responsibilities, matrix functions for synergies on core capabilities

Source Vivid Economics

- *Motive:* The motive of preserving service security subsequently led to Congress vetoing cost-cutting measures, such as ending Saturday deliveries and closing the least busy post offices, despite the obvious cost savings. Workers' benefits have also been given high priority over profitability, as evidenced by the $51.8 billion USPS has had to spend to pre-fund pensions for its future workforce. The price of postage is also regulated and is lower than all the major European postal companies, placing further pressure on USPS's margins.
- *Strategy response:* Combined, the above factors have resulted in USPS not adopting a clear strategic response to structurally lower letter volumes. USPS has remained in its core business areas, with falling letter volumes and strong competition in parcels continuing to place pressure on margins. Basic cost-cutting measures, such as reducing headcount, have been implemented, but have not been sufficient to offset the declining value in core businesses. However, supply security has been maintained and the price of postage remains low.
- *Organisational structure:* USPS has remained a government-controlled and regulated monopoly with a top-down hierarchical structure. This multi-layered structure provides strong oversight of operations and

drives uniform operational improvement, creating high levels of efficiency (USPS has half the workers per unit of mail compared with Deutsche Post), albeit within a declining market (Fig. 11).

(2) Royal Mail

The UK government's motive in 2011 was returning Royal Mail, which was balance sheet insolvent in 2011, to profitability. This involved the separation and privatisation of the postal service from the postal infrastructure of Royal Mail, which allowed for cost-cutting, delayering and a customer-orientated service culture that returned the postal service to profitability. Growth beyond its core areas has been limited by capital constraints, although Royal Mail has been exploring other markets.

- *Motive:* The UK government was focused on returning Royal Mail to profitability after four years of pre-tax losses from 2008–11 that made it a growing liability for the government. Royal Mail has had to contend with declining letter volumes since 2004, the loss of its monopoly in letters in 2006 and, more recently, intense competition in parcels from other European postal operators.
- *Context:* Four consecutive years of losses created internal capital constraints that were

Fig. 11 USPS remained under government control with a top-down regional structure that helped promote efficiency in a large operation. *Source* Vivid Economics

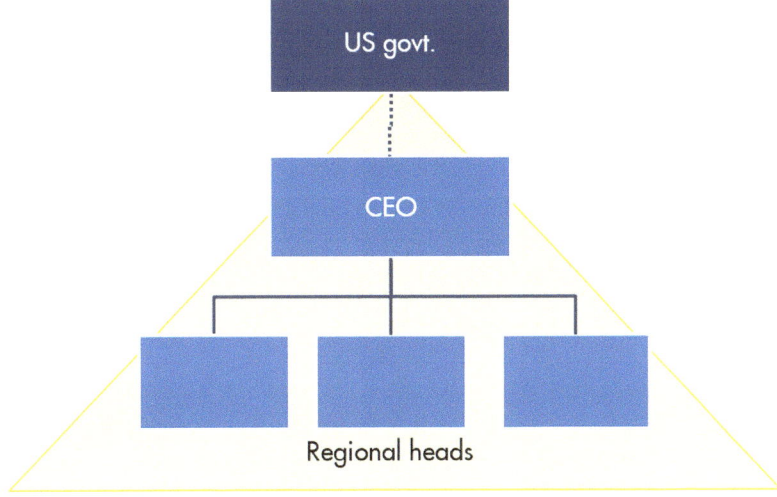

exacerbated by the stigma of using public funds for investment. As a government-owned company, Royal Mail was bound by strict regulation that made it difficult to renegotiate contracts or alter services. It maintained a universal service obligation that further limited its ability to diversify from its core letters business. It also had a pension plan that accrued large liabilities during the financial crisis of 2008, which contributed to making the firm insolvent.

- *Strategy response:* The response was the separation and privatisation of the postal service from the post offices, which remained under public ownership. The government had to take up the pension liabilities of Royal Mail to make its balance sheet solvent and enable privatisation. This enabled Royal Mail to reorient its core business, focusing solely on the service delivered to customers and implementing intense cost-cutting measures that resulted in the closure of a third of mail processing centres and a 10% reduction in headcount after privatisation. These measures were successful in returning the core business areas to profitability, despite minimal revenue growth from 2011 to 2017. However, Royal Mail has not yet been able to significantly diversify, and cost-cutting is unlikely to be a long-term strategy for generating profits. Royal Mail has begun to invest in other

geographies and logistic services. These investments have been small due to the constraints of its low capital reserves and dependence on external financing, which prevent large-scale acquisitions of the like Deutsche Post made in the early 2000s.

- *Organisational structure:* The separation of the postal service and infrastructure arms allowed for more distinct business models to be implemented. By the Post Office arm remaining in public hands, Royal Mail ensured it could act as network infrastructure for all postal companies, preventing wasted investment and the potential formation of a natural monopoly. The new, service-focused Royal Mail was able to take advantage of its less stringent regulatory environment to implement targeted cost-cutting and investment. To assist in this, Royal Mail brought in a CEO, with significant experience of the public-private transition, to reform the regulatory context and ultimately make the company more profitable (Fig. 12).

(3) Deutsche Post

Deutsche Post began a large programme of acquisitions to diversify its business after it became clear that its domestic market would be threatened by changing legislation. It benefited from favourable timing and had significant cash flows from its traditional business areas, allowing it to finance its

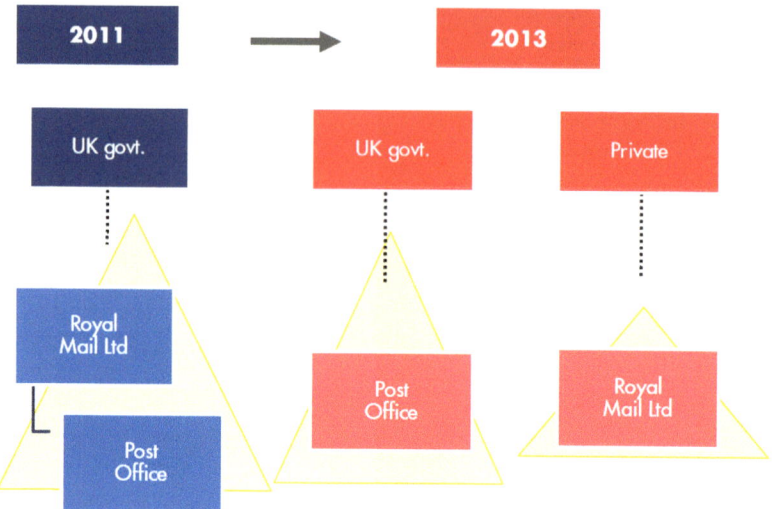

Fig. 12 Royal Mail was separated from the Post Office and privatised to facilitate intense cost-cutting. *Source* Vivid Economics

acquisitions internally. It also created divisions with a customer focus and increased divisional autonomy, while demonstrating organisational flexibility to maximise synergies across its increasingly broad areas of business.

- *Motive:* EU directives led to Germany passing new laws to promote greater competition in its domestic postal market in 1997, making Germany the leader in postal market liberalisation in Europe. At the time, domestic post accounted for more than 75% of Deutsche Post's revenues. Hence, Deutsche Post was motivated to expand into new value areas, both geographically and within logistic services, to preserve profits as domestic competition inevitably increased.
- *Context:* Deutsche Post benefited from favourable timing. Its transition was motivated by legislation, rather than by declining letter volumes. Consequently, it still had the large internal cash flows needed to undertake large acquisitions. Its privatisation in 2001 also afforded it the freedom to undertake radical strategies.
- *Strategic response:* Deutsche Post made several large acquisitions that transformed it into a major global logistics company. It made smaller acquisitions in the late 1990s (Danzas and Air Express International) before acquiring DHL and Excel in 2002 and 2005 respectively for more than €8 billion

following its privatisation in 2001. DHL had expertise in international express delivery and a developed postal network across the USA and Europe, while Excel made Deutsche Post a major player in supply-side logistics. However, such an extreme strategy comes with inherent risks, as evidenced by Deutsche Post discontinuing its domestic express service in the USA and incurring $3.9 billion in restructuring costs as a result in 2009.

- *Organisational structure:* Deutsche Post has demonstrated flexibility in its organisational structure as it has sought to increase the autonomy and customer responsiveness of its divisions. Similar divisions in new acquisitions were combined to create more efficient networks, and a shift from a three-tier to a two-tier management structure allowed for greater responsiveness to consumer demands. Deutsche Post also formed cross-cutting service divisions to maximise synergies across its new business areas, and a global services unit was introduced in 2006 to provide support across all divisions (Fig. 13 and Table 2).

(4) Lessons from the case studies—the utilities sector

(1) RWE

RWE had a large stock of lignite, coal and nuclear assets in 2004 that were low cost,

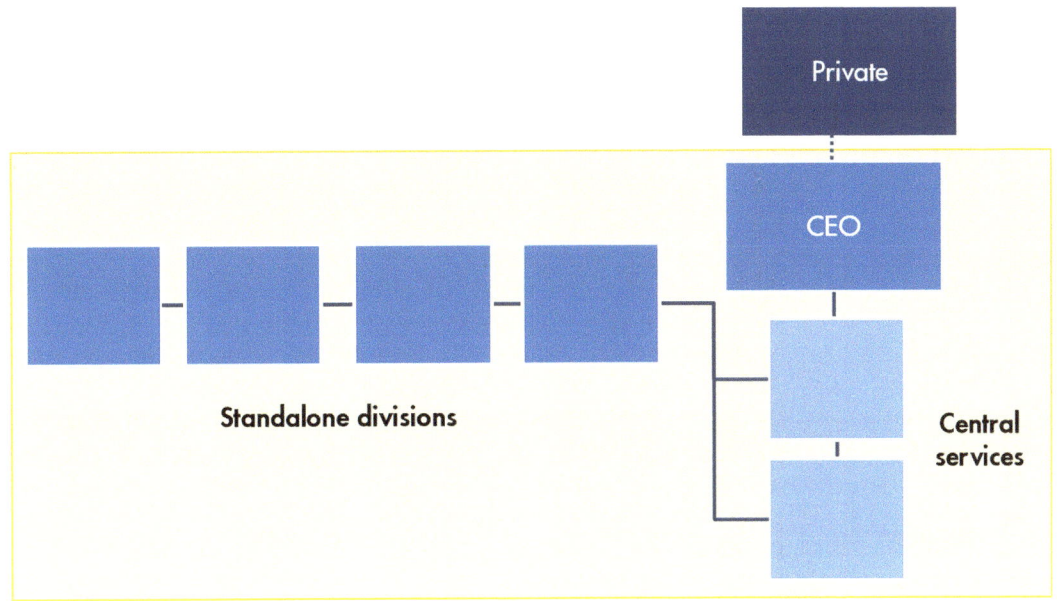

Fig. 13 Deutsche Post imposed a flatter organisational structure and central service divisions to maximise synergies. *Source* Vivid Economics

Table 2 There have been diverging responses among utility companies in response to a similar trend

FIRM	MOTIVE	CONTEXT	STRATEGIC RESPONSE	ORGANISATIONAL STRUCTURE
2004 RWE	**Harvest rents** Legacy assets provided high returns which made it difficult to justify alternative investments	**Monopoly culture** History of being a regional monopoly had made management less adaptive to change	**Hold firm** Little investment in alternative generation technologies while legacy assets were maintained and upgraded	**Little or no change** Assets and organisational structure largely remained the same up until the separation of Innogy
2016 innogy	**Pursue growth** Innogy split from RWE, with a mandate to invest in grids and infrastructure, retail and renewables	**Clear green trend** Anti-nuclear laws and general German policy fostered widespread trust in renewable energy	**Split and Reorient** Innogy pursues renewables and grid opportunities in Europe and abroad, while pursuing access to new finance	**Division and divisions** Split from "dirty" business as a separate entity, removing issues of cannibalisation and enabling a flatter, more agile portfolio of holdings
2009 DONG Ørsted	**Pursue growth** Subsidies created rents in offshore wind while conventional generation became less profitable	**Skill and certainty** Demand certainty and taxing offshore E&P expertise limited barriers to new investment	**Transformation** DONG (now Ørsted) moved steadily to completely shift its business from oil and natural gas to offshore wind	**Organic revamping** Wind was initially added as a separate division which then grew, drawing on and transforming, the existing organisation

Source Vivid Economics

delivered high returns and provided employment in key shareholder municipalities. Despite new legislation introduced in 2004 promoting renewables, RWE held firm and planned to increase its lignite generation capacity. Hence, RWE kept a centralised organisational structure

that could better focus on managing these concentrated, large-scale assets.

- *Motive:* In 2004, renewable capacity in Germany was noteworthy, but seemed unlikely to displace the extremely low-cost baseload power provided by lignite and nuclear generation, which had high returns. In addition, German municipalities owned 24% of RWE and relied on the local employment from lignite and coal assets. The employment from lignite generation was also a political tool, leading many politicians to support its use and protect its role in the German energy system, which further disincentivised RWE from diversifying into other technologies.

- *Context:* RWE had already invested in improving and expanding its traditional asset base, locking capital into assets with long lifespans. RWE's previous status as a regional monopoly also fostered a culture of inaction and risk aversion. This was further accentuated by RWE's history of large cashflows and dividends that investors were not prepared to compromise.

- *Strategic response:* RWE did not make any significant additions to its renewable capacity from 2004-10, instead it continued to focus on its legacy assets. In 2004, RWE considered Germany to be "at the beginning of a long-term investment cycle" and the plan was to replace old power stations with more efficient versions, rather than branch off into alternative generation methods. In its 2004 annual report, RWE mentioned the new renewable energy legislation solely in terms of the monetary burden it would place on the company, rather than the opportunities it offered. Consequently, in 2005 RWE announced plans to spend €3.5 billion on two projects to install 3.6 GW of new, optimised lignite generation capacity, which were among the largest projects ever planned in RWE's history. However, as renewable energy capacity increases at unprecedented rates, RWE has been forced to adopt a harvester mentality, with the aim to derive as much value as possible from its large legacy asset base that will be gradually phased out of Germany's power system.

- *Organisational structure:* RWE maintained the centralised management structure that oversaw its conglomerate of business areas. For its traditional large-scale assets, this allowed for efficient management, but it constrained the autonomy, flexibility and organisational development of its new energy areas and made investment and growth more difficult.

(2) Innogy

Innogy was separated from, but still majority-owned by, RWE in 2016. It contained the green assets of RWE, as well as the network and retail businesses. The motivation for this change was for Innogy to be able to pursue growth opportunities in renewables and other markets without being constrained by the growing liabilities of RWE's legacy assets or conflicts of interest. By splitting from RWE, Innogy was free to attract fresh investment and implement an organisational structure focused on energy services to better serve its new markets.

- *Motive:* Innogy was split from RWE in order to chase growth opportunities. RWE as a whole struggled to come to grips with the changes it was facing and lacked flexibility due to the capital it had invested in its legacy assets.

- *Context:* German renewable energy policy had motivated an unprecedented uptake of renewables that has completely altered the state of the energy system—it was clear that the future trend in Germany was one of mass renewable energy generation. In addition, the Fukushima nuclear incident in 2011 resulted in a moratorium on nuclear generation and even greater support for renewable power as a clean alternative.

- *Strategic response:* Innogy has marketed itself as a highly innovative company and is pursuing renewable energy opportunities internationally. It is attracting financing as a longer-term investment, which would not have happened if it were still bundled with the legacy assets of RWE, which represent more of a short- to medium-term investment.

- *Organisational structure:* Innogy was separated as a partially owned subsidiary of RWE. This removed any association it previously had with the declining legacy asset base of RWE, although 75% of the shares remain in RWE's hands. Separation afforded Innogy the

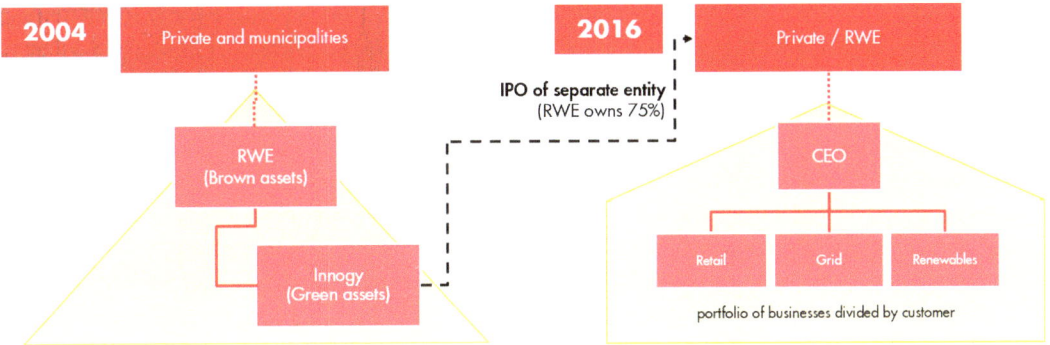

Fig. 14 Once Innogy was separated from RWE it adopted a consumer-focused structure, aligning its divisions with areas of the value chain. *Source* Vivid Economics

freedom to pursue its new energy service strategy and reorient the organisation accordingly. Innogy restructured itself in accordance with the different levels of the energy value chain to align with the demands of the end consumer rather than those of production. It also gave units a greater degree of autonomy (flattening the organisational hierarchy), while also pursuing a portfolio approach, whereby investment across those units could be ramped up or down depending on their relative success (Fig. 14).

(3) DONG Energy

DONG Energy (now Ørsted) is majority-owned by the Danish government. The company was motivated to pursue offshore wind technology as a reliable source of domestic clean energy to offset declining profits in conventional generation and also develop a new potential growth market.

DONG Energy's expertise in offshore oil and gas exploration and its experience in pilot offshore wind farms made it ideally placed to aggressively shift its business focus from oil and gas to offshore wind. The company had to change its organisational structure to accommodate this new business area: offshore wind was first added as a division under the CEO to develop, before later becoming the main focus of the company.

- *Motive:* Denmark has ample offshore wind reserves and has long promoted wind power with strong domestic subsidies. Declining electricity prices in the Nord Pool power market and fluctuating demand levels reduced thermal generation earnings, making the subsidies and secured earnings from wind generation more appealing.

- *Context:* DONG Energy had the necessary skills to drive the development of the offshore wind market. Its offshore oil and gas expertise was easily transferrable and it gained significant experience developing offshore wind farms from its merger with Elsam in 2006, mitigating the risks and barriers to investment.

- *Strategic response:* DONG Energy managed to completely transform its main area of focus from oil and gas production to offshore wind generation. Having already completed several medium-sized pilot projects in Denmark in the early 2000s, DONG Energy began seeking out larger opportunities that resulted in an agreement with Siemens in 2009 to buy 1.8 GW of wind turbines. The sheer size of this deal enabled economies of scale to develop in production and deployment and it marked the start of DONG Energy's commitment to pioneering offshore wind technology. Since 2009 DONG Energy has been involved in the largest offshore wind farms in Denmark (Anholt, 400 MW) and globally (London

Array, 640 MW) and has won the bid to deliver the world's first offshore wind farm of more than 1 GW (Hornsea Project One, 1.2 GW).

- *Organisational structure:* Offshore wind was formed as a special division under the CEO to prevent conflict of interest with other areas and to ensure that targets were set appropriately for a developing business area. At the same time, DONG Energy was able to use the existing organisational structures and capabilities from its oil and gas exploration business—scouting, construction and asset management—which were well suited for offshore wind projects. As the wind division expanded and traditional assets were divested, a more consolidated organisational structure emerged with a focus on the growth of the green businesses and an integrated approach to sharing functional expertise. The result (so far) has been a greater emphasis on the benefits of integration and focus, rather than autonomy and diversification (Fig. 15).

2.1.4 Conclusions and Implications for China

The oil and gas sector is facing structurally lower prices and more technology disruptions, which could challenge the long-standing business strategy and traditional, hierarchical organisational structure of companies.

Case studies from the postal and electric utility sectors show that companies may respond in different ways, depending on their motivation (the outcomes valued most by stakeholders) and context (the operational conditions that may constrain a company's strategic margin of manoeuvre).

- *Little change remains an option*—where stakeholders valued security of existing services and nearer-term profits, business strategy focused on harvesting the value of the existing business model, and organisational change was about making the existing structure leaner (USPS).
- *Splitting the old from the new*—where stakeholders valued longer-term profitability but were constrained by regulation, conflicting interests or capital (Royal Mail, RWE in 2004 and DONG Energy), business strategy focused on streamlining and redirecting resources to new capabilities, and organisational change focused first on splitting the old from the new.
- *Transforming with the trend*—where companies seeking new opportunities had relevant capabilities and sufficient capital (Deutsche Post, DONG Energy, and Innogy after 2016), and the strategic shift was more aggressive than for those acting later or facing resistance (Royal Mail and RWE in 2004).

Organisational change needs to align with the business model. Where new businesses were similar to old (Royal Mail and DONG Energy), the organisational change is more about refreshing the existing structure than radically restructuring it; while entirely new product or service

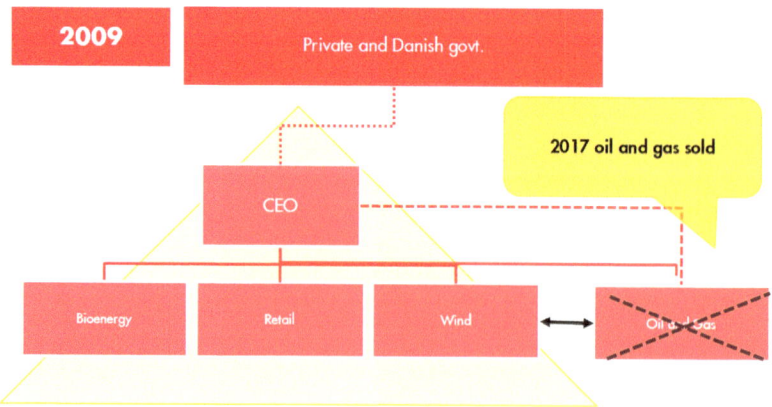

Fig. 15 DONG Energy reframed its organisational structure and applied it to offshore wind. *Source* Vivid Economics

models (Deutsche Post and Innogy) involved far-reaching change.

Governments can have a role to play in shaping the motives and context of companies to help guide outcomes, and they should consider what types of response current policies are encouraging. Through setting the conditions that influence strategic and organisational change, governments can determine the future role of the oil and gas industry. Current policies are setting the motive and context of companies that will encourage a particular response that may not be in line with the government's aims. Consideration should be given to how motives and context impact future responses and what the options are for a government to influence these factors.

(1) Response archetypes for oil and gas companies

There is no dominant or common response strategy to the prevailing future trends—the motivations and context behind each company will determine what types of strategic and organisational response they adopt. Figure 16 illustrates the generic range of motivation and context that oil and gas companies may potentially face. How a company responds to these factors will require the appropriate archetype of strategy response, which in turn will involve a different type of organisational structure.

When there is a focus on security and binding constraints exist, legacy assets are often maintained and harvested and their declining role in a changing market is accepted. Companies in this position will need to maintain their legacy assets to meet supply security or short-term employment targets but are unable to diversify given the restricting context that surrounds them. In such a situation their best response may be to take on a harvest mentality, using their legacy assets as much as possible during the transition, but with the knowledge that the value of these assets will be greatly diminished over time as the market changes. This strategy is facilitated by maintaining a top-down approach to keep central control, which is more efficient when dealing with a small number of high-value legacy assets. Central functions that aid asset management and organisational efficiency would also be useful to maximise the rents from harvesting.

This strategy is effectively what RWE and USPS employed, although they did so to a lesser extent. RWE has maintained its large capacity of lignite and nuclear generation and is earning what it can from those assets, while maintaining its hierarchical structure. USPS is not showing obvious indications of harvesting value, but the company is clearly bound by a service security commitment that is preventing it from adopting other responses.

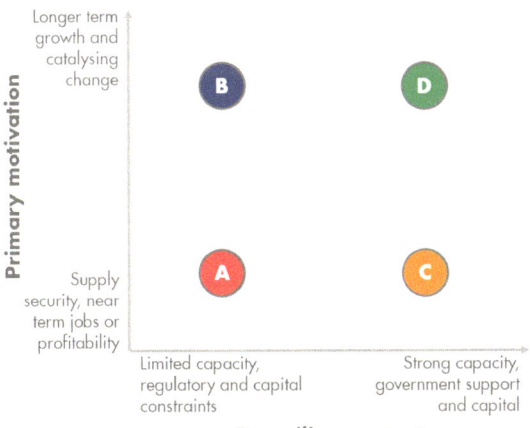

Four archetypes of strategy response arise:

A. Supply security focused harvesters:
  - Ensure supply security/near-term employment
  - Harvest value from legacy assets during the transition

B. Constrained niche growers:
  - Aspirations for growth, but context prevents moving beyond conventional oil and gas
  - Focus on streamlining their core functions and growing within niche markets

C. Dividing conquerors:
  - Split into two: one focused on harvesting value from legacy
  - The other focused on diversification for long term growth

D. Transformative diversifiers:
  - Move quickly to a new energy service model and divest hydrocarbon assets to finance the transition
  - Potential to deliver long-term growth but is more risky

Fig. 16 Motive and context will determine how Chinese oil and gas companies strategically respond to lower prices and new technology disruptions. *Source* Vivid Economics

When a similar focus on security exists but within a less stringent context, a company has the option to separate in order to provide the desired service security and seek growth opportunities elsewhere. When security and short-term gains are the priority, some continued use of legacy assets is unavoidable, as diversifying into new ventures will not immediately fulfil either of the two objectives. However, when the prevailing context is less restrictive companies should consider employing a divide and conquer strategy, separating themselves into two distinct parts, where one delivers the near-term supply security, dividends or employment targets and the other is free to pursue diversification for long-term growth, unhindered by the needs of the existing legacy assets. Organisationally, this requires a clear split between the two different parts of the company. Care must be taken to carry over any existing organisational strengths into the appropriate areas of the business.

Innogy was separated from RWE for these reasons. Although it became clear by 2011 that both lignite and nuclear generation were declining in relevance, RWE maintained them to preserve short-term cash flows and to try and recoup as much of its sunk investment in these assets as possible. Innogy was created as a separate subsidiary that was not burdened with any legacy assets and was free to innovate and invest in growing renewable markets.

A binding context can prevent a company from moving away from its traditional assets, even when the primary motivation is for long-term growth—in such scenarios, a focus on efficiency in core areas and expansion in niche markets is a sensible response. When regulatory or capital constraints are tight, companies may not be able to divest their legacy assets or effectively invest in new areas. To drive growth in this scenario, companies can focus on streamlining their core functions to maximise efficiency and squeeze up margins. Alternatively, they can aim to develop within smaller niche markets where lower capital investments may still be sufficient to generate decent returns. This type of strategy is implemented by forming agile, decentralised market units that can respond quickly to the unique and often changing circumstances of different niche markets and ensure that the core business stays as relevant and lean as possible. Portfolio optimisation is also an important step to ensure that all business areas are aligned, and central functions are implemented effectively.

Royal Mail found itself in this situation in 2013. It was limited by tight capital constraints after several years of losses and was seeking to generate long-term profit and growth. It began a programme of intense cost-cutting that returned the company to profitability despite little revenue growth, and it started to make small investments in other logistic areas and geographies to gain a foothold in potentially long-term growth markets.

If the motivation is for long-term growth and the context is non-restrictive, a transformative strategy can be adopted, although this comes with higher risk. When long-term growth is the primary motivation, investment into new areas becomes a priority as legacy assets will shrink considerably as structurally lower prices and greater technology disruptions set in. Companies facing this situation may seek to diversify aggressively and shift to new markets or incorporate new technologies. However, such extreme change comes with inherent risk, which makes the outcomes for supply security, employment and profitability uncertain. To adopt such a strategy effectively, companies should develop an organisational structure that is flatter and has more divisional autonomy to facilitate fast response to the needs of different business areas. It is also important to strengthen core capabilities across the group to maximise synergies wherever possible, especially if elements of the current business can provide a competitive advantage in new areas (Fig. 17).

Both Deutsche Post and DONG Energy are examples of companies that have undergone extreme transformations to preserve long-term profits. Deutsche Post used its large internal cash flows to invest heavily in other logistic areas and was flexible with its organisational structure to maximise synergies across its rapidly expanding business areas. DONG Energy transformed its

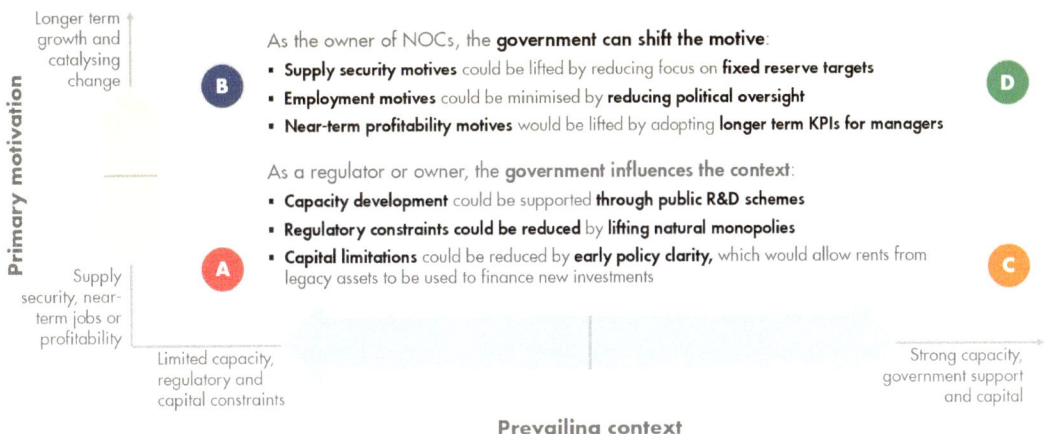

As the owner of NOCs, the **government can shift the motive**:
- **Supply security motives** could be lifted by reducing focus on **fixed reserve targets**
- **Employment motives** could be minimised by **reducing political oversight**
- **Near-term profitability motives** would be lifted by adopting **longer term KPIs for managers**

As a regulator or owner, the **government influences the context**:
- **Capacity development** could be supported **through public R&D schemes**
- **Regulatory constraints could be reduced** by **lifting natural monopolies**
- **Capital limitations** could be reduced by **early policy clarity,** which would allow rents from legacy assets to be used to finance new investments

Fig. 17   Motive and context are not set in stone but controlled by stakeholders—the Chinese government can therefore shape outcomes. *Source* Vivid Economics

core business from oil and gas to offshore wind. It was incentivised by consistent wind subsidies and implemented divisional autonomy for its offshore wind division to insulate it from competition from other divisions within the company, setting growth rather than profit targets until it reached maturity.

(2) Options for adjustment

Governments can play a role in shaping the future responses of companies by adjusting the motives and context through policy setting. Although strategy and organisational response choices are decentralised and made at the corporate level, there is potential for governments to intervene and guide the direction of future responses by setting the conditions that influence those responses. This can be done by setting different targets for government-owned companies or it can be implemented broadly with policy that affects all companies. These adjustments will be important to guide the oil and gas industry towards the role that the government envisages for it in the future, as structurally lower prices and increasing technology disruptions begin to take hold in the industry.

Government-owned oil and gas companies may be limited in their future response to prevailing trends by the current goals set for them. The government has a clear, direct role in setting the motive for companies that are under

government control and should be aware that some types of future response will not come about without the correct motivation. For example, a national oil company cannot transform its business model and shift into renewables, as DONG Energy has, if it is bound by an obligation to deliver fixed reserve targets. USPS was clearly limited in its ability to respond to the structurally lower demand it was facing by the government's (its owners) singular focus on service security. Similarly, employment goals and short-term profitability targets encourage continued use of legacy assets and will prevent large-scale diversification. This is an issue that affected RWE—employment and dividend requirements meant RWE prioritised its legacy assets over investment in new areas.

There is also the potential to collectively adjust the context across all companies by altering pertinent policies. As a regulator, the government can influence the context that all companies face. Public funding or R&D that supports new technologies or new capacity development can reduce the uncertainties that companies face when trying to enter new business areas. The Danish government's strong wind subsidies were key in encouraging DONG Energy to pursue offshore wind at a time when thermal generation earnings were volatile. Removing natural monopolies by allowing greater access to infrastructure networks can help companies to refocus on their core businesses. When Royal Mail was partly privatised in 2013, the Post Office arm was separated into an independent entity with a management structure centred on the postal service the company provided. Early policy clarity is the key to providing the robust signals needed for companies to optimally plan their longer-term strategy and organisational responses. Deutsche Post benefited from this level of clarity, as it was able to diversify into new markets and geographies before its domestic market declined, when it still had significant cash flows to invest. In contrast, Innogy, the green asset subsidiary of RWE, was separated from RWE in 2016, well after the German renewable energy revolution had cut into the earnings of RWE's legacy assets. The rapid escalation of renewable energy policies in Germany prevented a clear policy signal from forming, contributing to RWE's continued use of legacy assets and subsequent losses.

Appendix 1: Postal Companies: Responding to Lower for Longer Trends

Introduction

Since the early 2000s, the postal industry has faced two global trends: a decline in letter volumes from the spread of electronic communications; and a parallel, but smaller, increase in parcel volumes due to the rise in e-commerce. Combined, these two trends yield a fall in total volume of 1–2% per year, and global revenue growth of only 1.6%, significantly below the economy-wide average of 4.3% revenue growth. The effect on postal companies has clear parallels with oil and gas companies facing lower for longer hydrocarbon prices. Major postal companies have been hit particularly hard by the decline in letters, historically their main source of revenue and over which they had a monopoly—they were often state-owned and seen as delivering an essential service. Postal companies have revenues of the same magnitude as oil and gas companies.

On top of these two global trends, some countries have experienced deregulation and rising competition, while others have remained

closer to a regulated monopoly. Beginning in 1997, the European Commission has abolished national monopolies on mail in Europe. EU member states were required to allow competitors to enter their national postal services, at first only in certain product categories (such as parcels) and by 2012 across the full spectrum of postal services. In contrast, the USA allowed competition in parcels and express letters, but the United States Postal Service (USPS) continues to hold a legal monopoly on standard mail.

Postal companies have had a range of responses to the challenge of declining value in their main area of business—understanding the factors that led to these can help oil and gas companies plan their own transitions. The individual decision factors and circumstances facing each postal company dictated the way in which they responded to the lower for longer trend of falling letter volumes. Broadly speaking, the responses to lower for longer can be organised into three categories:

(1) *Inaction:* No organisational changes or divestments, with cost-cutting and efficiency improvements limited to a few areas due to restrictive legislation.

(2) *Divestment and cost-cutting:* Large efficiency gains from divesting inefficient, non-core areas and fully utilising cost-cutting opportunities. This has often been achieved through privatisation or reorganisation by separating mail operations from other parts of the business (such as pensions and post office operations).

(3) *Diversification:* In addition to separating certain parts of the business from core operations, companies can diversify into new geographies and new lines of business.

We have analysed the responses of three companies, each of which represents a different category of response. USPS, protected by a continuing monopoly on letters but prohibited from entering new non-postal markets and with limited cost-cutting options, represents the minimal response or inaction to industry-wide trends. Royal Mail, unbundled and privatised in 2013, has enacted a wide range of cost-cutting and

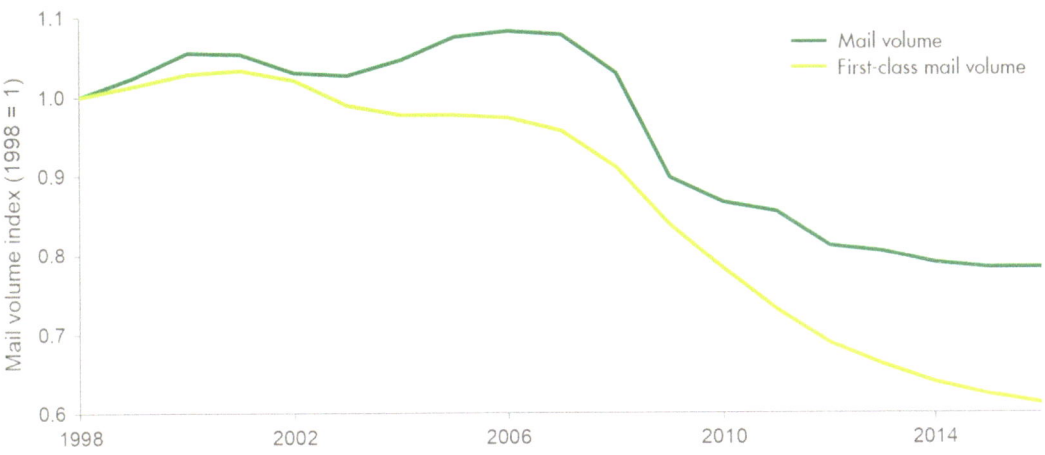

Fig. 18 Total mail volume did not begin to fall consistently until after 2006, although first-class mail volumes have been falling since 2001. *Source* Vivid Economics

modernisation measures that have significantly improved its recent financial performance. Deutsche Post, privatised in 2000, has taken similar steps as Royal Mail, but has gone faster and further, diversifying from its former core business, both geographically and in product offering, and growing substantially as a result.

The parallels with the oil and gas industry, both in terms of the lower for longer trend faced and the scale of the companies themselves, make the postal sector a good case study for oil and gas. Both have two core products with different outlooks: one that is facing challenges because of long-term falling demand (oil/letters) and another that is experiencing rising demand (gas/parcels). In addition, both industries have faced, or are facing, large regulatory change: deregulation and competition in the case of post and increasing climate change policy in oil and gas.

The large postal companies were similar to oil and gas companies: they were often nationalised, delivered an essential service and had huge revenues. The high capital and large economies of scale in both sectors mean that small falls in volumes can make large impacts on profitability. These similarities make the postal sector a good learning case for oil and gas companies.

USPS: An example of inaction

USPS has been facing a steady decline in the volume of first-class mail since 2001 and mail overall since 2006. Figure 18 shows the volume of first-class mail, which is the main source of revenue for USPS, peaked in 2001 and has been falling ever since. In contrast, overall mail volumes tended to rise until 2006, after which they fell sharply. The parcel market is dominated by multinational companies, like FedEx and UPS, that were quick to innovate and capture profitable delivery routes, leaving USPS with only an 8% share of the sector's main growth market.

From 2002–06, total mail volumes were rising and USPS posted a cumulative profit of $8.6 billion. Given the very small falls in first-class mail volumes from 2002–06 and the fact that total mail volumes continued to rise to 2006, it is unsurprising that USPS posted strong financial results pre-2006. Its average annual profit during those years was $1.7 billion, more than double that of the late 1990s, which averaged $726 million per year from 1997–99.

Since 2006, USPS has been forced into inaction, as legislation and Congress have prevented cost-cutting and diversification in response to mounting losses. The 2006 Postal Accountability and Enhancement Act was intended to modernise postal regulation that had been in place since 1971, reassessing the pricing of postal services and setting clearer barriers to entry for USPS in non-postal services. Given USPS's monopoly over non-express letters and its strong profits pre-2006, the concern was that USPS might use its monopoly profits from letters to unfairly subsidise its entry into new areas. Congress wants USPS to maintain a universal service and has continually blocked cost-cutting measures like stopping Saturday deliveries. Once the lower for longer trend set in after 2007, USPS had limited responses and had accumulated losses of $10.6 billion by 2016.

In 2006, letter volumes had yet to fall appreciably and with USPS posting profits in four consecutive years, there were over-optimistic expectations about the future of letter volumes and the payoff from inaction. The timing of the 2006 Postal Accountability and Enhancement Act was unfortunate in that it preceded the onset of falling letter volumes. It seems apparent that in 2006 expectations for the potential severity of the lower for longer trend were not accurately formed. This led to an overestimation of the potential payoff from adopting a strategy of inaction. The historical evidence suggested a mild lower for longer trend at the time (later forecasts in 2009 have proved more accurate in predicting letter volume falls). There was, therefore, little motivation to shift to a new strategy.

After the 2006 Postal Act, diversification was effectively no longer open to USPS—a prohibition that was criticised by USPS's management —limiting its strategy response options to either inaction or divestment.

The US government promises a universal service and responds strictly towards any action that may disrupt or threaten it. The government places a high value on security of service and benefits to employees, both of which contribute to USPS adopting a strategy of inaction. This resulted in Congress vetoing many cost-cutting measures, forcing USPS towards inaction by default. The large postal infrastructure system

that USPS operates requires economies of scale to be efficient—small falls in volume can quickly reduce margins and create large losses.

Once the lower for longer trend had taken hold after 2006, USPS had to accept its limited ability to respond strategically. Given the restrictions placed on it entering new non-postal services and the heavy competition it faces from other companies that specialise in parcel and express delivery, USPS has been limited to cutting costs and making efficiency improvements that address the decline of its main market.

Some cost-saving measures were implemented, mainly by reducing head count, although mandatory pre-payments of pension and security benefits for workers added huge liabilities. From 2006–14, the number of full-time employees at USPS fell by 30%, as it pursued efficiency gains to offset the sharp fall in letter volumes. However, these gains are small in comparison to pension pre-funding payments of more than $5 billion per year that USPS was forced to make between 2007 and 2016. While other postal companies have often had their pension obligations split off to make them solvent, USPS has had to bear the full cost. Once the pre-funding payments are separated from expenses, slight but persistent falls in operating expenses since 2007 are visible, despite a growing number of delivery points.

Other cost-cutting and divestment measures have been prevented by Congress to avoid potential disruptions to the universal service. USPS has pushed to stop Saturday deliveries since 2009, but Congress has vetoed it, most recently in 2013. Similarly, Congress vetoed a plan to close the 3,600 least busy post offices in 2012. There is also tight regulation around the pricing of letters, which closes another potential avenue to boost falling revenues. A price decrease of 5% was implemented in 2016, cutting deeper into margins already squeezed.

The result of this has been large cumulative losses by USPS over the past decade: $10.6 billion in total from 2007–16, excluding pension pre-funding payments. The case of USPS shows that if mandatory rules prevent a government agency from adjusting to long-term declines in

Fig. 19 USPS has suffered huge losses and depressed profits since 2006, illustrating the dangers of inaction. *Note* The profit/loss figures here exclude the security pre-payment obligations of USPS; all figures separate over 2012 and 2011 the double pre-payment made in 2012. *Source* Vivid Economics

its core markets, large losses are likely to accumulate. Figure 19 shows that expenses could not be reduced due to the restrictions on cost-cutting options and the large losses that resulted. In addition, USPS is prevented from seizing possible growth opportunities—by regulations, financial constraints and culture. However, social objectives (such as preserving a post office in nearly every town) and past commitments (such as the pension and healthcare benefits of postal workers) were honoured.

The clear parallel between USPS's experience and the oil and gas sector is the potential desire to maintain security, which can lead to large losses if it prevents a company from responding to trends. It is not difficult to envisage a nationalised oil and gas company being unable to reduce or diversify away from fossil fuel production in order to maintain domestic energy security, just as USPS was forced into inaction partly by the political desire to protect the USA's universal postal service. If a government objective like energy security or universal service is the goal, then policymakers should be aware that there is a significant possibility of the company accumulating loss, as new trends can rapidly change the market.

If energy security is a high priority, it is prudent to review policies regularly to account for shifting trends. Enacting policy that creates barriers to change is risky and possibly highly damaging for the companies affected. A balance needs to be reached by having policy that is strict enough to achieve an appropriate level of security, but which has the flexibility to allow the company to respond to trends when the negative effects become overwhelming.

Inaction may be an appropriate strategy in the short run, but change is likely to be required over the longer term; introducing independent scenario teams can help identify when a change in strategy is needed. Scenario teams that are not under the jurisdiction of any specific business unit can offer unbiased guidance of when inaction may no longer be appropriate. Given the structurally lower prices that are gradually taking hold, an alternative strategy to inaction will need to be considered in the longer term. Correctly timing this strategic shift will lead to better outcomes.

Royal Mail: An example of a divestment/cost-cutting strategy

The UK's Royal Mail was struggling with operational inefficiencies, declining letter volumes and poor financial results before its part-privatisation in 2013. UK letter volumes

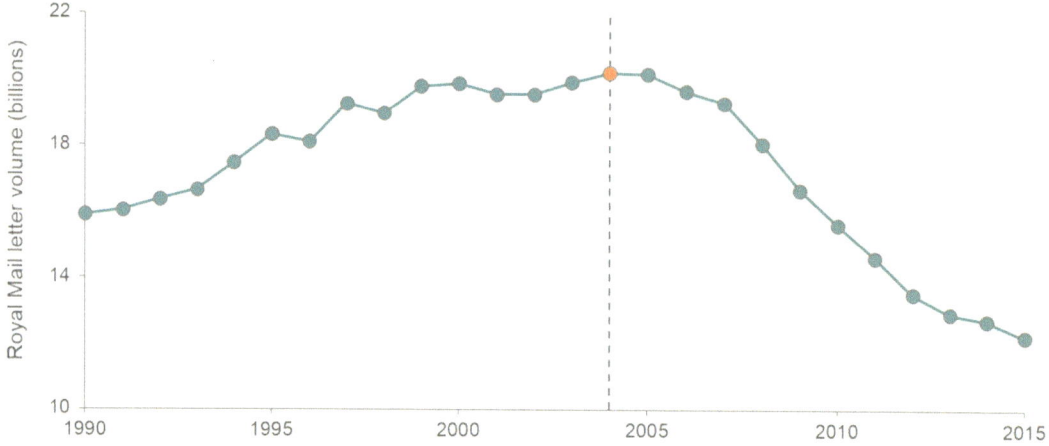

Fig. 20 Royal Mail letter volumes peaked in 2004, but experienced sharp falls after 2007. *Source* Vivid Economics

peaked in 2004 after steady growth since 1990. The initial decline was modest but gathered pace after 2007. Letters delivered fell by about 5% per year from 2007–15. In 2007–08, Royal Mail had the lowest operating margin of the 13 major western European postal companies and posted pre-tax losses each year from 2008–12. At the same time, the UK postal market was opened to competition from January 2006, causing a steady decline in Royal Mail's market share. Lastly, in the wake of the financial crisis of 2008, Royal Mail became balance sheet insolvent, as the asset value of its pension fund declined, with balance sheet net liabilities of more than £3 billion in 2011 (Fig. 20).

The British government decided to privatise the service arm of Royal Mail in 2011. This led to strong cost-cutting initiatives but limited diversification. The legislation that allowed for private control of Royal Mail was passed in 2011, with privatisation starting in 2013. Following privatisation, Royal Mail returned to profit in 2017 after five consecutive years of pre-tax losses, mainly driven by efficiency gains in its core domestic delivery service (profits rose without major increases in total revenue). Some diversification into international markets and vertically upwards into e-commerce has occurred, but these have been relatively minor compared to the strategies of other companies.

The positive examples of postal service privatisation in other European countries and clear financial struggles of Royal Mail since 2008 motivated a similar privatisation in the UK. By 2011, the decline in letter volumes had become a clear trend and the poor performance of Royal Mail showed that continued inaction would not be sustainable. Earlier examples of privatisation across Europe illustrated how postal companies could reform their operations and return to profit while maintaining service obligations. Hence, the expectations were that the payoff from inaction would be low due to the continued decline in letters and inefficiencies, while the payoff from divestment or diversification would be high given the evidence from similar strategies across Europe.

However, as a government-controlled corporation, Royal Mail was subject to tight regulations that created high barriers to change and prevented action pre-2013. Being under government control meant Royal Mail could not renegotiate contracts, access private capital, adjust its products or enter new markets without time-consuming approval processes. Consequently, even with expectations of high payoffs from divestment and diversification, Royal Mail was not able to respond in an effective manner to its declining financial performance, resulting in several consecutive years of pre-tax losses before

2013. Privatisation offered an easy avenue to reduce transformation costs and allow for new strategies to be adopted.

While the British government considers a universal service important, it has far more lenient price controls on postage than the USA. Post-privatisation, Royal Mail was still designated as the universal service provider, obligated to provide a nationwide service for a uniform price, six days a week. This designation meant that a strong focus was still needed on the core service of delivering letters, leading to a degree of aversion to change. Large diversification strategies inherently carried a great deal of risk and could disrupt this core business. Hence, the Royal Mail's preferences pushed it more towards cost-cutting strategies than diversification into new markets.

The British government split the historic post office into three parts in 2013 and only privatised the postal service arm, leaving the network of post offices under public ownership. These three parts were the letter and parcel service operations of Royal Mail, which were subsequently privatised in 2013; the network of post offices, which remains in public hands; and the net liabilities of the Royal Mail Pension Plan that were taken over by the government to make Royal Mail solvent again. In effect, the government bore a one-time cost to enable privatisation and reduce transformation costs. This unbundling was implemented to allow the post offices to be used by different postal service providers and avoid wasted spending on infrastructure.

Following privatisation in 2013, Royal Mail's financial situation improved dramatically, mainly through divestment in its core business area. While revenue growth was modest, with a compound annual growth rate of 1% between 2011 and 2017, profit levels increased substantially, suggesting the focus was on efficiency gains rather than entry into new areas. Between 2011 and 2016, Royal Mail achieved an overall headcount reduction of 9.7% and reduced the number of its mail processing centres from 57 to 38. This greatly improved the profitability of its core domestic letters and parcels business, reversing an operating loss of £120 million in 2011 to an operating profit of £411 million in 2017.

Although effective in the short run, this divestment strategy is unlikely to offer a long-term solution for generating profit growth. The short-term impacts of divestment have clearly been significant, reversing Royal Mail from losses into profit as seen in Fig. 21, mainly due to improvements in the core business. However, the future performance of this business

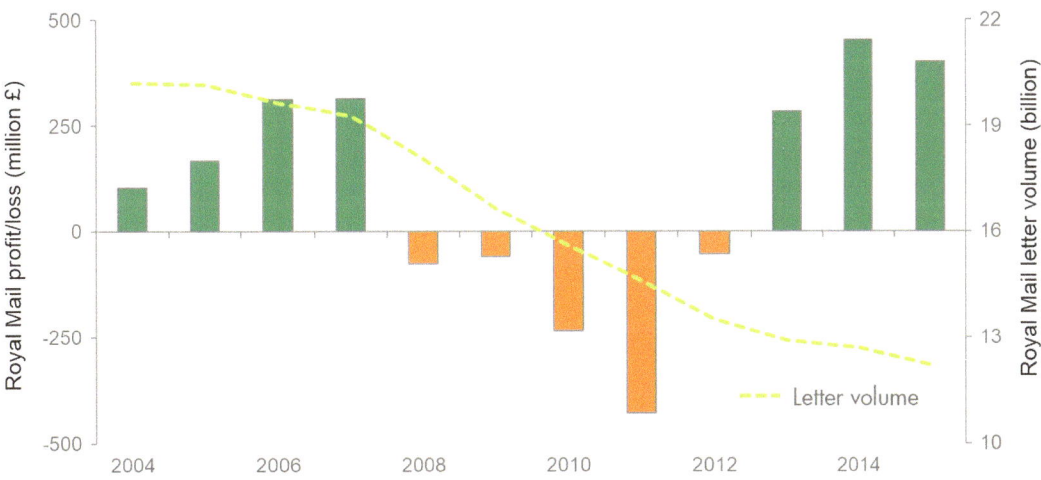

Fig. 21 Privatisation in 2013 led to a return to profitability, despite a continuous decline in letter volumes. *Source* Vivid Economics

is in doubt, as letter volumes continue their lower for longer trend (they are forecast to decline 4–6% annually). In addition, fierce competition from foreign operators in the parcel market has led to a 20% overcapacity, which prevents Royal Mail from staking a large claim in the main growth market. As efficiency gains are exhausted, further profit growth will likely have to come from parcels—where profit growth has been unimpressive so far—or diversification, which Royal Mail has only pursed at small scale and is losing ground to more aggressive competitors.

Royal Mail has begun to move vertically into e-commerce and logistics but is being outpaced by movement the other way, which is likely to increase pressure on margins. Royal Mail (like Deutsche Post and France's La Poste) is moving vertically up the value chain towards website development, digital marketing and parcel collection points. For now, Royal Mail appears to be progressing exclusively through mergers and acquisitions; no major organisational reshuffle—such as the creation of a new board position or business unit—has been announced. The deal volumes of its e-commerce-related acquisitions have not been disclosed, nor have revenue forecasts been given by Royal Mail, but it is estimated they will add only £100 million in revenue over the next two years. Amazon has started to build an in-house delivery network, expanding its operations from e-commerce to the underlying parcels and logistics business which, by comparison, reported revenues of £1.5 billion in 2016.

Royal Mail has acquired foreign postal operators to diversify into new international markets, but again these have been small. In the USA, it acquired Postal Express for $13 million in 2017 and Golden State Overnight for $90 million in 2016. In Europe, Royal Mail acquired ASM Transporte Urgente of Spain for €71 million in 2016. By comparison, Deutsche Post's acquisition of DHL in 2002 for around €2 billion and Exel for €5.6 billion in 2005 are of a different magnitude. Part of the reason for these smaller acquisitions by Royal Mail was capital limitations. There are benefits from taking a more measured, modular approach that does not require large-scale strategic and organisational shifts: the risk and costs are much lower. This is evidenced by the $3.9 billion in restructuring costs arising from Deutsche Post's expansion into the US market, followed by an exit from domestic US deliveries. But the reality is that a strong presence in new markets requires strong investment that Royal Mail is currently not delivering.

The government effectively paid a one-off fee to allow Royal Mail to break out of constraining monopoly regulation, which was crucial for Royal Mail to become profitable again. With its high transformation costs before privatisation, Royal Mail was unable to pursue any strategy other than inaction, despite expectations of a continuing lower for longer trend and a high payoff from divestment or diversification. By taking on the pension liabilities of Royal Mail, the government made privatisation possible, which proved to be net beneficial. For oil and gas in China, this could take the form of divesting network infrastructure (as a precondition to becoming a private company) and accessing private capital markets for growth opportunities.

With the right regulatory environment and leadership, stabilising the core business and maintaining service security is possible without major top-level organisational change. Except for the separation of the post office arm and the government taking on the pension liabilities, Royal Mail's turnaround has been achieved without making major changes to its organisational structure. However, as mentioned above, this turnaround was concentrated in Royal Mail's core business, and the company's long-term growth prospects are not yet certain.

Deutsche Post: An example of a transformative strategy

Deutsche Post's diversification strategy into global markets and logistic services began in the late 1990s, predating any fall in mail volumes. The decline of mail volumes in Germany began later than in both the USA and the UK, peaking in 2008 and falling only by 13% by 2015. In comparison, the falls in mail volumes in the USA

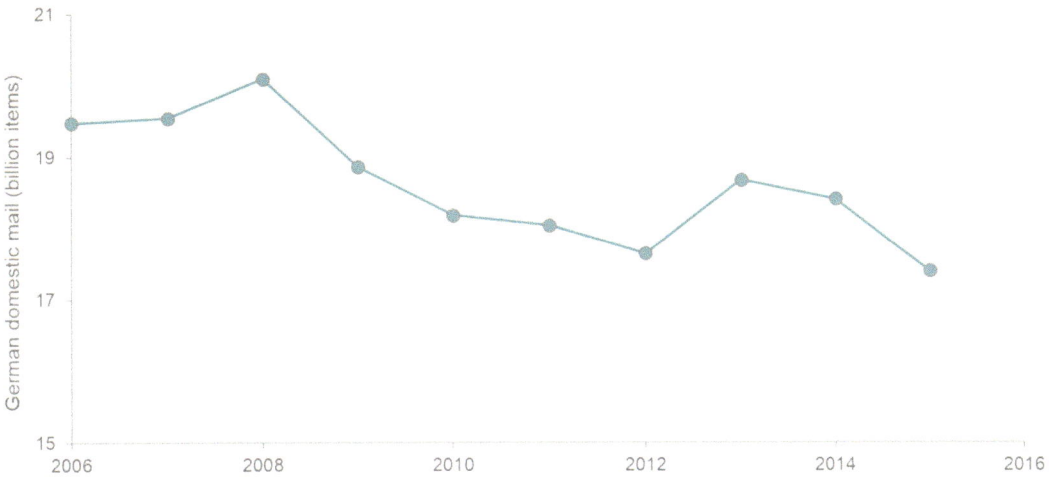

Fig. 22 The decline in Germany's postal market began in 2008, later than in the USA and the UK. *Source* Vivid Economics

and the UK from their respective peaks were 38% and 42% respectively (Fig. 22). By the time mail volumes began to fall appreciably, Deutsche Post was already well diversified into different geographies and markets—hence, lower for longer had a minimal overall impact on the revenue and earnings of Deutsche Post.

Since 2001, Deutsche Post has completed several large acquisitions and achieved consistently high profits, despite falls in domestic mail volumes. These large acquisitions include DHL in 2002 and Exel in 2005, although a global diversification strategy with smaller acquisitions has been implemented since 1998. As a result, Deutsche Post has taken early positions in growing markets and developed expertise, which has led to strong revenue growth (68% from 2001–15) from a wide range of ventures and little noticeable impact from lower letter volumes, as seen in Fig. 23.

Deutsche Post's motivation for diversification was triggered by the European Union postal reforms of 1997, which aimed to increase competition in domestic mail. The EU directives led to Germany passing new laws to guarantee postal service quality and promote greater competition in 1997, making Germany the forerunner in postal market liberalisation in Europe. This set expectations of a declining future share of the domestic postal market, which in 1997 represented more than 75% of Deutsche Post's revenues. The chosen response was to diversify into a global logistics company offering a full range of services—the payoffs from such a strategy were expected to be high given the rapidly growing courier, express and parcel markets in Europe and abroad, as well as the desire of business customers to have a one-stop shop for all logistics services.

Large cash flows and privatisation allowed diversification activities to occur with minimal internal barriers. Deutsche Post began to diversify well before lower for longer began to impact margins—the profit from its mail operations was around €2 billion in 2000. This allowed Deutsche Post to pursue diversification from internal financing rather than having to raise capital externally. Privatisation further facilitated this by streamlining decision-making, allowing much faster acquisitions. Systematic restructuring of the company in 1989, and again following German reunification up to 1997, provided valuable experience in assimilating and accommodating new assets.

Germany has had more flexible regulations for universal service provision than the USA and the UK, which has minimised the external barriers to diversification for Deutsche Post. Although

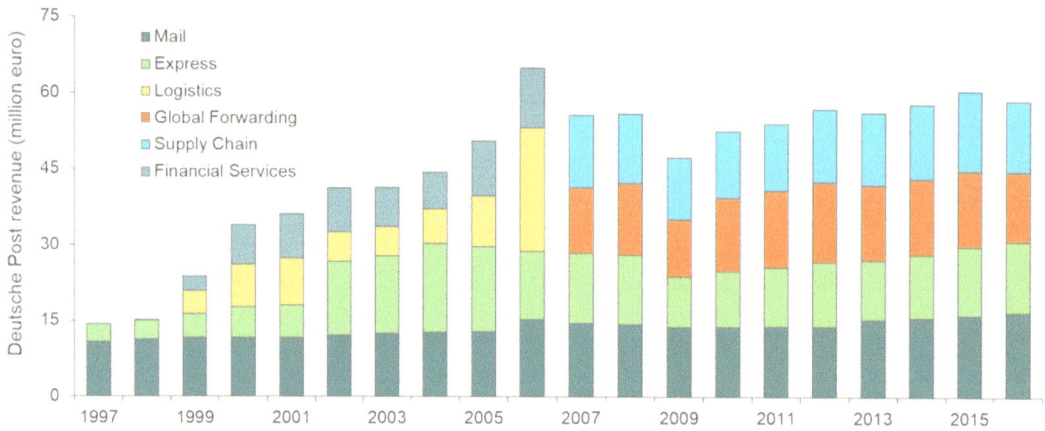

Fig. 23 Since the early 2000s, Deutsche Post's revenue has come from several areas; total revenue has not been appreciably reduced by declining domestic letter volumes. *Source* Vivid Economics

Deutsche Post has historically had a universal service obligation, as well as some exclusive rights, since 2008 the market has been fully liberalised, and Germany no longer formally designates one company to provide that universal service. Without this pressure, Deutsche Post has been free to continue exploring international markets and create new products, while willingly providing a universal service on business grounds.

Deutsche Post's path to privatisation in the 1990s was motivated by losses caused by an inefficient organisation and EU directives, rather than as a response to falling mail volumes. Before 1989, a state-controlled company, Deutsche Bundepost, was in control of all postal and telecommunications services. This company adopted the bureaucratic structure of its governing bodies that was inappropriate for a market-oriented company, leading to complex processes, no bookkeeping and no coherent strategy for marketing and sales. The result was heavy losses, with a deficit of €320 million in 1990 alone. The first round of postal reform in 1989 separated the different services and restructured them into corporate organisations, quickly reversing their fortunes. When EU directives on opening up postal markets to competition were released, Germany scheduled the privatisation of Deutsche Post, which began in 2000.

Deutsche Post stated its aim to become a globalised logistics company in 1997, as it became clear domestic markets would be exposed to increasing competition. The 1997 Postal Act laid out the regulations to induce greater competition in the German postal market. One of the main policies of this act was the removal by 2002 (later extended by six years in 2001) of Deutsche Post's exclusive licences on some postal products. Consequently, Deutsche Post began its globalisation strategy to become "the number-one global player", not only to escape a tightening domestic market, but also to access the rapidly growing express, courier and parcel markets in Europe and globally. This transition began with some smaller acquisitions and new international services, but large acquisitions did not occur until after privatisation in 2001.

Following privatisation, Deutsche Post launched an intensive programme of diversification through acquisitions and the divestment of non-core activities. Privatisation afforded Deutsche Post the freedom to use its strong internal cash flows to pursue larger acquisitions. DHL, a US express logistics company, was acquired in 2002 for €2.4 billion, just a year after Deutsche Post's privatisation; Exel, a UK-based logistics company, was acquired for €5.6 billion in 2005. These gave Deutsche Post an immediate foothold in new markets, transforming it from a mail-centred company into a global, full-spectrum logistics provider. Deutsche Post also made more than €8 billion from divestments in 2004-10, mainly from the sale of Postbank, raising funds for further acquisitions.

In more recent years Deutsche Post has focused on innovation, establishing innovation forums and creating a dedicated innovation unit within DHL. In 2010, Deutsche Post released E-Postbrief (E-mail letter) as a digital alternative to physical letters, although its popularity is hard to determine as usage figures have yet to be released. More recent innovations include drone deliveries, which are completing a third round of testing, electric postal delivery vehicles and a real-time supply chain management system. DHL Innovation Centers act as forums to encourage collaboration between customers and partners to develop new products and services.

Deutsche Post has made a remarkable transition that has allowed it to weather the decline in its previous core product (domestic letters) and is now positioned as a global leader in growing logistics markets. However, this strategy of rapid diversification is not without drawbacks; moving so rapidly into new ventures can lead to heavy losses. However, this should be accepted as a risk of such bold strategies. Deutsche Post discontinued its domestic express offering in the USA and incurred $3.9 billion in restructuring costs as a result in 2009.

2.2 Electricity Grids in Transition

2.2.1 China's Network Arrangements

(1) Overview

China's current network arrangements provide grid access to 100% of its vast and widely dispersed population. In 1990, 89% of the population had access to electricity; by 2014, 100% were connected. In the same period, electricity consumption per person increased eightfold, from 511 kilowatt-hours (kWh) to 4,047 kWh per person, which is half of the average OECD electricity consumption of 8,004 kWh per person. This vast increase in electricity access and consumption has been delivered through one of the largest and most reliable electricity networks in the world.

China has abundant energy reserves, such as hydropower and coal, to serve its rapidly growing electricity demand, but these lie far from large demand centres. The geographically uneven distribution of energy resources and demand centres limits the flexibility of the system to respond to imbalances in supply and demand. It also requires considerable investment in long-distance transmission networks.

The Chinese transmission system is organised on administrative lines and composed of provincial networks, with limited regional and national integration. Detailed planning, investment and operation are primarily coordinated at the provincial level, with a smaller degree of coordination at the level of the national network. The institutional framework of the Chinese electricity sector is complex, as shown in Fig. 24.

Network planning follows a top-down process. The National Development and Reform Commission (NDRC) specifies general network investment in five-year planning cycles, after which provincial governments and local NDRC branches finalise these plans for their administrative territory. The State Grid Corporation of China, the state-owned electric utility, invests in interconnectors between regions, while the regional grid companies invest in interconnectors between provinces. Provincial grid companies focus on the bulk transmission network within their provinces. Subsidiaries of grid companies at the prefectural and county levels are responsible for distribution networks.

Electricity pricing follows a similar top-down process. The regulatory bodies set the wholesale and retail prices of electricity, which can then be amended by provincial governments to achieve policy and economic development goals.

China explores alternative approaches for restructuring and marketisation of its electricity sector. In 2014, the NDRC launched a pilot project in Shenzhen to accumulate experience for wider adoption of performance-based regulation in China. The pilot project aimed to incentivise

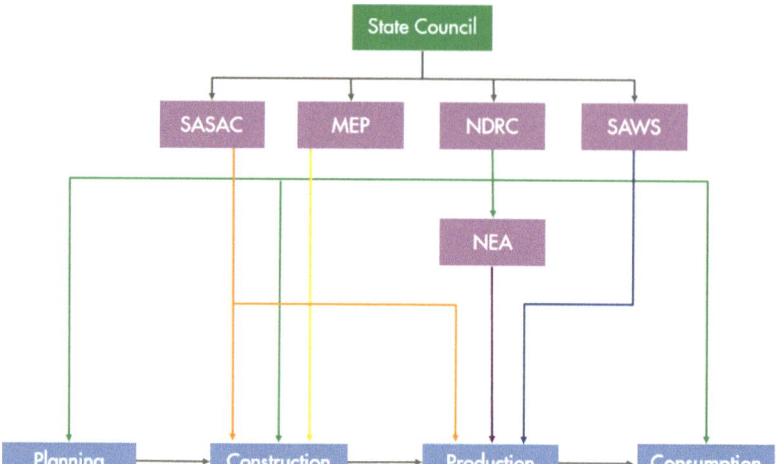

Fig. 24 The institutional framework of the Chinese electricity sector comprises several organisations with overlapping responsibilities. *Note* SASAC = State-owned Assets Supervision and Administration Commission; MEP = Ministry of Environmental Protection; NDRC = National Development and Reform Council; SAWS = State Administration of Work Safety; NEA = National Energy Association

grid companies to decrease their costs. It determined permitted costs and profits of grid companies and capped their total revenue. Following the success of the pilot project in Shenzhen, the NDRC decided to expand the reform step by step to other provinces and then nationwide.

(2) Performance of Chinese network arrangements

China's current network arrangements have delivered a large and stable grid with high levels of energy access, but the grid is fragmented geographically and has highly centralised management. As Chinese regulatory bodies are not independent, planning and pricing decisions at the central and local levels may be influenced by political objectives. The top-down planning approach and lack of coordination with provincial and regional grid companies may lead to inefficient investment decisions and poor coordination between generation and transmission investments. Such investment patterns create pockets of generation where electricity supply is abundant, but an inability to transmit to areas where it is scarce, due to limited transmission capacity. As a result, while China has significant

renewable generation capacity, such as solar and wind, large volumes of renewable energy are curtailed, increasing overall electricity costs and carbon emissions.

The transition to a low-carbon and decentralised electricity system is likely to worsen these inefficiencies and create new challenges. As the Chinese economy and energy sector decentralise, it will be increasingly difficult to maintain the electricity grid's high level of reliability at an affordable cost. Moreover, decentralisation of electricity resources requires significant investment in smart capabilities and creates challenges for the efficient planning and delivery of infrastructure across different networks.

Therefore, China may benefit from reforms to network arrangements, as have many other countries that have adopted international best practice for efficient network provision.

2.2.2 Key Principles of Efficient Network Provision

In order to realise the economic benefits of liberalised electricity markets, international best practice suggests a set of principles for efficient network provision:

(1) Proceed towards full liberalisation of the wider electricity system

Effective investment in, and operation of, the wider electricity system is a precondition for efficient electricity supply. This requires liberalisation of sectors suitable for competition (fuel production, generation, retail), use of markets to procure key services (capacity, balancing), and the pricing of externalities, such as air pollution and carbon emissions.

(2) Align incentives with public policy objectives

Make the incentives of network providers consistent with the provision of a reliable and affordable supply of electricity by controlling the monopoly behaviour of network companies and ensuring that prices reflect underlying costs:

- *Reform electricity network institutions.* Electricity networks are natural monopolies, with little scope for improvement through competitive markets. It is therefore critical that their incentives are aligned with public policy objectives. A monopoly faces incentives to underinvest in new infrastructure and to charge prices that are higher than its costs. A state-owned company may face incentives to prioritise short-term political objectives, rather than longer-term public policy objectives. These incentives can be mitigated through institutional reform of the electricity network. One option is to reform the network company's incentives through performance-based regulation enforced by an independent regulator. Another is to separate network operation and ownership through the creation of an independent system operator (ISO). The UK and most European countries currently use performance-based regulation, while the USA uses the ISO model across its transmission systems, for example, PJM Interconnection (the transmission system in north-eastern USA).
- *Consider the use of locational pricing.* Efficient network investment and operation make use of information on network congestion. If implemented, locational (nodal or zonal)

pricing can help reveal the costs of network congestion. Nodal pricing is used in several US states, Argentina, Chile, Ireland, New Zealand, Russia and Singapore, while zonal pricing has been adopted by most European countries and Australia. However, locational pricing has disadvantages as well as advantages. Importantly, locational pricing is most effective once time-of-use pricing is fully implemented across network users.

(3) Take further action to meet the challenges of a decarbonised system

The electrification of energy demand and improvements in the efficiency of electrical appliances will make the future volume and demand for transmission capacity more uncertain. Flexible resources such as electricity storage and demand response can substitute for new network investment, as long as sufficient investment incentives are present:

- *Designate strategic zones for transmission-scale renewable generation to reduce planning and investment uncertainty.* Renewable energy resources may be located far from demand centres and thus require large-scale transmission investment. Uncertainty over the volume and location of generation can be mitigated through zoning.
- *Ensure there are revenues available to encourage providers of flexible resources to offer a full range of system services.* The flexible resources needed for decarbonisation contribute several system services, such as balancing and frequency response, but there may be underinvestment if markets do not exist for these services. Several electricity markets in western China run demand curtailment markets, allowing flexible resources to generate revenues.

(4) Prepare for the development of a decentralised electricity system, and its associated digitalisation

By investing in the coordination of decentralised resources, their control, balancing, security and data flows:

- *Coordinate investment in decentralised resources.* A coordination problem arises when independent developers that lack information about the plans of other developers make similar investments, creating overinvestment or, if the developers are risk averse, underinvestment. Either way, the result can be inefficient. Solutions include formal processes for multilateral resource planning and the publication of current and consented resources, as used by the transmission and distribution system operators of Spain and Ireland.
- *Determine how decentralised resources will be controlled.* While distribution networks today are largely passive, an active network is capable of accommodating distributed resources. As the electricity system becomes more active and complex, a single system operator may start to rely on intermediaries (such as virtual power plants) and partners (such as distribution system operators) to assist with system balancing. New systems of control, with new computational requirements, administrative rules and institutional characteristics may then be employed, reflecting new operational vulnerabilities.
- *Balance data transparency with security.* As information communications technology infrastructure expands, the amount of data from the power system grows. The data systems need their own infrastructure, with public access to facilitate competition and optimise operations. Meanwhile, the distribution of data across resources creates new risks of cyberattack and privacy loss, which can be solved through adequate protocols.

2.2.3 Roadmap for Efficient Network Arrangements

This section presents a suggested roadmap for the development of network arrangements in China, as the country simultaneously carries out large-scale investment, market reforms and decarbonisation of its power system. These options are based on international best practice and leading thinking on future arrangements, as set out in Sects. 2.2.6 and 2.2.7.

The roadmap is based on the following guiding principles:

- *Strong markets need strong government.* Market-based solutions have the potential to identify and deliver cost-effective investment and operation of power systems. However, both markets and natural monopoly networks benefit from a strong government that takes an active role in ensuring institutional incentives are aligned with public policy objectives. Network incentives can be aligned through separation of roles (unbundling) or strong regulation.
- *The institutional framework can be developed progressively.* Wholesale institutional reform is challenging and disruptive. At the outset, small changes in current practice and small-scale pilots may provide proof-of-concept sufficient to build consensus for larger-scale reforms.

The roadmap suggests the following:

(1) Immediate actions

- *Continue the market liberalisation programme.* Phase 1 of the DRC-Shell cooperation suggested a programme of electricity market liberalisation; consistent with this, China's 13th Five-Year Plan (2016–20) aims to improve the systems by which markets play the decisive role in resource allocation. It will be important to continue with the market liberalisation programme to deliver a more advanced and efficient electricity system.
- *Rationalise investment planning.* Clearly defined metrics for reliability and economic efficiency help network planners to identify efficient investments. Meanwhile, the application of the "beneficiary pays principle" encourages investment that increases productivity and avoids diverting national resources to stimulate regional output. Together, these approaches, when applied at national and regional level, facilitate greater interconnection and sharing of generation services.
- *Implement a coordinated approach to investment.* The use of a common investment

framework enables coordinated planning of generation and network investment, harnessing both strategic decisions and market enterprise. For example, the framework might set out the role of generation zones for large-scale renewables alongside alternatives like small-scale distributed generation.

- *Implement smart system architecture.* Smart distributed resources are crucial to affordable decarbonisation. Before building the distributed resources, the system architecture to manage it should be laid down. At a minimum, this includes: deployment of smart meters to introduce time-of-use pricing to consumers in the distribution network; upstream information and communications architecture; and R&D to develop technical solutions for the smart grid.

(2) Move towards efficient pricing

- *Deregulate prices.* Cost-reflective pricing can signal investment and operational efficiency to decision makers. The deregulation of prices can proceed sequentially, with further price reform contingent on the success of previous reforms. Price reform could commence upstream and progress downstream, beginning with input fuels and progressing through generation, network access and ending with retail, with provision to protect retailers if wholesale prices rise above retail prices before deregulation
- is completed.
- *Create harmonised trading arrangements between transmission systems.* Use prices to determine interconnector flows between provincial and regional transmission systems, signalling which provinces or regions could benefit from new investment. The prices would stimulate lower-cost generators to respond to demand.
- *Implement time-of-use pricing.* Time-of-use pricing allows consumers and flexible resources to respond to variation in generation costs and demand. Again, it can proceed sequentially, starting with larger consumers, such as industrial facilities with flexible

production schedules, and ending with smart household appliances.

- *Consider locational pricing.* Similarly, locational pricing can signal investment and operational efficiency, but it brings disadvantages as well as advantages. China may consider locational pricing, after it has implemented time-of-use pricing, to signal geographical network constraints once demand peaks have been shaved. Zonal pricing is a potential intermediate step between uniform and full nodal pricing.
- *Protect end users.* The deregulation of retail electricity may result in rent-seeking by retailers, raising consumer prices. This can happen if consumers do not switch suppliers readily or for other reasons if competition is not effective. Policies to protect consumers could be developed alongside any deregulation of retail electricity prices.

(3) Begin market trials

- *Create small-scale trials.* Create small-scale trials to competitively procure new transmission investments, non-network alternatives to new transmission assets, ancillary services and so on. Use competitive tenders or auctions for the trials. These may provide proof-of-concept and experience with innovative, cost-effective solutions. To be successful, procurement must be open-access and transparent.
- *Progressively introduce market procurement.* If the competitive procurement trials are successful, they can be scaled up and wider market procurement progressively introduced, where appropriate, across each transmission system. Market procurement can reveal information about the relative cost and benefits of a range of technologies.

(4) Make institutional choices

- *Develop transmission network institutions.* Options include the status quo, an enhanced role for market procurement, and a regulated transmission system operator (TSO) or independent system operator (ISO). International experience of the regulated TSO or ISO model is yet to reveal the best performer of the two, so it matters more to adopt a good quality institutional model early than to choose between the options.

- *Select a model of control for decentralised resources.* Initially, when the number of resources is small, the TSO may be able to control them directly. However, as the number of resources increases, and as temporal and locational pricing become more sophisticated, the computational, commercial and contractual capacity of a single operator model may be exceeded, and new models of control may be needed.

2.2.4 An Introduction to Electricity Networks

Electricity networks transport electricity from generators to consumers through a combination of high voltage transmission networks and low voltage distribution networks. The system operator balances supply and demand on the network at all times, within the constraints of available network capacity.

Liberalisation of electricity markets can lead to more efficient electricity systems by replacing government control over markets with competition. However, government intervention remains necessary to correct market failures. Electricity networks are a natural monopoly and may be subject to regulation to ensure efficient investment and operational decisions and pricing.

(1) Basic concepts

Networks are used to transport electricity and keep costs down. First, power generators often operate under economies of scale—it is cheaper to build a small number of large power plants than a large number of small plants to serve a population. Second, generation may be located remotely, far from sources of demand (for example, due to environmental constraints). Third, networks can reduce redundancy in generation investment, where patterns of generation and demand vary geographically.

To reduce losses and unit costs, most networks combine high voltage transmission with low voltage distribution. Electrical losses are low when power is transported at high voltage, and high when transported at low voltage. To minimise losses, electricity networks use high voltage transmission to carry electricity over long distances, and low voltage distribution to deliver electricity to consumers. Figure 25 provides a stylised illustration of a conventional electricity network.

Transmission networks are typically meshed networks, while distribution networks are radial networks. Meshed networks are a complex arrangement of links, with multiple paths connecting different nodes. Meshed networks are more resilient: multiple links provide redundancy, such that if a single link fails, other paths remain

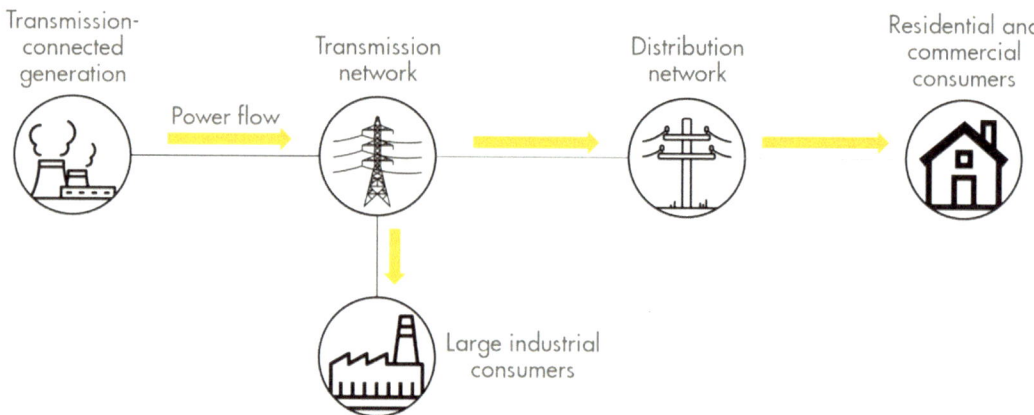

Fig. 25 Electricity networks transport electricity from generators to end users. *Source* Vivid Economics

intact. This redundancy raises the cost of the net-work. Furthermore, flows of electricity through a meshed network are governed by physical laws and are difficult to predict. The operation of a meshed network is, therefore, complex and com-putationally demanding. Radial networks are a simple arrangement of links, with single paths connecting nodes. Radial networks are less resi-lient, but also less costly and less complex.

The system operator balances supply and demand on the network at all times. Network voltage and frequency need to be maintained within precise limits to prevent damage to equipment and blackouts. The maintenance of stable voltage and frequency is known as system security. While generators and consumers can trade privately for electricity supplied through the network, system security benefits everyone equally. The system operator arranges with generators and consumers to adjust output or consumption to stabilise the network.

Transmission and distribution links have finite capacity, which leads to congestion. Congestion occurs if demand at a node is greater than the capacity of the links supplying that node. As demand is variable, it is usually not efficient to invest in sufficient capacity to meet peak demand at all times. The efficient level of capacity is lower, which gives rise to a degree of congestion. The system operator manages this congestion by balancing generation output and consumer demand across the network.

The system operator typically balances in one of two ways. Under a central dispatch system, the system operator observes potential generation and consumption on the network and optimises the pattern of generation and consumption within the network constraints. Under a bilateral trading system, the system operator observes contracted generation and consumption on the network and identifies areas of network congestion. The sys-tem operator then curtails some generators causing the congestion and arranges for alterna-tive generators on less congested lines to provide the missing generation.

(1) Balancing the electricity system

A simple example of balancing an electricity system under network constraints is shown in

Fig. 26. In this example, a system operator bal-ances a system of two cities interconnected with a capacity-constrained transmission line. The numbers in green, blue and purple are the inputs for the system operator's balancing problem. The numbers in red are the outputs.

In City A, generators can produce up to 150 megawatts (MW). Generation costs $10 per megawatt-hour (MWh). Consumers demand 50 MW. In City B, generators have 50 MW capacity. Generation costs $20/MWh. Con-sumers demand 90 MW, exceeding the local generation capacity. Consumer demand in both cities is constant and does not change with price. The interconnection between City A and City B can carry up to 80 MW.

To minimise the total cost of the system, the system operator first uses the cheaper generation in City A. Generators in City A serve local demand (50 MW) and export 80 MW to City B (in total 130 MW). Because the 80 MW capacity of the interconnection is fully utilised, the more expensive generators in City B serve the rest of City B's local demand (10 MW). As a result, the system price equals the cost of the generators in City B ($20/MWh), and the total cost is $2,800 (140 MW x $20/MWh).

(2) Electricity networks in a liberalised elec-tricity system

The liberalisation of electricity markets has a strong economic rationale: competition maximises effi-ciency and increases welfare. For this reason, many countries have gradually liberalised their electricity markets, reducing government intervention and increasing competition. Beginning in 1990 with the UK, liberalisation spread to Norway, Chile, Argentina, New Zealand and Australia in 1991, and began to spread across the USA from California in 1994. The European Commission published directives in 1996 that encouraged more countries across Europe to liberalise.

Liberalisation of electricity markets means the development of competitive markets, with mini-mal government control over the technologies and prices in these markets. Competitive markets operate in wholesale electricity, capacity and the procurement of balancing services. Historically,

Fig. 26 Example of system balancing with network constraints. *Source* Vivid Economics

this has been achieved through the unbundling of generation and retail from the natural monopolies of transmission and distribution. In parallel, a process of privatisation often occurred, particularly in the competitive parts of the supply chain.

However, market failures remain and government interventions are needed to correct them. As summarised in Fig. 27, measures include: a strong carbon price to incentivise low-carbon generation and reduce demand; a stable, predictable and credible policy setting to reduce policy risk and the cost of capital for low-carbon investments; and, the regulation of natural network monopolies.

2.2.5 Challenges in Network Provision
Network service providers face challenges as electricity systems decarbonise and decentralise.

(1) What are the challenges?
The challenges of efficient network provision encompass planning and delivery, operation and cost recovery. Challenges in planning and delivery

arise from uncertainty over future electricity demand and the difficulty of coordinating network investment with independent generators. Challenges in operation arise from the capacity constraints of the network, the complexity of the system, and the unpredictability of flows of electricity. Challenges in cost recovery arise from the natural monopoly characteristics of electricity networks, and the difficulties of mitigating monopoly behaviour with conventional regulation.

(1) Planning and delivery
Planning and delivery of electricity networks faces uncertainty over future electricity demand and difficulties in coordinating network investment with independent generators. First, the volume and location of future demand is uncertain and depends on population growth, changes in settlement patterns, and changes in industrial structure, technology and economic growth. Network planners judge where network assets will be required. Second, while vertically

Fig. 27 The objective of an electricity market is to supply services so that demand and supply can be balanced, while correcting market failures. *Source* Vivid Economics

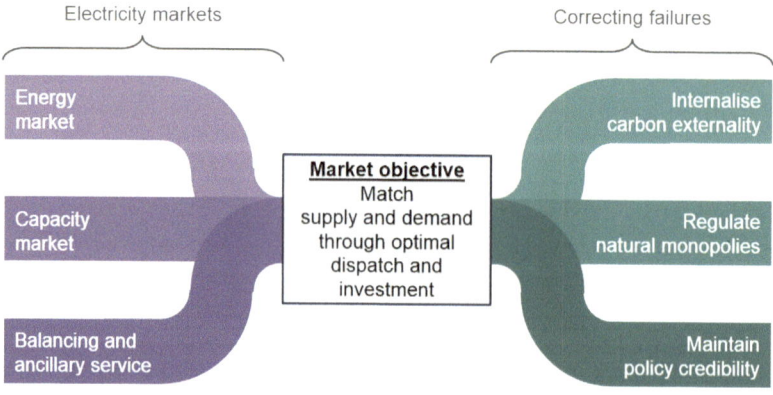

integrated utilities coordinate their plans for generation and network investment, in a liberalised electricity system, generation and network investment are carried out by separate organisations. Without coordination, generators face the risk that their revenues may be curtailed by network congestion, and networks face the risk that generators will underuse their assets.

(2) Operation

Balancing the electricity system is challenging due to system complexity and the unpredictability of electricity flows. With many sources of generation and consumption, as well as network constraints, the optimal level of production and consumption for each source is a complex calculation. Furthermore, due to the physical laws governing electricity networks, the precise flows of electricity through the network depend on the volumes of consumption and production of each generator and user and cannot be predicted in advance.

(3) Cost recovery

Electricity networks are natural monopolies. One of the roles of the system operator is to levy charges to pay the network owner. This is achieved by designing tariffs that recoup capital, operation and maintenance costs for the network owner, and passing these costs through to network users. When the network operator is also the network owner, it operates as a monopoly and is incentivised to underinvest in network infrastructure and charge high prices to consumers.

Innovative arrangements are needed to mitigate monopoly behaviour effectively. Electricity networks are characterised by high capital costs and economies of scale. For these reasons, electricity networks are natural monopolies, with a single network serving a given area. The conventional approach to mitigating monopolistic behaviour in a natural monopoly is regulation. However, regulators have imperfect information on current network costs and how these costs can be reduced over time as productivity improves. Depending on the type of regulation, network

companies may face incentives to overstate their costs or to overinvest.

(2) Future changes: decarbonisation and decentralisation

Potential changes in key characteristics of the electricity system are encompassed within two broader trends: decarbonisation of electricity and the wider energy system; and decentralisation of system resources, as summarised in Fig. 28. These changes, and their implications for the challenges of efficient network provision are described below in turn.

(1) Decarbonisation

Decarbonising an electricity system requires changes in generation technologies and far-reaching changes in patterns of electricity demand. Generation technologies will shift from fossil generation to low-carbon generation, that is, a mix of carbon capture and storage, nuclear, biomass and renewables. Electricity demand will be affected by increases in demand from electrification of end-use sectors, particularly heat and transport, as well as decreases in demand from greater efficiency of electrical appliances. There will also be a shift in the profile of demand, as low-carbon flexible resources (electricity storage and demand-side response) emerge to balance the relatively inflexible generation profile of nuclear and renewables.

These changes are likely to make harder the challenge of planning and delivering network infrastructure. Future volumes of demand will be more difficult to forecast, due to uncertainty over the level of electrification of end-use sectors and improvements in the efficiency of electrical appliances. Another element of uncertainty is the degree to which low-carbon flexible resources will reduce peak demand and, therefore, the level of network capacity.

Low-carbon flexible resources can be substituted for new network investments, thus reducing network costs. However, as these resources provide different system services (balancing, frequency response, network congestion mitigation), there may be underinvestment in

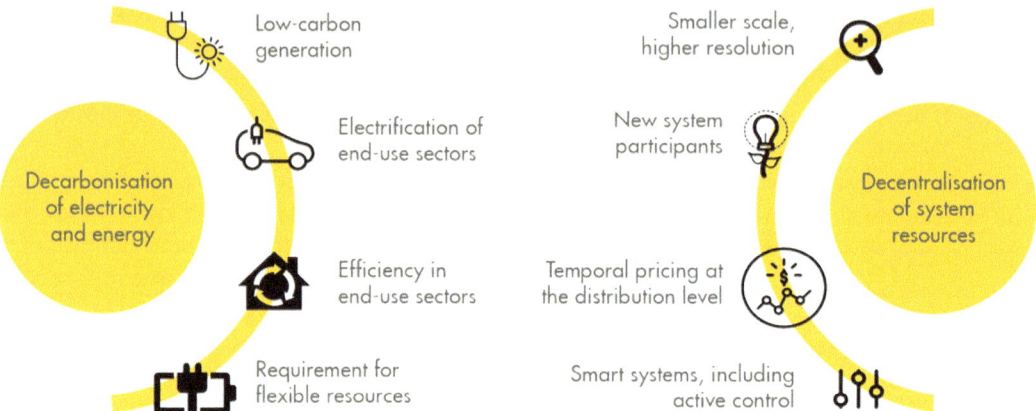

Fig. 28 Future electricity grids will be shaped by two broad trends: decarbonisation and decentralisation. *Source* Vivid Economics

them if markets do not exist for all the services the resources provide.

(2) Decentralisation

Decentralisation involves a shift in electricity resources from the transmission system to the distribution system. Decentralised electricity resources comprise generation, demand response and storage. Decentralised generation includes wind and solar, which are increasingly connected to the distribution network, including at the household level. Demand response is the flexible operation of electrical equipment in response to system conditions; electric vehicles are expected to significantly increase the potential for demand response as their batteries can be charged at times of high electricity generation and low demand. Decentralised storage may also increase, potentially even at the household level.

To unlock decentralised electricity resources requires a smart grid. A smart grid is characterised by the predominance of controllable electricity resources (generators, storage and appliances) throughout the electricity system, the ability for users to decide when to use their devices, the development of operating standards that allow resources to be operated in a coordinated way, sufficient development of digital technologies (communications bandwidth, data storage and computing power), and adequate data security and privacy protocols.

Decentralisation exacerbates the challenges to planning, delivery and operation:

- A coordination problem may arise between the transmission system and the various distribution systems. Most new investment occurs in the transmission system, where adequate information is available on current and planned resources to inform new investment decisions. A shift in investment to the distribution network, where adequate information is not typically available, will create a coordination problem, where investors will not have a clear understanding of system needs nor the potential returns on investment. This could result in overinvestment, underinvestment, a poor technology mix, or a poor spatial distribution of resources.

- The computational requirements of balancing a decentralised system will increase significantly. Optimising the operation of the entire system requires knowing the optimal volume of output and consumption of every resource in the system. As the number of controllable resources increases, from the limited set of large resources in a transmission system to the total set of resources across all distribution systems, the computational demands of this optimisation calculation also increase. If computing technology is not able to meet these computational demands, then intermediate levels of control are required, and only partial optimisation is possible.

Finally, decentralisation will require the risks to data privacy and cybersecurity to be effectively managed. Decentralisation will be accompanied by a very significant extension of digital technology across all distributed resources and will create risks to data privacy and cybersecurity. Protocols to manage data use and control these risks can be developed and implemented.

2.2.6 Network Arrangements to Address Current Challenges

Best practice arrangements are needed to ensure efficient network provision. Section 2.2.5 describes challenges in efficient network provision, and how these challenges will grow as power systems decarbonise and decentralise. International experience since the liberalisation of electricity markets has produced strong evidence—based on best practice in network arrangements—on how to meet the challenges of efficient network provision. Current best practice arrangements provide a foundation for new arrangements to meet future challenges arising from decarbonisation and decentralisation. These best practice arrangements comprise:

- an institutional model to align incentives with public policy objectives: an institutional model is needed that mitigates monopolistic behaviour and incentivises networks to invest in appropriate network infrastructure and operate it efficiently;
- strategic transmission planning: determining the appropriate profile of new transmission investment is a complex process that requires strategic planning;
- the appropriate level of locational pricing: investing in and operating networks efficiently requires an understanding of current network congestion; and
- a regime for merchant transmission investments: merchant transmission investors have the potential to deliver more adequate investment than a single-owner network.

(1) An institutional model to align incentives with public policy objectives

An institutional model is needed that mitigates monopolistic behaviour and provides networks with incentives to invest in appropriate network infrastructure and to operate that infrastructure efficiently. Two institutional models that can create an efficient regime are the transmission system operator (TSO) model with performance-based regulation, and the independent system operator (ISO) model. 0 highlights the key differences between these two models: a TSO both owns and operates the transmission system, requiring strong regulation; while an ISO is a system operator that is fully separated from ownership of all network resources. Several intermediate models also exist, for example, where the system operator and transmission owner are legally separate companies but owned by the same parent company (Fig. 29).

As shown in Fig. 30, most electricity systems have moved from vertically integrated monopoly utilities before liberalisation to a TSO or ISO model today. In 1985, Chile was the first country to adopt the ISO model. The UK shifted from vertical integration to a TSO with a performance-based regulation structure in 1990, with Germany following suit in 1998. Following the orders of FERC (the US electricity regulator), Pennsylvania-New Jersey-Maryland (PJM) and California (CAISO) transitioned from a vertically integrated structure to the ISO model in the late 1990s.

(1) TSO with performance-based regulation

A TSO is an entity that both owns and operates the transmission system; it therefore has incentives for monopolistic behaviour. The TSO owns all the network assets and is also responsible for planning, deployment and operation of the system. The TSO model is prevalent in most European countries.

Performance-based regulation is needed to align a TSO's incentives with public policy objectives. A TSO is difficult to regulate as it has better information than the regulator on the costs it faces. This gives rise to one of two problems. If the regulator tries to prevent monopoly

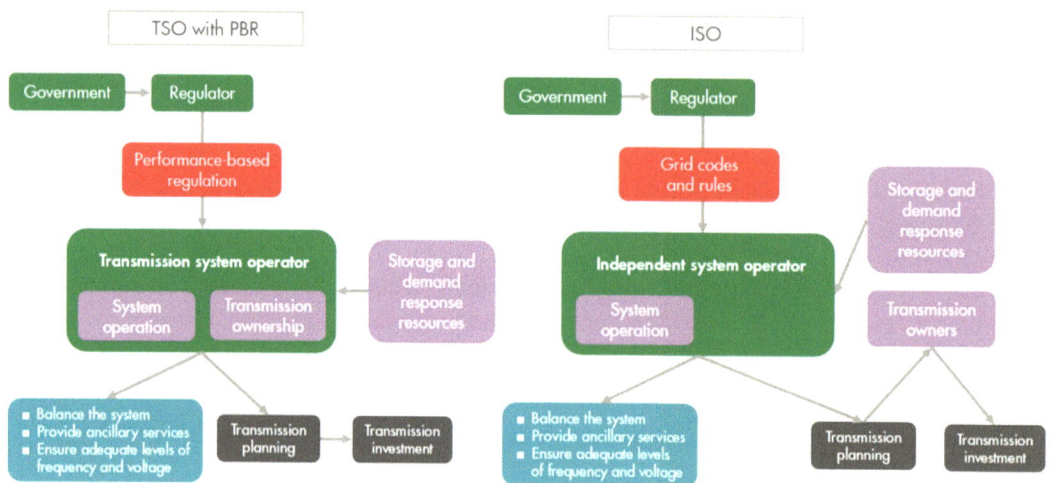

Fig. 29 The transmission system operator (TSO) and independent system operator (ISO) are the two main institutional models for system ownership and operation. *Source* Vivid Economics

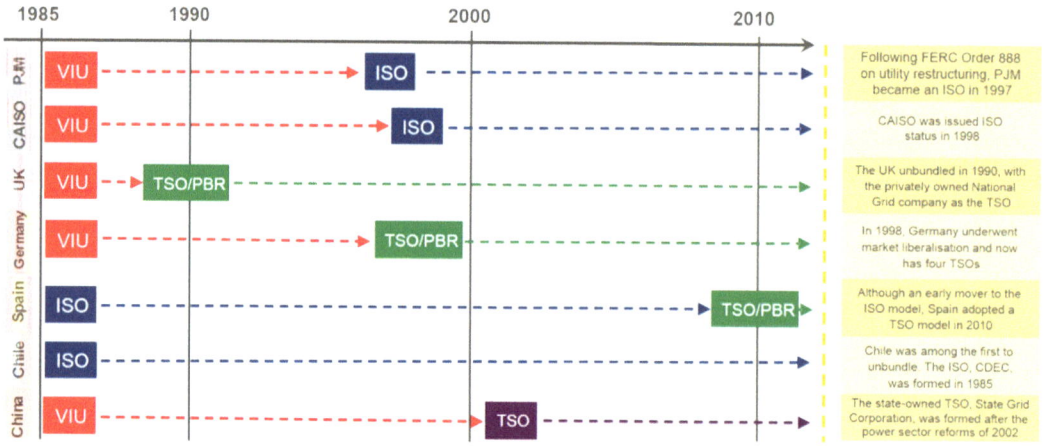

Fig. 30 Several countries shifted to the TSO and ISO models when they liberalised their electricity markets. *Note* VIU = vertically integrated utility; ISO = independent system operator; TSO = transmission system operator; TSO/PBR = transmission system operator with performance-based regulation. *Source* Chawla and Pollitt (2013)

behaviour by imposing a cap on prices, known as price regulation, it estimates the price level required for the TSO to recover its costs. The TSO then has incentives to exaggerate the costs it faces and secure price caps greater than its actual costs, a problem known as adverse selection. In this case, the TSO can underinvest and continue to set high prices. If instead the regulator tries to prevent monopolistic behaviour

by paying the TSO for its costs plus a regulated return (cost-of-service regulation), the TSO does not have any incentive to take necessary measures to decrease its costs, a problem known as moral hazard. In contrast to these approaches, performance-based regulation seeks to address both monopoly profits and underinvestment. An example of performance-based regulation is the imposition of a price (or revenue) cap, which is adjusted every year by the rate of inflation and target rate of productivity growth. The TSO is constrained in exaggerating its costs by the regulator conducting benchmark analysis and detailed studies of the TSO's historical accounts. The TSO has some incentive to lower its costs because it can retain the difference between the price cap and its actual cost as profit.

Performance-based regulation is still evolving and there is no consensus on the optimal practice. Several forms of performance-based regulation have been adopted, which continue to evolve. An example of their evolution is the introduction in the UK of the RPI-X (retail price index minus X) mechanism in 1992, and its eventual replacement by the RIIO (revenue = incentives + innovation + outputs) mechanism in 2013, as described below. A key challenge in performance-based regulation is its high information burden. The regulator has to review the TSO's accounts and business plans and undertake benchmark analysis.

Performance-based regulation in the UK
The UK introduced performance-based regulation in electricity networks in 1992 with the RPI-X mechanism, later replacing it with the more sophisticated RIIO mechanism in 2013:

- *RPI-X mechanism.* Ofgem and its predecessors, the electricity market regulator in the UK, used the RPI-X mechanism until 2013. Under this mechanism, Ofgem carried out cost forecasts to determine the base revenue required by the TSO to recover its costs. Based on this base revenue, Ofgem set a price cap and adjusted it each year for retail price inflation (RPI) and an assumed target rate of productivity growth calculated by statistical benchmark analysis. Ofgem reset the price cap at the end of the five-year price control period to ensure that the TSO's cost savings are passed on to end consumers. Under the RPI-X mechanism, the TSO had incentives to reduce its costs because it was able to retain the margin between the price cap and its own actual cost as profit, as shown in Fig. 31. However, the RPI-X mechanism did not

adequately incentivise service quality or innovation. It allowed cost savings to be achieved through decline in quality of service, and the five-year price control periods did not provide sufficient incentives to develop new technologies (such as smart meters) with longer investment cycles and the potential to provide cost savings over the longer term.

- *RIIO mechanism.* In 2013, Ofgem introduced the RIIO mechanism to address its concerns with RPI-X. The RIIO mechanism takes the elements of RPI-X that work well, such as determining base revenue, and adds extensive innovation and output targets to them. The RIIO mechanism determines the base revenue from output-led business plans with greater use of option analysis and scenario planning. In their business plans, companies compare the costs and benefits of options for delivering long-term outputs under various scenarios and assess the value of keeping these options open. Moreover, financial and reputational incentives strengthen the incentive structure. Ofgem rewards or penalises companies when they achieve or miss their output targets. Reputational incentives do not have a financial element, but they affect Ofgem's evaluation of base revenue in the next review periods. A longer control period, of eight years, encourages the TSO to focus on longer term investments, such as smart meters. RIIO provides a wider set of performance incentives than RPI-X, but at the cost of greater complexity and reduced transparency.

(2) Independent system operator
An independent system operator (ISO) is fully separated from ownership of all network resources. The ISO performs all system operation functions, including allocating network capacity among generators and consumers to respect the physical characteristics of the network, carrying out residual balancing to ensure electricity is delivered to where it is most valued, and maintaining the stability of the electricity system. In an ISO model, the transmission network

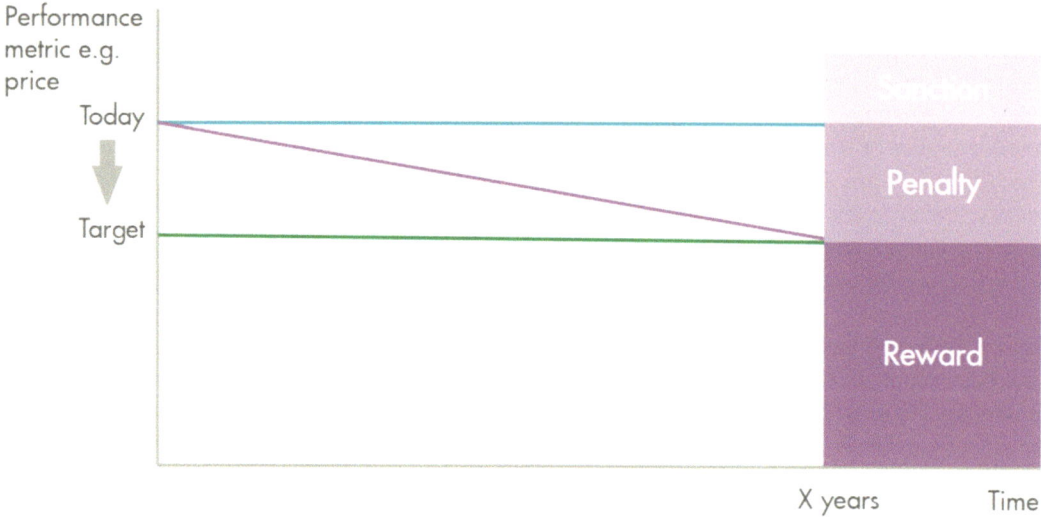

Fig. 31 Under the RPI-X mechanism, the TSO takes measures to decrease its costs because it retains revenue from cost savings. *Source* Vivid Economics

resources are owned by one or more transmission owners (TOs). The ISO levies charges from generators and/or consumers for use of the transmission system and pays these charges to the TOs. Typically, an ISO is also responsible for planning new transmission investments, either mandating TOs to make these investments or incentivise competitive tenders to facilitate delivery of investment. An ISO is usually a non-profit entity.

Separation of operation from ownership removes the system operator's incentives to charge monopoly prices. An ISO may be a non-profit organisation or may earn profits on revenues for system operation. As an ISO does not earn profits determined by revenues from use of the transmission system or by the cost of transmission investments, it has no incentive to charge tariffs that are higher than needed to recover the investment costs of the transmission network or underinvest in transmission assets. Unlike a TSO, therefore, an ISO does not need performance-based regulation.

The ISO's functions can be specified by its mandate and its behaviour can be governed by a set of rules. The mandate could be to minimise the total cost of meeting a given reliability standard. Rules could govern processes for transmission planning and investment; providing grid connections to new system resources; administering competitive tenders for new network assets; levying charges for network use; or monitoring market power in the electricity system.

Nevertheless, it is desirable to have in place a mechanism to incentivise the ISO. It is unlikely to be possible to specify a set of rules that perfectly incentivise the management to meet the ISO's mandate. That is, to encourage the ISO to carry out the planning, investment and operation of the transmission system to strike the best possible balance between reliability of electricity supply and economic efficiency of the network. Imposition of financial penalties on the ISO is likely to be a poor incentive mechanism as ISO revenues are likely to be small relative to the welfare losses arising from poor performance in operating the system. Instead, well-designed management incentives may be needed.

The ISO model is prevalent in North and South America. Chile, Argentina and Peru were early adopters of the ISO model. There are many ISOs in the USA, each covering a transmission network.

(2) Strategic transmission planning

The design of new transmission investment is a complex process and requires strategic planning. Specific challenges are: the uncertainty in volume and location of future demand, the coordination problem facing generation and network investors, and the large number of possible transmission investments. These challenges can be mitigated by strategic planning. Key characteristics of strategic planning include:

- *Define the objectives of new transmission investment.* Transmission planning is more effective when the objectives of new investment are clearly defined. These include reliability (security and adequacy of supply) and economic efficiency (reduction in the total cost per unit of output). Defining these criteria allows the benefits of new transmission infrastructure to be measured.
- *Estimate and compare the benefits of each proposed investment.* Cost-benefit analysis is a key planning tool to identify proposals that meet investment objectives. Reliability and economic efficiency can be assessed with electricity system modelling, and modelling of multiple scenarios can help identify the best possible infrastructure investments in conditions of uncertainty. If available, locational pricing provides a clear signal of congestion costs and can substantiate the economic benefits of new network infrastructure.
- *Consult all relevant stakeholders.* As the costs of investments are borne by network users, they have incentives to ensure that only the most valuable network infrastructure is developed. Stakeholder consultation can elicit views from generators, consumers (municipalities, consumer interest groups) and connected transmission and distribution systems to inform the cost-benefit analysis.

(3) Appropriate level of locational pricing

Investing in networks efficiently requires an understanding of current network congestion. New network investments that relieve significant network congestion are particularly valuable.

Many electricity systems operate a system of uniform transmission pricing, which does not signal network congestion. Under a system of uniform pricing, the price of electricity is the same at every network connection, regardless of the degree of congestion. It does not signal the need to invest in solutions to relieve network congestion—such as new network investment, generation or non-network alternatives to new transmission assets, such as electricity storage and demand-side response. While, in principle, the system operator can signal the need to invest through location-specific transmission charges, the rate of these charges is difficult to determine if the electricity price does not signal network congestion. Furthermore, uniform pricing results in redispatch costs, where a plant that is scheduled to generate is compensated for curtailment if the network is congested.

Nodal pricing can help signal network congestion. Nodal pricing, or locational marginal pricing, is a price mechanism that reflects the cost of supplying additional electricity at a specific network connection (node), given the demand for electricity, transmission constraints and options for local generation at that node. When there is no network congestion, overall demand is met at least cost and all nodal prices are the same. When network congestion occurs, demand is met by costlier local generation rather than cheap generation from another node, raising prices at congested nodes. Nodal pricing therefore signals the need to invest in solutions to relieve network congestion, such as new network investment, new local supply, and non-network alternatives to new transmission assets, such as electricity storage and demand-side response.

(1) Nodal pricing

Figure 32 provides a simple example of balancing an electricity system under nodal pricing. In this example, a system operator balances a system of two cities interconnected with a capacity-constrained transmission line under nodal pricing. The numbers in green, blue and purple are the inputs for the system operator's balancing problem. The numbers in red are the outputs.

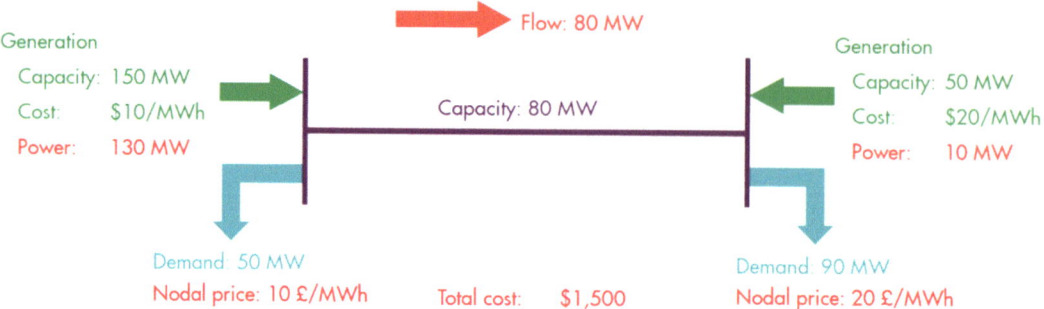

Fig. 32 Nodal pricing reflects the cost of supplying additional electricity at a given node. *Source* Vivid Economics

In City A, generators can produce up to 150 MW. Generation costs $10/MWh. Consumers demand 50 MW. In City B, generators have 50 MW capacity. Generation costs $20/MWh. Consumers demand 90 MW, exceeding the local generation capacity. Consumer demand in both cities is price inelastic (consumer demand is constant and does not change with price). The interconnection between City A and City B can carry up to 80 MW.

To minimise the total cost of the system, the system operator first uses the cheaper generation in City A. Generators in City A serve the local demand (50 MW). That is why the nodal price in City A equals the local generation cost ($10/MWh). Generators in City A also export 80 MW to City B. Since the capacity of the interconnection is fully used, expensive generators in City B supply the rest of local demand (10 MW). As a result, the nodal price in City B equals the local generation cost ($20/MWh). The total cost is $1,500 (= 140 MW x $10/MWh + 10 MW x $20/MWh).

However, nodal pricing has disadvantages as well as advantages. First, vulnerability to market power may arise because the segmentation of the electricity market into smaller locational markets increases the concentration of generators at each node with a supply deficit. Some authors challenge this view and argue that the network architecture is the main driver of market power, rather than the pricing mechanism. Nodal pricing can also reduce liquidity in long-term contracting, such as financial transmission rights and contracts for differences. This problem is addressed in the USA by averaging nodal prices into trading hub prices to provide liquidity to market participants.

Zonal pricing, another form of locational pricing, may provide a useful compromise between uniform and nodal pricing. Zonal pricing reduces the complexity of having large numbers of nodes by aggregating them into zones. Similar to dispatch under nodal pricing, the system operator first dispatches generation-given transmission constraints between zones. If transmission lines in a given zone are congested, the system operator has to redispatch generation in that zone to alleviate congestion. As a result, zonal pricing provides some of the benefits of nodal pricing in terms of signalling network congestion, but does not fully eliminate the redispatch costs associated with uniform pricing.

Locational pricing may be considered when time-of-use pricing is fully implemented. Wider changes to improve the flexibility of the electricity system through electricity storage and demand response are expected to reduce demand and generation peaks, which would automatically reduce network congestion relative to an inflexible system. These changes require time-of-use pricing to be fully implemented across all system resources (including end users) to be effective.

(4) A regime for merchant transmission investments

The entry of merchant transmission investors has the potential to deliver greater adequacy of investment than a single-owner network. Merchant transmission investors are third-party developers of transmission projects. If locational pricing is implemented, a merchant transmission investor has the incentive to invest in a new transmission link when the revenues from use of that link are greater than the investment cost. Therefore, in principle, merchant competition offers the potential to increase the adequacy of the transmission infrastructure by investing where an incumbent is not willing to do so. This might be the case if the incumbent is an unregulated monopoly or poorly regulated TSO.

Other attractive properties of the merchant model are the ability to include non-network alternatives to new transmission assets in planning processes, reduce the risk for consumers and minimise investment costs. In liberalised power markets, potential merchant transmission investors could invest in new transmission capacity or enter the generation market to supply local generation to a node that is served by a congested transmission link. Investment risk is transferred from regulated transmission owners and consumers to the merchant. As the merchant is the beneficiary of any cost saving, construction costs may also be minimised.

However, merchant investment alone is insufficient to ensure overall adequacy of the network, underscoring the importance of a well-designed institutional model. Transmission investments exhibit economies of scale, where large capacity investments carry only a small cost premium relative to small investments. As large capacity investments offer significant additional benefits at little additional cost, they are socially desirable; however, as these additional benefits are reflected in lower locational prices (due to lower congestion), they are less desirable for private investors. In this setting, merchants will tend to underinvest in new network capacity. Alternatively, with an ISO institutional model, the ISO can also ensure overall network adequacy by planning new network capacity and delivering new investment at minimal cost by running competitive tendering processes.

Merchant transmission investments have been implemented in the USA, Australia and Argentina. In the USA, merchant investment is promoted by the Federal Energy Regulatory Commission (FERC) Order 1000, and several projects are in progress or have been completed in recent years. Nearly all merchant-led investments have been on interconnectors, that is, links between separate networks. Here merchants alleviate coordination and cost allocation issues between different system operators.

2.2.7 Network Arrangements to Address Future Challenges

This section discusses new network arrangements to address the future changes brought on by decarbonisation and decentralisation. They are:

- *Strategic generation zones:* Strategic generation zones coordinate investment in transmission and generation assets and connect remote renewable energy resources to large population centres.
- *Markets for flexibility services:* Flexible resources, such as electricity storage and demand response, can be a substitute for new network investment, as well as providing a range of different system services. A simple set of markets for each system service can reward flexible resources and avoid underinvestment.
- *System for controlling decentralised resources:* While distribution networks today are largely passive (one-directional flow between the transmission system and the end user), they will need to become active (distributing power from various sources and bidirectional) to accommodate distributed resources. Distributed resources increase the complexity of the electricity system. If the system is too complex for a single system operator to balance, a hierarchy of resource control will be needed, with intermediaries such as virtual power plants and distribution system operators interacting with the transmission system operator. The hierarchy of resource control may reflect computational requirements, institutional characteristics or operational vulnerability.
- *Coordinated investment in decentralised resources:* Investment in generation and

storage resources should meet the needs of the whole electricity system. A decentralised electricity system may not be able to provide adequate information to investors on system needs, raising the risk of inefficient investment. A coordinated approach to investment can mitigate this risk.

- *Open-access, public data on system conditions and resources:* Market participants need access to market information to facilitate a level playing field for competition. A data exchange could be part of an efficiently functioning energy system, but would need to be both secure and accessible.

- *Accommodate future innovations:* Recent innovations in electricity networks include new network structures and peer-to-peer electricity trading, which may offer significant benefits. These innovations can be facilitated through pilots and early stage funding.

(1) Strategic generation zones

The decarbonisation of electricity generation and the wider energy system heightens the challenge of planning and delivering network infrastructure. The amount of new transmission investment needed will be more difficult to determine due to greater uncertainty over the level of total electricity demand. This uncertainty is due to: (a) the electrification of end-use sectors and improvements in the efficiency of electrical appliances; and (b) peak demand, as flexible resources contribute to smoother generation and consumption profiles. The degree to which generation will be centralised, that is, connected to the transmission network, will be difficult to forecast.

In many countries, renewable energy resources are located in areas that are distant from large population centres, and thus require large-scale transmission investment. For example, in the UK, most electricity demand is located in the south of England, while a large proportion of onshore wind resources are located in Scotland and offshore wind resources in the North Sea.

In a liberalised electricity system, investors in generation and network investment face a coordination problem. While vertically integrated utilities can plan generation and network investment simultaneously, in a liberalised electricity system, generation and network investment are carried out by different institutions. Generation investors face the risk that their revenues may be lower due to inadequate network investment, and network investors face the risk that generators will underuse their new network investments. This can lead to underinvestment.

This coordination problem can be mitigated with strategic generation zones. If a strategic decision is made to exploit a large renewable resource that is distant from large population centres, generation investors may be given incentives to invest there. This may require an overarching strategic plan to be developed by an institution, such as a government agency, with sufficient authority to determine the location of both transmission and generation investment. It may also require a credible, long-term regime for network connection to be developed to reduce stranded asset risks for generators. For example, in the UK, nine offshore wind farm zones of varying sizes with the capacity to deliver 33 GW were identified within British waters. The Crown Estate, the statutory owner of seabed rights, asked renewable energy developers to bid for exclusive rights to develop offshore wind farms within the zones. The Electricity Networks Strategy Group, a high-level forum of key stakeholders in electricity networks, including the Crown Estate, then identified the key transmission investments needed to meet future demand, given the expected location of future generation. The areas identified by this exercise are shown in Fig. 33.

(2) Markets for flexibility services

Decarbonisation will require flexible resources, electricity storage and demand response, which offer non-network alternatives to new transmission assets. Electricity networks are costly, long-lived assets. Investment in them is made with uncertainty over the future spatial and temporal profile of generation and demand. It will be increasingly valuable to substitute flexible resources like electricity storage and demand response for new network investments where possible, or defer new network investments until

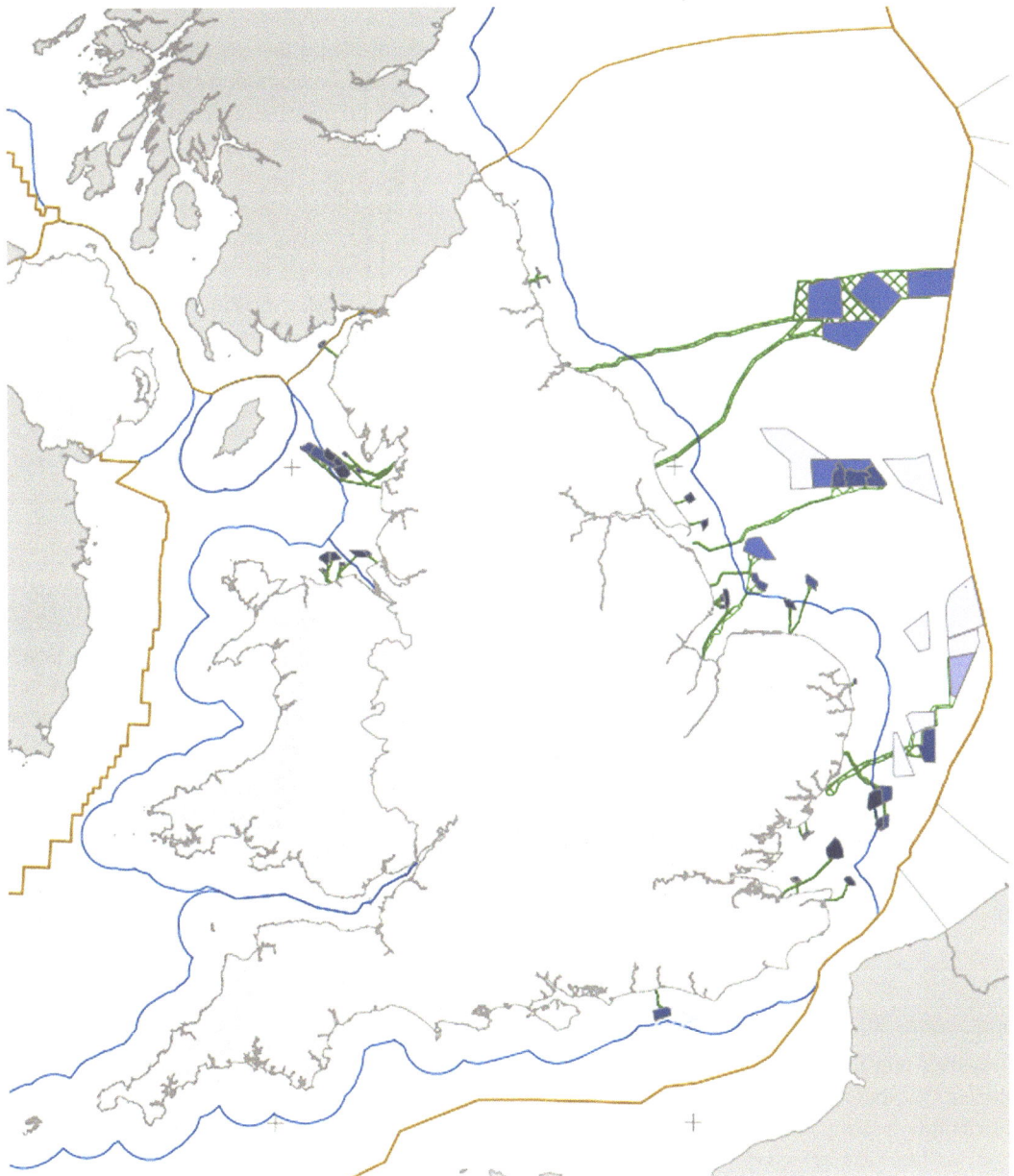

Fig. 33 UK offshore wind farm zones identified for the development of renewable energy. *Source* The Crown Estate (2017)

the future profiles of generation and demand are better understood. As discussed later in this report, nodal pricing provides a locational signal for non-network alternatives.

Flexible resources provide a range of system services. As well as substituting for and deferring new network investments, services provided by flexible resources include balancing and system stability. To provide balancing services, electricity storage can hold surplus energy until it is needed, and demand response can shift demand to when electricity is being generated. To provide system stability, electricity storage and demand response can adjust system voltage and frequency. For flexible resources to be deployed at volumes that bring the greatest benefits, mechanisms exist to reward them for the system services they provide.

A simple set of markets for each system service allocates existing flexible resources and signals investment need. Storage and demand response are already actively engaged in balancing supply and demand in several wholesale markets worldwide. However, markets for system stability are typically not developed enough to allow flexible resources to participate to their fullest extent. Procurement mechanisms for system services typically specify these services in terms of the properties of the thermal generators that historically provided them, such that these mechanisms are often closed to new, low-carbon flexible resources. Recent attempts to create procurement mechanisms for flexible resources have resulted in a patchwork of complex and mutually inconsistent mechanisms. For example, in the UK, electricity storage facilities providing a short-term ancillary service called enhanced frequency response are not allowed to participate in the capacity mechanism. A simple set of markets to reward all system services provided by flexible resources is needed to ensure adequate investment in non-network alternatives.

(3) Model for controlling decentralised resources

While distribution networks today are largely passive, they will need to become active to accommodate distributed resources. As demand on the distribution network is typically inflexible, generators for the transmission system operate flexibly to meet demand and maintain system security. As decentralised electricity resources (distributed generation, storage and demand response) are deployed in the distribution network, these resources will need to operate flexibly.

Distributed resources increase the complexity of the electricity system. Optimising the operation of the electricity system involves finding the optimal volume of output and consumption of every resource in the power system. A conventional, centralised electricity system typically includes a limited number of large transmission-connected generators, suppliers and large industrial consumers with flexible demand. However, a decentralised electricity system will include a very large number of small decentralised resources. The number of resources to be optimised might increase by a factor of several million.

A similar increase in computational demand would occur if the temporal resolution at which resources are controlled increases. For example, shifting from hourly to real-time (second-by-second) settlement of all dispatch and consumption actions would increase the computational demands of optimal system balancing by a factor of 3,600.

If the system is too complex for a single system operator to balance, a hierarchy of resource control will be needed. Complete optimisation of a decentralised electricity system would have very significant computational requirements. If computing technology is not able to meet these requirements, then control of the system will be distributed.

Virtual power plants (VPPs) and distribution system operators have a role to play in a hierarchy of resource coordination. VPPs, also known as aggregators, could coordinate (aggregate) decentralised resources and coordinate them individually to present the (transmission) system operator with a level of net generation (or consumption). If the system is complex, more than two levels of resource coordination might be needed, for example, with some VPPs coordinating the activity of smaller VPPs further down the hierarchy. Distribution system operators are VPPs that coordinate all resources in a given distribution system, either directly or via intermediate VPPs.

Hierarchies of resource control provide only partial optimisation of the electricity system. In order to optimise the whole electricity system, a single optimising agent must know the demand and supply curves for each system resource. Where no single agent has this information, only partial optimisation is possible, as groups of resources for which information is available must be optimised separately. Markets between groups of resources will be required to coordinate their operation to balance the whole system. However, supply and demand curves in electricity systems change in real time, while the process of price discovery in decentralised markets takes place over time through multiple iterative trades. Therefore, markets between multiple levels of resource coordination can provide only a partial optimisation of the electricity system.

A decision needs to be made about how distributed resources will be coordinated, with several possibilities available. Four possible models for operating a distribution system of distributed energy resources are currently being discussed in the literature, and are illustrated in Fig. 34. These models are:

- *Whole system operator.* This model involves only one level of control. The transmission system operator (TSO) carries out constrained dispatch of the whole electricity system. In other words, it carries out least-cost dispatch across all resources in the transmission and distribution systems, taking capacity constraints in both systems into account. In this model, distribution system operators retain their current, minimal functions of basic

planning and operation of the distribution system.

- *Whole system operator with distribution system operator.* This model has two levels of control. First, the TSO carries out economic dispatch of the whole electricity system, including constrained dispatch of the transmission system. In other words, it carries out least-cost dispatch across all resources in the transmission and distribution systems, but only takes capacity constraints in the transmission system into account. Second, distribution system operators (DSOs) modify the operation of distributed resources to allow for capacity constraints in the distribution system.
- *Virtual power plants (VPPs) with distribution system operator.* This model involves three levels of control. First, VPPs provide the transmission-level system operator with offer curves for generation or demand reduction for the resources they operate. Second, the transmission-level system operator carries out constrained dispatch of the transmission system. Third, DSOs modify the operation of distributed resources to take into account

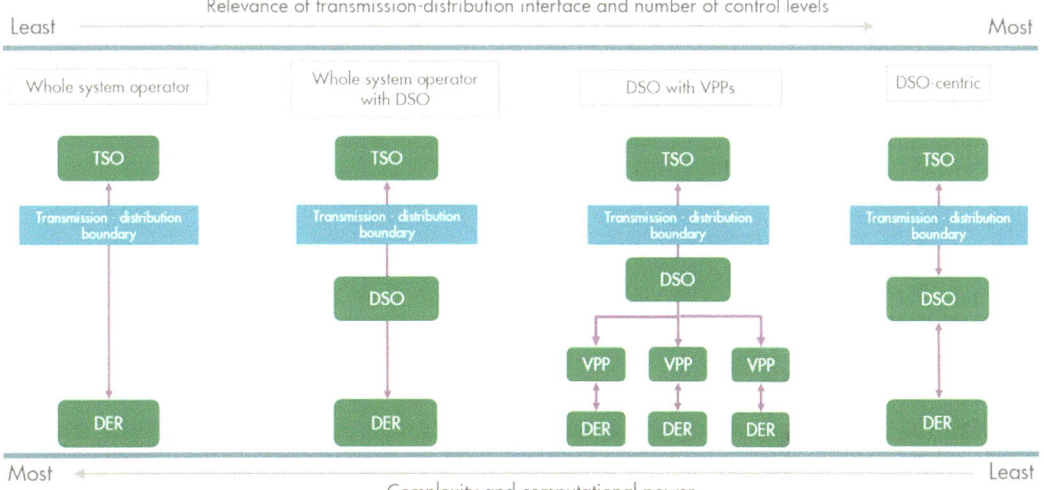

Fig. 34 There are several possible models for coordinating distributed energy resources. *Note* TSO = transmission system operator; DSO = distribution system operator;

DER = distributed energy resources; VPP = virtual power plant. *Source* Vivid Economics

capacity constraints in the distribution system.

- *Distribution system operator.* This model involves two levels of control. First, DSOs act as sole distribution-level VPPs, controlling all distribution system resources and carrying out constrained dispatch of the whole distribution system. DSOs provide the transmission-level system operator with bid or offer curves for the generation and demand resources they operate. Second, the transmission-level system operator carries out constrained dispatch of the transmission system.

Key criteria when choosing a model are the computational requirements, institutional characteristics and operational vulnerability of each model. As explained above, systems with high computational requirements or small improvements in computing technology will require more levels of resource coordination; while systems with low computational requirements or large improvements in computing technology will require fewer levels of resource coordination, and—with sufficiently developed computing technology—potentially a whole system operator. Preferences for specific institutional characteristics are also relevant. For example, the system operator and distribution system operator models involve coordination of resources by a single operator. Consumers may or may not have concerns about price, quality of service and privacy. If they have concerns, they might prefer a model involving control of resources by VPPs, which compete to meet customer requirements. Finally, the models may have different degrees of vulnerability to digital failures, such as those caused by cyberattack.

Hierarchies of resource control may be needed until digital capabilities are sufficiently developed. While in the near term a single operator may be able to optimise the electricity system with relatively low volumes of distributed resources, once sufficient volumes of distributed resources are deployed the task of optimisation may be too great for a single operator. Hierarchies of control may be established early to ensure increasing volumes of distributed resources can be accommodated if

improvements in computing technology fail to keep pace with increases in computational requirements. Even if computing technology improves to the point where a fully distributed system can be optimised by a single operator, an increase in the temporal resolution of system control (for example, from half-hourly settlement towards real-time settlement) would result in significant increases in computational requirements. A shift from hierarchies of control to a whole system operator model would only be viable if improvements in computing technology were sufficient to meet the requirements of optimising a fully distributed system in real time.

(4) Coordinated investment in decentralised resources

Electricity systems with largely centralised resources provide adequate information to investors on system needs. Electricity systems periodically require investment in new resources, such as new generation plants. In theory, developers invest in new resources in response to a price signal in the wholesale market. These new investments are typically large. In principle, this leads to a coordination problem, whereby either several investors might plan to develop a similar resource (overinvestment), or investors might not invest in required resources due to the risk that other investors might do so (underinvestment). In practice, these risks are minimised because the transmission system operator knows which resources are under development and awaiting grid connection and can make this information public.

However, a decentralised electricity system may not provide adequate information, risking inefficient investment. As resources shift to the distribution system, the same coordination problem may arise. This is because unless adequate procedures are introduced, no single market participant knows which resources are under development and awaiting grid connection across all electricity systems. The consequence is, again, inefficient investment: overinvestment, underinvestment, a poor technology mix, or a poor spatial distribution of resources.

The risk of inefficient investment can be mitigated through a coordinated approach to

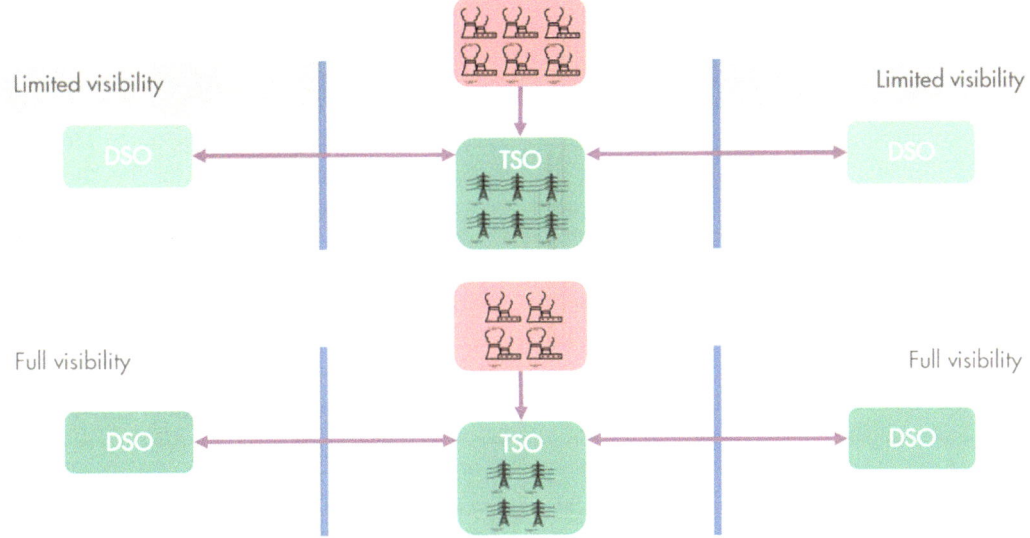

Fig. 35 A coordinated approach can ensure efficient investment in electricity system resources. *Note* TSO = transmission system operator; DSO = distribution system operator. *Source* Vivid Economics

planning new assets. As set out in Fig. 35, this could be achieved either through formal processes for resource planning and decision-making, or through provision of information:

- Formal processes for resource planning and decision-making have been introduced in Spain and Ireland by the system operators of the transmission and distribution systems. Processes include formal collaboration, with the TSO and DSOs planning infrastructure and generation investment together. For example, in Spain, some regional administrations formed evaluation boards. In these boards the administration, TSO, DSO and developers coordinate investment plans and grid connection requests. The TSO and DSO analyse and approve investment plans together, thereby minimising network development and project costs and risk. In Ireland, under the group processing approach, investment plans of developers are collected in batches and then submitted to the TSO and DSO for consideration. The TSO or DSO then processes the plans that are most suited to its system. This approach coordinates the development of the transmission and

distribution systems and efficiently allocates scarce capacity.

- An alternative approach is to provide adequate information. For example, compulsory registration in a publicly available database when applying for planning or grid connection consent would provide investors with an understanding of the pipeline of future resources and allow them to evaluate potential investments against the expected system requirements.

(5) Open-access, public data on system conditions and resources

The development of digital infrastructure is expanding the amount of data and information available. Smart meters in buildings track the profile of energy use every second, offering a new source of information on energy consumption and user behaviour. In the electricity grid, sensors and wide area networks monitor grid reliability, providing real-time information on network conditions.

Market participants need a degree of access to market information to facilitate a level playing field for competition. The Council of European Energy Regulators, for example, has identified limited access to information as a key barrier to entry for new market participants. Access to information about distributed energy resources and network conditions may also grow in importance in the future. This information may provide a foundation for new opportunities for system balancing, as new participants could enter the market and find more efficient balancing solutions.

A data exchange could be part of an efficiently functioning set of future energy networks but would need to be managed carefully to mitigate risks while being accessible. A data exchange is a secure store of data, for example, on customer use patterns, available distributed energy resources, local prices and network conditions. The availability of this data raises privacy concerns, so a balance would need to be struck between accessibility and protection. Use of data exchanges will require adequate institutional arrangements. For example, if DSOs, who participate in markets for electricity system services, ran data services to which they gave themselves preferential access. In the long term, digital developments may make the operation of data exchanges by centralised authorities unnecessary, particularly if electricity trading shifts towards peer-to-peer exchange.

(6) Accommodating future innovations

Innovations in electricity networks include new network structures and peer-to-peer electricity trading, which may offer significant benefits. New network structures, including microgrids and fractal grids (a system of multiple microgrids) have the potential to make electricity systems more resilient to failure (see Sect. 1 on new network architectures below). Peer-to-peer trading through a distributed data management platform, such as a blockchain, offers the potential to lower transaction costs and reduce the role of intermediaries in electricity markets (see Sect. 2 on peer-to-peer trading below).

These innovations can be facilitated by the necessary policies and models needed to deliver efficient networks today and in the future. Most fundamentally, liberalised electricity markets provide a supportive environment for the

development, demonstration and adoption of innovations. More specifically, an institutional model that aligns system operator incentives to public policy objectives will be needed to mitigate any incentive for incumbents to block the spread of innovations. A model for control of decentralised resources will also be needed to offer innovations, such as new network structures and peer-to-peer electricity trading, the opportunity to participate in electricity markets.

It is possible that over time these and other innovations will drive or enable larger changes that have the potential to restructure the electricity system more significantly. It will be worthwhile monitoring new technologies and business models, so that policy and regulation can respond appropriately, to realise value and address risks.

(1) New network architectures

Microgrids and fractal grids are innovative network architectures. Both architectures provide greater resilience than the radial links of conventional distribution networks. Microgrids achieve resilience through redundancy in generation, while fractal grids achieve resilience through redundancy in network infrastructure.

A microgrid is a small-scale, partially self-sufficient network, incorporating both generation and demand sources. Microgrids may be connected to the local distribution network, importing or exporting electricity according to system conditions, but may also disconnect from the distribution network and operate as an island. As a microgrid can meet some or all of its own demand, it is more resilient to wider system failures, caused by a fault or cyberattack, than a radial network, and is well suited for critical functions such as hospitals, military installations or data centres. As microgrids are self-sufficient, they may require more on-site generation than conventional networks. The deployment of on-site generation in microgrids may result in a larger volume of generation assets in the wider electricity system, implying a degree of asset redundancy and an increase in costs. In principle, the redundancy can be mitigated if sufficient generation is deployed to serve only essential loads when islanded. A microgrid can aggregate its resources

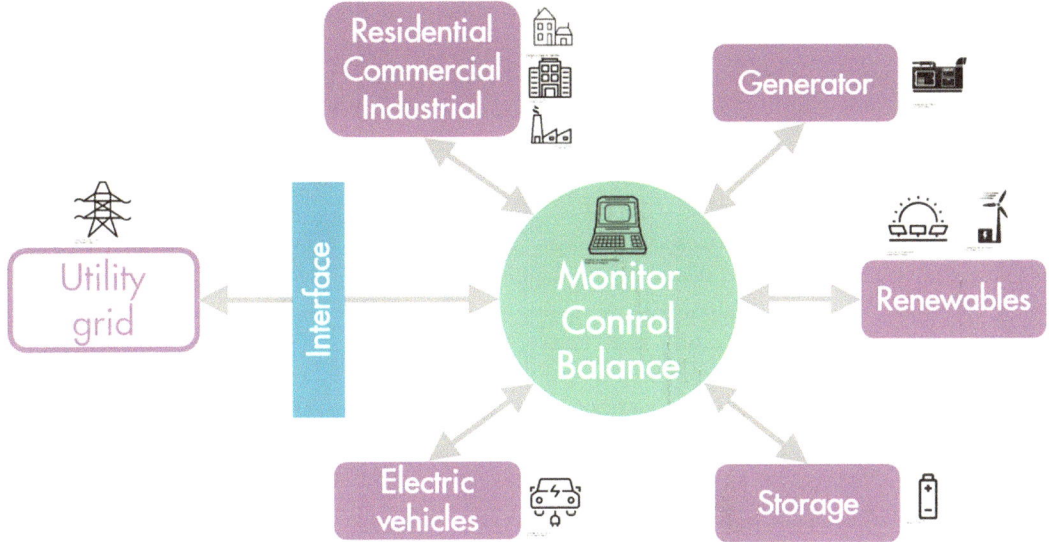

Fig. 36 A microgrid contains generation and flexible resources, as well as sources of electricity demand. *Source* Vivid Economics

like a virtual power plant to coordinate trading with the wider electricity system. This aggregation can be accomplished by a central controller, or potentially by peer-to-peer communication of individual microgrid resources, without a central controller. Figure 36 shows the structure of a microgrid with some of these features.

A fractal grid is a new network structure currently at the concept stage. A fractal grid combines the economy of radial networks, where redundancy is minimal, with the resilience of meshed networks, where nodes are connected by multiple links. The fractal grid achieves these properties through use of a fractal, or recursive pattern, where multiple sets of links with the same structure are connected in a parent-child relationship. Failure of a single link does not prevent power flow between nodes, and the fractal architecture can accommodate microgrids that are able to island themselves during wider system failure. Proponents of fractal grids note that most urban spatial areas already have a fractal structure, so a fractal grid system would be easy to develop in cities. There are several fractal grid demonstration projects ongoing. For example, CleanSpark's FractalGrid has a federated structure that connects microgrids in a parent-child relationship.

Microgrids in the FractalGrid can share their generation and services with other microgrids to shave peak demand and increase the reliability of the whole system, or they can island themselves to manage generation and load independently. The FractalGrid Demonstration at Camp Pendleton military camp in California and NRECA's Agile Fractal Grid are other examples of microgrids in a parent-child relationship.

(2) Peer-to-peer electricity trading

Peer-to-peer electricity trading could allow individual owners of small-scale generation, storage and demand resources to participate in electricity markets. Currently, electricity is traded between generators and large electricity suppliers. Virtual power plants are also likely to enter the electricity market. In addition, developments in peer-to-peer electricity trading could facilitate individual owners of small-scale generation, storage and demand resources to trade electricity. Peer-to-peer energy trading is being piloted by several small microgrids. For example, the Brooklyn Microgrid in New York connects generators, distribution lines, batteries and load sources, with trades and electricity flows tracked though a blockchain distributed ledger.

If peer-to-peer trading becomes widespread, it may reduce the role of intermediaries operating the electricity system. Some commentators suggest that blockchain could automate the active participation of large numbers ·of distributed energy resources, such that an intermediary like a virtual power plant is not required. In future, peer-to-peer electricity trading might take place not only within a microgrid, but between resources at the level of the distribution system, and potentially the transmission system. It is theoretically possible that sufficient automation might reduce the role of system operators in distribution or even transmission, though it is likely that their core roles, managing network constraints and maintaining system security, will remain.

2.2.8 Country Case Studies

This section presents six case studies of network arrangements across five countries: China, the USA, the UK, Germany and Australia. China's experience illustrates the challenges in providing efficient network infrastructure under complex institutional arrangements and limited use of strategic planning of infrastructure investments. The USA provides a contrasting example of a large territory, with multiple separate transmission systems, that has gone through the liberalisation process under the oversight of a central regulator. PJM, one of the larger US independent system operators, is considered the leading example of state-of-the-art network arrangements. The electricity transmission network in the UK is owned by three separate transmission companies, one of which (National Grid) is also the operator for the whole system. As a TSO, National Grid is subjected to innovative performance-based regulation, though the UK is now moving closer to the ISO model. Germany, like the USA, has several transmission systems, which participate in wider transmission planning, both nationally and with other European system operators. Australia has highly progressive markets for system services, facilitating electricity storage and enabling its use as an alternative to new transmission investments. Each section below discusses for each country the institutional arrangements, the processes for transmission

planning and delivery, the extent to which locational pricing has been implemented, and recent developments to prepare for the future challenges of decarbonisation and decentralisation.

(1) China

(1) Institutional arrangements

Liberalisation of China's electricity sector has progressed in three distinct phases. Before the start of the liberalisation process, the then Ministry of Electric Power owned and operated the generation, transmission and retail of electricity. In 1985, the generation market was opened to private investment to address severe power shortages. Then, in 1996, the ministry was abolished, the State Power Company was established, and the State Economic and Trade Committee took over the regulatory functions of the ministry. As a result, the electricity business was separated from government functions. In 2002, generation was unbundled from transmission and retail; two grid companies, State Grid Corporation of China and China Southern Power Grid, were created; and the State Electricity Regulatory Commission (SERC) was established. Taking the liberalised electricity markets as a model, the reform aimed to create competitive wholesale markets and regulated network tariffs.

After a decade of relatively limited reforms, a new phase of reform is now underway. SERC, which would lead the liberalisation of the electricity sector, was folded into the National Energy Administration in 2008, which is part of the National Development and Reform Commission (NDRC). As the leading planning agency under the Chinese State Council, NDRC defines policies for China's economic and social development. Transmission and retail of electricity has remained vertically integrated. Prices and network tariffs are still regulated centrally. In 2015, the Communist Party Central Committee and the China State Council published a statement committing to electricity market reform and the introduction of competitive wholesale and electricity markets.

The Chinese electricity sector has a complex institutional framework. Several institutions have overlapping authority over the sector, and there is

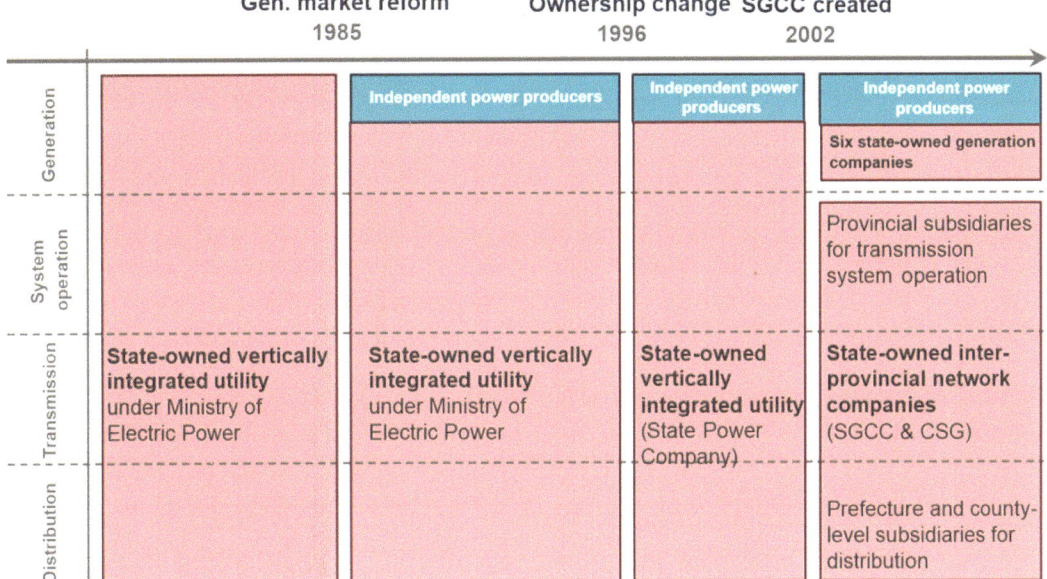

Fig. 37 Structure of the Chinese electricity system. *Note* State-owned companies are in red, privately-owned companies are in blue; SGCC is the State Grid Corporation of China and CSG is China Southern Power Grid. *Source* Vivid Economics

no mechanism to coordinate the actions of these institutions. All institutions in the electricity sector are overseen by central government.

The Chinese transmission system comprises a set of partially interconnected provincial transmission systems, with some additional interconnections at the regional level. State Grid Corporation of China and China Southern Power Grid are in charge of regional interconnectors, while the regional grid companies control interconnectors between provinces, and provincial grid companies manage the transmission network within their provinces.

The Chinese transmission grid is operated by provincial grid companies. Electricity dispatch and trading centres develop generation schedules and manage congestion. The provincial grid companies schedule annual and monthly generation and also manage ancillary services. Scheduling of generation is determined by provincial governments based on an allocation of operating hours, rather than on need. This causes significant curtailment of renewable electricity, if alternative generators are scheduled to operate at times of high renewable output. In 2015, total renewable curtailment

amounted to 1.6% of China's electricity demand. Figure 37 shows the structure of the Chinese power sector before and after the 2002 reform. The reform aimed to aggregate the provincial markets into six regional markets, but this is still in progress.

(2) Transmission investment

Network planning follows a top-down process. The National Development and Reform Commission's (NDRC) five-year plans define general network investment programmes. These programmes are aimed at driving economic growth, rather than meeting reliability or economic efficiency targets. Provincial governments and local NDRC branches refine these general investment programmes for their administrative territory, typically without stakeholder consultation.

In line with the multilevel structure of the grid companies, investment in the transmission network follows a multilevel structure. State Grid Corporation of China and China Southern Power Grid invest in interconnectors between regions, while the regional grid companies invest in interconnectors between provinces, and provincial grid companies are responsible for the transmission network within their provinces. Construction of transmission networks stretching through several regions requires the involvement of several regional grid companies. There is no merchant involvement in transmission investment.

(3) Level of locational pricing

There is no locational pricing and transmission congestion is not priced. The regulatory bodies, the NDRC and State Electricity Regulatory Commission set the wholesale and retail prices of electricity. Provincial governments amend the centrally set price to achieve local policy and economic development goals.

(4) Modernising network arrangements

China does not have specific procedures to ensure non-network alternatives to new transmission assets are assessed in new transmission planning and investment decisions, though investment in electricity storage is increasing. The Chinese government runs pilot projects and subsidises investment in private battery storage projects. According to Bloomberg New Energy Finance, around 180 MW of battery storage is under development in China. The government's Golden Sun programme subsidises investment in photovoltaic panels and battery storage.

(5) Summary of arrangements

See Table 3.

(2) USA

The US transmission network consists of interlinked regional transmission networks, each of which covers a large geographic area and has a different institutional model and set of arrangements to the others. Figure 38 presents the regional transmission networks of the USA. While some of them have adopted state-of-the-art approaches and are interconnected with their neighbours, others are isolated and remain vertically integrated from generation to retail. Moreover, five alternating current electricity grids (Eastern, Western, Quebec, Alaska and Texas interconnections) cover the USA and Canada and tie together regional transmission networks, and hence utilities, in their territory.

(1) Institutional arrangements

Two federal regulators, Federal Energy Regulatory Commission (FERC) and North American Electric Reliability Corporation (NERC), oversee the US transmission system. FERC regulates the transmission and wholesale sale of electricity in interstate commerce, reviews the siting application for transmission projects, and ensures reliability of the interstate transmission system by setting standards. NERC, overseen by FERC, develops reliability standards and ensures the reliability and security of the power system as a whole.

Before the liberalisation of the US electricity sector, vertically integrated utilities owned and operated generation, transmission and retail of electricity in their area. These utilities were operating in poorly interlinked regional transmission networks covering a single state or a group of states (region). Following large blackouts in the north-east in the 1960s, NERC promoted interconnection of neighbouring transmission networks, so they could exchange power and increase the reliability of the whole transmission system.

The Energy Policy Act of 1992 laid the ground for the liberalisation of the US electricity sector. It opened the electricity market to independent generators and gave FERC the authority to regulate all wholesale electricity transactions.

FERC issued several orders and proposed a standard market design for the US electricity sector. It promoted the unbundling of electricity generation, transmission and retail, with transmission arrangements following the independent system operator/regional transmission organisation (ISO/RTO) model. According to the standard market design, ISO/RTO ensures open access for all generators to the transmission system, manages a competitive wholesale spot market, and controls congestion using locational pricing. ISOs and RTOs have similar responsibilities, and in practice there is very little difference between the legal definition of each. FERC proposed to exercise jurisdiction over transmission owners and operators that had not adopted its standard market design. Many entities challenged FERC's proposal. Opponents argued that

Table 3 Summary of network arrangements in China

Institutional arrangements	Transmission planning and delivery	Network pricing	Modernising network arrangements
Institutional model: ■ Transmission system operators (TSOs) ■ State-owned inter-provincial network companies (SGCC and CSG) ■ Provincial subsidiaries of these companies own and operate each network as TSOs	**Planning:** ■ The NDRC sets five-year national investment plans to drive economic growth ■ Provincial governments and regional NDRC branches then adapt national plan targets to their territories **Delivery:** ■ Provincial grid owners are responsible for delivery in their province ■ Interconnection between provinces is delivered by reginal grid companies ■ SGCC and CSG invest in interconnection between regions **Investment regime:** ■ Currently, merchant transmission investment is not possible	■ No locational pricing in the grid ■ NDRC and SERC set central wholesale and retail power prices ■ Provincial governments can amend centrally set prices	**Readiness for decarbonisation:** ■ Under current network arrangements, planning processes do not incentivise non-network alternatives to new transmission assets ■ However, investment in storage is rising and the central government runs pilot projects and provides subsidies for storage, for instance, the Golden Sun programme

Note SGCC = State Grid Corporation of China; CSG = China Southern Power Grid.

Source Vivid Economics

one standard would not fit the needs of different regional transmission systems and asked for a voluntary regional approach, in contrast to FERC's mandatory approach. During the exchange of proposals between FERC and transmission system entities, several entities with high electricity prices, such as Midwest Independent Transmission System Operator (MISO) and Southwest Power Pool (SPP), adapted FERC's standard market design to their regional needs and introduced ISOs voluntarily. Following these adaptations, FERC gave up its attempt

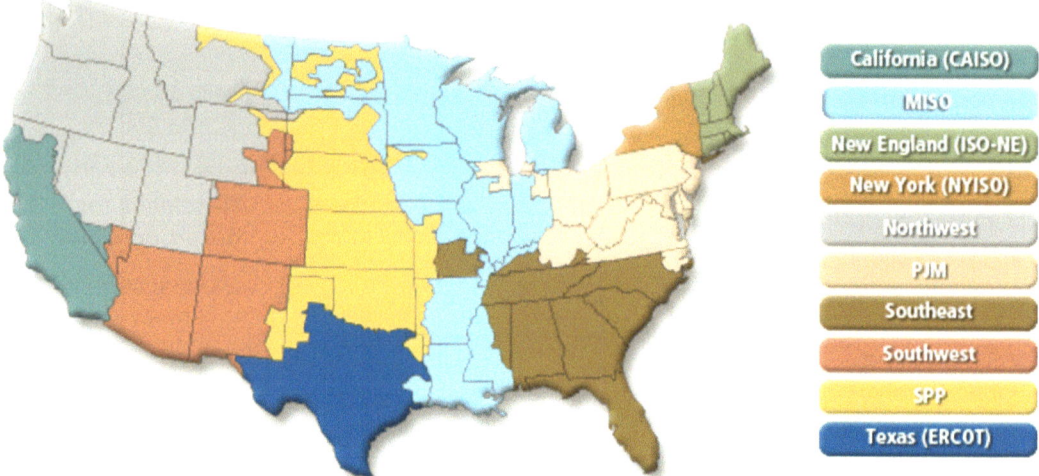

Fig. 38 The US transmission system is segmented into several regional transmission networks. *Source* Federal Energy Regulatory Commission (FERC) (2017)

to mandate its standard market design and pursued a voluntary regional approach.

FERC's push for a competitive electricity sector and standard market design were received differently by the regional transmission networks. Currently, the US power system consists of 10 regional transmission networks. Three of them (Southeast, Southwest and Northwest) have maintained their traditional vertically integrated utility structure and serve customers directly. The others—California, Midwest (MISO), New England, New York, Northeast (PJM), SPP and Texas—liberalised their markets and adopted FERC's standard market design. They unbundled generation and retail from their transmission systems and introduced competition in generation and retail. Figure 39 shows the key steps in the evolution of FERC's standard market design.

(2) Transmission investment

All ISOs assess transmission adequacy, in consultation with stakeholders. A typical ISO periodically reviews adequacy of current and planned transmission assets with a long-term horizon to identify reliability concerns and address public policy needs. The ISO also organises public stakeholder meetings to ask for comments from generators, transmission owners, retailers, end customers and other interested parties. NERC oversees the planning process.

ISOs and RTOs have adopted different approaches to transmission investment and merchant involvement in these investments. Some ISOs, for example New York ISO, identify investment needs, while transmission owners and merchants propose solutions and carry out the necessary planning and investment. Some ISOs, such as California and PJM, identify transmission investment needs and plan the investments to address those needs. Transmission owners, and sometimes merchant developers, then compete for aspects of transmission investment, such as land rights, operations and costs. The California ISO gives a role to merchant investors, whereas in PJM there is little merchant activity, despite provisions for it.

To improve reliability and increase trade volumes between the regional transmission networks, ISOs and RTOs coordinate their network planning and investment. Neighbouring ISOs and RTOs form committees to develop plans for their regions. Moreover, interconnections provide a platform for collaboration between system operators and coordinate development of interconnection-wide transmission plans and

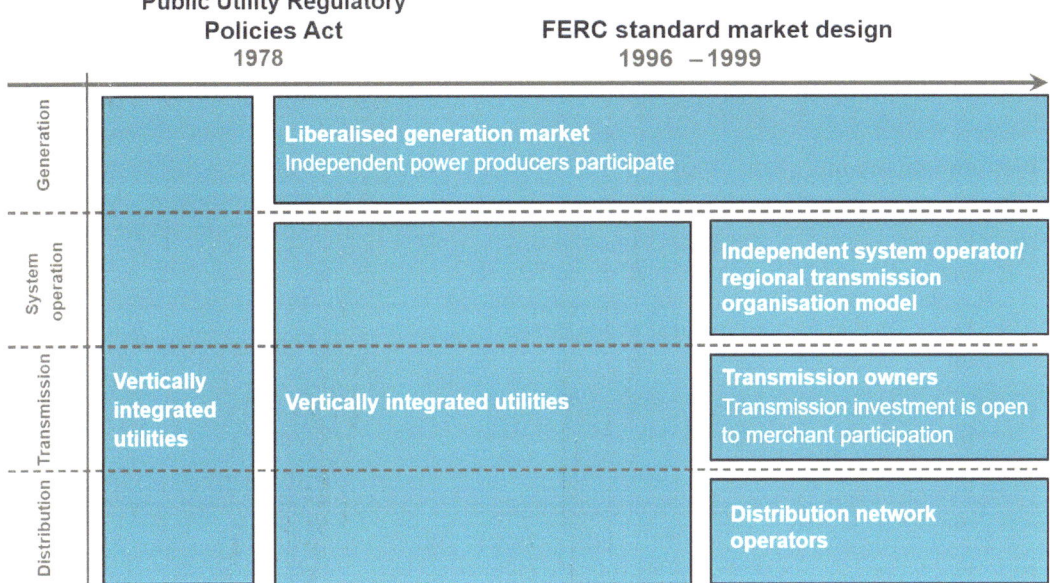

Fig. 39 Structure of FERC's standard market design. *Source* Vivid Economics

transmission investment. However, these collaborations are limited to interconnections and do not attempt to optimise the national transmission network as a whole.

(3) Level of locational pricing

Following FERC's order and standard market design, the liberalised electricity markets have adopted locational pricing at varying levels of granularity and apply different approaches to address disadvantages of nodal pricing. As an example, PJM adopted a highly granular nodal pricing mechanism, but it averages nodal prices into hub prices to provide liquidity to parties involved in long-term contracts. CAISO followed a less granular nodal pricing mechanism than PJM. It argues that high level granularity increases costs and system complexity, while not providing significant benefits to compensate for these drawbacks.

(4) Summary of arrangements

See Table 4.

(3) PJM

PJM is a regional transmission organisation in the USA. PJM (originally named after the Pennsylvania, New Jersey, Maryland Power

Pool), operates part of the Eastern Interconnection grid, serving all or parts of Delaware, Illinois, Indiana, Kentucky, Maryland, Michigan, New Jersey, North Carolina, Ohio, Pennsylvania, Tennessee, Virginia, West Virginia and the District of Columbia. PJM operates more than 82,000 miles (132,000 km) of transmission lines and coordinates 1,373 generating sources with 176,569 MW of generation capacity. Around 65 million people live in the PJM area. There are 20 transmission owners serving their respective transmission zones in PJM.

(1) Institutional arrangements

In 1927, three vertically integrated utilities interconnected their systems to exploit benefits and efficiencies from sharing their generation resources. In doing so they founded the world's first power pool, PJM. The pool carried out least-cost constrained dispatch to reduce costs for pool members. Later, other utilities joined the pool and extended its area of coverage. Until the beginning of the 1990s, member utilities took turns to operate PJM.

In the 1990s, following the liberalisation of the US electricity market and FERC's orders promoting a standard market design, PJM started its transition from a power pool to an ISO. In

Table 4 Summary of network arrangements in the USA

Institutional arrangements	Transmission planning and delivery	Network pricing
Institutional model: ■ FERC proposed best practice is the RTO model, which is functionally similar to the ISO model ■ RTO model has been adopted by seven out of ten interconnection regions ■ However, the Southeast, Southwest and Northwest pools retain a traditional vertically integrated utility structure	**Planning:** ■ In PJM and CAISO, the SO directly plans investment projects, while in regions such as NYISO, TOs or merchants plan and propose transmission solutions ■ All ISOs review current and planned TO investments for suitability against reliability and public policy needs ■ Upstream and downstream stakeholders are consulted during the review process, with NERC as the regulator **Delivery:** ■ In CAISO and PJM, the SO invites tenders for planned investment, with TOs and merchants competing for delivery ■ TOs and merchants propose and deliver their own solutions in NYISO **Investment regime:** ■ Merchants may make investments, although these have largely been limited to interconnections	■ Following FERC orders, liberalised markets adopted nodal pricing ■ However, ISOs apply this at different resolutions ■ PJM has highly granular prices, but aggregates to hub prices to ensure liquidity ■ Nodal prices in CAISO are less granular than in PJM

Note FERC = Federal Energy Regulatory Commission; CAISO = California Independent System Operator; NYISO = New York Independent System Operator; RTO = regional transmission organisation; ISO = independent system operator.

Source Vivid Economics

1993, the PJM member utilities formed the ISO PJM Interconnection Association, which operated the PJM power pool. In 1997, PJM opened its first bid-based electricity market and provided the underlying trading platform. Later that year, FERC approved PJM as the first ISO in the USA. In 2002, in line with FERC's standard market design, PJM became an RTO to operate the multi-state transmission network. Figure 40 shows changes in the market structure of PJM over time.

(2) Transmission investment

PJM carries out a planning process called regional transmission expansion planning (RTEP) to ensure future adequacy of the network. Under the RTEP process, PJM analyses several scenarios to assess the adequacy of network conditions over a time frame of 15 years. RTEP defines the need for and benefits of a transmission project, but review and approval are the responsibility of the member states where the project is located. The transmission investments identified are discussed publicly in stakeholder meetings with PJM members, generators, transmission owners, retailers, end customers and other interested parties.

To develop the best plans for the system as a whole, PJM coordinates its planning process with neighbouring system operators. These include Midcontinent Independent System Operator (MISO), ISO New England and New York ISO. This coordinated planning process seeks to ensure an efficient level of new generation resources and transmission lines across neighbouring systems.

PJM mandates transmission upgrades and extensions that are required to maintain the reliability standards of transmission owners. Under PJM rules, the cost of these projects is allocated to the respective transmission owners. Merchants can also explore interconnection business opportunities in the PJM region. Merchants' interconnection requests are subject to strict rules and procedures. PJM completes feasibility and system impact studies before approving the interconnection request.

(3) Level of locational pricing

In 1998, PJM replaced zonal pricing with nodal pricing to address inefficiencies caused by transmission congestion. Zonal pricing does not account for transmission constraints in a zone, hence market participants do not internalise transmission constraints in that zone. As a result, in the PJM region transmission congestion was under-priced, and widespread bilateral contracting—without considering network constraints—resulted in significant generation curtailment and redispatch costs. Under the nodal pricing scheme, the prices are discovered for around 2,000 nodes, accounting for transmission constraints between the nodes and minimising inefficiencies. To provide liquidity for hedging instruments, PJM averages nodal prices in given areas into hub prices.

(4) Modernising network arrangements

PJM has not developed specific procedures to ensure that non-network alternatives to new transmission assets are considered in new transmission planning and investment decisions. Nevertheless, it is actively encouraging flexible

resources to provide system services. PJM offers retail customers day-ahead and real-time options for emergency load response. Virtual power plants, known in PJM as curtailment service providers (CSPs), act as intermediaries between retail customers and PJM. CSPs help retail customers to reduce their demand in high price periods. PJM also operates a reliability pricing model capacity market. Both demand-response resources and energy efficiency resources can participate in these markets through CSPs and collect payments for reducing their demand and adopting efficiency measures. Demand resources can also participate in PJM's synchronised reserve, regulation and day-ahead scheduling reserve market.

(5) Summary of arrangements

Pennsylvania, New Jersey, Maryland (and other areas within the Eastern Interconnection) (Table 5).

(4) Great Britain

Great Britain comprises England, Scotland and Wales, whereas the UK comprises all three along with Northern Ireland. Britain has the larger of the UK's two electricity systems. The smaller is the Northern Ireland system, which is connected to the Republic of Ireland. It is owned by Northern Ireland Electricity Networks and operated by the Irish system operator EirGrid, as part of the island's single electricity market.

(1) Institutional arrangements

Before market reform, the vertically integrated Central Electricity Generating Board (CEGB) was the statutory monopoly provider of generation and transmission services in England and Wales. The CEGB was subject to cost-of-service regulation and its operations were characterised by high capital costs, low productivity growth, and a low return on assets. Distribution networks were operated by twelve regional monopolies, known as area electricity boards.

In 1990, the CEGB was unbundled and privatised into multiple generating companies and a single transmission company, the National Grid Company. The Electricity Act 1989 led to the creation of the Office of Electricity Regulation

Table 5 Summary of network arrangements in PJM

Institutional arrangements	Transmission planning and delivery	Network pricing	Modernising network arrangements
Institutional model: ■ PJM is the ISO/RTO for all or part of 13 US states and the District of Columbia	**Planning:** ■ PJM carries out a regional transmission expansion planning process to ensure the network meets future requirements ■ Review and approval decisions for projects are made by member states, following broad stakeholder involvement **Delivery:** ■ PJM mandates TOs to perform transmission upgrade and extension projects for security of supply **Investment regime:** ■ Merchants may invest in interconnection assets, although they are subject to strict regulatory oversight by PJM ■ PJM coordinates projects with RTOs in its region to ensure the Eastern Interconnection is efficiently coordinated, this includes NEISO, NYISO and MISO	■ PJM has high resolution nodal pricing, with around 2,000 nodes in its operational area ■ To ensure markets remain liquid, PJM averages nodal prices into lower resolution hub prices	**Readiness for decarbonisation:** ■ There is no specific procedure to ensure non-network solutions are considered ■ However, PJM does operate a reliability pricing capacity market which provides payments for load reduction ■ PJM currently has about 10 GW of demand-side response capacity **Readiness for decentralisation** ■ High-resolution nodal pricing provides investment incentives for distributed resources

Note NEISO = New England ISO; NYISO = New York ISO; MISO = Midcontinent ISO; RTO = regional transmission organisation; ISO = independent system operator; DSR = demand-side response.

Source Vivid Economics

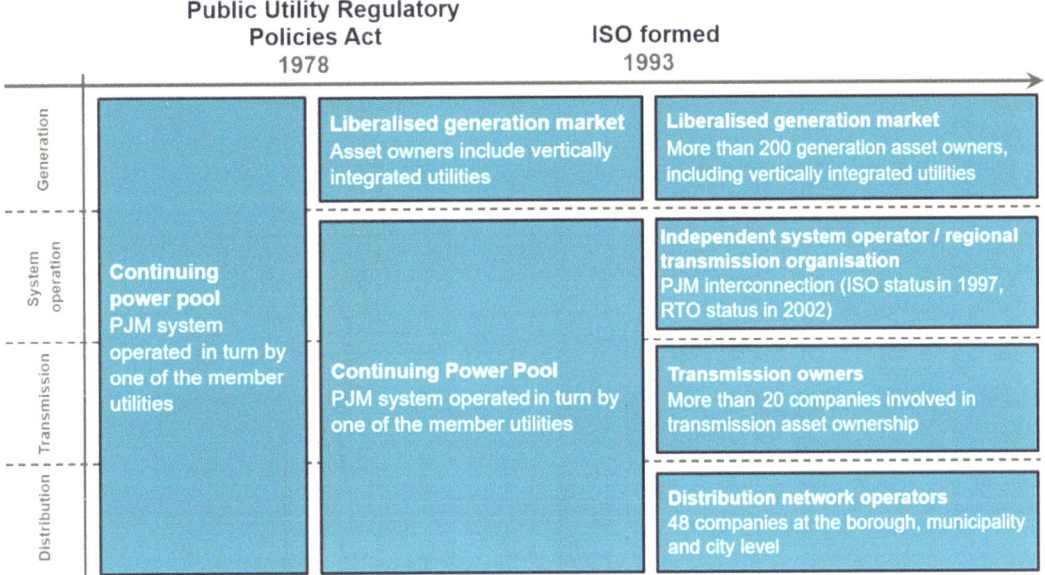

Fig. 40 Structure of PJM's electricity system. *Source* Vivid Economics

(OFFER) and the Director General of Electricity Supply (DGES). Their role was to regulate the transmission network monopoly of National Grid and the distribution network monopolies of the regional electricity companies and set price caps with periodic reviews. The act also established competition in generation, with the requirement that generators compete by selling power in a wholesale market.

Today, National Grid is the transmission system operator (TSO) for Great Britain. There are three onshore transmission owners (TOs) in Britain: National Grid Electricity Transmission for England and Wales, and Scottish Power Transmission and Scottish Hydro Electric Transmission for Scotland. Driven by European Commission directives on unbundling, offshore transmission assets are owned by offshore transmission owners (OFTOs), with owner-operators chosen by competitive tender run by the power sector regulatory body, the Office of Gas and Electricity Markets (Ofgem). Figure 41 shows the structure of key parts of the British power system before and after liberalisation in 1990.

National Grid's operations as transmission owner (TO) and system operator (SO) are subject to performance-based regulation by Ofgem. These functions are subject to separate incentive regimes. Transmission ownership is regulated through periodic price reviews, currently occurring every eight years. Following privatisation, this took the form of the RPI-X (retail price index minus x) framework, which has since been replaced by the RIIO (revenues = incentives + innovation + outputs) framework as described in 2.2.6 above. System operator regulation is revised every two years and is designed to incentivise efficient balancing, data provision and modelling.

As well as core system operator functions, National Grid also acts as the delivery body for the UK government's decarbonisation policies through electricity market reform. As an SO, it is required to contribute to the drive for competition in onshore transmission and act as the delivery body for new market arrangements, such as the capacity mechanism and contracts for difference schemes. The capacity mechanism is the UK's capacity market to ensure security of supply. Contracts for difference are feed-in tariffs that provide price support for newly contracted low-carbon generation.

The institutional model for the transmission network is moving closer to the independent system operator (ISO) model. In August 2017 Ofgem

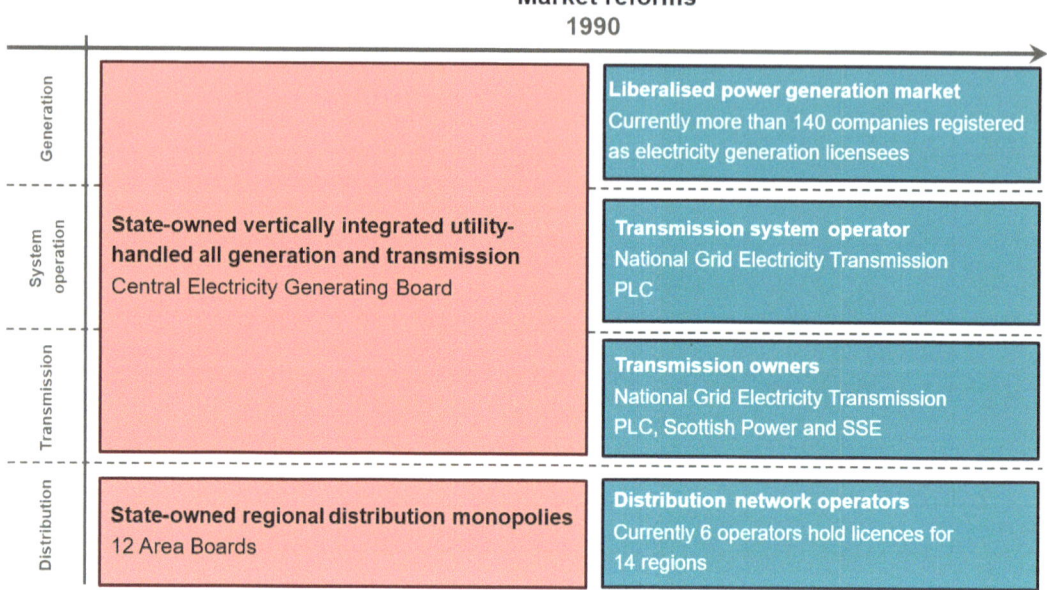

Fig. 41 Structure of the Great Britain electricity system. *Note* State-owned companies are in red, privately-owned companies are in blue. *Source* Vivid Economics

and National Grid confirmed the separation of National Grid's system operator business into a legally separate company within National Grid plc. This will take effect from 2019 onwards and will bring the British power grid's structure closer to the ISO model. The case for separation is based on a greater planning and delivery role for the SO, as well as removing the risk of conflict of interest between TO and SO functions going forward.

As an SO, National Grid is the delivery body for electricity market reform. Under Ofgem's new integrated transmission planning and regulation (ITPR) framework, National Grid is obliged to identify, plan and recommend transmission investment projects. The case is further strengthened by the underlying trend of decentralisation, which increases the need for coordination of the planning and operation of transmission and distribution systems by an SO.

(2) Transmission investment

Traditionally, planning, development and regulation of onshore transmission in the UK has been implemented by regulated monopoly TOs. In the future, transmission planning in Great

Britain will take place under Ofgem's integrated transmission planning and regulation (ITPR) model. Under this framework, the SO has new responsibilities to recommend transmission investments through the network options assessment (NOA) process. However, final implementation decisions remain in the hands of TOs.

National Grid currently owns all onshore transmission assets in England and Wales, with investment regulated through Ofgem's price control mechanism. It has traditionally held sole responsibility for the planning and delivery of investments in both countries, and occasionally coordinated with Scottish transmission owners on major projects with nationwide impact. This is set to change, with Ofgem developing a framework for competitive tendering in onshore transmission, although there has been no substantial non-incumbent transmission investment to date. Historically, the regulatory asset value (RAV) approach to TO revenue regulation has tended to favour capital expenditure-based solutions, with little incentive for TOs to present innovative, non-asset intensive solutions which would face greater regulatory scrutiny. The outcomes-based framework of Ofgem's RIIO regulatory model sought to address these issues.

Efficient offshore transmission will be vital to achieve decarbonisation in the UK, with investment requirements to 2030 estimated at £8-20 billion, perhaps even exceeding onshore investment needs. To achieve this, Ofgem developed a new offshore transmission investment regime. Windfarm developers plan and build offshore transmission assets, which they are then obliged to divest as part of the European Union's unbundling directives. Competitive auctions are then held for the rights to own and operate these offshore assets, known as offshore transmission licences. Licences are allocated to offshore transmission owners (OFTOs), who are required to be independent of both onshore TOs and offshore wind generation developers. Ofgem's decision not to extend National Grid's onshore transmission monopoly to offshore assets was based on a desire to promote competition and achieve more efficient outcomes. National Grid subsidiaries may participate in offshore transmission auctions but separated from National Grid's existing TSO business to avoid unfair advantages through its privileged position as system operator.

Interconnections with continental Europe and Ireland are largely owned by National Grid, alongside overseas partners. These assets are mainly governed by European Commission directives which determine operational and revenue models. As interconnectors are not classified as either demand or generation, they are exempt from transmission tariffs. This can lead to suboptimal siting signals and a lack of coordination between interconnection assets and onshore transmission planning and investment.

(3) Level of locational pricing

Transmission charges, known as transmission network use of system (TNUoS) tariffs, are set to provide efficient economic signals to grid users by reflecting the additional costs TOs incur in serving them. While electricity in England and Wales lacks full nodal pricing of the type seen in PJM and some other regions, TNUoS tariffs have a locational component which aims to reflect the

difference in cost impacts users have at different locations in the grid.

The location varying element of TNUoS tariffs aims to capture the investment, maintenance and operating costs of connecting different locations in the transmission grid. Tariffs are derived from the DC load flow investment cost-related pricing (DCLF ICRP) transport model. TNUoS tariffs also have a non-location varying or residual element, to recover the costs of historical investment.

The TNUoS pricing regime has been criticised for having a high degree of cost socialisation, with around 75% of system costs recovered from the flat residual tariffs. A high degree of cost socialisation leads to inefficient siting signals, as the TNUoS tariffs faced by generators do not adequately reflect the cost of providing them with transmission services.

(4) Modernising network arrangements

In July 2017, Ofgem published a plan setting out future actions needed to deliver a smarter, more flexible energy system. In this plan, Ofgem set out its intention to facilitate the participation of both storage and demand response in the electricity system:

- *Facilitating the participation of storage.* Ofgem commits to review network charges for electricity storage, which currently incur network and balancing charges for generators and consumers, despite its role in reducing the requirement for new network investments. Ofgem also commits to define storage in primary legislation and to clarify its regulatory status within the electricity system and planning regimes.
- *Facilitating the participation of demand response.* Ofgem commits to allow revenue stacking between the capacity market and ancillary services market, following concerns that market rules prevented demand-side response providers from participating in both markets; and to reform the balancing mechanisms (the procurement mechanism for residual balancing) to allow virtual power plants to participate directly in energy supply

and facilitate more demand-side response in residual balancing.

In line with these reforms, National Grid has also committed to reform the procurement processes for flexible energy resources. In June 2017, National Grid published its System Needs and Product Strategy to address key challenges to the deployment of flexible energy resources in Britain's electricity markets:

- National Grid identifies the number of products, lack of transparency over product specification and unclear assessment criteria for product parameters as key challenges to the deployment of flexible energy resources. First, the set of products offered for system services is large and complex (providers can choose from more than 20 different products, each with different technical requirements and routes to market). Second, there is a lack of transparency in product specifications, which vary with system conditions and which themselves are the result of underlying conditions. Third, assessment criteria are unclear, as the system operator does not specify the importance and value it attaches to key parameters, such as length of contract or speed of delivery, in the products it procures; and overlapping markets, where more than one product exist to solve the same system problems, each with different procurement processes.
- National Grid commits to simplify the products for system services through rationalisation, standardisation and improvement. National Grid commits to three actions. First, to carry out a review to reduce the suite of products that it procures, to remove products that are no longer required in their current form, or have been superseded by later products, and offer market-based alternatives where possible. Second, to standardise products within each service market, with standardised parameters such as contract length (e.g. 1 month, 6 months, 1 year, 2 years) and speed of delivery of reserve energy (e.g. 2, 5, 10 or 20 min). Third, to improve the products procured to better suit the technical abilities

and economic characteristics of the assets providing the services.

Alongside these reforms, the UK energy networks industry is planning the reforms required to deliver the smart grid. The Energy Networks Association, the industry body for owners and operators of gas and electricity networks in the UK, has set up the Open Networks Project, an initiative to reform the operation of electricity networks and underpin the delivery of the smart grid. The Open Networks Project's objectives include developing improved processes for transmission system operators (TSOs) and distribution system operators (DSOs) in connections, planning, shared TSO/DSO services and operation; and developing a more detailed view of the required transition from a distribution network operator (DNO) to a DSO, including the impacts on existing organisational capability.

(5) Summary of arrangements
See Table 6.

(5) Germany
(1) Institutional arrangements
Before liberalisation, the German electricity sector was characterised by regional monopolies at three levels. At the supra-regional level, eight network energy supply companies, each active in its own region, produced around 80% of all electricity. Four of these were vertically integrated from generation to retail, while the other four were vertically integrated across generation and transmission only. All eight companies also provided electricity to regional energy suppliers. At the regional level, around 80 energy supply companies generated electricity (10% of total electricity), managed the distribution network and supplied electricity to end consumers or municipal utilities. At the local level, around 900 municipal utilities generated the remaining 10% of total electricity, managed the municipal distribution networks and supplied electricity to end consumers.

The German electricity sector went through a structural change in the 1990s and 2000s. Territorial monopolies were abolished and generators gained access to end consumers in other

Table 6 Summary of network arrangements in Great Britain

Institutional arrangements	Transmission planning and delivery	Network pricing	Modernising network arrangements
Institutional model: ▪ NG is the TSO in England and Wales (sole TO) and an ISO in Scotland (where two other firms are TOs) ▪ Regulated under RIIO PBR (revenue = incentives + innovation + outputs performance-based regulation) ▪ Great Britain is moving towards the ISO model **Additional SO functions:** ▪ NG is a delivery body in CFDs and the capacity market as part of EMR	**Planning:** ▪ Historically, onshore investment has been planned by NG, with wind farm developers responsible for offshore planning ▪ Going forward, under Ofgem's ITPR, NG will plan and recommend investment projects in its role as SO **Delivery:** ▪ NG develops and operates new onshore assets; offshore assets are developed by generators, and divested to OFTOs ▪ NG is part of ENTSO-E, which coordinates planning across European TSOs **Investment regime:** ▪ Merchant involvement to date has been limited ▪ However, Ofgem is developing a competitive tendering framework which will include a larger role for merchants ▪ Interconnectors are co-owned by NG and overseas partners	▪ No nodal or zonal pricing ▪ TNUoS charges have a small locational component ▪ Wholesale electricity prices are updated on half-hourly basis	**Readiness for decarbonisation:** ▪ Strategic generation zones ensure coordinated investment in offshore generation and supporting onshore networks ▪ Reforms underway to reward flexible resources for system services **Readiness for decentralisation** ▪ ENA, the industry body for network owners and operators, started the Open Networks Project to support the DNO to DSO transition and better coordinate TSO and DSOs

Note NG = National Grid; Ofgem = the Office of Gas and Electricity Markets (UK regulatory body); TNUoS charges = transmission network use of service charges, ENA = Energy Networks Association.

Source Vivid Economics

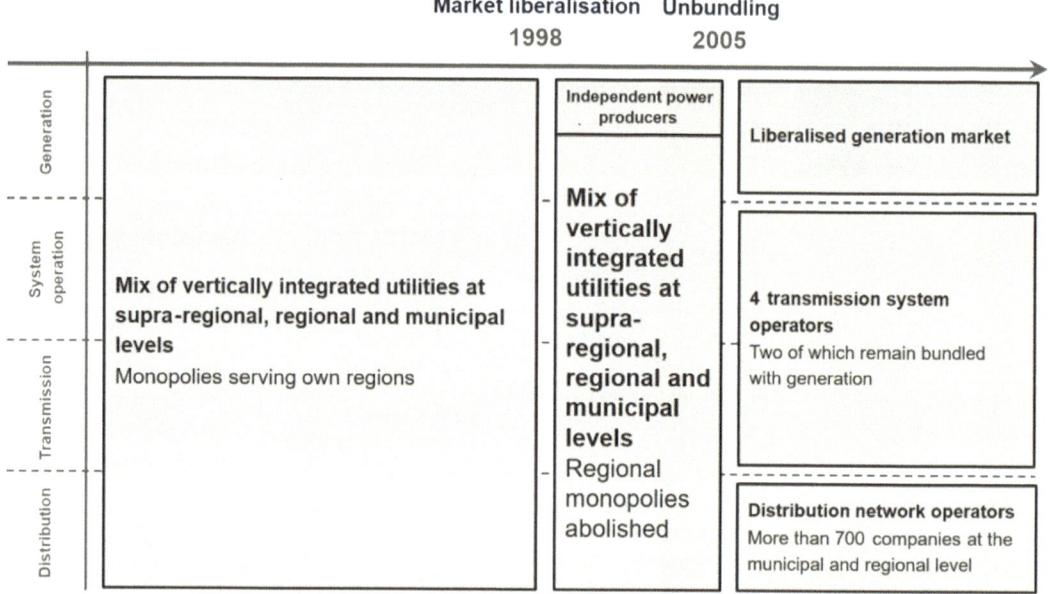

Fig. 42 Structure of the German electricity system. *Note* All segments of the German electricity system contain a mix of privately and publicly owned companies. *Source* Vivid Economics

territories. In the mid-2000s, following the EU's Electricity Market Directive, Germany unbundled generation, transmission, distribution and retail in its electricity sector, but allowed different degrees of unbundling. The levels of the supply chain could be separated completely, legally or functionally, resulting in a complex and heterogenous structure. Unbundling of transmission and generation took place in 2005, while unbundling of distribution and retail was completed in 2007.

Germany's transmission network now consists of four different transmission systems, with different institutional models. Two of these are based on the TSO model, with TenneT and 50 Hz owning and operating their networks. A further two retain a degree of vertical integration, with the generation companies RWE and EnBW continuing to own transmission assets, though these are regulated to mitigate conflicts of interest.

There are around 900 distribution operators serving 20,000 municipalities in Germany. The 900 distribution operators include four of the formerly vertically integrated supra-regional network

energy supply companies, a number of regional companies and around 700 municipality-owned utilities. Often municipality-owned utilities are vertically integrated from generation to retail.

At the national level, the federal network agency (Bundesnetzagentur) regulates the German electricity sector and has various responsibilities. It oversees competition in the electricity sector and unbundling of vertically integrated companies and ensures non-discriminatory access to the transmission and distribution networks. Bundesnetzagentur also regulates charges levied by transmission and distribution system operators. Since 2009, the TSOs are subject to regulation that caps grid tariffs, but which also provides incentives to increase efficiency and reduce costs. At the state level, 11 state regulatory authorities enforce regulations, while the other five states have delegated all their regulatory responsibility to the Bundesnetzagentur. Grids with more than 100,000 customers, or covering more than one state, are also overseen by the Bundesnetzagentur. Figure 42 shows the structure of the German power sector before and after liberalisation and unbundling.

(2) Transmission investment

The TSOs are responsible for transmission planning and investment. Together, they draft a framework that forecasts developments in the German electricity market over the next decade under different scenarios. After the federal network agency, Bundesnetzagentur, reviews and approves the framework, the TSOs define their transmission investment in line with the framework and draft a network development plan. If the investment project is located in one state only, it is submitted to the state government concerned for approval. If the investment project involves more than one state, the federal network agency makes the final approval decision.

The TSOs coordinate their operation and transmission planning with each other, with TSOs in other European countries and with power exchanges, in order to use existing generation and transmission capacities efficiently. The European TSOs participate in regional security coordination initiatives (RSCIs) to harmonise their electricity system operations. Developed voluntarily by the TSOs, RSCIs are service providers with no live system operation capabilities. They coordinate security analysis, short- and medium-term adequacy forecasts, capacity calculations and outage planning. The TSOs use services provided by RSCIs as input alongside with national factors for their final decision-making. Neighbouring TSOs collaborate to develop regional RSCIs, but they are currently working on a European-wide, central verification platform. The European TSOs will submit their planned energy exchange to this single central platform that will compare and coordinate the TSOs' actions.

The TSOs make investments in transmission assets. Merchants are involved in consortiums that invest in interconnectors that link Germany to its neighbours.

(3) Level of locational pricing

Germany does not have locational pricing on transmission and faces rising redispatch costs as a result. The federal network agency reported that in 2014 the TSOs intervened in generation and redispatched on 330 days. The intervention concerned 5,197 GWh of power and cost €186.7 million.

Redispatch costs are passed on to end consumers. In 2015, the costs increased to €402.5 million.

Transmission charges levied by the TSOs do not have any temporal (time-of-use) or locational component.

(4) Modernising network arrangements

Germany has an increasing share of intermittent renewable energy in its electricity mix. It aims to introduce flexible non-network services to integrate intermittency of renewable resources efficiently and thereby improve the reliability of its electricity sector and supply security.

TSOs allow retail end customers to auction for demand curtailment. The TSOs can enter into contracts with end consumers to curtail their demand at short notice when demand peaks. The TSOs compensate end consumers whose demand is curtailed with a fee. The level of the fee is set in weekly auctions. The auctions are designed for medium and large industrial end customers with high power consumption, but residential customers can also participate in the auctions through aggregators.

There are several pilot projects and public funding available for battery storage. For example, Younicos built a battery park in Schwering to assist the distribution grid with frequency regulation and integration of wind energy. Similarly, ENERCON provides primary control services to Feldheim, whose energy mix is 100% renewables. Since 2013 the German government provides grants to incentivise battery storage of energy generated by photovoltaic panels.

(5) Summary of arrangements
See Table 7.

(6) Australia
(1) Institutional arrangements
Before market reform, Australia's power market comprised vertically integrated state-owned monopolies. Examples include the State Electricity Commission of Victoria and the Electricity Commission of New South Wales. State governments appointed boards of commissioners which were responsible for the operations of the companies. This structure was also present in

Table 7 Summary of network arrangements in Germany

Institutional arrangements	Transmission planning and delivery	Network pricing	Modernising network arrangements
Institutional model: ■ Germany has four transmission networks ■ Two of them are operated by TSOs ■ The other two are vertically integrated, with the generation companies RWE and EnBW owning transmission assets	**Planning:** ■ TSOs are responsible for setting transmission development frameworks which are then reviewed by the federal network agency ■ TSOs then define network investment plans in line with these frameworks ■ Depending on project scope, state governments or the federal network agency then provide final approval for projects **Delivery:** ■ TSOs are responsible for delivering their own projects **Investment regime:** ■ The four German TSOs coordinate their system operation and planning ■ German TSOs are part of ENTSO-E, which coordinates network development and planning across European TSOs ■ There is no merchant transmission investment in Germany	■ Uniform pricing	**Readiness for decarbonisation:** ■ TSOs run auctions for demand curtailment services ■ These auctions are tailored towards medium- to large-sized industrial customers with high consumption ■ Residential end users can participate indirectly through aggregators ■ Several pilot projects and government funding for storage

Note ENTSO-E = European Network of Transmission System Operators for Electricity.
Source Vivid Economics

Australia's other regulated industries and was characterised by low productivity growth and inefficient performance.

The electricity sector underwent market reform in the 1990s, with a move towards private ownership that began with the breakup of vertically integrated monopolies into generation, transmission, distribution and retail components.

In the late 1990s, generation and retail were largely privatised, with some transmission and distribution companies following suit. The National Electricity Market (NEM) began operations in 1998 as Australia's first wholesale spot market. To facilitate the new power pool, market reforms were accompanied by significant investment in interconnection capacity, making

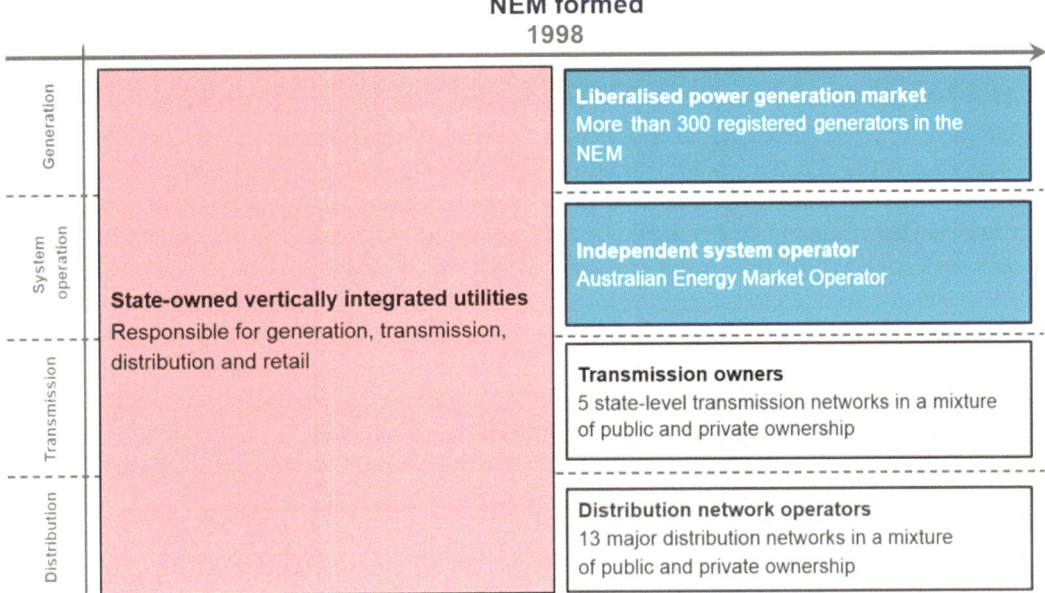

Fig. 43 Structure of the Australian electricity system. *Note* State-owned companies are in red, privately-owned companies are in blue. Segments containing a mix of privately and publicly owned companies are in grey. *Source* Vivid Economics

NEM one of the world's longest interconnected systems.

NEM covers the five interconnected Australian states of Queensland, New South Wales, South Australia, Victoria and Tasmania, and represents 89% of the country's generation capacity. The Australian Energy Market Operator (AEMO) is the ISO. Alongside traditional SO functions, AEMO publishes a long-term transmission programme known as the National Transmission Network Development Plan (NTNDP) and makes regional demand forecasts as part of its role of national transmission planner. AEMO is also directly responsible for transmission planning in the state of Victoria. Transmission services are provided by a transmission network service provider (TSNP) for each state and which are jointly state- and privately owned. There are 13 major distribution networks, each of which is a monopoly provider in its designated region. The market is regulated by the Australian Energy Regulator (AER). The structure of the Australian power sector before and after the formation of the NEM is shown in Fig. 43.

(2) Transmission planning

While AEMO conducts transmission planning in Victoria, TOs in other states are responsible for their own investment planning. TOs are required to publish annual planning reports (APRs) analysing their proposed network investments over the next five years. TOs have to take into account AEMO's National Transmission Network Development Plan (NTNDP) or risk financial penalties. They coordinate their plans with those of the distribution companies in their networks. While TOs are obliged to take the NTNDP into account, they retain autonomy over final planning decisions.

The NTNDP sets out AEMO's strategic national vision for transmission planning over a period of 20 years. Consistent with requirements for TOs, AEMO takes recent APRs into account when preparing its national transmission development plan. The aim of the NTNDP is to provide a long-term focus for investment planning and inform TOs of likely future developments, rather than influence individual investment decisions. The entire planning process is designed to create a positive feedback loop between AEMO, TOs and distribution companies, which may lead to better coordinated and more efficient transmission planning.

When considering major new transmission investments, TOs are required to perform a cost-benefit analysis known as the regulatory investment test for transmission (RIT-T). The process involves compiling a complete list of network and non-network investment solutions, with the help of stakeholder engagement, and choosing the solution with the highest expected return. Cost categories considered include construction and other asset provisions, operation and maintenance, and regulatory compliance. Benefit categories include more efficient dispatch, increased security of supply, reduced need for other investments, lower network losses or ancillary service costs, and contribution to renewable generation targets.

AEMO provides only a monitoring role in this process, ensuring that TOs comply with the RIT-T protocol. The procedure applies only to projects involving network enhancement, not to maintenance, and only when an option's costs exceed AUD 5 million. In Victoria, AEMO is directly involved in transmission planning and operates a competitive tendering process for projects that do not affect the assets of the incumbent TO, AusNet Services.

The Australian Energy Market Commission (AEMC) recommended enhancing AEMO's role in developing national transmission plans and in overseeing TO planning and investment tests. The commission also found that Victoria had similar levels of reliability and service, but paid less in achieving this, suggesting that AEMO-led investment planning is more efficient than incumbent TO planning. This could be due to the regulatory asset value (RAV) approach, which can lead to conflicts of interest and incentives for gold-plating (the more a TO spends on infrastructure, the more it gets paid by the government).

(3) Level of locational pricing

The NEM uses a hybrid of full nodal and zonal pricing, in which the price at each connection point is determined relative to the price at a common regional reference node. The spot price at each network connection point is calculated as the regional reference node price multiplied by a factor that accounts for intra-regional losses associated with that connection point. Under this system, costs of supply are generally higher for loads further from the regional reference node, reflecting higher losses. When capacity constraints are non-binding, prices across regions will vary based on network losses only. By contrast, when congestion occurs, differences in regional prices will depend on marginal generation costs in each region.

Spot prices in each reference node are calculated as the time-weighted average of dispatch prices and vary every 30 min. Dispatch prices are determined through central dispatch at each node and vary every five minutes.

(4) Modernising network arrangements

A set of market-based arrangements for frequency control, known as frequency control ancillary services (FCAS), has been in place since 2001. FCAS provides eight separate real-time spot markets for frequency control. These markets are considered highly progressive: they are market-based with standardised parameters. For example, products to raise or lower frequency are specified for 6-second, 60-second and 5-minute response times. This contrasts sharply with markets for ancillary services in other countries, such as the UK, where products do not have standardised parameters and are not open to providers across all technologies.

Automaker and energy company Tesla built the world's largest electricity storage facility in South Australia in 2017 to provide frequency response under FCAS:

- In 2016, Southern Australia experienced a catastrophic state-wide power outage (black system event), in which more than 800,000 customers lost power supply. In the initial stages of the system failure, high wind speeds damaged transmission lines, causing sequential faults and a dip in voltage. This led to an interconnector failure, islanding of Southern Australia from the remainder of the NEM, and supply failure as the system could not be balanced.

- A lack of inertia caused by high volumes of renewable generation in the system was found to be a key factor in the outage. South Australia has high levels of wind generation, which accounted for more than 40% of its power in 2016. Traditionally, fossil-fired thermal plants have played a key role in managing such events, acting as synchronous generators that provide real-time frequency response. Wind turbines lack this capability, increasing the risk of entire power network failure in the event of asset loss.

- Utility-scale storage units have been developed in Southern Australia by Tesla to address these issues. Tesla built a 100 MW battery installation in 2017 that is capable of providing power to 30,000 homes in the event of a blackout. The facility was funded by South Australia's $150 million Renewable Technology Fund that supports renewable energy projects. The facility adjoins the 100 MW Hornsdale 2 wind farm, allowing the batteries and wind farm to provide fast response services together in the case of a network fault, and to time-shift wind capacity and help meet peak demand periods.

(5) Summary of arrangements Australia (National Electricity Market, NEM) (Table 8)

3 Drivers, Supporting Conditions and Pathways for China's Energy Revolution

The history of human society is closely linked with energy supply changes. Fossil fuels like coal, oil and gas have driven the industrialisation and modernisation of global society. However, after more than two centuries of industrialisation, the energy system dominated by fossil fuels has brought severe environmental pollution and greenhouse gas emissions. Safeguarding the environment and combating climate change are perhaps the two biggest challenges facing the world, and the transition to a low-carbon global energy system is now an irreversible trend. Human society is entering a new era of energy transition from high carbon to low carbon, from low density to high density, and from black to green.

3.1 New Features in Energy Development Are the Foundation of the Energy Revolution

From 2012, China's economic development entered a new normal of slowing growth, structural optimisation and a shift in drivers. Understanding, adapting to and leading the new normal is a major issue in China, both now and in the near future. Meanwhile, developments in the energy sector are also changing, sharing some of the features of the new normal: slowing growth in energy demand, a gradual shift of growth drivers from manufacturing to services and households, the emergence of new business models and smart energy, an increase in renewable energy and the continuous optimisation of the energy mix, and success in supply-side reform.

3.1.1 Slowing Growth in Energy Demand

In the new normal, energy consumption shows a clear slowdown in growth. In this century, China's total energy consumption has almost tripled from 1.47 billion tonnes of coal equivalent (Btce) in 2000 to 4.3 Btce in 2015. This average annual growth in energy consumption of 7.42% produced an average annual economic growth of 9.6%. Although high, the growth rate has in fact been slowing down. In 2006–10, the average annual growth rate in energy consumption was 6.65%, down 5.55 percentage points from 2001–05. In 2011–15 it dropped to 3.58%, down 3.07 percentage points from 2006–10. In general, the current slowdown in energy growth matches that of the economy, thanks to proactive industrial

Table 8 Summary of network arrangements in Australia

Institutional arrangements	Transmission planning and delivery	Network pricing	Modernising network arrangements
Institutional model: ■ AEMO is the ISO	**Planning:** ■ TOs are responsible for investment planning, except in Victoria (AEMO) ■ AEMO sets out annual NTNDPs detailing long-term network needs ■ TOs publish APRs detailing plans over a five-year horizon ■ TOs and AEMO must take each other's plans into account, creating a feedback loop leading to coordinated national plans **Delivery:** ■ TOs are responsible for delivery of assets outside of Victoria, and can use tendering processes for third-party delivery ■ In Victoria, AEMO runs competitive tenders for projects which do not affect the assets of the incumbent TO, AusNet Services **Investment regime:** ■ TOs must engage stakeholders when optioneering for solutions ■ TOs required to perform cost-benefit analysis (RIT-T) for potential solutions ■ Third parties may participate in the delivery of assets	■ Pricing in NEM is a hybrid between nodal and zonal pricing ■ Each state has a reference node price ■ Connection point prices include locational adjustments based on network losses	**Readiness for decarbonisation:** ■ RIT-T process for transmission planning is designed to consider both network and non-network investment solutions ■ World's largest battery facility provides frequency response in South Australia in conjunction with AEMO

Note AEMO = Australian Energy Market Operator; NTNDP = National Transmission Network Development Plans; RIT-T = regulatory investment test for transmission; NEM = National Electricity Market.

Source Vivid Economics

restructuring, continuous efforts to improve energy efficiency and reduce emissions across all industries. These are the principal features of China's new normal and the foundation of the energy revolution.

3.1.2 Energy Growth Drivers Are Shifting from Manufacturing to Services and Households

Traditionally, as drivers of energy use weaken, new ones gradually arise to take their place. Take power consumption, for example. Average annual growth in power consumption in the tertiary (service) industries and in households during the 12th Five-Year Plan (2011–15) was 4.8 and 2.4 percentage points higher respectively than in the secondary (manufacturing) industries. In 2016, power consumption in the service industries and households grew 11.2% and 10.8% respectively, much faster than the growth in manufacturing (2.9%). This shows that the major driver for power consumption growth is shifting from high energy-consuming industries like manufacturing to the service industries and households.

3.1.3 Growth of New Energy Business Models Represented by Smart Energy and Internet+

A third driver of the energy revolution is the integration of energy with digital and Internet technologies. The rapid development of smart energy and Internet + (Energy Internet) is expected to diversify energy production and supply and stimulate new business models. Energy carriers will shift from single supply mode to a diversified one, in which conventional coal businesses, utilities and oil and gas companies transform themselves into integrated energy suppliers with multiple types of energy sources and the flexibility to provide various energy services that meet different user requirements. Internet + smart energy technologies can coordinate and optimise the control of energy storage equipment and controllable loads by: (1) building information interconnections between distributed

energy systems and users and between various local energy networks; these will leverage the spatio-temporal complementarity of distributed power supply systems in a wide area network and the system control capacity between energy storage equipment and demand-side controllable resources; and (2) enabling "horizontally complementary energy sources and networks and the vertically coordinated development of energy sources, networks, loads and energy storage". These measures will help control the impacts of intermittence on local grids and are a feasible way to connect distributed renewable energy at scale, thus substantially increasing the share of clean electricity in the energy mix.[6]

As the Internet + smart energy sector develops, its most striking feature is the rapid growth of electric vehicles (EVs). On the one hand, EVs are electrifying the transport system, which reduces oil dependency in socioeconomic development. On the other hand, EVs are the only way to fully connect the transport sector with Internet + and build a new and modern intelligent transport system enabled by artificial intelligence technologies, including remote control and unmanned driving. By 2016, China's EV ownership exceeded 1 million cars; there were more than 150,000 public charging stations and more than 200,000 private recharging points.

3.1.4 More Clean Energy and Optimisation of the Energy Mix

Global energy systems are increasingly turning to clean electrification.[7] They are investing in and developing clean power generation and new technologies like smart energy and the Energy Internet to create a new power supply system of integrated distributed and centralised energy. Such a system will increase electrification levels in manufacturing and living. And they are decarbonising non-electricity energy applications

---

[6]Gao Shiji and Guo Jiaofeng et al., Energy Internet Boosts Energy Transformation and Institutional Innovation in China, China Development Press, 2017, pp.11–13.

[7]The Energy Transitions Commission, Better Energy, Great Prosperity: Achievable Pathways to a Low-carbon Energy System, 2017.

in industry and transport by replacing conventional fossil fuels with biomass and hydrogen and by developing and deploying carbon capture, utilisation and storage technologies.

Currently, China is substituting oil and gas for coal, and non-fossil fuels for fossil fuels. In 2016, coal's share of China's total energy consumption was 62%, down 6.5 percentage points from 2000; whereas the share of natural gas and fossil fuels was 6.4% and 13.3% respectively, up 4.2 and 6 percentage points from 2000. In particular, the slowdown in energy consumption growth, which began in 2013, provided an opportunity to optimise the energy mix. The share of coal consumption in 2016 was 5.4 percentage points lower than in 2013, and the share of fossil fuels and natural gas consumption in 2016 was 3.1 and 1.1 percentage points higher respectively than in 2013 (Table 9).

3.1.5 Early Successes in Energy Supply Reform

As China's economic development entered the new normal, its rapidly growing energy industry —driven by the traditional modes of supply and demand—began to show inadaptability and regional and structural overcapacity. In response, the national government implemented energy supply-side reforms and reported its initial success in 2016. First, coal overcapacity was cut by 250 Mt annually. Second, the installed base of power generation became cleaner: more than 200 GW of coal-fired generating units were upgraded and made more energy efficient, more than 100 GW of coal-fired generation was upgraded to ultra-low emission standards, and the installed capacity of non-fossil energy rose to 36.1% of the total. Third, substituting clean energy for fossil fuels was vigorously encouraged. This included the installation of electric heating in homes and electric boilers in manufacturing facilities, as well as clean heating demonstration projects using renewable energy in provinces such as Inner Mongolia, Hebei and Jilin. Fourth, the NDRC's Opinions on Accelerating the Use of Natural Gas was introduced to drive fast and coordinated development along the natural gas value chain—upstream, midstream and downstream.

Table 9 Total energy consumption and energy consumption structure, 2000–15

Year	Total energy consumption		Energy consumption by type (%)			
	Total	Growth (%)	Coal	Oil	Natural gas	Non-fossil energy
2000	14.70	4.55	68.5	22.0	2.2	7.3
2001	15.55	5.84	68.0	21.2	2.4	8.4
2002	16.96	9.02	68.5	21.0	2.3	8.2
2003	19.71	16.22	70.2	20.1	2.3	7.4
2004	23.03	16.84	70.2	19.9	2.3	7.6
2005	26.14	13.50	72.4	17.8	2.4	7.4
2006	28.65	9.60	72.4	17.5	2.7	7.4
2007	31.14	8.72	72.5	17.0	3.0	7.5
2008	32.06	2.94	71.5	16.7	3.4	8.4
2009	33.61	4.84	71.6	16.4	3.5	8.5
2010	36.06	7.30	69.2	17.4	4.0	9.4
2011	38.70	7.32	70.2	16.8	4.6	8.4
2012	40.21	3.90	68.5	17.0	4.8	9.7
2013	41.69	3.67	67.4	17.1	5.3	10.2
2014	42.58	2.13	65.6	17.4	5.7	11.3
2015	43.00	0.99	63.7	18.3	5.9	12.1
2016	43.60	1.4	62.0	18.3	6.4	13.3

Source China Statistical Yearbook 2017

However, to transform the traditional energy system into a clean, low-carbon, secure and efficient modern energy system requires some deeply seated conflicts and issues to be addressed urgently. First, the excessive production and use of coal remains unsolved. The measures to cut overcapacity, implemented in 2016, haven't resolved oversupply. Cutting overcapacity will remain the industry's main focus over the next 3–5 years. Second, the slowdown in growth in power demand is at odds with the rapidly increasing installed capacity of new power generation. As it takes time to improve peak-shaving capacity, there are still difficulties in connecting renewable energy to the grid. Wind, solar and hydro curtailments will remain for some time. Third, as the coal-fired generating units under construction in the latter part of the 12th Five-year Plan (2011–15) are put into operation, the total number of operating hours of coal-fired generating units are predicted to fall to around 4,100 in 2017. Meanwhile, as the price of coal returns to a reasonable level, the risk of operating at a loss increases across the entire coal-fired power generation sector. Fourth, the development of natural gas is still restricted by its comparatively expensive price. There are numerous difficulties in deregulating the natural gas end-use pricing system, which are slowing the growth of downstream gas consumption.

3.2 Five Drivers of the Energy Revolution

3.2.1 Changes and Diversification in International Energy Supply

According to the forecasts of energy companies like BP and agencies like the U.S. Energy Information Administration (EIA), global energy demand will continue to grow slowly. By 2020, total global energy demand will reach 14.6 billion barrels of oil equivalent (boe), and the growth rate will decrease from 2.0% in 2010 to 1.3% in 2020. By 2030, the growth rate will decline to about 1.0%, with global energy demand at 15.4 billion boe.

Meanwhile, with the rise of unconventional energy like shale oil and shale gas, supply is becoming increasingly diversified. Influenced by the shale gas revolution in the USA, other countries in the Americas—including Argentina, Brazil, Canada and Venezuela—are exploiting their rich resources. As the development of unconventional oil and gas increasingly matures, the Americas are expected to become the second Middle East.

According to the EIA, US oil imports dropped to 24% in 2015, in sharp contrast to 60.3% in 2005. The USA is very likely to become a net oil exporter by 2020. Canada's oil production, according to the IEA's predictions, will reach 30–60 million barrels per day by 2030. With abundant conventional oil and gas resources, both onshore and offshore, Canada is expected to develop into an energy superpower in the coming years.

As clean energy, renewables will play an increasingly important role in diversifying energy supply. According to the IEA, renewable energy (including hydropower) will account for half of newly added global power output and almost a third of global power generating capacity by 2035, making it the dominant power source.

3.2.2 Stable Economic Development Is a Solid Foundation for the Energy Revolution

Needless to say, China's socioeconomic development will face unprecedented difficulties and challenges in the future. These include a fall in the working age population, severe overcapacity in traditional industries like steel and mining, low participation in high value-added segments, and increasing exposure to environmental problems caused by intensive industrial development. However, China still holds huge development potential and resilience. It has fully developed industries, rich human resources and rising innovation capability, which provide a solid foundation for future development.

In October 2015, the Fifth Plenary Session of 18th CPC Central Committee adopted the CCP Central Committee Proposals for the Formulation of the 13th Five-Year Plan for Economic and Social Development. Guided by the concepts of "innovative, coordinated, green, open and sharing", China is expected to make continuous progress in economic restructuring and transformation, and deliver stable and sustainable medium-to-high growth. This will help ensure that the Chinese dream of national rejuvenation will be fulfilled by mid-century. China's GDP per capita will approach $40,000 (at 2015 prices) by 2050,[8] an increase by a factor of five on 2015. As a strategic requirement of future socioeconomic development, this increase in GDP per capita will become a solid foundation for the revolutions in energy production, consumption and supply.

3.2.3 Combating Climate Change and Protecting the Environment Are Key Drivers of the Energy Revolution

According to the Fifth Assessment Report (AR5) of the Intergovernmental Panel on Climate Change (IPCC), the rise in global average surface temperature between 1951 and 2012 was 0.72°C, almost double that of 1880. Global warming has become an indisputable fact. Almost 200 ratifying countries agreed on the Paris Agreement in 2015. They identified the long-term goal of keeping the increase in global average temperature to well below 2°C, and even limiting the increase to 1.5°C by 2100. China has also developed action plans to combat climate change, such as setting the target of reaching peak level CO_2 by 2030 or earlier and achieving a 60-65% decrease in CO_2 emissions/GDP per capita by 2030, compared to the level in 2015. However, long-term extensive economic development impacts the environment negatively and significantly. In response, China issued the Air Pollution Prevention and Control Action Plan in 2013. Currently, China is making

vigorous efforts to control haze and is implementing effective measures to reduce emissions of air pollutants. These include shutting down small coal-fired industrial boilers, substituting electricity or gas for scattered coal, replacing the internal combustion engine with electric vehicles, and using oil of higher quality.

The pressure to combat climate change and control haze is forcing China's energy industry to develop a clean and low-carbon energy system that will gradually deliver sustainable socioeconomic development.

Substituting electricity and gas for fossil fuels like coal and oil is an important way to optimise the energy system and achieve energy efficiency and emissions reduction. However, electricity and gas are both constrained by system factors: gas by price and the need to overhaul the transmission system; and electricity by costs, and a lack of interconnected nationwide infrastructure and critical technologies. During the ongoing 13th Five-Year Plan (2016–20) and beyond in the medium and long terms, optimisation is urgently needed to improve the level of coordinated energy development.

3.2.4 Innovation Is an Important Support for the Energy Revolution

New business models like the Energy Internet are evolving rapidly and driving the entire energy technology innovation value chain. Taking the overall energy industry and its long-term development requirements into consideration, the Energy Internet, enabled by smart technologies, achieves deep integration of energy and information and new technologies and business models. As the new driver of a rejuvenated energy industry, the Energy Internet will support the steady implementation of the energy production and consumption revolution.

China's Energy Internet (Internet +) will be achieved through a three-stage strategy:

- In 2017–20: (i) distributed power generation and storage technologies will be deployed at scale to allow flexible grid connection of various types of distributed energy; (ii) an

[8]The State Information Center of China. 2016. Internal Research Report.

Internet-based multi-energy trading system will go live; and iii) demonstration projects of multiple interconnected energy networks and energy sources will trigger the extensive use of Energy Internet technologies.

- In 2021–25: (i) intelligent scheduling between diversified energy carriers will be possible and distributed power generation and storage systems widespread among end users; and (ii) smart and diversified urban energy networks will be established to allow accurate supply scheduling from different sources to meet fluctuations in demand.
- In 2026–30: (i) new microgrids will be developed nationwide and an interconnecting network of non-fossil energy will be built to help reduce the share of non-fossil energy in primary energy to the target of 20%; and (ii) an open and sharing Energy Internet ecosystem will be formed to significantly improve energy efficiency.
- After 2030: building on the achievements of the Energy Internet, the use of renewable energy will span many sectors—agriculture, industry, transport, commerce and households. The ecosystem supporting the fast and sound development of renewable energy will continue to improve and push renewable energy development firmly into the fast lane.

3.2.5 Natural Gas Should Play a Dominant Role in the Energy Revolution

According to China's National Hydrocarbon Resources Assessment (2015), China's geological gas resources are:

- conventional gas, including tight gas: 90.3 trillion cubic metres, including 50.1 trillion cubic metres of recoverable reserves;
- shallow shale gas within a depth of 4,500 m: 121.8 trillion cubic metres, including 21.8 trillion cubic metres of recoverable reserves; and
- shallow coalbed methane within a depth of 2,000 metres: 30.1 trillion cubic metres, including 12.5 trillion cubic metres of recoverable reserves.

By the end of 2016, the cumulative proven geological gas resources were:

- conventional gas, including tight gas:[9] 11.7 trillion cubic metres, including 5.2 trillion cubic metres of recoverable reserves;
- coalbed methane: 692.83 billion cubic metres, including 334.40 billion cubic metres of recoverable reserves; and
- shale gas: 544.13 billion cubic metres, including 122.41 billion cubic metres of recoverable reserves.

Currently, the conversion rate of natural gas resources to reserves and recovery rate of proven reserves in China are all relatively low. The proven resource rate of conventional gas, coalbed methane and shale gas are 13.0%, 2.3% and 0.4% respectively, so there is a huge potential to recover more resources through technological innovation. In addition, China has made great progress in exploring for offshore natural gas hydrates, of which there is huge resource potential to be unlocked.

Before non-fossil energy technologies mature, natural gas (the cleanest burning fossil fuel) is the best replacement for coal and oil to reduce pollution and greenhouse gas emissions from energy consumption. Even after the technologies mature, natural gas—thanks to its flexibility as an energy carrier—still has great potential for extensive deployment. As unconventional gas resources are still being discovered, it makes sense to make natural gas a dominant energy carrier after coal and oil. This is also a key part of the energy revolution.

3.3 Analysis of Strategic Pathways for the Energy Revolution

3.3.1 Scenario Setting

(1) Strategic pathway design should follow the principle of "letting the targets be the guide"
China's Strategy of Energy Production and Consumption Revolution is a clear strategic plan

---

[9]Conventional gas here refers to gas field gas, excluding solution gas.

for energy development between 2016-30. Using the plan as a guide, this report looks at possible strategic pathways to implement the plan and achieve China's energy revolution. The report is not about making predictions in multiple scenarios. Rather it focuses on the following three goals identified by China: first, aligning medium- and long-term economic and social development with the goal of building China into a great modern socialist country, as stated by President Xi in his address to the 19th National Congress of the Communist Party of China in 2017. Second, controlling greenhouse gas emissions and air pollution. And third, guaranteeing energy security.

(2) Strategic pathway drivers—economic development, energy security and environmental protection

Research on China's Medium- and Long-Term Energy Development Strategy[10] summarises the main drivers of China's energy development in the future—economic development, energy security and environmental protection—and proposes optimal energy development pathways by balancing the goals of those three drivers. In the traditional economy and energy system, the three drivers are independent rather than integrated: (i) rapid economic growth means more energy consumption, which traditionally means more fossil fuels—this leads to worsening environmental pollution and puts energy security at risk from greater dependence on oil and gas; (ii) the environment-first approach tends to slow down economic growth; and (iii) the security-first approach slows down economic growth and reduces demand.

However, in the new normal of China's slowing economic growth, and as new energy technologies and business models appear, the integration of economic development, energy

security and environmental protection significantly strengthens. In economic development, the traditional growth drivers of real estate and energy-intensive industries gradually slow down, and the new drivers—primarily the high-tech and service industries—begin to take over. This changeover gradually loosens the rigid coupling between economic growth and energy demand. As energy demand growth will mainly come from the service industries and households, the demand for cleaner energy and supply flexibility will grow. The more developed the economy, the greater the demand for clean energy and supply flexibility. This in itself represents a new challenge.

In energy security, the national government has identified electric vehicles (EVs) as the main means of road transport in the future. EV manufacturing and its upstream and downstream industries will see rapid growth. It is possible that the replacement of conventional petrol and diesel vehicles with EVs may escalate at a speed beyond expectation, triggering a rapid shift from oil to electricity as the predominant energy for transport. This will mitigate the long-existing energy security risk of oil import dependency and facilitate the switch to renewable energy in China's power grids. China is also diversifying into natural gas, thanks to rapid growth in domestic shale gas production and successful pilot exploration of natural gas hydrates. This gradual strengthening of domestic supply capacity will support the substitution of gas for scattered coal at scale, and increase the capacity to use low-cost, high-quality natural gas resources in international markets.

In environmental protection, haze prevention is a major concern. China vigorously promotes measures such as replacing coal with electricity and gas and shutting down small coal-fired boilers. This not only reduces coal-induced pollution, it helps to unlock the market potential for cleaner alternatives like renewable energy, geothermal power and natural gas. Greater use of clean energy creates opportunities for the deployment of Energy Internet technologies, which spurs innovation and encourages new business models.

---

[10]Development Research Center of the State Council and Shell International Limited. 2013.

(3) Strategic pathway scenarios

Based on the above analysis, the scenarios developed in this study are as follows:

- First, in alignment with the strategic goals set out in President Xi's report to the 19th National Congress of the Communist Party of China in 2017, China's socioeconomic development scenario for 2050 forecasts the economic aggregates, industry structure and development of major industries in 2020, 2030 and 2050, which are then used as external input variables to analyse China's energy supply pathways.

- Second, in light of China's socioeconomic development trends, and in accordance with the Strategy of Energy Production and Consumption Revolution (2016–30), the recommended pathway scenario for China's future end-use energy development is described, based on the present analysis.

- Third, we present our research on the impact of scattered coal governance and EV development on end-use energy demand. Currently, haze prevention and control is the biggest uncertainty influencing China's energy supply development pathway. Substituting electricity and gas for scattered coal (SEGFSC) and replacing the internal combustion engine (ICE) with EVs are the two most important ways to prevent and control air pollution. The level of success in both efforts will directly affect China's future energy supply system. This report analyses both measures and uses the energy system analysis model we developed to conduct sensitivity research on their impacts on the energy system. We developed three scenarios for substituting electricity and gas for scattered coal—High, Medium and Low—and two scenarios for replacing ICE vehicles with EVs—the Recommended and Extreme.

- Fourth, based on research into end-use energy demand, and in accordance with the need for low-carbon energy development, China's future primary energy supply is forecast to follow the Recommended scenario for the energy supply revolution.

3.3.2 Socioeconomic Development and Demand for Energy Services

Research on China's medium- and long-term macroeconomic development draws different conclusions. O'Neill and Stupnytska (2009) estimate that China's GDP growth rate in 2011–20, 2021–30, 2031–40 and 2041–50 will be 7.9%, 5.7%, 4.4% and 3.6% respectively. In the rapid economic growth scenario where supply-side structural reform is implemented, Li and Lou (2016) believe that the potential average economic growth rate in the 13th Five-Year Plan (2016–20) and 14th Five-Year Plan (2021–25)—neither of which had been published at the time—would be 6.5% and 5.8% respectively. According to Xiao Lin (2016), China's long-term economic growth will gradually slow down and approach the world average (currently around 3.5%) before settling at 3–4% in 2050. Based on these projections, this report forecasts population trends, growth drivers and development trends for major industries. It uses the SICGE model to make structural predictions and ensure there is consistency in macroeconomic trends, changes in energy demand and shifts in industry structure. It also provides input variables for the energy-environment system analysis model.

(1) China's population trend

According to the latest statistics of the National Population and Family Planning Commission (NPFPC),[11] China's population will peak at 1.45 billion in 2030, 30 million more than the 2015 forecast of the United Nations. At the same time, the number of people aged 65 or older will rise to 350 million in 2050, 2.4 times that of 2015. Those aged 65 or older made up 11% of the population in 2015, compared to a projected 25% in 2050. Correspondingly, future labour supply will decline absolutely, dropping to 830 million in 2050, down 170 million from 2015 (Fig. 44). Population and aging trends are crucial for China's future economic development.

---

[11]http://www.nhfpc.gov.cn/xcs/s3574/201510/b03bbb9da18044c299f673f0b84eeab1.shtml.

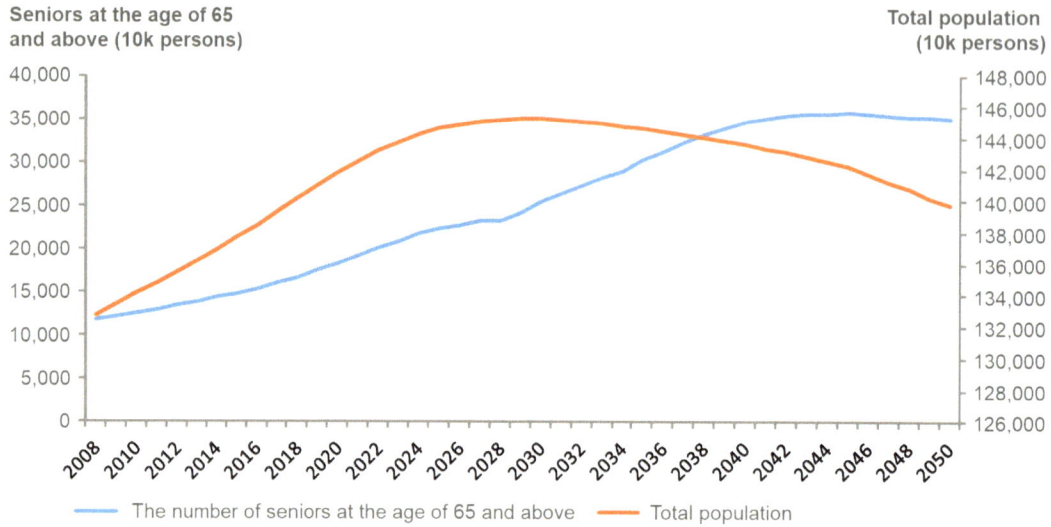

Fig. 44 Forecast trend of China's total population and population aging. *Source* Extension of the predictions of the National Population and Family Planning Commission

(2) China's growth in floor area

According to research by the Ministry of Housing and Urban-Rural Development, China's floor space stock in 2015 was around 60 billion square metres. It comprised 17.6 billion square metres of urban residential floor space, 27.6 billion square metres of rural residential floor space, and 14.0 billion square metres of urban commercial floor space. In the experience of developed countries (see Fig. 45), when China's floor space peaks urban residential floor space per person will be around 40 square metres and public floor space per person around 20 square metres (based on a conversion factor of 0.8 with the floor space per person in developed countries). To ensure steady and sustainable development of the real estate industry, this report suggests that the time-to-peak should be postponed to around 2040, in accordance with the population forecast for that year. Based on the projected population in 2040, the floor space peak will be about 92 billion square metres (see Fig. 46). In this pathway, the rapid expansion of the real estate industry has ended and the annual floor space will decrease from 3.2 billion square metres in the 13th Five-Year Plan (2016–20) to 2.3 billion square metres in the 14th Five-Year Plan (2021–25) and 1.7 billion square metres in the 15th Five-Year Plan (2026–30). As a result, the ability

of the real estate industry to stimulate economic growth will continue to weaken.

(3) Trends for major industrial products

(1) Vehicle ownership is expected to increase

In the experience of developed countries, car ownership goes through three stages—slow growth, explosive growth and saturation—as GDP per capita increases. Currently, China's GDP per capita exceeds $8,000. Judging by the growth in passenger car ownership in recent years, China's car industry has entered a fast growth stage, which is expected to last until 2035. After which, it will gradually reach saturation and a sharp slowdown in growth. It is projected that China's vehicle ownership will exceed 500 million in total by 2050 and 350 units per 1,000 people.

(2) Steel output is predicted to decline

Currently, more than 50% of crude steel in China is used in the building sector. As future annual average floor space continues to decline, the demand for crude steel will gradually decrease. It is predicted that crude steel demand will decline to 730 Mt in 2020, 600 Mt in 2030 and 200–250 Mt in 2040. The steel will be used mainly to make mechanical equipment, household appliances and cars. As China's stock of steel products (equipment,

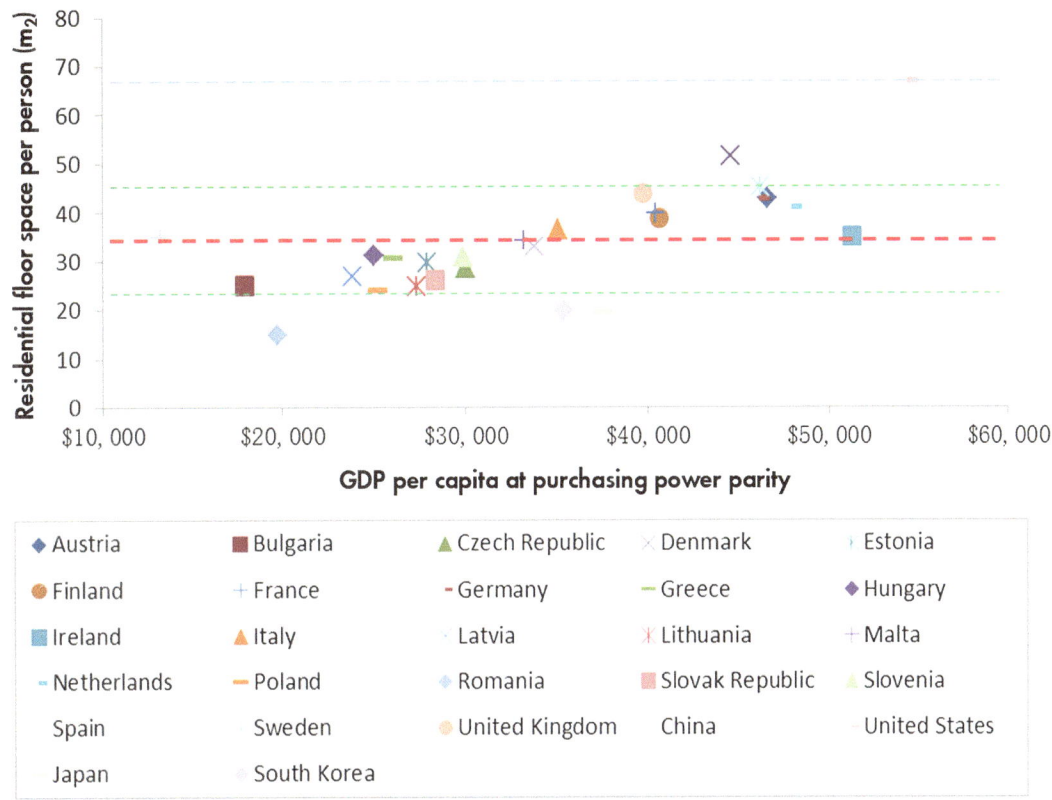

Fig. 45 Comparison of residential floor space per person in selected countries. *Source* The Standard Quota Department, Ministry of Housing and Urban-Rural Development (MOHURD) (2016)

appliances and cars) gradually increases, the recycling rate of scrap steel will also rise and is expected to reach 20% by 2025, 25% by 2030 (the same as in developed countries) and 40–70% by 2050.

(3) Energy use by the nonferrous metal sector is expected to grow steadily until 2030

An important basic raw material, nonferrous metals are widely used to make equipment and products for many industries: power, transport, machinery, electronics and aerospace. As the Made in China 2025 strategic plan is implemented, the nonferrous metal sector is expected to grow annually by around 3% until 2030, which will cause its energy consumption to continuously rise. After 2030, as the stock of equipment and products made of nonferrous metals increases and ore smelting is gradually

replaced by recycled scrap metal, the sector's energy consumption will fall sharply.

(4) Demand for traditional building materials will gradually decrease, while demand for new building materials will significantly increase

As demand diversifies and the low-carbon economy develops, the growth space for traditional building materials (bricks, wood, plaster and stone) gradually narrows. Demand for cement and wall materials is expected to peak in the present 13th Five-Year Plan (2016–20), before contracting. Cement output will gradually decrease from 2.35 Bt in 2015 to 1.8 Bt in 2030, and to 1 Bt in 2050. These traditional materials will be mainly used for the construction and maintenance of public infrastructure and buildings. As floor space continues to grow to 2040,

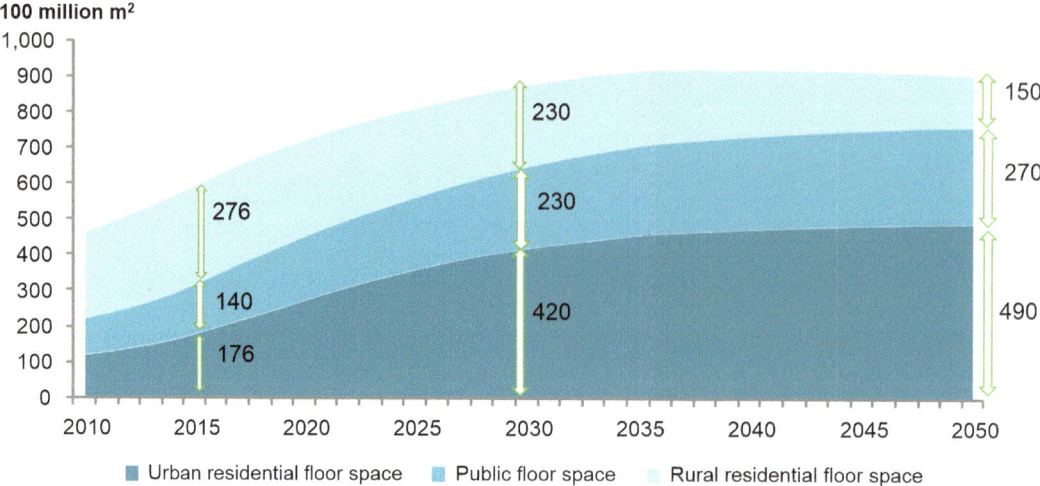

Fig. 46 China's floor space predictions. *Source* The Standard Quota Department, Ministry of Housing and Urban-Rural Development (MOHURD) (2016)

the annual need for maintenance and decoration will also increase, which means growing demand for glass, ceramics, plastics and other new building materials.

(4) China's economic development to 2050

The 19th National Congress of the Communist Party of China in 2017 identified the strategic goal of achieving the country's modernisation by 2035 and developing China into a great modern socialist country by 2050. Compared with the previous goal of achieving modernisation by mid-century, the revised target is 15 years ahead of the original timeline. This shows that China has made greater progress than expected and that there is still huge development potential in the long term. By 2035, China's GDP will reach $33.8 trillion (at 2015 prices), passing the USA to become the world's largest economy. GDP per capita will be more than $20,000, rising to around $40,000 in 2050, which is above that of moderately developed countries. China's share of global GDP will exceed 20%.

By 2050, China will have shifted from a production-led to a consumption-driven economy. Manufacturing's share of the economy will continue to decrease, while that of the service sector will rise from 50.2% in 2015 to 58% in 2035 and around 70% by 2050 (at 2015 prices). In

industry, equipment manufacturing and light industry's share of the economy will increase from 46% in 2015 to 61% in 2035 and 65% in 2050.

3.3.3 Energy Demand to 2050, as Forecast by the Strategy of Energy Production and Consumption Revolution (2016–30)

The Strategy of Energy Production and Consumption Revolution (2016-30) uses the same socioeconomic development analysis as above to forecast China's energy demand pathway to 2050. It is important to note that this is not a business as usual scenario, but an integrated one that takes such factors as the potential for energy efficiency in various industries, clean energy requirements and economic feasibility into consideration. The energy supply revolution should be carried out on the premise of meeting end-use energy demand (Fig. 47).

(1) Agriculture

Agriculture is a low energy-consuming sector. Its share of total end-use energy consumption in 2015 was only 2% and energy consumption per unit of added value of RMB 10,000 ($1,600) was only 0.1 tonnes of coal equivalent (tce). China's agricultural growth will remain stable for a long time. The one major factor that influences energy

RMB 100 million

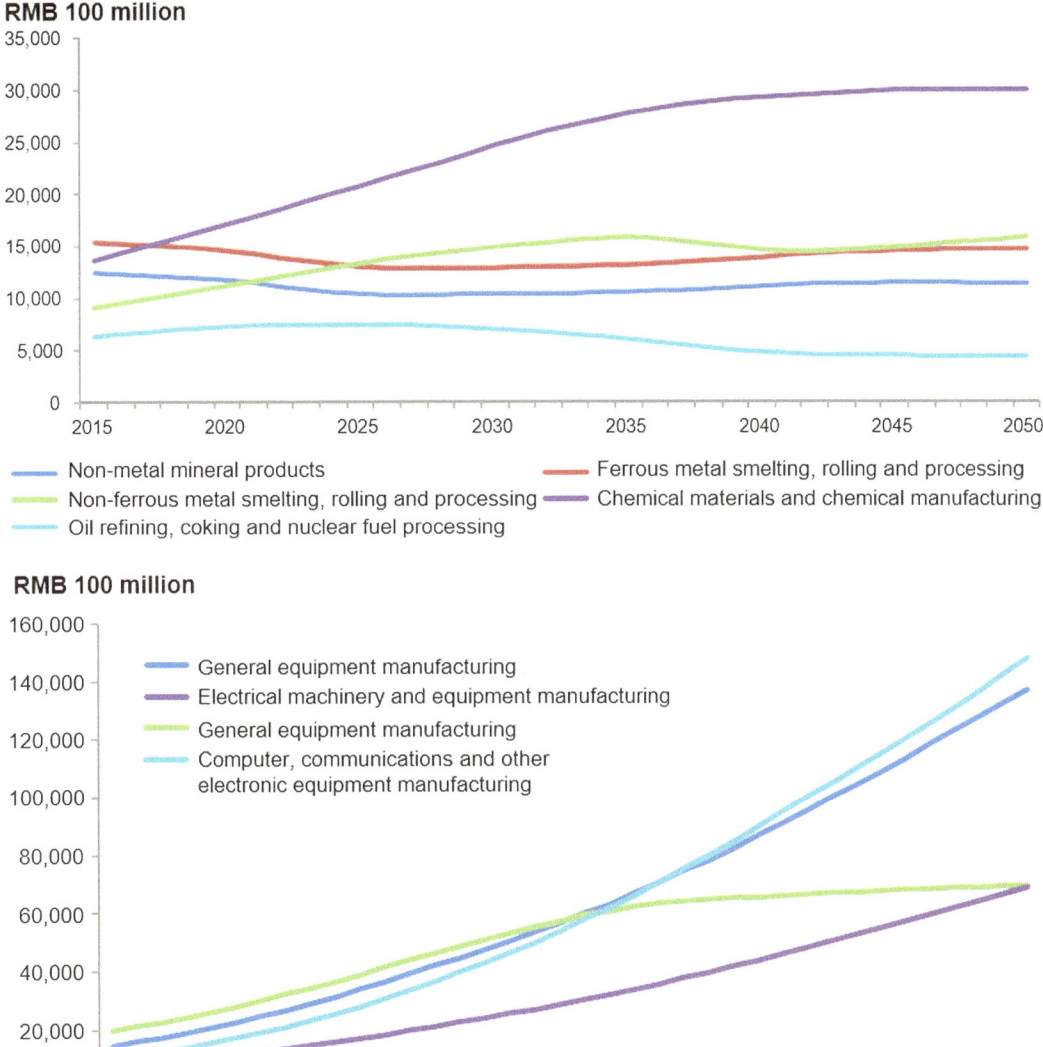

Fig. 47 Trends in China's major manufacturing sectors

consumption in agriculture is mechanisation, which began to improve in 2005. By 2015, agricultural machinery used 1,120 GW of electricity, double that of 2005. As mechanisation and energy consumption rise, cumulative energy consumption per unit of added value falls—by 17% in the past 10 years. In the future, as agricultural mechanisation becomes saturated, modernisation will gradually shift to the use of

biotechnology and information technology (IT), which will further reduce energy consumption per unit of added value. When it was adopted in 2016 the 13th Five-Year Plan (2016–20) predicted that agricultural mechanisation would continue to grow and that the amount of energy used in agriculture would approach 66 Mtce by 2020. As the efficiency of agricultural machinery improves and the energy efficiency benefits from

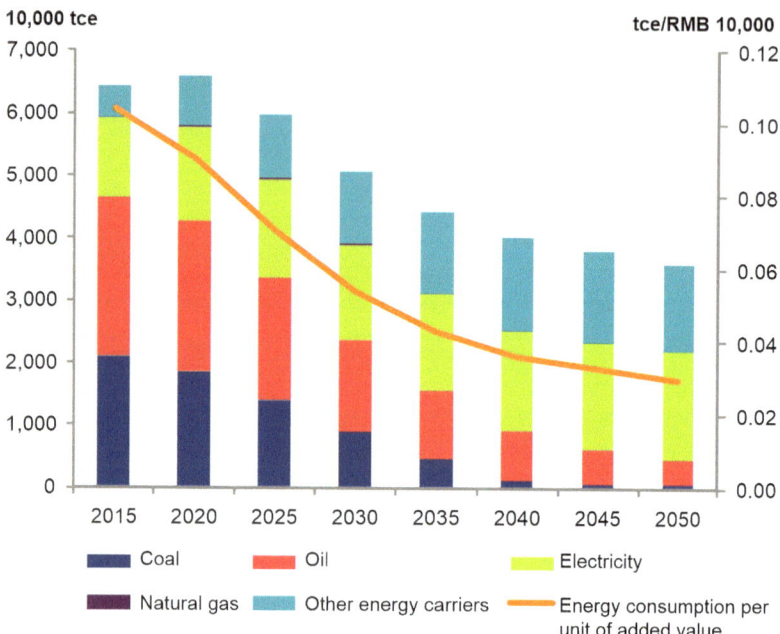

Fig. 48 Energy consumption in Chinese agriculture

biotechnologies and IT become evident, energy consumption in agriculture is expected to decrease to 51 Mtce in 2030, 43 Mtce in 2035 and 36 Mtce in 2050.

Eventually, the use of scattered coal in agriculture will completely disappear, and oil-fuelled machinery will be replaced by electric and biomass fuelled machines. The use of electricity and renewable energy will gradually grow, rising from 28% in 2015 to 54% in 2030 and 87% in 2050 (Fig. 48).

(2) End-use energy consumption in industry and buildings

In 2015, the industrial and building sectors consumed 88% of coal, 36% of oil, 55% of natural gas and 71% of electricity and heat. In this report, we assess the future energy consumption of industry and buildings. Our assessment is based on trends in high energy-consuming industries—including iron and steel, cement, glass, aluminium, ammonia, ethylene and methanol—and the future industrial development plans outlined in Made in China 2025. As shown in Fig. 49, China's total energy use in industry and buildings is expected to

peak in 2025-30 at about 2.3 Btce, slightly higher than the current level. It is then predicted to steadily decrease to 2.2 Btce in 2035 and 1.9 Btce in 2050. Energy consumption per unit of added value in industry and buildings will steadily decline— by 2050, it is expected to be 80% lower than in 2015.

In our scenarios, the assumption is that as the stock of scrap steel increases and the mainstream technologies in iron and steel gradually shift from predominantly long processes (steelmaking starts with iron ore and coke) to equal focus on long and short processes (electric furnace steelmaking starts with scrap steel), the share of short-flow steelmaking technologies will gradually increase to 60% by 2050. The replacement of small industrial coal-fired boilers and kilns will reduce the use of industrial scattered coal. However, as living standards improve and demand for consumer products increases, the use of chemicals in manufacturing will be the only high point in otherwise decreasing demand for coal and oil. Coal's share of the energy mix will fall sharply from 55% in 2015 to 45% in 2030 and 29% in 2050. Correspondingly, the share of

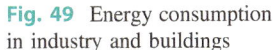

Fig. 49 Energy consumption in industry and buildings

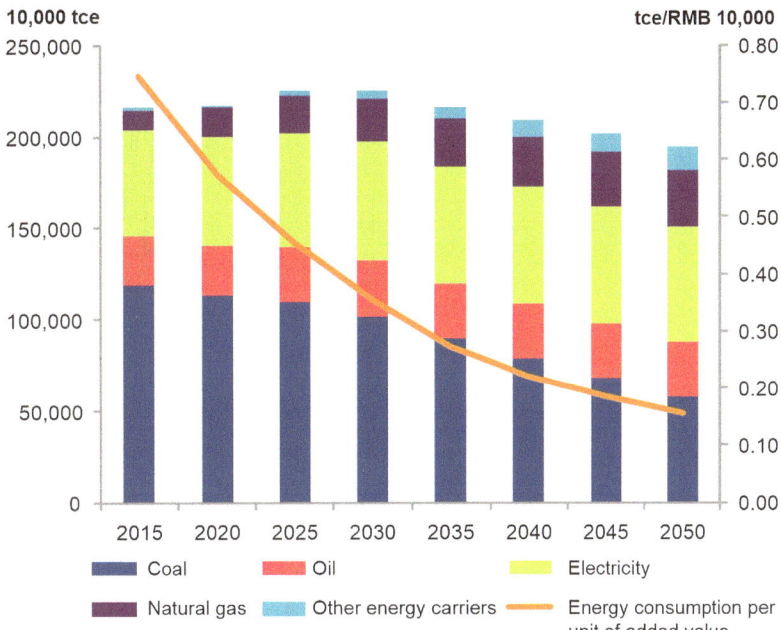

natural gas, electricity and heat will increase from 32% in 2015 to 55% in 2050.

(3) Transport

The energy consumption of the transport sector includes petrol and diesel used by households, wholesale and retail businesses and public services. Some of the petrol used by industry and buildings should also be included. In the future, as more and more families buy vehicles and air travel becomes more popular, energy consumption in the transport sector will grow. However, fuel economy is improving and public transport becoming more prevalent, which means energy consumption per unit will decrease. In our scenarios we have plotted vehicle ownership and unit energy consumption of road, rail, waterborne and air transport to forecast future energy demand. The most important results are as follows: by 2030, due to population growth and higher vehicle uptake, energy consumption in transport will continue to increase, peaking at around 700 Mtce in 2035. It will then slowly decline, reaching 640 Mtce in 2050. Petroleum-based fuels like petrol, diesel and kerosene will still play a dominant role. Demand for these fuels will be 410 Mtce in 2050, but their share of the transport fuel mix will

gradually decline from 87% in 2015 to 65% at mid-century. In contrast, electricity and the natural gas and biofuels that replace petroleum will increase significantly to 200 Mtce in 2050, which is 35% of the sector's fuel mix. Based on China's medium- and long-term targets for electric vehicles (EVs), it is predicted that EV ownership will reach 3 million units in 2020, 80 million units in 2030 and 270 million units in 2050. The share of EVs in total vehicle ownership will exceed 50% by 2050, with annual electricity consumption reaching 400,000 GWh (Fig. 50).

(4) Services (excluding transport)

The development of the service industries will play an important role in optimising China's industrial structure and shifting its growth drivers. As the service industries expand, so too will their energy consumption, although this will be constrained by the floor space they occupy. We estimate that energy consumption in services will increase to 350 Mtce in 2035 and 470 Mtce in 2050. Due to modernisation and the use of electricity and gas instead of scattered coal, coal's share of energy consumption in the sector will rapidly decrease, falling to only 2% by 2030. Internationally, power demand in services is

Fig. 50 China's energy consumption in transport

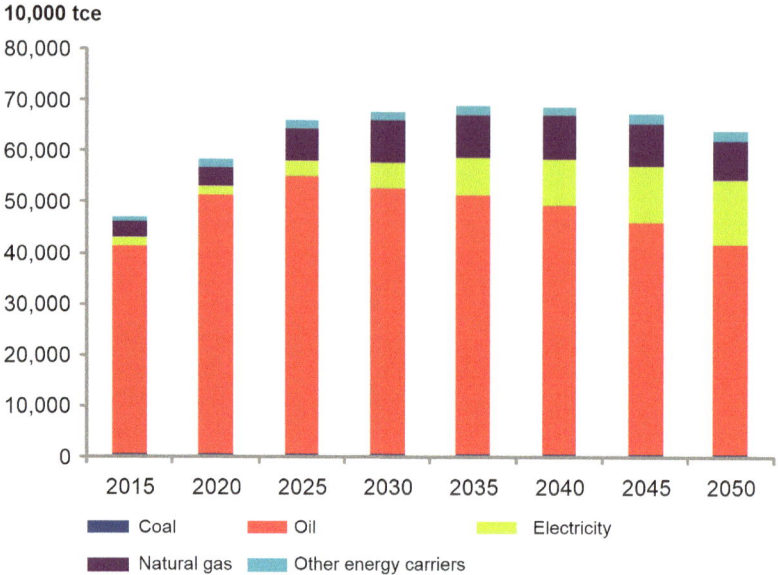

Fig. 51 China's energy consumption in the service sector

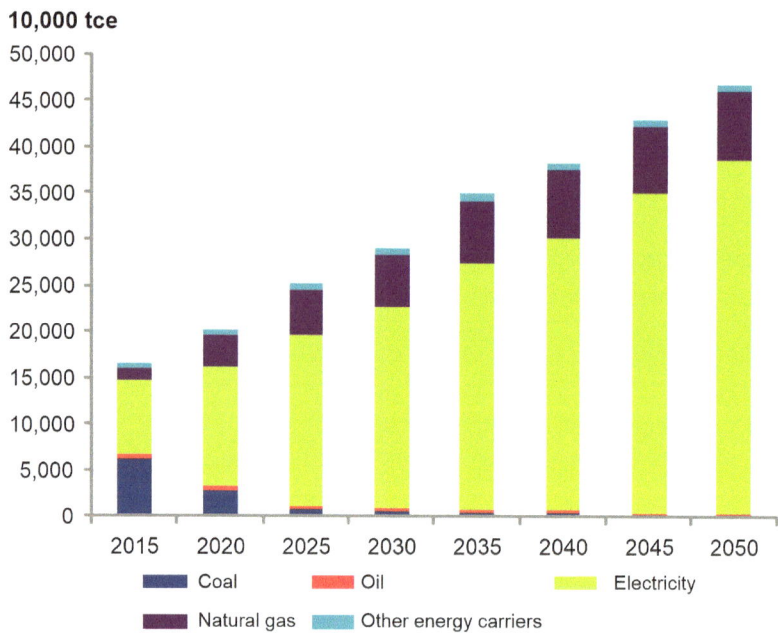

similar to that of households. As such, power consumption in China's service industries is projected to reach 2.4 trillion kWh by 2050, which is 82% of the sector's total energy use. As the natural gas distribution system develops, gas demand from large buildings is expected to rise steadily. By 2050, gas consumption in the service sector will reach 75 Mtce, 16% of its total energy use (Fig. 51).

(5) Households

As living standards improve, households use more energy. Household energy consumption is expected to reach 520 Mtce in 2035 and 660 Mtce in 2050. By replacing scattered coal with electricity and gas, coal consumption will rapidly decline, from 23% of the household energy mix in 2015 to 4% in 2030 and only 1% in 2050. Liquefied petroleum gas will be gradually replaced by

Fig. 52 China's energy consumption by households

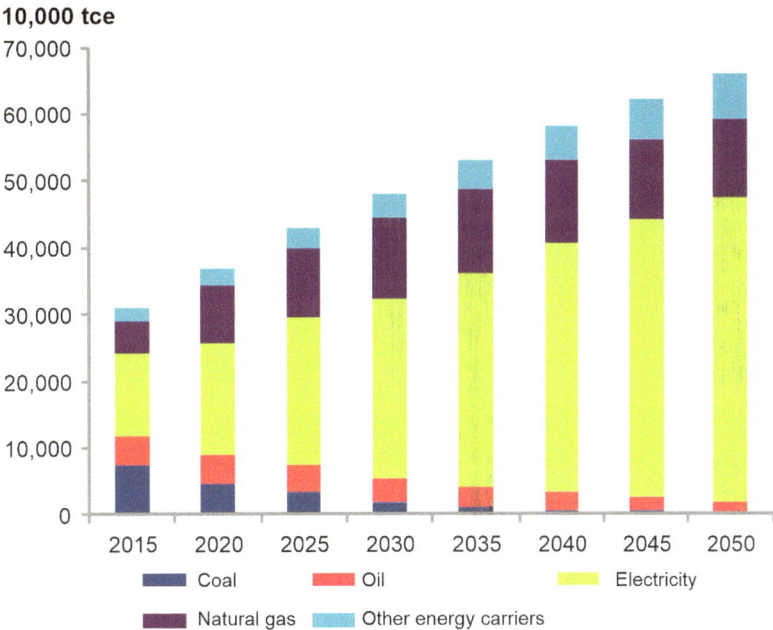

10,000 tce

natural gas, decreasing from 44 Mtce in 2015 to 14 Mtce in 2050, which is 14% and 2% respectively of the household energy mix. China's household power consumption per capita will gradually increase from 540 kWh in 2015 to 1,100 kWh in 2030 and to 2,000 kWh in 2050. Electricity and heat's share of the household energy mix will rise from 40% in 2015 to 69% in 2050. Natural gas use will reach 120 Mtce in 2050, which is 17% of household energy demand.

(6) Total end-use energy demand

In summary, China's future energy demand in agriculture, industry and buildings, transport, services and households shows the following characteristics: (i) China's energy consumption in these sectors will peak around 2040 at 3.8 Btce; (ii) energy demand in agriculture, industry and buildings, and transport will peak around 2020, 2025–30 and 2030–40 respectively, but not until 2050 in services and households; and (iii) the share of coal and oil in the sectors' energy mix will decrease from 42% and 24% respectively in 2015 to 15% and 20% respectively in 2050. Use of electricity and heat will almost double, from 810 Mtce in 2015 to 1.61 Btce in 2050, and their share of the sectors'

energy mix will increase from 26 to 43%. Electricity demand is expected to reach 7 trillion kWh in 2020, 9 trillion kWh in 2030 and 11.7 trillion kWh in 2050, equivalent to 25.6%, 30 and 38% of the sectors' energy use (Figs. 52 and 53).

3.3.4 Analysis of Three SEGFSC Scenarios

The China Energy Statistical Yearbook 2016 estimates that China's scattered coal consumption in 2015 was 617 Mt: including 120 Mt in coal mining, 93 Mt in households, 90 Mt in chemical production and 260 Mt in services and light industries like food, textiles and equipment manufacturing. Scattered coal accounts for 50–70% of the air pollutants emitted by coal in China (the emissions from burning 1 tonne of scattered coal are equivalent to those from burning 5–10 tonnes of thermal coal in a power plant).[12] Substituting electricity and gas for scattered coal (SEGFSC) is, therefore, the first measure to prevent and control air pollution. According to China's Action Plan for the Prevention and Control of Air Pollution and

---

[12]http://www.chinanews.com/ny/2016/10-21/8038629.
shtml.

Fig. 53 China's overall
energy use by sector and type

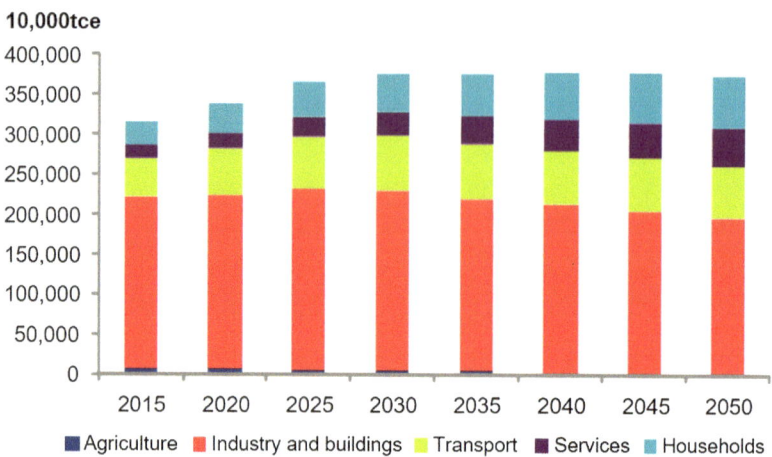

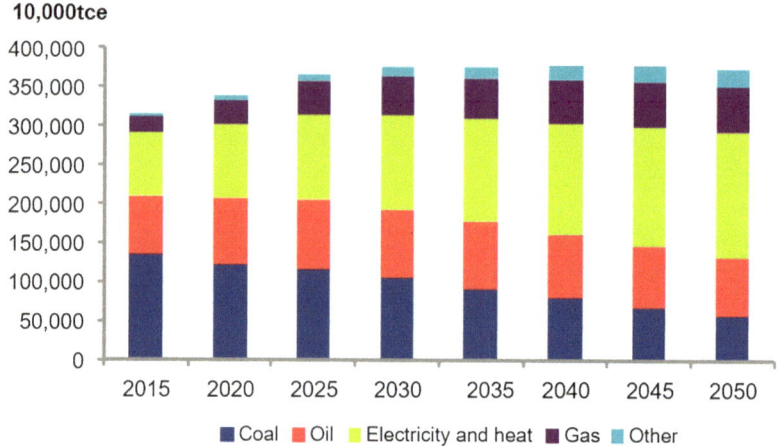

other relevant local policies, the measures for reducing scattered coal use in residential heating include central heating with gas-fired boilers, geothermal heating or waste heat recovery; and using electricity instead of scattered coal in areas where central heating is not possible. The measures for replacing scattered coal with electricity and gas in industry and commerce include shutting down small coal-fired boilers and using waste heat recovery and other energy-efficient technologies (Fig. 54).

Given the gap between China's Class 2 air quality (35 parts per million, ppm) and the current average air quality in major cities (50 ppm), this report predicts that more than 35% of scattered coal will be displaced by 2020 and about 70% by 2030.

To understand the impact of varying degrees of SEGFSC on primary energy, we made the following sensitivity analysis in three different scenarios: in the High SEGFSC scenario, if scattered coal is to be completely replaced by 2030, more than 35% and about 70% will have to be substituted by 2020 and by 2030 respectively. In the Low SEGFSC scenario, and if there are no additional policy measures after 2017, 20% and 50% of scattered coal should be substituted by 2020 and by 2030 respectively.

Scattered coal consumption in the High, Medium and Low SEGFSC scenarios is shown in Fig. 55: (i) in the Medium SEGFSC scenario, scattered coal use will decrease to 400 Mt in 2020, 180 Mt in 2030 and 60 Mt in 2050; (ii) in the High SEGFSC scenario, scattered coal

10,000 t

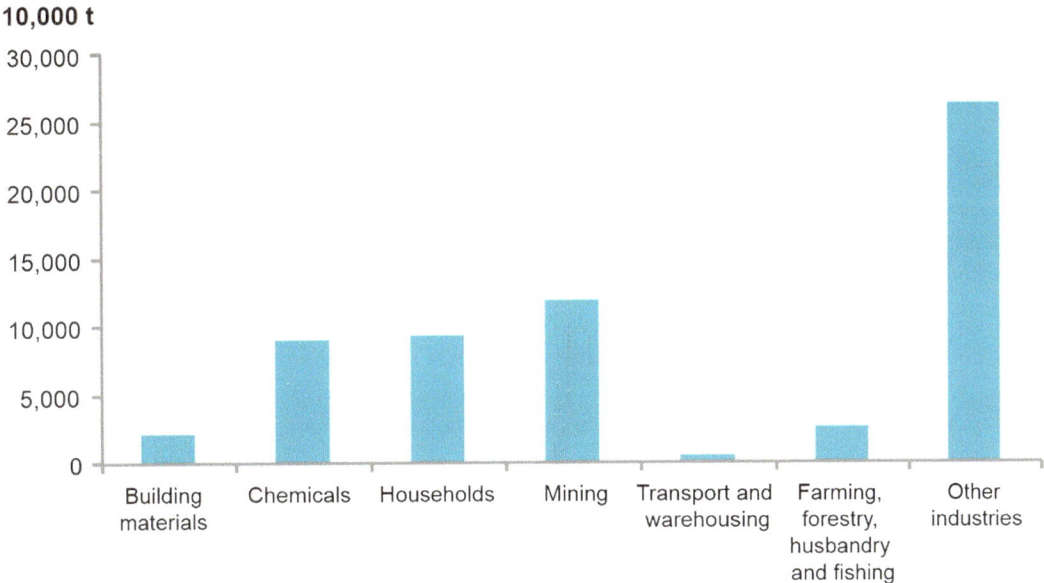

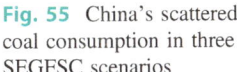

Fig. 54 China's use of scattered coal by sector in 2015

Fig. 55 China's scattered coal consumption in three SEGFSC scenarios

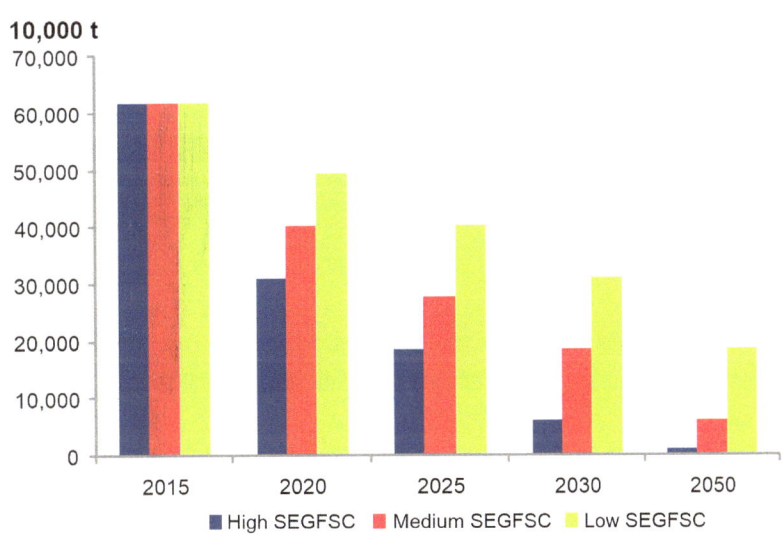

consumption will decline to 300 Mt in 2020, 60 Mt in 2030, and if the almost complete replacement of scattered coal is achieved, to 10 Mt in 2050; and (iii) in the Low SEGFSC scenario, scattered coal consumption will decrease gradually to 500 Mt in 2020, 300 Mt in 2030 and 200 Mt in 2050.

Based on the assumption that electricity and natural gas each have a 50% share in SEGFSC, and that the efficiency of gas-fired boilers is 90% and the energy efficiency ratio of electric heating is 3.2, demand for electricity and natural gas in the three SEGFSC scenarios is calculated as shown in (Table 10).

Table 10 Electricity and natural gas demand in three SEGFSC scenarios

			2020	2025	2030	2050
Low SEGFSC	Natural gas	Billion cubic metres	23.96	41.93	59.90	83.86
	Electricity	GWh	79,370	138,900	198,430	277,800
Medium SEGFSC	Natural gas	Billion cubic metres	41.93	65.89	83.86	107.82
	Electricity	GWh	138,900	218,270	277,800	357,170
High SEGFSC	Natural gas	Billion cubic metres	59.90	83.86	107.82	117.86
	Electricity	GWh	198,430	277,800	357,170	390,430

3.3.5 Analysis of Two EV Development Scenarios

The development of electric vehicles (EVs) is of great importance for China. First, EVs can help reduce fossil fuel consumption in transport and lower CO_2 emissions and environmental pollution; second, EVs can alleviate oil supply and demand pressure, reduce China's dependency on oil imports and improve energy security; and third, as one of the strategic emerging industries, EVs are expected to become a pillar of China's future economic development.

According to the Strategy of Energy Production and Consumption Revolution (2016–30), China's EV ownership will reach 3 million units in 2020, 80 million units in 2030 and 270 million units in 2050. From the current industry development trend, however, the possibility of accelerated EV development cannot be ignored:

(i) there is a global boom in EV R&D—mainly around such critical technologies as batteries, automated driving systems, charging infrastructure and support systems—and in issues like leasing, financial services and shared business models. All these factors can speed up the current development trend even more;

(ii) as artificial intelligence technology evolves, automated driving and EVs could become a standard means of future transport, extending vehicles from a travel tool to a home on wheels. Such new travel experiences can help speed up the transition from conventional vehicles to EVs in the medium and long terms; and

(iii) EVs can serve as both a means of transport and a distributed energy storage facility linked to smart grid technologies. They can play an important role in making the most of installed renewable energy capacity and supporting China's long-term development of renewable energy.

In addition to the Recommended scenario, our report uses an Extreme scenario based on accelerated EV development and the impacts it has on China's energy supply system (Table 11).

The two EV development scenarios are illustrated in Table 11.

If vehicle ownership remains unchanged and EVs replace mainly diesel and petrol vehicles but not gas-powered vehicles, vehicle energy demand in the Extreme scenario is estimated to be as follows (see Fig. 56): annual consumption of petrol and diesel will decrease by 100 Mt by 2030 and by 130 Mt by 2050, whereas annual electricity demand will increase by 200,0000 GWh by 2030 and by 340,000 GWh by 2050, compared with the Recommended scenario.

Table 11 Two EV development scenarios

		2020	2030	2050
Vehicle ownership		27,000	45,000	54,000
Recommended scenario	Ownership (million)	5.24	83	270
	Share (%)	1.9	18.4	50.0
Extreme scenario	Ownership (million)	5.24	200	500
	Share (%)	2	44	93

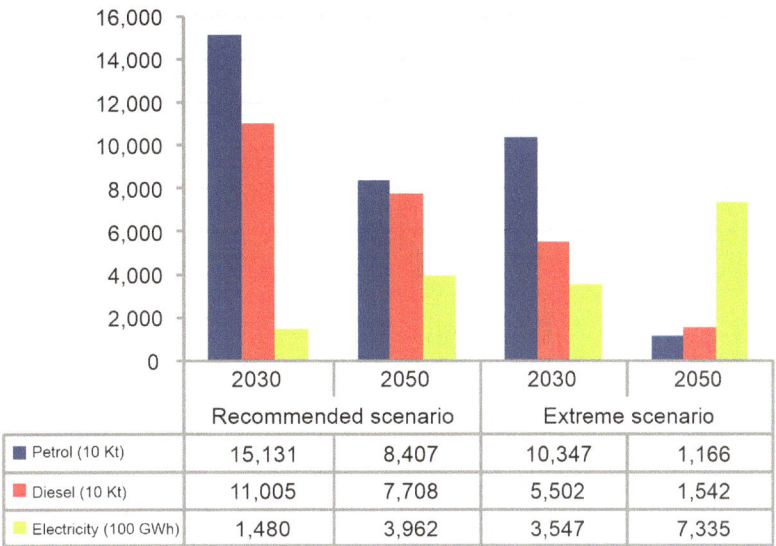

Fig. 56 China's total energy use by vehicles in two EV development scenarios

	Recommended scenario		Extreme scenario	
	2030	2050	2030	2050
■ Petrol (10 Kt)	15,131	8,407	10,347	1,166
■ Diesel (10 Kt)	11,005	7,708	5,502	1,542
■ Electricity (100 GWh)	1,480	3,962	3,547	7,335

3.3.6 Strategic Pathways for China's Energy Revolution

(1) Total end-use energy demand

Based on the analysis of China's energy demand above, our predictions for China's future end-use energy demand in the SEGFSC, Recommended and EV Extreme scenarios are in Table 12.

In the Recommended scenario, the share of electricity in end-use energy demand will steadily increase from 23% in 2015 to 26% in 2020, 29% in 2030, 32% in 2035 and 38% in 2050. The share of clean energy, including electricity and natural gas, will increase from 30% in 2015 to 50% in 2030, 54% in 2035 and 62% in 2050.

Thanks to the higher efficiency of using electricity and gas instead of coal and oil, end-use energy demand will decrease in the SEGFSC scenarios, though not drastically. Even by 2050, the largest decline will not exceed 4%.

Although faster replacement of scattered coal with electricity and gas will increase demand for both alternatives, the pressure this will create in the energy system is not excessive. In the High SEGFSC scenario, demand for natural gas and electricity in 2030 will be 24 billion cubic metres and 80,000 GWh higher than in the Recommended scenario. Such increases make up only 4% of end-use natural gas demand and 0.9% of end-use electricity demand in 2030, so the pressure they create is relatively small. However, regional and peak-hour supply shortages in the colder months and regions, caused by greater use of electricity and gas instead of scattered coal, should not be ignored.

In contrast, the rapid development of EVs will lead to significant growth in electricity demand. In the Extreme scenario, electricity demand in 2050 will be 0.34 trillion kWh higher than in the Recommended scenario, which increases total electricity demand from 11.7 trillion kWh to 12 trillion kWh. Electricity's share of end-use energy demand is expected to rise to 40.5%, 2.7 percentage points higher than in the Recommended scenario. Such an increase in electricity demand will require the power system to have sufficient backup capacity or better demand-side management to address the uncertainties in demand from electric vehicles.

(2) Electricity supply

In the Recommended scenario, electricity demand is divided into peak load and ancillary services, intermediate load, base load and distributed load. To ensure the future power supply system is optimised, the following principles should be observed: (i) non-fossil energy power generation should be prioritised in scheduling, but wind power should not be used to address peak load; (ii) gas power generation should

Table 12 Comparison of China's future end-use energy demand in four different scenarios (Mtce)

		2015	2020	2030	2035	2050
Recommended scenario	Coal	1,360	1,150	960	840	630
	Oil	770	900	940	910	840
	Natural gas	250	460	800	860	900
	Electricity	700	860	1,120	1,230	1,440
	Total	3,070	3,370	3,820	3,850	3,800
High SEGFSC	Coal	1,360	1,080	870	800	590
	Oil	770	900	940	910	840
	Natural gas	250	490	840	860	910
	Electricity	700	870	1,130	1,240	1,440
	Total	3,070	3,340	3,770	3,830	3,780
Low SEGFSC	Coal	1,360	1,220	1,050	930	710
	Oil	770	900	940	910	840
	Natural gas	250	440	770	850	860
	Electricity	700	860	1,110	1,230	1,430
	Total	3,070	3,410	3,860	3,920	3.850
EV development extreme scenario	Coal	1,360	1,150	960	840	630
	Oil	770	900	790	720	650
	Natural gas	250	460	800	860	900
	Electricity	700	860	1,140	1,270	1,480
	Total	3,070	3,370	3,700	3,700	3,650

Note Heat is included in coal, natural gas and electricity consumption

mainly be used for peak shaving and for building distributed energy stations; and (iii) coal power should be considered the last resort in the entire power supply system and gradually shift to flexible sources of supply.

As shown in Table 12, the share of clean power (i.e. not coal- and oil-fired generation) will gradually increase from 30% in 2015 to 38% in 2020, 52% in 2030 and to around 80% in 2050. Of this, the share of wind and solar power is expected to increase from 4.9% in 2015 to about 15% in 2030 and more than 25% in 2050. The share of nuclear power will increase steadily from 3% in 2015 to more than 10% in 2030 and more than 20% in 2050. The share of gas-fired power generation will increase from 3% in 2015 to 9% in 2030 and around 10% in 2050. Despite continuous growth in hydropower capacity, its share will gradually decrease from 17% in 2015 to 14% in 2030 and 13.5% in 2050.

In terms of installed capacity, coal-fired power generation in 2020 will be 1,100 GW, after which there will be a flat-growth period until 2030. Installed capacity will gradually decrease to 800 GW in 2050, as decommissioning of coal-fired generating units progresses. Average operating hours of coal-fired generating units will also gradually decrease from about 4,400 h per year in 2015 to 4,000 h in 2020-30, and 3,000 h in 2050. The installed capacity of hydropower (excluding pumped storage) will reach 300 GW in 2020, after which growth will slow down, reaching 350 GW in 2030 and 450 GW in 2050. Gas-fired power generation and nuclear, wind and solar power will grow rapidly. Their installed capacity will reach 220, 136, 420 and 400 GW respectively in 2030, which equates to new installed capacity of 10, 7, 19 and 24 GW annually. By 2050, their installed capacity will have grown to 340, 350, 1,000 and 1,000 GW

Table 13 Output of power generation types in the Recommended scenario (GWh)

	2015	2020	2030	2035	2050
Hydropower	991,800	1,050,000	1,290,000	1,370,000	1,580,000
Coal-fired	4,006,100	4,380,000	4,400,000	4,100,000	2,470,000
Oil-fired	5,000	5,000	5,000	5,000	5,000
Gas-fired	195,800	390,000	840,000	1,000,000	1,240,000
Nuclear	196,400	410,000	980,000	1,400,000	2,500,000
Wind	223,100	460,000	780,000	1,000,000	1,720,000
Solar	55,900	200,000	520,000	700,000	1,330,000
Biomass	47,400	100,000	220,000	300,000	590,000
Others	20,100	55,000	85,000	100,000	305,000
Total	5,741,600	7,050,000	9,120,000	9,975,000	11,740,000

Table 14 Installed capacity of power generation types in the Recommended scenario (GW)

	2015	2020	2030	2040	2050
Hydropower	296.66	299.00	355.00	380.00	455.00
Coal-fired	884.19	1,100.00	1,100.00	1,100.00	800.00
Oil-fired	6.00	6.00	6.00	6.00	6.00
Gas-fired	66.37	110.00	220.00	270.00	340.00
Nuclear	26.08	58.00	136.00	197.00	350.00
Wind	129.34	247.00	420.00	560.00	1,000.00
Solar	43.18	146.00	400.00	550.00	1,000.00
Biomass	10.00	18.00	46.00	60.00	120.00
Total	1,461.74	1,984.00	2,683.00	3,133.00	4,071.00
Effective installed capacity	1,166.78	1,507.08	1,846.77	2,050.00	2,245.65

respectively, which requires new installed capacity of 6, 10, 30 and 30 GW respectively per year (Tables 13 and 14).

In the SEGFSC and accelerated EV development (Extreme) scenarios, China needs to increase the deployment of new renewable energy capacity. In the Extreme scenario, we estimate that 340,000 GWh of additional power will be needed by 2050. Although such an amount could be supplied by an additional 425 operating hours of 800 GW of coal-fired power generating units, this contradicts the goal of promoting renewable energy. If the installed capacity and operating hours of coal-fired and gas-fired power generating units remain unchanged, an additional 260 GW of solar power or 190 GW of wind power will be needed by 2050 to support the development of EVs (Fig. 57).

(3) Demand and supply of primary energy

China's total demand for primary energy in the Recommended scenario is shown in Table 15. The country's total primary energy demand is expected to increase continuously from about 4.8 Btce in 2020 to around 5.4 Btce in 2030 and some 5.8 Btce in 2050. Coal demand will flatten by 2020 and decrease thereafter. The share of non-coal fuels is expected to increase gradually from 35.7% in 2015 to 55% in 2030 and 73% in 2050. Oil demand growth will flatten by 2030,

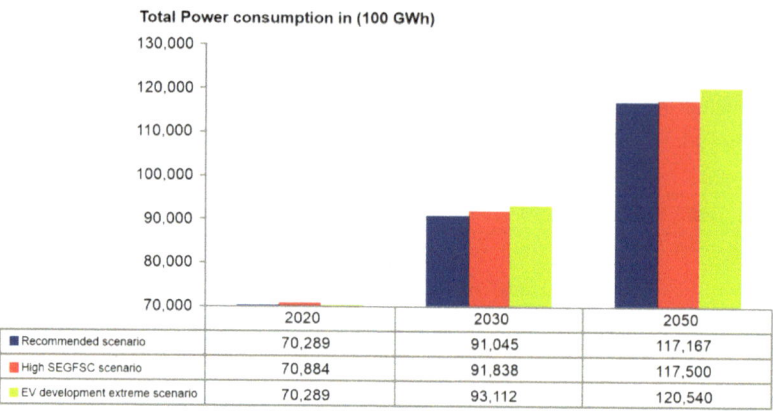

Fig. 57 China's total power consumption in three scenarios

Total Power consumption in (100 GWh)

	2020	2030	2050
■ Recommended scenario	70,289	91,045	117,167
■ High SEGFSC scenario	70,884	91,838	117,500
□ EV development extreme scenario	70,289	93,112	120,540

Table 15 Total primary energy demand in the Recommended scenario

		2015	2020	2030	2035	2050
Total primary energy demand (Mtce)	Coal	2,750	2,670	2,430	2,230	1,570
	Oil	770	900	940	910	840
	Natural gas	250	460	800	860	900
	Non-fossil fuels	510	750	1,210	1,540	2,420
Total		4,280	4,780	5,380	5,550	5,730
Share of non-coal energy (%)		35.7	44.2	54.8	59.9	72.6
Share of non-fossil energy (%)		11.8	15.7	22.5	27.8	42.3
Primary energy for power generation	Mtce	1,750	2,100	2,610	2,830	3,140
	Share (%)	40.9	44.1	48.5	51.0	54.8

Note According to China's current method for calculating primary energy demand, when non-fossil fuel power generation is included in primary energy, it should be converted to coal consumption equivalent

and gradually decrease as EVs replace the internal combustion engine. Clean energy will play an increasingly important role in meeting energy demand. The share of non-fossil energy will increase from 11.8% in 2015 to 22.5% in 2030 and to more than 40% in 2050. In addition, increases in electrification will raise power generation's share of energy supply from 40.9% in 2015 to 48.5% in 2030 and 54.8% in 2050.

Compared to the Recommended scenario, in 2030 total primary energy consumption will be 30 Mtce lower in the SEGFSC scenario and 90 Mtce lower in the accelerated EV development (Extreme) scenario. By 2050 the decrease will be more gradual: 10 Mtce in the SEGFSC scenario

and 100 tce in the accelerated EV development scenario. The share of non-fossil energy power generation will increase to 43% in the SEGFSC scenario and 45% in the accelerated EV development scenario in 2050 (Fig. 58).

(4) Fossil energy supply by energy source
Total energy supply by energy source (coal, oil and natural gas) in the three scenarios is as follows:

(1) Coal supply
Coal demand shows a clear decline in all three scenarios, dropping to 3.5 Bt in 2030 and 2.2 Bt in 2050. Correspondingly, coal supply is expected to reach 3.8 Bt in 2030 and 2.4 Bt in 2050. If China's

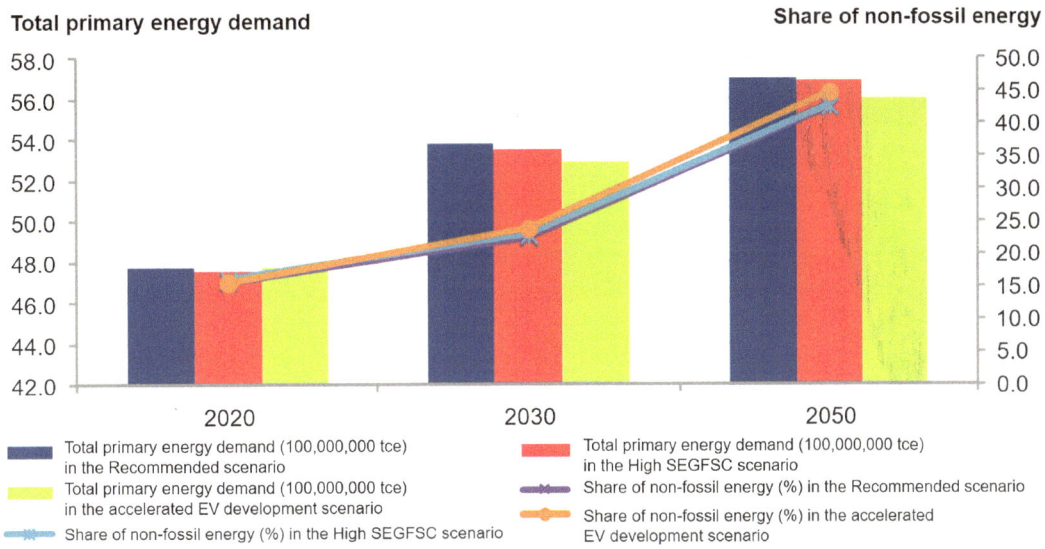

Fig. 58 China's total primary energy demand and share of non-fossil energy in three scenarios

annual coal imports are 200 Mt, its domestic coal capacity will be 3.6 Bt in 2030 and 2.2 Bt in 2050. As China's current coal capacity exceeds 5.0 Bt, the major task facing its coal sector is steady capacity cutting. Meanwhile, coal mining safety and production efficiency should be improved.

(2) Oil supply

Although EVs can significantly reduce future oil demand, in the Recommended scenario China's oil demand and oil processing capacity peak at more than 650 Mt and 720 Mt respectively in 2030. Even in 2050, China's oil demand will still be around 650 Mt. Demand control and structural adjustment are the major tasks ahead, in particular: improving chemical refining capacity to speed up the shift from oil refining to chemical production.

(3) Natural gas supply

In 2016, China's domestic natural gas, conventional natural gas and shale gas production was 136.9 billion cubic metres, 123.0 billion cubic metres and 7.9 billion cubic metres respectively, and the volume of coalbed methane surface extraction was about 4.4 billion cubic metres. Natural gas imports in 2016 were 72.1 billion cubic metres, which is 35% of the country's total

gas consumption. China's natural gas demand is expected to grow rapidly, largely from replacing scattered coal with gas. To ensure that demand can be met, supply capacity needs to improve significantly. If China's import dependency is consistently controlled at around 35%, then 260 billion cubic metres, 450 billion cubic metres and 490 billion cubic metres of natural gas will be needed in 2020, 2030 and 2050 respectively. Therefore, China still needs to vigorously explore and exploit conventional gas, tight gas, shale gas and coalbed methane. It also needs to drive research on how to exploit natural gas hydrates (Table 16).

4 China's Energy Revolution: Changes in Primary Energy

4.1 Characteristics of Fossil Energy Supply

4.1.1 Coal: Secure, Clean and Efficient

For a long time, coal has been the dominant source of energy supply in China. Coal makes up more than 70% of China's total primary energy production and more than 70% of its total power output. However, extensive development of coal

Table 16 China's fossil energy demand and capacity forecasts

			2020	2030	2035	2050
Coal (Mt)	Coal demand	Recommended scenario	3,807.43	3,471.83	3,181.26	2,210.36
		High SEGFSC scenario	3,714.78	3,348.29	3,129.49	2,158.59
		EV development extreme scenario	3,807.43	3,471.83	3,181.26	2,210.36
	Coal mining capacity		4,188.17	3,819.02	3,499.38	2,431.40
Oil (Mt)	Oil demand	Recommended scenario	628.42	654.77	639.65	585.76
		High SEGFSC scenario	628.42	654.77	639.65	585.76
		EV development extreme scenario	628.42	551.91	505.58	451.69
	Refining and supply capacity		691.26	720.25	703.61	644.34
Natural gas (billion cubic metres)	Natural gas demand	Recommended scenario	347.6	604.8	648.3	673.0
		High SEGFSC scenario	365.6	628.8	658.3	683.1
		EV development extreme scenario	347.6	604.8	648.3	673.0
	Domestic exploitation capacity		261.4	449.6	470.7	488.4

resources over a long time has caused severe pollution and environmental damage, and the major coal-producing regions are faced with daunting challenges. Although China's coal demand is about to peak—in some regions it is already in decline—in the long term, coal will remain a major source of energy. Without a revolution in coal production, the energy revolution will be nothing but empty talk.

Developing coal into a secure, economic, clean and efficient resource is, therefore, an effective way to drive the energy revolution.

(1) Effective control of coal capacity

Low-carbon energy and decarbonisation are the megatrends in energy development worldwide. As the long-term extensive use of coal has caused severe pollution, environmental damage and climate change, a reduction in coal use is inevitable. This will present coal-producing regions with a formidable challenge. The fundamental idea behind China's coal production revolution is to minimise production. In the next

decade, China's coal capacity will be effectively controlled to no more than 4.5 Bt and 4.2 Bt respectively in 2020 and 2030.

(2) Significant increase in scientific coal

To solve the environmental problems caused by coal extraction and use, coal mining needs to take into consideration production conditions in the various coal-producing regions to enable scientific coal production. Geology, coal quality, reserves, water resources and ecology in each coal-producing region should be assessed. From this, we expect a third of China's coal mines to meet the criteria for scientific coal production. A further third will fail to meet the criteria but will be upgraded, and a third will be too far behind criteria requirements to warrant upgrading and will gradually be shut down. As a result of these measures, scientific coal capacity will increase. By 2020, forecasts expect scientific coal capacity to reach 3.5 Bt, which is more than 80% of total coal production capacity. By 2030, scientific coal capacity will be 4.2 Bt, or 100% of total coal production capacity. By 2050, coal production will decline, but scientific coal will still make up 100% of total coal production capacity.

(3) Mechanised, smart coal mines will minimise environmental impact and safety incidents

By 2020, the rehabilitation rate of old incidents of collapsed land caused by coal mining in China will be more than 80%, and that of recent ground collapses will be more than 90%, which is more than in developed economies. China's mechanisation of coal mining will reach 90% and the reuse of coal mine gas and mine water will reach 85% and 100% respectively. The reuse of currently stockpiled and newly added coal gangue (solid waste) will be 90% and of associated minerals 70%. China's raw coal washing rate will be 90%. The mortality rate per million tonnes of coal mined will be 0.03, which is close to that of developed economies. The number of major accidents will be less than five and the number of fatalities below 100, which although high is considerably below current levels.

By 2030, the rehabilitation rate of all collapsed land from coal mining in China will be 100%. Mechanisation will approach 100% and in large- and medium-sized mines reach 100%. Reuse of coal mine gas and mine water will be 100%, as will the reuse of currently stockpiled and newly added coal gangue and associated minerals. China's raw coal washing rate will also be 100%. The mortality rate per million tonnes of coal mined will be less than 0.02, the most advanced level internationally, and annual fatalities will not exceed 60.

By 2050, all of China's coal mines will be automated and smart. The fatality rate per million tonnes of coal mined will be less than 0.01, and there will be no major accidents during mining. Coal mining will be safe and green, with near-zero environmental impact and casualties.

(4) New business models will transform coal-dependent cities

New business models are continuously emerging. Cross-industry combinations in the form of mergers and acquisitions between coal and power companies will spread.

Traditional coal companies and coal-dependent cities in mining regions will successfully transform. As mechanisation and mine intelligence improve, the number of coal mine workers will rapidly decrease, especially those working in the mines. Many coal-dependent cities will make vigorous efforts to implement innovative reforms and successfully complete their transformation. Through the coordinated development of strong industries, support infrastructure and social programmes, these once coal-dependent cities will evolve into integrated regional centres that are modern, sustainable, liveable and business-favourable.

4.1.2 Oil and Gas

China's focus on oil and gas exploration and production will shift from predominantly conventional resources to an equal mix of both conventional and unconventional resources—onshore and offshore, and at different depths, from shallow to ultra-deep.

Increasing its exploration and extraction of oil and gas resources will strengthen supply capacity and help safeguard national energy security. Following long-term intensive development, China sees limited production growth potential in onshore conventional oil and gas. However, China has abundant unconventional oil and gas resources, including 19–22 billion tonnes of recoverable unconventional oil resources and nearly 40 trillion cubic metres of recoverable unconventional gas. China's production of deep-sea oil and gas resources is low. In recent years, China's exploration of deep and ultra-deep oil and gas has made continuous breakthroughs. China must continue to make major adjustments in oil and gas exploration and production to achieve the equal mix of conventional and unconventional resources it strives for to achieve energy security. By 2030, China's unconventional crude oil output will be 50 Mt, which is 20% of its total oil production. By 2030, natural gas production will quadruple, with unconventional gas (including tight gas) increasing to two-thirds of total production. In addition, offshore oil and gas production will exceed 100 Mt and dependence on oil and gas imports will fall to 40–45% of total consumption.

(1) Unconventional oil and gas is the key to increased production and reserves

Tight oil and gas is an effective supplement to conventional oil and gas. China will focus on Ordos, Sichuan and the Tarim Basin to ensure speedy scale development of medium- and low-grade tight gas resources and difficult-to-recover oil reserves. By 2030, the output of tight oil and tight gas will be 50 Mt and 100 million cubic metres respectively.

As the exploration and development of shale gas resources accelerate, the technologies and equipment needed to exploit them will improve. Seismic measurement and interpretation during drilling, rotary steering drilling, hydraulic fracture monitoring, horizontal well drilling and downhole micro-seismic monitoring will all progress. Breakthroughs will be made in the exploration and development of organic-rich marine shale in south China and organic-rich mudstone in the north. By 2030, China's shale gas production will exceed 100 billion cubic metres.

Coalbed methane (CBM) production will grow. Technologies like CBM reservoir engineering and dynamic evaluation, underbalanced drilling, fracturing stimulation and extraction, and the recycling of low-concentration coal mine gas will all advance as CBM extraction shifts from high-quality reservoirs to deep and complex strata. By 2030, China's CBM production will reach around 50 billion cubic metres.

(2) Unconventional oil and gas will replace conventional

As breakthroughs are made in exploration and extraction technologies, the challenges in exploiting China's oil and gas resources in reservoirs deeper than 3,500 m onshore and 500 m offshore will be addressed, enabling production at scale. In deep onshore shale, technical difficulties such as identifying deep-strata targets and fracturing at high earth temperatures, at high pressure and in complex lithology will be solved. Innovation will enable oil and gas resources to be exploited in deep carbonate rock, clastic rock and volcanic rock. This in turn will allow China to develop at scale its deep shale oil and gas

resources in the Tarim Basin, Sichuan, Ordos, Bohai Bay, Songliao Basin and Junggar Basin. Offshore, breakthroughs in deep-water technologies like deep-water drilling, subsea processing, subsea pipeline transmission and semisubmersible production will accelerate exploration and development in the East China Sea, the South China Sea and the Huanghai Sea.

(3) The shift to clean and green production

Oil and gas fields should be developed in an optimal and integrated way to ensure that production, refining and waste disposal are centralised. This will minimise the production footprint and oil and gas losses. Drilling technologies like multilateral well, horizontal well, slimhole drilling and air drilling will be widely deployed and the waste produced from oil and gas field construction effectively controlled. The use of non-toxic or slightly toxic chemicals in oil and gas fields will increase significantly, and oil spills will be avoided to ensure maximum recovery and safeguard the environment. The well liquid recycling rate will exceed 95%. The wastewater produced in the drilling process will be completely recycled and acid residual liquids, fracturing residual fluids and flow-back fluids fully recycled or disposed of in a harmless way. When a field is decommissioned, the production platform and infrastructure will be completely recycled. Potential environmental risks in shale gas development will be identified and avoided, and the whole-process supervision system will continuously improve to ensure environmentally friendly development.

4.2 Characteristics of Non-fossil Energy Supply

4.2.1 Hydropower: Continuous Growth in Capacity, Within Ecological Red-Line Regulations

Large hydropower projects will continue. By 2020, the construction of large hydropower plants in the middle and lower reaches of the Jinsha, Yalung, Dadu and Lancang rivers, and the upper reaches of the Yellow River, will be under way. Total hydropower capacity will reach 370 GW in 2020, which is more than 50% of China's clean and low-carbon power generation capacity. By 2030, hydropower development will focus on the upper reaches of Jinsha River, the upper reaches of Lancang River, and Nujiang River. Newly constructed hydropower projects will reach 80–100 GW in 2020–30 and 400–450 GW in 2050.

Hydropower development and ecological protection will be taken into equal consideration to minimise the species affected. Fish pass structures will be developed, demonstrated and implemented, and rare animals, plants and ancient trees will be well protected. The land requisition compensation system will be continuously improved to ensure reasonable compensation for those displaced by hydropower projects. Centralised and decentralised relocation will be combined to provide an optimal solution to any resettlement or employment problems facing those required to move.

4.2.2 Renewable Energy: Increasing Share of the Energy Mix

(1) Wind power: Equal focus on onshore and offshore to ensure coordinated development
By 2020, onshore and offshore wind power will play a dominant and complementary role. Wind power projects in the north, north-east, north-west, eastern coastal China and areas with medium and low wind speed, will be developed. By 2020, cumulative installed wind power capacity will reach 250 GW, which is 12% of China's total installed power generation capacity. Wind power will meet 6.5% of electricity demand, at parity with conventional coal-fired generation.

In 2020–30, equal focus will be put on onshore and offshore wind power development, with new installed capacity close to 20 GW annually. In 2030, cumulative installed wind power capacity will exceed 400 GW, which will be 8.6% of China's total power output. The share of wind power in the energy source mix will increase to about 15%.

In 2030–50, offshore wind power development close to load centres will take off. In 2050, cumulative installed wind power capacity will exceed 1,000 GW, or 14.6% of China's total power output. The share of wind power in the energy source mix will increase to about 25%.

(2) Solar power: Deploy the right solar technology for local conditions
In desert areas with abundant sun like west and north China, Gansu, Qinghai and Xinjiang, solar power will be prioritised. In medium and large cities in east and central China, building-integrated PV systems will rapidly develop and be widely deployed. In regions that feature good sunshine conditions, large areas of available land and abundant water resources, solar thermal pilot projects will be built to test different solar thermal technologies. Installed solar power capacity will reach 140 GW in 2020, 400 GW in 2030 and 1,000 GW in 2050.

Solar power will speed up the process of addressing access to electricity issues and poverty alleviation in areas that large grids don't reach. Poverty alleviation projects will be started in those areas to help provide people in need with a stable source of income.

(3) Liquid biofuels: A substitute for oil
In the heavy-duty truck, aviation and shipping sectors, non-food liquid biofuels will be gradually commercialised and used at scale. In the short and medium terms, the production cost of non-food liquid biofuels will drop sharply, with production reaching 10 Mt in 2020. As breakthroughs are made in second and third generation biomass technologies—like cellulosic ethanol and algae-based biofuel—liquid biofuels will gradually replace petrol, diesel and kerosene in heavy-duty trucks, aviation and shipping. The production of liquid biofuels for transport will reach 20 Mt and 60 Mt in 2030 and 2050 respectively, replacing about 50 Mt of petroleum-based fuels.

(4) Renewable heat will grow extensively
Solar thermal energy will play an important role in heating and cooling buildings and heating water for homes, businesses and industry. By 2020, solar-powered water heating systems will be extensively deployed to ensure that the heating capacity of solar thermal applications reaches 800 million cubic metres, and that solar thermal

energy replaces 87 Mtce of fossil fuels annually. In 2020–30, solar thermal capacity will be 1.2 billion cubic metres, replacing 120 Mtce of fossil fuels per year. In 2030–50, as medium- and high-temperature solar application technologies are developed, solar thermal will extend to heating and cooling medium- and high-temperature commercial and industrial applications. The production capacity of solar thermal applications will grow to 1.4 billion cubic metres, replacing 150 Mtce of fossil fuels annually.

Biogas and biomass briquettes will help meet heat demand for cooking and boilers. By 2030, biogas purification technology will be close to maturity, and the industrial biogas production sector will begin to take shape. Intensive agriculture and compact biomass briquette technology will be developed in suitable areas at scale to achieve an annual production target of 30 Mt of briquettes. Centralised geothermal heating will become predominant in the north, north-east, east and south-west, and ground source heat pumps will be deployed at scale in regions that are hot in summer and cold in winter, like the middle and lower reaches of the Yangtze River. In 2030–50, biogas technology will be widely used in urban and rural areas, and annual production of biogas will reach 50 billion cubic metres. As the reliability and lifespan of biomass briquette production equipment improves, markets will be extensively developed and suitable areas will see the large-scale development of biomass briquette manufacturing. As geothermal heat technologies mature, applications will be promoted extensively in China. By 2050, geothermal energy will heat or cool 1 billion square metres of floor space, replacing 60 Mtce of fossil fuels per year.

4.2.3 Nuclear Power: Safe Development and Larger Share of China's Power Output

(1) Installed nuclear power capacity will significantly increase with third-generation pressurised water reactor (PWR) technology

A well-proven and safe technology, third-generation PWR is the mainstream technology in nuclear power plants across the world and for China's future nuclear power development. China will increase its investment in nuclear power plants in eastern coastal areas and build a nuclear power belt in east and central China, all using third-generation PWR. In 2020, the installed capacity of nuclear power plants, built or under construction, will be 58 GW and 30 GW respectively, and nuclear power's share of China's total power output will be more than 5%. In 2030, the installed capacity of nuclear power plants built or under construction will grow to 120–150 GW and 30 GW respectively, which will increase nuclear's share of China's total power output beyond 10%. In 2050, the installed capacity of nuclear power plants built will reach 350 GW, and nuclear's share of China's total power output will top 20%, becoming one of the main pillars of the power sector.

(2) China will become a nuclear power hub

Alongside nuclear power development in east and central areas of the country, China's nuclear power equipment manufacturing, and project construction capability will significantly improve. The CAP1400 and Hualong 1 nuclear power demonstration projects will become a national brand, enabling China's nuclear power technologies to go global. By 2020, China's nuclear power technology and equipment export framework will have taken shape. By 2030, the level of exports will be an important indicator of China's international competitiveness in the nuclear power sector, helping China to become a nuclear power equipment manufacturing hub for the world.

(3) Nuclear power development will be based on informed decision-making and public acceptability

Acceptability has restricted nuclear power development. In the future, the decision-making process will be more scientific and consistent to ensure nuclear development strategies and plans are clearly defined and implemented. Laws and regulations will be improved to ensure all activities have a legal basis. Positive publicity and public guidance will ensure better information and improved public acceptability of nuclear power.

4.3 Potential Energy Production Technologies

4.3.1 Natural Gas Hydrate: A Strategic Alternative to Natural Gas

China has abundant natural gas hydrate (NGH) resources, with reserves estimated at 83.7 trillion cubic metres. There are large NGH deposits in polar tundra sandstone in the Qinghai-Tibet Plateau and in seabed sandstone in the South China Sea. However, large-scale NGH extraction technology has yet to be developed, as there are environmental and safety production risks like methane leakage and submarine landslides.

China has made considerable breakthroughs in NGH development. Bluewhale 1, the world's most advanced semi-submersible drilling rig, has been charted to explore for NGH off the China coast. China has also innovated 20 critical NGH technologies and conducted safe and controlled trial mining of muddy silt-type combustible ice, which is a world-first. For the past two decades, the USA has led the world's shale oil and gas revolution through wide deployment of horizontal drilling and hydraulic fracturing technologies. Like the USA, China will also rely on continuously improving technologies and equipment to make extraction of NGH commercially viable. China should also strengthen research so that it can assemble a complete range of support technologies for NGH exploration and production. These should include predicting, targeting and evaluating NGH deposits; drilling and wellbores; well pattern design; environmental impact assessment of extraction; and safety. In addition, China will accelerate its NGH surveys and screening of potentially profitable exploration areas. After 2030, as commercial mining of NGH becomes possible, China will strive to sharpen its international competitiveness in NGH exploration and production, to lead a new revolution in oil and gas production.

4.3.2 Advanced Nuclear Power Technologies: Research on Fourth-Generation Reactor and Fusion Technologies

China's research and development of fourth-generation reactor technology will drive the nuclear power revolution. China will conduct experimental research into the technology's application and will complete demonstration projects of high-temperature gas-cooled reactors as soon as possible. China's short-term and long-term targets for outlet temperature are 700–950°C and above 1,000°C respectively. Research on critical technologies will be completed by 2025 and pilot demonstrations by 2030.

China will strive to speed up the design and certification of small reactors by 2020 to enable wide deployment in applications such as floating nuclear power plants; naval vessels; neutron source; power supply for remote areas, households and industrial heating; offshore oil exploitation; and sea water desalination. In addition, following the fast reactor development strategic pathway of "test-demonstrate-commercialise", China will strive to build fast-reactor demonstration projects by 2025 and large commercial fast-reactor demonstration projects by 2030.

China will also focus on basic research and design concepts for such advanced technologies as nuclear fusion reactors, very-high-temperature reactors, supercritical water reactors, thorium molten salt reactors and travelling wave reactors. China will also carry out long-term R&D and demonstration projects on basic and critical technologies to enable the wide deployment of mature technologies after 2030.

4.3.3 Marine Energy: Research for the Future

Marine energy refers to the energy generated by tides, waves, currents and temperature differences. It is renewable, abundant and environmentally friendly. The world's oceans contain the energy equivalent of double the current global energy consumption. But ocean energy has disadvantages as well, such as instability, low energy density, harsh operating conditions and poor commercial viability.

China has a vast coastline, which means abundant marine energy. Currently, marine power generation technology is still under research and experiment, in particular how to transform marine energy into reliable and easily accessible energy carriers like electricity. As marine energy technologies progress, it is expected to become an important component of China's future energy system.

4.3.4 Space-Based Solar Power: Exploration

The sun is the original source of most types of energy developed and used by humans. Solar energy radiates from the sun into space, mainly in the form of electromagnetic waves. Only a very small fraction (1/2,200,000,000) of this energy reaches Earth. If the solar power in space can be harnessed, the world could have an inexhaustible supply of energy. Space-based solar power has many benefits, including 24-hour uninterrupted supply and the possibility of transferring the energy to ground-receiving stations on Earth, which could then distribute it to where it is needed.

Space-based solar power is still at the design and exploration stage. Building solar power stations in space will present many challenges, including the development of large-scale space delivery systems and special materials for solar panels and other equipment, performing maintenance while in orbit, and avoiding other spacecraft and space debris.

With space exploration increasing, aerospace technologies advancing, and major innovations in new forms of energy and materials taking place, space-based solar power may soon become reality. When it does, it will be a new and better way to harness solar energy.

5 China's Energy Revolution: The Transition to Secondary Energy

5.1 Characteristics of Secondary Energy Supply

5.1.1 Electricity

(1) Electricity is becoming more important

Economic development increases the need for energy and changes the end-user demand structure. High-quality energy carriers like electricity will gradually replace high-pollution, non-renewable fossil fuels. According to BP Energy Outlook 2017, by 2035 about two-thirds of new global energy consumption will be power generation. Electricity will also play a major role in China's future energy system. Electricity's share of China's end-user energy demand is forecast to increase from 20% in 2015 to 25% in 2020, 32% in 2030 and more than 50% in 2050.

Replacing coal and oil with electricity and increasing the capacity of long-distance power transmission are two important changes in China's future energy system. Electricity will replace coal fuel in industry, agriculture, buildings and homes, specifically in industrial boilers and kilns and in household heating and cooking. This will reduce the use of scattered coal across sectors and significantly increase the conversion of coal into electricity. The electrification of vehicles, rail and irrigation will be vigorously developed to reduce oil dependency. Ultra-high voltage power transmission lines will be built to transmit thermal, wind and solar power in west and north China, and hydropower from south-west China, to east and central China at scale. This will help optimise the energy mix, improve power availability across regions, control thermal power construction and pollution emissions in east and central China, and reduce haze in eastern parts of the country.

(2) Clean and efficient coal-fired power generation

The share of installed coal power capacity and coal power output will continue to fall. Taking into account China's economic and social development, energy source endowment, and the requirements of the energy revolution, China's power and coal power sectors still hold great growth potential. Installed power capacity is predicted to reach 2,000 GW and 3,170 GW in 2020 and 2030 respectively, and power output will be 8.5 trillion kWh and 12 trillion kWh respectively. Installed coal power capacity will reach 1,100 GW in 2020, up 33% from 2014, and 1,450 GW in 2030, up 76% from 2014. Coal power's share of installed capacity is expected to decline from 60.7% in 2014 to 55% in 2020 and 46% in 2030, and coal power output to decrease to 66% and 56% in 2020 and 2030 respectively.

The efficiency of coal-fuelled power generation will improve. The amount of coal consumed per unit of electricity will fall sharply, thanks to energy-efficiency technologies, upgrading existing plants to ultra-low emission status, and shutting down outdated units that do not conform with mandatory standards. According to the Action Plan on Upgrading and Reconstructing Coal-Fired Power Plants for Energy Conservation and Emissions Reduction (2014–20), in 2020 the amount of coal consumed per unit of electricity should be: (a) lower than 300 grams coal equivalent per kilowatt-hour (gce/kWh) in newly built coal-fired power generating units; (b) lower than 310 gce/kWh in upgraded coal-fired power generating units; and (c) lower than 300 gce/kWh in upgraded coal-fired generating units of 600 MW or more.

Pollutants and greenhouse gas emissions from coal-fired power plants will gradually decrease. The key targets for coal-fired power generation include clean production, reducing emissions of soot, CO_2 and NO_x, and controlling wastewater discharge. By 2030, emissions of soot, CO_2 and NO_x per kWh will decrease to 0.04 g, 0.15 g and 0.2 g or lower respectively. China has set the goal of reaching a peak in CO_2 emissions by around 2030, which makes carbon emissions reduction the biggest challenge facing coal-fired power generation. Deployment of carbon capture, utilisation and storage (CCUS) technology at scale will become an important option for reducing greenhouse gas emissions.

Coal-fired power generation technologies and equipment will significantly improve. China will increase support for the development of emission reduction technologies for coal-fired power plants. It will back the development of advanced dedusting, desulphurisation and denitration processes and innovate new ways to save energy, water and land. Advanced coal power technologies like ultra-supercritical will be deployed to achieve the highest efficiency and lowest emissions. The combination of clean coal gasification and combined cycle power generation in integrated gasification combined cycle (IGCC) technology turns coal into pressurised gas that can be used to generate power with lower emissions, thus opening another way to use clean coal efficiently.

(3) Distributed energy resources encourage local consumption

Distributed solar photovoltaic (PV) systems. Distributed solar PV power systems could become the main component in distributed renewable energy. According to China's 13th Five-Year Plan for Energy Development (2016–20), installed solar power capacity is expected to be more than 110 GW in 2020. This will include 60 GW of distributed solar power, which is more than the 45 GW installed capacity of centralised solar power plants. The installed capacity of distributed solar power is expected to make up 56% of total installed solar power capacity in 2020. In accordance with China's 13th Five-Year Plan for Solar Power Development, 100 distributed PV demonstration plants will be built by 2020.

Gas-fired distributed generation. Gas-fired distributed generation is an important part of China's energy strategy. According to China's 13th Five-Year Plan for Energy Development (2016–20), China will vigorously promote gas-fired distributed generation, focusing on the development of combined cooling, heat and power (CCHP) plants and include gas-fired distributed generation in national energy development strategies. China's installed capacity of gas-fired distributed generation will be more than 110 GW in 2020, including 15 GW of CCHP plants. According to the 13th Five-Year Plan for Natural Gas Development, China will encourage the development of efficient gas projects like gas-fired distributed generation, constructing peak-shaving gas-fired power plants and adapting CCHP plants to local conditions.

Distributed wind power. According to China's 13th Five-Year Plan for Wind Power, China will proactively explore business models suitable for the development and deployment of distributed wind. China will also promote the development of microgrids that combine wind, solar and energy storage. Small- and medium-sized wind power systems have unique advantages. Northern China has excellent wind resources, as do other parts of the country. These resources are often more suitable for small- and medium-sized wind power solutions than for large wind parks. Distributed wind power provides local grid connections, eliminating the need for long-distance transmission. Local distributed wind systems can also relieve power supply pressure, especially in load centres in east China.

5.1.2 Heating and Cooling
(1) Growing demand for heating
According to forecasts in Market and Policy Research on Renewable Heat, total demand for heating will reach around 1.67 Btce in 2020 and about 2.24 Btce in 2030, an increase of 34%. There are four main types of demand in the market: household water heating, building heating, building cooling and industrial heat. The combined share of building heating and cooling will exceed 70% in 2020 and 2030.

Household water heating. In 2020, the energy used to heat household water will be around 152 Mtce. In 2030, the energy needed to heat public buildings will be about 189 Mtce. Energy consumption for residential and public building water heating will be about 145 Mtce and 7 Mtce respectively in 2020, and 178 Mtce and 11 Mtce respectively in 2030. Household water is a major application for renewable heat, with thermal the biggest application for solar, geothermal and biomass water heating technologies.

Building heating. Buildings are the largest and most important application in heating. Heat generated for buildings will reach 631 Btce in 2020 and 794 Btce in 2030, which equates to 38% and 36% respectively of the total heat market. Rural building heating in cold regions will account for 40–45% of total building heating demand in 2030. Currently, building heating in rural areas relies mainly on scattered coal, and the coal-fired heaters used are often inefficient and pollute severely. Renewable heat is, therefore, a good clean energy alternative to coal-fired heating.

Building cooling. Demand for building cooling will reach 612 Btce in 2020 and 903 Btce in 2030, which is roughly the same as for building heating. Electric air conditioning systems are the main cooling technology for buildings now and will remain so in the future. In renewable cooling, ground source and water source heat pump cooling technologies are well proven. Solar air conditioning is expected to deliver technical breakthroughs by 2020.

Industrial water heating. Industrial heat demand is expected to reach 274 Mtce in 2020 and 355 Mtce in 2030. Biomass-fired boilers can replace coal-fired boilers to heat industrial water, but other renewable heat technologies need to operate with conventional heating technologies to meet industrial water heating demand. Renewable heat technologies can replace fossil energy in certain parts of the heating process, for instance in preheating and heat tracing, helping to increase the use of clean energy in industrial water heating.

(2) Strong potential for renewable heat
Renewable heat is generated mainly by solar, biomass and geothermal energy. Renewable heat production is forecast to reach 1.84 Btce in 2020 and 3.59 Btce in 2030. Given the huge potential of solar and geothermal energy, renewable heat could grow even faster than forecast, which would increase the share of clean and low-carbon energy in the heating market.

Solar heating. According to China's 13th Five-Year Plan for Solar Energy, the surface area of solar thermal collectors will reach 800 million square metres in 2020. Converted into tce/m^2, solar thermal applications could replace 96 Mtce of energy. As medium- and high-temperature solar technologies mature, the growth potential of solar heating increases. If solar heating maintains the same growth rate as envisaged in the 13th Five-Year Plan, the surface area of solar thermal collectors will reach 1.2 billion square metres by 2030, replacing 144 Mtce of energy.

Biomass heating. The use of agricultural and forestry waste will gradually shift from power generation to biomass-fired combined heat and power (CHP) and biomass heating. Biomass heating will be fuelled mainly by biomass briquette-fired boilers, thereby helping to displace industrial coal-fired boilers. According to the China Biomass Energy Technology Development Roadmap 2050, China's installed capacity of power generation from agricultural and forestry biomass will reach about 5 GW in 2020 and around 8 GW in 2030, equivalent to 4.2 Mtce and 8.96 Mtce respectively.[13] The heat supplied from biomass briquette-fired boilers will be 600 petajoules (PJ) in 2020 and 1,000 PJ in 2030, enough to replace 20.5 Mtce and 34.2 Mtce of conventional energy in the heating market. Based on the above, biomass heating will amount to 24.7 Mtce in 2020 and 43.16 Mtce in 2030.

Geothermal heating. Given the current state of geothermal energy and technology development, China's future geothermal heating strategy will focus on shallow geothermal energy (the upper few metres of the ground), which can be developed to heat the cold areas of north China. If geothermal heating maintains an annual growth rate of 12% in 2016-20, the surface area heated by geothermal energy will approach 900 million square metres in 2020, replacing 13.2 Mtce of conventional energy. If it maintains annual growth of 15% in 2021-30, the surface area heated will be close to 3.6 billion square metres in 2030, replacing 53.5 Mtce of conventional energy.

5.1.3 Oil Products

(1) Oil product marketisation will accelerate

Marketisation of oil products will accelerate in the future, resulting in the following changes in policy: (i) the oil product import and export market will be deregulated; (ii) oil product pricing will be deregulated; (iii) mixed ownership will be initiated to encourage and attract private capital in the upstream, midstream and downstream oil and gas industry, thereby promoting diversification of market participants; and iv) the introduction of oil product futures will be accelerated. The aim is first to meet demand in the oil refining sector before deregulating the oil product import and export markets. This will benefit downstream users and facilitate the orderly development of crude oil and oil product markets. Full oil product marketisation will be possible by 2030.

(2) Demand growth for petrol and diesel will gradually slow down

Petrol and diesel in China are mainly used by motor vehicles, so vehicle ownership numbers directly determine the amount of petrol and diesel consumed. The CNPC Petrochemical Research Institute has forecast that vehicle ownership will maintain an annual growth rate of about 12% to 2020. The number of vehicles owned is expected to rise from 173 million in 2015 to 266 million in 2020 and 300 million in 2030. By 2050, the share of EVs in passenger vehicle ownership will exceed a third.

(3) Diesel demand will flatten

China is still a developing economy. As such, its economy is still growing rapidly and so too is its need for energy. Diesel consumption in China has yet to peak. According to the scenarios of Zhang Hailing of the Institute of Policy and Management of the Chinese Academy of Sciences, in the High Regular Consumption scenario,[14] diesel use will peak in 2020–25 at 174 Mt and decline slowly

---

[13]If the share of CHP in China's total power output from burning agricultural and forestry biomass were to be 60% in 2020 and 80% in 2030, the installed capacity of biomass-fired CHP systems would reach 3 GW and 6.4 GW in 2020 and 2030 respectively. If these CHP systems operate 6,000 h annually, they would provide enough heat for 150 days.

[14]In this scenario, the economy and population grow rapidly. Economic development and income growth are the priorities. Lifestyle trends are consumption-oriented. As oil supply security and environmental problems become increasingly severe and oil use becomes more efficient, the saturation point of commercial vehicle ownership per thousand people is 400 units.

thereafter. Alternative energy carriers like electric and hydrogen vehicles will develop slowly. In the Low Energy Revolution scenario[15]—characterised by strong economic and social development, optimised industrial structures, improved vehicle fuel economy and fast-growing alternative fuels—diesel consumption will peak at 170 Mt in 2015–20. After 2020, as vehicle ownership slows down and alternative energy carriers grow, the decline in diesel consumption will speed up.

(4) Kerosene demand will grow rapidly
One of the effects of higher income per capita is growth in air travel and air cargo, which means greater demand for aviation fuel. In 2016–20, the annual average growth rate in aviation fuel will be 9.2%. Use of kerosene-based aviation fuel is expected to reach 40 Mt in 2020, after which growth will slow down, reaching 53 Mt in 2025 and 64 Mt in 2030. Air cargo will grow at a similar rate as air travel, but will gradually slow down after 2030. The average flight length in air travel will increase from 1,669 km today to 1,900 km in 2050. That of air cargo will increase from 3,312 to 4,200 km. Longer aircraft range and better fuel economy will cause growth in kerosene demand to decline. It will increase from 64 Mt in 2030 to 72 Mt in 2040, and then flatten after 2040.

(5) Liquefied gas use will be driven by industrial demand
In recent years, industry's increasing use of liquefied gas is the main reason for the fuel's rapid growth. Growth in household liquefied gas demand will slow down, although not in rural areas where growth potential remains. Industry's share of total liquefied gas demand will continuously increase. Use of liquefied gas in transport will also grow, due to it being more economical than petrol. End-user liquefied gas demand is forecast to reach 44 Mt in 2020 and 50 Mt in 2030. After 2030, household demand for liquefied gas will remain flat before gradually

declining, due to natural gas replacing fossil energy and slower growth in demand in rural areas. The production capacity of propane dehydrogenation plants, where liquefied gas is made, will be saturated by 2030. As the future deployment rate of new technologies and plants isn't clear, liquefied gas demand in the chemical sector is expected to remain stable after 2030.

5.1.4 Hydrogen Energy
Hydrogen has been included in China's energy plan as a strategic option to optimise energy consumption and safeguard national energy security. China produces hydrogen energy from a variety of sources, including fossil fuels, renewables and industrial gases. Hydrogen production technologies and processes, like water electrolysis and hydrogen purification by pressure swing adsorption, are mature. Hydrogen safety technologies in China are almost in line with internationally advanced levels, and breakthroughs have been made in developing safe high-pressure hydrogen cylinders and tanks. China's hydrogen fuel infrastructure is behind that of developed economies like the USA, Japan and Germany, but is growing fast. In addition, China has completed a standards system for hydrogen and fuel cell technologies. Based on international standards, the system will play an increasingly important role in driving the development of the hydrogen energy industry.

(1) The hydrogen energy industry will grow rapidly
According to the report on China's Hydrogen Energy Industry Infrastructure Development, China will make major progress in hydrogen energy infrastructure development by 2020. Hydrogen energy production capacity will reach 72 billion cubic metres, supplying 100 hydrogen refuelling stations, 10,000 fuel cell vehicles and 50 hydrogen-powered railway trains by the end of this decade. The gross output value of the hydrogen energy industry will amount to RMB 300 billion ($43.5 billion).

By 2030, the hydrogen energy industry will be an integral part of China's new energy strategy, and its output value will exceed RMB 1 trillion. There will be 1,000 hydrogen refuelling

---

[15]In this scenario, the population grows continuously. The economy successfully transitions to low-carbon energy and oil use improves. As public transport and rail transport are on the rise, the saturation point of vehicle ownership per thousand people is 350 units.

stations serving 2 million fuel cell vehicles, supported by 3,000 km of high-pressure hydrogen pipelines. In addition, China will develop a technical standards system for hydrogen energy infrastructure to match those in developed economies. And it will innovate hydrogen and fuel cell technologies and services to support the continued development of the hydrogen energy industry.

In the next 3–5 years, the hydrogen energy industry will see massive growth thanks to extensive production, safe and large-scale storage, continuous supply, long-distance transmission and fast replenishment. Hydrogen fuel cells and lithium-ion batteries are the two pillars of the new energy vehicle industry. Fuel cells are being vigorously promoted across China. For instance, Shanghai released a development plan for fuel cell vehicles in September 2017, which forecast that the city will have more than 100 fuel cell vehicle related companies by 2020 and build 50 hydrogen refuelling stations by 2025. Fuel cell vehicle technology and manufacturing will match international levels by 2030, and the annual output value of Shanghai's fuel cell vehicle industry will exceed RMB 300 billion ($43.5 billion).

(2) Coal-to-hydrogen and carbon capture and storage will become mainstream

China has abundant and relatively inexpensive coal resources, so coal-to-hydrogen will probably become the main method for large-scale hydrogen production. However, the coal-to-hydrogen process emits large amounts of CO_2, so the use of carbon capture and storage (CCS) technology is necessary to reduce emissions. Currently, CCS technology is mainly deployed in thermal power generation and chemical production. According to the U.S. Environmental Protection Agency (EPA), CCS technology can help reduce 80–90% of CO_2 emissions in thermal power plants. China Shenhua Group began development of a CCS project in Ordos in 2009 and has successfully developed 300,000 tonne CCS demonstration projects. As CCS technology gradually matures, the coal-to-hydrogen plus CCS method of hydrogen production will predominate, providing sufficient hydrogen for the medium- and long-term development of China's hydrogen economy.

5.2 Secondary Energy Infrastructure

Energy infrastructure is the core of China's energy transition. It is where change in the energy system is perhaps at its greatest, driven by deep integration, digitalisation and Internet technologies. China's energy system will focus equally on centralised and distributed energy production. Energy sources, networks, loads and storage will be integrated into a single system using multiple types of energy, which will gradually evolve into a low-carbon, smart, secure and efficient national energy system.

5.2.1 General Characteristics
(1) A combination of centralised and distributed energy systems

China's future power system will comprise centralised smart grids and distributed low-carbon energy networks. Historically, centralised production has been the dominant energy supply mode in China. By optimising resource allocation and improving energy use efficiency, centralised supply will still play an important role in driving China's energy system and socioeconomic development. However, as distributed energy systems have distinct advantages like low transmission losses, efficiency and environmental protection, China's energy supply mode is shifting from a predominantly centralised mode to a combination of centralised and distributed energy supply. This is an irreversible trend in China's energy system, and the only way to ensure an optimised and clean energy system.

Centralised smart grids will be the platform for optimising power resource allocation in the long term. From the beginning, China developed large centralised power generation facilities, connecting them with end users through transmission and distribution systems. Thanks to economies of scale and supply reliability, centralised power systems generally succeeded. However, as China's energy sources are often located far from load centres, smart grids are needed to optimise supply and demand. According to China's Strategic Action Plan for Energy Development (2014–20), 14 large 100 Mt coal bases and nine large 10 GW coal power bases will be formed, mainly in the north, north-west and south-west of the country; and nine large wind power bases will be created, mainly in the north, north-east and north-west. The installed capacity of solar power in the north, north-east and north-west will reach 37 GW, 37% of China's total installed capacity in solar. Given the distance between energy production in the north and west and load centres in the east and south, China's priorities include the continued development of long-distance high-capacity power transmission, expansion of the West-to-East power transmission project, and implementation of the North-to-South power transmission project.

Secure, clean and efficient distributed energy is an important part of China's future energy system. Distributed energy is a network that connects low-carbon energy producers with consumers. It is capable of effectively integrating different types of energy and making energy use more efficient. As such, it is an important tool to counter energy shortage and environmental pollution. A distributed energy network can comprise multiple renewables like building-integrated photovoltaics, rooftop solar, waste-to-energy, biogas, geothermal, energy storage, and small wind farms. Each network can operate as an independent microgrid serving a village, organisation or community, or be connected to the centralised grid for regional integration, leveraging technologies such as artificial intelligence and big data.

Distributed energy offers efficient and flexible grid connection. This enables the grid to operate optimally by combining centralised and distributed energy resources. Generally, distributed energy networks need to be connected to the grid to import or export power during periods of surplus production or deficiency or network outages. During normal operation, the grid can provide distributed energy networks with services like voltage and frequency support and system backup. These can help optimise power distribution and safeguard grid operation.

Distributed energy will be mainly connected to power distribution systems and microgrid/micro energy networks, resulting in multi-directional energy flows. This is detrimental to network protection and control. To offset such negative effects, active power distribution and smart microgrid/energy network technologies can be developed to ensure grid connection of all distributed energy and safe and stable grid operation.

(2) A digital platform for smart energy services

Digitalisation, the Internet and advanced information software are changing the way energy is produced, distributed and used. Together, they are refashioning the energy landscape, creating new business models and driving economic and social development.

Smart, digitalised energy infrastructure creates a platform for sharing energy and information. Whereas the traditional grid is a pure power transmission and distribution system, the smart grid is an integrated energy and information carrier. This enables the grid to function as a central platform for wide-area optimisation and to coordinate distributed energy infrastructure. In time the strength of the grid's information and communication resources can help build a public data-sharing platform that can improve energy production and consumption, and better serve people's needs.

The integration of energy networks with the Internet triggers the emergence of new business models and energy service players. It connects decentralised users, diversified energy carriers and different service providers in the energy system, expanding the scope and frequency of market interaction and lowering transaction costs. Business model innovations will initially target power trading and new energy vehicles, and then extend to other parts of the energy system like energy demand consulting, energy efficiency products and online-to-offline services. Many innovative energy service companies will also start up, driving the development of the energy Internet industry into value-added services.

(3) Complementary energy systems from multiple sources

A smart energy system comprises multiple energy sources. It consists of various energy conversion technologies that enable subsystems like power, heating, cooling, gas, oil and transport to deliver complementary benefits along the entire value chain, from production and transmission to conversion and use. In this way, the entire energy system is interconnected and complementary. In addition, coordinated planning and development of the energy subsystems—power, heating, etc.—can reduce resource waste and improve the economic benefits of the entire energy system.

Large-scale transmission, energy storage, wind and solar, and power-to-hydrogen and power-to-methane projects will be developed to build an energy network of complementary sources that reduce volatility in renewable energy and allow it to be used efficiently.[16] Energy bases should identify a suitable multi-energy complementary model for their region based on local energy sources, land resources, transmission corridors, grid strength and other variables. For example, hydropower complemented by wind and/or solar power in an energy base focused on hydropower and new energy; or wind, solar and energy storage complemented by thermal power in a base integrating new energy and thermal power. Energy bases of this kind can help increase the use of renewable energy.

Multiple types of complementary energy provide end users with options, enabling them to customise their energy supply in power, heating, cooling and gas. This can be achieved with natural gas-fired combined cooling, heat and power plants, distributed renewable energy systems and smart microgrids. Complementary energy systems enable energy demand-side management and improve energy efficiency.

(4) Coordinated development of energy sources, networks, loads and storage

Energy supply loads and energy storage will be integrated to form an open and coordinated platform that connects generation, transmission and distribution with demand and consumption. In this way, deep integration and the bi-directional flow of power and information will improve efficiency and the share of clean energy in end-user consumption.

The coordinated development of energy sources, networks, loads and storage is not limited to one energy carrier, it applies to the entire energy system. Energy sources include oil, natural gas, electricity and other carriers. Networks means oil pipeline networks, heating networks, power grids and other energy networks. Loads refers to power load and other energy demands of users. Energy storage includes batteries, pumped storage, thermal (molten salt) storage and others.

Various energy conversion, information and integration technologies will be deployed to coordinate the development and production of energy sources and their transport or transmission in the energy system. First, energy sources will be combined in a flexible and efficient way to ensure coordination between new types of power generation and the grid. This will improve the grid's self-regulating capacity and reduce the impact of intermittent renewable energy on grid stability. Second, networks, loads and energy storage will be integrated, enabling multidirectional interaction between the grid, energy storage and demand-side resources like energy efficiency and load management. This will connect energy storage with grid control and allow operators to track fluctuations in renewable energy output and respond with orderly (smart) charging and discharging of energy storage resources, thus strengthening grid security and stability.

User demand can be connected to the system to enable integrated energy demand management across the entire energy system. Users' overall energy demand and energy storage resources are controllable, allowing information and energy to flow between demand, storage and supply. In this way, a highly automated and controllable energy supply and demand system will take shape, improving the efficiency, security and stability of the entire energy system.

---

[16]Zhou Xiaoxin, Development of Next Generation Energy System, in Electric Age. Issue 1, 2017.

5.2.2 The Power Grid of the Future

According to Zhou Xiaoxin, China's renowned power system expert and his research team, China's grid will evolve from the second to the third generation by 2050. Building on more than 100 years of first- and second-generation grid technologies, the third-generation grid supports large-scale new energy power generation, significantly reduces grid security risks and integrates a wide range of information and communications technologies. It is the smart, sustainable power grid of the future. The overall trend of grid development is to build a flexible, efficient and integrated smart energy system comprising a network of national transmission systems, local transmission and distribution systems, microgrids with distributed energy resources, energy storage facilities and energy utility systems.

The power grid of the future will be a combination of conventional grids and microgrids. China's West-to-East and North-to-South power transmission projects will remain the backbone of the system, transporting a mix of hydropower, coal power, large-scale wind power and desert-based solar power to load centres.[17] Distributed power generation systems that use local energy sources and regional grids and microgrids will be developed, and the number of electric vehicles and other energy storage facilities will substantially increase. The future power distribution system will probably be divided into several independently operating zones, which can connect with virtual power plants and microgrids of varying sizes.[18]

AC and DC power links will form the spine of the national transmission system. Ultra-high voltage AC (UHVAC) transmission systems are synchronous, providing instantaneous response and adjustment to regional changes in demand. Ultra-high voltage DC (UHVDC) point-to-point transmission systems transfer large amounts of power at very high voltages over very long distances, with low electrical losses. UHVDC and UHVAC systems complement each other, and a mix of the two is where the future of UHV transmission in China lies. Flexible HVDC and DC power systems will be effective supplements, delivering technical and cost-efficient advantages. As the manufacturing technologies of cut-off devices and DC cables continuously improve, flexible HVDC will become the most important transmission mode in DC power systems.

Advanced information and communications technologies will be deployed across the entire power system value chain to increase automation, digitalisation and intelligence and provide deeper and better information on system operations. Artificial intelligence, the Internet of things, big data, cloud computing, deep learning and blockchain will be deployed in asset management, system control, and energy management and trading to improve security, cost-effectiveness and reliability.

Diverse sources of energy will be integrated to enable interactive supply and demand. The power grid of the future will support large volumes of renewable energy and enable smart operation and integrated information management. It will become an Energy Internet that integrates energy, information and business to balance supply and demand, provide users with flexible options and the ability to conduct real-time transactions. The Energy Internet will dynamically manage and optimise multiple energy sources, delivering grid reliability and maximising financial returns.

5.2.3 Distributed Energy

China's structural transformation of the economy and energy system is accompanied by cheaper distributed energy and digitalisation. China's distributed energy prospects are highly promising. According to research on China's energy system evolution by Professor Li Liying of South China University of Technology and his team, the pathways of distributed energy development are as follows: first, in the short term (up to 2020), centralised energy supply will remain

---

[17]Zhou Xiaoxin, et al. Development Mode and Critical Technologies for China's Future Grid, Proceedings of the CSEE, Issue 25, 2014.

[18]Ma Zhao, et al. Form and Trends of Future Power Distribution Systems, Proceedings of the CSEE, Issue 6, 2015.

dominant, supplemented by distributed energy. Natural gas-fired combined cooling, heat and power (CCHP) systems will be extensively deployed, and the development and use of renewables, like wind and solar, will switch from centralised to distributed mode. Second, in the medium term (by 2030), the share of distributed energy in the energy system will increase, as will energy diversification. Third, in the long term (by 2050), distributed energy will be everywhere and comprise a major component of the energy supply system. Energy producers and consumers will make free transactions via Internet-based energy trading platforms. The International Energy Agency, in its Prospects for Distributed Energy Systems in China, explains the trends driving distributed energy development. These include making consumers an active part of the energy system, providing innovative services, and Internet-based business models.

Making consumers an active part of the energy system. In distributed energy systems, consumers have the opportunity to play an active role in the energy system. They can have their own generator, energy storage device and smart equipment, enabling them to be both an energy consumer and a producer. More and more house owners, businesses and organisations like local government can all be part of distributed energy solutions. This provides a way for them to express their values, such as using renewable energy only or a desire to use smart technology.

Integrating multiple energy sources to provide innovative services. Deep integration of distributed and centralised energy with intelligent control, information management and user terminals enables energy providers to offer innovative services. First, they can provide energy optimisation services for end users. The production and consumption of different energy sources like heat, gas and electricity will be automatically measured and stored, providing useful information to help optimise energy production and consumption. Big data, data mining, and the analysis of users' energy consumption and production data will provide a deep understanding of user characteristics. Second, they provide value-added services within the energy system. Most distributed energy systems will be connected to centralised power distribution systems, exchanging energy and information. Distributed energy systems can provide value-added services for centralised systems, including peak shaving, frequency adjustment, system backup and better power quality.

Internet-based development of distributed energy business models. Innovative technologies like information and communications, cloud computing and big data analytics allow digital technologies to penetrate each part of the energy system and connect them all in an intelligent way. For instance, by analysing the operating hours of different distributed energy systems and the data from demand-side response and energy storage systems, supply and demand can be coordinated to reduce costs and consumer prices. Second, by using energy storage and demand-side response, generation and consumption in centralised and distributed energy systems can be coordinated to deliver an optimal energy system. Microgrids will be developed in remote areas to reduce the cost of constructing power transmission and distribution systems.

5.2.4 Energy Storage

Energy storage is of strategic importance for the energy transition and for changing the way power is generated and consumed. It can boost supply at peak times and improve grid operations and the use of grid equipment. Energy storage holds great potential for the future. According to the International Energy Agency, by 2050 the global installed capacity of energy storage will exceed 800 GW (equal to 10–15% of total installed power capacity) and be worth several trillion dollars. China's installed capacity of energy storage will reach 200 GW in 2050 and be valued at more than RMB 2 trillion. This shows that China has massive and urgent demand. In the context of the Energy Internet, energy storage technologies like electrochemical, thermal, hydrogen and electric vehicles, will be gradually deployed at scale, accelerating the development of centralised renewable energy, distributed power generation and microgrids.

Provides grid support and makes large-scale centralised renewable energy more efficient. Energy storage allows renewable energy to have the same attributes as conventional energy, making it schedulable, predictable and controllable. Large-scale centralised renewable energy storage shows two key trends: the first is to enable local renewable energy to be used as a reliable source of power in a national or regional grid. Energy storage can offset the effects of intermittency by releasing energy when needed, increasing the acceptability of renewables as an important contributor to grid operations. The second is a shift from single-point and single-type energy storage to multi-point and multi-type storage. As energy storage evolves into a system, it can control, schedule and optimise multiple types of energy storage from different grid connection points in a holistic manner.

Energy storage drives the development of distributed renewable energy. Energy storage is a critical support technology for distributed generation and smart microgrids. It plays an increasingly important role in integrating, coordinating and managing multiple energy sources. Energy storage units are becoming more compact, modular and fast responding. As energy demand diversifies, and users of AC and DC power at different voltages coexist, energy storage will be the mainstream technology to enable end users to switch roles from power consumer to producer. Energy storage will be a buffer and an enabler of flexible and smart interactions.

Coordinated development of energy storage and electric vehicles. Electric vehicles (EVs) will be part of the future grid as an important means of energy storage. EVs have two main energy storage applications—they can transfer power from a secondary battery located outside of the vehicle, or the vehicle can act as an energy storage unit itself. As sales of EVs increase, the energy storage potential of a secondary battery will remain at around 15% of the capacity of the on-board battery. By 2020, the total energy storage capacity of EVs in China is expected to reach 3.766 GW/13.749 GWh, with 2.1 GW/10.4 GWh from secondary batteries.[19] In the future, the interaction between EVs and the grid will be two-way instead of one-way. EVs will feed energy into the grid and also function as a distributed energy storage source. In addition, they will provide many ancillary grid services, help consume distributed new energy and improve the cost effectiveness and security of grid operations.

The cloud will become a new type of energy storage in the future. According to research by scientists such as Kang Chongqing, cloud energy storage will be a grid-based service that allows consumers to use a shared energy storage pool at anytime and anywhere. It would lower the cost of providing energy storage services sharply. Cloud energy storage is part of the sharing economy in which assets, resources and services are shared by a group of individuals or companies. Cloud energy storage can be either a grid-scale centralised facility or a fleet of smaller distributed units, operated and managed by cloud energy storage providers.

6   The Implications of China's Energy Revolution

Vigorous development of clean energy and building a diversified supply system are the foundations and goals of the energy revolution. International experience shows that changes in energy supply require policy support, innovation in production and supply technologies, the transformation of industry, and optimised allocation of investment and employment resources. This section of the report will analyse the implications of China's energy supply revolution in four areas: energy supply, the transformation of industry and energy capacity, investment, and employment.

---

[19]Sun Wei, Li Jianlin and Wang Mingwang, et al. *Business Operation Models of Energy Storage Systems and Analysis of Typical Cases*, China Electric Power Press, Beijing, 2017.

6.1   Energy Supply

As a result of the energy revolution, China's future energy supply system will be diversified, comprising the clean and efficient use of coal and the robust development of other types of energy —oil, gas, nuclear power, new and renewable energy—as well as stronger transmission and distribution networks and the widespread use of energy storage. This will require changes in the way energy is produced and innovation in energy supply technologies.

6.1.1   Energy Supply Structure: The Energy Supply System Will Prioritise Non-fossil Energy and Conversion to Electricity

In the Recommended scenario of the energy revolution, fossil fuels' share of primary energy will decrease annually, and non-fossil energy will gradually replace oil and coal to become the number-one energy source. Coal consumption will decrease sharply, and its share will drop to 27% in 2050 from 64% in 2015. Oil consumption will rise and then decline, although its share will remain steady at around 16% through to 2050. Natural gas, a cleaner energy carrier, will grow rapidly in the medium term, reaching 15%

of the primary energy mix in 2030 and remaining stable thereafter. Non-fossil energy will increase substantially. In 2030, its share of the energy mix will exceed 20%, making it the second largest energy source after coal. In 2050, its share will be more than 40%, replacing coal as the number-one energy source (Fig. 59).

In the end-user energy supply system, electricity's share will gradually increase from 22% in 2015 to 30% in 2030 and 38% in 2050. This will be made possible by the rapid development of clean energy and the use of clean coal. The share of renewable energy (wind, solar and biomass) and nuclear power will significantly increase. In 2030, the share of non-fossil energy will increase to 50.5%, exceeding that of fossil energy and becoming the main power source. By 2050, the share of non-fossil energy (installed capacity) will be more than 70% and the share of non-fossil energy output will be 66%. In the medium term, the total consumption of coal as primary energy will not change greatly; but its share of end-user energy supply will fall, even though installed coal power capacity will slightly increase. This shows that the energy supply revolution shifts from direct coal use to the conversion of clean and efficient coal into electricity (Fig. 60).

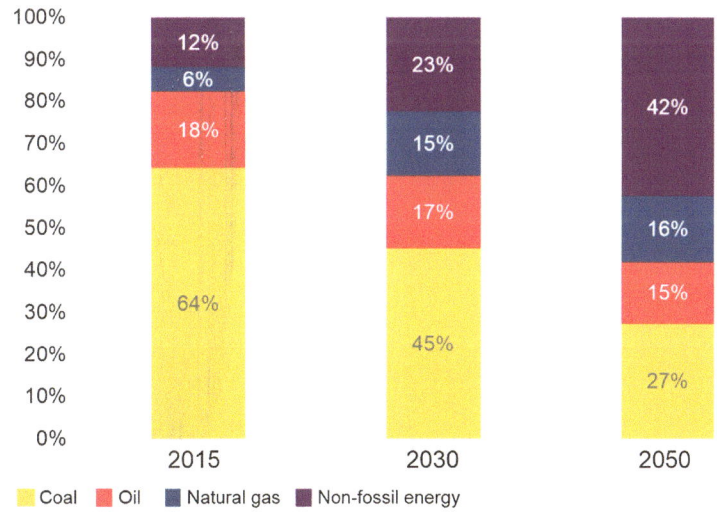

Fig. 59 China's primary energy supply system in the Recommended scenario

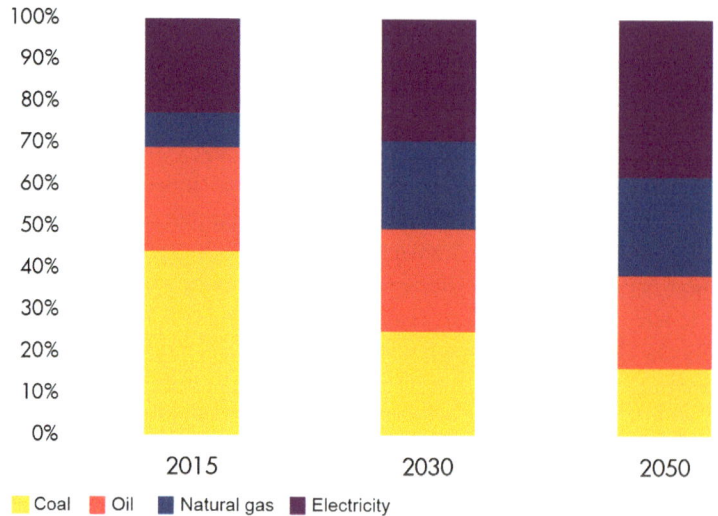

Fig. 60 China's end-user energy system in the Recommended scenario

6.1.2 Energy Production: A Combination of Centralised and Distributed Clean and Conventional Energy

One of the main goals of the energy revolution is to increase the use of clean, renewable energy. China's clean energy sources and load centres are separated, in west and east China respectively. To achieve the goal of the energy revolution, the development of clean energy in these two regions needs to be coordinated. Coordination will result in an optimised energy system comprising large centralised renewable energy plants and outgoing transmission systems in west China and distributed energy systems in east and central China. These smaller distributed energy systems will consist of small wind farms, solar power, gas-fired combined cooling, heat and power (CCHP) and ground-source heat pumps (GSHP) connected to microgrids or local power distribution systems.

Coal will remain the main energy source in China in the longer term. To meet the targets of the energy revolution, coal needs to follow the principles of safe production, efficient development and ecological priority. At the exploitation stage (as primary energy), coal will change from extensive development to intensive and green production. In conversion (as secondary energy),

coal-fired plants will be upgraded to ultra-low emission units using clean coal and high efficiency generating processes.

6.1.3 Energy Supply Technology: The Energy Internet and Demand-Side Response Will Optimise Supply from Multiple Energy Sources

Lateral multi-energy interconnection and vertical supply- and demand-side coordination will be an important way to implement the energy revolution. To integrate and coordinate multiple energy sources, networks, loads and energy storage, new energy supply technologies like the Energy Internet and demand response are needed.

The Energy Internet will enable deep integration between different energy sectors (coal, gas, renewables, etc.) and between the energy industry and other industries. It will also improve fossil energy use efficiency across all industries through digitalisation and the use of big data. Moreover, the Energy Internet will harness the full potential of the power system by increasing the share of clean energy in end-user energy consumption. This represents a shift in China's long-established way of supplying and using energy, and a transformation of the link between socioeconomic development and energy consumption.

The Energy Internet can incorporate the different requirements of energy users and extend power demand-side management into integrated energy demand management, spanning the entire energy industry. It enables bidirectional flow and allows energy consumers to become energy producers, transforming the traditional way of meeting growing demand with increased energy supply.

6.2 Industry and Energy Capacity Transformation

In essence, the energy revolution is about changing the energy system, enabling it to shift from quantity-based to quality-based supply. It includes the development of low-carbon energy from high-carbon sources and the green development of black coal and oil, made possible by energy system reform and technology innovation. The energy revolution is continuously fostering new types of energy and making conventional types cleaner and greener. To slow down energy consumption growth, China's energy industry needs to switch from extensive mode to efficient, high-quality development.

6.2.1 Industry Trends

The goal of energy supply is to build a secure, efficient, cost-effective, clean and low-carbon energy production and transmission/distribution system. To achieve this goal, energy sectors like coal, oil, natural gas and renewable energy need to urgently transform their current development mode into one that is green, efficient and cost-effective.

(1) Clean, efficient coal

Coal will remain dominant in China's energy mix for a long time to come. Clean and efficient coal use is, therefore, an indisputable requirement for the coal sector. China identifies four priorities in coal: (i) shift development from resource-driven to innovation-driven; (ii) shift coal use from a fuel-only focus to one centred equally on fuel and the quality of the raw material; (iii) shift coal production from extensive to intensive green production; and (iv) shift coal utilisation from high-emission to low-emission, clean and efficient use.

(2) Green development of conventional and unconventional oil and gas resources

China's onshore and conventional oil and gas production holds limited growth potential, and its exploitation of deep-sea oil and gas is low. However, China has abundant unconventional oil and gas resources. It should therefore adjust its oil and gas development strategies, switching focus from: conventional resources to an equal focus on conventional and unconventional resources; from onshore to an equal focus on onshore and offshore; from shallow reservoirs to middle and deep reservoirs; and to proactively develop clean and green oil and gas.

(3) Low-cost distributed clean energy

China's clean energy development of wind, solar and biomass is still at an early stage. Renewables do not have a cost advantage over conventional energy, and their development relies on national policy support and price subsidies. To deliver the energy revolution goal of more clean energy, while remaining competitive and relieving the fiscal pressure on national government, China urgently needs to increase investment in clean energy technologies, reduce clean energy costs and encourage the development of distributed systems to use clean energy efficiently.

6.2.2 Energy Capacity Transformation

Replacing fossil energy with clean energy and substituting electricity for coal and oil is the key to addressing energy security, environmental pollution and climate change, and to achieving the energy revolution. In primary energy, the goal is to shift from predominantly fossil energy to predominantly clean energy. In secondary level conversion, the objective is to replace coal and oil with electricity and promote the deployment of electric boilers, electric heating systems and e-mobility to increase the share of electricity in end-user energy consumption and reduce fossil energy use and environmental pollution.

Table 17 Fossil energy capacity changes in the Recommended scenario

Sector	Unit of measurement	2015	2030	New capacity in 2015–2030	2050	New capacity in 2030–2050
Coal	Mt	5,351	4,200.92	−1,150.08	2,674.54	−1,526.38
Oil	Mt	236.22	250.00	13.78	250.00	0
	Mtce	337.46	357.15	19.69	357.15	0
Natural gas	Billion cubic metres	139.9	330.0	190.1	462.0	132.0
	Mtce	186.01	438.90	252.89	614.46	175.56
Total	Mtce	6,124.67	5,279.97	−844.69	3,942.35	−1,337.62

(1) The transformation of primary energy—substituting clean energy for fossil energy

This transformation is about substituting non-fossil energy for fossil fuels in the primary energy system. It requires replacing coal with cleaner natural gas in the fossil energy supply system, using clean coal in the coal supply system and accelerating the shift in coal from its extensive use as an all-purpose fuel to an emphasis on clean coal and raw material quality. The energy revolution will reduce coal production capacity by 115 Mt in 2030 and 153 Mt in 2050, a drop of more than 50% compared to 2015. Greater use of natural gas can offset 16% of this reduction, with the remainder met by non-fossil energy (Table 17).

(2) Secondary energy—substituting electricity for coal and oil

In the energy revolution, secondary energy conversion has two objectives: the first is to substitute electricity for coal and oil to increase the share of electricity in end-user energy consumption; and the second is to replace coal in power generation with clean energy like natural gas, wind, solar, hydropower and biomass. As the revolution progresses, new clean thermal power generating units will be built and existing outdated ones shut down. Some thermal power generating units will be transformed to gas-fired ones, and others upgraded to ultra-low emission status to reduce emissions of air pollutants. This will help achieve the shift to efficient, clean coal. As renewable energy expands, thermal power will be used increasingly for peak shaving.

Gas-fired generating units will be used mainly for heating and as peak shaving for renewable energy (Table 18).

6.3 Investment

To deliver the energy revolution goal of transforming the energy supply system and greatly increasing the share of clean energy, China's energy industry needs to invest more in both clean energy and the clean use of fossil fuels. In recent years, China's investment in clean energy has prioritised renewables. However, with the energy revolution, China will also invest in building clean energy transmission systems, R&D and skills development. As the marketisation of the energy industry improves, its investment targets will diversify and more private capital will finance transmission system construction and equipment manufacturing, where previously state funding was predominant.

6.3.1 Diversification of Investment Priorities

In the energy revolution, investment will focus on clean energy, energy infrastructure, primary energy exploitation and secondary energy conversion. To deliver the energy system described in the Recommended scenario, by 2030 new investments in primary energy exploitation, installed power capacity and energy transmission systems need to be RMB 10.9 trillion in total. In 2030–50, new investments in the same three categories need to reach RMB 35.37 billion, RMB 12,194.87 million and RMB 4,057.3

Table 18 Secondary energy changes in the Recommended scenario

Sector		2015	2030	New capacity in 2015–2030	2050	New capacity in 2030–2050
Fossil energy	Coal power	884.19	1,100.00	215.81	800.00	−300.00
	Oil-fired power generation	6.00	6.00	0	6.00	0
	Gas-fired power generation	66.37	220.00	153.63	340.00	120.00
	Subtotal	956.56	1,326.00	369.44	1,146.00	−180.00
Nuclear power		26.08	136.00	109.92	350.00	214.00
Renewable energy	Hydropower	296.66	355.00	58.34	455.00	100.00
	Wind power	129.34	420.00	290.66	1,000.00	580.00
	Solar power	43.18	400.00	356.82	1,000.00	600.00
	Biomass	10.00	46.00	36.00	120.00	74.00
	Subtotal	479.18	1,326.00	846.82	2,120.00	794.00
Total		1,461.82	2,788.00	1,326.18	3,616.00	828.00

Note Unit of measurement is megawatts (MW)

Table 19 New capacity and transmission capacity demand in the Recommended scenario

Field	Sector	Unit of measurement	2015	New capacity in 2015–2030	New capacity in 2030–2050
Primary energy exploitation	Oil	Mt	236.2162	13.7838	
	Natural gas	Billion cubic metres	139.854	190.146	103.04
Secondary energy conversion	Hydropower	GW	296.66	58.34	77.2764
	Coal power	GW	884.19	215.81	
	Gas-fired power generation	GW	66.37	153.63	94.7061
	Nuclear power	GW	26.08	109.92	177.7453
	Wind power	GW	129.34	290.66	478.2806
	Solar power	GW	43.18	356.82	497.0266
	Biomass	GW	10.00	36.00	61.5393
Energy transmission systems	Power transmission lines	km	609,100	657,200	1,743,700
	Substations	Billion kVA	3.366	7.312	34.079
	Crude oil pipelines	km	21,000	12,700	9,100
	Product oil pipelines	km	27,000	44,400	46,200
	Natural gas pipelines	km	64,000	196,100	358,200

billion respectively, amounting to RMB 16,287.55 billion in total. In the medium and long terms, the energy revolution will stimulate a total of RMB 27.2 trillion of investments, of which renewable energy like wind and solar power will take the largest share (Tables 19, 20, 21 and 22).

Table 20 Decrease in unit investment in energy projects in one cycle (one cycle = 5 years)

Field	Sector	Decrease in unit investment in 2015–2030 (5%)	Decrease in unit investment in 2030–2050 (%)
Primary energy exploitation	Oil	2%	2%
	Natural gas	2%	2%
Secondary energy conversion	Hydropower	2%	2%
	Coal power	2%	2%
	Gas-fired power generation	2%	2%
	Nuclear power	5%	2%
	Wind power	5%	2%
	Solar power	10%	2%
	Biomass	10%	2%
Energy transmission systems	Power transmission lines	2%	2%
	Substations	2%	2%
	Crude oil pipelines	5%	2%
	Product oil pipelines	5%	2%
	Natural gas pipelines	5%	2%

6.3.2 Diversified Investment Sources

In accordance with China's Strategy of Energy Production and Consumption Revolution (2016–30) issued by the State Council, China will reform the administrative approval system for the energy industry. It will also revise the negative list restricting foreign investment and market access and encourage and guide market participants to legally and equally invest and operate in those energy fields not on the negative list.

Currently, state-owned capital dominates investment in China's energy industry, although there is some private capital invested in renewable energy like wind and solar. However, most investment in distributed energy resources is by users. As development of renewable energy, distributed energy-based microgrids and the Energy Internet gathers pace, the amount of private capital invested in these areas will increase.

Marketisation of the energy industry will also diversify investment sources. Currently, the new round of power market reforms has deregulated the additional power distribution and sales businesses, allowing private capital to take part. Oil and gas market reforms have deregulated unconventional oil and gas exploitation, oil and gas pipelines and crude oil reserve storage. As energy sectors are further deregulated, investors in energy production, transmission system construction and equipment manufacturing will also diversify.

6.4 Employment

Currently, China's energy supply-side reform focuses on five issues: cutting overcapacity, reducing excess inventory, deleveraging,

Table 21 Forecast unit investment in energy projects

Field	Sector	Unit of measurement	2015	2030	2050
Primary energy exploitation	Oil	RMB/t	2,100	1,977	
	Natural gas	RMB/cubic metres	0.3000	0.2824	0.2604
Secondary energy conversion	Hydropower	RMB/kW	13,000	12,235	11,286
	Coal power	RMB/kW	3,700	3,482	
	Gas-fired power generation	RMB/kW	6,500	6,118	5,643
	Nuclear power	RMB/kW	13,500	11,575	10,676
	Wind power	RMB/kW	8,000	6,859	6,327
	Solar power	RMB/kW	9,000	6,561	6,052
	Biomass	RMB/kW	10,000	7,290	6,724
Energy transmission systems	Power transmission lines	RMB/km	1,080	1,016	938
	Substations	RMB/kVA	0.2000	0.1882	0.1737
	Crude oil pipelines	RMB/km	10,000,000	8,573,750	7,908,154
	Product oil pipelines	RMB/km	10,000,000	8,573,750	7,908,154
	Natural gas pipelines	RMB/km	10,000,000	8,573,750	7,908,154

Table 22 Forecast new investments in the Recommended scenario

Field	Sector	2030		2050		New investment in 2015–2050
		New investment	Subtotal	New investment	Subtotal	
Primary energy exploitation	Oil	27.796	82.381	–	35.369	117.75
	Natural gas	54.584		35.369		
Secondary energy conversion	Hydropower	727.899	8,607.348	1,161.763	12,194.873	20,802.221
	Coal power	766.23		–		
	Gas-fired power generation	954.157		696.239		
	Nuclear power	1,316.442		2,341.736		
	Wind power	2,072.966		3,762.819		
	Solar power	2,486.942		3,722.349		
	Biomass	282.712		509.968		
Energy transmission systems	Power transmission lines	0.68	2,257.411	2.012	4,057.307	6,314.718
	Substations	1.397		6.932		
	Crude oil pipelines	113.953		95.081		
	Product oil pipelines	396.892		467.691		
	Natural gas pipelines	1,744.489		3,485.591		
Total			10,947.14		16,287.549	27,234.689

Note Unit of measurement: RMB billion

lowering costs and strengthening weaknesses. Cutting overcapacity will result in large-scale outflows of personnel from conventional energy sectors like coal, causing severe unemployment and reemployment difficulties. On the other hand, the rapidly developing clean energy sector urgently needs a large number of highly qualified people, creating considerable employment opportunities for those with the right skills.

As the energy revolution deepens, China's modernised and technologically more advanced energy industry will place higher demands on the labour force. In short, more quality and less quantity will be required. Moreover, industrial, regional and structural imbalances in the work-force will become evident.

6.4.1 Industrial Structure of Employment

Cutting overcapacity will have a negative impact on employment in conventional energy sectors like coal. According to the State Council, in 3–5 years from 2016, about 500 Mt of coal capacity will be shut down and 500,000 employees (about 10% of the workforce in the coal sector) will require reemployment. In 2015–50, the number of workers employed in the coal sector will fall by more than 2 million, compared to the scenario where energy supply revolution measures are not taken (Tables 23 and 24).

Coal aside, the energy revolution will not lead to large-scale personnel outflows from the energy industry. On the contrary, it will create more jobs.

Table 23 Average employment in the energy sectors in the Recommended scenario

Sector		UoM	2015	2030	2050
Coal	Output	Mt	3,750.00	3,819.02	2,431.40
	Average employment	Persons/Kt	1.17	0.9	0.80
Oil and gas	Output	Mtce	486.07	723.68	883.28
	Average employment	Persons/Ktce	1.693	1.44	1.15
Thermal power	Installed capacity	MW	9,565.6	13,260.0	11,460.0
	Average employment	Persons/MW	0.40	0.34	0.27
Nuclear power	Installed capacity	MW	260.8	1,360.0	3,500.0
	Average employment	Persons/MW	4.00	3.4	2.72
Renewable energy	Installed capacity	MW	4,791.8	12,210.0	25,750.0
	Average employment	Persons/MW	6.41	5.45	4.36

Note When automation is taken into account, average employment in each energy sector will decrease by 5% every five years. UoM = Unit of measurement

Table 24 New employment in the energy sectors in the Recommended scenario

Sector	2015	2030	New employment in 2015–2030	2050	New employment in 2030–2050	Total new employment
Coal	4,424,000	3,798,000	−626,000	1,934,000	−1,864,00,0	−2,490,00,0
Oil and gas	823,000	1,041,000	219,000	1,017,000	−25,000	194,000
Thermal power	383,000	451,000	68,000	312,000	−139,000	−71,000
Nuclear power	104,000	462,000	358,000	952,000	490,000	848,000
Renewable energy	3,073,000	6,655,000	3,583,000	11,229,000	4,573,000	8,156,000
Total	8,807,000	12,408,000	3,601,000	15,444,000	3,035,000	6,637,000

Note Unit of measurement = persons

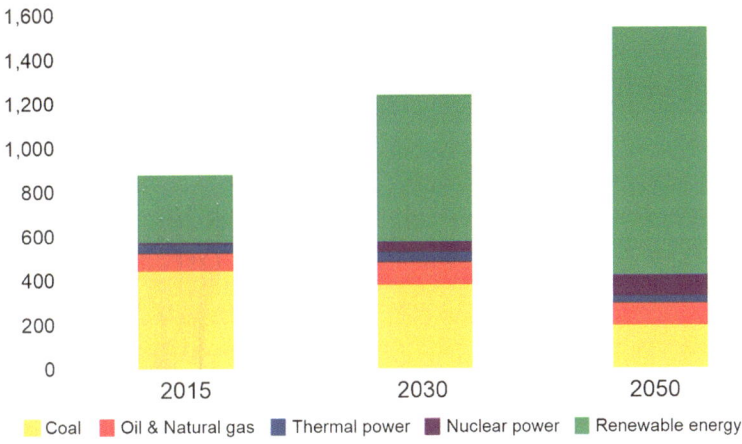

Fig. 61 Employment trend in the energy industry in the Recommended scenario (in 100,000)

Coal ◼ Oil & Natural gas ◼ Thermal power ◼ Nuclear power ◼ Renewable energy

More than 6 million employment opportunities will be added to the energy industry by 2050, thanks mainly to the energy transition. Clean energy sectors like wind, solar, nuclear and biomass will develop rapidly, creating a vast number of employment opportunities in technology development, equipment manufacturing, installation and maintenance. It is projected that the new energy sector will create 4 million new jobs for the entire energy value chain by 2030, and more than 8 million new jobs overall by 2050. The new energy industry will play the role of stabiliser by absorbing surplus workers from conventional energy sectors. For example, those workers from the coal sector with less expertise and fewer skills could work in infrastructure construction. The development and deployment of energy-efficient technologies will also create employment in the technology and service fields (Fig. 61).

6.4.2 Regional Structure of Employment

China's energy industry is faced with two contradictory challenges: excess supply of inadequately skilled labour and a shortage of qualified people with the right skills in renewable energy, energy transmission systems, and emerging services like demand-side response, energy efficiency and integrated energy management. China's energy revolution will, therefore, impact different provinces in different ways, depending on the structure of the energy industry within their domain.

In provinces such as Zhejiang, where the new energy and energy service sectors are developing well, the energy revolution will have only a slight impact on the provincial employment structure. Thanks to higher revenues and the creation of numerous employment opportunities by the new energy industry, surplus employees from capacity-cutting sectors can be repositioned within those provinces.

In provinces like Shanxi and Inner Mongolia, where conventional energy sectors like coal dominate, the energy revolution will have a greater impact on the employment structure. On the one hand, there is high unemployment in capacity-cutting sectors like coal. On the other hand, these provinces have relatively abundant new energy sources. Yet, despite large amounts of new energy capacity, wind and solar curtailment is severe. The new energy sector in these provinces has, therefore, limited development potential in the near future and cannot absorb surplus employees from the conventional energy sectors. High unemployment and economic downturn will form a vicious circle, resulting in outflows of skilled young residents. This will in turn adversely impact the provinces' future economic development.

6.4.3 Employment Supply Structure

The workforce in China's energy industry encompasses rural migrant workers, graduates from vocational schools, and graduates from colleges and universities. As there is a relatively large number of rural migrant workers in primary energy exploitation (mainly coal), capacity-cutting will hit those workers hard. Booming clean energy fields like wind, solar and nuclear power require skilled people. They need employees with expertise in the relevant technologies. Increasingly, graduates from vocational schools and colleges and universities will be attracted to these sectors.

Graduates from vocational schools are generally more educated than rural migrant workers and more skilled than college graduates. They also have a comparative advantage in labour cost than rural migrant workers and the average college graduate. In the future, they might be employed in clean energy equipment manufacturing, installation and maintenance.

Relative to the lower-wage requirement of rural migrant workers and the higher reward/price ratio of graduates from vocational schools, college graduates with better skills are mainly working in technical development in the energy industry. However, the number of college and university graduates is growing rapidly. In 2015, a record 7.49 million students graduated from colleges and universities in China. In addition, as university college reform progresses slowly, and graduate qualifications deviate to some extent from societal requirements, the employment situation in China, especially in those provinces with a large number of graduates, looks grim.

7 China's Energy Revolution: Measures, Policies and Adjustments

The low-carbon and efficiency requirements of the energy revolution drive the transformation and diversification of the energy supply system. To deliver a successful energy revolution, measures need to be taken to cut capacity in, and shift to the clean production of, primary energy, and expand secondary energy applications. Policies and proposals need to be developed to guide investment in the relevant energy sectors and solve the reemployment problem of surplus workers laid off by conventional energy companies. This section will, therefore, analyse the measures, policies and proposals for China's energy revolution from the following three perspectives: the primary energy revolution, the secondary energy revolution, and policies and proposals for energy supply-side investment and reemployment.

7.1 Measures for the Primary Energy Revolution

Excessive energy consumption and environmental degradation urgently require the energy revolution to make progress. The goals of the primary energy revolution include: (i) substituting clean energy for fossil energy, shutting down outdated capacity and shifting to the green production of fossil energy; (ii) the rapid development and efficient use of renewable energy; and (iii) the development of new energy production technologies, exploration, and the use of this new energy.

7.1.1 Fossil Energy Supply Revolution

The fossil energy revolution mainly concerns coal, oil and gas. In terms of coal, the revolution is about the safe, cost-effective, green and efficient development of coal resources. In terms of oil and gas, it requires an equal focus on conventional and unconventional resources—onshore and offshore and at all depths, from shallow to ultra-deep.

(1) Coal supply revolution

(1) Keep the exploitation of coal resources within acceptable environmental limits to ensure reasonable and scientific capacity

Coal production should be reduced and the low-carbon and zero-carbon world energy trends fully understood to ensure that the coal industry is aligned with scientific capacity requirements.

(2) Environmentally friendly coal production
The principle of clean coal should be followed and new technologies like water-preserved mining and cut-and-fill mining promoted to reduce land collapse in worked-out sections. Land rehabilitation efforts should be increased and ecological recovery and governance technologies deployed. Quality coal processing technologies should be widely used to improve coal washing and dressing.

(3) Improve integrated coal production
To make coal production efficient, significant improvements should be made in: (i) the recovery rate of integrated coal resources with gas and water extraction; (ii) the comprehensive reuse of coal gangue and mine water; and (iii) the coalbed methane recovery rate. Efficient coal mining technologies should be deployed to raise the level of mechanisation and automation. Gas, water and other resources should be extracted during mining to make coal recovery cost-effective and sustainable.

(2) Oil and gas revolution
(1) Unplug technical bottlenecks in the exploration and development of unconventional oil and gas
Priority should be given to accelerating the exploration and development of tight oil and gas and to making breakthroughs in super-lateral well drilling and multistage fracturing technologies. Coalbed methane (CBM) exploration and development should be promoted and advances in various emerging technologies made to enable CBM extraction to shift from high-quality reservoirs to deeper, more complex strata.

(2) Develop deep and ultra-deep offshore oil and gas at scale
Strong efforts should be made to solve such difficult challenges as identifying deep reservoir targets, working at high temperatures and in complex lithology, and extracting oil and gas from deep carbonate, clastic and volcanic rock. Offshore oil and gas engineering equipment and technologies should be developed, as should semi-submersible drilling and production platforms and subsea production systems to speed up the exploration and extraction of deep-sea oil and gas.
(3) Promote clean oil and gas development

A blueprint for oil and gas field construction should be drawn up to minimise field footprint and reduce oil and gas losses. Waste collection, treatment and disposal should be centralised. Environmental impact assessments (EIA) should be carried out when decommissioning oil and gas fields, and ecological recovery made in accordance with the EIA results. The sealing processes in oil and gas collection and transmission should be strengthened to reduce gas emissions. An EIA and information disclosure system should be created to strengthen supervision and ensure environmentally friendly oil and gas development.

7.1.2 Non-fossil Energy Revolution
The non-fossil energy revolution refers mainly to renewable energy and nuclear power. For renewable energy, it means increasing the share of renewables in the energy mix. For nuclear power, it means safety and substituting nuclear for fossil energy.

(1) Renewable energy revolution
(1) Protect the environment and deliver the hydropower development goal
Large hydropower stations should be built to deliver the hydropower development goal, on the understanding that strict environmental protection requirements are met and displaced people relocated correctly.

(2) Construct support facilities to develop wind power in an orderly way
At the introductory stage of wind power development, onshore and offshore wind power should be complementary. Priority should be given to wind power development in north, north-east and north-west China, eastern coastal China and areas with medium and low wind speed. At the mid-developmental stage, equal focus should be placed on developing onshore and offshore resources. And at the mature stage, offshore wind power should be fed into load centres.

(3) Use different solar technologies to ensure solar power generation is optimised for local conditions
In desert areas with abundant solar energy in west and north China, solar power development

should be given priority. In medium-sized and large cities in east and central China, building-integrated photovoltaics should be encouraged and grid-connected rooftop PV systems widely deployed. In regions that feature good sunshine conditions, large areas of available land and abundant water resources, solar thermal pilot projects should be tested. In this way, different solar power technologies will be combined to deliver complementary benefits.

(4) Develop liquid biofuels and substitute them for oil at scale
In the short and medium terms, more effort should be made to develop non-grain liquid biofuel technologies to lower the production cost of non-grain liquid biofuels. As breakthroughs are made in second- and third-generation biomass technologies like cellulosic ethanol and algae-based fuel, liquid biofuels should gradually replace petrol, diesel and kerosene in heavy-duty road transport, aviation and shipping.

(5) Use renewable heat at scale
Solar thermal energy can play an important role in heating or cooling water, homes, industry and commerce. In both rural and urban areas, biogas and biomass briquettes should be promoted to fuel cooking and boilers.

(2) Nuclear power revolution
(1) Significantly increase nuclear power capacity
Third-generation pressurised water reactor technology should be deployed to increase investment in nuclear power plants in eastern coastal areas. Effective measures should be taken to protect plant sites and strengthen nuclear power development in central China, building a nuclear power belt across eastern and central regions.

(2) Become a hub for the global nuclear power industry
China's nuclear power equipment manufacturing and project construction capability should be enhanced to complement nuclear power

development in east and central China. Strong efforts should be made to actively create national brands for China's future nuclear power technologies and equipment, enabling them to go global.

(3) Foster an atmosphere of acceptability for nuclear power development
A science-based decision-making system should be established and long-term development strategy drawn up to bolster the growth of nuclear power. More effort should be made to improve laws, regulations and supervision, ensure consistent decision-making, create positive publicity and guidance, and improve information disclosure, thus building up public confidence in and acceptance of nuclear power.

7.1.3 Promote New Energy Production Technologies
New energy production technologies could potentially provide a richer range of energy sources for production and living. These mainly include the exploration and production of natural gas hydrates (NGH), advanced nuclear power and marine energy.

(1) Research NGH exploration and exploitation
China has abundant NGH resources in polar tundra sandstone and seabed sandstone reservoirs. It should strengthen basic research to deliver a complete range of support technologies for NGH exploration and production. China should accelerate NGH resource surveys and screen potentially profitable exploration areas to implement drilling and trial extraction of NGHs. The aim should be to achieve earliest commercial extraction, thus sharpening China's international competitiveness in NGH exploration and exploitation.

(2) Explore and develop more advanced nuclear power technologies
Guided by the inherent safety concept, China should research and develop fourth-generation safe reactor technology to continue the nuclear

power revolution. International collaboration and capital investment should be strengthened to explore and develop advanced nuclear technologies like fusion power.

(3) Forward-looking research on marine energy

China has a vast coastline, which means it has abundant marine energy. R&D is pivotal to finding ways to convert marine energy into reliable and accessible energy carriers like electricity. As these technologies progress, marine energy is expected to become an important part of China's future energy system.

7.2 Measures for the Secondary Energy Revolution

Electricity-centred energy supply becomes increasingly important in the energy revolution. It is an efficient and easily accessible secondary energy carrier, with the potential to be clean and carbon-free. The strategy of replacing coal and oil with electricity and long-distance clean power transmission is a priority goal of China's secondary energy supply revolution.

7.2.1 Combine Centralised and Distributed Energy Optimally

Centralised power generation and supply have dominated China's energy system and driven China's socioeconomic development forward. However, because distributed energy systems have lower electrical losses, are more efficient and more environmentally friendly, China's energy system is shifting towards a combination of centralised and distributed energy supply. This is an irreversible trend in China, and the only way to ensure an optimised and clean energy system.

7.2.2 Replace Fossil Energy with Electricity

Replacing fossil energy with electricity means coal, oil, gas and direct fossil fuel burning. Electrical equipment should be extensively deployed to reduce fossil energy consumption.

This conserves energy resources and reduces wastewater, residuals and exhaust gas, thus helping to deliver China's energy efficiency and emissions reduction strategy.

7.2.3 Increase Demand-Side Response

Reasonable pricing and incentive mechanisms should be developed, and flexible demand-side response resources coordinated—such as distributed energy systems, energy storage and electric vehicles—to drive users to implement demand-side response. This will help reduce peak loads and peak-valley differences. It will lessen the need for new power generating capacity and for more power transmission and distribution systems. The efficiency and reliability of existing systems will improve, as will the use of critical resources like land, capital and human resources.

7.3 Policies and Proposals for Supply-Side Investment and Reemployment

Replacing fossil primary and secondary energy with cleaner alternatives and electricity respectively will shift China's energy supply system towards a low-carbon future. New investment opportunities will arise. There will, however, be costs. Shutting down outdated capacity will result in unemployment in conventional energy companies. Investment in clean energy needs to be guided and conventional energy companies helped to reposition their surplus workers. Both factors are important to a successful delivery of the energy revolution.

7.3.1 Supply-Side Investment Guidance

Development of clean energy is one of the major tasks of the energy revolution. Investment guidance is therefore important. In recent years, several problems have been exposed, such as: increasingly severe wind and solar curtailment, despite a sharp increase in installed capacity; and weak market competitiveness and low marketisation due to instability in clean power generation. Policies and suggestions for supply-side investment are necessary.

(1) Increase investment in clean energy R&D
Government should play a leading role in encouraging investment in clean energy research and development, providing companies with the policy support they need. Reducing production costs through technical innovation is the key to clean energy development. Clean energy R&D should deliver technologies and products with proprietary intellectual property rights. It should feature collaboration with industry, universities and research institutions, and support the development of a clean energy industry.

(2) Encourage investment in energy storage and support facilities
Policies encouraging investment in energy storage will help further clean energy development and grid connection. Government authorities can introduce a compensation mechanism for energy storage system operators to provide ancillary services and encourage clean energy providers to install energy storage systems. This will better facilitate grid connection and maximise use of clean energy generation.

(3) Implement environment-related tax incentive policies
Levying an environmental tax is a relatively common practice to promote clean energy development. It enables government to discourage emissions and discharges through taxes. A pollution tax could be levied on companies that discharge pollutants during production. It would dissuade high energy-consuming and high-pollution companies from using conventional energy. This will reduce energy consumption, encourage the use of clean energy, and help protect the environment. A proportion of the taxes levied could be used for pollution control and part of them to fund clean energy development.

(4) Improve clean energy grid connection and consumption
Implementing the policy that prioritises clean energy for grid connection and consumption plays an important role in clean energy development. The policy is recommended for implementation at local level. Local energy planning

authorities can measure their power consumption and peak shaving capability and then identify the clean energy quota that can be connected to the grid and consumed through bidding.

7.3.2 Supply-Side Reemployment
The energy revolution will improve the overall employment structure, but it will also generate unemployment. Given the multiple effects of the energy revolution on employment, policies and proposals on how companies should reposition surplus workers and government solve unemployment are necessary.

(1) Companies should create channels to reposition surplus workers
In supply-side reforms, capacity cutting will inevitably cause unemployment. The relevant enterprises should think about how to actively and properly reposition surplus workers through several channels, thus gradually alleviating unemployment by region and sector. The repositioning of surplus workers in companies facing overcapacity should begin by safeguarding workers' rights and interests. The channels for repositioning surplus workers should be expanded to include internal job transfer, starting a business, early retirement and public-service jobs. Government will provide the relevant support policies for companies that hire surplus workers laid off by capacity cutting. Surplus workers who want to start a business should be offered the chance to enter a business incubator. For surplus workers who are difficult to reposition, public-service jobs will be a priority.

In the context of the energy revolution, the channels to address surplus worker reemployment are shown in Fig. 62.

(2) How government should address reemployment
In the short term, supply-side reforms will inevitably impact adversely workers in some regions and sectors. In the long term, a structural imbalance in employment will be more evident than an imbalance between supply and demand. To ensure a smooth and orderly employment market, labour demand should be optimised in

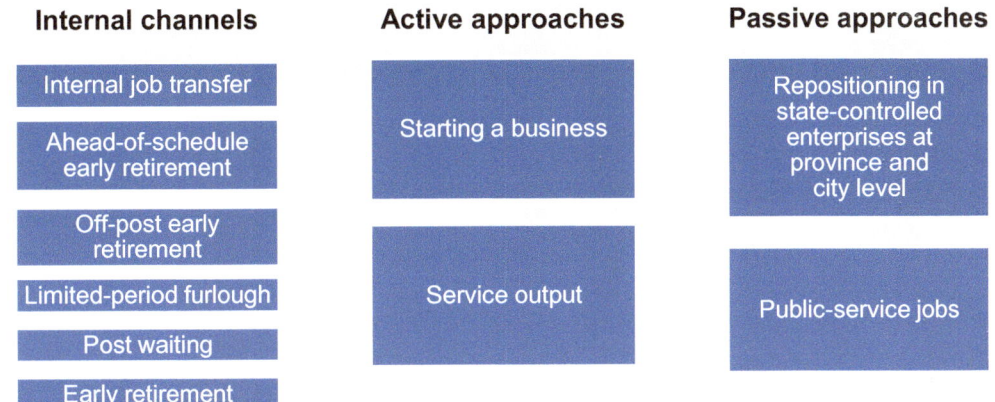

Fig. 62 Channels to address surplus worker reemployment

the short term and labour supply improved in the long term. The public employment service and social security systems should also be improved to match labour supply with demand and provide basic job security.

(1) Short-term optimisation of labour demand
(i) The structure of the energy industry should be adjusted and emerging businesses with a strong job-creating capability developed. The service industry has stronger job-creating potential than manufacturing, so its share of employment should increase.
(ii) Adjustment of the corporate ownership structure should be speeded up, and medium, small and micro-sized private enterprises prioritised. Private enterprise has become the main channel to address the employment challenge. Relevant support policies need to be implemented to encourage and grow more private enterprises, especially those in technology.

(2) Long-term improvement of labour supply
(i) Adjustment of academic disciplines should be accelerated to effectively lead and match societal requirements. Universities and colleges should adjust their academic disciplines and programmes to correspond with socioeconomic developments. They should channel more effort

into developing talent and play an effective role in leading and supporting industry advancement.
(ii) Adjustment of the educational structure should be faster and vocational education strongly promoted. Vocational education should be included in the national education plan and given priority. Investment in vocational education should be increased. Vocational schools should be guided to collaborate with companies, to develop the talent and skills needed by society.

(3) Improvement of the public employment service and social security systems
(i) The social security system should be improved and include a basic government guarantee. The share of commercial insurance in healthcare and the pension system should be gradually increased. Cross-provincial commercial insurance should be improved to reduce labour transfer costs and integrate labour markets across China. This will facilitate cross-provincial coordination of employment.
(ii) The public employment service should be improved to match labour supply with demand. Existing employment and training policies should be optimised to improve quality and provide the skills required for job transfers and reemployment, thus creating an effective employment training system.

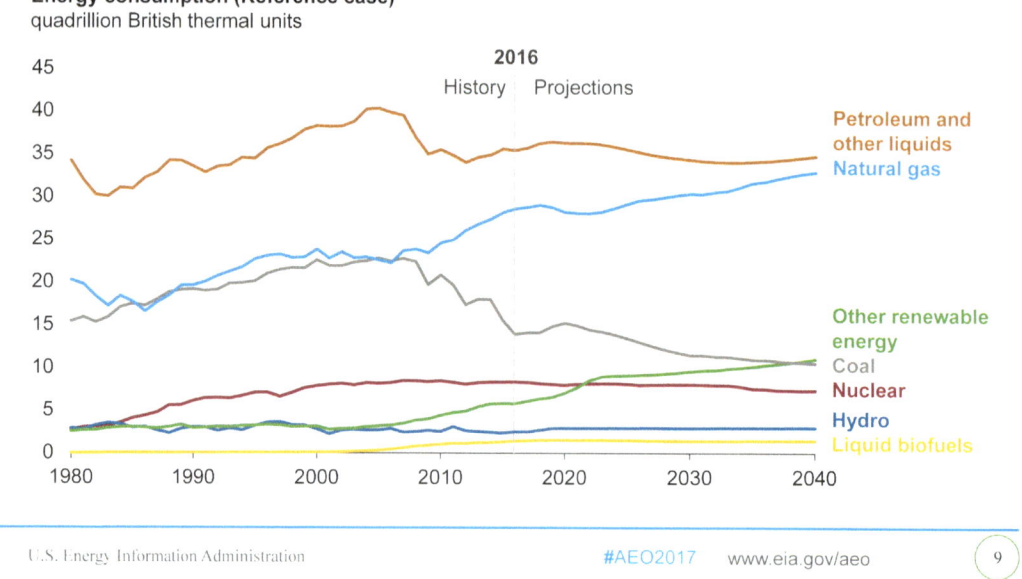

Fig. 63 Energy consumption (Reference case) in quadrillion British thermal units. *Source* EIA (2017)

Appendix 2: United States Energy Transitions to 2050: Trends, Challenges and Targets

This appendix summarises the slide deck created by University of California, Davis, for Shell on United States transitions to 2050: Trends, challenges and opportunities, and serves as a companion to that presentation. It covers recent research reports on US energy futures to 2050 and draws conclusions on where the US energy system is headed, and how policies might change this future. It also relates the US situation to that of China and draws lessons that may be of interest to the Chinese context.

We relied on several major reports, including those of the U.S. Energy Information Administration (EIA 2017), the U.S. Department of Energy (DOE 2016), the International Energy Agency (IEA 2017) and BP (BP 2017), as well as other reports for specific topics. These are listed at the end of the appendix.

1. Summary of our findings

Our key findings include:

- given current trends, the USA will reach flat energy use and marginally declining oil and

  CO_2 emissions after 2030; there will be no deep reductions without major new policies;

- natural gas rises to match oil as the biggest US energy source after 2030; renewables remain far below;

- however, renewables for power generation rise dramatically to 2040, nearly matching gas; both rise at the expense of coal;

- wind power becomes the top renewable power source after 2020 (passing hydro); solar passes hydro a decade later;

- CO_2 emissions decline little overall, as decreases in transport and power generation are partially offset by rising industrial emissions; and

- the potential for large oil reductions mainly resides in light-duty vehicle efficiency and electric vehicles, but EVs have little effect until after 2035. Efficiency improvements are 10 times more important to 2030.

The following figures tell much of the story. The EIA Reference case projection of energy use is shown in Fig. 63. From now to 2040, the EIA projects that total US energy demand will be roughly flat, with a rise in natural gas and non-hydro renewables offsetting a decline in coal, with nuclear, hydro and oil use remaining

Energy production (Reference case)
quadrillion British thermal units

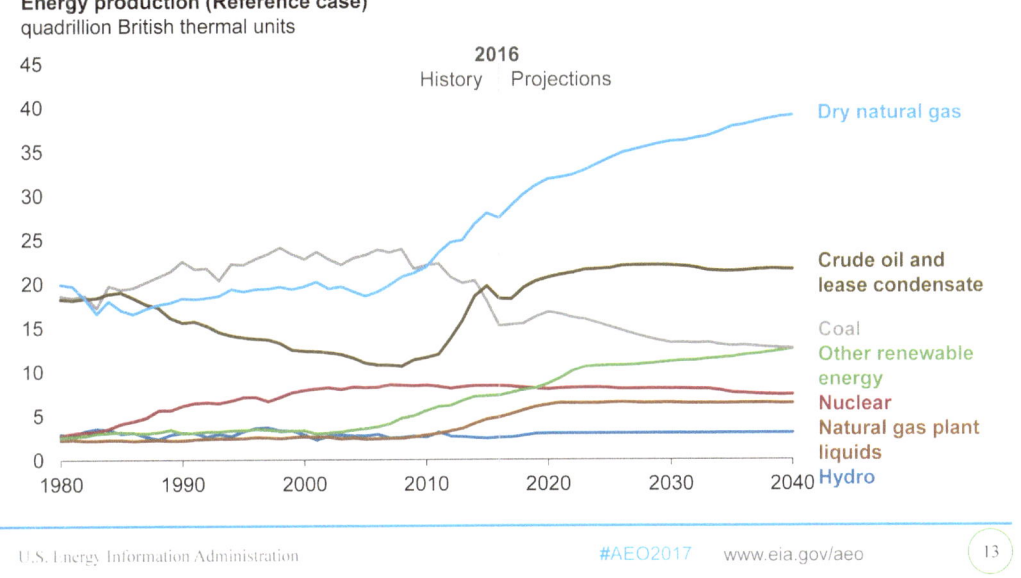

Fig. 64 Energy production (Reference case) in quadrillion British thermal units. *Source* EIA (2017)

fairly constant. The lack of decline in oil use relates to ongoing increases in travel in the USA, barely offset by increases in energy efficiency and fuel-switching in that sector.

In energy production (Fig. 64), natural gas will continue to rise steadily, keeping prices low, while oil will plateau but avoid a decline as new capacity comes online. The steady rise in renewable energy production is enough to compensate for the decline in coal production by 2040, but remains far below oil and gas production.

In terms of electricity production, renewables —specifically wind and solar power—are expected to rise dramatically (Fig. 65), with wind passing hydro as the leading source by 2020 and remaining the top source to 2040, despite plateauing in 2025. Solar rises steadily, with both utility-scale and end-use applications growing at a rapid rate over the timeframe.

Even EIA "side cases", such as the high oil price case and the low economic growth case, show relatively little decline in CO_2 over the timeframe (Fig. 66).

The U.S. Department of Energy released a very different vision of the future at the end of the Obama Administration in 2016: The United States Mid-Century Strategy for Deep Decarbonization. This featured a major shift to electricity across the energy economy, with deep decarbonisation of electricity largely due to faster renewables growth than in the EIA Reference case. It also included very strong increases in transport efficiency and slower demand growth, coupled with a shift away from oil towards electricity and biofuels. This combination, with some supporting measures in other sectors, provides a pathway for an 80% reduction in CO_2 by 2050, but achieving this would require much stronger policies than exist today (Fig. 67).

2. The USA and China

Our key findings on the USA that have implications for China include:

- the rise in US shale oil (horizontal fracking) production has greatly increased domestic oil supply and lowered natural gas prices, both of which are absent in China;
- the increase in natural gas and renewable electricity generation (and slower demand growth) has enabled a more rapid decline in coal use than has been possible in China;

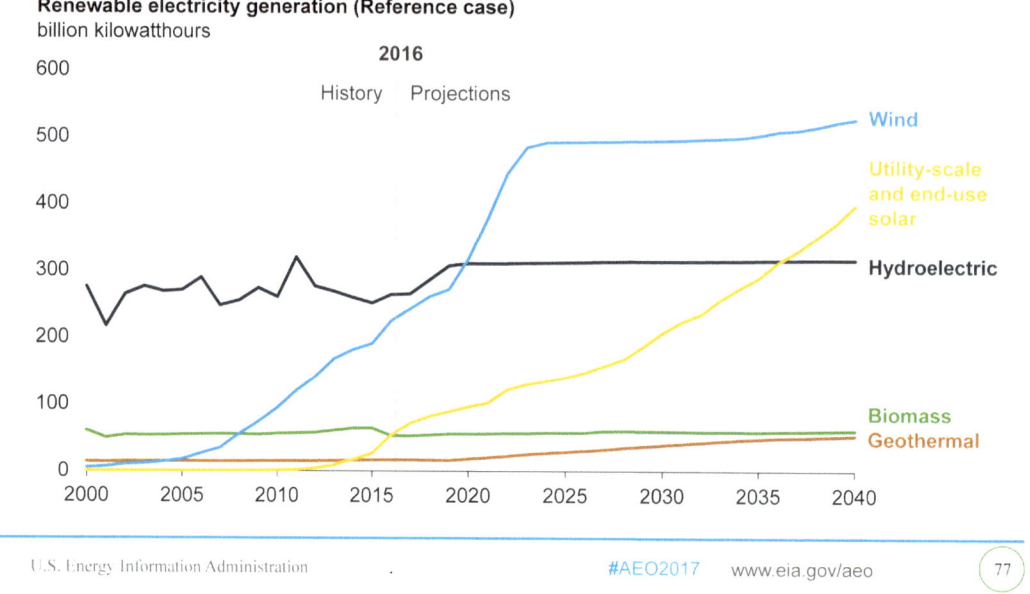

Fig. 65 Renewable electricity generation (Reference case) in billion kilowatt-hours. *Source* EIA (2017)

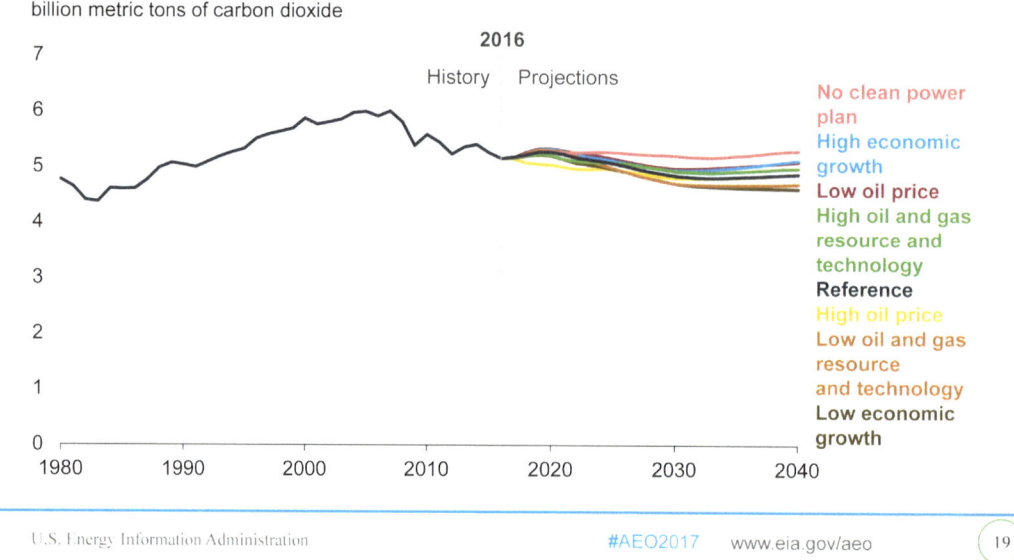

Fig. 66 Energy-related CO_2 emissions in billion metric tonnes of CO_2. *Source* EIA (2017)

- although renewables (solar photovoltaic and wind) are rising rapidly in the USA, the pace is behind China and the absolute value is far behind;
- electric vehicle (EV) sales growth in the USA is projected to be similar to China, although China's very recent adoption of a new energy vehicle credit policy will likely accelerate growth; the overall size of the EV market appears much larger in China, where the government has announced aggressive EV sales targets of 20% of new vehicle sales by 2025;
- CO_2 emissions decline marginally in the USA, while they rise in China; both nations will need

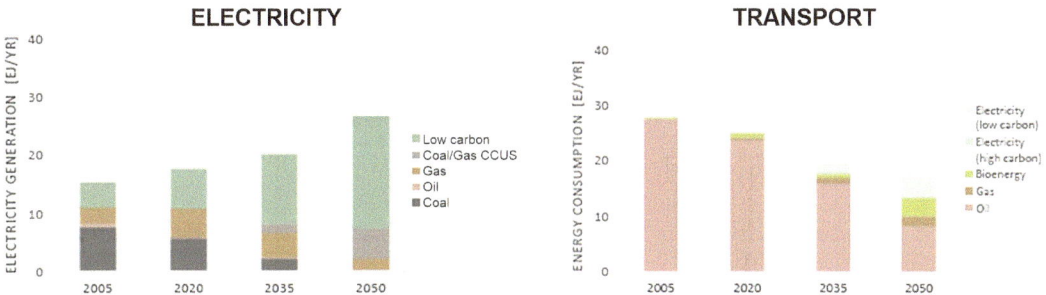

Fig. 67 U.S. DOE Mid-Century Strategy scenario, electricity and transport energy projections. *Note* CCUS = carbon capture, utilisation and storage. *Source* U.S. Department of Energy (2016)

Growing oil demand in emerging economies...

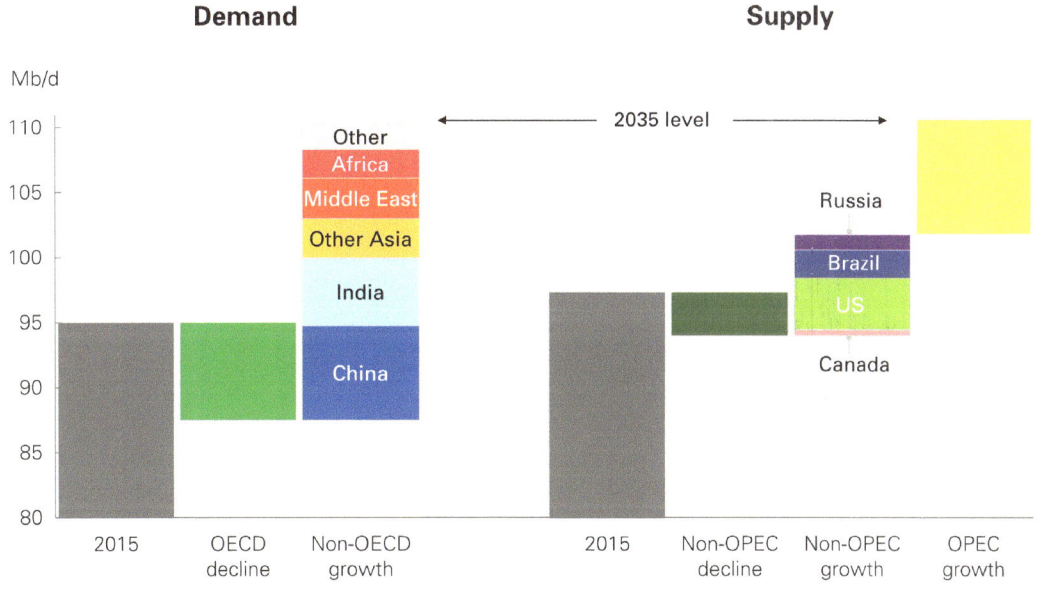

Fig. 68 Growing oil demand in emerging economies. *Source* BP Energy Outlook 2017

to turn a corner to achieve outright reductions; renewables growth, along with nuclear growth in China, will be the key to this in both cases; and

- the potential for deep oil demand reductions mainly reside in light-duty vehicle efficiency and EVs, but EVs have little effect until after 2035. Efficiency improvements are 10 times more important to 2030.

Considering both oil supply and demand, BP Energy Outlook 2017 shows China with major oil demand growth to 2040, and the USA important in supply growth (Fig. 68).

BP also expects a very similar percentage growth in renewables for power generation to 2040, though with much higher absolute levels (and thus share of total global growth) in China.

Renewables continue to grow rapidly...

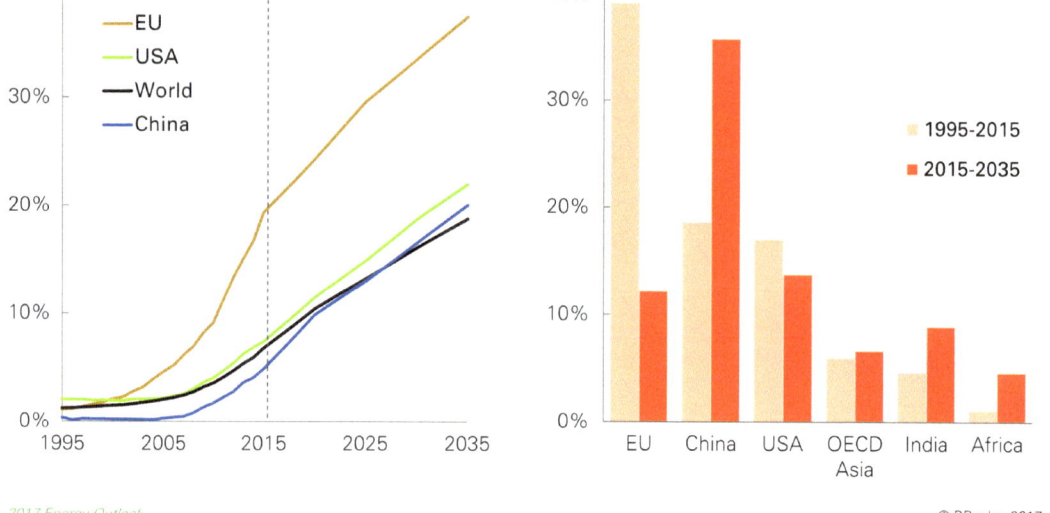

Fig. 69 Renewables share of power generation and by region. *Source* BP Energy Outlook 2017

This reflects China's rapidly growing electricity market, among other things. The EU will remain the highest in terms of renewables' share of electricity (Fig. 69).

BP estimates that wind costs are already lower in North America (mainly the USA) and China than coal (Fig. 70); solar and wind become the cheapest options in both countries by 2035. These low costs will drive both the rise in these renewables and an eventual decline in coal power in both countries.

In the light-duty vehicle sector, both the USA and China are expected to experience rapid increases in electric vehicle market penetration. Bloomberg (2017) projects that EVs will reach almost 50% of new car sales by 2040 in both countries (Fig. 71). China's slightly slower start in EV sales has been offset by very rapid increases over the past two years, and this is expected to continue. The introduction of new models in both the USA and China, along with

incentives and the provision of recharging infrastructure, is driving growth.

The International Energy Agency (IEA 2017) projects that, under their respective Paris Agreement nationally determined contributions, US CO_2 per capita would drop dramatically by 2030 while China's would be flat. Although China (and most other countries) starts from a much lower position and would remain lower than the USA (Fig. 72).

3. Technology trends

A range of technology trends in the USA and beyond was identified in the IEA 2017 report, including:

- solar photovoltaic (PV) costs decline by 50% from 2010–25 and become competitive with natural gas; wind power is often competitive now;

Renewables growth is driven by increasing competitiveness...

Cost of power generation from new-build plants*

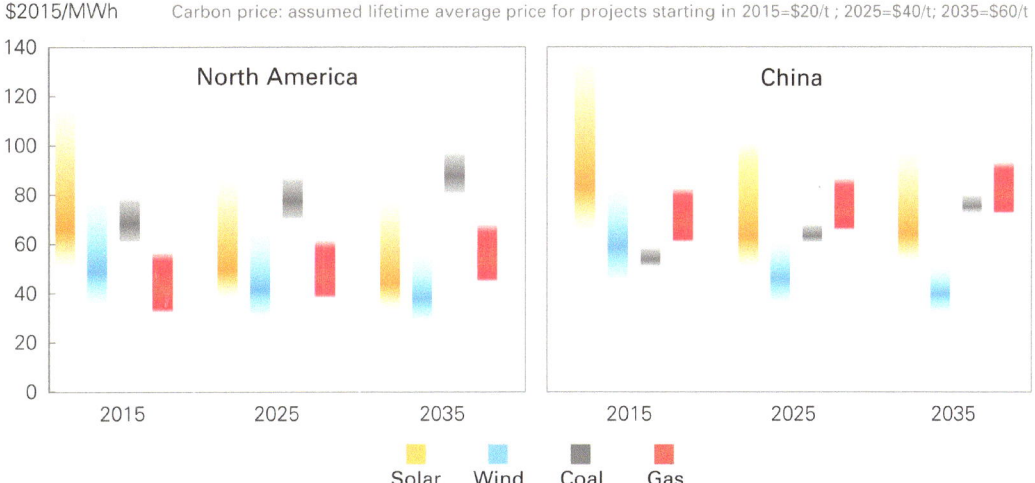

Fig. 70 Renewables growth is driven by increasing competitiveness. *Source* BP Energy Outlook 2017

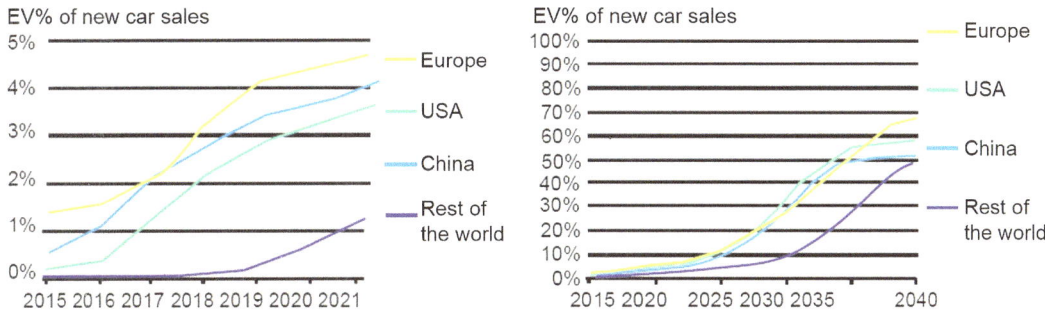

Fig. 71 Electric vehicle sales projections. *Source* Bloomberg New Energy Finance (2017)

- fracking has led to both an increase in oil supply (and projected future supply) and low-cost natural gas; this is expected to continue for at least the next decade;
- tight/shale oil production rises marginally from 2020–25, then flattens to 2040—but there is no significant decline;
- electric vehicle growth is largely technology-driven; future projections to as high as 30% of vehicle sales by 2030 depend on reductions in battery costs and performance improvements;
- there is no major revival of nuclear—generation declines slowly to 2040; and
- a deeper reduction in CO_2 emissions in the future is partly dependent on greater use of solar PV and wind, and on battery costs continuing to decline.

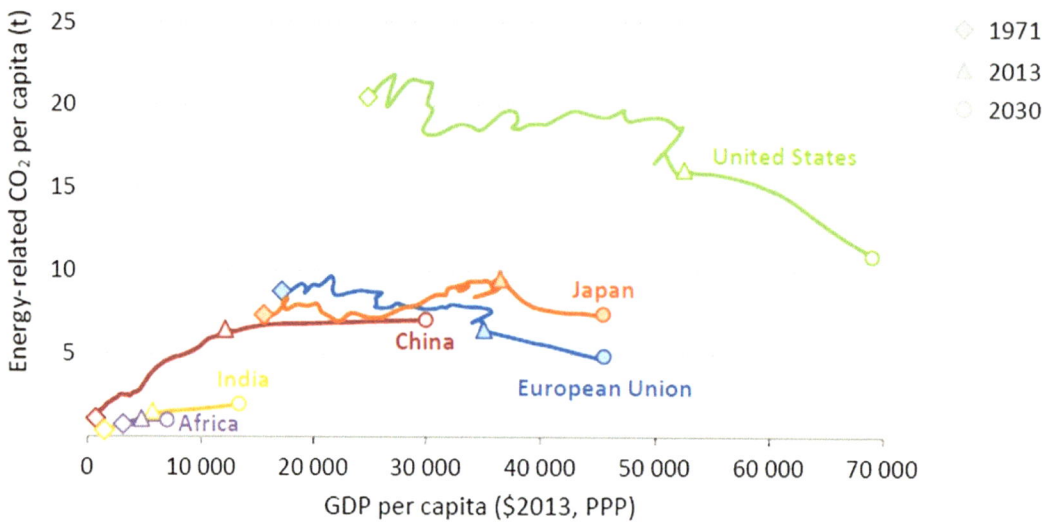

Fig. 72 CO$_2$ emissions for GDP/per capita for different countries. *Note* PPP = purchase power parity. *Source* IEA (2017)

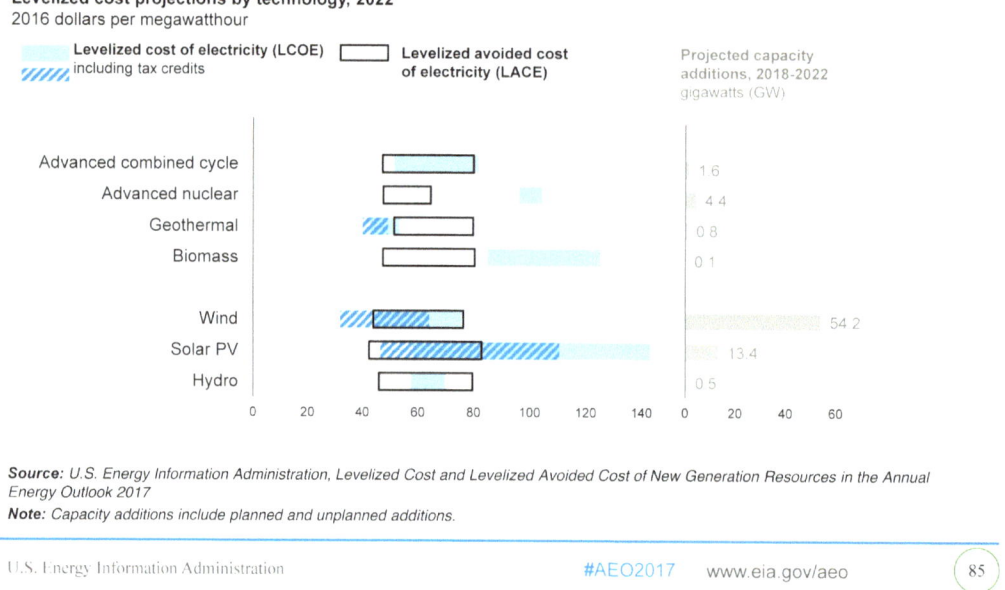

Fig. 73 Levelised cost projections by technology, 2022 (in 2016 $ per megawatt-hour). *Source* EIA (2017)

The example of declining solar PV costs driving growth is highlighted in the EIA data in Fig. 73. As PV costs have dropped over the past decade, installations have grown rapidly. While cost reductions will slow, the impacts will continue to be felt in rapid PV growth in the USA.

PV technology penetration will be driven largely by on-site installations, especially for residential use. The EIA projects this to reach 100 GW by 2040 (Fig. 74).

The EIA projects electric vehicles in the USA to approach 2 million vehicles by 2040, or about

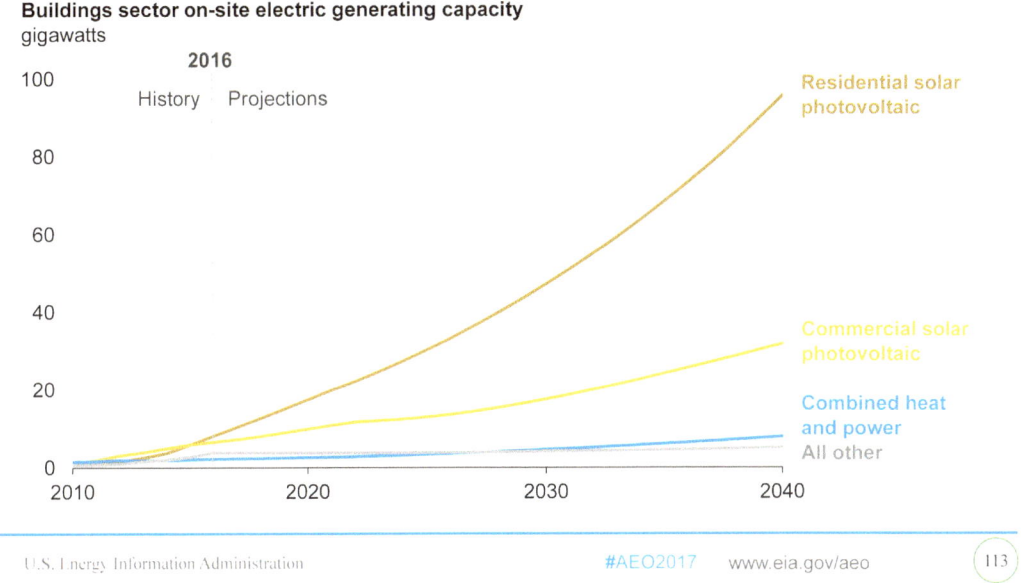

Fig. 74 Building sector on-site generating capacity in gigawatts. *Source* EIA (2017)

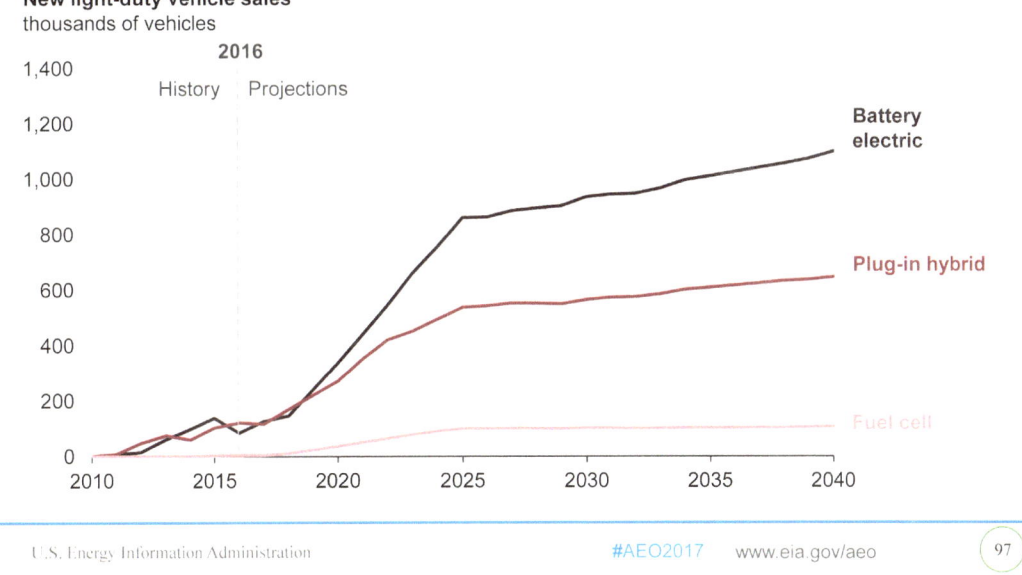

Fig. 75 Projected sales of new light-duty electric vehicles in thousands of vehicles. *Source* EIA (2017)

10% of total light-duty vehicle sales (Fig. 75). Other sources project higher levels (Fig. 76). In any case, the cost and performance of batteries, along with the availability of recharging infrastructure, will likely determine how far EVs grow.

Most EV projections to 2030 and beyond are far higher than the EIA's, and have increased from 2016 to 2017, reflecting a growing optimism that the EV's time is coming (Fig. 76; note that Fig. 76 shows stocks while Fig. 75 shows sales).

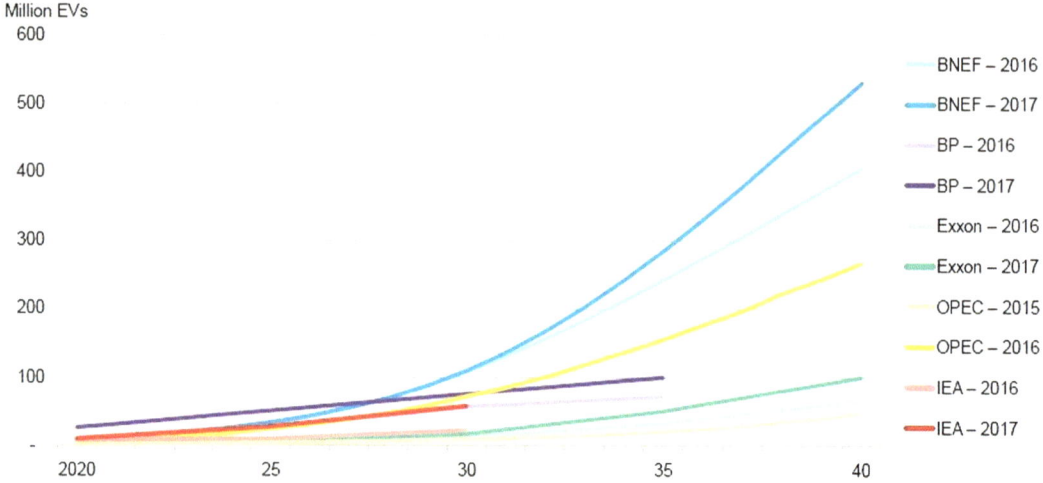

Fig. 76 Projected stock of electric vehicles. *Note* BNEF = Bloomberg New Energy Finance; Exxon = ExxonMobil; OPEC = Organization of the Petroleum Exporting Countries; IEA = International Energy Agency. *Sources* Bloomberg New Energy Finance, New Energy Outlook 2017. BP Energy Outlook 2017. International Energy Agency, World Energy Outlook Special Report: Energy and Climate Change, 2015. International Energy Agency, World Energy Outlook 2016. Natural Resources Defense Council, Americas Clean Energy Frontier: The Pathway to a Safer Climate Future, 2017. U.S. Energy Information Administration, Annual Energy Outlook 2017. U.S. Department of Energy: United States Mid-Century Strategy for Deep Decarbonization, 2016. University of California, Davis, Three Revolutions in Urban Transportation, 2017

Appendix 3: Lessons from Supply Revolutions in the Nordic Countries

1. Summary of our findings

Supply revolutions fundamentally change the energy mix to meet the changing goals of society. Momentum for a supply revolution can gather pace as economies develop and demand new energy services or combat pollution, or discover new energy resources and invent new technologies, or face external supply shocks. This involves the simultaneous uptake of new fuels and the abandoning of existing fuels, or at least an end to the growth of existing fuels. The goals of China's energy system are becoming increasingly important: to improve energy quality and access, meet growing energy demand, drive investment into new technology and maintain energy security. They are likely to necessitate a shift from coal to new energy sources, such as gas, nuclear and renewable power.

These goals cannot be met without the cooperation of stakeholders in the energy system. The key stakeholders are consumers, who will ultimately pay for a transition; companies, who will implement the transition; and energy workers, whose livelihoods will depend on the transition. These stakeholders must be considered by policymakers when planning a supply revolution, as either their cooperation is needed, or their opposition avoided.

The Nordic countries, in particular Denmark and Norway, are at the frontier of reducing the environmental impact of their energy systems. Denmark is unique in having integrated very high levels of variable wind power into its electricity system, while maintaining energy supply security. Norway is a pioneer in adopting electric vehicles and decarbonising its transport sector. Despite their relatively small size, the energy transitions of both Denmark and Norway offer interesting lessons for China, as the Nordic countries are global leaders in the energy transition.

These two case studies, on Denmark and Norway, extend a previous report, which considered the supply revolutions in Germany, France, Japan and the UK.

(1) Consumers

The case studies suggest that consumers will participate in the energy transition if it is subsidised, as was the case in Norway, and they may even pay for the transition if they are convinced of the environmental benefits, as in Denmark. However, both the Norwegian electric vehicle (EV) subsidies and the Danish energy taxation system that finances the deployment of offshore wind suffer from design problems that Chinese policymakers can learn from. Specifically:

- China may wish to align carbon costs by implementing a universal carbon price and support R&D in green technologies to reduce the cost of transition. Both the Danish energy taxation system and the Norwegian EV subsidies failed to align carbon costs with mitigation opportunities; this is likely to have increased the total cost of the Scandinavian transitions, as the most expensive mitigation options were encouraged by policy. Norwegian EV subsidies have been an expensive way to reduce carbon, and cheaper reductions could likely have been made in other sectors of the economy or by investing in R&D to reduce EV costs. Likewise, Danish taxes were unequally distributed across energy carriers and sectors. As a result, industry has had insufficient incentives to reduce energy consumption and emissions. This is unfortunate because similar decarbonisation levels could have been achieved at a lower cost if policies had been technology-neutral and all sectors and carriers had been treated equally.
- China may wish to further evaluate the distributional consequences of planned energy policies. Both the Danish and Norwegian energy transitions have had unforeseen distributional effects. Norwegian EV subsidies have favoured city dwellers who experienced

greater benefit from in-kind subsidies, such as free parking and the use of bus lanes during traffic congestion, than rural citizens. This is good from an environmental efficiency perspective as air pollution is more problematic in cities. However, the system favours richer citizens as they tend to live in cities and the policy might therefore widen inequality. Likewise, the burdens of Danish energy taxation have been unequal, with residential consumers and small and medium-sized enterprises (SMEs) paying for the transition. This has protected heavy industries but has put significant pressure on household energy bills, which hit poorer households disproportionally hard.

(2) Companies

The Danish and Norwegian cases illustrate that companies can both finance and benefit from the energy transition. National oil companies (NOCs) have been a key part of the transition, both in Norway and in Denmark. In Norway, Equinor (previously Statoil) and many international oil companies (IOCs) have helped finance the transition through their petroleum taxes. In Denmark, DONG Energy has transformed from being a conventional NOC into a largely green energy service company that develops Danish offshore wind resources. To reflect this transition, DONG, which was short for Danish Oil and Natural Gas, renamed to Ørsted in October 2017, after a Danish scientific innovator, given the company no longer operates in oil and gas. Both cases illustrate how companies, and in particular NOCs, can be the key to the transition. However, the Scandinavian cases highlight that a series of supporting conditions must be in place if the transition is to be successful. Specifically, two lessons can be learned:

- China may wish to develop credible and long-term strategies for its energy transition, as it gradually moves towards non-subsidised systems and as technology costs are reduced. The Nordic countries are characterised by strong institutions and widespread public-private partnerships, which have enabled the energy transitions of Norway and Denmark. Norwegian petroleum tax revenues, which have helped finance its energy transition, are the result of attracting investments in domestic oil and gas exploration and development. This has been possible through a petroleum tax system that is credible and shares risks between government and companies. For example, the government of Norway shares exploration risk with oil and gas companies through tax exemptions and the ability to deduct losses. Likewise, Danish government subsidies for offshore wind and its support for R&D have been long term and credible. This has allowed developers to reduce deployment costs and finance their investment through pension funds and other private investors. Both the Norwegian and Danish cases highlight the importance of credible policy that includes an element of risk sharing between the government and the private sector.

- China may wish to integrate different energy resources—such as nuclear, wind, solar and hydro—and provide public oversight of infrastructure development to reduce system integration costs. Government provision of public goods, such as system integration and pipeline infrastructure, has been key to both the Norwegian and Danish energy transitions. In Norway, the petroleum tax base rests on a publicly managed and regulated infrastructure system for oil and gas transport. Likewise, the Danish wind transition rests on public action to improve infrastructure and reduce system integration costs. High levels of offshore wind integration have only been possible due to a combination of interconnections with neighbouring electricity markets and incentives to make conventional power plants more flexible. These public initiatives have reduced system integration costs and made the Danish energy system capable of integrating significant quantities of renewables.

(3) Energy workers

The Danish and Norwegian energy transitions have had very different employment outcomes. In Norway, there were almost no green manufacturing jobs created, as the country did not have an indigenous automotive industry. In contrast, the Danish wind revolution resulted in the development of a domestic industry, which is today internationally competitive and supports a significant number of green manufacturing jobs. The Scandinavian cases present three important insights for China:

- China may wish to take advantage of early adoption to gain comparative advantage and ease the transition for its energy workers. Denmark benefitted from being an early mover in offshore wind, providing many of the initial technology advances, and therefore became home to a fast-growing industry. As a result, the transition created a significant number of green manufacturing jobs in Denmark. In contrast, Norway did not see a similar benefit from its EV transition because it did not host a domestic automobile industry. Instead, it had to import vehicles and pay high subsidies because EV costs had not yet come down when it began its transition.
- China may wish to combine its renewable energy policy with a regional development strategy to maximise the benefits for its energy workers. The energy transition has benefitted poorer rural areas of Denmark where wind power has been developed and deployed. A similar pattern is seen in other countries, such as the UK and the USA, as wind resources are often located far from conventional centres of economic activity. The spatially dependent characteristics of many renewables, such as solar and wind, can therefore be seen as an advantage and can be integrated into a wider regional development strategy.
- China may wish to reskill hydrocarbon workers and establish educational hubs that

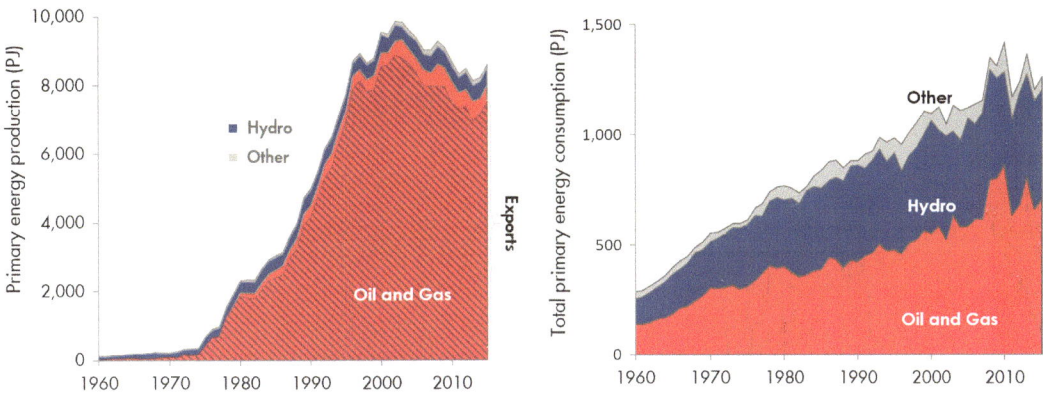

Fig. 77 Norway is one of the largest producers of hydrocarbons, yet consumes green energy and is a leader in dealing with climate change. *Source* IEA (2016)

can educate the workers required for the energy transition. The case of Denmark illustrates how workers can be reskilled and transferred from oil and gas production to the renewables industry through targeted educational programmes. A specific example is the Port of Esbjerg, which used to be the primary base for servicing Danish oil and gas production in the North Sea but is increasingly becoming a base for offshore wind operations and maintenance. This demonstrates how workers can be transferred from developing hydrocarbons to renewable energy resources.

2. Overview of supply revolutions

(1) Norway

Norway has paradoxically emerged as a world leader in climate action, despite being one of the largest producers of hydrocarbons. Oil and gas dominate domestic energy production, representing about 94% of total Norwegian energy production in 2014. However, more than 90% of this production was exported. In contrast, hydropower provides most of Norway's domestic electricity consumption, making it one of the cleanest energy systems in the world (Fig. 77).

Norway's commitment towards domestic decarbonisation has achieved remarkable reductions in transport emissions thanks to its adoption of electric vehicles. Since 2010, average CO_2 emissions per kilometre from passenger cars have fallen by 9% in Norway, while the USA has seen a small increase in its emissions. The fall in transport emissions has been driven by an unrivalled uptake of fully battery-powered electric vehicles that run on low-carbon hydroelectricity. Norway has achieved the most successful deployment in EVs globally, with a market share of around 28% of all new vehicles in 2016.

Government policies on tax exemptions and in-kind subsidies for electric vehicles have contributed to the increase in EV sales. EVs in Norway are exempted from import duties, a one-time purchase tax and 25% VAT on sale. EV users also benefit from low annual road tax, free toll-road use, free municipal parking, and access to bus lanes. These subsidies have made EVs cost-competitive with comparable internal combustion engine vehicles. However, government policies are becoming increasingly expensive and have unequally favoured rich urban citizens, for whom in-kind subsidies have had the highest value (Fig. 78).

For China, Norway's EV transition provides lessons on how to increase market penetration of electric vehicles, but warns of using coal-based electricity to power them. The key findings from Norway's EV revolution are:

- tax exemptions and in-kind subsidies can increase the uptake of EVs. However, subsidies are expensive. If China waits for EV

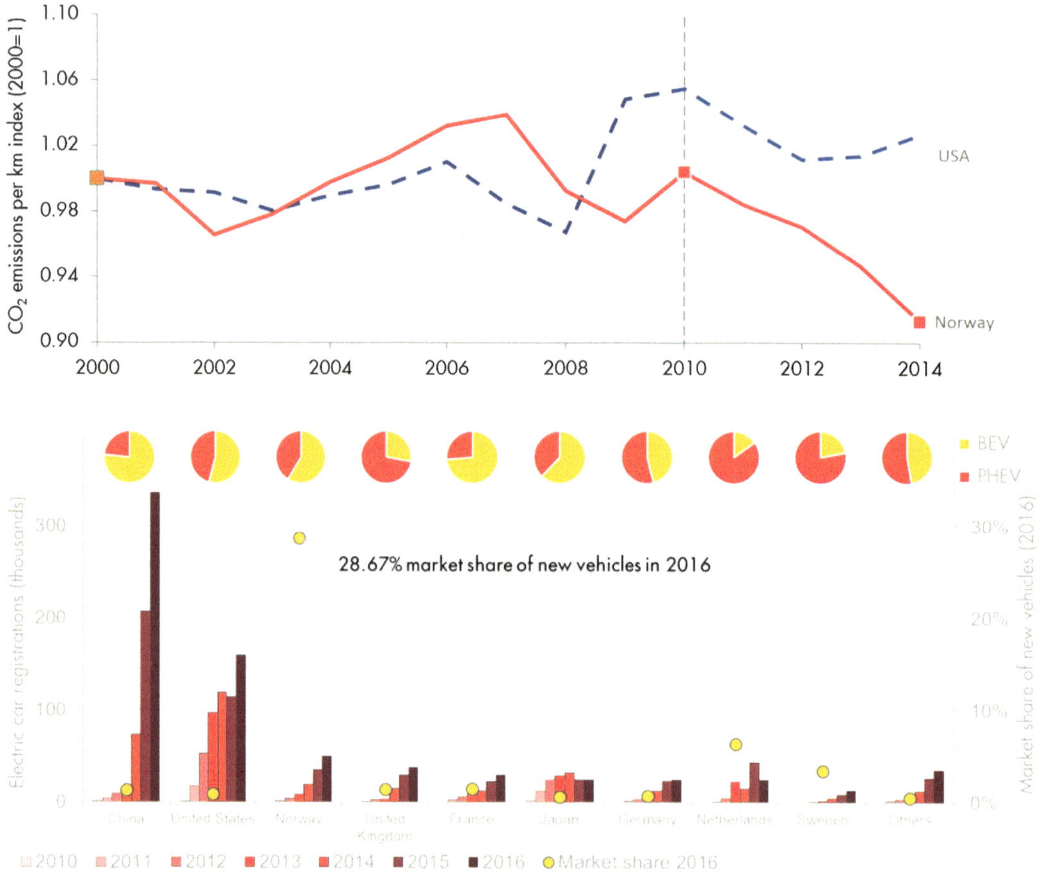

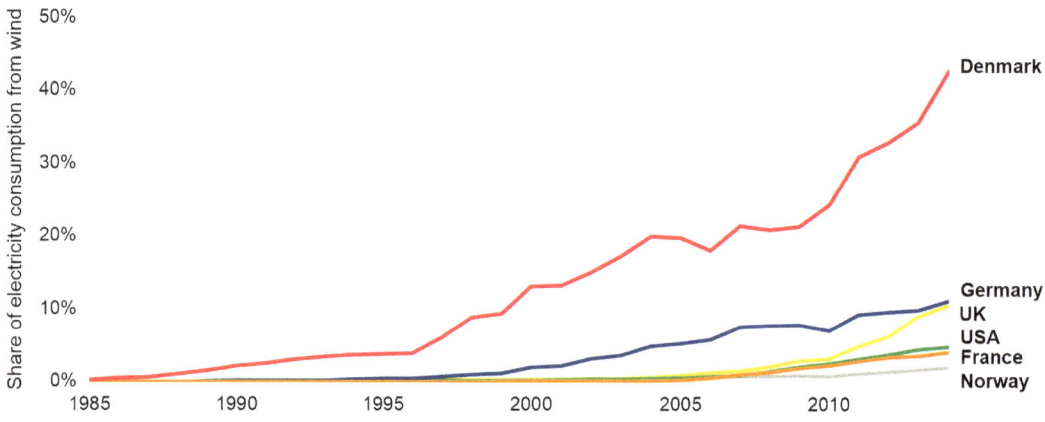

Fig. 78 The most recent example of Norway's commitment to decarbonisation is its reduction of transport emissions, driven by an unrivalled uptake of fully battery-powered electric vehicles. *Source* Vivid Economics

Fig. 79 Denmark is a world leader in integrating offshore wind into its electricity system. *Source* Vivid Economics

costs to decline, as they are expected to, subsidies to support EVs could be lower;

- EV subsidies have disproportionally favoured city dwellers but have reduced air pollution in cities. China could use similar measures to bring down air pollution in its urban areas. This could, however, raise equality concerns;

- a combination of a neutral tax system, public infrastructure and R&D support has kept investors interested in Norwegian oil and gas, revenues from which have financed the EV transition. This model of financing is now being challenged by low oil prices. China may need to consider sustainable ways of financing the energy transition; and

- Norway's EV transition has failed to translate into manufacturing jobs due to a highly underdeveloped automotive industry. China is better situated to capture the supply side benefits of a changing vehicle stock.

(2) Denmark

Denmark is a world leader in developing and integrating offshore wind into its electricity system. Wind power's share of domestic electricity consumption was more than 40% in 2015, as shown in Fig. 79. Denmark's wind power capacity has increased five times since 1997, reaching 5,227 MW in 2016.

Interconnectors with neighbouring countries have enabled high uptake of offshore wind, while maintaining energy supply security. Denmark is interconnected with Norway, Sweden and Germany, a result of the 2013 System Operation Agreement in the Nordic countries. By integrating its electricity market with those of its neighbours, Denmark can draw on a portfolio of otherwise inaccessible energy resources. Norway and Sweden have a hydro- and nuclear-based electricity system respectively, while Germany has a large electricity market that is based on a range of generation technologies. Denmark has benefitted from interconnecting with these markets because they offer supply security when the wind does not blow and potential offtake markets when excessive wind energy is generated. Interconnection has helped solve the intermittency

problem and limited the otherwise large system integration costs associated with variable wind power.

Interconnection has also reduced the need for investment in energy storage facilities. Energy storage, such as batteries and hydro-pumped storage are often considered vital to integrating non-dispatchable renewables. However, by interconnecting with neighbouring countries and, more importantly, with a range of generation technologies, Denmark has avoided the significant costs associated with building storage facilities (Fig. 80).

For China, Denmark's wind transition provides lessons on how to increase wind integration into the electricity system while fostering employment opportunities for energy workers. The key findings from Denmark's supply revolution are:

- Denmark's energy transition was financed by taxing residential consumers and small and medium-sized companies, while exempting heavy industries. This taxation system is inefficient from an environmental perspective as it does not align carbon costs, although it is politically expedient. China may wish to consider more direct carbon pricing to minimise the cost of its energy transition;

- Danish wind integration and deployment was achieved through subsidies, supported by initiatives such as interconnection and policies to increase the flexibility of conventional power plants. Similarly, China may wish to support its renewable subsidy regime by integrating its different energy generation technologies through intra-country interconnections;

- Ørsted (formerly DONG Energy), originally an NOC, has played an important role in the Danish wind transition and China's NOCs may wish to consider whether to follow its example; and

- Denmark's early entry into wind power has created a domestic industry that is globally competitive and sustains high-value jobs in rural areas. Similar employment opportunities exist in China. Energy workers could be

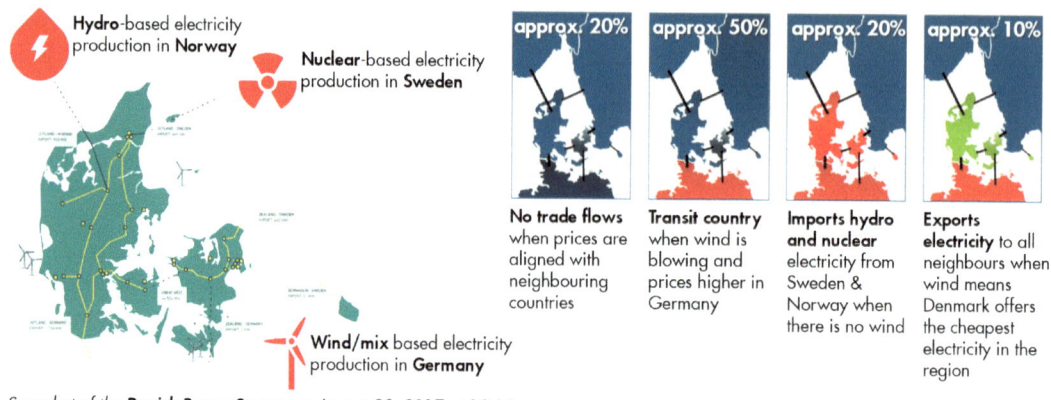

*Snapshot of the **Danish Power System** on August 29, 2017 at 19:15*

Fig. 80 Interconnectors with neighbouring countries provide energy security supply while reducing the need for investment in energy storage facilities. *Source* Vivid Economics

reskilled through targeted training programmes to avoid job losses associated with the transition away from fossil fuel extraction.

3. Consumers

(1) Summary

Consumers will participate in the energy transition if it is subsidised, as was the case in Norway. They will even pay for the transition if they are convinced of the environmental benefits, as happened in Denmark. However, each alternative suffers from design problems.

Both the Danish energy taxation system that finances the wind transition and the Norwegian EV subsidies failed to align carbon cost, which is likely to have increased the total cost of the Scandinavian transitions. Norwegian EV subsidies have been an expensive way to reduce carbon; cheaper reductions could likely have been made in other sectors of the economy or by investing in R&D to bring down EV costs. Likewise, Danish taxes were unequally distributed across energy carriers and sectors. As a result, industry has had insufficient incentives to reduce energy consumption and emissions. This is unfortunate because similar decarbonisation levels could have been achieved at lower cost if policies had been technology-neutral and all sectors and carriers had been treated equally.

Both the Danish and Norwegian energy transitions have had unforeseen distributional effects. Norwegian EV subsidies have favoured city dwellers who gained greater benefit from in-kind subsidies, such as free parking and the use of bus lanes during periods of traffic congestion, than rural citizens. This is good from an environmental efficiency perspective as air pollution is more problematic in cities. However, the system favours richer citizens as they tend to live in cities and the policy might therefore widen inequality. Likewise, the burdens of Danish energy taxation have been unequal, with residential consumers and small and medium-sized enterprises paying for the transition. This has protected heavy industries but has put significant pressure on household energy bills, which hit poorer households disproportionally hard.

(2) Norway

Generous tax exemptions and in-kind subsidies for electric vehicles have triggered a surge of demand for EV vehicles in Norway. Since the early 1990s, various initiatives in the form of purchase tax exemptions, use-tax exemptions and in-kind subsidies have been gradually introduced. Electric cars in Norway are exempt from import duties, one-time purchase tax as well as 25% VAT on sale. Additionally, EV users benefit from low annual road tax, free toll, free

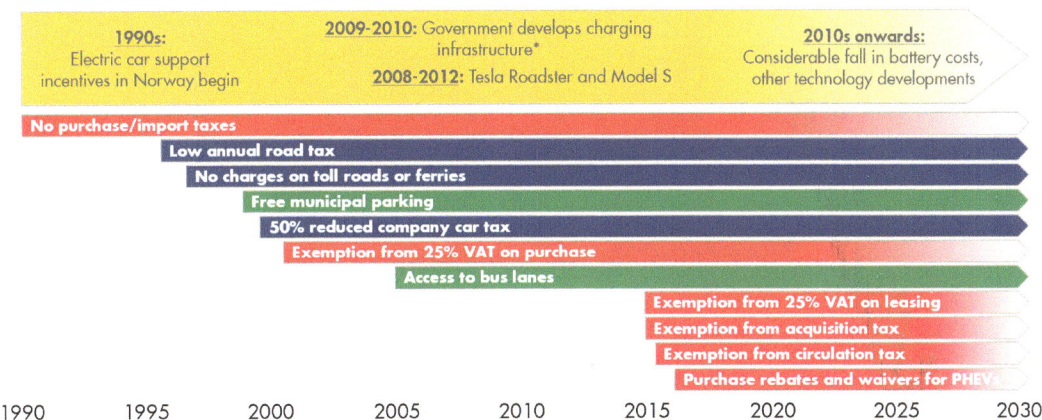

Fig. 81 Technological advances and government initiatives have made electric vehicles ever more attractive since the 1990s. *Note* Norwegian government entity Enova provides private actors and municipalities with public funding for fast-charging stations every 50 km (on average) on main roads. *Source* Vivid Economics

municipal parking and access to bus lanes. This series of policies has led the way for the uptake of electric vehicles (Fig. 81).

Technological advances along with government initiatives on EV charging infrastructure were enabling conditions for the success of EVs in Norway. The Tesla Roadster and Tesla Model S were launched in the years 2008–12, marking the emergence of consumer-friendly electric vehicles. To reduce the barriers for electric mobility, the Norwegian government simultaneously established Transnova (now part of Enova) to develop an extensive charging infrastructure. Transnova was a public funding

body which gave financial support to private actors and municipalities to build charging stations. Subsequently, about 1,800 standard charging stations were established all over the country through an earmarked allocation of around $6 million in 2009, and 70 fast-charging stations were built in 2011. This cooperation between private actors and the government established a solid support infrastructure for EVs in Norway (Fig. 82).

Generous government tax exemptions mean that the total ownership cost of electric vehicles is considerably lower than the total cost of comparable internal combustion engine vehicles.

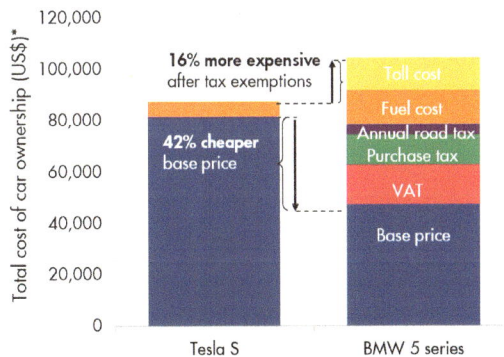
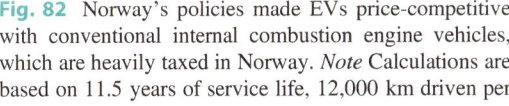
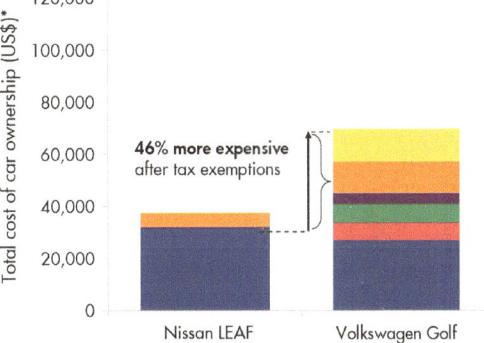

Fig. 82 Norway's policies made EVs price-competitive with conventional internal combustion engine vehicles, which are heavily taxed in Norway. *Note* Calculations are based on 11.5 years of service life, 12,000 km driven per year and five days a week access to toll roads. In-kind subsidies such as free parking and access to bus lanes are not included in the cost calculations. *Source* Vivid Economics

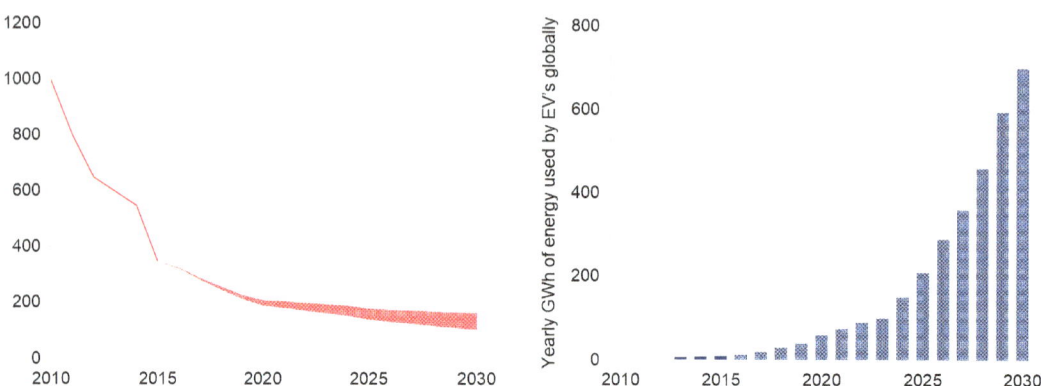

Fig. 83 Falling battery costs reduce the need for subsidies, which increases demand as batteries make up one-third of EV costs. *Source* Vivid Economics

Cars are traditionally taxed heavily in Norway and tax exemptions therefore work as a significant subsidy to EV consumers. For example, when the total ownership cost of a Tesla S is compared with a BMW 5 Series, which has a 42% cheaper base price, the BMW turns out to be 16% more expensive when taxes are included. In addition, fuel costs over the lifetime of a BMW 5 are more than double those of a Tesla S. The same pattern is seen with smaller EVs where ownership cost differentials are even higher. For example, an otherwise cheaper Volkswagen Golf becomes 46% more expensive than a Nissan LEAF when taxes are included (Fig. 83).

The cost of subsidising EVs in Norway has been high, but is expected to fall with technological developments in EV batteries. Battery costs currently make up a third of the cost of an EV. However, new developments in battery technology are expected to make them dramatically cheaper. Falling costs will not only reflect improvements in battery chemistry and in manufacturing processes, but also in economies of scale as the industry grows. These developments have the potential to make non-subsidised electric vehicles cost-competitive with their fuel counterparts, reducing the need for and cost of subsidies. The need for Norwegian EV subsidies might therefore be smaller in the future. More importantly, Norway could potentially have achieved the transition at a lower cost if it had waited for battery costs to come down or, if

instead of subsidising vehicle ownership, it had invested the money in R&D to reduce EV production costs (Fig. 84).

Uptake of EVs in Norway has reduced carbon emissions because of a low-carbon, hydropower-based electricity system, but this will not necessarily be the case in China. The source of the electricity used to power electric vehicles is an important concern when evaluating net emissions from transport. An electric vehicle in Norway, say a Tesla S, emits almost zero emissions as the electricity it uses comes from Norwegian low-carbon, hydropower-based generation. However, net emissions from the same Tesla S in China could potentially be 16% higher than a comparable combustion vehicle, due to the largely coal-based electricity system in China. Transforming the vehicle stock alone will therefore not help, but rather add to, the carbon emissions of the Chinese transport sector. To reduce carbon emissions, China must also decarbonise its electricity generation. That said, electrifying transport in China might still have local environmental benefits, such as reduced particle pollution in cities (Fig. 85).

Despite their success in motivating the uptake of EVs, Norwegian policies have unequally favoured urban citizens, spurring higher than average uptake in cities. Policies such as free municipal parking, toll-road charge exemptions and access to bus lanes are more practical and valuable in urban areas due to higher congestion.

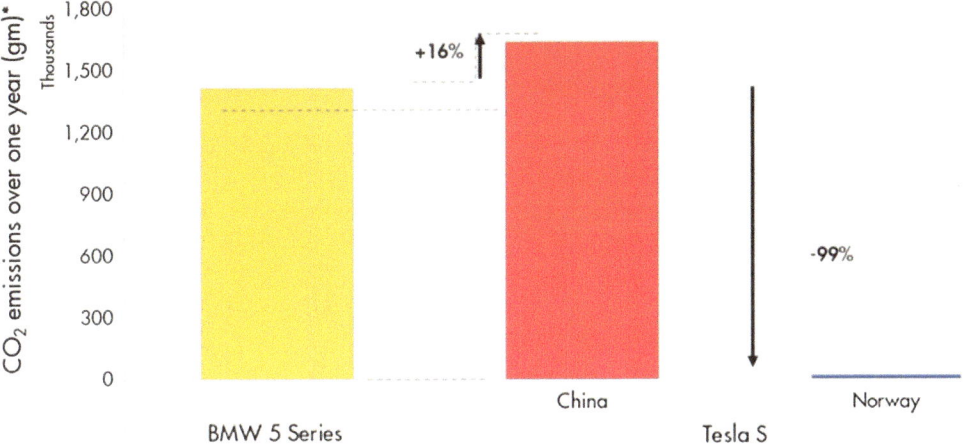

Fig. 84 Uptake of EVs has reduced carbon emissions because Norway has a low-carbon, hydropower-based electricity system

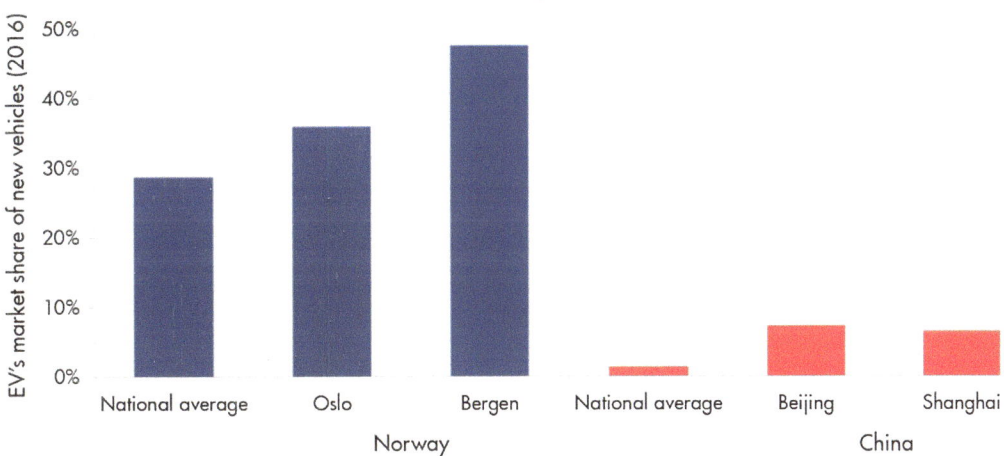

Fig. 85 Norwegian policies have unequally favoured cities, where EVs are more practical and in-kind subsidies are more valuable. *Source* IEA (2017)

As a result, cities like Oslo and Bergen have seen a higher uptake of EVs than the national average. These cities are also richer than the national average, and Norway's EV policy has therefore been regressive from a social perspective. That said, the health benefits of EVs are also larger in cities where air pollution problems are more prevalent. As China suffers from serious air pollution and congestion on its urban roads, policymakers could be inspired by the Norwegian policies to reduce existing problems. For example, in-kind subsidies such as access to bus

lanes or free parking could be effective in increasing demand for EVs in major Chinese cities such as Shanghai and Beijing.

(3) Denmark

Denmark's move into offshore wind was initiated by public concerns about energy security, specifically the desire to reduce dependence on imported hydrocarbons following the 1973 oil crisis. As a response, the Danish government began to implement high energy taxes in order to reduce oil dependence and finance public

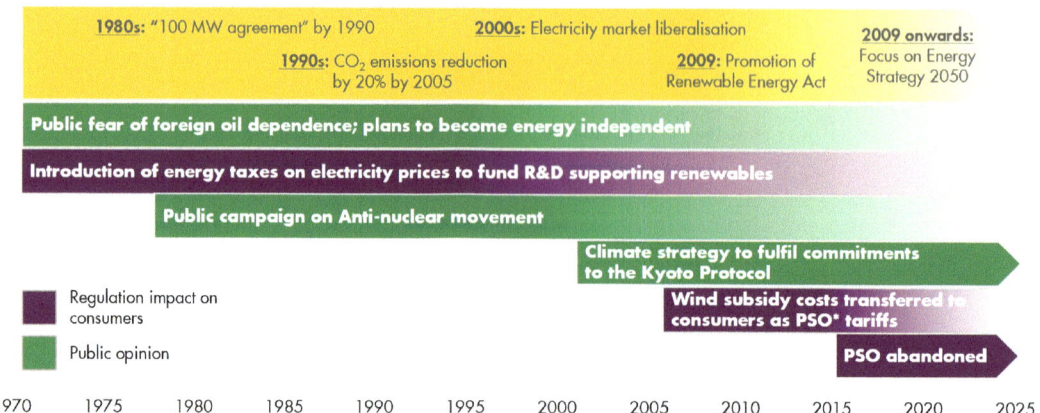

Fig. 86 Denmark's wind transition has been financed by taxing energy, but is supported by security concerns and a pro-climate movement. *Note* Public service obligation (PSO) tariff is a levy on all electricity consumption, revenues from which finance the subsidies paid to wind, combined heat and power plants and R&D in the renewable energy sector. *Source* Vivid Economics

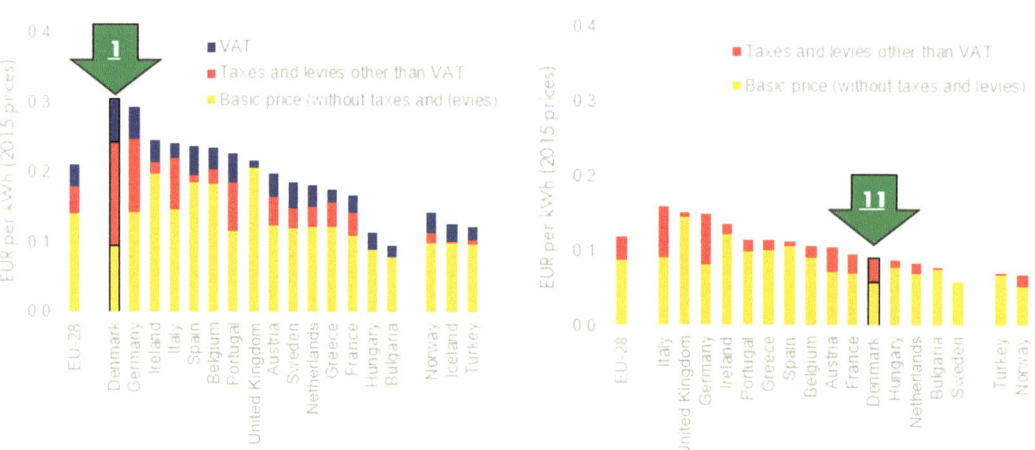

Fig. 87 The tax burden has been uneven, with residential consumers and small and medium-sized businesses paying a higher price than industrial consumers. *Source* Vivid Economics

research into alternative energy sources. In contrast to neighbouring Sweden, Denmark quickly abandoned nuclear power as a part of its future energy system. This was largely because a strong anti-nuclear movement fostered public concern about the safety of nuclear power. Due to public sentiment, the Danish government decided to exclude nuclear and include wind in its future energy planning (Fig. 86).

Denmark's wind transition has been financed by taxing energy consumers. In addition to existing high energy taxes, the public service obligation

(PSO) tariff was introduced in 2005, which is a levy on all electricity consumption. It is collected by an independent enterprise owned by the Danish Ministry of Climate, Energy and Building called Energinet. The revenues from PSO have been used to finance the energy transition, including development of renewable energy, support for decentralised combined heat and power plants, energy efficiency, R&D, and other expenses related to energy security (Fig. 87).

However, the burden sharing has been uneven, with residential consumers and small

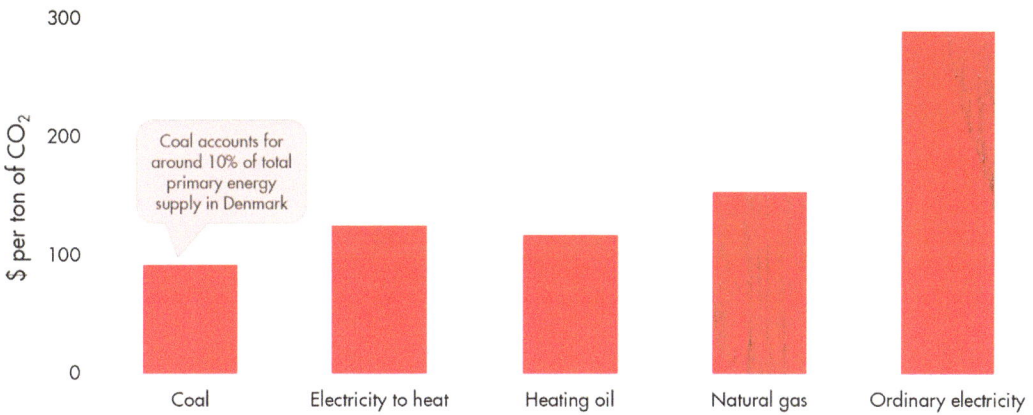

Fig. 88 Energy taxes do not align with carbon costs across sectors and energy carriers. *Source* Vivid Economics

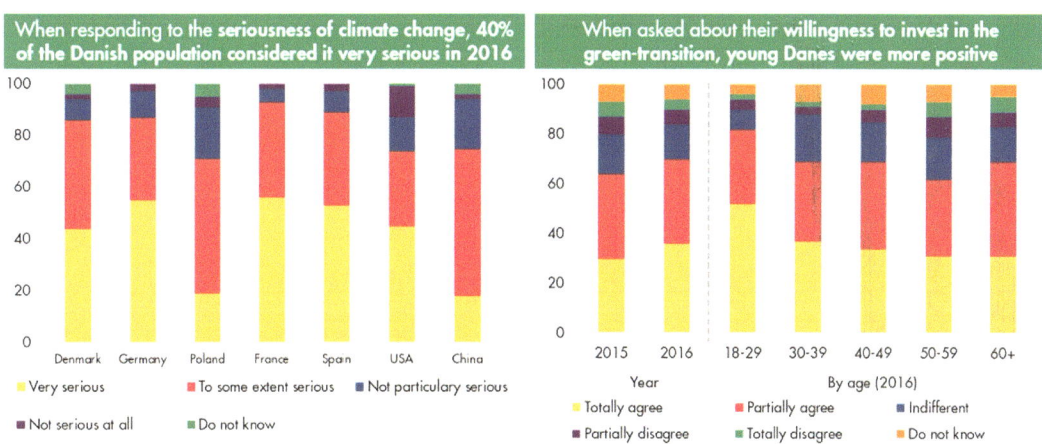

Fig. 89 Public willingness to participate in the transition remains high, despite expensive and inefficient energy taxes. *Source* Vivid Economics

and medium-sized enterprises (SMEs) paying a much higher tax than industrial consumers. Residential and SMEs in Sweden pay the highest price for electricity in the EU. The cost of electricity represents only around 30% of the overall price and energy taxes are almost three times the European average. In contrast, industrial consumers are not subject to the same high energy taxes. This is primarily because the government has sought to ensure global competitiveness for energy-intensive industries (Fig. 88).

The result is an inefficient tax system which fails to align carbon costs across sectors and energy carriers. Electricity is taxed by far the most out of all energy carriers, while natural gas

and coal are taxed less heavily. The result is that energy taxes do not align with the carbon content of the carrier and the tax system fails to satisfy the polluter pays principle. Furthermore, as the least taxed fuels are mainly used for high heat processes, industries are insufficiently incentivised to increase their energy efficiency. The exclusion of key emission contributors from the energy transition has likely meant that Denmark's energy transition has been more expensive than necessary (Fig. 89).

Despite expensive and inefficient energy taxes, public willingness to participate in the transition has remained high. A 2016 survey conducted by the green think tank CONCITO found that the Danes consider climate change a more serious issue than in many comparable countries. More importantly, a significant majority of Danes say they are willing to pay for the transition to a cleaner energy system. This sentiment is gaining momentum, as seen by the rise in willingness to pay between 2015 and 2016 in Fig. 85. Finally, it should be noted that younger people were more positive about contributing to the country's energy transition than older citizens.

4. Companies

(1) Summary

The Nordic cases illustrate that companies can both finance and benefit from the energy transition. National oil companies (NOCs) have been a key part of the transition in both Norway and Denmark. In Norway, Equinor and several international oil companies (IOCs) have helped finance the transition through their petroleum taxes. In Denmark, DONG Energy has transformed from being a conventional NOC into a largely green energy service company that develops offshore wind resources under the new name of Ørsted. Both cases illustrate how companies, especially NOCs, can be key to the energy transition. However, the Danish and Norwegian cases also highlight that a series of framework conditions must be in place if the transition is to be successful.

Both Norway and Denmark illustrate the importance of credible policies that include an element of risk sharing. The Nordic countries are characterised by strong institutions and widespread public-private partnerships, which have enabled the energy transitions of both Norway and Denmark. This is illustrated by the Norwegian petroleum tax system which, by incorporating an element of public-private risk sharing, has helped attract investment in oil and gas exploration and development. For example, the government of Norway shares exploration risk with oil and gas companies through tax exemptions and loss deductions. Likewise, the Danish government's subsidies for offshore wind and

support for R&D have been long term and credible. This has allowed developers to reduce deployment costs and finance their investments through pension funds and other private investors.

Government provision of public goods, such as system integration and pipeline infrastructure, has been key to both the Norwegian and Danish energy transitions. In Norway, the petroleum tax base rests on a publicly managed and regulated infrastructure system for oil and gas transport. Without this infrastructure, companies would be less inclined to invest in Norwegian oil and gas exploration and there would be less of a tax base to finance the EV transition. Likewise, the Danish wind transition rests on public action to improve infrastructure and reduce system integration costs. High levels of offshore wind integration have only been possible due to a combination of interconnections with neighbouring electricity markets and incentives to make conventional power plants more flexible. Both public initiatives have reduced system integration costs and made the Danish energy system capable of integrating significant quantities of renewables (Fig. 90).

(2) Norway

Government EV subsidies have resulted in falling tax revenues from vehicles in Norway. Vehicle taxes include a one-time purchase tax, registration fee, annual road tax, and fuel and CO_2 taxes, which represent on average 12% of all taxes in Norway. Since the inception of EV subsidies, these revenues have been decreasing as the sale of conventional vehicles declines. This trend will continue and the success of the EV policy is therefore threatening to eliminate vehicle tax revenues, which is an important source of government income in Norway. For example, if all new vehicles are zero emission by 2025 and current subsidies are continued, vehicle tax revenues could be halved. EV subsidies are thus becoming increasingly expensive, which puts pressure on government revenues (Fig. 91).

Electrification of transport has so far been financed by revenues from Norway's oil and gas

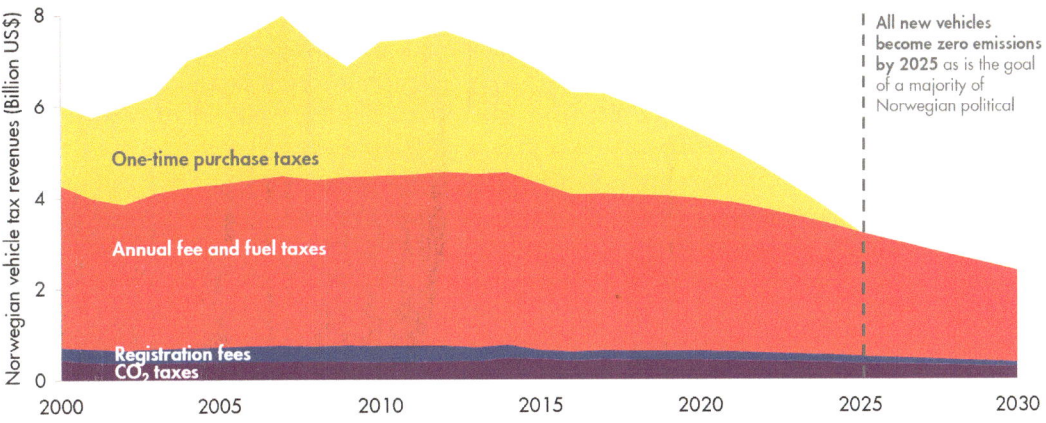

Fig. 90 Car tax revenues could fall by up to 50% by 2025 if all new vehicles are zero emission and current subsidies are continued. *Source* Vivid Economics

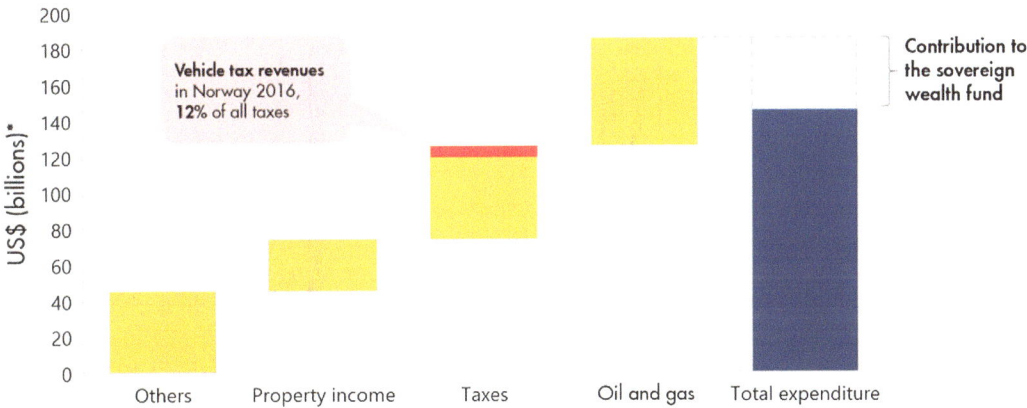

Fig. 91 Oil and gas revenues have enabled high levels of government spending and reduced dependence on taxes

production, reducing the need to raise taxes and reform EV policies. Around 30% of government revenues come from oil and gas, which contribute to a significant government surplus. The Norwegian sovereign wealth fund, also known as the oil fund, was established in 1990 and invests the Norwegian state's petroleum revenues. In September 2017, it was worth $192,307 per Norwegian citizen. Norway is thus uniquely rich in natural resources and this has helped finance the decarbonisation of its transport sector without taxing consumers.

The government has developed the petroleum tax base by establishing a sophisticated tax system which reduces entry barriers and shares risks between government and the private sector to encourage oil and gas exploration. The petroleum taxation system is intended to be neutral, under which only a company's net profits are taxable and losses may be carried forward with interest. Furthermore, a reimbursement system for exploration costs is offered: if a company incurs losses, it has the option to request an immediate refund of the tax value of exploration costs or carry the losses forward to future years. The taxation system is also flexible as it allows for consolidation between fields. This means the exploration costs can be written off against income from operations elsewhere on the Norwegian continental shelf.

Government support for oil and gas transport systems provides cost-effective infrastructure for oil and gas investors. The Norwegian gas transport system is highly regulated to avoid high tariffs associated with natural monopolies. Gassled, a joint venture, owns most of the gas infrastructure, while Gassco, a 100% state-owned neutral and independent operator, ensures equal access for all. Gassco's duties include administering system capacity, coordinating and managing gas streams, and running the infrastructure in accordance with regulations under the Petroleum Act. Oil transport infrastructure is less regulated, as it makes up a smaller part of the value chain in the oil industry. The infrastructure is independently owned, where owners and users negotiate agreements on access to pipelines governed by regulations.

In addition, R&D support has been vital to the competitiveness and innovation of the Norwegian petroleum industry. The Ministry of Petroleum and Energy established Oil and Gas in the 21st Century (OG21) in 2001, which brought together oil companies, research institutions and suppliers to agree on a joint national strategy for the country's oil and gas sector. The government encourages R&D through legislation or direct allocations to the Research Council of Norway, which funds the PETROMAKS 2 and DEMO 2000 research programmes. PETROMAKS 2 promotes long-term research and competence-building, while DEMO 2000 supports pilot and demonstration projects in the industry.

Resource rents from Equinor, in which the Norwegian state has a 67% holding, and IOCs contribute to the sovereign wealth fund. Equinor (then Statoil) was established in 1972, with the state as the sole owner, but was partially privatised in 2001. The Norwegian government now receives state's direct financial interest (SDFI) from Equinor and from all other offshore operators. The government covers its share of costs and investments in the Norwegian continental shelf and receives a corresponding share of income as SDFI. The state also receives taxes from more than 50 international companies involved in exploration, production and infrastructure off the Norwegian coast (Fig. 92).

The combination of a sophisticated tax system, well-functioning infrastructure and R&D support makes Norway attractive for oil and gas production, which provides the tax revenues used to finance the energy transition. However, Norway is exceptionally resource-rich and this system of financing is unlikely to work in other countries (Fig. 93).

(3) Denmark

Government grants and subsidies have given wind energy producers the financial support they needed to develop and integrate wind into the electricity system. In the early years of wind development, the Danish government provided wind energy producers with capital grants, up to 30% of their installation costs. The grants were progressively phased out as wind installations became cost-effective. In the 1990s, several measures were initiated to support wind projects for the first five years of their operations. These included a fixed feed-in tariff, where the price paid for the electricity generated from wind was set at 85% of the utility's production and distribution costs. Wind projects also received refunds from carbon and energy taxes. More recently, under the Promotion of Renewable Energy Act 2009, wind producers receive an environmental premium added to the market price, along with an additional compensation for balancing costs. Falling development costs mean that wind energy is now close to becoming cost-competitive with conventional fossil-fuel plants without subsidies. The Danish government therefore hopes to phase out wind energy subsidies soon (Fig. 94).

Government support has led Danish firms to become world leaders in the manufacture and deployment of offshore wind turbines. Although only 10% of Europe's total wind capacity is installed in Denmark, Danish companies manufacture—either wholly or partly,—more than 90% of wind turbines deployed in Europe. This demonstrates the strength and global competitiveness of the Danish turbine manufacturing industry. In addition, Danish companies are also

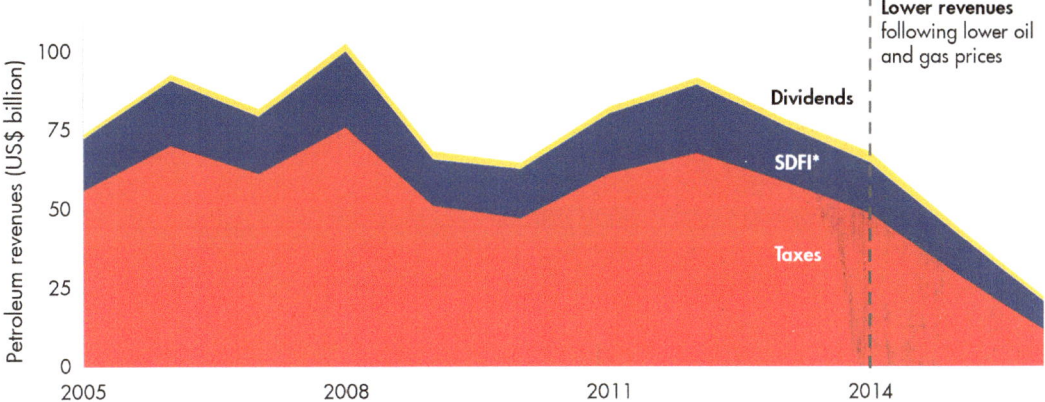

Fig. 92 Resource rents have been collected mostly by taxing Equinor and IOCs. *Note* State's Direct Financial Interest (SDFI) is the Norwegian government's directly owned exploration and production licenses for petroleum and natural gas on the Norwegian continental shelf. It includes pipelines and land facilities. All revenue from SDFI is transferred to the sovereign wealth fund. *Source* Vivid Economics

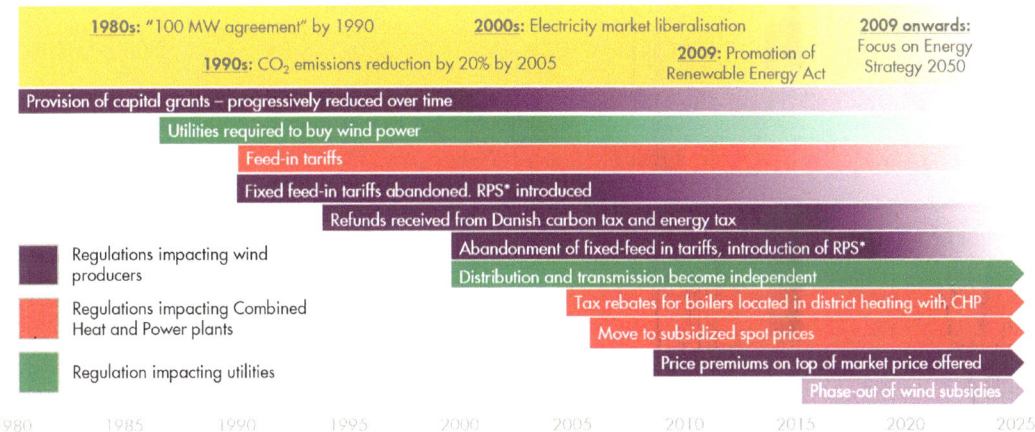

Fig. 93 Government grants and subsidies have given wind energy producers the financial support they needed to flourish.   *Note* Renewable portfolio standard (RPS) includes CO_2 quotas, tradable emission allowances and renewable energy certificates for a green electricity market. *Source* Vivid Economics

big owners of wind farms, with DONG Energy owning 16% of total capacity in Europe. Denmark's early mover advantage in offshore wind is still seen in current deployment. For example, almost 20% of offshore wind installations in Europe were developed by DONG Energy in 2016. These statistics show that Danish companies continue to play a leading role in the global offshore wind industry (Fig. 95).

DONG Energy is an excellent example of how a conventional national oil company can be transformed into a renewable energy services company. In 2005, DONG Energy committed itself to supplying energy that was green, smart and sustainable. To achieve its transformation from a conventional power plant operator into a green energy leader, DONG signed a pivotal agreement in 2010 to deliver 500 offshore wind turbines with a total capacity of 1,800 MW, in collaboration with wind turbine manufacturer Siemens. As a result, DONG has managed to achieve more than a 50% reduction in its emissions and to more than double its renewable generation since 2006. In addition, DONG's

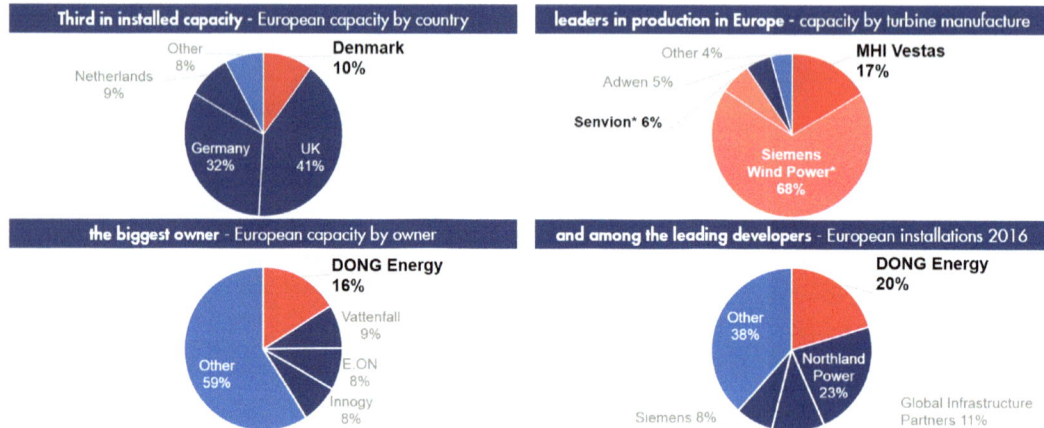

Fig. 94 Government subsidies have supported Danish companies to become leaders in the manufacture and deployment of offshore wind. *Note* Siemens Wind Power was formally Danish company Danregn, and Senvion was formally Danish company Jacobs (both acquired in 2004). The figures in the pie charts are for 2016. *Source* Vivid Economics

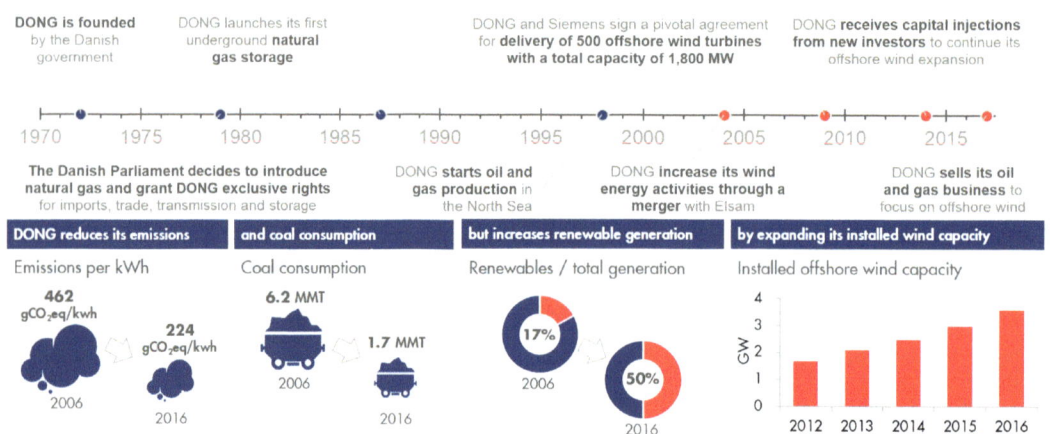

Fig. 95 DONG Energy illustrates how a conventional national oil company can be transformed into a renewable energy services company. *Note* DONG Energy changed its name to Ørsted in 2017 to signal its transformation from a black to a green energy company. DONG originally stood for Danish Oil and Natural Gas. *Source* Vivid Economics

average return on capital (ROC) has been higher than the average ROC for oil and gas companies over the past 10 years. The transition away from conventional hydrocarbon production was concluded in 2016 when DONG Energy sold its remaining North Sea oil and gas business. The company renamed itself Ørsted in October 2017 (Fig. 96).

Denmark's geographic location has been key to DONG's success in offshore wind. Consisting of the peninsula of Jutland and 443 relatively small islands, Denmark has a relatively high population density, which prevents the large-scale deployment of onshore wind. However, the country has large areas of sea territory with comparatively shallow waters and

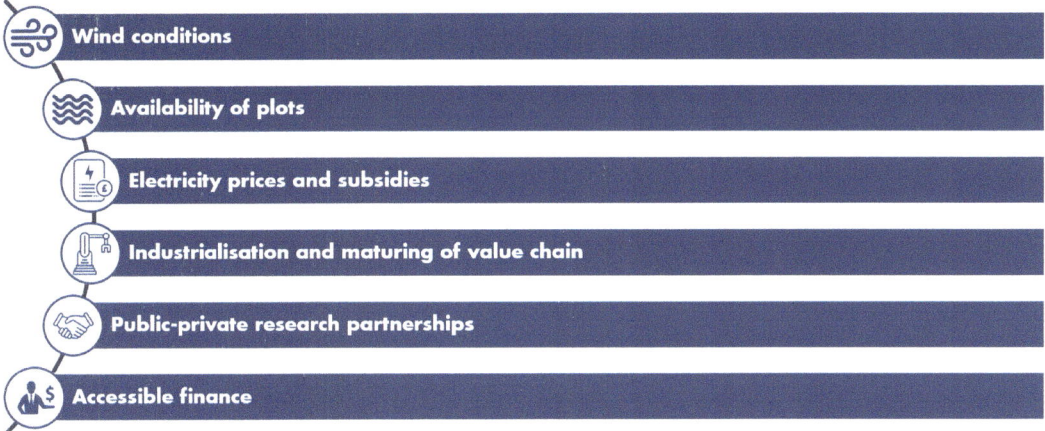

Fig. 96 Government support, together with favourable geographic and investment conditions, enabled DONG's transformation. *Source* Vivid Economics

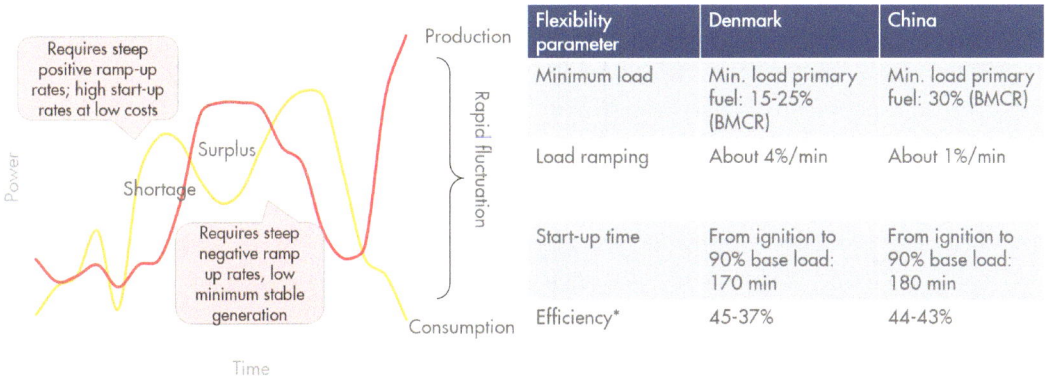

Fig. 97 Wind power growth is enabled by utilities enhancing the operational flexibility of conventional power plants. *Note* BMCR=boiler maximum continuous rating. *Source* Vivid Economics

significant offshore wind resources with high wind speeds of 8.5–9.0 metres per second at 50 metres height. Denmark therefore offered the ideal testing ground for deploying offshore wind, and DONG Energy utilised this.

Government support together with favourable investment conditions provided DONG Energy with the financial resources needed for its transformation. In addition to government subsidies, the significant volume of government tenders and large test facilities reduced offshore wind deployment costs. DONG also collaborated closely with publicly financed research institutions to ensure certification, testing and standards, which later contributed to its international competitiveness. Finally, financially stable returns and credible policies have made offshore wind energy as attractive as some financial bonds. This has allowed DONG to convince institutional investors, such as pension funds, to help finance their wind investments at relatively low cost (Fig. 97).

Enhanced operational flexibility of conventional power plants has helped reduce the cost of integrating non-dispatchable wind into the electricity system. The challenge to variable renewable electricity production is its unpredictable and inconsistent nature. To address the issue of load fluctuations, Denmark has focused on enhancing the operational flexibility of its

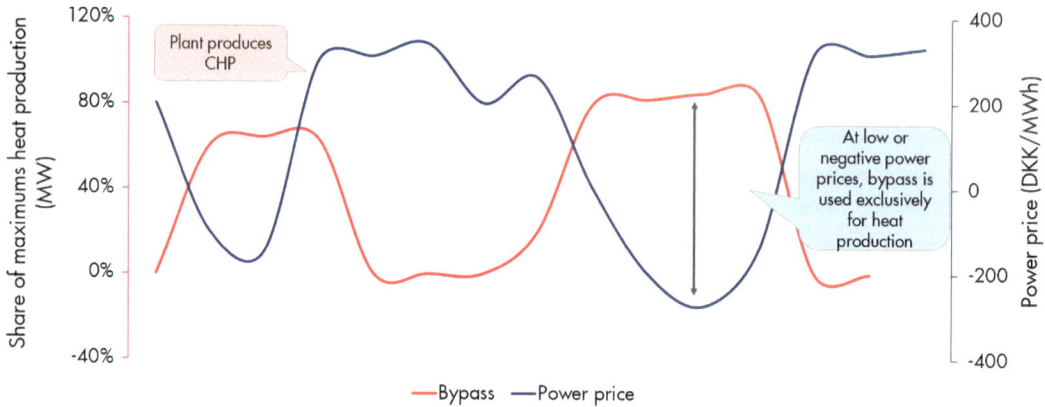

Fig. 98 CHP plants are incentivised to produce only heat when electricity prices are low, providing greater flexibility for wind integration. *Note* Large Danish CHP plants can let steam bypass the turbines to produce only heat. *Source* Vivid Economics

conventional power plants for the past 15 years or more. As a result, Danish coal-fired power plants have been optimised to have shorter start-up rates, lower minimum loads and steeper ramp-up rates than similar installations in other countries.

To further address the intermittency problem, combined heat and power plants (CHP) are incentivised to produce only heat when the wind is blowing and electricity prices are low. In an attempt to increase wind feed-in during hours of greater wind power generation, regulation around CHP plants was changed in 2005 to offer discounts on boilers in CHP facilities. This has incentivised CHP plants to produce only heat at times of high wind power generation, which in turn reduces system integration costs (Fig. 98).

5. Energy workers

(1) Summary

The Danish and Norwegian energy transitions had very different employment outcomes. In Norway, there were almost no green manufacturing jobs created as the country did not have a domestic automotive industry. In contrast, the Danish wind revolution resulted in the development of a domestic industry which is today internationally competitive and supports a significant number of green manufacturing jobs.

The energy transition has benefitted poorer rural areas of Denmark where wind energy is developed. A similar pattern is seen in other countries, such as the UK and the USA, where wind resources are often far from conventional centres of economic activity.

The spatially dependent characteristics of many renewables, such as solar and wind, can thus be seen as an advantage and can be integrated into a wider regional development strategy.

The case of Denmark illustrates how workers can be reskilled and transferred from oil and gas production to the renewables industry through targeted educational programmes. One example is the Port of Esbjerg, which used to be the primary base for servicing Danish oil and gas production in the North Sea. Today, it is a base for offshore wind operations and maintenance. This demonstrates how workers can be transferred from developing hydrocarbons to renewable energy resources.

(2) Norway

Rising EV demand has failed to translate into Norwegian manufacturing jobs. Almost all electric vehicles in Norway are imported from the USA, China and other EV manufacturing countries due to a highly underdeveloped domestic automotive industry. Almost 43% of EV production worldwide was produced by Chinese manufacturers in 2016. China has the potential to be a leader in the supply of electric vehicles. In contrast to Norway, China can create significant green jobs and opportunities for its energy workers by transforming the existing capital stock in the transport sector (Figs. 99 and 100).

The lack of green manufacturing jobs in Norway is problematic because the recent fall in oil prices has resulted in declining oil and gas employment in Norway. Historically, employment in the oil and gas sector was increasing. However, global markets and oil price trends determine employment in the Norwegian oil and gas industry. The recent fall in oil prices has led to job losses in both service and operations in the Norwegian oil and gas sector. This trend is likely to continue in the future as global climate policies and initiatives, such as the Paris Agreement, reduce demand for oil and gas. The lack of green manufacturing jobs might therefore prove to be a long-term challenge for the Norwegian economy.

(3) Denmark

Denmark's early adoption of offshore wind has made its domestic industry globally competitive and created a green manufacturing sector. Denmark's wind-related products and service exports account for around 7% of its total exports. This is by far the largest share of wind exports to total exports among the 28 European Union countries.

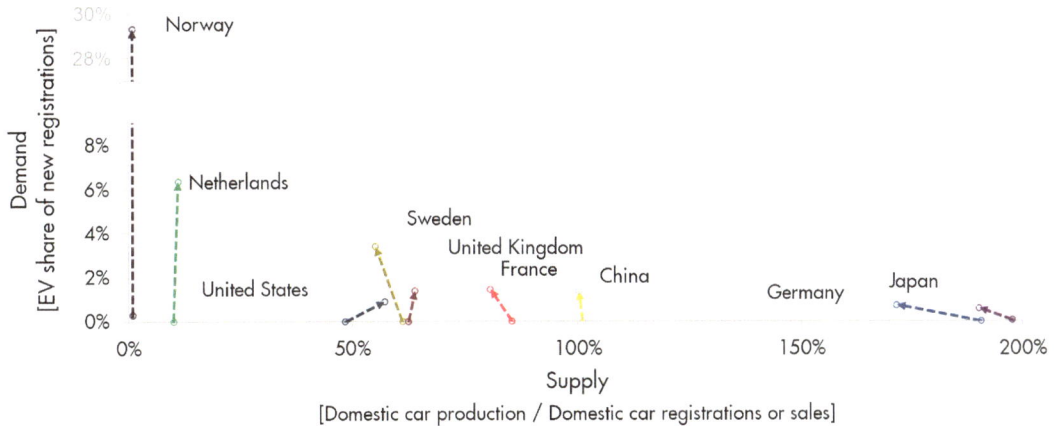

Fig. 99 eV demand has not translated into Norwegian manufacturing jobs, but China already has an automotive industry to take advantage of. *Source* Vivid Economics

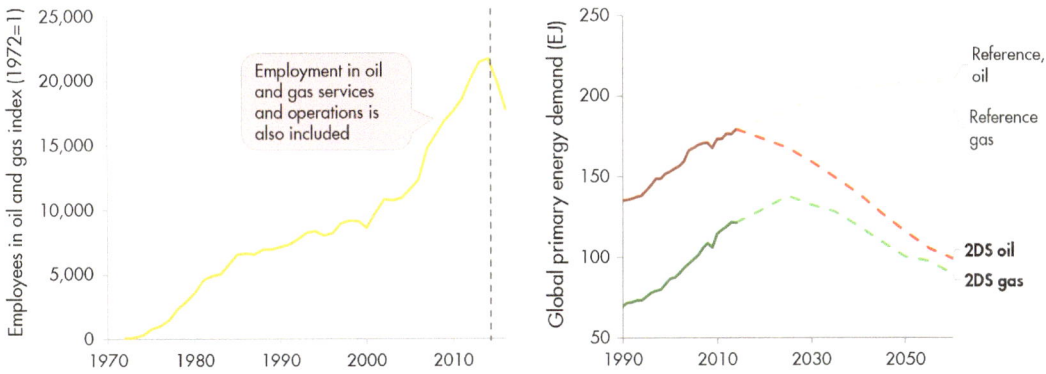

Fig. 100 Oil and gas employment is falling and is expected to continue to do so. *Note* 2DS=IEA's 2°C Scenario. *Source* Vivid Economics

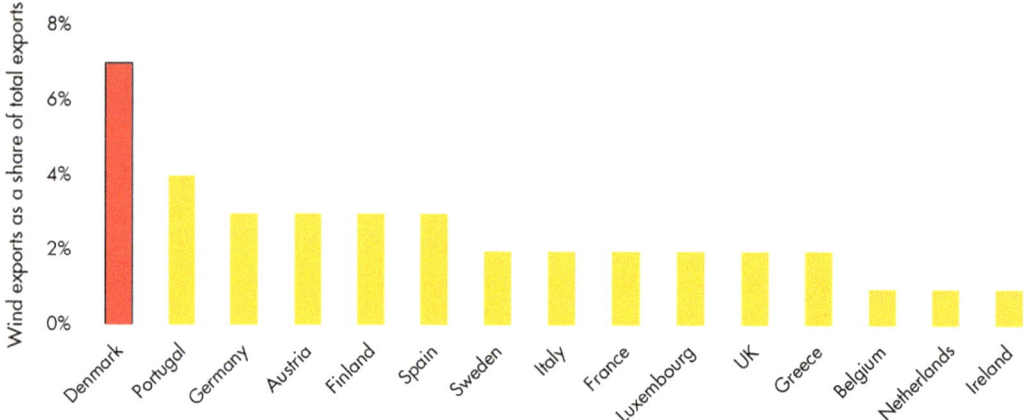

Fig. 101 Denmark's early adoption of offshore wind has made its domestic industry globally competitive, creating more jobs. *Source* Vivid Economics

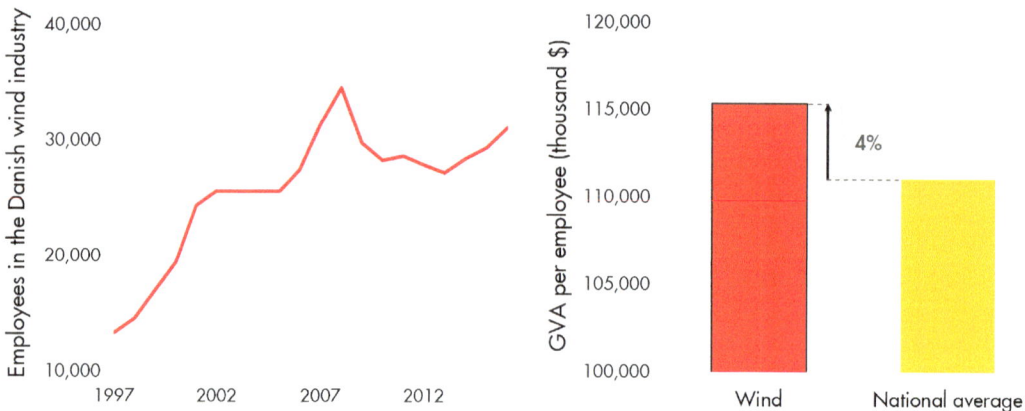

Fig. 102 The development of wind energy has created sustainable and high-value jobs in Denmark. *Note* GVA=gross value added. *Source* Vivid Economics

The success is attributed to Denmark's early adoption of wind technology, which gives it a competitive advantage globally.

Development of wind power has created sustainable and high-value jobs in Denmark. Employment in the Danish wind industry has been rising significantly since the 1990s. These jobs are not only green but also high value in terms of productivity. In 2016, the gross value added (GVA) per employee in the wind industry was higher than the national average by 4 percentage points. Notably, the national average for GVA per employee is already high in Denmark (Figs. 101, 102, and 103).

Most wind jobs are concentrated in rural areas and have been supported by tailored education programmes. 22% of the Danish population lives in the capital, Copenhagen. However, most wind

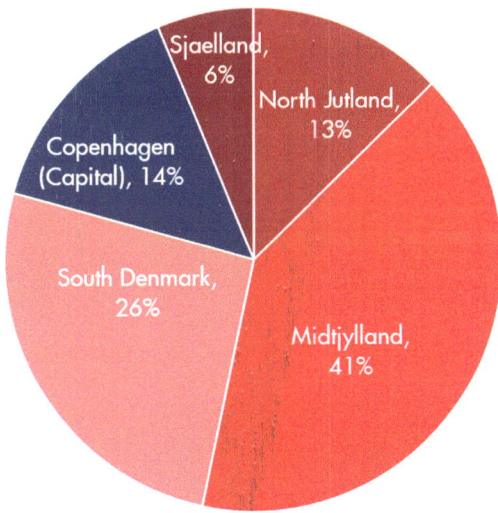

jobs are concentrated outside this economic hub. The development of offshore wind has provided new jobs and new sources of income for otherwise less affluent rural areas. These new job opportunities have been complemented with programmes for wind energy education to provide the training required for such jobs. The Danish Wind Power Academy, located in southern Denmark, provides customised courses for owners and operators of wind turbines, while the Technical University of Denmark near Copenhagen offers master's programmes on wind technology engineering (Fig. 104).

The Port of Esbjerg has been the cornerstone for the development of Denmark's offshore industry. Esbjerg has been the primary base for oil and gas activity in the Danish North Sea since extraction began in the 1970s. Its services for oil

Fig. 103 Wind jobs are concentrated in less-affluent rural areas. *Source* Vivid Economics

Fig. 104 The Port of Esbjerg has been dynamic in exploring synergies between conventional and renewable energy systems

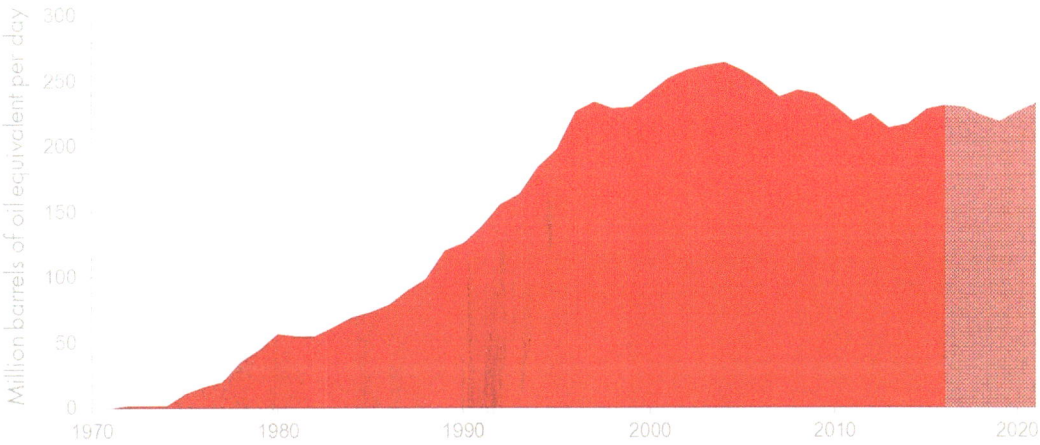

Fig. 105 Norway's oil and gas production is projected to be stable until 2020; longer-term forecasts are conditional on climate change. *Source* Vivid Economics

and gas include operations and maintenance of existing platforms, establishing new fields, security training and decommissioning. As energy production has shifted away from oil and gas, the port has increasingly transformed itself to service the offshore wind industry. Its services for offshore wind include pre-assembly, installation and maintenance of wind turbines. The port is an excellent example of how to exploit synergies between conventional and renewable energy services by reskilling energy workers.

Additional figures

See Figs. 105, 106, 107 and 108.

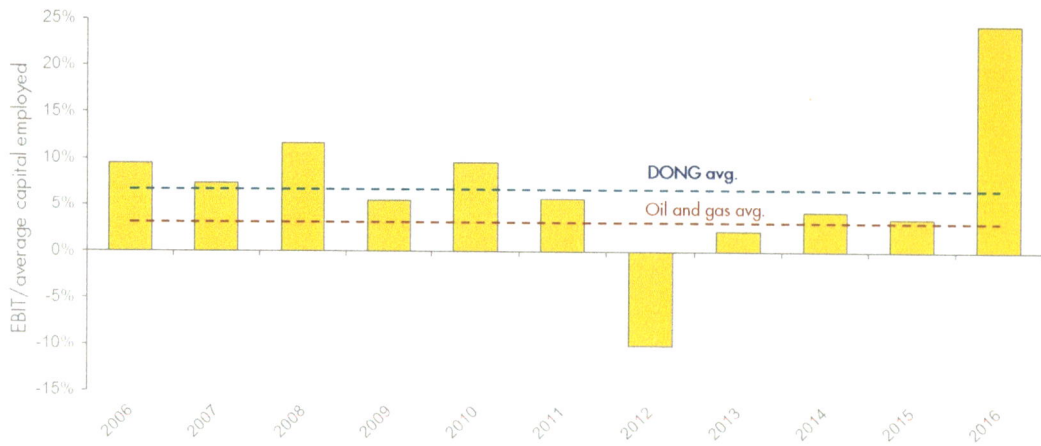

Fig. 106 DONG Energy's average return on capital (ROC) has been higher than the average ROC in oil and gas over the past 10 years. *Source* Vivid Economics

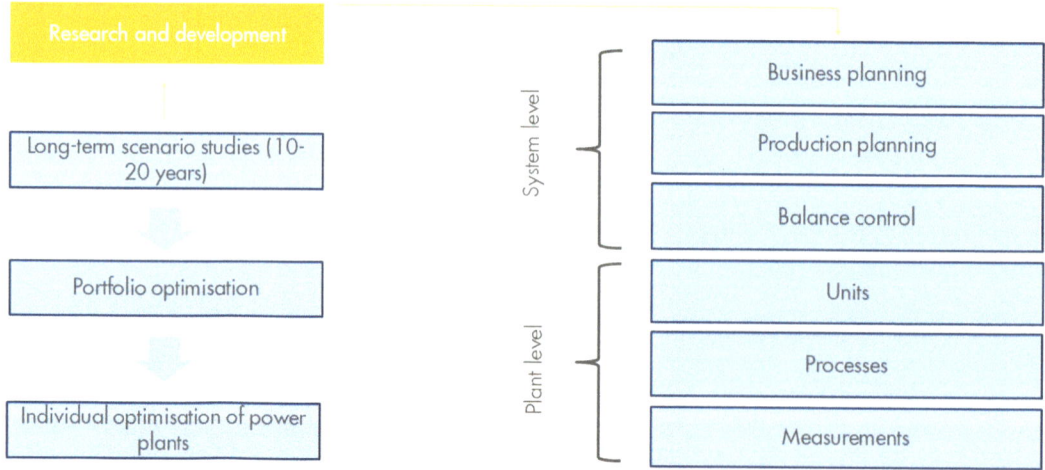

Fig. 107 Improving the operational flexibility of conventional power plants has been in focus for 20 years in Denmark. *Source* Vivid Economics

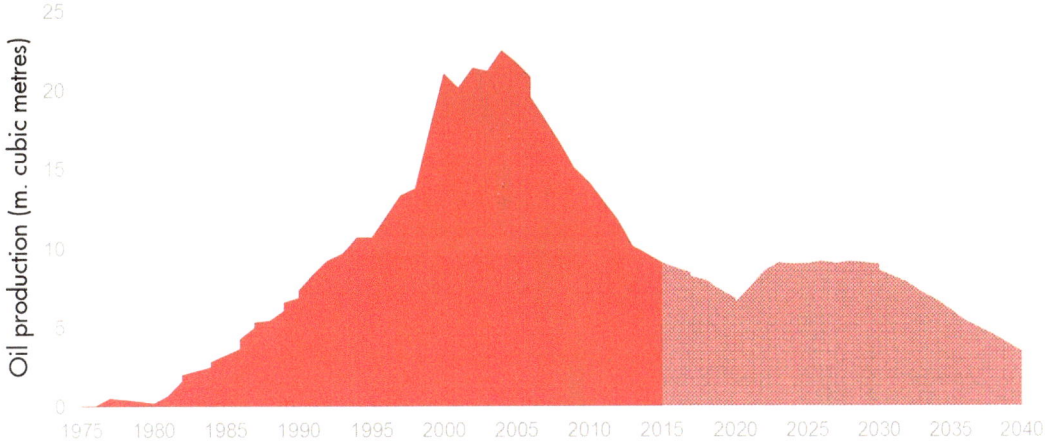

Fig. 108 Denmark's oil production is projected to continue to decline. *Source* Vivid Economics

Special Report 2: Research on China's Energy Demand Revolution

Yang Jianlong and Martin Haigh

1 From Quantity to Quality: International Experience of Energy Demand Revolutions

International experience suggests that China's next energy revolution will focus on energy quality instead of quantity. China's rapid economic growth from the 1980s has been partly driven by substantial expansion of the energy system. However, even if China's economy con-

tinues to grow, the country's energy demand will be unlikely to maintain the same growth rate in the future. In international experience, energy consumption per capita tends to be stable when GDP per capita stands at a higher level, as shown in Fig. 1. Types of energy use and fuel choice tend to change with economic growth when energy demand is stable. If China follows a path similar to international experience, the country's energy demand will not grow in quantity, but shift focus to energy use and the quality of the fuels chosen.

Energy demand tends to be stable when GDP stands at a higher level, which can be explained by the changes in how energy is used. Specifically, these patterns may result from changes in energy service demand (for example, travel mileage per capita) and/or improved conversion efficiency from energy to energy services. To illustrate this point, it is necessary to break down the energy intensity of GDP as follows:

Economic service intensity

Economic service intensity is the result of complex interactions between exogenous (uncontrollable) and endogenous (controllable) factors. As income rises, consumers demand increasingly more energy services (such as kilometres travelled, tonnes of steel produced, quality of lighting, etc.), which are largely related to specific needs or situations. However, even at a given income level, the travel mileage of rural residents tends to be

DRC Team Lead of Special Report 2:
Yang Jianlong from the Research Department of Industrial Economy, DRC of the State Council of China.

Shell Team Lead of Special Report 2:
Martin Haigh, Senior Energy and Climate Change Advisor, Shell International B.V.

Contributors:
Philip Gradwell and Cameron Hepburn from Vivid Economics; Ren Xianfang from Shell China; Duan Hongbo and Liu Ying from the University of Chinese Academy of Sciences; Mo Jianlei and Ji Qiang from the Institutes of Science and Development, Chinese Academy of Sciences; and Li Zhenyu from the Petrochemical Research Institute, CNPC.

Y. Jianlong (✉)
Research Department of Industrial Economy, DRC of the State Council of China, Beijing, China
e-mail: yangjl@drc.gov.cn

M. Haigh
Shell International B.V., The Hague, the Netherlands

© The Authors(s) 2020
Shell International B.V. and the Development Research Center (DRC) of the State Council of the People's Republic of China (eds.), *China's Energy Revolution in the Context of the Global Energy Transition*, Advances in Oil and Gas Exploration & Production, https://doi.org/10.1007/978-3-030-40154-2_3

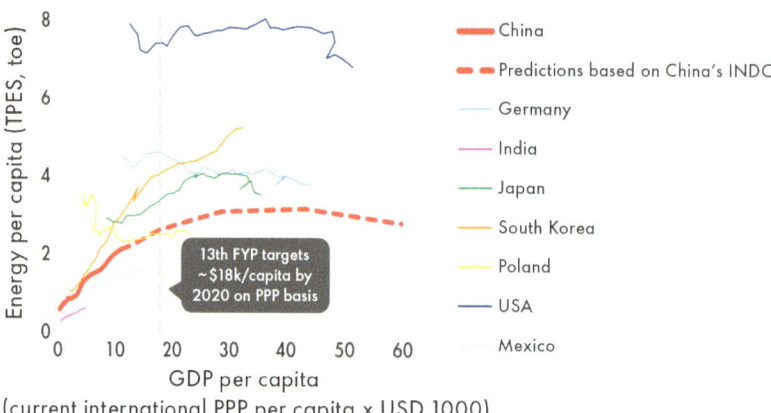

Fig. 1 No need for energy demand growth in quantity when GDP per capita reaches a moderately prosperous level. *Note* 13th FYP = 13th Five-Year Plan (2016–20); PPP = purchasing power parity; INDC = intended nationally determined contributions. *Source* International Energy Agency, World Energy Balances (2016)

longer than that of urban residents. Hence, if social development is accompanied by accelerated urbanisation, then even if the economy grows, the total demand of transport services may fall. This shows that by influencing endogenous factors, policymakers can guide service demand and ultimately steer the development of energy demand. However, other exogenous factors, such as geography, climate or culture, can also influence service demand, but these cannot be controlled by policymakers. It is therefore important to identify the exogenous and endogenous factors of energy service demand, for they allow policymakers to better predict and guide how energy demand develops with economic growth.

Energy conversion efficiency

Energy conversion efficiency is mainly determined by technology, and technology development is influenced by the price of energy and new equipment, regulatory standards, consumer behaviour and underlying trends in technology improvement. By breaking down energy intensity into energy services and conversion efficiency, the analysis can focus on the pure demand-side impacts of service demand before considering subsequent dynamic effects between technological progress and changing demand. Meanwhile, there are fuel-specific characteristics, which are irrelevant to service intensity and conversion efficiency but are needed by consumers. Although all fuels are energy carriers, they can be differentiated by the two dimensions of quality—cleanliness and flexibility.

Economic growth enables our society to prioritise energy cleanliness and flexibility, thus shaping the trend to high-quality energy carriers. In some cases (transport for example), this may result in improved conversion efficiency. If this trend in primary energy is not given sufficient attention, energy demand may be distorted, causing imbalance in the energy system.

If a study fails to consider the evolution from quantity to quality in energy service and carriers, forecasting errors may occur, which may lead to misleading policies. The Annual Energy Outlook published by the U.S. Energy Information Administration, gives a good example of such forecasting errors. The predictions in 1990 overestimated not only the future energy consumption level, but also the use of coal, a high-pollution fuel. Identification of the changing relationship between income and energy, and the preference of consumers for high-quality energy carriers, can help reduce these errors and improve decision-making (Fig. 2).

International experience suggests that China's future energy demand may change by between +25% and −33% against the level of international experience as a result of the energy service

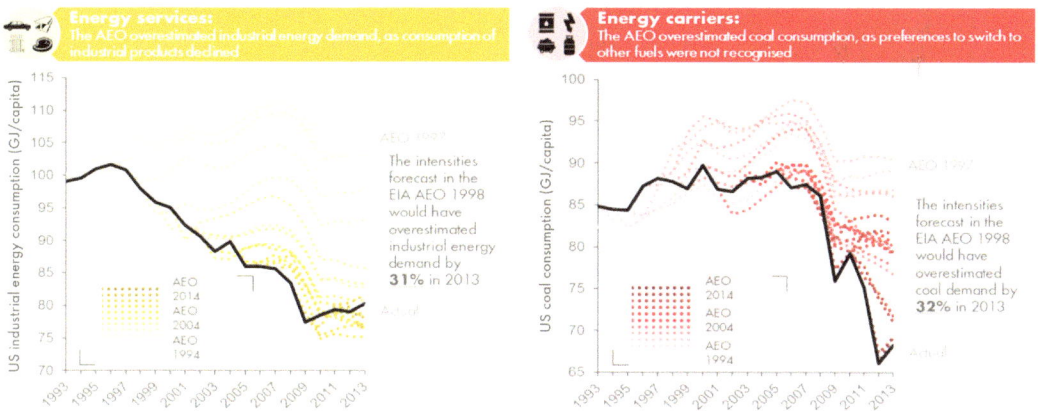

Fig. 2 Predictions of US energy demand. *Note* The dark line represents the predictions provided in the later versions of AEO. *Source* U.S. EIA, Annual Energy Outlook (1994–2014)

demand intensity paths it follows. Assuming that conversion technologies are constant, by 2030, if the conventional energy service demand mode is followed, China's energy demand will increase by 75% from the current level, as shown in Fig. 3. However, if the low service intensity path of international experience is taken, China's energy demand will fall by 33%. In contrast, if the high service intensity path of international experience is followed, China's energy demand will increase by 25%. This explains the important role of service demand in identifying China's future energy demand path, and it highlights the possible impacts of policy interventions on energy demand.

Policy plays an important role in affecting service demand, especially for China's future economic development. Impacts that can be changed by policy include those of population density on transport services, gross capital formation on industrial services, income inequality on building services, and urbanisation on agriculture energy services. These factors can explain why the difference in energy demand in the low and high service intensity paths is significant, when compared with China's expected income level in 2030. This highlights the importance of current policies and their ability to impact long-term energy demand.

When both energy cleanliness and flexibility are considered, the fuel mix in buildings, for example, will change rapidly. As income increases, the buildings sector shifts rapidly from coal and biomass to electricity and natural gas, because these energy carriers are easy to use and do not cause local pollution. In other sectors, however, cleanliness and flexibility cannot be provided by the same energy carrier. In transport, the cleanest fuel today (electricity) has a lower driving range than molecular-based fuels (oil or in future, hydrogen) making it less flexible. Due to the absence of a dominant fuel, the energy carrier structure remains constant, unless new technologies eliminate the trade-off between cleanliness and flexibility, as shown in Fig. 4.

- in buildings, it is possible to improve quickly both the flexibility and cleanliness of energy carriers;
- in power generation, adopting gas and nuclear improves cleanliness without lowering flexibility, but renewable technologies require a trade-off between cleanliness and flexibility;
- carriers in transport cannot improve simultaneously in flexibility and cleanliness with current technologies—future technologies may do this and eliminate the trade-off; and

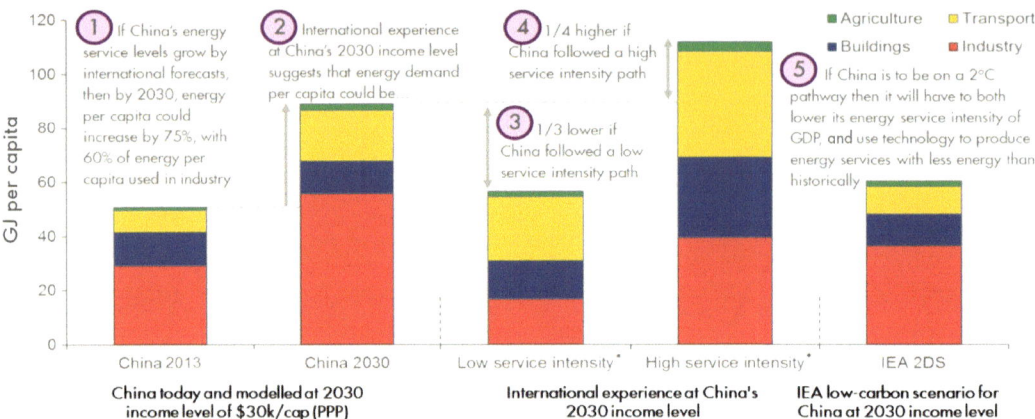

Fig. 3 Changes in China's 2030 energy demand. *Source* Vivid Economics; International Energy Agency, Energy Technology Perspectives 2016

Fig. 4 Not all sectors can see improvement in both the flexibility and cleanliness of energy carriers. *Source* Vivid Economics

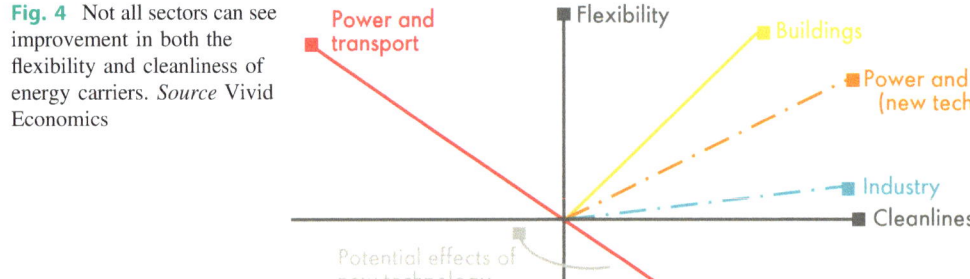

- in industry, the merit order of carriers varies by subsector—a single, representative line cannot be plotted.

Policymakers should understand these pattern changes and guide the formation of an energy system that accurately meets consumer demand, enables the efficient use of resources and restricts externalities like local air pollution or national energy security concerns. Lack of understanding of how energy demand will change and the factors causing these changes may result in an inefficient energy system that features oversupply, poor-quality fuels and an inability to achieve climate goals.

China's transition to the new normal of slower economic growth will impact energy service and demand pathways across the entire economy, especially in industrial energy demand. Lower capital investment and less reliance on heavy industry, which previously drove economic growth, will sharply reduce the demand for industrial energy services like steel consumption, and potentially bring them back in line with international experience. However, it remains to be seen how much effort China will make to drive this change. Even if no specific policies are implemented, as China's economy becomes more service-oriented and tangible investments peak, some regions may replicate Shanghai's experience and see a natural peaking of industrial energy demand. The balance between industry and service-oriented economic development, and the proximity of regions to their peak capital stock (assets, plant and equipment that help with production), will determine the future level of China's steel consumption, and therefore its level of industrial energy demand. In addition, China has experienced uneven development between its urban and rural areas. This imbalance is itself a critical determinant for energy service demand in buildings, especially at China's current income levels, and should be understood by policymakers.

In the buildings sector, smaller housing and increasing urbanisation will drive the transition to higher-quality energy carriers. China's high urban population density and relatively small housing floor areas will drive the transition to cleaner and more flexible fuels in buildings. As more and more rural areas are urbanised and energy access channels are expanded, the demand for *high-quality* fuels rather than *high-quantity* fuels may surge. Policymakers should ensure that investments in energy carrier supply networks match future consumer demand rather than current income levels, or their investments may restrict consumers' energy choices or result in energy waste.

1.1   Energy Services

In this special report, energy service is defined as the final output used by consumers, such as lighting, steel products or travel mileage. By analysing the demand modes of these energy services instead of energy consumption itself, the changes in potential demand modes and efficiency can be separated.

Energy service demand may vary with income and structural factors. Some structural factors may be affected by policies, for example, capital investment and urbanisation; other structural factors, such as climate and culture, are the fixed endowments of a country. In general, energy service demand grows as consumer incomes increase, but this relationship is not necessarily stable and may even be reversed.

If structural factors in energy service demand are not considered, the predictions for energy demand may be inaccurate. Without understanding the changes in service demand modes and the drivers behind such changes, the forecasts and the policies based on them may rapidly lose their relevance.

Using historical data of international service demand, this study identifies common development trends, classifies them into high and low demand pathways, and identifies their potential drivers. The aim is to: (i) provide insights into how energy service demand changes as incomes increase; (ii) identify the range of demand in international experience; and (iii) determine how policies can impact future demand.

The methodology of this study includes the following steps:

(1) collect global panel data (multidimensional measurements over time) on energy service demand, GDP and structural factors;
(2) plot the mean line into the third-degree line of best fit through all data points (across all countries and periods);
(3) split all data points into $5,000 GDP per capita tranches ($0–5,000, $5,000–10,000, etc.);
(4) identify the high and low service demand pathways as the lines of best fit through the top and bottom quartiles of service demand across each $5,000 GDP per capita tranche;
(5) highlight China's current values (by province or city, if applicable) and third-party forecasts of China's energy service demand in 2030;
(6) identify the explanatory drivers of the variations in service demand; and
(7) plot the lines of best fit through the data points that make up the top and bottom quartiles of each explanatory driver in each $5,000 GDP per capita tranche.

As indicated by the data analysis in international experience, energy demand comes mainly from four sectors. This study analyses the key service in each sector as a proxy indicator of the sector as a whole.

Transport: 28% of global final energy demand —excluding chemical feedstock.

- Proxy indicator: Road passenger and freight make up 77% of transport energy demand (22% of total energy demand).

Industry: 35% of global final energy demand —excluding chemical feedstock.

- Proxy indicator: Iron and steel production accounts for 28% of industrial energy demand (10% of total energy demand).

Buildings: 34% of global final energy demand —excluding chemical feedstock.

- Proxy indicator: Lighting emissions are an effective instrument for measuring building energy services.

Agriculture: 3% of global final energy demand—excluding chemical feedstock.

- Proxy indicator: Beef, pork and poultry account for 60% of agricultural energy demand (2% of total energy demand).

1.1.1 Transport

China needs policies that guide transport service demand. Overall, China's transport service demand is very close to the average seen in international experience. Given this, and the fact that transport service demand, according to international experience, does not plateau until very high income levels are reached, it is likely that the expected 2030 demand level will not occur without guiding policies, as shown in Fig. 5.

Population density is a key factor in travel mileage, so urban planning could be a means to limit transport service demand, without constraining people's choices. In international experience, more densely populated countries have lower levels of road travel than scarcely populated countries at the same level of income. As Fig. 6 shows, transport service demand in South Korea and Finland differ greatly, despite their similar income levels. Should China wish to limit the growth of transport service demand, it could do so by increasing population density and

urbanisation. Given current urbanisation patterns in Shanghai and Beijing, for example, this should be possible.

1.1.2 Industry

China's demand for energy services in industry is higher than international levels, but there is uncertainty as to whether this trend will continue. China's current steel demand is double that of the high path level of international experience. Given this fact, China needs to make great changes to get to the low path level. Even if no change is made, China's steel demand will remain at the high end of international experience in 2030. However, steel demand in some provinces (Shanghai, for example) has peaked and started to decline. If such a trend is replicated in other provinces, China's future steel demand may stabilise early and converge with international experience by 2030.

Gross capital formation can explain the difference between the high and low paths in international experience at the GDP per capital level of $30,000. This shows that China's steel demand and related energy consumption may decrease with slowing investment in fixed assets —a process that may occur naturally as capital stock reaches saturation. Much of China's infrastructure has already been constructed, so the need for more investment may decline in the coming years, as shown in Fig. 7.

Another important factor is whether China's goal is to maintain its leadership in exporting heavy industry products or transition to a service-oriented economy. Just like China, South Korea's economy is dominated by heavy manufacturing. South Korea has the world's highest steel consumption per capita, as shown in Fig. 8. However, if China decides to build a service-oriented economy, as the UK did in the 1990s, demand for industrial services will decrease sharply. Such a shift in China's industrial structure may impact future industrial energy demand. Some provinces in China may retain their export-oriented heavy industry, while others may switch to a service-oriented economy. How to balance these development options and how close each province is to its peak capital

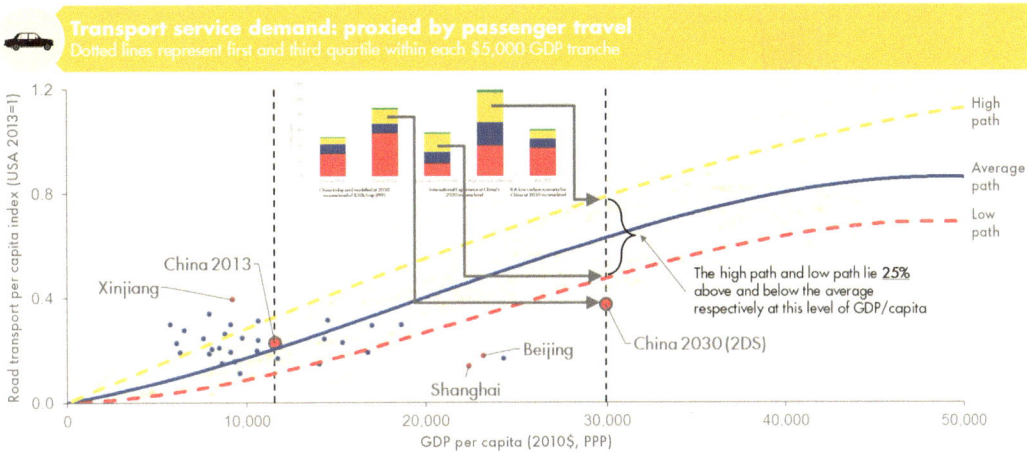

Fig. 5 Comparison of China's transport model with international experience. *Note* 2DS = the IEA's pathway representing a 2°C goal for limiting global warming; PPP = purchasing power parity. *Source* OECD; IEA, Energy Technology Perspectives; National Bureau of Statistics of China

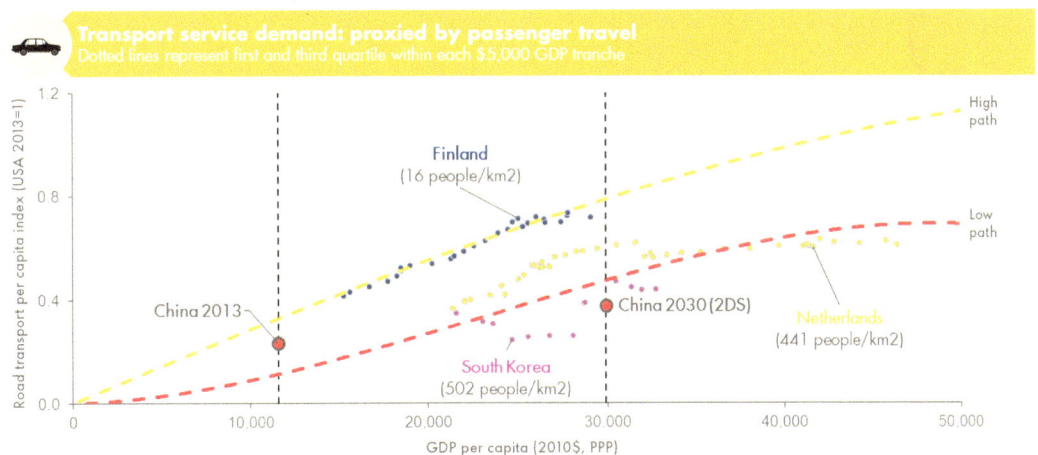

Fig. 6 Higher population density can restrict transport service demand. *Note* PPP = purchasing power parity. *Source* OECD; IEA

stock will determine China's future steel demand and industrial demand for energy.

1.1.3 Buildings

The range of energy service demand for China's buildings is wide even in 2030. In the low case, based on IEA's 2DS scenario, China's energy service demand for buildings may remain almost constant to 2030, which is not in line with historical experience. As most Chinese cities lie within general international experience, the deviation required for the 2DS case suggests changes in efficiency or economic activity. For example, in Alxa, Inner Mongolia, high secondary industry output causes GDP per capita to be extremely high, while building energy services remain low as income largely does not remain in the city. This is in contrast to Sanya, Hainan Island, where the nature of the tourism-dominated economy causes building

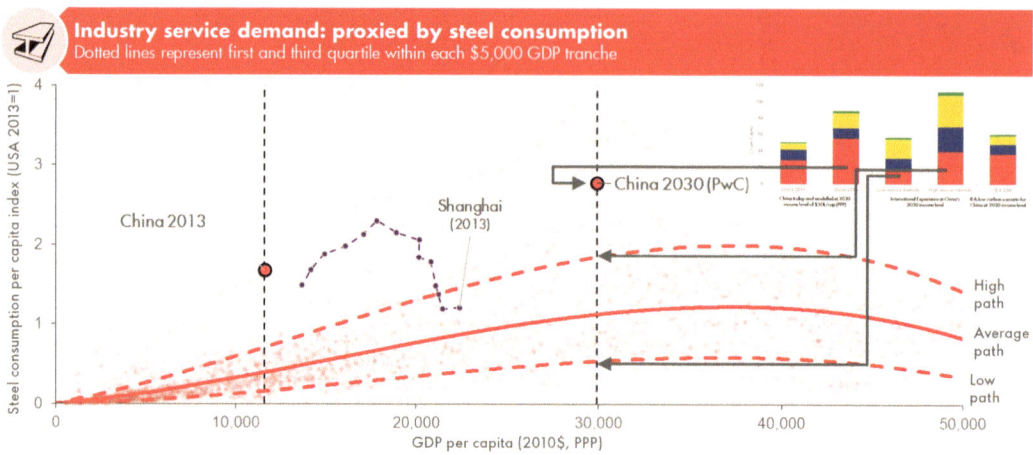

Fig. 7 China's steel demand exceeds that of most countries. *Note* Province-level steel demand is measured by gross fixed asset formation; PPP = purchasing power parity. *Source* World Steel Association; IEA; National Bureau of Statistics of China; PricewaterhouseCoopers (PwC)

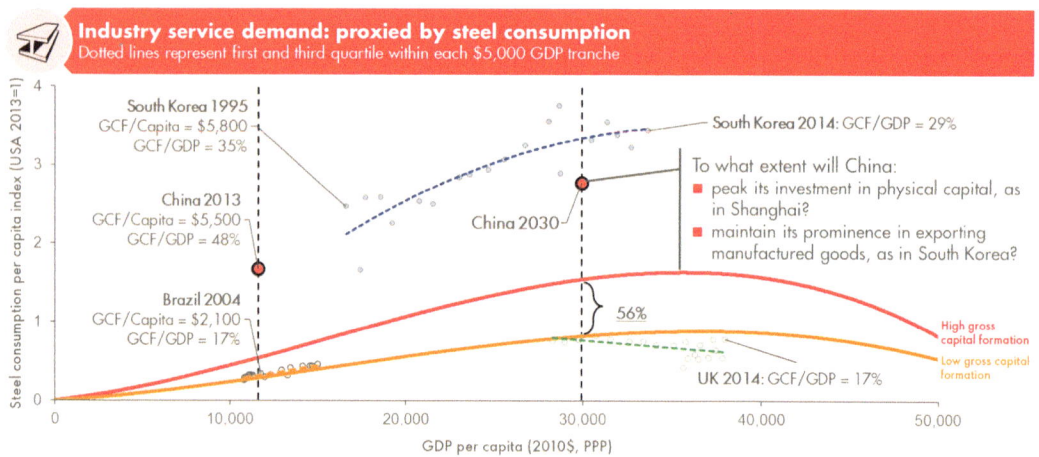

Fig. 8 Physical capital investment is a key factor in China's industrial energy demand pathway. *Note* GCF (gross capital formation) = change in gross economic fixed assets; PPP = purchasing power parity. *Source* World Steel Association; IEA; World Bank

energy services to be outside the range of most international experience at that income level (Fig. 9).

Income inequality can restrict energy access and result in low energy service demand at a higher income level. In international experience, income inequality plays an important role in explaining the variation between the high and low pathways at a given income level. This is especially the case at China's 2013 income level,

where income inequality explains the 53% variation in high and low demand for energy services in buildings. If China's development is accompanied by an increase in income equality, a higher level of building energy services should be provided than if income growth was unequal.

1.1.4 Agriculture

China's agriculture energy service demand is consistent with that of most countries, with only

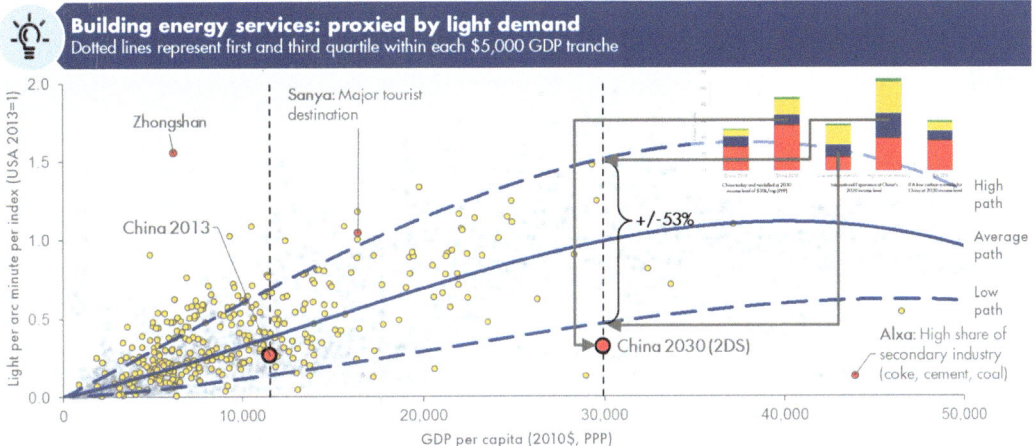

Fig. 9 China's current building energy service demand is consistent with international experience. *Note* Light per arc minute index = light emissions per arc minute/(population density) ^0.5; China 2030, based on lighting energy consumption forecasts in IEA Energy Technology Perspectives; major oil producers are excluded due to flare; 2DS = the 2°C goal for limiting global warming; PPP = purchasing power parity. *Source* National Oceanic and Atmospheric Administration (NOAA); IEA; Socioeconomic Data and Applications Center (SEDAC); Beijing City Lab (BCL); Vivid Economics

very slight growth expected. The difference between the high and low paths in agricultural services is largely due to strong cultural factors—predominantly vegetarian countries such as India are unlikely to ever have high meat consumption, while in most South American countries meat makes up a high percentage of the diet. In absolute terms, China consumes a large quantity of meat (28% of global meat and 50% of all pork), but its meat consumption per capita is still lower than that in the high path of international experience. As predicted by the Food and Agriculture Organization of the United Nations (UNFAO), meat consumption will not change significantly by 2030, which may be due to a change in meat-eating behaviour. This can be seen in the recently revised dietary guidelines issued by the Chinese Nutrition Society to tackle the societal and health effects of excessive meat-eating.[1]

Urbanisation enables substantial economies of scale for cold-storage supply chains, lowering prices for meat and increasing demand. Urbanisation helps centralise market demand, justifying

the fixed capital expense of large cold-storage facilities that reduce transport costs for meat producers. The effect of urbanisation is more pronounced at lower levels of income: when disposable income is low, meat is a luxury and differences in price equate to large changes in consumption. At higher income levels, meat becomes a staple and is less sensitive to price movements. When GDP per capita is $30,000, the difference in meat demand between high and low levels of urbanisation accounts for only 17% of the variation between the high and low demand paths, whereas it accounts for 39% at China's current income levels. It is important for China to understand the potential impacts on meat demand of further increases to its high urbanisation rate, but also the possible effects of rising income on attitudes towards meat-eating (Figs. 10 and 11).

1.1.5 Lessons from International Experience

As suggested by international experience, China's energy service demand tends to go through structural change at its current level of GDP per capita. Figure 12 clearly shows that energy service demand changes over time: China

[1]http://dg.cnsoc.org/article/04/
8a2389fd54b964c80154c1d781d90197.html.

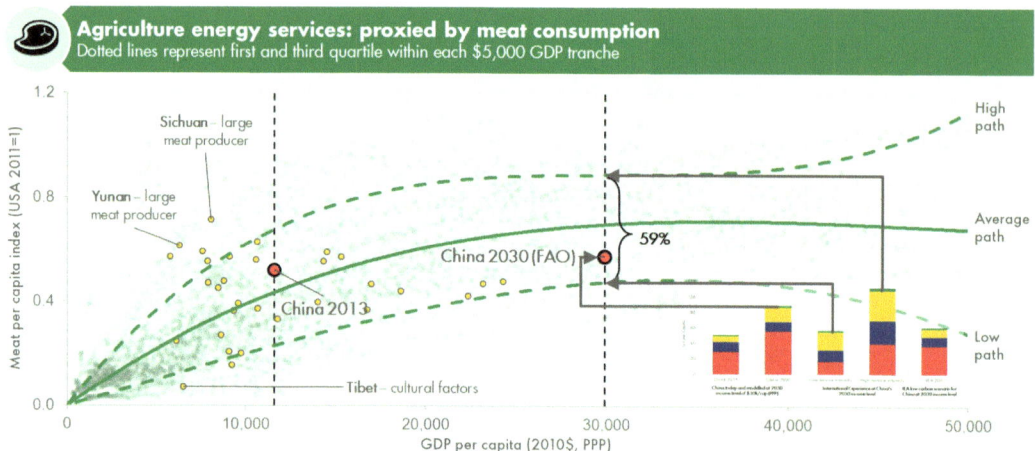

Fig. 10 China's meat consumption is consistent with international experience. *Note* Province-level meat consumption based on consumption data in 2011; PPP = purchasing power parity. *Source* Food and Agriculture Organization of the United Nations (FAO); IEA; National Bureau of Statistics of China; National Grains and Oil Information Center (NGOIC)

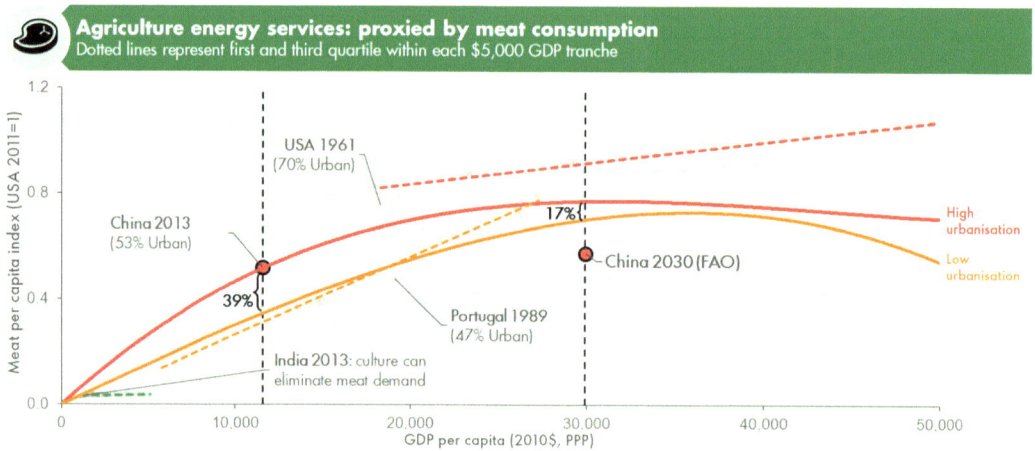

Fig. 11 Urbanisation and accompanying cold-storage supply chains are two key drivers of meat consumption. *Note* PPP = purchasing power parity. *Source* FAO; IEA; World Bank

is approaching levels of income where the relationship between GDP per capita and service demand begins to plateau. By identifying structural change and understanding the impact of policy on service demand, China's decision makers can develop policies to guide service and energy demand, thus maximising benefits and avoiding wasted investment.

Transport demand tends to plateau at a very high level of GDP per capita, but large infrastructure projects like high-speed rail may enable China to follow a completely different pathway. International experience suggests that transport service demand increases almost linearly with income. Before applying this trend to China, it is necessary to understand China's unique context. For instance, China has developed its high-speed rail network faster than any other country. Before April 2007, China had no high-speed rail; now it has more kilometres of

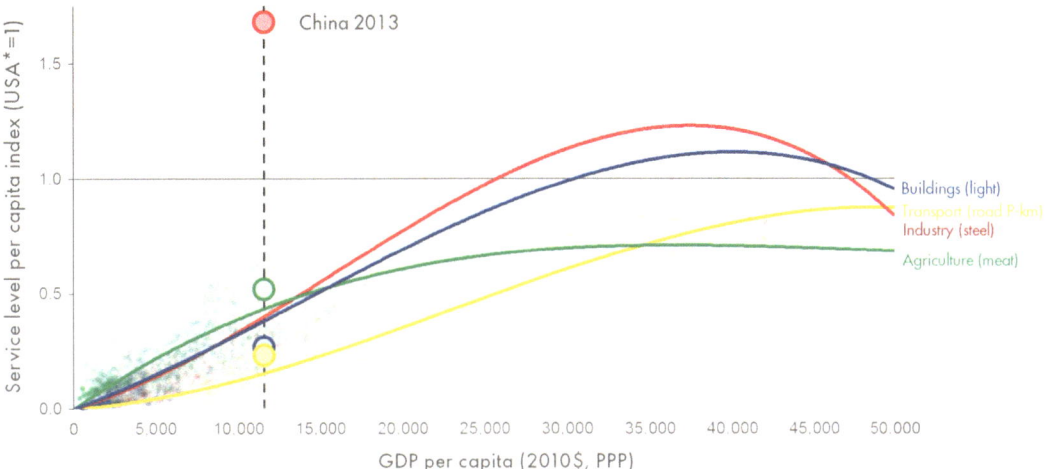

Fig. 12 Demand for key services develops differently with economic growth. *Note* For buildings, transport and industry: 1 = service level per capita in the USA 2013; for agriculture, 1 = service level per capita in the USA 2011; P-km = passenger travel in km; PPP = purchasing power parity. *Source* IEA; National Oceanic and Atmospheric Administration; Food and Agriculture Organization of the United Nations; OECD; World Steel Association

high-speed rail line than the rest of the world combined. Such a high level of infrastructure development may increase transport service demand but may also decrease road transport demand and improve energy efficiency in the sector. It is therefore important to take China's infrastructure into account when considering which service demand path to follow in the 2°C scenario. In addition, China's historical experience suggests that large-scale deployment of infrastructure investment within a short time can gradually change service demand. For example, China has built many airports, creating a market for energy-intensive air travel. International experience might therefore not be a good predictor for future Chinese transport pathways, and policymakers should continually update their expectations as the infrastructure landscape changes.

China's industrial energy demand is very high relative to international experience. Whether this trend will continue in the context of China's new normal of slower economic growth is unclear. China's current high levels of industrial service demand are largely driven by its heavy-manufacturing sector and significant levels of fixed capital investment. However, this growth model is widely believed to be unbalanced and

unsustainable, so China's new normal economy is shifting investment away from heavy industry to high value-added services. This indicates that industrial service demand will not continue on its previous growth trend. Shanghai's transition to a service-oriented economy, and the associated fall in service demand, can act as a case study for other provinces in China. However, this process will not be uniform across the country. Some provinces may retain their export-oriented heavy industry, while others may move towards a service economy. How to balance these development options, and how close each province is to its capital stock peak, will determine China's future steel consumption and its level of industrial energy demand.

China's somewhat imbalanced urban-rural growth, due to the *hukou* residential registration system as well as relatively low expenditure on social services, has limited the demand for building energy services. China's new normal economy highlights higher quality growth, which covers social equality to some extent. China's rural areas will likely see the greatest change in development, as there is clearly capacity for welfare improvement and investment. This will not only increase social wealth, but also reduce the economic imbalance between urban and rural areas. The positive correlation between building energy service demand and the urban-rural imbalance indicates that China's building energy demand may grow sharply.

Agricultural energy service demand will peak early, but it only accounts for a very small proportion of total energy use. China is at the income level where, in international experience, meat consumption growth begins to slow down. This, combined with the small share of global energy that agriculture accounts for, makes this sector a lower priority for policymakers. However, as urbanisation increases and the cold-storage supply chain it drives matures, meat consumption in rural areas will increase and counteractive policies will be needed. Cultural attitudes towards different meats may also bring unexpected changes to the energy system. Demand for beef, which accounts for only 8% of meat consumption in China, is currently being met by imports rather than domestic supply. Given China's goal of reducing meat consumption and CO_2 and methane emissions from the livestock industry, the supply and demand sides of meat may need to be addressed in the future.

1.2 Energy Carriers

Energy carriers are fuels that can be converted to provide a useful service, such as light or heat, or drive a physical process like powering a car. Common energy carriers are fossil fuels, electricity and biofuels such as wood. The quality of an energy carrier can be defined along two dimensions:

- cleanliness: the ability to provide energy without producing local pollutants, measured in particulate matter (PM10) emissions; and
- flexibility: the convenience and effectiveness of using the energy (in joules) carried by an energy carrier in a specific sector.

In international experience, as income increases consumers demand not only more energy but higher quality energy carriers. While relative prices are undoubtedly important in determining the carrier mix, even countries with large endowments of cheap, low-quality fuels, such as coal, tend to switch to higher-quality carriers as they develop. This indicates that rising incomes allow consumers to prioritise and pay for quality characteristics such as cleanliness and flexibility.

For some sectors, there is no single carrier that offers improvements along both quality dimensions. This has limited their transition to new carriers. In sectors where a single carrier offers improvements in both cleanliness and flexibility, such as electricity in the buildings sector, transitions occur quickly. In contrast, change is insignificant in sectors, such as transport, where, for all carriers, there are trade-offs between cleanliness and flexibility. However, it should be acknowledged that future technologies may eliminate existing trade-offs and so trigger future energy carrier transitions.

This section uses international experience to illustrate how shifts to high-quality carriers have occurred in different sectors and discusses potential pathways that China could take. The analysis shows that Chinese consumers are approaching income levels where cleanliness and flexibility are valued more than energy at the lowest possible cost. Policymakers should recognise these patterns of carrier demand. Many fuels have expensive supply chains and countries should avoid locking themselves into a dirty and inflexible carrier mix if consumer demand is about to shift to cleaner and more flexible fuels.

This study uses a consistent methodology to analyse the development of international energy carriers in four fields: buildings, power generation, transport, and iron and steel. The methodology comprises the following steps:

(1) collect global panel data (multidimensional data on measurements over time) on energy carriers in different sectors;
(2) for each sector, choose an appropriate definition of flexibility and identify fixed factors for cleanliness and flexibility for each energy carrier per joule of energy—as factors are fixed, technology is assumed to remain constant across time and countries;

(3) calculate the values of cleanliness and flexibility for each sector, country and observation in time, and normalise them to the sample average value;

(4) plot, as a third-degree line of best fit through all data points (across all countries and all time periods), the average line in a sector for each dimension of quality;

(5) divide the data into $5,000 tranches and identify the top and bottom quartile data points for each dimension of quality;

(6) plot the high and low path lines of best fit through the top and bottom quartile data points of energy carrier quality across each $5,000 income tranche; and

(7) highlight China's current values and third-party forecasts of China's energy demand in 2030.

1.2.1 Buildings

In the buildings sector, excess air ratio is a proxy for the flexibility of an energy carrier. Excess air ratio is the amount of air above the stoichiometric air quantity ratio (the absolute minimum amount of oxygen required for complete combustion of the fuel) that is required for a fuel to burn effectively. It is used as a proxy indicator for the size of equipment needed to convert the carrier fuel into a useful energy service, such as heat. The most flexible energy carrier is the one that can be used without large equipment, as space is limited in buildings. According to this definition, electricity is flexible, while coal generally needs a large amount of excess air to combust to avoid emitting harmful carbon monoxide, making it difficult to use in buildings.

Because it is possible to improve both cleanliness and flexibility simultaneously in buildings, changes in the carrier mix have occurred rapidly. International experience shows that buildings have shifted quickly towards electricity and gas as incomes have risen. This is because electricity and gas are cleaner and more flexible than the other available energy carriers. The income-driven improvements in both aspects of carrier quality are clearly evident in Fig. 15. Belgium and the UK are examples of countries

that have made rapid transitions to higher-quality fuels in this sector.

1.2.2 Power

Flexibility in the power sector is dependent on the capacity factor and the cost of transmission infrastructure for different methods of generation. A composite factor accounting for both the temporal and spatial flexibility of different generation methods is used as a proxy for the flexibility of carriers in the power sector:

$$Flexibility = [(1 - capacity\,factor) \\ *LCOE\,of\,gas\,generation] + system\,cost$$

The capacity factor is the ratio of a power plant's actual output relative to its potential output if it were to run all the time. It captures temporal flexibility and is low if the installation cannot generate electricity consistently over time. The capacity factor is therefore small for variable renewables, such as solar and wind, but large for dispatchable technologies, such as gas-powered generation. To capture the value of having power when needed, missing capacity is priced at the lowest levelised cost of dispatchable electricity, which here is assumed to be gas. System costs represent the spatial flexibility of carriers and are proxied by how costly it is to move power generated from the plant to the transmission grid. Renewable technologies tend to perform relatively badly in terms of system costs, as they are spatially distributed and can require significant grid investments to distribute power. In contrast, most dispatchable technologies have small system costs.

In power generation, cleanliness can be improved without compromising flexibility, but it is impossible to improve the two quality dimensions simultaneously. As shown in Fig. 13, as incomes increase, the cleanliness of power generation gradually improves. This is mostly driven by the uptake of gas and nuclear generation capacity, which emit only a fraction of the pollutants of oil and coal generation. However, these improvements in cleanliness are not as fast-paced as in the buildings sector, as switching

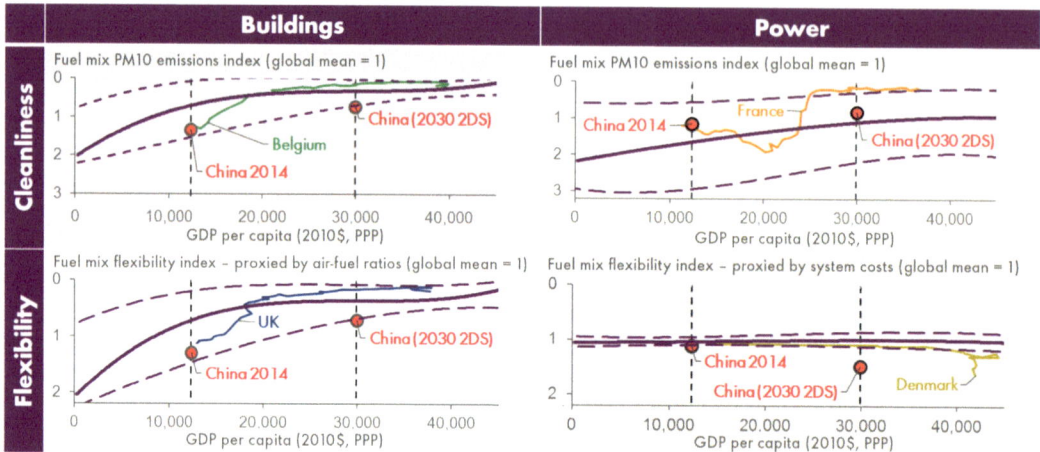

Fig. 13 Rising income can drive improvements in fuel quality when there are no trade-offs. *Note* The lower the value, the higher the cleanliness and flexibility—approaching 0 implies improvement. 2DS = the 2°C goal for limiting global warming; PPP = purchasing power parity. *Source* IEA; UK National Atmospheric Emission Inventory

to cleaner carriers does not carry similar flexibility benefits. Even so, there are examples of very fast transitions. The adoption of the Messmer Plan in France in the 1970s and its rapid deployment of nuclear capacity is an example of such a rapid transition towards cleaner fuels.

There is no clear relationship between power generation flexibility and income, but increased uptake of renewable energy or improvements in technology may change this. International experience shows that the flexibility of the carrier mix in power generation remains relatively stable as incomes grow. This is mainly because most energy systems are simply scaled up as incomes rise and because all dispatchable technologies offer similar flexibility. However, we do see a sudden drop in flexibility among countries that have adopted large volumes of renewable generation. This is the case with Denmark, which adopted significant amounts of wind generation in response to the oil crisis in 197 The variable nature of renewable power makes it costly to integrate into the power system, and consequently it has a flexibility factor that is an order of magnitude worse than that of dispatchable generation. This highlights the trade-off between cleanliness and flexibility that has limited the global uptake of renewables to date. However,

should battery technology improve and eliminate trade-offs, it may lead to a rapid uptake of renewables by removing the flexibility issues of renewable generation. In contrast, if trade-offs remain but countries decide to take up renewables to improve cleanliness regardless, then flexibility may begin to trend downwards with increasing incomes, at a cost to consumers.

1.2.3 Transport

In transport, there is a clear trade-off between cleanliness and flexibility, which prevents shifts to other fuels. Fuel shifting has been almost non-existent in the transport sector since the 1960s, and petroleum continues to dominate transport energy use in all countries. While there has been a recent push towards cleaner energy carriers, particularly electricity, it cannot yet match the flexibility of petroleum with current technology. The best, modern electric cars can travel up to 350 km on a single charge, while the best diesel cars can reach more than 1,200 km on a single tank of fuel. Although electricity has a better conversion efficiency rate than any other carrier, the low energy density of a battery gives it one-tenth the flexibility of diesel, as shown in Fig. 14. The lack of recharging stations, compared to petrol stations, puts an additional

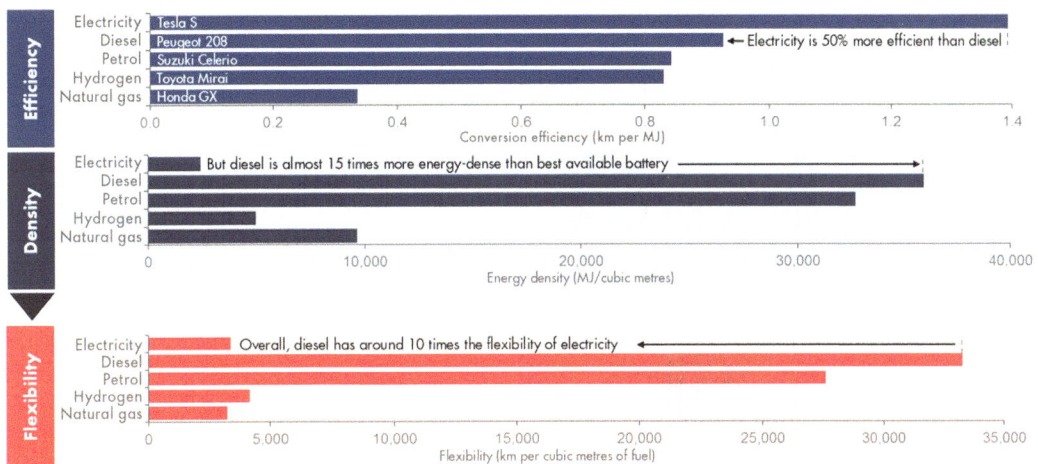

Fig. 14 Electric transport is less flexible than internal combustion engines. *Source* Next Green Car Ltd.; University of Birmingham (fuel data sheet) Staffell

constraint on undertaking long journeys with electric vehicles. Until technology improves to the point where this trade-off between different aspects of quality no longer has to be made, transport will likely continue to be a mono-fuel sector.

Flexibility is difficult to define for the industrial sector, as different subsectors use fuels in different ways for different purposes. We derived flexibility factors in the industrial sector econometrically, using fixed-effect regression to compute the gain in gross value-added for each subsector from a 1% shift from coal to other energy carriers. However, the range of processes within each industrial subsector, and the potential for an energy carrier to be used as a feedstock rather than as a fuel source, means that a singular merit order cannot be identified for industry as a whole.

1.2.4   Iron and Steel

In iron and steel manufacturing, the largest industrial subsector in China, electric arc furnaces can, to some extent, improve cleanliness without sacrificing flexibility. Currently, there are two major steelmaking technologies: basic oxygen furnace (BOF), which uses molten iron as the raw material; and electric arc furnaces, which use scrap steel. If electricity is provided by coal-fired power plants, electric arc furnaces use

around one-fifth of the coal of the BOF process to produce the same amount of steel. However, this assumes that scrap steel comes at no energy cost, whereas molten iron needs to be extracted using coke. In addition, the BOF process produces a large amount of particulate matter pollution, which needs careful collection and management. For primary steel production there is a trade-off between cleanliness and flexibility, whereas electricity can offer improvements along both quality dimensions for scrap steel.

1.2.5   Lessons from International Experience

International experience suggests that consumers switch to higher-quality energy carriers as incomes rise, and that technology improvements may trigger further shifts. With this in mind, it is possible to tailor the energy system to the preferences of future consumers, rather than locking it into a low-quality carrier that will ultimately reduce benefits. However, policymakers should understand the unique trade-offs for each sector, as these will likely lead to different rates of demand-driven carrier changes.

China is currently near the low path of both quality dimensions within the buildings sector, but the relatively high density of its urban areas may cause carrier mix changes to occur faster than international experience suggests. Both

Average new residential area

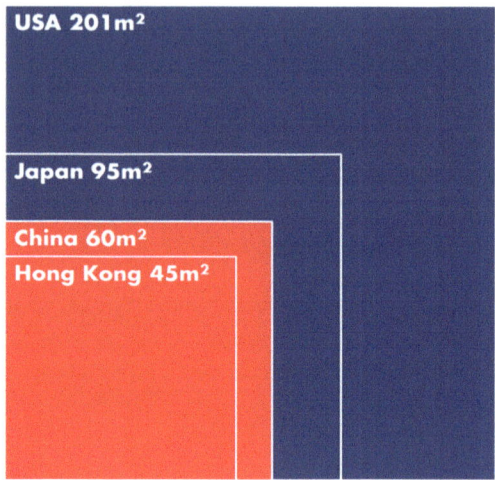

Fig. 15 China's average home size indicates higher population density

cleanliness and flexibility become more important in buildings as population density increases. As local pollution is a negative externality upon an entire area, a highly dense population might be willing to pay a higher price for clean energy carriers than rural societies. The high cost of land and relatively small living space, as shown in Fig. 15, will increase the value of a flexible fuel that can be more easily used in confined spaces. It is difficult to imagine a coal or wood heating system being used in the confines of future small urban households. The public demand for cleaner and more flexible carriers in buildings is, therefore, likely to increase as China continues its urbanisation, and policymakers should prepare for this.

Iron and steel production in China uses increasing amounts of electricity. The pressure this places on the power system may require planning. The large, dynamic loads of electric arc furnaces can reduce the quality of power for other consumers in the locality. To manage this, power stations should be located as close as possible to the electric arc furnace to minimise disturbances such as flickering. The recent uptake of electricity in iron and steel manufacturing has been fast, doubling between 2006 and 2014. It now accounts for 18% of total energy use in the steelmaking sector. If this trend continues, careful consideration will be needed on how to adapt the power grid to accommodate this increasing industrial demand for electricity.

China's power generation is dominated by coal, but higher generating capacity and expertise in solar generation may lead to a cleaner but less flexible system. China is currently better than the average path of international experience in cleanliness within the power sector. Hydropower makes up 19% of generation, and while 73% of China's electricity generation is currently provided by coal, it is still a cleaner carrier than heavy fuel oil which is used for power generation in some countries. China is expected to continue this improvement in cleanliness up to 2030, although seemingly at the expense of flexibility if scenarios by the International Energy Agency are to be believed.

Despite China's deployment of high-speed rail, road vehicles still dominate energy consumption in transport and almost exclusively use oil. Until technology develops to the extent that long-distance journeys can be reliably made with alternatively fuelled vehicles, there is unlikely to be any large energy carrier transition in this sector. While China accounted for 46% of all electric vehicle sales in 2016, the sector has seen disappointing growth and represents only around 1.3% of total car sales.

1.3   Conclusions

The energy system is driven by energy demand, and energy demand itself is driven by consumer demand. For policies to remain effective, the shifting nature of consumer demand must be recognised and catered for. Investing in energy that is not demanded is a misuse of resources and may actively reduce welfare by limiting future consumer choices if networks are locked in patterns of demand.

Consumer demand is more complex than simply demand for the greatest quantity of energy at the lowest cost. International experience shows that rising incomes will change consumer demand for energy services and allow consumers to pay for higher-quality carriers that are cleaner and more flexible than existing fuels such as coal.

Consumer demand for energy tends to shift from quantity to quality as incomes increase, and China is likely to cross this inflection point in the near future. The shift from quantity to quality involves a plateauing of service demands and increasing use of higher-quality fuel carriers. Rapid changes are possible should the drivers of service demand undergo a large shift, or if technology developments facilitate universal shifts to higher-quality energy carriers within a sector.

Hence, future energy demand cannot be projected on the basis of one country's historical patterns of consumption, but should be based on international experience, taking into account local characteristics. Historical experience is useful for identifying a baseline trend and a set of possible outcomes, but it should not be taken as an accurate predictor of the future. How local characteristics naturally develop or react to new policies will largely determine future consumer demand. China's shift towards the new normal of slower economic growth will, for example, reduce the share of energy-intensive heavy industry and increase the rate of urbanisation in rural areas.

However, these demands can also be shaped by policy to ensure resources and energy are used efficiently or to limit externalities. Many factors can play a role in changing consumer demand and preferences. They include economic structure and investment, population density, equality of access, support networks for clean and flexible fuels, and support technology that enables universal improvements in energy quality. Nudging energy service demand can help mitigate long-term negative externalities such as climate change and enable a more desirable use of resources to improve efficiency without jeopardising benefits.

An efficient system can be adopted by taking into account changing service demand trends and the shift towards higher-quality fuels, and by using policy to shape demand. Policymakers should aim to design an energy system that will meet future consumer demand. However, they should also take advantage of the ways they can shape it to avoid irrational energy consumption, achieve climate targets and other long-term goals, and improve resource-use efficiency.

2 Model Building

2.1 Model Description

3E models provide an integrated view on energy, the economy and the environment. They combine an optimal growth model based on neoclassical economic theory and an infinitely lived agent (ILA) model that follows the Ramsey Rule. 3E models describe economic operations through investment, consumption and the accumulation of capital. They use the traditional top-down approach, while providing richer technical content.

Our 3E model examines how fossil and new energy fuels evolve into dominant fuels in a sequential manner, using a built-in logistic sub-model of policy. In addition to fossil energy, the technical objects studied include seven low-carbon or zero-carbon energy sources: nuclear, biomass, hydropower, solar photovoltaic, wind, geothermal and marine energy. This enables our model to cover more bottom-up modelling characteristics and explore the role of zero-carbon energy technologies in reducing carbon emissions.

Another advantage of this richer technical approach is that it helps endogenise energy technologies and describe dynamic technological progress with an empirical curve based on learning by doing. This will significantly reduce the uncertainty in model results caused by exogenous technological progress, thus improving the robustness of the model's calculations. It must be noted that the model built for this section

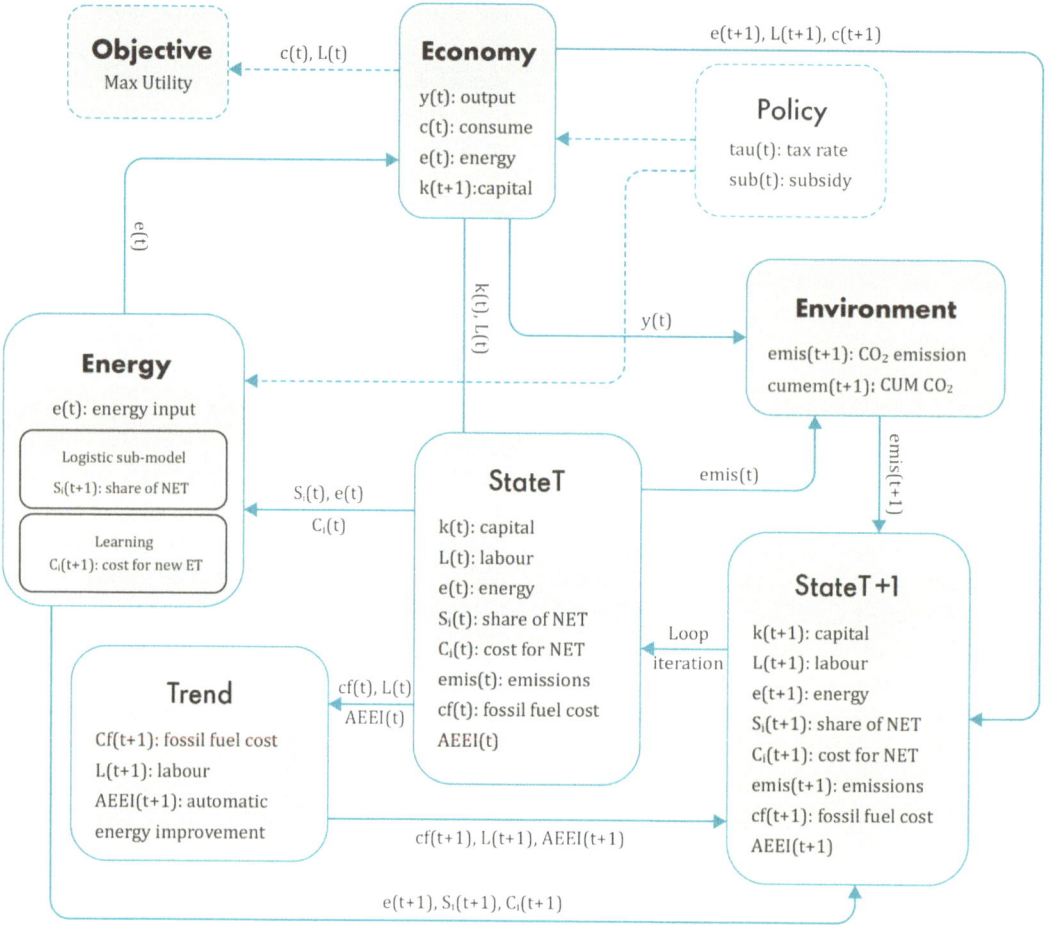

Fig. 16 Block diagram of the optimal 3E integrated system model for growth. *Note* CUM = cumulative carbon emissions; ET = energy technology; NET = new energy technology

is a regional 3E integrated system model for a single sector—the energy industry. The specific model architecture and operation routes are shown in Fig. 16.

The model assumes the existence of a forward-looking central planner who maximises society-wide utility by choosing the optimal path of decision variables, including investment and consumption. Utility here is measured mainly by consumption per capita (c), so the objective function of this model can be expressed as:

$$Max \sum_t \left(L(t) \cdot \log\left(\frac{c(t)}{L(t)}\right) \prod_{v=0}^{t} (1 + \sigma(v))^{-1} \right) \tag{1}$$

where:

L is population, and the utility discount factor is measured by time preference rate σ, and

$$\sigma(t) = \sigma_0 \cdot e^{-d_\sigma \cdot t} \tag{2}$$

d_σ is the annual decrease factor.

2.1.1 Economic Sector

The production process of the model is mainly described by the Cobb-Douglas production function, constant elasticity of substitution (CES). The input factors include capital (k_t), labour (L_t), and energy (e_t), i.e.:

$$y(t) = \left(\alpha(t) \left(k(t)^{\gamma} \cdot L(t)^{1-\gamma} \right)^{\rho} + \beta(t) e(t)^{\rho} \right)^{1/\rho}$$

$$(3)$$

where:

$y(t)$ represents output and α represents the level of technological progress achieved by the combination of capital and labour. β describes automatic progress in energy technologies, including the improvement in energy efficiency from non-price factors. γ and ρ respectively represent the constant elasticity of substitution between the share of capital and energy, and the capital-labour combination and energy. The capital stock in a new period is expressed as the capital discount stock in the previous period plus new investments in the current period (i):

$$k(t+1) = (1 - \delta)k(t) + i(t+1) \qquad (4)$$

To enable the model to describe gross domestic product (GDP) as a two-way relationship between energy input and economic output, the study defines GDP as the difference between output and energy cost:

$$gdp(t) = y(t) - ec(t) \qquad (5)$$

Energy cost is expressed as the product of energy input and composite energy price (pe), i.e.

$$ec(t) = e(t)pe(t) \qquad (6)$$

In addition, the allocation flows of GDP mainly include investment, consumption, and imports and exports (when energy R&D is considered, total R&D expenditure should also be included):

$$gdp(t) = i(t) + c(t) + x(t) - m(t) \qquad (7)$$

where:

x and m respectively represent imports and exports. In light of China's historical import and export conditions, the model also sets the lower limit of the share of exports in GDP (θ_1) and the upper limit of the share of imports in GDP (θ_2):

$$x(t) \geq \theta_1 gdp(t) \qquad (8)$$

$$m(t) \leq \theta_1 gdp(t) \qquad (9)$$

2.1.2 Energy Sector

The multiple logistic curves of policy intervention represent the core part of the energy module in the model. By building the logistic curves into the model, the study can enrich technical details of the traditional energy-economy endogenous growth model. It facilitates bottom-up analysis of how the substitution of new energy technologies for fossil fuels evolves. It also calculates the impacts of environment policies, like carbon tax and renewable energy subsidies, on economic and energy systems. The classic logistic model can be expressed as:

$$\frac{ds_i(t)}{dt} = a_i s_i(t) \left(1 - \sum_i s_i(t) \right) \qquad (10)$$

where:

s_i is the share of energy technologies in the market and a_i is the substitution parameter. Obviously, the model above cannot take into account the potential of various energy sources and the impact of policy incentives on the evolution of energy technologies. Hence, the model is further improved as:

$$\frac{ds_i(t)}{dp_i(t)} = a_i s_i(t) \left(\widehat{s_i} \left(1 + s_i(t) - \sum_i s_i(t) \right) - s_i(t) \right)$$

$$(11)$$

and

$$p_i(t) = \frac{c_f(t)(1 + \tau(t))}{c_i(t)(1 - \eta_i(t))} \qquad (12)$$

On the left side of the model, the relationship between share and time is modified as the relationship between share and relative price p_i, and p_i is expressed by the price ratio between the reference technology (fossil energy technologies, such as coal) and the new energy technology. Such policy variables as the ad valorem rate of carbon tax τ and renewable energy subsidy rate η_i are also introduced to ensure that the price ratio has reflected the impact of environmental policies on change of relative price. In addition to the impacts of price changes, the price ratio will also be affected by carbon tax and subsidy policies—when the carbon tax rate or subsidy rate increases, the relative price ratio will also increase, driving substitution of zero-carbon new energy for conventional fossil energy. Moreover, the market potential of various energy technologies can be expressed with the parameter S_i, and therefore $0 \leq s_i(t) \leq S_i \leq 1$.

The finite difference equation tends to be easier for working out the numerical iterative solution than the continuous differential equation, so the study differentiates Eq. (11) into the following form.

$$s_i(t+1) = s_i(t) + a_i s_i(t) \left(\widehat{s_i} \left(1 + s_i(t) - \sum_i s_i(t) \right) \right.$$
$$\left. - s_i(t) \right) (p_i(t+1) - p_i(t)) \qquad (13)$$

The new energy cost in the model is mainly described by the empirical curve of learning by doing, i.e. the unit cost of each technology will decrease with its increased accumulated installed capacity:

$$c_i(t) = c_i(0) \left(\frac{kdg_i(t)}{kdg_i(0)} \right)^{-b_i} \qquad (14)$$

where:

$c_i(0)$ is the initial energy cost and kdg_i is the knowledge stock of technology i which is generally characterised by the cumulative installed capacity, i.e. the knowledge stock in a new period is the knowledge stock in the previous period minus outdated knowledge factors, plus the new installed capacity of the technology in the current period:

$$kdg_i(t+1) = (1 - \psi)kdg_i(t) + s_i(t+1)e(t+1) \qquad (15)$$

where:

ψ is the knowledge depreciation rate (also known as out-of-date rate). In Eq. (14), Parameter b_i is the learning index, and its relationship with the learning rate is:

$$1 - lr_i = 2^{-b_i} \qquad (16)$$

Learning rate is generally defined as the rate of cost reduction induced by a doubling of cumulative production or installed capacity. In addition, the model's treatment of fossil energy price change is relatively simple—it assumes that the future price of fossil energy will rise due to resource scarcity and an increase in uncertainty in energy imports; it then sets the annual average growth rate to define the fossil energy price change exogenously. Therefore, the composite price of each energy source can be obtained by weighting the share of the energy source in primary energy consumption, i.e.

$$pe(t) = cf \left(1 - \sum_i s_i(t) \right)(1 + \tau(t)) + \sum_i s_i(t)c_i(t)(1 - \eta_i(t)) \qquad (17)$$

Climate change is a global challenge, and there are many uncertainties in defining the impacts of regional carbon emissions on atmospheric concentrations and rising global temperature. What's more, the environmental losses (especially non-market losses) are very difficult to measure accurately. Therefore, the model simplifies the environment module in the traditional integrated assessment model (IAM)—it only takes into consideration the man-made CO_2 emissions from production activities and the cumulative emissions from natural net emission factors, excluding the feedback into production activities from the greenhouse effect caused by emissions. The emission equation is given below:

$$emis(t) = \xi_f s_f e(t) + natem(t) \qquad (18)$$

where:

$emis(t)$ is man-made emissions, i.e. the product of fossil energy consumption and carbon emission factor ξ_f, and $s_f(t)$ is the share of fossil energy consumption:

$$s_f(t) = 1 - \sum_i s_i(t) \qquad (19)$$

where:

$natem(t)$ represents the annual natural emissions from China's landmass and ocean. Annually cumulative carbon emissions $cumem(t)$ can be expressed as:

$$cumem(t+1) = (1 - sr)cumem(t) \\ + emis(t+1) \qquad (20)$$

where:

parameter sr is the natural sinking rate of CO_2.

2.2 Data Processing and Parameter Estimation

The model built in this section is a cross-period dynamic optimisation model for China. Starting from 2010, the model defines every five years as a period, and looks at the policies for 2015–50. According to the latest data from the National Bureau of Statistics (NBS) of China, by the end of 2010, China's total population was 1.341 billion. Moreover, based on research on China's future population by Men Kepei et al.[2] and predictions by the World Bank, China's population is assumed to peak at 1.47 billion. Other key macroeconomic initial values and parameter values are given in Table 1.

In addition to such fossil energy sources as coal, oil and natural gas, the model also looks at seven non-fossil energy sources: biomass, nuclear, hydropower, geothermal, solar photovoltaic, wind and marine energy. The consumption of each energy source in the base year uses the coal equivalent calculation data in the China Statistical Yearbook 2011. See Table 2 for details.

It is difficult to obtain the initial energy cost. For easy quantitative calculation by the model, the study chooses the unit end-use cost of each energy source (RMB/tce). Fossil energy cost is the average of the coal price in China, the international crude oil price and the price of imported natural gas. The cost of using new energy technologies varies greatly due to their varying installed capacity, technological level and method of utilisation. Based on the estimated cost fluctuation of various renewable and new energy technologies by Anderson et al.[3] and Gerlagh et al. and in the Energy Report 2050 by the China Energy and Carbon Emissions Workstream, the study calculates the initial energy cost of each energy source as shown in Table 2.

In this model, wind energy is considered for power generation, without distinguishing between onshore and offshore wind power. Solar photovoltaic (PV) and marine energy are also considered for power generation only, but biomass and geothermal energy are considered for both power generation and non-electricity uses, with the focus on the latter. In addition, due to lack of official data on the systematic introduction of renewable energy, the data used in this study is estimated, based on the China Statistical Yearbook 2011, the Annual Report on Electricity Regulation 2010 and World Energy Resources 2010 published by the World Energy Council. As the share of renewable energy in China's primary energy consumption is very small, the model results are insensitive to these initial data.

---

[2]Kepei Men, Lianyu Jiang and Hongting Zhu, China Population Projection Based on Two New Grey Models. Economic Geography, Volume 27 (6): pp. 45–49 (2007).

[3]Anderson, D. and Winne, S., Innovation and Threshold Effects in Technology Responses to Climate Change. Working Paper 43, Tyndall Centre for Climate Change Research, (2003).

Table 1 Assumptions of key macroeconomic initial values and parameter values

Name	Value	Source
GDP (gdp)	40.12	China Statistical Yearbook 2011, UoM: RMB trillion
Investment (i)	27.71	
Consumption (c)	13.33	
Exports (x)	10.70	
Imports (m)	9.47	
Initial time preference rate (σ)	0.03	Refer to the settings of the world's average level in the DICE and RICE models (Nordhaus, 2007[a])
Annual decrease rate of time preference (d_σ)	0.3%	
Capital depreciation rate (δ)	5%	Set at 7% by Nordhaus and Yang (1996), Gerlagh et al. (2004) and Popp (2004), and at 3% by Kumbaroglu et al. (2008[b]). The annual capital depreciation rate is assumed at 5%
Share of capital (γ)	0.31	Refer to Nordhaus (2007), Gerlagh et al. (2004)
Elasticity of substitution (ρ)	0.40	
Initial AEEI	0.70	Refer to the settings of the world's average level in the DICE and RICE models (Nordhaus, 2007)
Annual decrease rate of AEEI	0.2%	
Lower limit of the share of exports in GDP (θ_1)	0.40	Based on the share of China's actual imports and exports in 1995–2011 (China Statistical Yearbook, 1995–2011)
Upper limit of the share of imports in GDP (θ_2)	0.30	
Carbon emissions factor (ξ)[2]	0.645	The calculation method in IPCC Guidelines for National Greenhouse Gas Inventories

[a]Nordhaus, W., The Challenge of Global Warming: Economic Models and Environmental Policy. (New Haven, USA, 2007)

[b]Nordhaus, W.D. and Yang, Z., A Regional Dynamic General Equilibrium Model of Alternative Climate Change Strategies. American Economic Review 86, pp. 741–765, (1996). Gerlagh, R., van der Zwaan, B., A sensitivity Analysis of Timing and Costs of Greenhouse Gas Emission Reductions. Climatic Change 65, pp. 39–71, (2004). Popp, D., ENTICE: Endogenous Technological Change in the DICE Model of Global Warming. Journal of Environmental Economics and Management 48, pp. 742–768, (2004). Kumbaroğlu, G., Karali, N. and Arıkan, Y., CO$_2$, GDP and RET: An Aggregate Economic Equilibrium Analysis for Turkey. Energy Policy 36, pp. 2,694–2,708, (2008)

The substitution parameters and technical learning rate in the logistic model of policy are provided in Table 2. The substitution parameter in this model refers to the key parameters that affect substitution between energy sources. For the substitution parameter value, this study refers mainly to the research by Anderson et al. (2003) which discusses the non-linear behaviour of this model, and the relationship between the substitution parameter and the variance of relative price distribution, as well as the basis for the reference value. The substitution relationship between energy structure and technology is shown in Fig. 17. Learning rate is an important parameter representing new energy technology

progress and cost evolution. McDonald and Schrattenholzer[4] made initial estimates of the technological progress rate of various energy technologies. Later, Kumbaroglu et al., Rout et al. and Rubin et al.[5] reviewed the estimates in

[4]McDonald, A., Schrattenholzer. L., Learning Rates for Energy Technologies. Energy Policy 29, pp. 255–261, (2001).

[5]Rout, U.K., Blesl, M., Fahl, U., Remme, U. and Voß, A,. Uncertainty in the Learning Rates of Energy Technologies: An Experiment in a Global Multi-regional Energy System Model. Energy Policy 37, pp. 4,927–4,942, (2009).Rubin, E. S., Azevedo, I. M. L., Jaramillo, P. and Yeh, S., A review of Learning Rates for Electricity Supply Technologies. Energy Policy 86, pp. 198–218, (2015).

Table 2 Initial share of technology, energy cost, rate of technical substitution and learning rate

	Share of energy technology (%)	Initial cost RMB/tce	Technical substitution capability coefficient	Rate of new energy technology progress (%)
Coal	66.32	2,034.17	–	–
Oil	19.1	3,336.05	9.5	–
Natural gas	5.9	1,505.29	9.0	–
Hydropower	7.10	1,627.34	7.5	0.970
Nuclear power	0.73	4,068.35	9.0	0.940
Wind power	0.19	4,475.18	6.3	0.885
Solar PV	2.73 E−4	16,273.39	5.5	0.785
Biomass	0.57	8,136.70	4.3	0.920
Marine energy	2.60 E−5	9,764.04	4.0	0.795
Geothermal energy	0.0797	8,950.37	4.0	0.870

Fig. 17 Relationship between energy structure and technical substitution

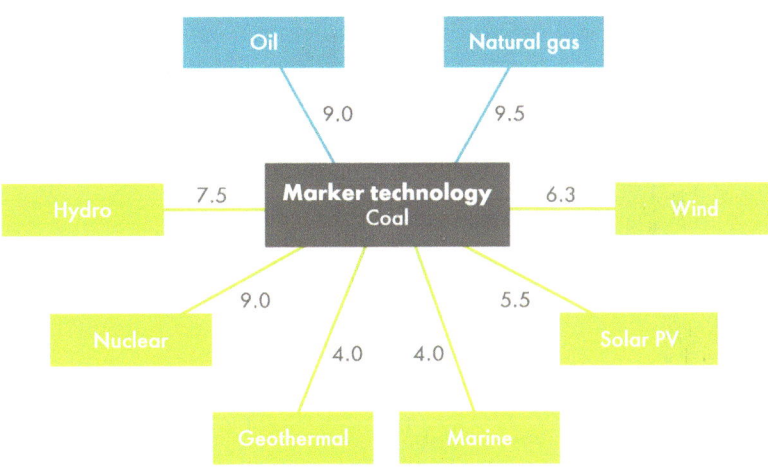

their more recent research. In general, new technologies have greater potential to drive technological progress, and thus have a higher learning rate of 12.9–18.7%. The relatively mature technologies have a slightly lower learning rate of 9.8–12.9%. The basically proven technologies have relatively small potential for technological progress and accordingly have a lower learning rate of about 7%.

2.3 Baseline Results

From the simulation results, China's economy will continue to grow in the future despite its slowing growth rate. Specifically, China's economic aggregate will increase from $8.7 trillion in 2015 to $20.5 trillion in 2030 and about $43 trillion by 2050 (Fig. 18). However, it will be difficult for China to maintain its current growth

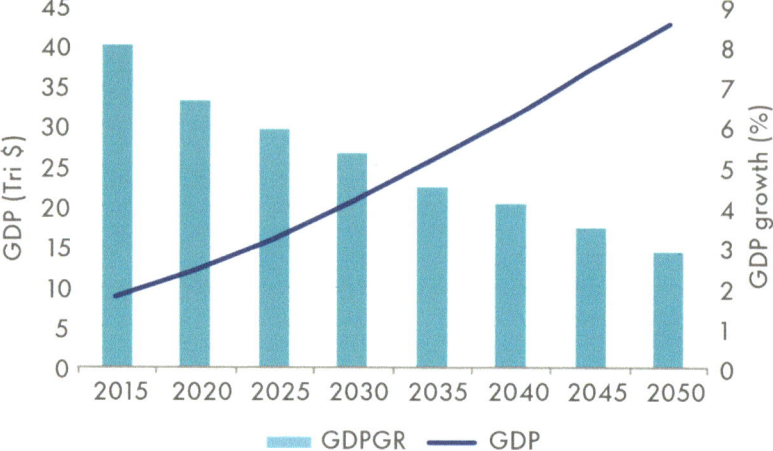

Fig. 18 China's macroeconomic development in the Baseline scenario. *Note* GDPGR = GDP growth

rate. In fact, compared with the 8% average growth rate during the 12th Five-Year Plan (2011–15), average economic growth during the 13th Five-Year Plan (2016–20) is expected to decrease to around 6.6%, and decline further to 4% and 2.86% respectively in 2040 and 2050.

To verify the reasonableness of the estimated economic development of this model, the study compares the relevant results with the predictions for China's future economic development by research institutions, as shown in Fig. 19. Compared with the economic growth estimates provided by these research institutions, the model result stays around the midpoint. In particular, the Energy Research Institute of the National Development and Reform Commission (NDRC) provides a relatively optimistic prediction of China's future economic development—it believes that China's macroeconomy still holds a steady growth potential of 8% by 2020, and even in 2020–30, the average economic growth rate is expected to be 7.1%. Lawrence Berkeley National Laboratory also believes that China's economy will maintain strong momentum by 2020, with an average growth rate of up to 7.8%. In contrast, the economic growth expectations of

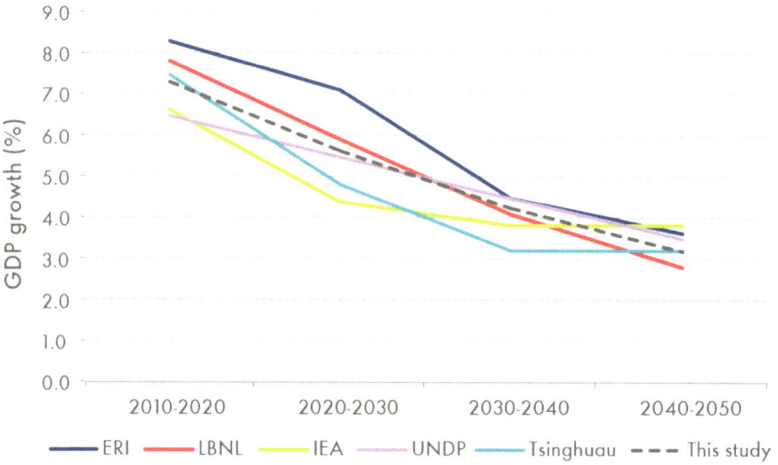

Fig. 19 Cross-study comparison of China's economic expectations by 2050. *Note* ERI = Energy Research Institute of the National Development and Reform Commission (NDRC); LBNL = Lionel Berkeley National Laboratory; IEA = International Energy Agency; Tsinghua University

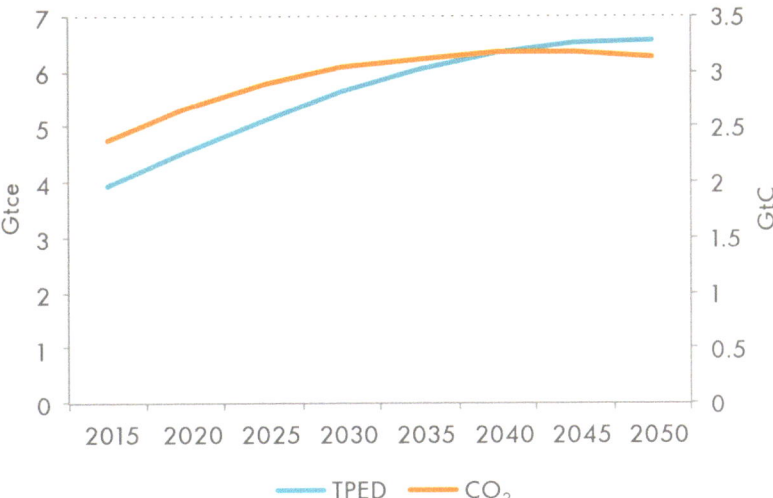

Fig. 20 Energy demand and carbon emissions path in the Baseline scenario. *Note* Gtce = gigatonnes of carbon emissions; TPED = total primary energy demand

the International Energy Agency and Tsinghua University are lower—they predict China's average GDP growth rate to be 5.5% and 6.2% respectively by 2030. Tsinghua University forecasts China's economic growth expectation in three scenarios—Optimistic, Moderate and Pessimistic—in its research report. In this study, only the forecast in the Moderate scenario is used for comparison.

Figure 20 shows China's future energy consumption and expected carbon emissions in our Baseline scenario. China's total primary energy demand (TPED) will stay in a growth trajectory in the long term, although the rate of growth is slowing down. Energy demand will be 4.53 Btce in 2020 and 6.35 Btce in 2040, the latter double that of 2015. This is consistent with the research conclusion of He Jiankun (2013). China's future CO_2 emissions will peak at 3.17 gigatonnes of carbon (GtC), equivalent to 11.6 billion tonnes of CO_2, by around 2040 in the Baseline scenario. This indicates that without policy interventions to reduce emissions, such as a carbon tax or carbon emissions trading, China's goal of reaching a carbon emissions peak by 2030 will be difficult to achieve.

2.4 Evolution of the Energy Structure and Development of Non-fossil Energy

Adjustments to the conventional energy system and the substitution of new energy technologies for fossil fuels depend largely on policy incentives. Our model, therefore, uses three incentive policy scenarios to encourage a shift to non-fossil energy technologies—Conservative, Moderate and Optimistic. This section will focus on the dynamic evolution of China's future energy structure in those three scenarios, focusing on the long-term development path of the various non-fossil energy technologies.

Figure 21 shows the dynamic evolution of the energy structure in the Moderate scenario, which is expressed by demand for different energy sources. It is easy to see that coal remains the dominant energy source in 2050, despite its decreasing share of total primary energy demand. Even in 2050, coal's share of total primary energy demand will be as much as 41.2%. The share of oil remains relatively stable—on the one hand, diminishing oil reserves determine its limited growth potential; on the other, the

Fig. 21 Energy structure in the Moderate scenario. *Note* GtC = gigatonnes of carbon

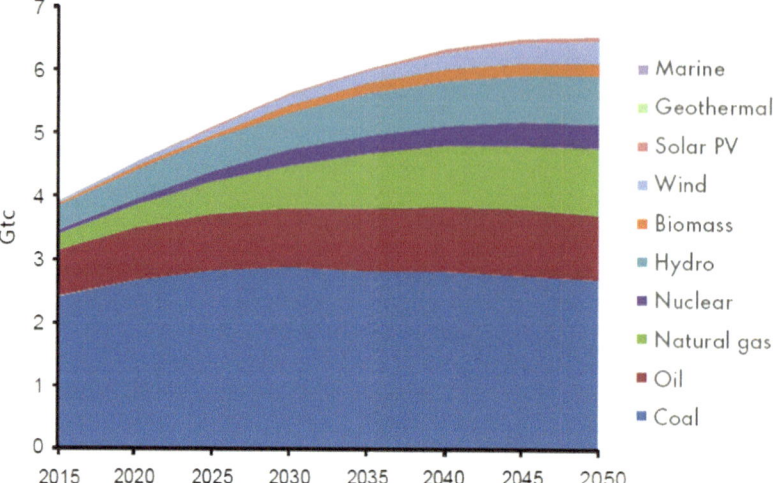

Fig. 22 Energy technology share path in the Moderate scenario

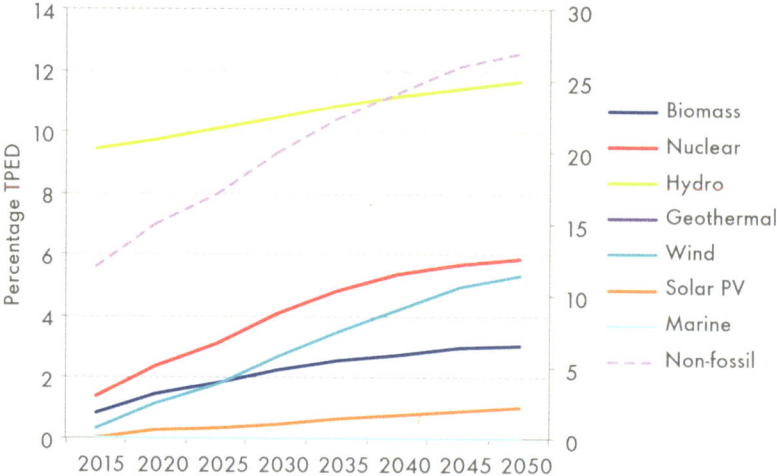

dependency of economic development on oil and the inertia of related industries prevent oil's share of demand from being squeezed by new energy technologies. By 2050, the share of oil in energy demand will decrease by only 3.3% from 2015. Natural gas demand will grow substantially, from 260 Mtce in 2015 to 670 Mtce in 2030, and its share of energy demand will increase from less than 6–12%.

In non-fossil energy technologies, hydro-power and nuclear power will retain their dominance in China's future non-fossil energy demand. By mid-century, the share of hydro and nuclear in total primary energy demand will reach 10.5% and 4.1% respectively (Fig. 22).

Wind and solar PV will also achieve significant growth. In particular, wind power will gain strong momentum after 2025, with demand reaching 349 Mtce, or 5.3% of total primary energy demand (TPED), by 2050 (Fig. 22). Non-fossil energy's share of TPED reaches 20% in the Moderate scenario, which suggests that the government can deliver its goal of non-fossil energy development by 2030 on schedule. By 2050, the share of non-fossil energy will rise to 27%. It should be noted, however, that the rapid increase in non-fossil energy is mainly due to conventional technologies like hydro and nuclear power, and that the combined contributions of non-hydro renewable energy remain low. For

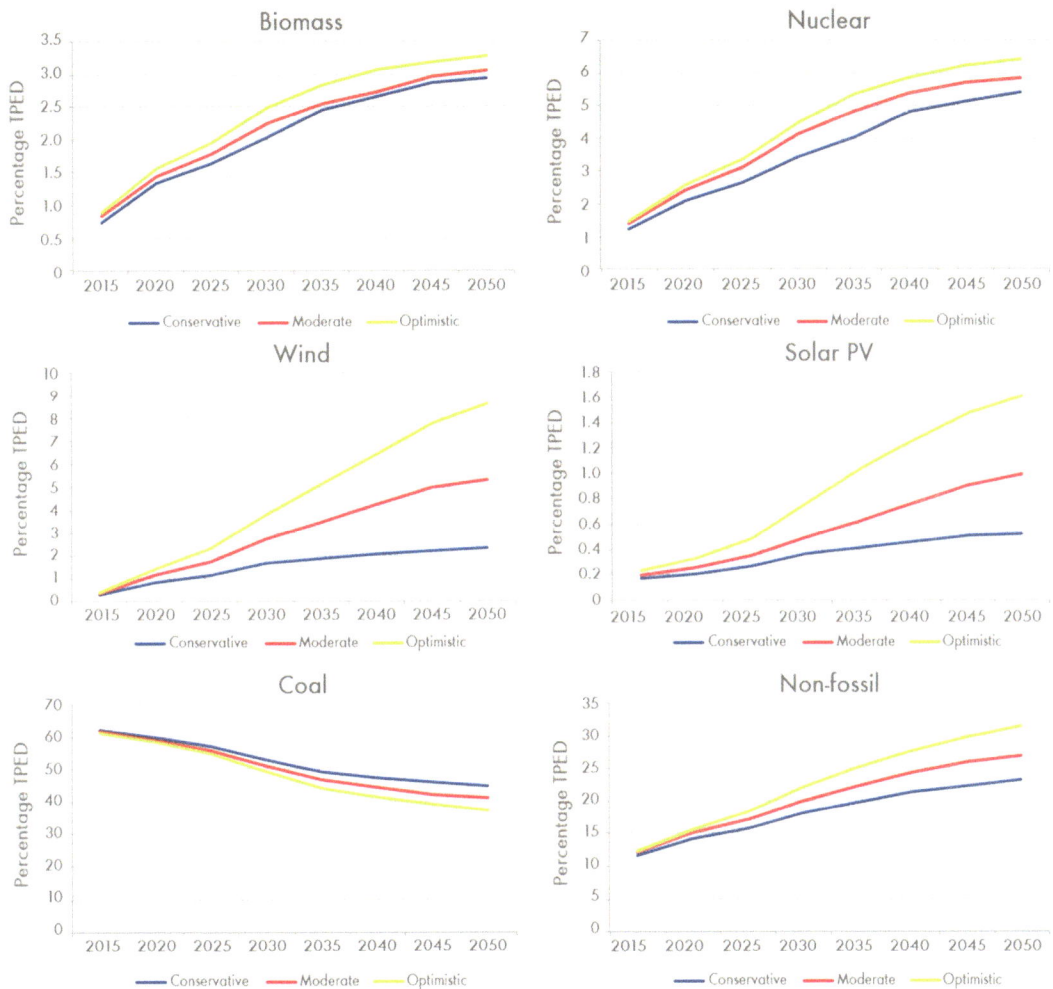

Fig. 23 Impact of different incentive polices on the evolution of energy technologies. *Note* TPED = total primary energy demand

instance, by 2050, the share of non-hydro renewable energy in TPED will be 15.2%, and the share of non-hydro and non-nuclear renewable energy in TPED will be even lower at 9.4%.

The development of each major energy technology (in terms of growth in its share of TPED) in different incentive policy scenarios is shown in Fig. 23. Wind and solar PV are the most sensitive to incentive policy, which suggests they have greater development potential in the future. In the prediction results, the share of wind power demand is 8.6% in the Optimistic scenario, 6.3% higher than in the Conservative scenario. Similarly, solar PV shows remarkable development in

the Optimistic scenario, reaching 0.75% in 2030 and 1.59% in 2050. In contrast, the share of solar PV will be only 0.51% in 2050 in the Conservative scenario, a difference of more than 1% compared with the Optimistic scenario.

Total non-fossil energy demand will grow rapidly. In the Moderate scenario, China's goal of achieving a 20% share of non-fossil energy by 2030 will be delivered on schedule, and in the Optimistic scenario the figure will reach 22%. By 2050, the share of non-fossil energy in the Optimistic scenario will be 31.2%, about 4% and 8% higher than the Moderate and Conservative scenarios respectively. This is basically

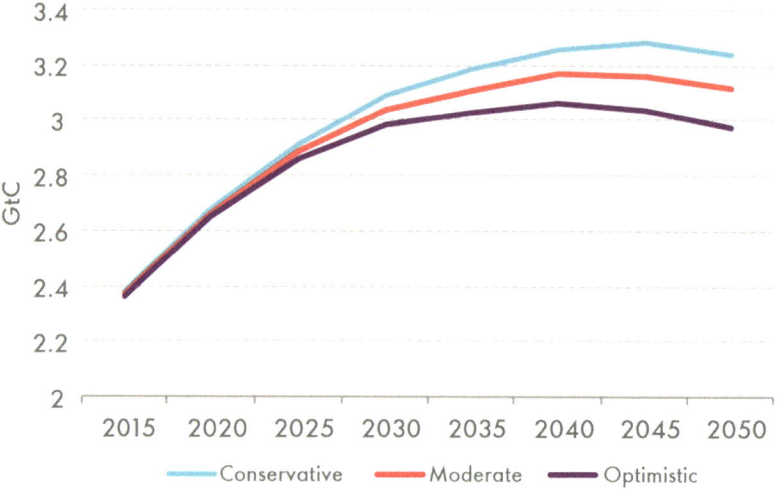

Fig. 24 Impact of non-fossil energy technology on carbon emissions. *Note* GtC = gigatonnes of carbon

consistent with the predictions of He Jiankun.[6] Due to rising demand for non-fossil energy technologies, coal's share of TPED will be continuously squeezed. Even so, coal will still make up more than 50% of demand in 2030 in the Conservative and Moderate scenarios. In the Optimistic scenario, coal's share of demand will decrease to 49% in 2030 and 37.1% in 2050, which echoes mainstream opinion about the long-term decline in coal demand.

Finally, the study briefly analyses the impact of energy technology development on future carbon emissions, as shown in Fig. 24. Policy stimulates technology development, which in turn can generate significant impacts on carbon emissions. In the Optimistic scenario, technology development reduces total carbon emissions, especially in the later maturing stages of technologies. On the other hand, technology can help carbon emissions to peak early. Figure 24 shows that in the Optimistic scenario, although China's goal of attaining a peak in carbon emissions by 2030 is not realised, the peak will be achieved around 2035 (at 3.06 GtC, equivalent to 11.2 Bt of CO_2), which is earlier than the Conservative and Moderate scenarios.

2.5 Energy Development and Climate Change: Optimisation and Choice of Policy

The significance of using policy to achieve specific energy and emission reduction goals has aroused extensive interest among researchers, both in China and abroad, especially with regard to scenarios on policy choice and policy cost evaluation. He Jiankun[7] established the low-carbon scenario indicator system to determine if China's CO_2 emissions would peak on schedule in 2030. He identified two preconditions for delivering the peak carbon emissions goal on time. Policy scenario-makers have since identified other key factors relevant to achieving the peak CO_2 goal. These include the transition from a high-growth to a slower growth economy, improvements in energy efficiency, progress in non-fossil energy technologies (nuclear power and renewables), deployment of carbon capture and storage (CCS), and the shift to a low-carbon lifestyle. With proactive policy packages, China's carbon emissions from energy-related activities could peak by 2025 or even earlier. However, current emission reduction efforts are not enough to achieve the goals of peak carbon

---

[6]Jiankun He, China's Energy Development and Response to Climate Change. China Population, Resource and Environment, Volume 21 (10): pp. 40–48, (2011).

[7]Jiankun He, CO_2 Emission Peak Analysis: China's Emission Reduction Goals and Policies. China Population, Resource and Environment, Volume 23 (12): pp. 1–9, (2013).

emissions and a 20% share of non-fossil energy by 2030. It is therefore necessary to introduce stronger policies and incentives to improve energy efficiency and renewable energy uptake and reduce carbon emissions.

2.5.1 Policy System Design and Basic Assumptions

Carbon pricing is a widely used policy across the globe to address the challenge of reducing carbon emissions. It is also an important policy option for China to help achieve the goal of peak carbon emissions by 2030. Carbon pricing includes carbon emissions permit trading via cap control and a carbon tax levied by price regulation. Theoretically, a balanced carbon market price is equivalent to the optimal carbon tax level, which means that carbon tax and carbon emissions trading can deliver the same policy effect under given conditions. Based on this assumption, carbon emissions are mainly controlled by introducing an endogenous (controllable) carbon tax in the model simulations of this report. A carbon emissions reduction policy alone provides limited incentives, especially in the short and medium terms. A relevant subsidy (ad valorem) is an essential policy option for driving the diffusion of new energy technologies. This study therefore considers a non-fossil energy subsidy as the second endogenous variable (after carbon tax) in the model optimisation process. These two variables allow us to analyse the impacts of coordinated and optimised policies on China's intended nationally determined contributions target for the 2015 Paris Agreement.

The optimised endogenous (controllable) carbon tax path requires setting an exogenous (uncontrollable) carbon emissions cap (CAP). The CAP used in this study is set by mainly referring to the additional carbon emissions that could be allocated to China under the scenario of limiting global temperature rise to below 2°C. Raupach et al.[8] provide the carbon emission space allocation plan for all countries and regions based on the principles of fairness, historical emission inertia and mixing under the 2°C scenario. China's cumulative carbon emissions will reach 105.55 billion tonnes by 2050 under the representative grandfather clause. Therefore, this study sets the estimated value as the exogenous CAP.

In the process of model optimisation, this study assumes that carbon tax revenues are always sufficient to compensate for the cost of subsidies, and that different policy mix options are realised by adjusting the ratio between cumulative carbon tax and subsidies throughout the entire simulation period (2010–50). When calculating the cumulative carbon tax and subsidies, this study follows international estimation practice and uses a discount rate of 5%, which is consistent with the capital depreciation rate used in the model. The cumulative carbon tax is the sum of the carbon tax on three fossil fuels (coal, oil and natural gas), and the cumulative subsidy is the sum of the subsidies for seven non-fossil energy technologies: biofuel, nuclear, hydro, geothermal, wind, solar and marine.

2.5.2 Analysis of INDC Target and Policy Optimisation

The policy optimisation results mainly reflect the impacts of optimised policy on the 2030 peak carbon goal and the non-fossil energy development goal.

When policy is optimised around the peak carbon and non-fossil energy development goals, the tax revenue from carbon pricing policy is far higher than the cost of non-fossil energy subsidies. As shown in Fig. 25, to achieve the 2030 peak carbon emissions goal, the ratio between cumulative carbon tax and subsidies needs to be more than 4:1. On the one hand, the policy does not need funding, other than the carbon pricing revenue to meet the cost of subsidies. On the other hand, the high ratio between cumulative carbon pricing revenue and subsidy costs is related to the 2030 peak carbon emissions goal. Generally, the higher the ratio between carbon pricing policy and subsidy policy, the greater the possibility of achieving peak carbon emissions

---

[8]Raupach, M. R., Davis, S. J., Peters, G. P. and R. W. Andrew, et al. Sharing a quota on cumulative carbon emissions. Nature Climate Change, 4: pp. 873–879, (2014).

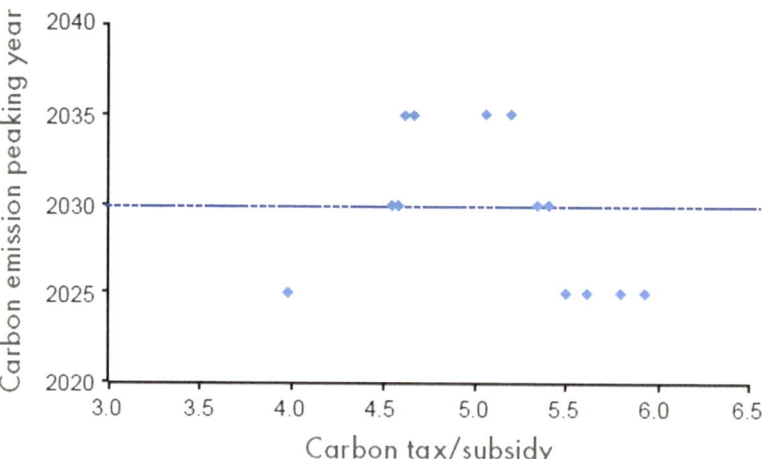

Fig. 25 The relationship between policy optimisation and peak carbon emissions

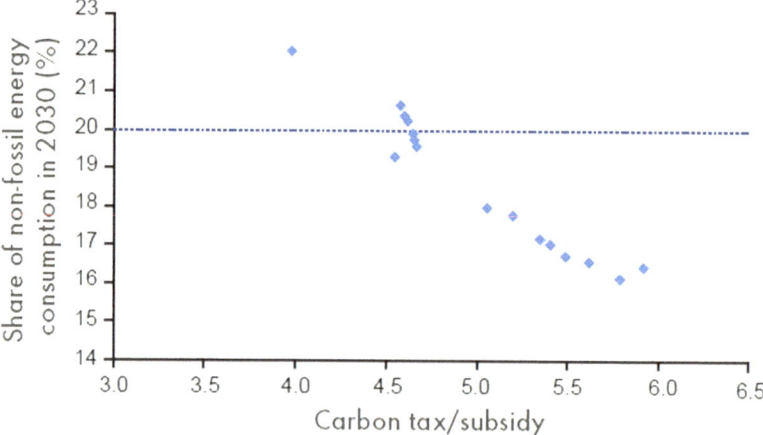

Fig. 26 The relationship between policy optimisation and the non-fossil energy development goal

early. For example, to reach peak carbon around 2025, the ratio between cumulative carbon pricing revenue and subsidies will be more than 5.5:1. Figure 25 shows that under the High and Low policy mix options, the goal of reaching peak carbon emissions by 2030 is possible—the difference is mainly in the peak level of emissions. In general, the stronger the carbon pricing policy, the lower the corresponding carbon emissions peak level. For instance, when the ratio between carbon tax and subsidies is 4.5:1 and 5.4:1, carbon emissions can peak around 2030 at 10.3 billion tonnes and 10 billion tonnes of CO_2 respectively. Therefore, when assessing how to achieve peak carbon, optimal policies and the difference in peak level need to be taken into account.

Comparatively speaking, the development of non-fossil energy is more significantly affected by subsidy policy than by carbon tax. As indicated in this study, the higher the share of subsidies in the policy mix, the faster the development of non-fossil energy technologies. As the carbon tax/subsidy ratio decreases, the share of non-fossil energy consumption steadily increases (Fig. 26). The figure shows that when the tax/subsidy ratio is higher than 5.5:1, the share of non-fossil energy in total primary energy demand is around 17% in China, but that when the ratio approaches 4.5:1, the share rises to around 20%. In particular, when subsidy policy strengthens and the ratio between cumulative carbon tax and subsidies is below 4:1, the share of non-fossil energy consumption will be higher

Fig. 27 The relationship between the peak carbon emission and non-fossil energy development goals

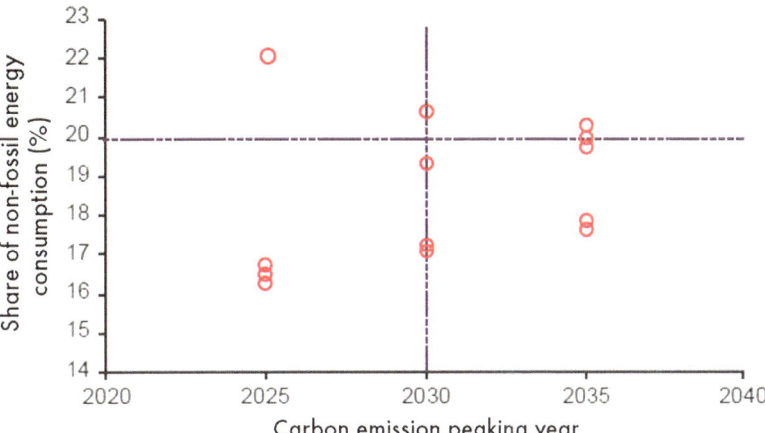

than 22%. This indicates that to achieve the 2030 peak carbon emissions and non-fossil energy development goals at the same time, China should take carbon pricing and subsidy policies into account and coordinate delivery of the two goals.

In the context of policy optimisation, the relationship between the peak carbon emission and non-fossil energy development goals is shown in Fig. 27. This relationship can be analysed across two dimensions. On the one hand, the two goals conflict in most cases, i.e. the looser the controls on carbon emissions, the greater the possibility of delivering the non-fossil energy development goal, and vice versa. This is because loose carbon emissions control reduces the need for carbon pricing in the optimisation process, which strengthens the role of subsidies for the development of non-fossil energy in the policy mix. On the other hand, there is a potential synergy between the peak carbon and non-fossil energy development goals, in that the two targets can be achieved at the same time under certain policy mix options. In particular, when subsidy policy plays a sufficiently significant role, it not only drives non-fossil energy technology development and delivers the targeted share of non-fossil energy, it also compensates for the emissions reduction effect of carbon pricing policy, thus achieving an early peak in carbon emissions. Figure 27 shows that when the ratio between cumulative pricing and subsidy is low at

3.9:1, China's CO_2 emissions will peak early around 2025, while the share of non-fossil energy consumption will reach 22%.

2.5.3 Analysis of Policy Choice and Macroeconomic Costs

The macroeconomic cost of energy and climate policies correlates significantly with the role of carbon pricing in the policy mix. As shown in Fig. 28, as the carbon tax/subsidy ratio increases, cumulative policy costs rise substantially (if the discount rate is 5%). When the ratio between cumulative carbon tax and subsidy is around 5:1, the cost of the policy mix is only 0.19% of GDP, but when the ratio is around 6:1, the cost rises to 0.77%.

However, when the role of subsidies in the policy mix increases (i.e. the carbon tax/subsidy ratio decreases), cumulative policy costs decrease sharply. In particular, when the ratio is lower than a certain threshold (4.7:1 for example), a policy mix of carbon pricing and subsidies will not be detrimental to China's macroeconomic growth. As indicated in Fig. 28, when the ratio between cumulative carbon tax and subsidies is lowered to 4.66:1, the overall gains from that policy mix will amount to 0.27% of GDP, and when the ratio further decreases to less than 4:1, the gains will amount to 0.75% of GDP.

The reduction in CO_2 emissions achieved by the carbon pricing policy comes mainly from a reduction in fossil energy consumption. This will

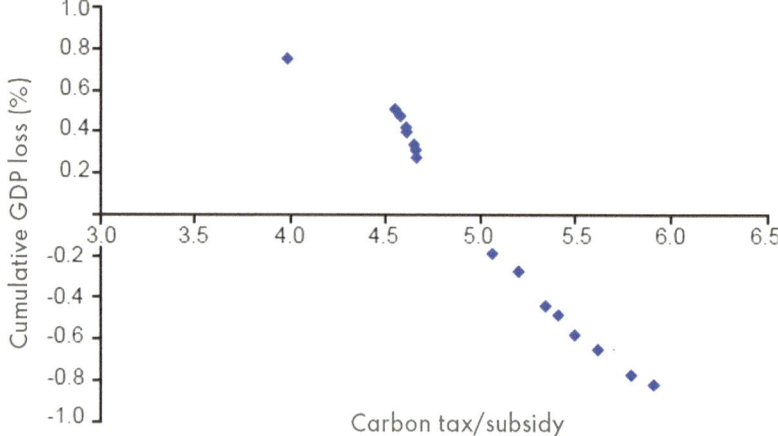

Fig. 28 Impact of policy choice on China's macroeconomy under the energy and climate policy goals

undoubtedly have a negative impact on energy-driven economic growth, especially as fossil fuels dominate total energy consumption. However, as the subsidy policy takes hold and the incentive effects become increasingly evident, renewable energy technologies like wind and solar will mature and be deployed at scale, replacing fossil energy technologies to support and drive macroeconomic growth.

2.5.4 Proposals for Policy Choice

In the process of delivering the non-fossil energy development and peak carbon emission goals, conflicts and synergies coexist, depending on the policy mix options adopted. The stricter the emission controls, the more likely peak carbon emissions will be achieved on time. When carbon pricing plays a prominent role in the policy mix, the non-fossil energy development goal is more difficult to achieve. When subsidies play a large enough role in the policy mix, the non-fossil energy development and peak carbon emission goals can be realised on schedule at minimal cost. When the carbon emission reduction policy is well designed, the decarbonisation goal can be achieved completely at near-zero cost.

To ensure smooth delivery of the non-fossil energy and peak carbon emission goals by 2030, the Chinese government should take the interactive relationship between the two goals into consideration. The government should focus on choosing and optimising a policy mix of carbon pricing and subsidies. Understanding how the two goals interact will help determine the optimal policy mix and strategy to smoothly deliver the two goals. In so doing, any conflicts between the two goals should be minimised or avoided and their synergies harnessed. This requires the government to focus on the following two factors when developing support policies: first, a policy mix—in which carbon pricing and subsidies play a dominant and an ancillary role respectively—should be adopted; and second, the policy mix should be continuously refined until the optimal combination point is found, i.e. increasing subsidy policies without changing the dominant ancillary relationship between carbon pricing and subsidies. It is, therefore, inappropriate to reduce or cancel technology subsidies and rely only on carbon pricing to deliver China's energy and climate goals when renewable energy technologies have not been commercially deployed at scale. Only when the carbon pricing and subsidy mix is fully optimised can the two goals be realised on schedule, and policy costs minimised and the benefits of the policy mix strategy harvested.

3 The Effects of the Energy Pricing Mechanism on China's Medium- and Long-Term Energy Demand

3.1 The Evolution of Fossil Energy Prices in the Medium and Long Terms

3.1.1 Coal Market

As the most important primary energy source in China, coal plays a dominant role in the entire energy demand structure. As shown in Fig. 29, despite the reduction in coal consumption over the past five years and a decline in absolute terms in 2015 of 1.5%, China's coal consumption in 2015 reached 1,920.4 Mtoe, which is 63.7% of the country's total energy consumption. China is the world's largest coal consumer, accounting for half of global demand in 2015. China's coal market is, therefore, not limited to its own supply and demand, but affects the international coal market as a whole.

In light of current medium- and long-term economic development trends in China, both supply and demand in China's coal market may undergo major transformations and be significantly affected by policy factors.

On the demand side, China's economic development is now undergoing a major transition: economic growth is shifting from high speed to a new normal of medium to high speed, and future growth in total energy demand will slow down as the transition progresses. China is also under great pressure to address haze and other environmental problems, and this pressure will continue in the short and medium terms. China will also struggle to achieve its energy transition goals away from coal for 2020 and 2030. There is, therefore, limited potential for growth in coal demand in the medium and long terms.

On the supply side, there is severe overcapacity. In China's ongoing structural reform of energy supply, a top priority is to cut excess capacity, mainly in the coal sector. As production is reduced, China's short-term supply of coal will be tight, and prices are likely to rise. This was evident already in 2016. In the last quarter of that

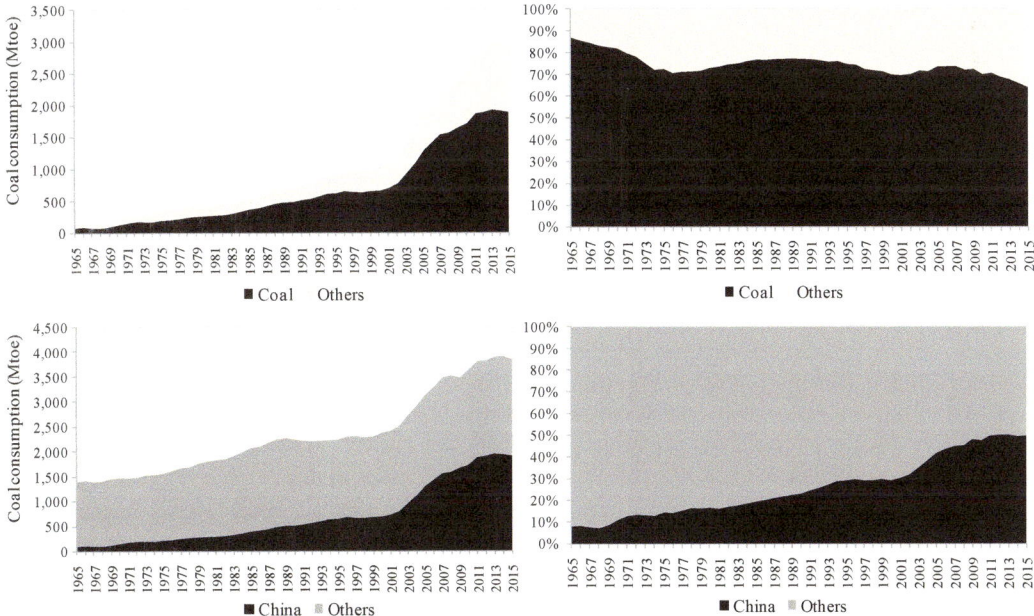

Fig. 29 Evolution of China's total coal consumption. *Source* BP Statistical Review of World Energy (2016)

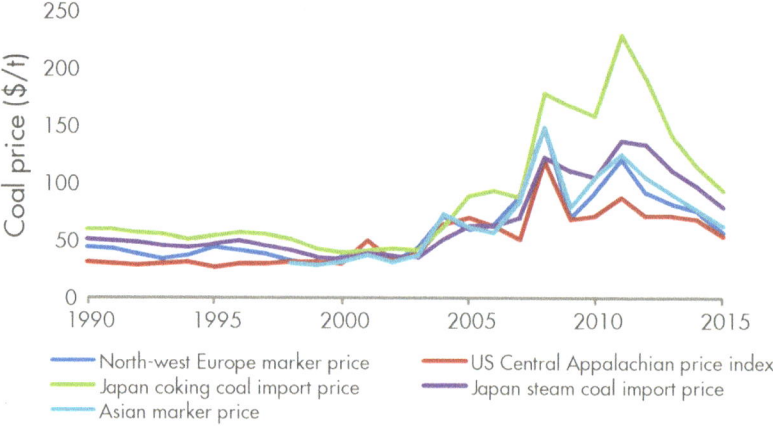

Fig. 30 Price movements in the coal market of major countries and regions. *Source* BP Statistical Review of World Energy (2016)

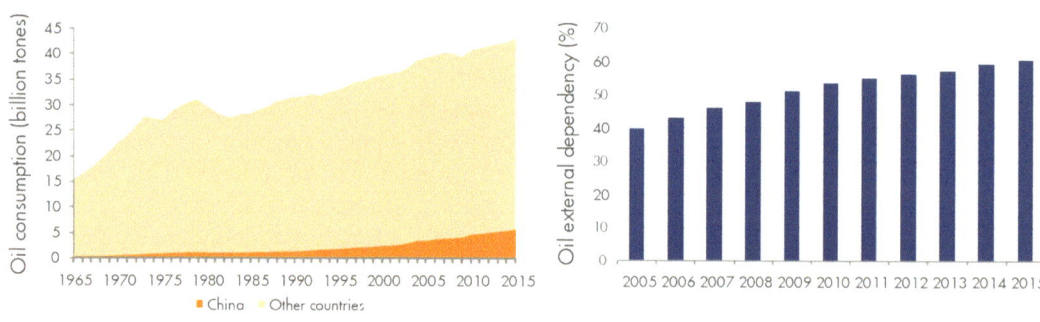

Fig. 31 Evolution of China's oil consumption and dependence on oil imports. *Source* BP Statistical Review of World Energy (2016)

year, coal prices rose rapidly: the price of thermal coal doubled over the year as a whole, that of coking coal almost tripled, and coke prices quadrupled.

In summary, after four years of declining prices (Fig. 30), China's coal prices in the short and medium terms are likely to rise as production is reduced to tackle overcapacity. In the medium and long terms, due to the transformation of China's economy (the new normal of slowing growth), environmental problems and constraints on alternative energy sources, coal prices are not likely to grow significantly.

3.1.2 Crude Oil Market

Due to China's fast economic growth in recent years, the country's oil demand has grown rapidly from 224 Mt in 2000 to 559 Mt in 2015, an average annual increase of 6.3%. China's oil

consumption now accounts for around 13% of the world total, making it the world's second largest oil consumer (Fig. 31). Owing to the limitations of China's resource endowments, the country's own oil production can hardly meet the needs of its rapid economic growth. With the growth in oil consumption, China's dependence on oil imports has been increasing year by year. In 2015, China's net oil imports were 328 Mt and its dependence on oil imports rose to 60.6% (Fig. 31). China's future oil price movements will be closely linked, therefore, with world crude oil price trends.

World crude oil prices have soared since 2000, rising from $30/barrel (bbl) to $100/bbl in 2008. After the global financial crisis in 2008, crude oil prices plunged to $60/bbl, then quickly rebounded to a peak of $110/bbl in 2010 and remained high and volatile in 2013–14. In 2015,

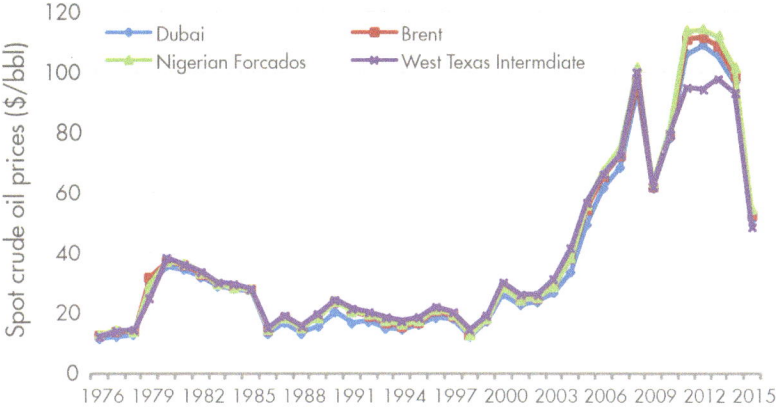

Fig. 32 Crude oil spot prices in major countries and regions. *Source* BP Statistical Review of World Energy (2016)

world crude oil supply and demand underwent profound changes, leading to plummeting crude oil prices in the short term (Fig. 32).

On the supply side, breakthroughs in unconventional oil and gas extraction technology (shale oil and gas) in the USA have greatly increased the country's crude oil production, making it the world's largest oil producer. Second, attempts by OPEC to limit production have had little impact on global oil supply, as its member states tend to focus on their own interests rather than those of the OPEC collective. As a result, oil-producing countries have maintained their production capacity, greatly driving up the world's crude oil supply.

On the demand side, after 2014, three factors combined to make global oil demand lower than expected: China's economic transformation, the weak economic recovery of Europe, and the failure of Japan's economic stimulus policy to make a difference. In the end, as the US dollar became stronger, international oil prices are expected to drop in the medium and long terms.

To sum up, in the medium and long terms, world crude oil prices will remain low within a certain range, but they will be affected by changes in supply and demand and other uncertainties. In particular, the status of the world's major economies (China, the EU and Japan) will determine future oil demand, while changes in US energy policies under the present administration, the production-limiting attempts of OPEC, and the complex geopolitical situation in

the Middle East, will have a significant impact on future oil supply.

3.2 Technology Evolution and Cost Reduction in New and Renewable Energy

New and renewable energy has significant advantages over conventional fossil energy in reducing pollutants and greenhouse gas emissions, while increasing the diversity of energy supply and improving energy security. Therefore, all countries around the world are pinning great hopes on the development of new and renewable energy. Over the past 20 years, especially in the past decade, the world has seen considerable progress in renewable energy, particularly in solar and wind power, the installed capacity of which reached 2,300 GW and 4,350 GW respectively in 2015. In the past decade, China has made remarkable achievements in new energy development. The country's installed capacity of wind and solar power was 440 GW and 1,450 GW respectively in 2015, accounting for 18.9 and 33.3% of the global total, making China the world leader in terms of installed capacity of wind and solar power (Fig. 33).

Without taking the influence of other external factors into account, the cost evolution of the various energy technologies is the fundamental factor that determines energy demand. Other important factors are the relative competitiveness

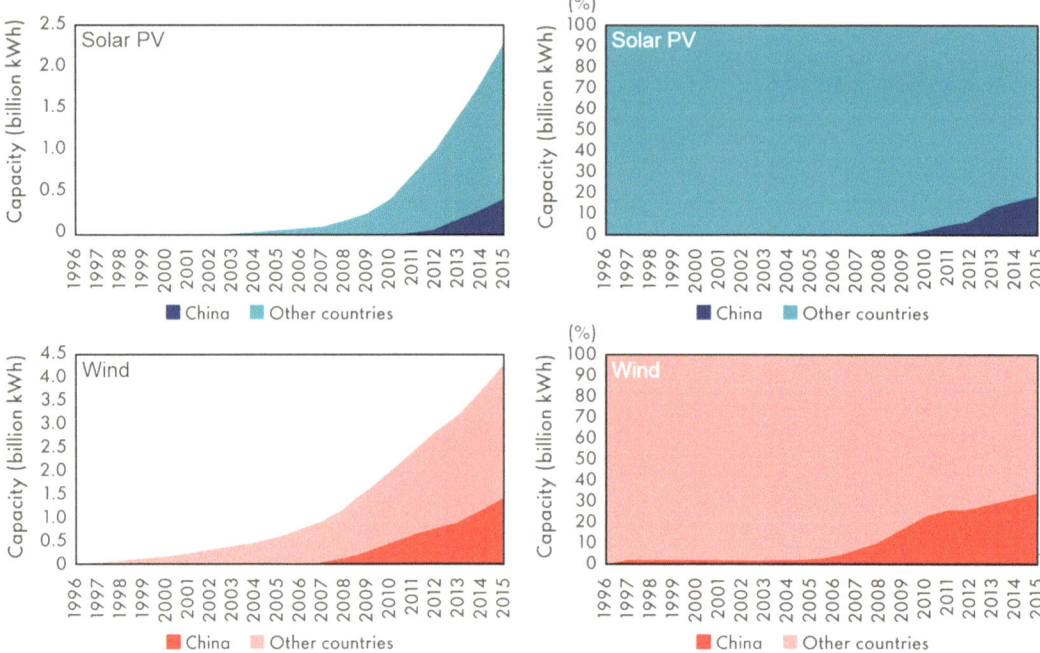

Fig. 33 Development paths for installed capacity of renewable energy in China and the world. *Source* BP Statistical Review of World Energy (2016)

of energy sources and the evolution of the energy demand structure in the medium and long terms. Technology evolution and the cost reduction potential of a mature energy like conventional fossil fuels are relatively limited. Whereas new energy like renewables, especially wind and solar, which are in the early stages of development compared to fossil energy, still have great potential for technology evolution. The rapid development of renewable energy over the past decade has made cost reductions possible, which is evident in China and all over the world. In Fig. 34, the evolution of wind and solar PV generation shows that although costs differ across the world, this is due mainly to differences in investment costs, resource conditions and other factors. The average cost of onshore wind power generation decreased from \$0.2/kWh in 1995 to \$0.10/kWh in 2005 and \$0.07/kWh in 2015, with an average cost reduction of about 65% over the 20-year period. The cost of utility-scale solar PV generation dropped more significantly, from \$0.31/kWh in 2010 to \$0.13/kWh in 2015,

giving an average cost reduction of 58% over the five years.

The technological advances and cost reductions achieved in renewable energy over the past decade are mostly due to significantly lower investment costs, made possible by the learning effects of growing installed capacity. Future reductions in the cost of renewable energy generation will come from lower equipment and component costs, improvements in operating efficiency, and lower operations and maintenance costs. According to the International Renewable Energy Agency (IRENA), the potential for future reductions in renewable energy costs will remain significant, with a projected 26 and 35% reduction in onshore and offshore wind power levelised cost of energy (LCOE) by 2025. The reductions in solar power will be even greater: the LCOE of solar photovoltaic and solar thermal will fall by 43% and 59% respectively. Other new and renewable types of energy, including nuclear, biomass, geothermal, hydro and marine energy, also have potential for cost reductions.

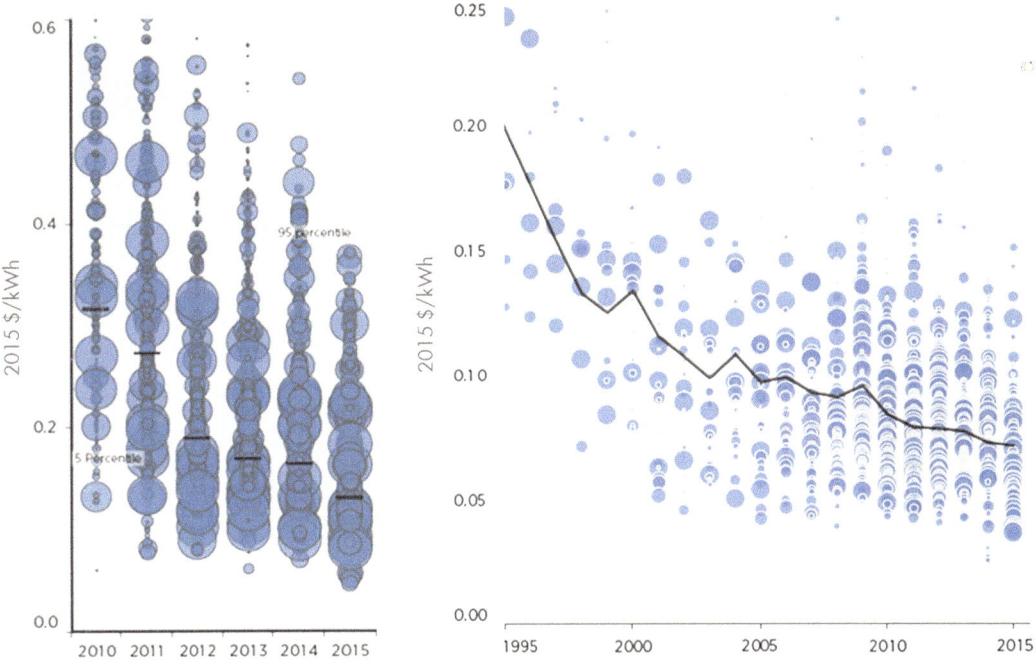

Fig. 34 Evolution of solar PV and wind power generation costs (LCOE). *Note* LCOE = levelised cost of energy. *Source* International Renewable Energy Agency (2016)

3.3 Policy Factors Affecting Energy Prices

3.3.1 Renewable Energy Policy

Although China has made significant progress in non-fossil energy over the past decade, renewable energy is still not cost-competitive because the current pricing mechanism does not include the external costs of fossil energy use (resource depletion and environmental impact). In order to support the development of renewable energy, China has adopted several policy measures over the past decade. The Renewable Energy Law, introduced in 2005, has laid the foundation for the rapid development of renewable energy in China. Similarly, the renewable energy feed-in-tariff has improved the price competitiveness of renewable energy and increased the deployment of renewable energy technologies. As shown in Fig. 35, renewable energy consumption has increased rapidly since 2005. By 2015, the consumption of wind and solar power was 185.1 terrawatt-hours (TWh) and 39.2 TWh respectively, and the proportion of non-fossil

energy in China's total energy demand system reached about 12%.

Under the current renewable energy policy mechanism, demand for renewable energy by consumers and for renewable energy subsidies by generators is growing. The subsidy deficit has now become a major obstacle to the sustainable development of renewable energy in China. To solve this problem, the government has announced several cuts in renewable energy feed-in tariffs (Table 3). At the same time, China has raised the feed-in tariff surcharge level successively from RMB 0.002/kWh in 2006 to RMB 0.019/kWh in 2016 (Table 4). Nevertheless, renewable energy subsidies are still far from meeting real needs. By the end of 2016, the cumulative gap in renewable energy subsidies was close to RMB 60 billion, and the sustainability of the original renewable energy subsidy policy faced challenges. Therefore, China's current and future renewable energy subsidy policies will undergo important transformations: (i) subsidies will be further reduced until their cancellation, while the transformation of the subsidy

Fig. 35 Evolution of wind and solar power consumption in China. *Source* BP Statistical Review of World Energy (2016)

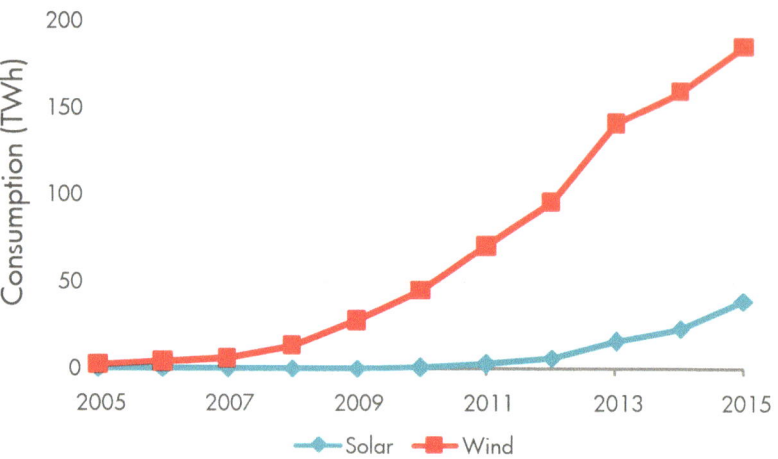

Table 3 Feed-in tariff adjustment for solar and wind power in China

	Year	Power on-grid price (RMB/kWh)
Solar	2009	Golden Sun programme tender price (first batch): 1.0928
	2010	Golden Sun programme tender price (second batch): 0.7288–0.9907 on-grid power price: 1.15
	2011	1
	2012	1
	2013	0.9, 0.95, 1 for three categories
	2015	0.8, 0.88, 0.98 for three categories
	2017	0.65, 0.75, 0.85 for three categories
Wind	2009	0.51, 0.54, 0.58, 0.61 for four categories
	2014	0.49, 0.52, 0.56, 0.61 for four categories
	2016	0.47, 0.50, 0.54, 0.60 for four categories
	2018	0.40, 0.45, 0.49, 0.57 for four categories

Source Ministry of Finance; National Development and Reform Commission; National Energy Administration

Table 4 Adjustments to the feed-in-tariff surcharge for renewable energy in China

Year	FIT surcharge for renewable energy (RMB/kWh)
2006	0.002
2009	0.004
2012	0.008
2013	0.015
2016	0.019

policy will be closely related to the technology and cost evolution of renewable energy in the future; and (ii) the future renewable energy quota trading mechanism and carbon trading scheme should boost future renewable energy development and become an important supplement to the renewable energy subsidy policy.

In addition, due to China's power market system, the characteristics of renewable energy technologies and other factors, wind and solar curtailment remains a tough challenge in China. For example, in 2015, the rate of wind and solar curtailment was 15% and 12.5% respectively. Due to the economic slowdown and the decline in electricity demand growth, wind and solar curtailment in the first half of 2016 increased to 38.9% and 19.7% respectively. Wind and solar curtailment greatly reduces the economic benefits of renewable energy investment and weakens the sustainable development of renewable energy. Therefore, addressing wind and solar curtailment has been a top priority of renewable energy development in the 13th Five-Year Plan (2016–20) and will remain so in the medium and long terms.

3.3.2 Carbon Pricing Policy

Due to the strong connection between energy demand and greenhouse gas emissions, China's climate policy will have a profound impact on the country's energy demand. On December 12, 2015, 195 countries entered into an historic agreement on global climate change at the United Nations Framework Convention on Climate Change (UNFCCC) in Paris. Signatories to the Paris Agreement committed to strengthen the global response to climate change by keeping the increase in global average temperature to well below 2°C above pre-industrial levels and by striving to limit the increase to 1.5°C. On November 4, 2016, the Paris Agreement came into effect. It is the second legally binding climate agreement following the Kyoto Protocol of 1997, and provides an institutional basis for the global response to climate change after 2020.

China is the world's largest emitter of greenhouse gases and plays a key role in combating climate change across the world. As a responsible developing country, China has made a positive contribution to the adoption and implementation of the Paris Agreement. Within the framework of the agreement, China submitted its ambitious intended nationally determined contributions (INDC). By 2030, China will: (i) reach peak CO_2 emissions and make its best effort to achieve peak carbon earlier; (ii) lower CO_2 emissions per unit of GDP by 60–65% from the 2005 level; (iii) increase the share of non-fossil fuels in primary energy demand to around 20%; and (iv) increase its forest stock volume by around 4.5 billion cubic metres from the 2005 level. This is the first time the Chinese government has set goals for its total carbon emissions, which is of great significance to advance the global response to climate change.

The carbon pricing mechanism is a market-based policy tool for reducing greenhouse gas emissions. It is a major institutional innovation in combating climate change that has been used in recent years by more and more countries and regions in their emission reduction practices. It is cost-efficient, environmentally effective and politically feasible to enact. According to the World Bank, by 2015, some 40 countries and more than 20 regions had adopted or planned to adopt carbon pricing instruments, including carbon trading schemes, covering 12% of global carbon emissions. Carbon pricing instruments are valued at close to $50 billion, and their scope and scale are expected to expand.

As a key measure to tackle climate change, China started to roll out a national carbon trading market in 2017 on the basis of seven pilot regional carbon trading schemes. The carbon market will be built in three phases. In phase 1, the IT system for data reporting, registry and trading will be built, which will take about a year. In phase 2, which will take another year or so, trial operation will start for the power generation industry. In phase 3, power generators will be able to transfer and trade carbon credits. The scope of participation will gradually extend to beyond the power generation sector, with more trading products and transaction types introduced.

The carbon pricing mechanism will influence future fossil energy demand—and thus the energy mix—by affecting the relative price competitiveness of fossil energy. For China, where coal consumption plays a dominant role, the implementation of carbon pricing is of great significance for the country's future energy demand. Even if China's adoption of clean coal technologies leads to fewer constraints on haze and other pollutants, in the present and in the medium terms—due to the various obstacles facing carbon capture and storage (CCS) technology—China is unlikely to have effective carbon solutions for coal at the end-use point. Therefore, when established, China's national carbon market is likely to reduce the country's consumption of fossil fuels, especially coal. The

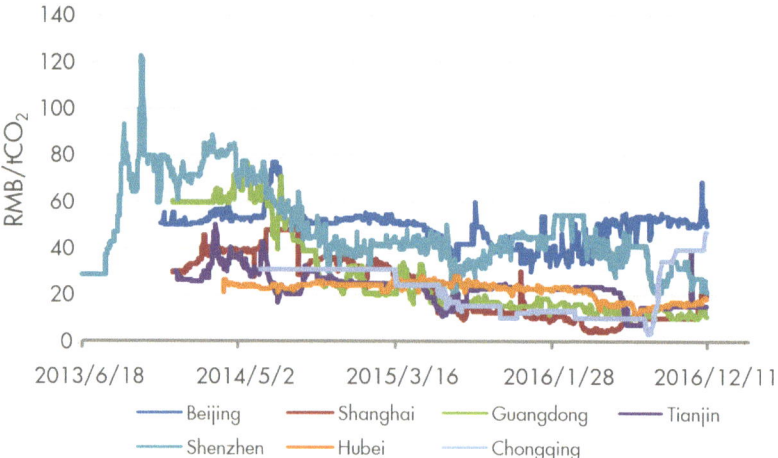

Fig. 36 Price trends in China's seven pilot carbon markets. *Source* China Carbon Trading Network

Table 5 Trading status in China's pilot carbon markets

	Starting time	Accumulated allowance volumes traded (Mt)	Accumulated trading volumes (million RMB)	Average price (RMB/tCO$_2$)
Beijing	2013-11-28	4.67	236.79	50.70
Shanghai	2013-11-26	7.6	150.58	19.81
Guangdong	2013-12-19	25.13	394.63	15.70
Tianjin	2013-12-26	2.42	39.86	16.47
Shenzhen	2013-06-18	17.76	587.22	33.06
Hubei	2014-04-02	34.05	730.52	21.45
Chongqing	2014-06-19	0.42	7.48	17.81

Source China Carbon trading Network

size of that reduction will be closely related to price trends in the carbon market.

The price trend and trading status of the seven pilot carbon trading markets are shown in Fig. 36 and Table 5 respectively. Due to their differences in economic development, carbon emission reduction goals and carbon market rules, the seven pilot schemes vary greatly in price, ranging from RMB 15 per tonne of CO$_2$ (t/CO$_2$) to RMB 60/tCO$_2$. At the same time, the market price fluctuates significantly in some pilot markets, which to some extent reflects the difficulty of reducing emissions and the market participants' varying predictions of what their future reductions will be. But on the whole, the current price level in the pilot carbon markets is generally low, and the impact on businesses' emissions reduction is not very significant.

It is likely that in the early years of the national carbon market, for the purpose of giving businesses time to learn and adapt to market rules, emission allowances will be relatively loose and the starting level of carbon prices will not be too high. However, as carbon emission reduction goals increase and the carbon market system improves, the carbon price level will rise, giving full play to the guiding and incentivising role of carbon pricing. According to McKinsey, Mo and Zhu[9] and estimates by the National Development and Reform Commission (NDRC), only a carbon price level of RMB 200–300/tCO$_2$ can play a significant role in the development of low-carbon technologies and the transition to a

---

[9]McKinsey & Company, Pathways to a low-carbon economy: Version 2 of the global greenhouse gas abatement cost curve. New York, McKinsey & Company, (2009). Jian-Lei Mo and Zhu, L., Using floor price mechanisms to promote carbon capture and storage (CCS) investment and CO$_2$ abatement. Energy and Environment 25(3/4), pp. 687–707, (2014).

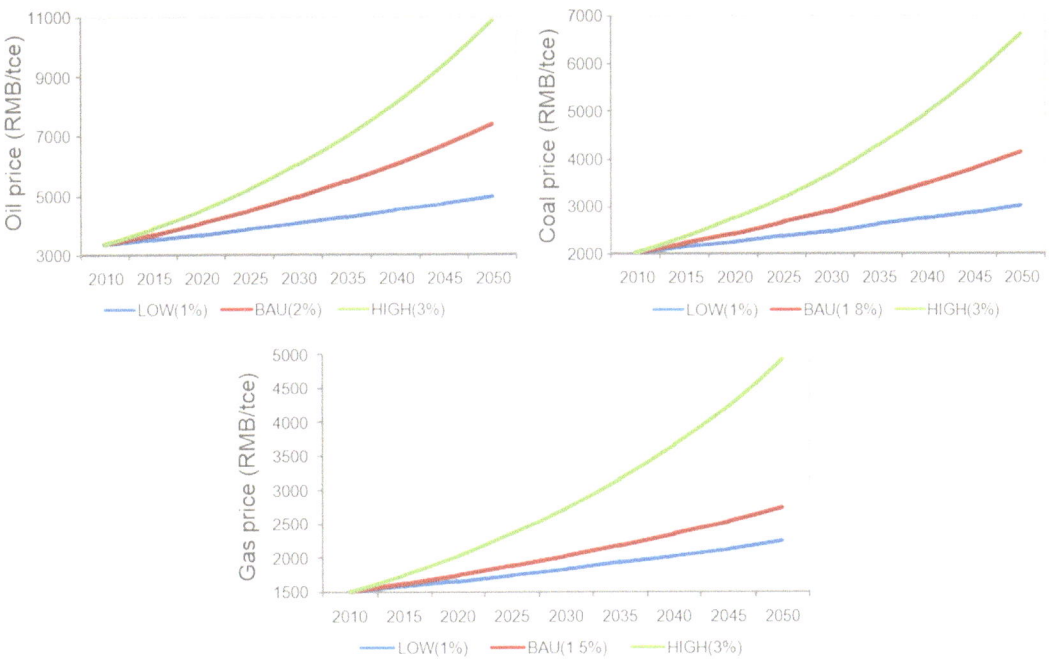

Fig. 37 Evolution of fossil energy prices in three scenarios

low-carbon economy. National carbon market prices are, therefore, expected to rise after 2020 to enable carbon emissions to peak by 2030.

3.4 Effects of the Pricing Mechanism on Medium- and Long-Term Energy Demand

3.4.1 Effects of Future Fossil Energy Prices on Energy Demand

Based on the historical trend of fossil energy prices and the current status of the fossil energy market, we designed three scenarios—High, Business as usual (BAU) and Low—to analyse future fossil energy price trends in China's

energy, economy and environment (3E) model. The average annual growth rate of future coal, oil and gas prices will be 2%, 2.3% and 1.7% respectively in the BAU scenario, 3% for all three fuels in the high-price scenario, and 1% for all three fuels in the low-price scenario. The price trends in the medium and long terms are shown in Fig. 37 (Table 6).

Figure 38 shows that, the evolution trend of fossil energy prices has a very significant impact on future total fossil energy demand. By 2030, energy demand in the BAU, low-price and high-price scenarios will be 77%, 98% and 48% higher respectively than in 2010. By 2050, the difference will be much greater: 1.67, 2.16 and 2.65 times respectively than in 2010. In the

Table 6 Future fossil energy price scenarios		Average annual increase rate (%)		
		High	BAU	Low
	Coal price	3	2	1
	Oil price	3	2.3	1
	Gas price	3	1.7	1

Note BAU = Business as usual

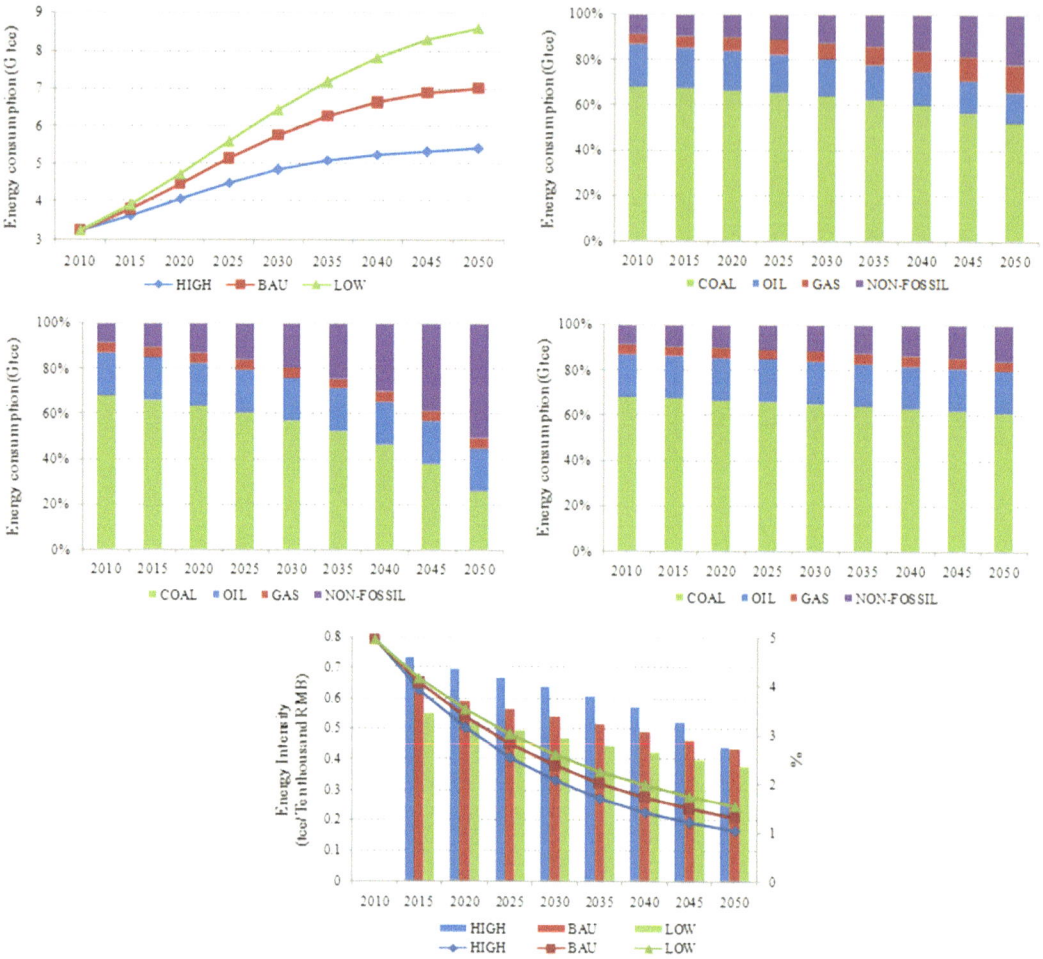

Fig. 38 Evolution of total energy demand, energy mix and energy intensity for different fossil energy price paths

high-price scenario, the total energy demand growth trajectory will shift down significantly compared with the base case, with demand starting to plateau around 2040.

The price of fossil fuels can have either a significant or insignificant impact on the energy mix. In the BAU scenario, the share of non-fossil energy will increase from 8.6% in 2010 to 12.4% in 2030 and reach only 21.9% in 2050, which is a relatively insignificant transformation of the energy mix. In the low-price scenario, the substitution of non-fossil energy for fossil energy is small and therefore insignificant in terms of energy mix transformation. In the high-price scenario, however, the share of non-fossil energy will reach 19.7% by 2030, close to the Chinese

government's 2030 INDC goal, and 50.6% by 2050. This means that the price of fossil energy in this scenario does have a significant impact on demand and the energy mix in China. Reducing subsidies for fossil energy and maintaining high fossil energy prices can limit energy demand and help optimise and transform the energy mix in the medium and long terms.

China still has great potential for energy intensity reduction in the future. In the BAU scenario, energy intensity in 2030 will be 51.8% lower than in 2010, and the average annual rate of decline during 2025–30 will be 3.3%. In the high fossil energy price scenario, energy intensity in 2030 will be 58.2% lower than in 2010. Even in the low fossil energy price scenario, energy intensity in 2030 will be 46.8% lower than in 2010. This is due mainly to the improvement in energy efficiency caused by non-price factors, such as economic restructuring and technology advances.

3.4.2 The Effects of Non-fossil Energy Technology on Energy Demand

In addition to fossil energy prices, technology advances and the cost of non-fossil energy can impact future energy demand significantly. The cost reduction potential of new and renewable energy is closely related to progress in technology. Based on previous research, this study summarises progress in several power generation technologies, as shown in Table 7. The various studies identify some differences in learning parameters within the same technology; but between different technologies, the differences in learning are more significant. Overall, the learning potential of technologies like solar, wind, marine and geothermal is huge, while that of hydropower, nuclear power and biomass is small. Based on previous studies and analyses in the energy, economy and environment model, this report sets three non-fossil energy technology evolution scenarios (High, BAU and Low): the high technology evolution scenario implies that non-fossil energy technologies have great learning potential and their future cost reductions will be significant with their diffusion. The low scenario implies that non-fossil energy technologies are relatively mature and their learning potential and cost reductions are small.

Figure 39 shows the evolution of total energy demand and the energy mix in different non-fossil energy technology scenarios. For total energy demand in the short term, the changes in the three scenarios will be basically the same. In the medium and long terms, differences in the three scenarios will gradually materialise but will not be significant. Although there is great uncertainty in the evolution of renewable energy technologies, the simulation results show that even in the most optimistic and pessimistic technology evolution scenarios, the differences in energy demand are not significant, indicating technology's limited impact on future energy demand. There are two possible reasons for this: (i) due to technological inertia, the learning effects of renewable energy usually take a long time to materialise, so their impact on energy demand is not evident in the short term; and (ii) since the share of non-fossil energy in China is relatively low, even if great progress has been made in non-fossil energy technologies, their impact on the cost of the entire energy system is small—it is therefore impossible for them to lead the evolution of the entire energy system in the short term.

In terms of the energy mix, the share of non-fossil energy in the three technology evolution scenarios does not differ greatly in the short term. In 2030, the share of non-fossil energy in the high, BAU and low scenarios will be 12.5%, 12.3% and 11.8% respectively. In 2050, the share will be 23.1%, 21.9% and 18.2% respectively, which is still a minor difference. The substitution of non-fossil energy for fossil energy will not be significant in the short term and its impact on the energy mix will be small. Similar results apply to energy intensity.

To summarise, in the absence of external policies, the impact of non-fossil energy technology advances on the evolution of the entire energy system will take a long time to materialise,

Table 7 Learning parameters of non-fossil energy technologies		GEO	SOL PV	WIND	MAR	BIO	NUC	HYD
	High	0.82	0.72	0.81	0.73	0.89	0.91	0.95
	BAU	0.87	0.79	0.89	0.80	0.92	0.94	0.97
	Low	0.92	0.85	0.96	0.86	0.95	0.97	0.99

Note Learning parameter (technical progress rate) = 1-learning rate = −2-learning index. GEO = geothermal; SOL PV = solar photovoltaic; WIND = wind; MAR = marine energy; BIO = biomass; NUC = nuclear power; HYD = hydropower

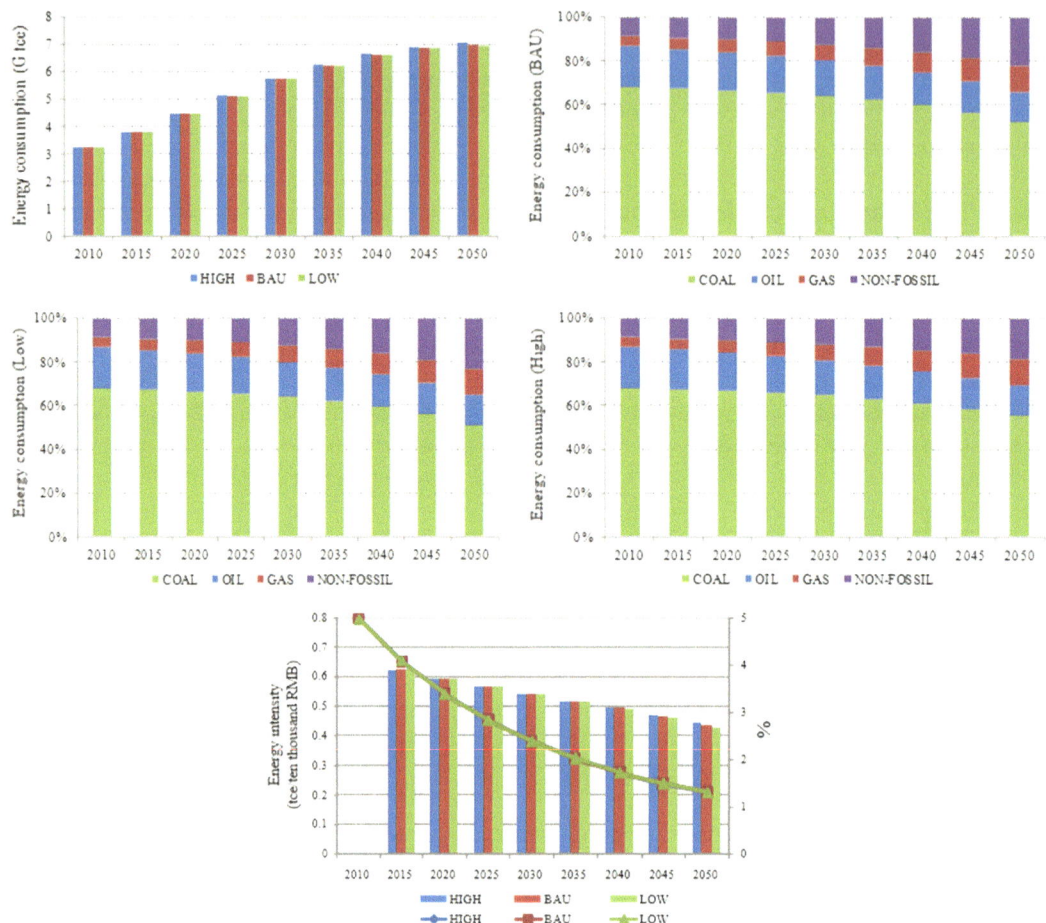

Fig. 39 Total energy demand, energy mix and energy intensity in different non-fossil energy technology evolution scenarios

and their impact in the short term will be limited. In the context of energy saving, emissions reduction and combating climate change, introducing additional policy mechanisms to regulate the future evolution of the energy system would be a logical step to achieving the energy transition and energy revolution goals.

3.4.3 The Effects of Energy Price Policy on Energy Demand

The above analysis shows that policy mechanisms are essential to control total energy demand and adjust the energy mix. This study introduces two policy mechanisms that affect energy prices in the energy, economy and environment (3E) model: carbon pricing (for fossil energy

emissions) and non-fossil energy subsidies. Carbon pricing determines the price of fossil energy emissions and turns them into a cost, thereby reducing the price competitiveness of fossil fuels. Subsidies, on the other hand, increase the relative price competitiveness of non-fossil energy.

This study assumes that China's national carbon pricing mechanism was introduced in 2015. We set three carbon price scenarios (high, medium and low: $90 per tonne of carbon (tC), $60/tC and $30/tC respectively), and assumed an average annual growth rate of 5% for carbon prices in the future. We based non-fossil energy subsidies on China's current renewable energy subsidy policy. We set two scenarios (a price subsidy of 20% and 30% respectively) and

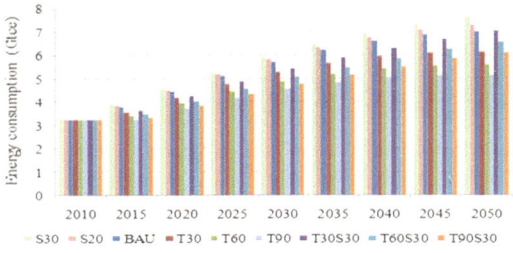 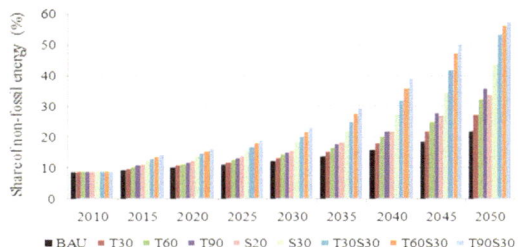

Fig. 40 The effects of carbon pricing and non-fossil energy subsidies on total energy demand and the energy mix. *Note* S30 = subsidy 30%; BAU = business as usual; T30 = tax 30%; etc.

assumed that the two policies were introduced in 2015. The future evolution of total energy demand, the energy mix and energy intensity are shown in Fig. 40.

The implementation of carbon pricing policy will have a significantly negative impact on the future trend of total energy demand. As the initial carbon pricing level increases, total energy demand will decline. Specifically, by 2030, in the policy scenarios of $30/tC, $60/tC and $90/tC, total energy demand will be 5.3 gigatonnes of coal equivalent (Gtce), 4.9 Gtce and 4.6 Gtce respectively, which is 7.4%, 14.0% and 19.4% lower than the 5.7 Gtce in the BAU scenario. By 2050, in the policy scenarios of $30/tC, $60/tC and $90/tC, total energy demand will be 6.2 Gtce, 5.6 Gtce and 5.2 Gtce respectively, which is 11.9, 20.0 and 25.8% lower than the 7.0 Gtce in the BAU scenario.

It can therefore be seen that the effects of carbon pricing policies materialise in the short term and become significant over time. Unlike carbon pricing policies, renewable energy subsidies will increase total energy demand. However, comparison of the two policy mechanisms indicates that subsidy policies will have a less significant impact on total energy demand than carbon pricing policies. If both policies are introduced, total energy demand will be higher than in the pure carbon pricing scenario and lower than in the subsidy scenario. An important conclusion can therefore be drawn: If policymakers are concerned that the introduction of carbon pricing could result in significant negative

impacts on energy demand, non-fossil energy subsidy policies can be introduced at the same time. This would increase demand for non-fossil energy and reduce that for fossil energy, ensuring smooth control of energy demand.

In terms of the energy mix, with the introduction of carbon pricing and non-fossil energy subsidy policies, the share of non-fossil energy will rise. Non-fossil energy subsidies will have a more significant impact on the energy system shift than carbon pricing. This is because the cost reduction of non-fossil energy technologies will be slow without proper policy support. While carbon pricing policies can inhibit fossil energy demand, the uptake of non-fossil energy can still be very challenging if no subsidy is introduced to reduce the cost of the technology. With a policy combining carbon pricing and new energy subsidies, fossil energy demand can be inhibited while non-fossil energy can be developed faster.

When both policy mechanisms are adopted, the impact on the energy mix is significant. The simulation results show that in the single policy scenarios, neither carbon pricing at $90/tC nor a subsidy of 30% is enough to achieve the 20% share of non-fossil energy goal by 2030. Whereas in the combined policy scenario, the joint effects of a carbon price of $30/tC and a subsidy of 30% puts the 20% share goal within reach by 2030. In the medium and long terms, the share of non-fossil energy can rise to 50% by 2050 in the combined policy scenario. This indicates that a combined policy mechanism is effective and necessary to achieve the energy revolution.

3.4.4 Analysis of China's INDC

China's intended nationally determined contributions (INDC) reflect its goal of reducing its carbon emission intensity by 60–65% by 2030. This study simulates the evolution trends of China's future carbon emission intensity in different policy scenarios, as shown in Fig. 41.

In the BAU scenario, where the emission reduction and energy efficiency efforts implemented through to the 12th Five-Year Plan (2011–15) are continued, without any additional policy measures, China's future carbon emission intensity will decline by 63.4% between 2005 and 2030. Figure 41 shows that by continuing existing energy saving and emission reduction efforts, China will be able to reach the lower limit of the carbon intensity reduction goal by 2030. But additional emission reduction efforts will be needed to reach the upper limit of the carbon intensity reduction goal.

In the single policy scenario, where a non-fossil energy price subsidy of 20% is introduced, carbon emission intensity will decrease by 64.5% between 2005 and 2030. If the subsidy level is increased to 30%, carbon emission intensity will decline by 65.3%, reaching the upper limit of the carbon emission intensity reduction goal.

In terms of carbon pricing policy, when the carbon price level is $30/tC, carbon emission intensity will decrease to 66.7% in 2030, which also reaches the upper limit of the carbon emission intensity reduction goal. Carbon pricing policy will, therefore, play a more significant role in reducing future carbon intensity.

Achieving peak carbon emissions by 2030 is one of China's INDC goals. To forecast the path to achieving this goal, this study simulated the evolution of China's total future carbon emissions in different policy scenarios, as shown in Fig. 42.

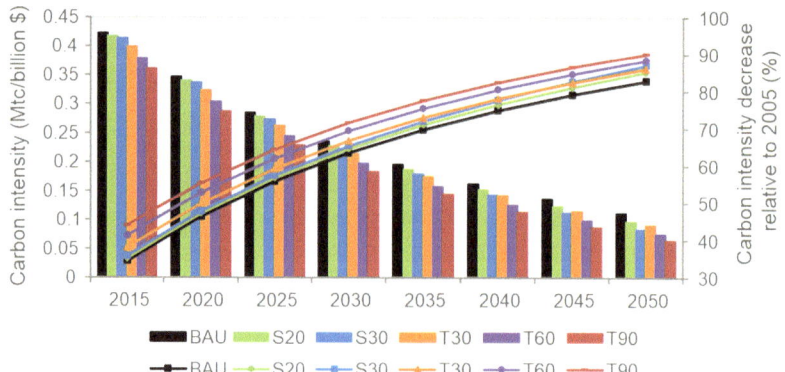

Fig. 41 Evolution of China's future carbon emissions intensity in different policy scenarios. *Note* BAU = business as usual; S20 = subsidy 20%; T30 = tax 30%, etc.

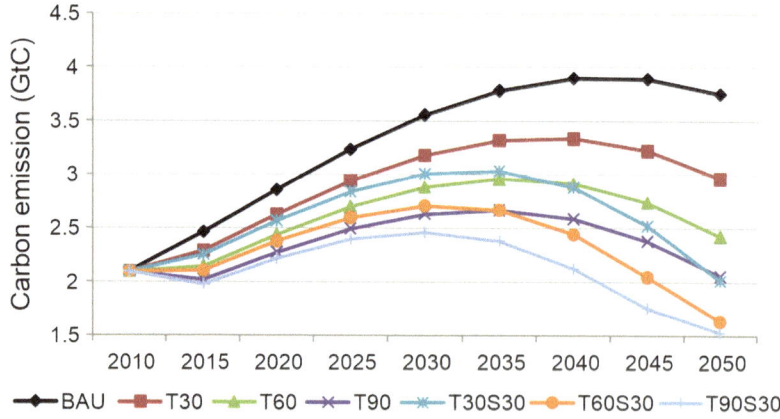

Fig. 42 Evolution of China's future carbon emission paths in different policy scenarios. *Note* BAU = business as usual; T30 = tax 30%; S30 = subsidy 30%; etc.

In the BAU scenario, where no additional policies are introduced, China's carbon emissions growth will continue to 2040 and decrease slowly thereafter. In the single carbon pricing scenario, China's carbon emissions curve will gradually decline. When the initial carbon price level is set at $30/tC, China's carbon emissions will peak in 2035, remain stable for five years, before decreasing after 2040. When the carbon price level stands at $60/tC, China's carbon emissions peak will not arrive significantly earlier, although the emissions path will be adjusted noticeably compared to the $30/tC scenario. Cumulative carbon emissions will also decline significantly by 2030 and 2050. When the carbon price is increased to $90/tC, China's carbon emissions will peak around 2030 and maintain that level until 2035, and then decrease.

In the combined policy scenario of carbon pricing and non-fossil energy subsidies, the evolution of carbon emissions shows a significant difference compared to that of the single policy scenario. First, in the combined policy scenario, a lower carbon price level can make an early carbon emissions peak possible. Specifically, with a carbon price level of $30/tC and a non-fossil energy price subsidy of 30%, China's carbon emissions peak in 2030 and maintain that level to 2035, before decreasing. Moreover, the carbon emissions curve in the combined policy scenario shows little difference to that in the single carbon pricing policy scenario in the short

term. In the medium and long terms, however, a gap develops: carbon emissions in the combined policy scenario are significantly lower than those in the single carbon pricing policy scenario. This is due to less renewable energy in the short term and the path dependence effect (once a path is entered it is difficult to leave).

Finally, we can conclude from the policy scenarios the following: Given that China's carbon policy goal is to reach peak carbon emissions by 2030 (with no specific peaking level target), China could have various carbon emission paths to choose from. Which carbon pathway China is to pick will to a large extend dictate the policy pathway in the future.

A larger share of the energy mix for non-fossil energy is an important component of China's intended nationally determined contributions (INDC). This study plots the evolution of China's future non-fossil energy share in different policy scenarios, as shown in Fig. 43.

In the BAU scenario where carbon pricing and non-fossil energy subsidy policies are absent, the evolution of the non-fossil share of the energy mix is stable—even in 2045–50, its share barely reaches 20%. With the introduction of the carbon pricing and non-fossil energy subsidy policies, the share of non-fossil energy increases. This study finds that the impact of non-fossil energy subsidies on the energy mix is greater than that of carbon pricing. The study also finds that in the combined policy scenario, the cost of

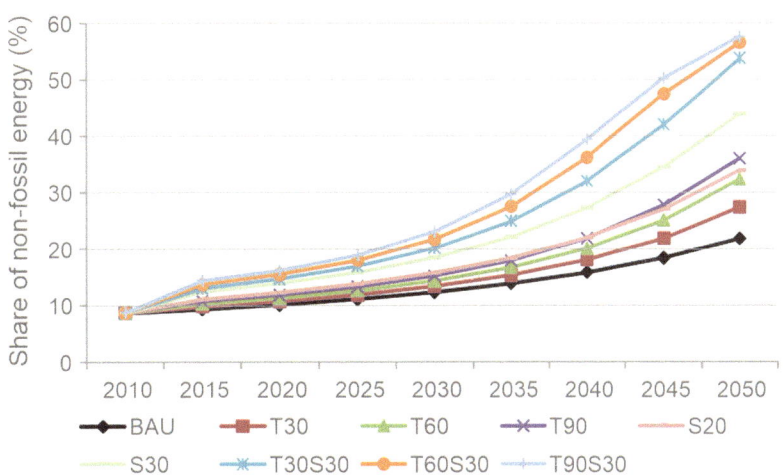

Fig. 43 Evolution of China's non-fossil energy development in different policy scenarios. *Note* BAU = business as usual; T30 = tax 30%; S20 = subsidy 20%; etc.

fossil energy use is pushed up by the carbon pricing policy, which limits fossil energy demand; on the other hand, the cost of consuming non-fossil energy is lowered by the subsidy policy, which increases non-fossil energy demand. The simulation results show that in the single scenario of $90/tC carbon pricing and the single scenario of 30% non-fossil energy subsidy, the share of non-fossil energy is 15.2% and 18.4% respectively, neither of which is sufficient to reach the goal of a 20% non-fossil energy share by 2030. However, in the combined policy scenario of $30/tC carbon pricing and a 30% non-fossil energy subsidy, the share of non-fossil energy reaches 20.1% by 2030, thus achieving the goal. When the carbon price level is increased to $60/tC and $90/tC, the share of non-fossil energy will rise to 21.7% and 23.1% respectively. In the single carbon pricing scenario, the share of non-fossil energy will be around 30% by 2050, while in the single non-fossil energy subsidy scenario, the figure will reach 40%. In the combined policy scenario, the share of non-fossil energy will exceed 50% and even reach 57.6% in the strictest combined scenario of carbon pricing and non-fossil energy subsidy.

3.4.5 Comparison of the Three INDC Goals and Analysis of Their Relationships

This section briefly summarises how China can achieve its three INDC goals of reducing carbon intensity, reaching peak carbon emissions, and increasing the share of non-fossil energy in the energy mix to 20%, all by 2030.

In terms of carbon intensity, no additional policy efforts are needed to reduce carbon intensity by 60% by 2030. However, to ensure carbon intensity reduction by 65%, additional policy efforts are required, such as a low carbon-pricing scenario ($30/tC) or a single policy of 30% non-fossil energy subsidy.

The goal of reaching peak carbon emissions by 2030 is not possible in the single non-fossil energy subsidy or low carbon pricing scenario. It can only be delivered on schedule in the high carbon pricing scenario ($90/tC) or the combined

policy scenario of carbon pricing and non-fossil energy subsidy.

In terms of non-fossil energy development, even if the high carbon pricing policy of $90/tC or high price subsidy policy of 30% is implemented, the share of non-fossil energy in primary energy demand will still be lower than 20% by 2030. The goal of 20% non-fossil energy share can only be achieved in combined policy scenarios (for example, a combination of a $30/tC carbon price and a 30% subsidy).

The analysis above shows that the carbon intensity goal requires the least additional policy efforts and is easiest to achieve, whereas the goal of 20% non-fossil energy share requires the greatest effort and is the most difficult.

4 The Impact of Information Technology on Energy Demand

4.1 The Evolution of IT and Its Impact on Energy Demand

From the 1990s on, continuous innovation in information technology (IT) and rapid, sustained development of the IT industry drove the world into the information age and, lately, the era of digitalisation. IT and digitalisation will inevitably trigger great changes in energy supply and in the demand patterns of businesses and people.

4.1.1 Evolution of IT

Informatisation is the extent to which an economy or society becomes information-based. It is an evolutionary process. In industrial society, the creativity of individuals, efficiency of businesses and organisations and the competitiveness of countries were restricted in both time and space. In the age of information, technological innovation continuously improves information infrastructure and increasingly refines production and management in business.

Evolution from the Internet to big data is inevitable. Information technology created the Internet, breaking the temporal and spatial boundaries between production, living and exchange. The emergence of desktop and laptop computers makes the computer an indispensable tool in business and in people's lives. With the rise of smartphones and tablets, mobility expands the Internet, providing a broader and more convenient channel for people to obtain information. The Internet of things (IoT) greatly accelerates the informatisation process with its application in multiple fields, including intelligent transport, environmental protection, government, public security, disaster forecasting, smart homes, personal health monitoring, lighting control and intelligence gathering. These advances push the world into the era of big data. A digital world parallel to the physical world is created around the behaviour of individuals and organisations. It is a world in which information is stored and analysed and information-based predictions made. This capability will further reshape how we live, produce and consume.

4.1.2 The Impact of IT on Energy Demand

IT has many impacts on energy demand.

First, IT affects energy demand. As big data scales up, more and more energy-hungry data centres are needed to process and store the data. To address the high energy consumption of big data management systems, new energy-efficient hardware, powered by new and renewable energy technologies, is needed.

Second, the impacts and opportunities generated by the deep integration of new-generation IT and energy technologies enable the Energy Internet: IoT, cloud computing, big data and the blockchain merge with renewable power generation, decentralised energy resources and the smart grid. By connecting various systems like transport, power and natural gas with the Internet information system, the Energy Internet can connect multiple energy carriers, such as electricity, heating, cooling and gas across the entire value chain, from production, transmission and storage to consumption. This will help increase the share of renewable energy production and supply, and eventually enable a harmonious energy supply ecosystem based on renewables and driven by IT. Some developed economies in

Europe and the Americas have already prioritised the Energy Internet. For example, Germany has developed the E-Energy programme to create a new energy network of digital interconnection, computer control and monitoring across the entire energy supply system. Blockchain technology will also drive development of the Energy Internet, opening up new applications and business models in energy generation, transmission, distribution, use and storage, and reshaping energy consumption patterns.

4.2 The Impact of IT on Household Energy Demand Modes

Household energy consumption accounts for 10% of China's total energy, the second largest category after industry. Residential energy consumption affects not only people's lives and well-being, but the environment and China's carbon emissions as well. With household energy consumption rising, especially in urban areas, and energy infrastructure improving, IT can play a crucial role in changing household patterns of energy consumption.

4.2.1 Changing Patterns in Household Energy Demand

The ubiquity of the Internet makes it easier for people to consume energy and spend more time at home. The more they are at home, the more energy they consume—for household appliances, heating or cooling, or for surfing the Internet to buy non-energy-related products and services.

In the Internet age, people can pay their energy bills and buy almost any product online, which makes consumption easier. In transport, the emergence of ride-hailing apps like Didi Chuxing makes it easier for people to get around.

If the Internet stage of informatisation makes people's energy consumption more convenient, then the next stage—big data—makes it intelligent. Household energy management is easier and more efficient, and energy consumption greener.

The Energy Internet centres on the large-scale deployment of renewable energy, especially distributed renewable energy. Big data analysis and cloud storage are crucial support technologies that can accelerate the development of the Energy Internet. In recent years, tech companies like Google and Alibaba have made inroads into the intelligent building and smart home industry. The smart home industry is likely to undergo rapid growth in the coming years.

As the Energy Internet and smart home systems evolve, the problems that face conventional energy networks—such as overproduction, low scheduling accuracy, high energy transmission losses and poor grid connections for new energy —will be solved, thus optimising the energy system, saving energy and reducing environmental impact.

By harnessing energy demand and supply information, energy providers can implement a scheduling strategy to make energy flow efficiently. They can also personalise household energy demand based on the energy consumption information they have collected. Households in turn can gain a deeper understanding of their consumption behaviour from the data.

The Energy Internet will revolutionise energy demand and make household energy consumption intelligent. When Internet titans can extract more accurate and richer information from big data and accurately push that information to individuals in a relevant manner, the changes in people's indirect energy consumption will no longer be limited simply to convenience.

4.2.2 Changing Patterns in Household Energy Consumption

In the era of big data, one-way energy consumption will change. The role of residents will shift from traditional energy consumer to energy consumer and provider, or prosumer. Energy consumption will not be a one-way relationship, dominated by energy companies, but a two-way interaction.

IT triggers change on a comprehensive scale. The energy supply, transmission and consumption system—supported by technologies like sensors, high-speed networks, mobile Internet, smart terminals, cloud platforms, big data processing and geographic information systems (GIS)—has

evolved into a new IT-driven energy supply and retail ecosystem. In the existing energy supply and retail system, energy production and consumption are strictly separated. Energy transmission companies (State Grid, for example), are responsible for supply and demand adjustments through the one-way transmission-consumption approach. The Internet and big data extend the concept of the consumer beyond a mere end user to a partner across the entire value chain. As the Energy Internet evolves and the cost of information falls, both suppliers and consumers will have access to market information like supply and demand and the price of raw materials via the Internet. This opens the possibility of decentralised trading, driving the shift from centralised to decentralised resource optimisation and allocation. With continuous deregulation of the power retail sector, there surely will be more and more trading parties in the energy supply and retail ecosystem. In addition, the combination of different trading parties, decentralised users and diversified energy sources will also drive the emergence of new business models.

It will be increasingly difficult to differentiate producers from consumers in the energy demand network. Power retail companies, industrial parks, buildings and even individual residents will have access to this ecosystem, directly generating two-way and even multiple-route transmission in energy consumption. This will strengthen the position of people as energy consumers. With the Energy Internet, households can choose from different energy consumption options to achieve the same effect. They can build the smallest energy supply and demand subnetworks through information sharing to lower their energy costs. They can build supply and demand models, based on energy prices and their own patterns of demand. Broadly speaking, households will not only benefit from the effects of IT on their energy consumption, they will impact the entire energy ecosystem themselves. For example, an electric vehicle is an interface with the energy ecosystem. It can be used as a means of transport or as an energy storage device to counter energy fluctuations. In Germany, surplus power from rooftop solar PV systems can be fed into the grid.

It is evident that IT has and will continue to change household energy demand patterns. Currently, IT has made household energy use increasingly convenient and diversified. In the future, as information technologies like mobile Internet, the Internet of things, cloud computing, big data and blockchain are effectively integrated, household energy consumption will evolve from convenient to green and intelligent, and its role will shift from passive one-way consumer to active two-way prosumer. IT brings infinite possibilities to household energy consumption.

4.3 The Impact of IT on Business Energy Consumption

IT applications show that extensive deployment of modern information technologies can substantially improve the efficiency of capital, technology and manpower, reduce energy consumption and costs, generate new business models and drive business transformation, thus changing energy production and consumption patterns.

4.3.1 Changes in Business Energy Use in the Age of IT

In the initial stage of IT in business, IT's impact on business energy consumption is mainly confined to energy management information systems to improve energy use and make energy savings. Its role is to improve the business efficiency of individual companies.

As energy resources decline and energy demand grows, the share of energy in a business's operating costs gradually increases. To lower operating costs and sharpen competitiveness, businesses have taken numerous measures to make their use of energy more efficient. However, before IT was widely deployed by businesses in China, the country was confronted by various challenges, including very high economic growth, poor management, low use of IT software and automation technologies, and inefficient energy use. Poor management means incomplete energy saving-related measurement, statistics and assessment; low IT use; and severe loss and waste

of energy. Businesses were therefore in urgent need of an integrated energy management and control system. The energy management information system that evolved with Internet technology met this need. Its continuous improvement and wide deployment are an effective means to improve businesses' IT-based energy management and energy efficiency.

An IT-based energy management information system provides monitoring, analysis and control of energy use—water, electricity gas, etc. With such a system, businesses can: (i) develop an energy procurement strategy to meet their production needs efficiently and cost-effectively; (ii) monitor and analyse energy consumption in their production processes; (iii) make timely scheduling of water, electricity, gas and fuel based on changing production and operational parameters; and (iv) unlock energy saving potential based on energy consumption data and monitoring. According to Chen Yanfei of the China Petroleum Planning and Engineering Institute, with an energy efficiency management system, energy-intensive industries like petrochemicals can monitor, analyse and assess plant energy consumption. The system provides IT-based visual management of energy use, optimises energy consumption in the production process and improves plant energy efficiency.

For energy-intensive industries like iron and steel, non-ferrous metals, chemicals and cement, energy management information systems are especially important. Before these systems existed, businesses had to record energy use manually, which often led to incomplete and missing data. Data plays a vital role in delivering energy savings, so data errors or incompleteness cause many problems. Despite their desire to save energy, businesses often attempt to do so without an energy-saving strategy. Without such a strategy, there is a risk of investing in popular energy-efficient technologies without considering their suitability. With an energy management information system, businesses gain a better understanding of their energy consumption and can develop an energy-saving programme for the weaknesses identified, thus improving energy efficiency.

4.3.2 Changes in Business Energy Consumption in the Age of Data Technology

As information technology evolves, the Internet changes how people live, work and think, profoundly. In this process, the role of data evolves from simple recording and analysis to advanced applications. It gradually becomes the second language of humankind, driving civilisation to evolve from the era of information technology to the age of data technology (cloud computing, IoT, big data and mobile Internet). The effects of data technology on businesses' energy consumption include intelligent energy management, IT-based energy efficiency and improvement of the entire energy use value chain.

Data technology enables intelligent energy management in business. The Internet, big data and cloud computing allow companies to set energy efficiency benchmarks for their operations and equipment, Businesses can analyse and compare their energy use data, automatically evaluate their energy efficiency and generate energy consumption reports. Functionalities like real-time monitoring, timely adjustment of parameters and automatic alarms signalling faults or excessive energy use can be set up along the value chain, from procurement to consumption, making smart energy management possible. Through its online energy consumption monitoring system, CNPC Offshore Engineering Company Limited (CPOE) has practical experience of the operation and benefits of smart energy management. Li Peng, head of CPOE, said, "With the online energy consumption monitoring system, we can understand our energy use level in real time, effectively controlling our total energy consumption and achieving cost reductions. It also helps us take economic and social benefits into consideration when making decisions."

The age of data technology makes IT-based energy efficiency possible. Intelligent automation systems enable businesses to control their energy use and flexibility by using the right type of energy at the right time at a given site. For instance, in buildings, the central air conditioning system accounts for more than half of total building energy consumption. As buildings become bigger and standards improve, the number of mechanical and electrical devices increases sharply. This equipment is deployed throughout the building and on all floors. A decentralised management system, requiring local monitoring and operation on each floor, would require a lot of human operators. If modern computer technology and network systems are used to create a centralised energy management system to continuously monitor all mechanical and electrical equipment, energy consumption and staffing can be reduced by about 25% and 50% respectively.

The age of data technology can improve the entire value chain of an industry. By integrating information technologies like smart detection of operational problems, advanced data display graphics, wireless communications and network control, IT can optimise resource allocation and make energy scheduling between businesses possible. In the power industry, for example, IT can: (i) optimise combustion to reduce fuel consumption in conventional power plants; (ii) improve scheduling efficiency through automated dispatch in power transmission; (iii) ensure grid reliability and power quality by enabling life cycle management of power metering devices in distribution networks; and (iv) prevent power shortages by adopting time-of-use (TOU) tariffs based on power demand patterns in retail.

4.3.3 IT Optimises Business Energy Demand

The Energy Internet is a network of equal exchange between distributors and consumers that enables power to flow in two or multiple directions between participants. It uses advanced power electronics, IT and smart management to integrate distributed energy resources, energy storage facilities and loads of various types into a single system. The Energy Internet can effectively address the geographical mismatch between energy production and consumption in China, reduce renewable energy curtailment, and improve businesses' energy use.

The Energy Internet can improve the efficiency of new power generation hugely. When operating separately, the output from distributed energy resources is random and intermittent. When fluctuating distributed generation is connected to a conventional grid, it threatens grid security and reliability. To eliminate this threat and safeguard the grid, the amount of intermittent solar and wind power in the grid has to be limited, which results in severe wind and solar curtailments (i.e. a large proportion of the energy generated is not used). The Energy Internet solves this problem by enabling small distributed energy networks and microgrids to: (i) meet local consumption needs first; and (ii) export surplus energy to the larger grid. This significantly improves the use and efficiency of renewable energy.

The Energy Internet also plays a critical role in smart energy storage. When there is a renewable energy surplus, pumped storage power plants and electric vehicles can store the surplus and release it when needed. Smart household appliances like washing machines, dishwashers and water heaters can be programmed to consume energy when demand and tariffs are low. These energy storage facilities and smart household appliances can form a virtual power plant to address peak demand by releasing more energy into the network and by consuming it at the optimal time.

4.4 IT Drives Change in Energy Consumption

Currently, China is faced with unreasonably high energy use and low energy use efficiency. The Energy Internet is highly promising and will be of great significance to change China's energy consumption patterns, especially by making the shift from centralised to distributed energy production possible.

4.4.1 The Rapidly Evolving Energy Internet

First, the fast and large-scale development of cities, industrial parks, high energy-consuming companies and green buildings means there is an urgent need for the development of the Energy Internet.

Second, China is confronted with such challenges as unsustainable energy production, low energy use efficiency and conservative thinking in the energy industry. The Energy Internet can reduce wind and solar power curtailment and improve power output and asset utilisation through big data-based life cycle management and interaction between multiple energy sources and loads. By integrating big data analysis on user energy consumption, energy management, demand response and smart homes that can sell surplus power, energy use efficiency can be significantly improved. Ultimately, more renewables in the energy mix and better energy use efficiency will drive change in the energy system.

4.4.2 From Centralised to Distributed Energy Trading

According to Professor Zeng Ming, Director of the Research Center of Energy and Electricity Economy, North China Electric Power University, electricity will eventually become a traded commodity. In the age of the Energy Internet, the traditional consumer is both a producer and a consumer (prosumer). Consumers can produce power with their own distributed renewable energy system. More importantly, they can provide user-side load resources to participate in demand response through smart energy solutions. Consumers can be active in community demand-side response projects, and also be part of a virtual power plant. In addition, consumers can also sell power to the grid from their electric vehicle and energy storage facilities.

5 Changes in Conventional Fossil Energy Demand

5.1 Coal

5.1.1 Current Status and Trends in Coal Demand

According to the official website of the Ministry of Land and Resources of the PRC, by the end of 2014, China's proven reserves of coal were 1,531.7 billion tonnes (Bt), third largest after the USA and Russia. In 2015, China's coal production was 3.75 Bt, down 3.3% from 2014, and 47% of the world's total coal production. China's

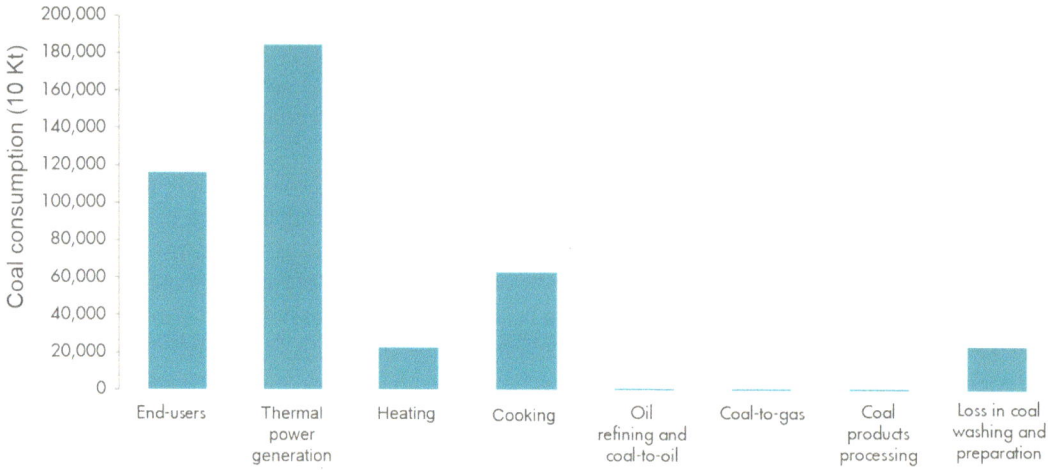

Fig. 44 China's coal consumption in 2014. *Source* National Bureau of Statistics of China

coal consumption in 2015 was 3.965 Bt, down 3.7% from 2014, and 50% of the world's total coal consumption. Coal's share of China's primary energy use was 64% in 2015, far higher than the world average of 30%.

In 2014, China's total coal consumption was 4.116 Bt. Coal consumption in major industries and processes is shown in Fig. 44.

In 2014, thermal power generation was China's largest consumer of coal at 1,845.25 Mt, 44.82% of the total. Coal consumption by end users and in coking was 1,160.44 Mt (28.19%) and 628.94 Mt (15.28%) respectively, while that used for heating and in oil refining and coal chemicals was 224.45 Mt (5.45%) and 23.30 Mt (0.57%) respectively. Losses from coal washing and preparation were 233.75 Mt, which is 5.68% of total coal consumption.

China's annual coal consumption from 2000 to 2015 is shown in Fig. 45. Consumption increased year by year, rising from 1.357 Bt in 2000 to 4.244 Bt in 2013, an annual growth rate of 9.5%. Consumption then dropped to 4.116 Bt in 2014 and to 3.965 Bt in 2015. Due to China's slowing economic growth and increasingly strict environmental policies, controlling coal consumption has become an important means to adjust primary energy demand. There is, therefore, only a slim possibility that coal consumption will increase in the future. China's coal

consumption is forecast to decline to 4.08 Bt in 2020 and to 3.6 Bt in 2030. This indicates that China's coal consumption peaked in 2013.

In the future, China's coal sector will make vigorous efforts to achieve technological breakthroughs and implement successful demonstration projects in safe, efficient and intelligent coal mining and in clean, efficient and intensive coal use, thus improving the sustainability of the coal sector. In the 13th Five-Year Plan (2016–20), China placed total energy consumption high on the agenda. As a priority in controlling total energy consumption, coal's share of primary energy is targeted to decline below 60%, while research on the development of commercial coal and clean coal use standards will be accelerated.

China is the world's largest consumer of coal, followed by the USA, India, the EU and Japan. In 2014, China's share of world coal consumption was 50.6%, while that of the USA, India, EU and Japan was 11.7%, 9.3%, 7.0% and 3.3% respectively. According to the BP Statistical Review of World Energy 2016, world total coal consumption dropped by 1.8% in 2015, compared to an average growth level of 2.1% in 2005–14, and coal's share of world primary energy consumption dropped to 29.2%, the lowest since 2005. In 2015, coal accounted for 64% of China's primary energy consumption, far higher than the world average of 29.2%. In

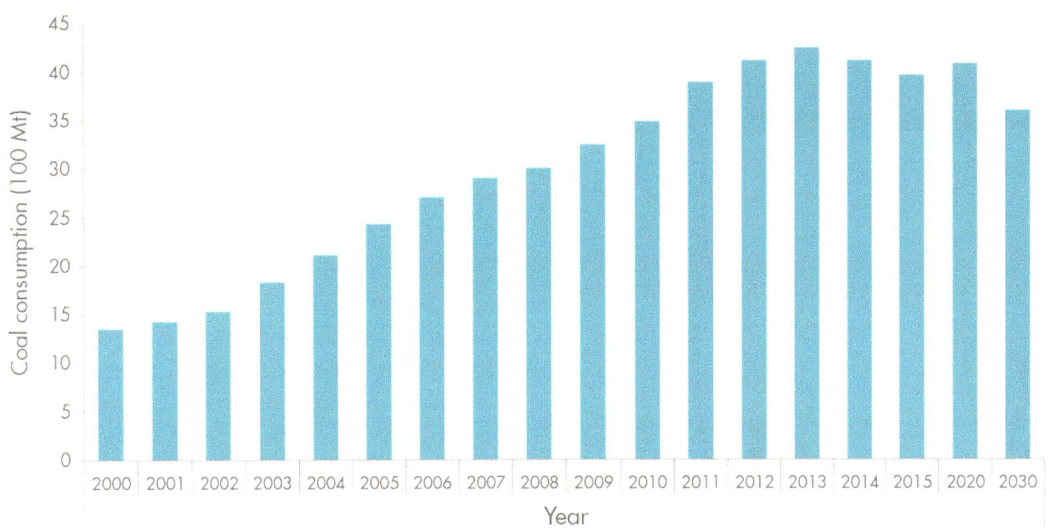

Fig. 45 China's annual coal consumption, trend and forecast. *Source* National Bureau of Statistics of China

addition, the net decrease in world coal consumption in 2015 came from the USA and China, declining by 12.7% and 1.5% respectively.

5.1.2 Opportunities in Coal Demand

(1) Huge potential in clean coal power generation and combined heat and power
Strong demand for power and great prospects for clean coal power generation. Clean coal power generation is the safest, most cost-effective and environmentally friendly way to use coal. Electricity will be the fastest-growing energy source in the next two decades, in China and in other countries. China's annual demand for power is forecast to grow by 4.9% and 2.3% respectively in 2010–20 and 2020–30. In 2013, China's power demand per capita (3,936 kWh/person) was higher than the world average (3,293 kWh/person). A year later it had risen to 4,047 kWh/person. By 2020, China's power demand per capita and household power demand per capita will reach 6,200 Wh/person and 1,400 kWh/person respectively. Compared with gas-fired power generation, nuclear and wind, coal-fired power generation is safer and more cost-effective. Thanks to continuous technological innovation in recent years, near-zero emissions from coal-fired power generation are possible in China. Coal therefore has great prospects as

China's dominant energy source in clean coal power generation.

Development of combined heat and power (CHP) will be accelerated in order to increase the share of centralised coal burning in total coal consumption Currently, China's centralised coal burning accounts for 48% of total coal consumption. The extremely large number of coal-fired facilities that are not equipped with pollution control technologies is a key contributor to air pollution. Industry experts have forecast that by substituting electricity for coal, PM2.5 emissions in east and central China will decrease by about 12%, 20% and 28% respectively in 2015, 2017 and 2020, compared to 2010. In contrast, 99% of coal consumption in the USA is used to generate power, which makes centralised pollution control of coal burning possible. In the future, China needs to take several measures to significantly increase centralised coal burning, thus reducing emissions from coal burning.

(2) Modern coal chemicals will be a breakthrough in future coal consumption
Modern coal chemical technologies include coal-to-oil, coal-to-olefin and aromatics, coal-to-ethylene glycol, and coal-to-gas. Compared with traditional coal chemicals, modern coal chemical processes are more scientific and

environmentally friendly and make higher-quality and higher value-added products. As a complement to petrochemicals, modern coal chemicals can bring the benefits of China's coal resources into play, reduce dependency on oil imports and strengthen China's energy security. China's modern coal chemical industry has made impressive progress since the 1990s. Technologies have been successfully tested in demonstration projects and upgraded to industrial-scale production facilities. Major breakthroughs have been made in advanced coal chemical synthesis, domestic coal chemical technologies have been commercialised and catalysts developed—all in China. All these advances have laid strong foundations for a clean, efficient and economically viable modern coal chemical industry.

5.1.3 Optimal Paths for Coal Consumption

The cleanest and most efficient ways to use coal are in power generation, coal-to-oil and coal-to-gas. In terms of energy efficiency, using coal-fired electricity to power electric vehicles is more efficient (28.6%) than using coal-to-oil for oil-fuelled vehicles (19.2%) and coal gas for gas-fuelled vehicles (13.3%). In terms of vehicle operating costs, if the oil products, natural gas and electricity are priced at RMB 7.72 per litre, RMB 3.1 per cubic metre and RMB 0.68 per kWh respectively, the operating cost of oil-fuelled buses is the highest at RMB 524,100/year, followed by buses fuelled with compressed natural gas at RMB 355,600/year and electric buses at RMB 352,000/year. As China's natural gas industry becomes increasingly market-oriented, higher gas prices will be inevitable. Therefore, using coal-fired electricity to power electric vehicles is the most cost-competitive option. In terms of safety, oil and natural gas pose higher safety risks in storage, transport and use, while power generation, transmission, distribution and consumption are relatively safe. This makes coal-fired electricity for electric vehicles the safest option.

Coal-to-electricity is the safest and most cost-effective and environmentally friendly way to use coal. To achieve the goals of the energy revolution, China should follow the coal-based and electricity-centred guiding principles of substituting electricity for poor quality coal and using higher quality coal cleanly and efficiently. Meanwhile, China should continue to develop modern coal chemical technologies, including coal-to-oil, coal-to-gas and coal-to-olefins.

Use efficient and ultra-low-emission coal-fired power generating plants at scale. China should speed up development of large coal power bases, especially 100 Mt coal bases and 10 GW coal-fired power generation bases, and long-distance large-capacity power transmission systems (the West-East and North-South power transmission projects). China should also allow private enterprise to develop large, clean coal-fired power plants in major load centres like the Beijing-Tianjin-Hebei region, the Yangtze River Delta and the Pearl River Delta in accordance with the environmental requirements of ultra-low emissions.

Build combined heat and power (CHP) and combined cooling, heat and power (CCHP) plants. CHP and CCHP plants achieve two objectives: (i) they significantly improve coal use efficiency and the economy of thermal power plants; and (ii) they can reduce emissions and pollution by replacing a large number of small coal-fired boilers used for heating. The non-power sectors should reduce their use of scattered coal and increase the amount of electricity in their energy consumption. The use of coal-fired industrial boilers in the iron and steel, chemical and building material sectors should be strictly controlled. Secondary energy should be supplied by centralised power plants. CHP and CCHP should be used in heating and cooling load centres like urban centralised heating areas and industrial parks.

After years of development, China's modern coal chemical industry has made major progress in four areas. First, independent coal gasification technologies have been widely deployed. More than 100 locally manufactured gasifiers are in operation, including opposed multi-burner gasifiers, HT-L pulverised coal gasifiers and two-stage dry pulverised coal gasifiers. Second, significant breakthroughs have been made in advanced coal chemical synthesis. Specifically, many locally developed technologies have been commercialised, including direct coal liquefaction and catalysts, high/low temperature Fischer-Tropsch synthesis and catalysts, methanol-to-low-carbon-olefins and catalysts, and fluidised bed methanol-to-aromatics and catalysts. Third, critical equipment for the production of coal chemicals has been locally manufactured in China, including large air separators, methanol-to-olefins reactors and the main pump valves. Fourth, the development and deployment of treatment technologies for waste gas, wastewater and solid waste in coal chemical production has made great progress, with near-zero discharge of wastewater now possible.

There are also many obstacles to overcome in the modern coal chemical industry. These include high consumption of resources, especially water; high capital investment in plants; difficulties in, and the high cost of, treating waste gas, wastewater and solid waste; and high CO_2 emissions. In recent years, sharply declining international oil prices have lowered the price of oil-based chemicals, severely restricting growth for modern coal chemical companies. In the context of China's slower economic growth, resource conservation and environmental awareness, the Chinese government has introduced increasingly strict policies for the coal chemical industry to raise the access threshold. It should be remembered that China is still modernising its coal chemical industry. In-depth research and prudence are needed.

5.2 Oil

5.2.1 Current Trends in Oil Demand

China's remaining oil resources are still abundant —the geological reserves of new proven oil resources were 1.118 Bt in 2015 and have exceeded 1 Bt for nine successive years. By the end of 2015, China's cumulative geological reserves of proven oil resources reached 37.176 Bt, including economically recoverable reserves of 2.569 Bt, with the reserves to

production ratio at 11.9. China sustained oil production at a level above 200 Mt for six successive years; oil production reached 215 Mt in 2015, increasing by 1.5% compared with the previous year. Oil consumption grows continuously—in 2015, China's oil consumption reached 550 Mt, rising by 6.1% compared with the previous year, and dependency on oil imports stood at 60.6%. The share of oil in China's primary energy consumption is also increasing—in 2015, oil accounted for 18.1%, rising by 0.7% compared with the previous year, yet far below the world average of 32.9%.

In terms of oil demand, petrol, kerosene and diesel consumption was 97.76 Mt, 23.35 Mt and 171.65 Mt respectively in 2014. Petrol and kerosene use increased by 4.4% and 7.9% respectively from the previous year, while diesel consumption grew by 0.14 Mt. Figure 46 shows the production and consumption data of oil and major oil products in China from 2000 to 2015. In that period, China's oil consumption grew rapidly. Compared with 2000, oil consumption in 2015 increased by 142%, but oil production grew only by 31.6%. The production of major oil products increases on a yearly basis. Kerosene production in 2015 was 17.35 Mt higher than in 2010, reflecting an average annual growth rate of 11.31% over the six years. The average annual growth rate of petrol was 8.52%, while that of diesel was much lower at 3.18%. The consumption of oil products also shows a yearly growth trend—from 2010, the average annual growth rate of petrol, kerosene and diesel was 7.04%, 5.76% and 3.15% respectively. As the gap between oil consumption and production expanded, China's dependency on oil imports reached 60.6% in 2015, which posed huge challenges in terms of oil supply security. Diesel production was slightly higher than consumption, and their growth rates were basically the same. In 2014, the production of petrol and kerosene was 12.83 and 31.95% higher than consumption, which is consistent with China's current diesel-petrol ratio of consumption. The diesel-petrol ratio peaked in 2005 and 2007 at 2.26:1, dropping to 1.76:1 in 2014.

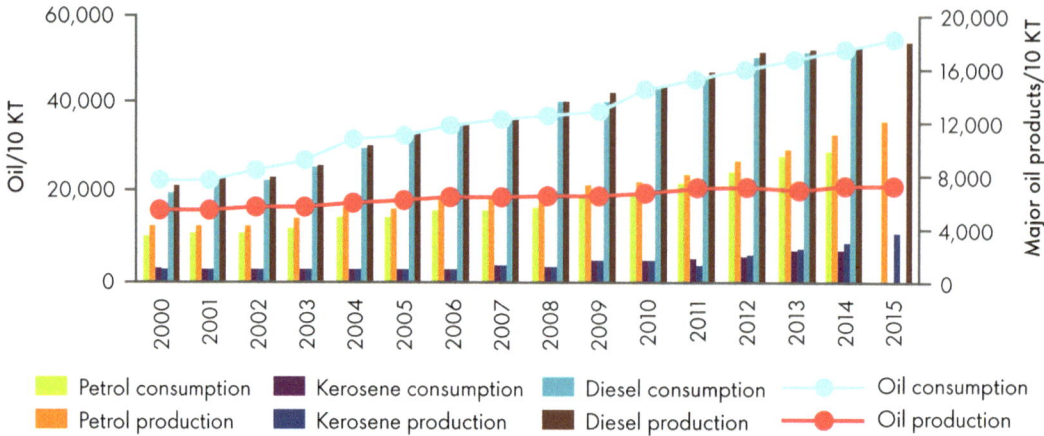

Fig. 46 China's production and consumption of oil and major oil products, 2000–15. *Source* Ministry of Land and Resources of the People's Republic of China

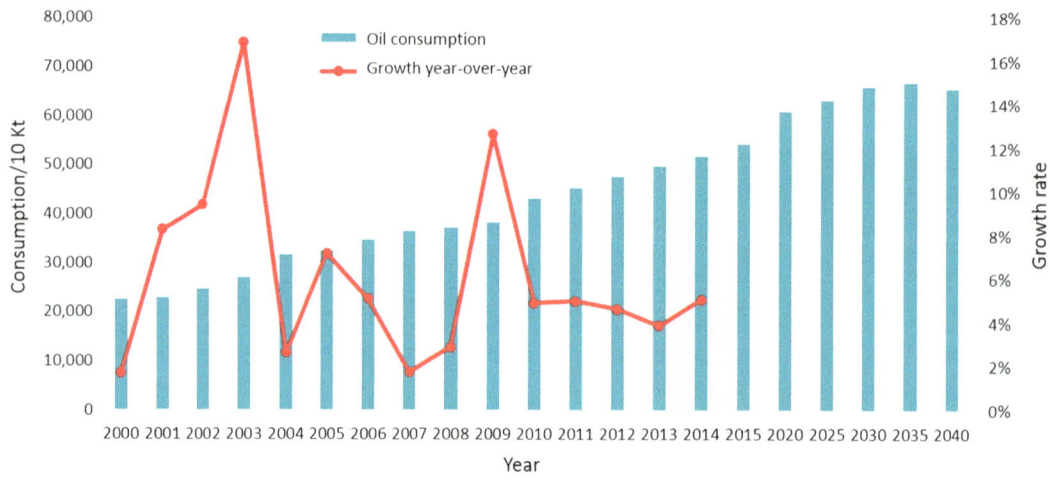

Fig. 47 China's oil consumption growth rate

As China's industrialisation accelerated from 2000 and vehicle sales increased rapidly, oil consumption maintained a strong average annual growth rate of more than 5.6%. After 2009, as a result of China's lower economic growth, industrial restructuring and the development of oil alternatives, oil consumption growth slowed down. The average annual growth rate of oil consumption in 2010–15 was 3.9%, with volumes at 544 Mt in 2015 (Fig. 47). This pattern of ongoing, but decreasing, growth will continue. During the 13th Five-Year Plan (2016–20), the average annual growth rate of China's oil demand

is expected to decrease to around 2%. After 2020, as China enters the later stages of industrialisation, its oil demand growth will slow even more, dropping to around 0.8% between 2020 and 30, with volumes rising slowly to 610 Mt in 2020. In general, China's oil consumption follows an S-shaped growth trend that is expected to peak around 670 Mt in 2025–30, then stay relatively steady, before declining in 2035–40.

In the forecast period, there are some uncertainties in China's oil demand, especially in the development of alternative fuels. Once major breakthroughs are made in electric vehicle technologies, the development of the EV industry will exceed expectations, which may result in oil consumption peaking earlier than predicted.

As China's vehicle ownership and air transport sector grow, demand for oil products also increases. Oil will remain the dominant transport fuel for at least the next 20–30 years. Demand for diesel in China is historically weak and will peak around 170 Mt in 2015–20. Demand for petrol will grow rapidly and is expected to reach around 170 Mt in 2030. The rapid development of air transport will increase demand for kerosene, which is predicted to rise to 36 Mt and 58 Mt in 2020 and 2030 respectively. The diesel-petrol ratio of consumption will inevitably fall, to 1.3:1 in 2020 and 1.1:1 in 2030.

As reported in the BP Statistical Review of World Energy 2016, global oil consumption grew by 94.61 Mt in 2015, up 1.9% from 2014 and far higher than the average annual growth rate of 1% in the previous decade. The share of oil in global primary energy consumption was 32.9%, an increase of 0.3% on 2014. China is the world's second largest oil consumer after the USA, accounting for 12.9% of the world total in 2015. India replaced Japan as the world's third largest oil consumer, with an oil consumption of 207 Mt. China ranked first in terms of incremental oil demand, followed by India, at 38.342 Mt and 15.437 Mt respectively. India has the world's highest year-on-year growth rate for oil demand at 8.1%, 1.8% higher than China and far higher than the USA at 1.6% and the EU at 1.5%. Japan registered the world's largest decline in incremental oil demand growth at 3.9%, equivalent to a drop of 7.967 Mt.

The growth rate of global oil production exceeded that of global oil consumption for two consecutive years after 2014. Global oil production growth stood at 139 Mt in 2015, up 3.2% on the previous year, and the fastest annual growth rate since 2004. Oil production in Iraq and Saudi Arabia hit record highs, and oil production in OPEC countries rose to 1.9 Bt in 2015, exceeding the previous record high of 2012. The world's largest oil producer, the USA, also achieved record growth in oil production in 2015—an increase of 49.795 Mt on the previous

year. Although China's oil production was 4.9% of the world total in 2015, this was an increase of 1.5% on 2014.

5.2.2 Opportunities in Oil Demand

Improvements in fuel quality accelerate. China implemented the China IV petrol quality standard in 2014 and the China IV diesel quality standard in 2015. Provinces including Beijing, Shanghai, Jiangsu and part of Guangdong, have implemented the China V fuel quality standards ahead of schedule. As of January 2016, China supplied China V petrol and diesel in 11 provinces and cities in north-east China, with supply to be extended across the country from January 2017. China's National Energy Administration released a draft of the China VI petrol and diesel quality standard in June 2016, which was enforced across China on January 1, 2019. China has completed the fuel specification upgrade from China II to China VI In 14 years. But further upgrades are expected, with China VI (B) for petrol scheduled to take effect in 2023, which will possibly lead to further reductions in olefins and increases in octane number.

China needs to reduce its diesel-petrol ratio urgently. After 2000, China's fixed investment in refineries maintained strong momentum. As diesel demand grew, the diesel-petrol ratio of consumption increased correspondingly, reaching a record high of 2.26:1 in 2005 and 2007. As a result, the diesel-petrol ratio of production was higher in most of China's refineries. However, the diesel-petrol ratio of consumption declined from 2008, falling to 1.76 in 2014. Due to weak diesel demand and strong petrol demand, the diesel-petrol ratio of consumption is expected to drop continuously, making it harder to reduce the diesel production ratio. To this end, China's oil refineries must take comprehensive measures to restructure, thus improving fuel quality and efficiency and optimising production.

The fuel saving potential of vehicles is huge. China has been striving to improve vehicle fuel economy to reduce fuel consumption. As a result, the vehicle fuel economy indicator has been significantly improved. China began implementing the passenger vehicle fuel economy standard in July 2005. Initially, a single vehicle fuel consumption limit was adopted. Now, there are two indicators of fuel consumption limit: by vehicle model and by the corporate average fuel consumption (CAFC)/Target CAFC (TCAFC). In accordance with the Passenger Vehicle Fuel Consumption Limits and the Evaluation Methods and Indicators for Passenger Vehicle Fuel Consumption, effective as of January 1, 2016, the average fuel consumption of new vehicles is to be reduced to 5 l/100 km by 2020. In addition, energy efficiency and new energy vehicles are clearly identified as a priority in the Made in China 2025 strategy, which states that the fuel consumption of new passenger vehicles (new energy passenger vehicles included) should be reduced to 4 l/100 km by 2025. In 2015, the average fuel consumption of China's passenger vehicles was 7.97 l/100 km. There is, therefore, huge potential to reduce vehicle fuel consumption to achieve the targets.

5.2.3 Optimal Paths for Oil Consumption

Optimal paths for upgrading of fuel quality. The major difficulty in upgrading fuel quality from China IV to China V lay in reducing sulphur content from 50 to 10 μg/g, which could be effectively addressed through petrol and diesel hydrodesulphurisation or adsorptive desulphurisation. This ensured full implementation of the China V standard as scheduled on January 1, 2017. The major difficulty in upgrading fuel quality from China V to China VI lies in reducing olefins without changing the octane number. Meanwhile, the declining diesel-petrol ratio and growing demand for petrol should also be taken into consideration. Firstly, breakthroughs in series alkylation technologies should be made. Alkylation is the principal method for upgrading fuel quality from China V to China VI. To upgrade fuel quality, research on series alkylation technologies, including solid acid alkylation and ionic liquid alkylation, should be increased to secure early breakthroughs. Secondly, catalytic cracking should be given priority and advances in catalytic cracking technologies made to improve petrol yield,

reduce olefins, increase octane number and boost propylene production. Through isomerisation and etherification of olefins in catalytic cracking light petrol, reduction of olefins while increasing octane number can be made possible. Thirdly, optimisation of refining-chemical integration should be enhanced and optimisation techniques like hydrocracking and catalytic reforming adopted to increase yields of high-octane petrol and low-cost raw materials for ethylene and aromatics. Fourthly, hydrogen costs should be lowered, as this largely determines petrol and diesel production costs.

The most important ways to reduce the diesel-petrol ratio include optimising oil refining units, improving production technologies, reducing diesel output and increasing the production of petrol. Specifically, the production of alkylated and isomerised fuels can be increased to improve the octane number and increase petrol output. Poor-quality diesel fuels like catalytic cracking diesel and recycled diesel can be converted into petrol components of high-octane number or aromatic hydrocarbon products, thus reducing diesel production. Integrated refining and chemical companies should reduce straight-run naphtha and properly increase straight-run diesel as the raw material for ethylene. In addition, diesel exports can be increased to release diesel overcapacity, and a national fuel pricing mechanism should be leveraged to gradually reduce the diesel-petrol ratio of production and promote diesel sales.

For example, a company with an integrated 10 Mt oil refining plant and 1 Mt ethylene plant, faced with reducing its diesel-petrol ratio and coping with slowing growth in demand for oil products, would need to shift production from diesel and petrol to high-grade petrol, aviation kerosene, clean diesel and low-cost chemicals. That is, it would need to shift its production focus from fuels to chemicals. A change like this takes time. Structural adjustments need to be speeded up during the 13th Five-Year Plan (2016–20) to allow more low-quality raw materials to enter ethylene cracking plants.

Given the current situation in China, the paths for vehicle energy saving can be categorised as technological and non-technological. Technological paths include improving the lubrication and cylinder fuel injection systems to raise the conversion efficiency of the internal combustion engine. The engine ignition system should be improved to boost engine stability, component service life should be increased to reduce the need for spare parts, and design modifications made to enhance fuel economy. Advanced energy-efficient technologies should be developed to improve the performance and fuel economy of hybrid electric vehicles, petrol and diesel engines, heavy-duty vehicles and engines, and diesel-fuelled saloon and light-duty vehicles. Non-technological paths include the provision of sufficient highway and transport facilities, on-demand production of vehicle models and oil products, and fuel-efficient driving, all of which can significantly reduce vehicle energy consumption.

5.3 Natural Gas

5.3.1 Current Trends in Natural Gas Demand

According to the Ministry of Land and Resources of the PRC, by the end of 2014, China's proven reserves of natural gas were 4.9 trillion cubic metres. In the same year, China imported 33 billion cubic metres of gas, accounting for 55.5% of total gas imports. Most of these gas imports came from Turkmenistan, with a small amount from Uzbekistan, Kazakhstan and Myanmar. Imported liquefied natural gas (LNG) was 18.93 Mt (equivalent to 26.5 billion cubic metres), accounting for 44.5% of total gas imports. These LNG imports came mainly from Australia, Indonesia, Malaysia and Qatar.

In 2014, China's apparent natural gas consumption[10] stood at 186.89 billion cubic metres, including domestic gas of 130.16 billion cubic metres and imported gas of 59.13 billion cubic metres. China's gas exports were 2.61 billion cubic metres. Total gas consumption by industry;

households; transport, warehousing and postal services; wholesale, retail, accommodation and catering; and other sectors was 122.13 billion cubic metres (65%), 34.2 billion cubic metres (18%), 21.44 billion cubic metres (12%), 4.66 billion cubic metres (3%) and 4.4 billion cubic metres (2%) respectively (Fig. 48).

China's natural gas consumption was 46.7 billion cubic metres in 2005 and 106.9 billion cubic metres in 2010, jumping to 192 billion cubic metres in 2015. As China increases its efforts to reduce carbon emissions and control pollution, during a period of growing urbanisation, its demand for natural gas will gradually increase. China's natural gas consumption is predicted to reach 290 billion cubic metres in 2020, equal to an average annual growth rate of 7.5%, and 480 billion cubic metres in 2030. The share of natural gas in China's total energy consumption will increase from 5% in 2015 to 12% in 2030.

By sector, gas demand in mining will peak and then decline, while that of manufacturing will grow, sourced mainly from coal-to-gas. Gas demand in power generation and heating holds huge potential and could increase substantially in the future. Increased gas demand in transport will come mainly from vehicles fuelled by compressed or liquefied natural gas. Gas demand by households will maintain steady growth. Figure 49 shows the trends and forecasts for China's natural gas demand over 30 years.

China is one of the world's largest natural gas consumers in terms of total consumption, but its gas consumption per capita is lower than comparable countries. In 2014, China's gas consumption (187 billion cubic metres) ranked third in the world, but was only 25% that of the USA (759.4 billion cubic metres) and 46% that of Russia (409.2 billion cubic metres). However, China's gas consumption per capita (137 m^3/person) was only 5.8% that of the USA (2,382 m^3/person), 4.8% that of Russia (2,845 m^3/person) and 29.3% of the world average (467 m^3/person). Even if China's total gas consumption reaches 480 billion cubic metres and gas consumption per capita registers 350 m^3/person in 2030, it would be no more than

---

[10]Apparent gas consumption is a country's dry natural gas production plus imports minus exports.

Fig. 48 China's natural gas consumption in 2014. *Source* National Bureau of Statistics of China

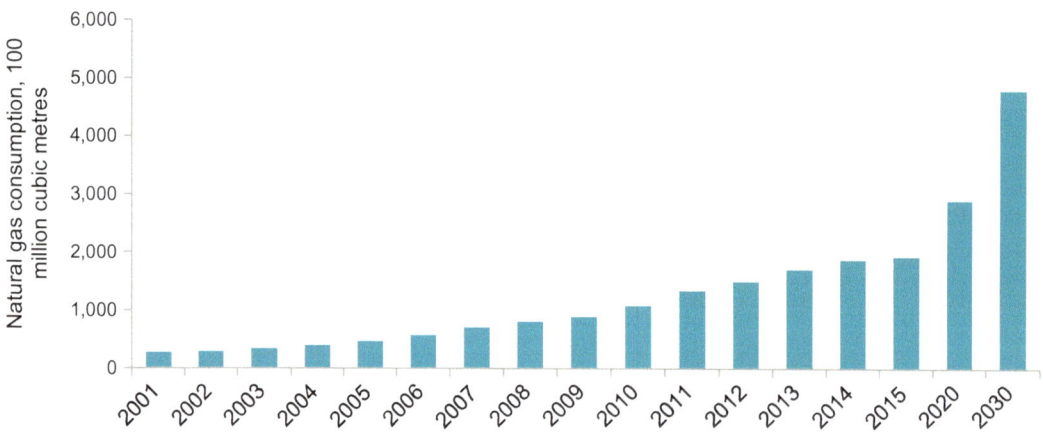

Fig. 49 China's natural gas consumption, 2001–30. *Source* National Bureau of Statistics of China

75% of world average gas consumption per capita in 2014.

5.3.2 Opportunities in Natural Gas Demand

(1) High growth potential in transport
Compared with oil products, natural gas as a vehicle fuel delivers many benefits. It is safe, efficient, cleaner and low carbon. In 2000–14, natural gas consumption in China's transport,

warehousing and postal services sector increased from 880 to 21,440 million cubic metres, which is an average annual growth rate of 25.6%, a figure far higher than that of petrol (8.5%) and diesel (9.5%) in transport in the same period. The sector's share of China's total gas consumption increased from 3.6 to 11.5%.

In recent years, gas-fuelled vehicles have developed rapidly in China, mainly in compressed and liquefied natural gas (CNG and

LNG). In 2000, there were only around 6,000 gas-fuelled vehicles registered in China. In 2010, there were 1.11 million units (1.10 million fuelled by CNG and 10,000 by LNG). By 2014, the figures had escalated to 4.595 million units (4.411 million fuelled by CNG and 184,000 by LNG), making China the leader in gas-fuelled vehicles with 20% of the world total. From the second half of 2014, the market turned due to China's slowing economic growth, weak demand in transport, sharp decline in international oil prices and high natural gas prices. As China increases its efforts to prevent and control air pollution, development of gas-fuelled vehicles, especially freight vehicles and urban buses, will be fast-tracked. China's gas demand in transport is expected to rise to 37 billion cubic metres in 2020 and 52.3 billion cubic metres in 2030.

(2) Rising gas demand in power generation

Compared to coal, gas-fired power generation delivers many benefits, including high energy conversion efficiency, strong peak shaving capacity, small gas turbine footprint, fast start and stop, large single unit capacity, safe and stable operation, and low emissions of pollutants and CO_2. By the end of 2015, China's installed capacity of gas-fired power generation was 66.37 gigawatts, accounting for only 4.4% of China's total installed power generating capacity and 20% of the world average.

Affected by such factors as low levels of indigenously manufactured equipment, high procurement and maintenance costs, slow progress in power market reform and high gas prices, China's gas-fired power generation is less cost-effective than coal, hydropower and nuclear power. Gas-fired power generation projects in China make only small profits at best. Others are loss-making, so the industry takes a wait-and-see attitude to assess those projects under construction or planned. In 2014, power generation accounted for 14.1% of China's total gas demand, far below that of the USA (30.4%), the UK (38%), Germany (36%) and South Korea (44%). Given the environmental benefits of gas-fired power generation, the segment holds strong growth potential.

(3) Steady growth in household gas demand

In 2000–14, China's gas pipeline capacity increased from 33,700 to 434,600 km, an average annual growth rate of 18%. As China's long-distance gas transmission and urban gas distribution pipeline networks grew, gas demand by households rose just as rapidly. In 2000–14, household gas demand increased from 3.23 billion cubic metres to 34.26 billion cubic metres, averaging 18.4% annually, according to the National Bureau of Statistics of China. Gas demand per capita increased from 2.6 to 25 m^3/person, with urban gas demand per capita rising from 7.0 to 45.7 m^3/person. At the end of 2015, the total population of mainland China was 1.374 billion, of which 771 million were urban residents, 56% of the total. As more people move into cities, China's urbanisation rate will reach 60% in 2020 and 70% in 2030, and the number of residents using gas will also grow. Household gas demand in China is forecast to rise to 48 billion cubic metres in 2020 and 62 billion cubic metres in 2030, about 13% of total gas demand in 2030.

5.3.3 Optimal Paths for Natural Gas Demand

(1) Optimal paths for gas-fuelled vehicles

By 2020, the number of gas-fuelled vehicles in China is expected to reach 10.5–11 million units, including 400,000–500,000 vehicles fuelled by LNG. Together, they will account for more than 5% of China's total vehicle stock. The number of LNG service stations will amount to 4,500–5,000.

Gas-fuelled vehicles have significant environmental and cost advantages over petrol and diesel. The operating cost of natural gas transport fuel is lower than that of petrol and diesel. Compared with electric and other new energy vehicles, gas is a more mature technology. Gas-fuelled vehicles perform better, have a longer driving range and are safer.

Small gas-fuelled passenger vehicles like urban taxis and family saloons have lower operating costs, according to the China Automotive Technology and Research Center. The fuel cost of a CNG-fuelled taxi driving 350 km a day is only 45% that of a petrol-driven taxi (measured at the Beijing price level in March 2016). This amounts to a lower daily fuel cost of RMB 85, or RMB 30,000 per year over 350 days. Even though it costs up to RMB 10,000 to convert a small taxi from petrol to CNG, operating costs are much lower than those of petrol taxis.

LNG offers a long driving range and is more suitable for large trucks and urban buses. The price of an LNG-fuelled heavy-duty truck is RMB 80,000–100,000 higher than a diesel truck of the same horsepower. Given the price of LNG is only 50–70% that of #0 diesel of the same calorific value, the fuel cost of an LNG-fuelled heavy truck travelling 150,000 km per year is around RMB 140,000 lower than diesel, enabling the difference in purchase price to be recovered in less than a year.

(2) Optimal paths for gas-fired power generation

Constraints are holding back the development of gas-fired power generation in China. These include barriers to collaboration on critical technologies between China's major gas-fired equipment manufacturers and overseas companies, a lack of locally developed gas-fired alternative technologies, uncoordinated gas supply and power generation, incomplete support policies and standards, and rising gas prices. And power plant design, construction, operation and maintenance, and safety management need to be improved.

To make gas-fired power generation sustainable, China should: (i) increase investment in innovation and develop core technologies, including gas-fired power generation equipment; (ii) unplug the management bottlenecks in the power and gas industries; and (iii) deepen power system and gas pricing reform, and allow the market to truly play the decisive role in resource allocation.

To summarise, due to technology and pricing constraints, there are still big uncertainties in China's gas-fired power generation sector. However, with sufficient support from the government and breakthroughs in local development

of core technologies, the sector has great growth potential in the future.

5.4 Accelerate the Integration of New Energy with Conventional Fossil Energy

In the vast country of China, the geographical distribution of fossil energy and new energy sources like solar, hydro, wind and biomass is uneven. Due to unbalanced economic development in east, central and west China, energy supply and demand in these regions vary greatly. Energy development should therefore be adapted to local conditions and local energy resources. National energy revolution policies should be implemented to diversify energy demand. And, energy consumption should consistently follow resource conservation and be clean, sustainable and efficient.

An integrated development system and cooperation platform for conventional and new energy should be established. Currently, conventional energy and new energy develop separately. A cooperation platform that gives play to their complementary advantages and delivers synergies is absent. For instance, the long-distance transmission of solar photovoltaic and wind power needs the infrastructure of coal power transmission networks. But due to higher energy costs and unstable output, as well as constraints from the power industry, it is difficult to connect new energy to the grid and deliver it to consumers. China, therefore, needs to build a flexible cooperation platform for different types of energy demand to speed up the integration of conventional and new energy.

Subsidy policies should be improved to drive the sustainable development of new energy. It will take a long time for new energy to replace fossil fuels, but the shift is already under way. The initial stages of new energy development are characterised by high capital investment, low returns and high production and operating costs, so national subsidy policies are needed. When new energy technologies mature and their costs fall to an acceptable level, new energy can be fully integrated in the energy market. New energy subsidy policies are indispensable, but they should be proper and reasonable, thus encouraging technological innovation and the sound development of the new energy sector.

6 Electrification of Energy Demand

International experience suggests there are three key factors that have historically driven electrification rates: technological progress, increased income and structural change.

Technological progress has been the most important driver of electrification, particularly in buildings. Assuming a continuation of historical technology trends, this driver could increase China's electrification rate by 6 percentage points by 2050.

Income increases could raise China's electrification rate by more than 5 percentage points by 2050, as consumers demand a higher quality energy carrier in terms of flexibility and cleanliness.

Given structural change affects electrification, a move to less power-intensive sectors implies lower electrification rates: an increased share of transport in energy use in China is expected to reduce the electrification rate by 3 percentage points by 2050.

In the future, decarbonisation will be a key driver of electrification, as the power sector can be decarbonised relatively easily and electricity used across end uses. Decarbonisation will increase electrification of the transport and buildings sectors, principally due to the uptake of electric vehicles and heat pumps. However, further electrification of industry is challenging due to technical constraints (Fig. 50).

China's electrification rate could increase from 23% today to 32% by 2050 on the basis of historical experience of macroeconomic drivers, or to 40–48% indicatively as a result of future trends in decarbonisation. To achieve these high rates of electrification, policymakers may wish to implement a range of policy packages to overcome increased peak load variability, underdeveloped

networks, investment and behavioural barriers, and innovation constraints.

6.1 Macroeconomic Drivers of Electrification

Technology, income and economic structure drive changes in electrification rates.

Technology effect: Common technology trends across countries over time have led to increased levels of electrification. A country today will be more electrified than a country in the 1990s with the same level of income. Technology could contribute 6 percentage points to China's electrification rate by 2050.

Income effect: Level of electrification increases with income. A country with a higher income level will be more electrified than a country with a lower income level in the same period, all things being equal. Income could contribute 5 percentage points to China's electrification rate by 2050.

Structural change: Differences in economic structure lead to different levels of electrification. A country with more energy use in an electricity-intensive sector, such as buildings, will have greater levels of electrification. Structural change will lower China's electrification rate by 3 percentage points by 2050.

To estimate this, we use a fixed effect (within) model. This uses the panel dimension of the dataset to account for both entity and time fixed effects. Geology is an example of an entity fixed effect that varies across countries but not across time; and general electrification due to technological progress is an example of a time fixed effect that varies across time but not across countries. The time trend chart is generated by storing and plotting the time fixed effects. It shows the common electrification trend across time, controlling for income and entity fixed effects (i.e. country-specific factors such as latitude and geology).

Technological progress includes the introduction of new electricity-powered services, as well as the substitution of electricity for other

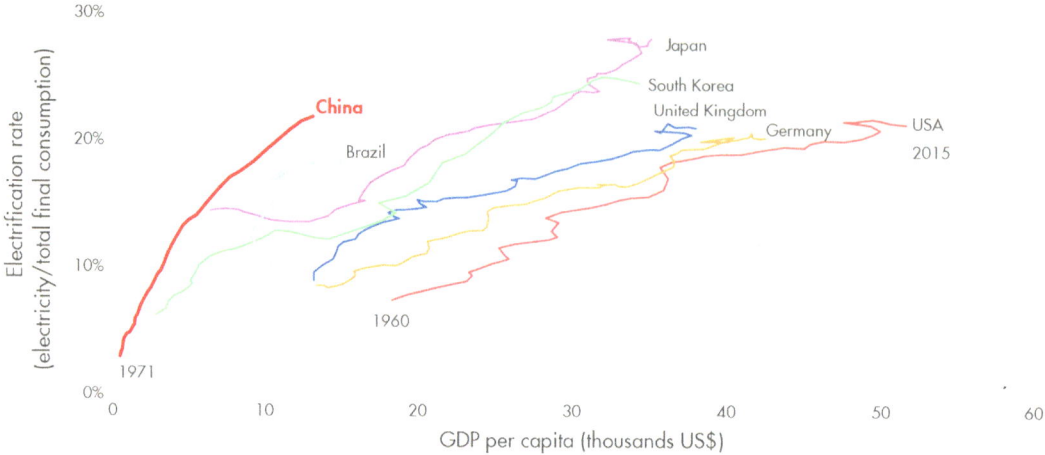

Fig. 50 Electrification rates tend to increase with income, with China achieving rapid electrification over a short period of time. *Source* International Energy Agency World Energy Balances

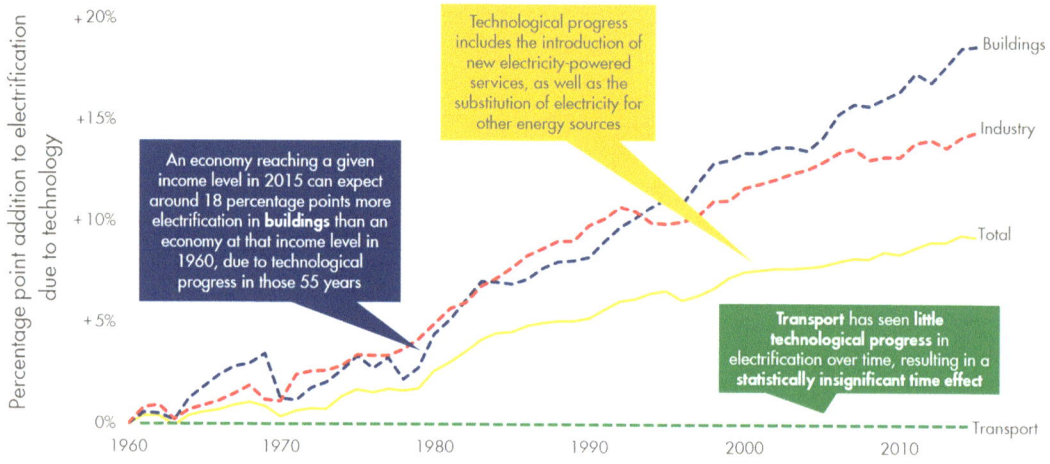

Fig. 51 Technological progress has historically been the most important driver of electrification. *Source* Vivid Economics

energy sources. An economy reaching a given income level in 2015 can expect around 18 percentage points more electrification in buildings than an economy at that income level in 1960, due to technological progress in those 55 years. Transport has seen little technological progress in electrification over time, resulting in a statistically insignificant time effect.

Assuming historical trends in electrification from the 1960s will continue, and based on the China GDP forecast for 2050, China's 2050 electrification rate will be similar to the electrification rate of select developed countries, but the electrification level of sectors will vary greatly (Figs. 51 and 52).

Structural change tends to shift activity to less power-intensive sectors, such as transport, as is likely to occur in China. There are two different electrification paths: one is for economic structure with a high share of industry, as in South Korea; and the other is for economic structure with a low share of industry, as in the UK.

In addition to technology, income and structure, a set of country-specific characteristics

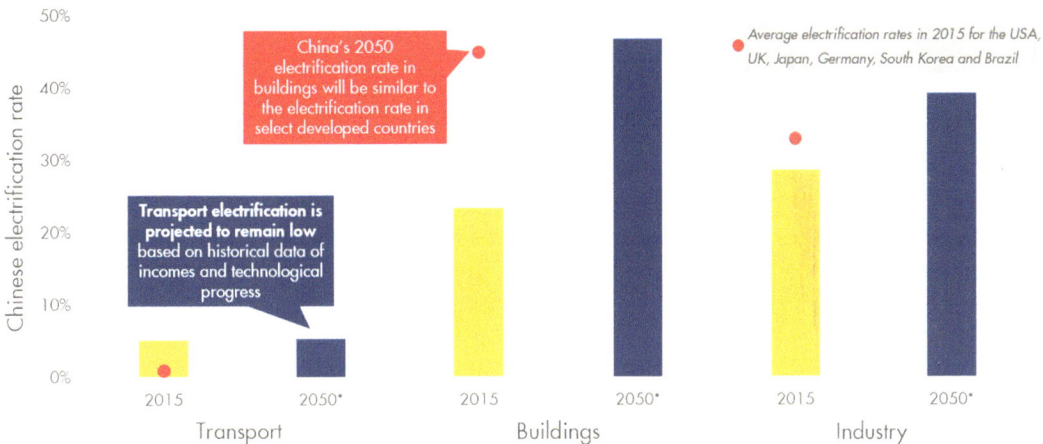

Fig. 52 Sectors have different power intensities, which are expected to increase over time with new technology and higher incomes. *Source* Vivid Economics

determines the potential for electrification, including urbanisation level, share of heavy industry, climate and resource endowment. For example, a highly urbanised and heavy-industry oriented country typically has a higher electrification level.

After controlling for technological progress and country-specific effects, the income effect on electrification is significantly reduced. An economy growing from $15,000 per capita to $40,000 per capita can expect to increase electrification by around 4 percentage points, when controlled for technological progress and country-specific effects (Figs. 53, 54 and 55).

China's electrification rate could increase from 23 to 32% by 2050, based on historical international experience of macroeconomic drivers. Increase in income leads to an increase in electrification of 5 percentage points, and technological progress adds another 6 percentage points. Structural change decreases electrification by 3 percentage points, as the share of energy use for transport (currently the least electrified) rises by 7 percentage points, whereas the share of industry (currently the most electrified) falls by 6 percentage points (Fig. 56).

6.2 Decarbonisation Drivers of Electrification

Decarbonisation is expected to accelerate electrification and lead to somewhat higher overall rates. For example, the median of low-carbon scenarios analysed by the Intergovernmental Panel on Climate Change puts global electrification at 35% in 2050, compared to 30% for the median of high-carbon scenarios. This is because electricity can be decarbonised with relative ease compared to other energy carriers, and there are electricity alternatives to conventional technologies in transport, buildings and industry.

Transport

Transport has the greatest scope for electrification, with just 1% of transport energy powered by electricity in OECD countries. Infrastructure is key for light-duty vehicle electrification, while freight and air electrification are limited by technical constraints.

Light-duty vehicles: High costs to consumers and development of charging infrastructure, particularly fast charging stations in high-density urban areas, constrain uptake. Range anxiety remains a behavioural barrier. However, these

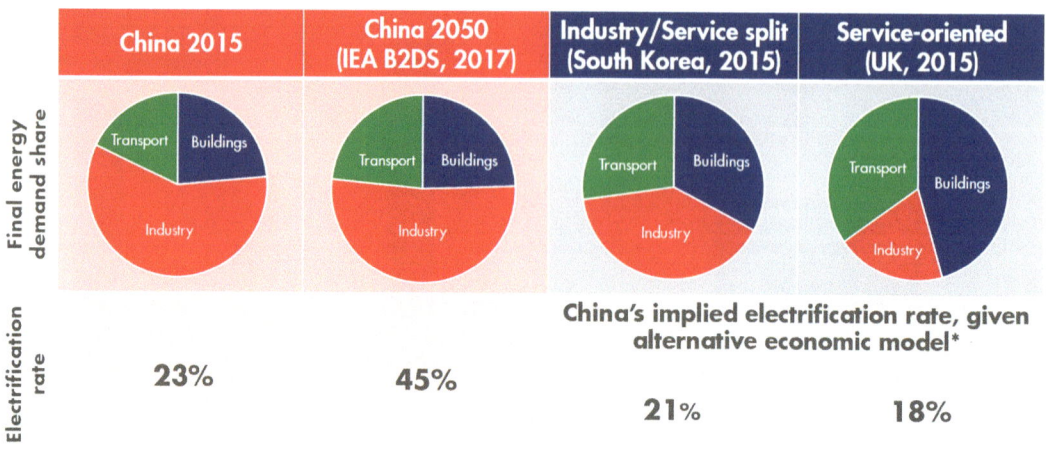

Fig. 53 Structural change tends to shift activity to less power-intensive sectors. *Note* IEA 2017 B2DS = International Energy Agency, Beyond 2°C Scenario (2017). *Source* Vivid Economics

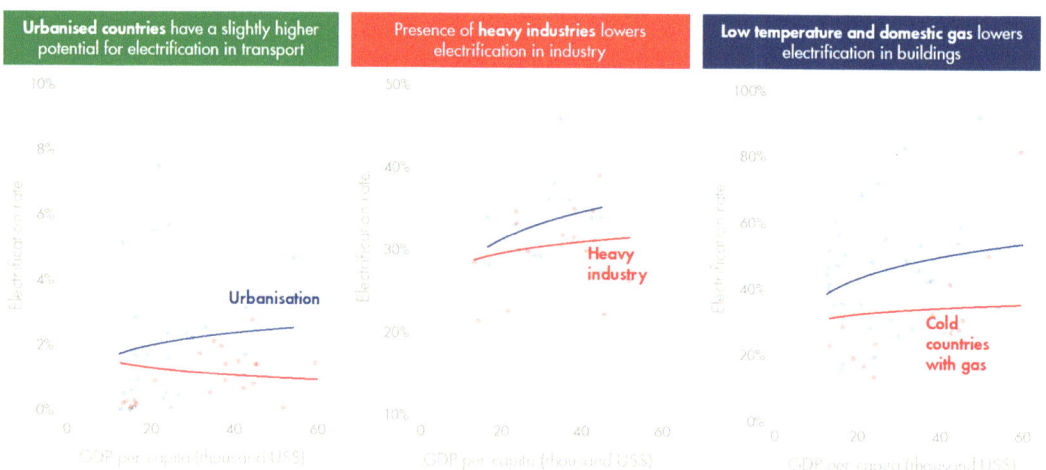

Fig. 54 Country-specific characteristics affect the potential for electrification. *Source* Vivid Economics

constraints are expected to be lowered in the future.

Heavy-duty vehicles: Electrification requires major investment in either catenary lanes, costing $2 million per kilometre, or in expensive inductive charging infrastructure, which is lower efficiency and requires vast changes to the existing infrastructure.

High-speed rail: Limited to specific regions of high-population density; it also requires

significant infrastructure investment in electric trains and train lines. Modal shift and behavioural barriers are key to unlocking the benefits of rail electrification.

Shipping: Limited potential for electrification, possibly limited to coastal shipping due to shorter distances.

Air: Limited potential in electrifying air transport unless major technology breakthroughs occur (Fig. 57).

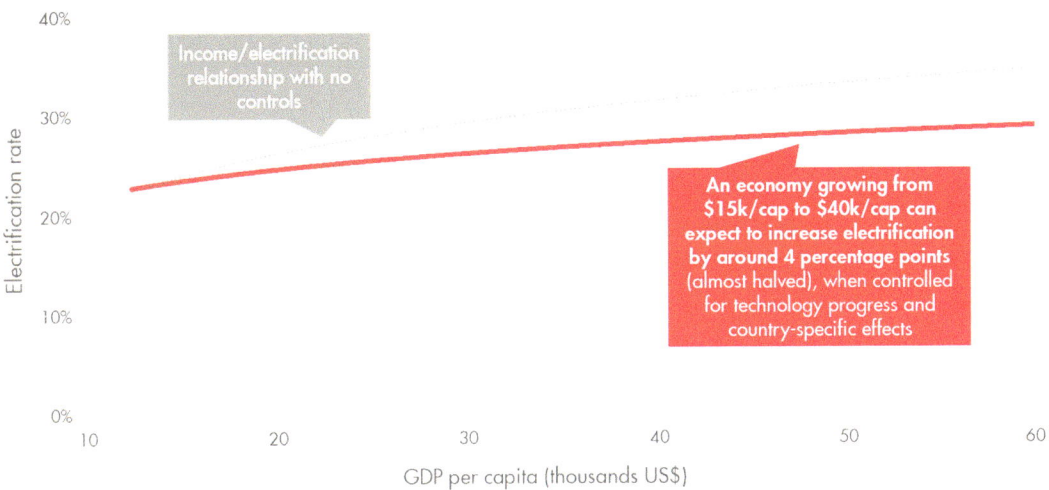

Fig. 55 Income effect on electrification. *Source* Vivid Economics

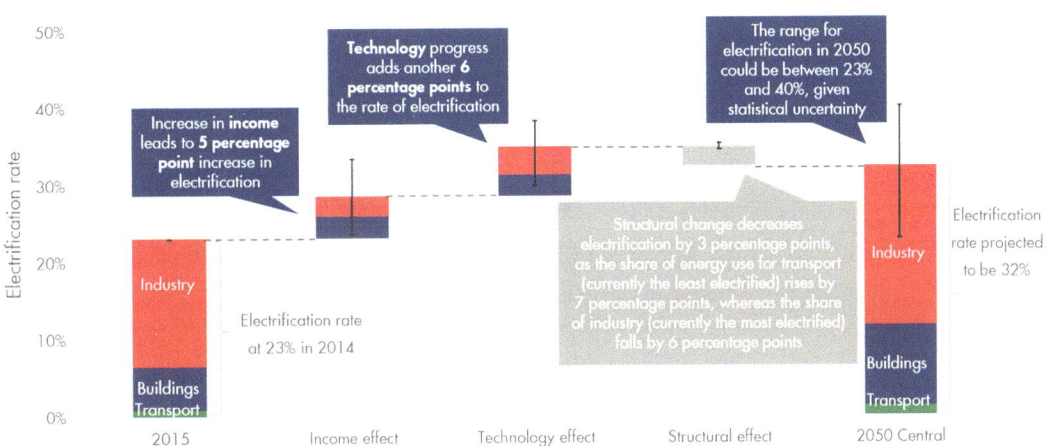

Fig. 56 China's electrification rate could increase from 23 to 32% by 2050. *Source* Vivid Economics

6.3 Buildings

Water and space heating have the greatest potential for increased electrification, although they face cost barriers.

Lighting and appliances: There are no technical limits to electrification; both are already electrified in developed economies.

Air conditioning: There are few barriers to electrification; 75% of air conditioning in developed countries is electrified. Electricity demand is expected to increase with rising demand for air conditioning as incomes increase in hot, humid areas.

Cooking: Electric options exist with moderate barriers to further electrification. In rural areas biomass dominates cooking energy use; in non-rural areas electrification rates depend on the presence of competing gas networks.

Water heating: Key electrification technologies are electric water boilers and heat pumps, though high capital costs and competing gas networks keep uptake rates low.

Space heating: Electric options exist with high barriers to increased electrification. Technologies such as heat pumps are limited by latitude, existing consumer attitudes, available

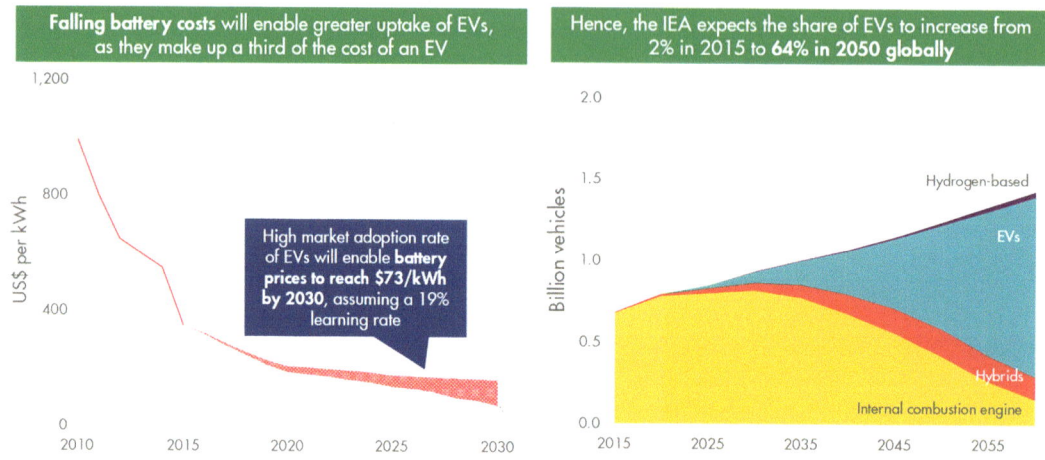

Fig. 57 The share of passenger electric vehicles is expected to be more than 60% by 2050, due to falling battery costs. *Source* International Energy Agency, Energy Technology Perspectives (2017), Bloomberg New Energy Finance (2017)

space, building envelope efficiency and large switching costs (Fig. 58).

Industry

Electrification options exist for many medium- and high-grade industrial heat processes, though relative costs remain high.

Iron and steel: Electric arc furnaces hold significant potential for scrap-based crude steel production. In iron production, electrowinning (extracting iron from the ore through electrolysis), is still in its early stages of development.

Chemicals and petrochemicals: Electrothermal furnaces could potentially be used for a variety of petrochemical cracking processes, though it remains a marginal technology. Ammonia production through grid electrolysis is technically feasible.

Cement: Electrothermal dryers can be used for clinker calcination. While technically feasible, the extremely high heat requirements of clinker production make electrification of the process prohibitively high cost.

Pulp and paper: Electrothermal technologies, such as induction or electric arc furnaces, can be used in medium-heat pulp and paper drying processes. These technologies are increasingly cost-competitive.

Aluminium: Electrothermal dryers can be used for the bauxite reduction process, though technical availability is low. Production of fused aluminium oxides in electric arc furnaces by electrothermal fusion is commercially feasible, though it remains a marginal process.

Decarbonisation will raise electrification of transport and buildings due to electric vehicle and heat pump uptake, but industry options are limited.

In summary:

Transport

- Passenger transport electrification will be driven by falling battery costs and increasing electric vehicle cost competitiveness.
- Freight, shipping and air transport energy end uses are more difficult to electrify.

Buildings

- Building electrification could increase from less than a quarter to more than half globally in a decarbonising world by 2050.
- Progress depends on the diffusion of heat pumps and electric boilers for space and water heating and behavioural change away from fossil fuels.

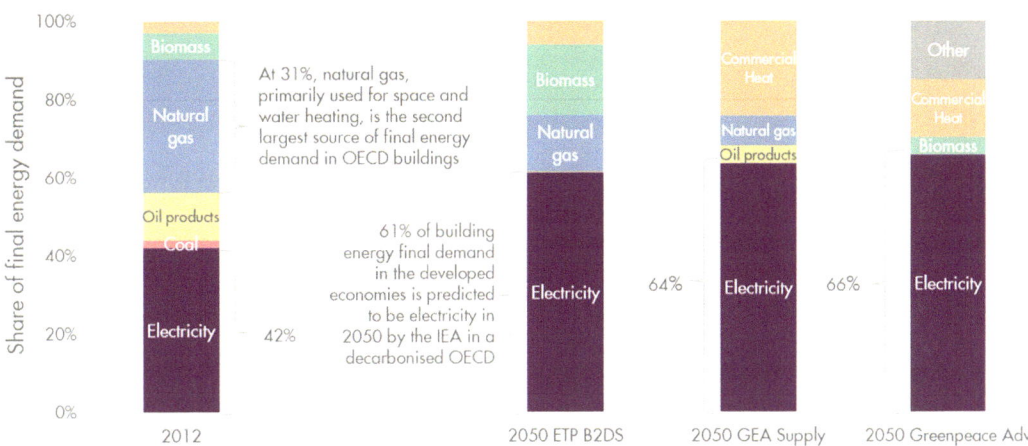

Fig. 58 Several decarbonisation scenarios suggest up to two-thirds of energy demand in OECD buildings could be electrified by 2050. *Source* IEA ETP 2017; GEA Energy pathways for sustainable development (2014); Greenpeace Energy Outlook 2015

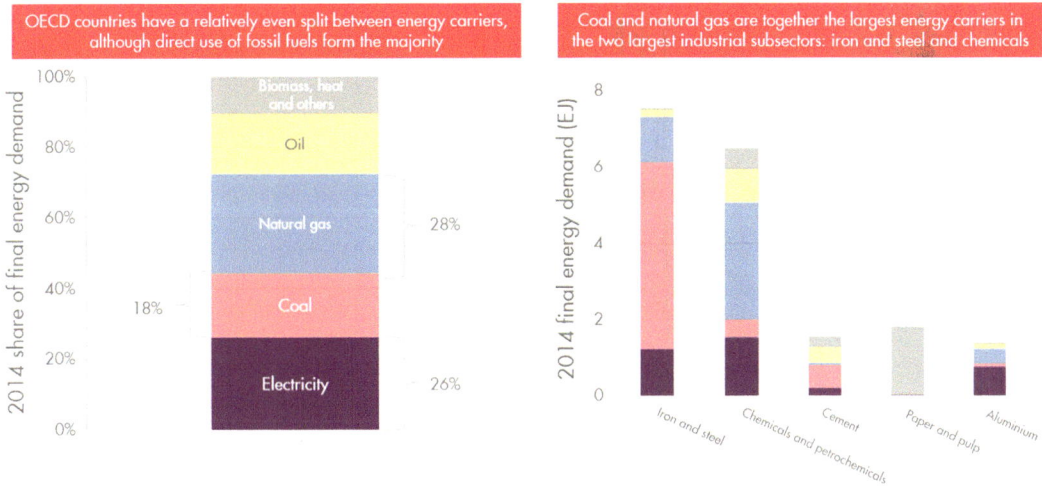

Fig. 59 Less than a third of OECD industry energy demand will be electrified in 2050, according to the IEA Beyond 2°C Scenario. *Source* International Energy Agency, Energy Technology Perspectives, Beyond 2°C Scenario (2017)

Industry

- Industry has limited scope for electrification.
- Electric arc furnaces hold promise in steel-making, but further R&D is necessary to support deployment.
- However, technical limitations and low capital turnover rates limit scope elsewhere (Fig. 59).

6.4   China's Electrification Potential

China's electrification rate could rise to 32% by 2050 given macroeconomic drivers, or to 40–48% given decarbonisation drivers.

- Income effect: +5.5%
- Technology: +6.4%
- Structural change: −2.6%

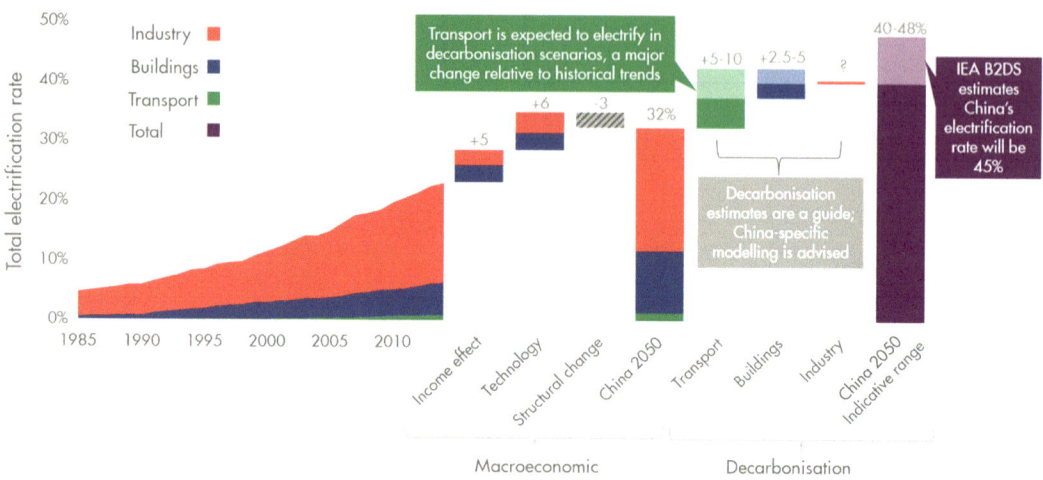

Fig. 60 China's electrification rate could rise to 32% by 2050 given macroeconomic drivers, or to 40–48% given decarbonisation drivers. *Note* IEA 2BDS = International Energy Agency, Beyond 2°C Scenario. *Source* Shell International

Total electrification rate under macroeconomic drivers: 32.4%.

- Decarbonisation in transport: +5–10%
- Decarbonisation in buildings: +2.5–5%
- Decarbonisation in industry: +?%

Total electrification rate under decarbonisation drivers: 40–48% (Fig. 60).

6.5 Constraints and Policy Suggestions

Policymakers must consider a range of systemic and sector-specific constraints, such as increased peak load variability, underdeveloped networks, investment and behavioural barriers, and hurdles preventing innovation.

Seasonal and daily fluctuations in energy demand will increase in proportion to the uptake of electric vehicles, electric heating and renewable generation. Heightened peak load variability can be overcome with more grid interconnections and flexible markets. The policy responses include:

- increase investment in energy storage technologies, such as batteries (although this is expensive and only technically feasible for short-term energy storage);

- promote grid interconnections to facilitate a more flexible energy distribution system;
- increase reserve electricity generation capacity to meet unpredictable daily demand peaks; and
- adopt demand-tiered electricity pricing to discourage electricity use during peak periods and spur consumption during off-peak times.

A lack of support infrastructure for electrified technologies will depress the uptake of electrified products and reduce learning-by-doing cost reductions. Policy responses to network constraints include investment in charging networks and efficiency mandates:

- direct support across the entire electric vehicle ecosystem, including maintenance, repairs and charging stations;
- mandate electrified technologies in public buildings to build up technical know-how and facilitate the creation of a support network; and
- implement building, industry and transport efficiency standards to ensure product uptake across the broadest set of consumers and businesses, which will reduce overall electricity demand and ease network pressure.

Investment and behavioural barriers: Investors may be unwilling to invest in electrified technologies due to uncertainty over investment returns on electrified products. Consumers may prefer gas or biomass options. Investment and behavioural barriers can be overcome through the following policies:

- apply stable and consistent tax benefits and subsidies to consumers and suppliers across energy carriers;
- maintain a credible electrification policy to mitigate investor uncertainty and spur investment;
- build product networks to encourage the formation of liquid secondary markets for assets associated with electrified processes to reduce asset resale uncertainty; and
- increase support for energy service companies to encourage heat pump uptake during building and industry retrofits.

In terms of innovation, private companies and individuals are often unwilling to invest resources in underdeveloped technologies. Innovation support is required to reduce technology costs and increase electrification in more challenging subsectors:

- increase tax benefits and direct expenditure for research, innovation and demonstration;
- develop R&D ecosystems and support learning by doing to facilitate technology

transition from laboratories to product commercialisation; and
- focus innovation support on areas with substantial electrification potential (shipping, heavy-duty trucks, aviation).

6.6 Conclusion

Electricity is almost a quarter of China's final energy demand. China has already proved its capacity to electrify rapidly, and electricity could meet almost half of final energy demand by 2050. China's electrification rates are comparable to many OECD economies.

As the country grows, with consumers demanding higher quality energy carriers, and technology and structural shifts drive further changes, historical experience suggests that driven by these macroeconomic factors, China's electrification rate could increase from 23% today to 32% in 2050.

Decarbonisation will further drive the process. A high uptake of electric vehicles in transport and heat pumps in buildings could increase China's electrification rate to 40–48% in 2050.

To achieve this, policymakers may wish to implement a range of policy packages to overcome increased peak load variability, underdeveloped networks, investment and behavioural barriers and innovation constraints.

Special Report 3: A Study of China's Technology Revolution

Song Zifeng and Nigel Dickens

Energy is a key driver of social development. The energy transition correlates strongly with social and economic reform. Among the many drivers of the energy transition, a revolution in energy technology is undoubtedly one of the most critical.

DRC Team Lead of Special Report 3:
Song Zifeng from the Research Department of Industrial Economy, DRC of the State Council of China.

Shell Team Lead of Special Report 3:
Nigel Dickens, Economics and Analysis Manager—New Fuels, Shell International Petroleum Company Ltd.

Contributors:
Philip Gradwell from Vivid Economics; Cameron Hepburn, Huang Yanghua from the Chinese Academy of Social Sciences (CASS); Kang Xiaowan from the Energy Research Institute, National Development Reform Commission (NDRC); Huang Bibin from the State Grid Energy Research Institute Co., Ltd.; Tu Junming from China National Travel Service Group Corporation Limited; Xing Lu from China National Petroleum Corporation (CNPC); and Zhang Jun from the State Grid Energy Research Institute Co., Ltd.

S. Zifeng (✉)
Research Department of Industrial Economy, DRC of the State Council of China, Beijing, China

N. Dickens
Economics and Analysis Manager—New Fuels, Shell International Petroleum Company Ltd., London, UK
e-mail: Nigel.Dickens@shell.com

To build modern energy systems that are affordable, secure and sustainable, countries need to push technological innovation. Although innovation in energy is at a relatively high level, it is unlikely to be high enough to achieve the needed revolution in energy technology. The International Energy Agency's 2017 Tracking Clean Energy Progress scorecard of 26 technologies finds that only three are on track for wide-scale deployment. To deliver a low-carbon energy system, more effort is needed in 15 technologies and eight are not on track. As many of the current set of new energy technologies are based on a pipeline of innovation that was started in the 1970s, it is important to understand the energy technology issues of the current era.

1 The Implications of Energy Technology Revolution

It is extremely rare for innovations to be evenly distributed across time and space. Instead, innovations tend to occur in a concentrated manner, which is likely to trigger a technology revolution. Technology revolution usually refers to the transition process, in which a technology or technologies is/are replaced by another technology or technologies in the short term. This process—from emergence to deployment and diffusion—will eventually impact socioeconomic development. Up to now, countless energy-related

Shell International B.V. and the Development Research Center (DRC) of the State Council of the People's Republic of China (eds.), *China's Energy Revolution in the Context of the Global Energy Transition*, Advances in Oil and Gas Exploration & Production, https://doi.org/10.1007/978-3-030-40154-2_4

technologies have been invented, but few qualify as an energy technology revolution. To understand the deep implications of energy technology revolution, multiple dimensions and perspectives must be taken into account.

1.1 Energy Technology Revolution Is a Long-Term Process

The history of technology shows that it often takes several decades for new technologies to flourish. It can take as long as 30 years for a new technology to gain a 1% market share, and even longer for it to be deployed at scale. There are many such examples: the steam engine appeared in the late 18th century, but the number of workers employed by steam engine-powered factories and manual workshops did not pass 50% of the total until 1880[1]; the history of electric vehicles (EVs) can be traced back to the 1830s, with the world's first commercially operated EV unveiled in New York in 1897, although little progress was made in the next 100 years; the modern oil industry took shape in 1859, but didn't take off until the early 20th century; fracturing and horizontal well drilling technologies have a history of more than 50 and 30 years respectively, but the shale gas revolution did not occur until after the global financial crisis in 2008.

For those technologies that qualify as revolutionary, the deployment life cycle of their core technology is often far more than 100 years. Rail, electricity and motor vehicles are typical examples.[2] This tells us that energy technology revolution is typically a long-term process.

1.2 Energy Technology Revolution Is Strongly Correlated with Industrial Revolution

Human society has witnessed two energy revolutions—from wood and biomass to coal, and from coal to oil and gas. These two revolutions spawned or were accompanied by technological or industrial revolutions, including the steam engine, the internal combustion engine and electrification. Together, they drove productivity and enabled civilisation to leap forward. As summarised by Freeman and Soete,[3] over the past 200 years, the energy system has undergone several transitions—from hydraulic power to steam, to electricity, oil, and oil and gas. These transitions underpin the Kondratiev waves of economic and technological cycles. The rationale behind these waves is that a rapid drop in the price of critical energy resources is enabled only by a technology revolution, such as steam, that eventually boosts productivity significantly (Table 1).

1.3 Energy Technology Revolution Triggers New and Important Energy Sectors

There is a strong correlation between energy technology revolution and industrial revolution. Every energy technology revolution in history has enabled new and important energy sectors to emerge. These include coal and oil and gas in primary energy, and electricity in secondary energy, all three of which are still significant pillar industries for economic growth.

For example, the oil and power industries that emerged in the second half of the 19th century still play an important role in driving economic growth in the USA. Measured by industry R&D investment per capita and the proportion of employees with an education in the STEM disciplines (science, technology, engineering and

[1]Vaclav Smil, Made in the USA: The Rise and Retreat of American Manufacturing, The MIT Press, 2015.

[2]Jan Fagerberg, David C. Mowery and Richard R. Nelson, The Oxford Handbook of Innovation, Oxford University Press, 2006.

[3]Chris Freeman, and Luc Soete, The Economics of Industrial Innovation, Routledge, 1997.

Table 1 Consecutive waves of technology change

Long wave or period		Major characteristics of the base structure			
Period	Kondratiev wave	Science, technology, education and training	Transport	Energy system	Common and inexpensive critical elements
First, 1780–1840	Industrial revolution: industrialised production of textiles	Apprenticeship, learning-by-doing, schools and scientific associations with different opinions	Canals, roads	Hydraulic power	Cotton
Second, 1840–1890	Steam power and rail	Professional mechanical and civil engineers, technical colleges and public entry-level education	Rail (iron), telegraph	Steam	Coal and iron
Third, 1890–1940	Electricity and steel	Industrial R&D laboratories, national chemical and electrification laboratories, and standard laboratories	Rail (steel), telephone	Electricity	Steel
Fourth, 1940–1990	Mass production of vehicles and synthetic materials (Fordism)	R&D in mass production industries and government institutions, widespread access to higher education	Motorways, radio and television, air routes	Oil	Oil and plastics
Fifth, 1990–?	Microelectronics and computer networks	Data networks, global R&D networks, life-long education and training	Information highway and digital networks	Oil and gas	Microelectronics

Source Freeman and Soete (1997)

mathematics), the USA currently has 50 industries that contribute nearly a quarter of its total employment opportunities (direct and indirect) and 17% of its GDP. They also account for 90% of R&D investment in the private sector, 85% of US patents, and 60% of the country's exports.[4] Among the 50 industries, power and oil and gas refining rank second and third respectively in terms of industry scale, with their combined gross added value (GVA) accounting for 16.1% of the total of all 50 industries. This shows that every energy technology revolution triggers the emergence of important energy sectors that hold long-term development potential.

Based on the above, this study argues that energy technology revolutions generally appear in a concentrated manner within a short time-frame. They generate long-term and significant impacts, including industrial revolution and important new energy sectors.

2 Energy Technology Innovation and Development

Technology readiness is a necessary but insufficient condition for an energy revolution. Technologies that achieve high levels of deployment benefit from a supporting set of factors, in addition to their technological development. These support factors consist of demand for the services the technology provides, such as clean or secure energy; supply of the input the technology requires, such as the components of the technology or primary fuels; and markets that incentivise the deployment of the new technology, such as a newly liberalised market that favours a new lower-cost technology. If the

[4]Mark Muro, Jonathan Rothwell, Scott Andes, Kenan Fikri and Siddharth Kulkarni, America's Advanced Industries: What They Are, Where They Are, and Why They Matter, Brookings Institution, 2015.

supply, demand and market factors are well-facilitated in the innovation ecosystem, the prospect of triggering an energy technology revolution becomes likelier, and thus the marginal impact of innovation is greater.

2.1 Innovation Policy: Basic Findings

It is hard to predict which technologies will win through, given that adoption is driven by a mix of demand, supply and market factors, in addition to the technology's characteristics. As a result, there is a tension between the need to fund a large pipeline of innovation, and the uncertainty about which technologies will be adopted. Innovation policy is ultimately about resolving this tension.

2.1.1 Prioritise Financial Support

Government funding of innovation should focus on the following three aspects.

First, on the early stages of research, development and deployment, rather than on the later stages of niche and wider market deployment. At these early stages the scale of financing is often relatively small, so limited government budgets can fund a broader range of innovations. The risk is high, but government can bear it and share losses across society, which is fair as gains will be shared by society if the innovation is successful.

Second, innovations that create new classes of technology, rather than innovations for specific technologies.

Third, technologies with high capital intensity, longevity and requiring new networks, but only if alternatives with lower capital intensity and transition costs have been explored first. Technologies with high capital intensity and longevity are riskier and require public support further into the deployment phase when they do not plug and play in the current energy system.

2.1.2 Encourage and Coordinate Diversity

Government institutional support for innovation should focus on the following three aspects.

First, it should support an ecosystem of innovators. Innovation has several development stages. Different skills, resources and types of organisation are needed at each stage—from universities for basic research, to entrepreneurs for entry into niche markets, and large companies for wide deployment of the technology. Government should acknowledge that this diversity is needed, rather than trying to deliver the entire innovation pathway through one organisation.

Second, it should create links between stages of innovation. Government can support collaboration by distributing information and by supporting networks of innovators. Government can also step in if part of the ecosystem is weak; for example, by running demonstration projects, as long as the findings are then passed on to organisations in the next stage of the innovation pathway.

Third, providing standards for and access to infrastructure.

2.1.3 Continuously Evaluate and Adapt

Innovation is an uncertain and dynamic process. So, the question of whether to fund an innovation should be regularly evaluated to take into account new findings and changes in circumstances. Government is not always best placed to evaluate the success or failure of an innovation. All decision makers have biases, but government can be too biased as its decision-making power is often concentrated. Also, a policymaker's view of a technology may differ from that of end users. Ultimately, a technology will need to succeed in a market. So, as innovations mature, they should be increasingly exposed to market competition.

2.1.4 Be Able to Fail Fast

Government can often face a tension in that while it is economically best placed to bear the losses from risky innovations, it can face greater institutional barriers to accepting failure than the private sector. However, as prolonging failure comes at an increasing economic cost, it is best to fail fast before rising costs make failure inevitable. This requires more than constant evaluation. It also requires a political understanding that innovation is a dynamic, uncertain process, where failure can be a good outcome, and it may be best to fail fast.

2.2 Innovation Policy: Case Study

Many of the technologies leading the low-carbon transition were initially developed more than 40 years ago and may still require years to change the energy system materially. For example, mentions of solar energy peaked in 1982, but solar energy provided only 1.1% of global electricity in 2015 (Fig. 1).

To better illustrate the process of innovation and the associated challenges, four case studies will be referred to throughout. These case studies cover a range of innovations, including synthetic fuels in the USA in 1979; the Human Genome Project (HGP) in the USA in the 1990s; early wind turbine development in Europe and the USA in the 1970s; and bioethanol fuel in Brazil in 1975 (Table 2).

2.2.1 Innovation Rate

Innovation is a process of experimentation, so the rate of innovation depends on the nature and number of experiments that can be undertaken within the research budget. In broad terms, four factors characterise the process of experimentation and, therefore, the rate of innovation. These can be divided into two categories: frictions and capital characteristics. Frictions include: (i) complex processes: progressing through the many stages of innovation and experimentation, from R&D to market deployment, is an uncertain and lengthy process; and (ii) collaboration: combining interrelated technologies (clustering) or adjacent technologies (spillovers) requires collaboration, which can be difficult to achieve. Second, capital characteristics include: (i) capital intensity: the high upfront costs of minimum viable units limit the number of experiments that can take place within a budget; and (ii) capital longevity: a technology with a long lifespan cannot be repeated as quickly as one with a short lifespan.

(1) Frictions

The process of innovation is complex and uncertain. The innovation process comprises several stages, each with different characteristics. The outcome of innovation is very rarely known at the start of the process, and surprise is a

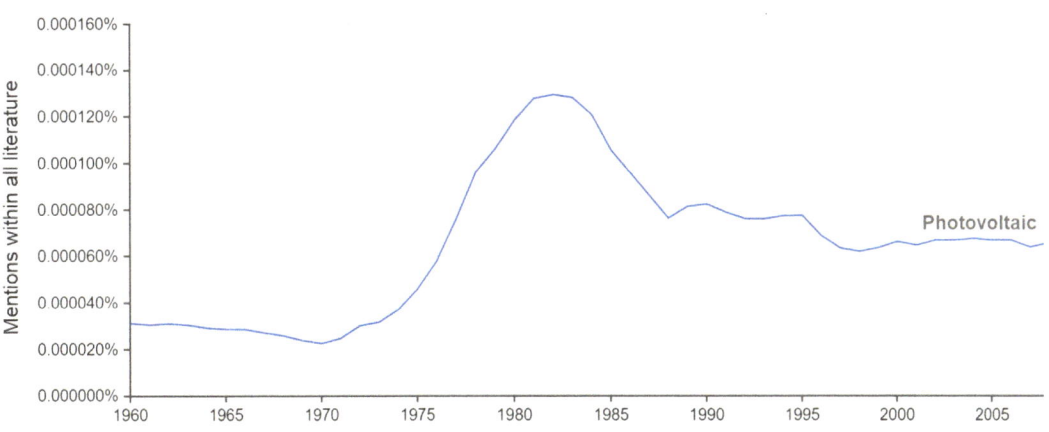

Fig. 1 Technology innovation can have long lead times. *Source* Google Ngram

Table 2 The chosen case studies cover a range of interventions and learnings

	Motivation	Intervention	Lessons
Synthetic Fuels USA (1979)	Supported due to high predicted oil prices, energy security benefits and high capital costs ■ Synfuels are created by liquefying coal and act as a oil substitute ■ Post 1979, oil prices were predicted to increase rapidly and make synfuels cost competitive ■ The process is expensive and gains were only expected in the long run – this hindered private investment	R&D support followed by large, incentivised niche market deployment ■ Government provided research funding in the 1950s-70s, but low oil prices limited development ■ 1980-1986: large niche market deployment supported by government subsidies ■ Goal of 0.5 million barrels by 1986 and $12.2 billion (1980$) initial budget	Shows the need for flexible goals and policy in response to new circumstances ■ Oil prices already falling when niche market deployment just begun ■ 1985 – projects produce only 2% of 1986 targets ■ 1986 – Project cancelled, total cost of $4.5 billion (2010$)
Human Genome Project USA (1990)	Access to the genome sequence would be restricted by patents if a private firm accomplished the sequencing first ■ 1986 – invention of automated sequencing machines made deciphering the human genome feasible ■ The data is of huge value to drug producers and medical research ■ Threat of a private company patenting sections of the genome accelerated intervention	A government programme raced against a private venture to complete the sequencing ■ Government funding estimated at $5.6 billion (2010$) ■ 1990 – Project officially begins with completion date of 2005 ■ 1998 – Celera Genomics begins sequencing, public effort intensified ■ 2000 – Drafts announced, public programme is ahead by 3 days ■ 2003 – Full release	There can be value in government replacing private sector funding, but the benefit to society must be clear ■ The economic impact of the HGP to 2012 has been valued at $965 billion ■ For this style of intervention to be justifiable the project must have unequivocal benefits ■ Argument that competition helped HGP come under budget and 2 years ahead of schedule
Wind Turbines Global (1970s)	Was a clean, alternative power source but was uncompetitive in the 1970s ■ Interest in wind power grew after the oil crises highlighted energy issues ■ Lack of efficient turbines, low lifetime of generation assets and limited appreciation of pollution externalities meant private investment in this technology were limited	Different approaches by different countries – Denmark regarded as the most successful ■ US: large subsidy programme delivered capacity, but unreliable and led to market crash ■ Denmark: focus on supporting small scale turbines, knowledge sharing and market support ■ Germany: R&D focus on large turbines, but no significant market support, hence little demand for unproven and unreliable technology	There needs to be support across stages and communication between agents for effective deployment ■ Germany spent 5 times more on R&D than Denmark up to 1990 but achieved no notable wind capacity. ■ The Netherlands created a competitive market that discouraged knowledge sharing ■ The US had a booming market driven by large subsidies, but a lack of standards hampered reliability and led a market crash
Bio-ethanol Fuel Brazil (1975)	Ethanol is easily produced in Brazil and can both regulate sugar prices as well as increase energy security ■ 1975 – Oil and sugar price shocks in the 1970s led to the "ProAlcool" programme ■ Innovation in upscaling ethanol production and increasing usage rather than producing ethanol itself ■ Support for development and deployment of ethanol only vehicles and "flex-fuel" vehicles	Ethanol subsidies, agricultural R&D funds and support for ethanol vehicles at varying levels over time ■ Mandatory fuel blending of ethanol with gasoline creates demand ■ Incentivised supply with low interest loans and guaranteed prices ■ Grants provided for agricultural research to increase crop yields ■ 1986 – Guaranteed ethanol prices reduced below average cost of production due to oil price collapse	Exposure to market forces drives cost reductions and supporting technologies can be essential for success ■ 1980 – beliefs of increasing oil prices meant no concentrated effort was made to improve efficiency ■ Post 1986 – subsidies were rolled back, leading to cost reductions across the production chain ■ Post 2004 – introduction of flex-fuel cars re-energises the programme

Source Vivid Economics

fundamental characteristic of innovation. These are common issues that all technologies face, and there are various frictions. Misaligned economic incentives between investors and society give rise to additional issues. This occurs most prominently when externalities exist. Whenever an innovation has a positive externality that benefits a third party or creates a common good, it is unlikely that it will be fully appreciated by the private sector, as the social gains cannot be captured, leading to a suboptimal level of investment.

Governments have a key role in correcting these frictions and ensuring an optimal outcome is realised. The issues of misaligned incentives, the high risk of innovation, and changing parameters of investment along the innovation pathway are all frictions that can result in socially beneficial innovations not reaching market deployment. It is in governments' interests to help ease these frictions and help innovation deliver the best societal outcome possible.

(2) Capital characteristics

The capital intensity of an asset is the capital cost of developing a minimum viable commercial energy business. This will tend to be low for modular technologies, such as solar photovoltaic cells, and higher for technologies that need to be deployed at scale, such as nuclear power plants. High capital intensity leads to greater financial risk and a longer payback time. This can reduce demand; without sufficient demand an innovation will not be profitable, even if it is successfully developed. This in turn leads to less investment in innovation and slower innovation overall. Capital longevity is the useful lifetime of an asset—it can also reduce innovation by causing lower turnover rates, creating fewer opportunities to learn by doing, and generating less demand for new capital assets.

Power generation technologies generally have high capital intensity and long capital longevity, limiting the potential for experimentation and slowing deployment rates. Government can provide support to overcome these barriers. The case study of wind turbine development in the 1970s provides an example of this. In Germany, the emphasis was on developing very large turbines, which had a high capital intensity but strong economic potential. In Denmark, the government wanted to accelerate innovation in wind turbines and actively supported the development of smaller turbines. These smaller turbines were less complex than the large German turbines and the capital cost was lower. This lower complexity and greater number of opportunities to experiment reduced the risks that demonstration

turbines would fail and accelerated the learning rate, which led to Denmark's faster deployment of wind power, establishing the industry rapidly with the minimum viable product.

Another key trend in energy is the transition towards decentralised energy, where technologies have smaller unit sizes. Since 2009, a period characterised by increased small-scale investment, even the major markets have generally seen increasing levels of investment. The lower capital intensity of decentralised energy is opening new demand avenues that increase adoption rates, creating the necessary signals for new innovation (Table 3 and Fig. 2).

2.2.2 Innovation Pathway

There are four stages that an innovation must pass through before it can be marketed: R&D, demonstration, introduction into niche markets, and deployment. Different actors are present and the relevant types of support vary at each stage. In the early stages, funding and favourable loans are the most important factors. When a technology moves beyond demonstration, market creation policies, such as support infrastructure and public procurement, become the more common options for policy interventions (Fig. 3).

(1) Challenges along the innovation pathway

The changing characteristics of each of the four stages and fundamental market failures create challenges along the innovation process. The two main challenges that can hinder innovation are misaligned incentives between actors at each stage of innovation, and the increases in capital requirements and risk as projects advance along the innovation pathway.

First, misaligned incentives. Misaligned incentives can result in innovations with positive externalities being under-supported and socially inefficient technologies surviving. Misaligned incentives can lead to imitation, where near-identical products are deployed in the market, which is wasteful for society. Imitations arise when competing products are not interoperable because they follow different standards or formats. This can be less efficient than a common,

Table 3 The example of renewable energy generation

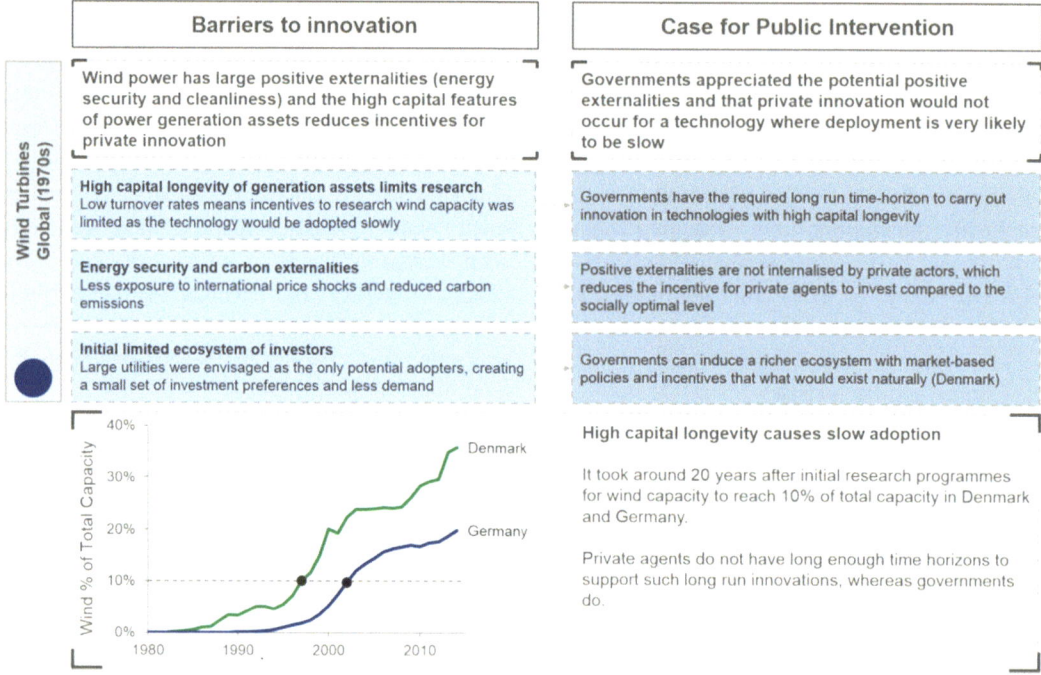

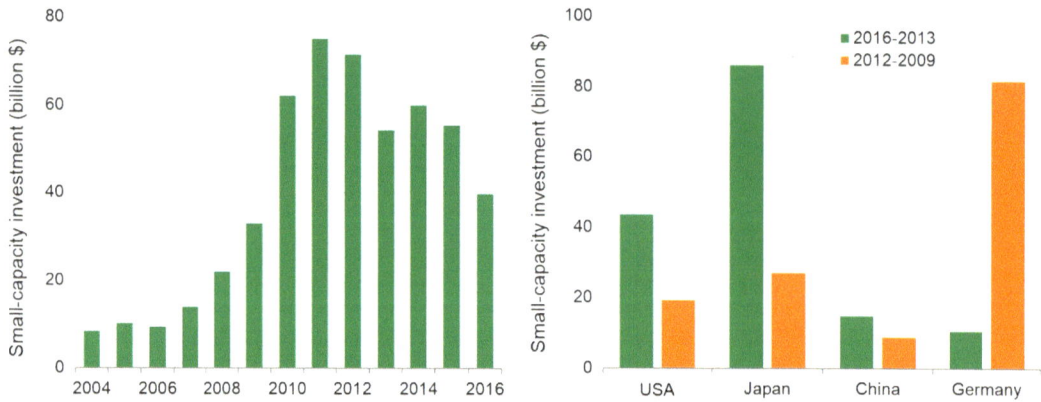

Fig. 2 Small-capacity investments. *Note* Small capacity is defined as roof-mounted solar PV cells with a total capacity of under 1 MW

interoperable system. The diversity of electric vehicle plug formats is an example.

However, active collaboration can arise should synergies exist. Ethanol fuel is a typical example. Ethanol fuel was heavily supported in Brazil, despite being more costly than imported oil due its positive externalities of better energy security and its potential to help regulate the domestic sugar market. This was achieved by creating demand through ethanol-petrol blending mandates, providing ethanol distribution infrastructure and supporting flexible fuel innovations (Table 4).

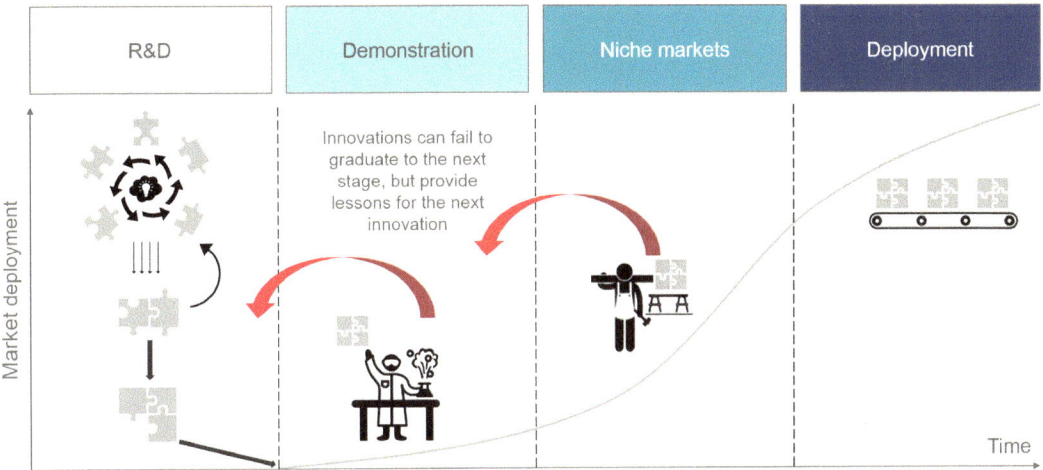

Fig. 3 Four stages of innovation pathway. *Source* Vivid Economics

Table 4 The example of ethanol fuel

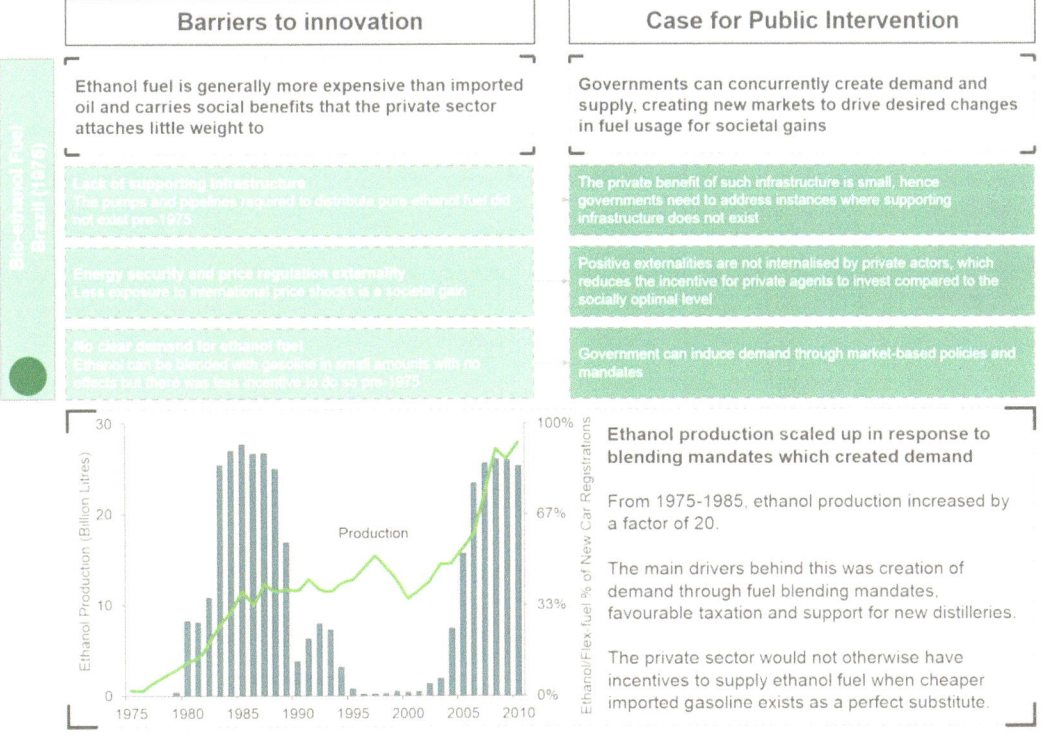

Second, changing capital requirements. Each stage of innovation is distinct and has its own level of risk and capital requirements. A broad range of investors, with different risk preferences and capital availability, will ensure there is adequate support at all stages of innovation. When there are no investors for all stages, innovations can fail to progress and be left stranded.

Take the Human Genome Project (HGP), for example. The level of funding increased over time as the project advanced. In 1988, $54 million was provided, much of it from public funds,

Table 5 The example of the Human Genome Project

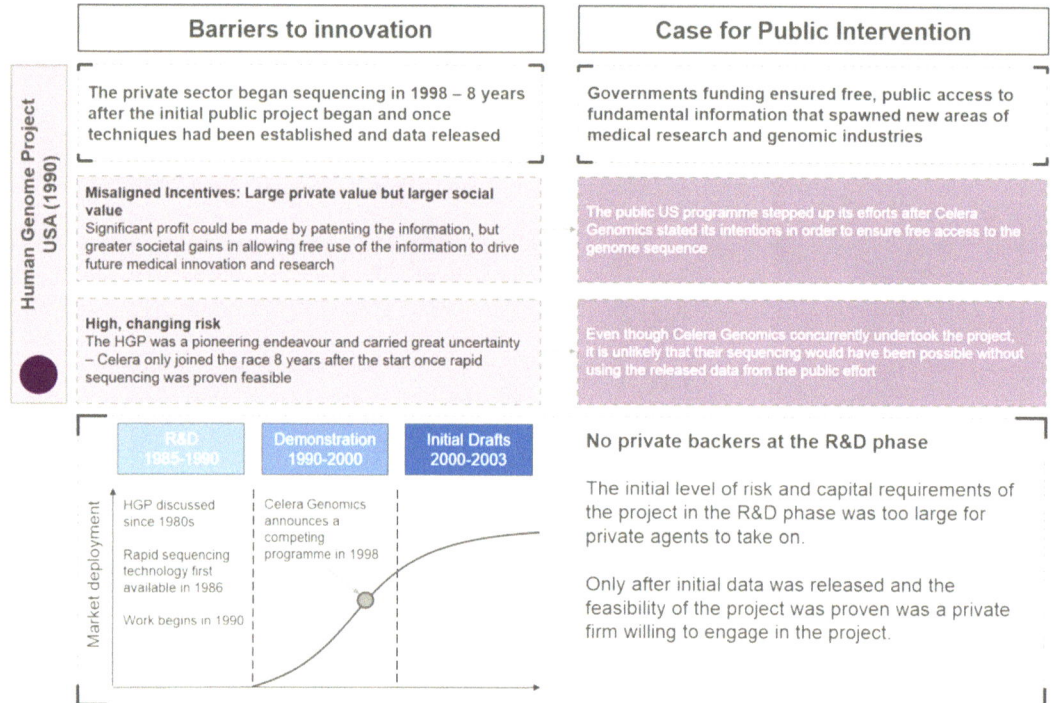

which rose to $290 million in 1992. In the final two years of the project (2002–03), average yearly funding stood at $550 million, 10 times that of 1988. A private company, Celera Genomics, announced its intention to decode the genome in 1998. By that time data had already been released and the feasibility of the project proven. The lower risk attracted private investment that was not available during the initial, higher risk phase of the project (Table 5).

(2) Risk and reward along the innovation pathway

A project having high risk or requiring high capital is not immediately a sign of a bad investment if the expected return is high—the balance between risk, capital and return should be considered. Increasing capital requirements and increasing risk are both undesirable from the point of view of an investor. Hence, for a certain level of risk and capital investment, an appropriate level of return is expected.

Take synthetic fuels (synfuels) in the USA, for example. The expected private return of a synfuel plant is tied to the expected future oil price, as the two commodities are near-perfect substitutes. Given the volatility of oil prices and the generally poor accuracy of previous price forecasts, synfuels carry a degree of risk that, when coupled with the high level of capital at stake, is unattractive for private investors. Public agents have longer time horizons. This, and the desire to increase energy security played a large role in convincing the US government to support a large-scale synfuel project (Table 6).

(3) Summary

The role of a government or public body is to minimise these challenges of misaligned incentives and changing capital requirements via monetary and non-monetary interventions. Governments have a range of policy options available to them to address these challenges and improve the social outcomes of innovation.

Table 6 The example of synfuels

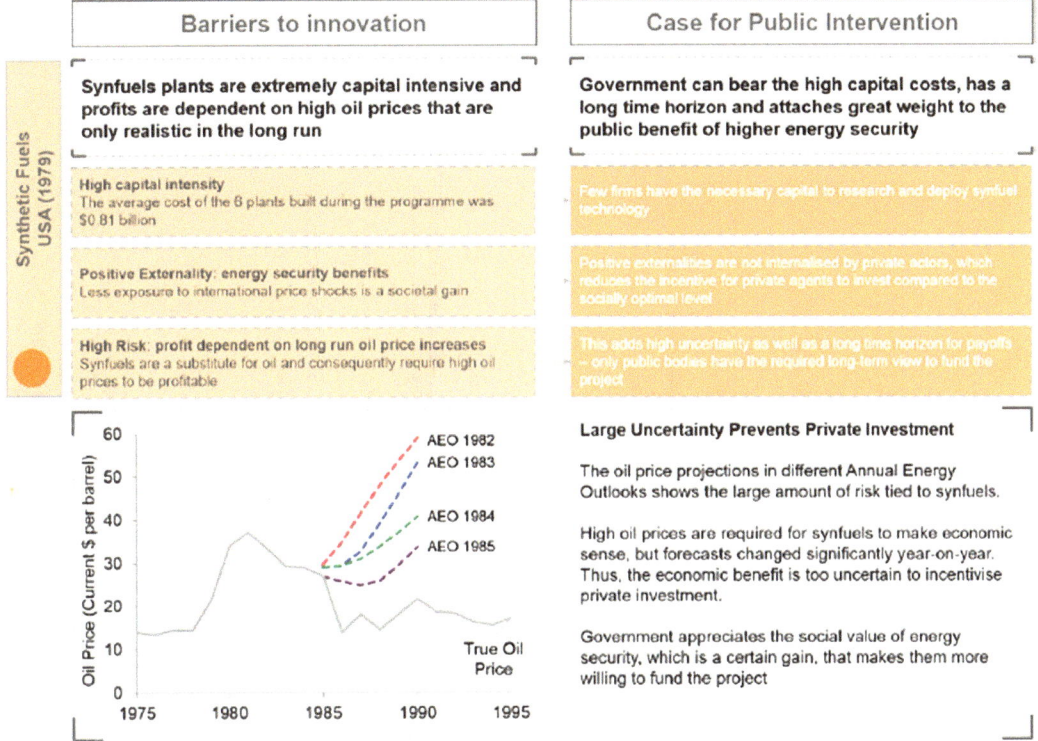

Note AEO = U.S. Energy Information Administration (EIA), Annual Energy Outlook.

These can range from acting directly as an investor with monetary interventions to compensate for: (i) a limited ecosystem of investors or; (ii) addressing misaligned incentives that prevent investment in socially beneficial projects. They can also include non-monetary interventions, such as fostering a culture of innovation investment, ensuring strong links exist between all relevant agents in the ecosystem, and providing market support via mid-stream investment choices and operating standards.

2.2.3 Intervention Policy

(1) Defining success and assessing policy interventions

It is clear that the inefficiencies within innovation create a role for public intervention. However, what defines the success or failure of such intervention is less clear. Innovation is an experimental

process: while discovering viable technologies is one side of this, the unpredictability of the outcome means that it is not certain that the choice of support technologies is correct. There are also valuable spillovers and lessons learned, even when technologies do not reach the general market. Consequently, defining success purely on the final, direct output or widespread use of a new technology does not create the optimal environment for successful innovation. Hence, exploring both the dynamic pathway of innovation, as well as the final outcome, is the key to understanding and assessing public intervention.

First, assessing innovation with hindsight. Once an innovation project has been completed, its final outcome is defined by the realised social value of the innovation and whether the technology reached deployment. Hindsight removes uncertainty and reveals both the true social value of the innovation project and the extent to which it was supported. High social value innovations

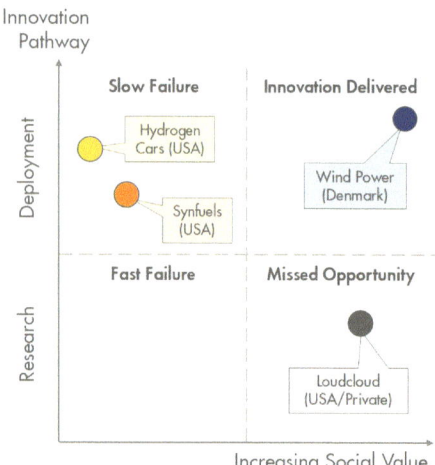

Innovation Pathway

Deployment

Research

Slow Failure — Hydrogen Cars (USA)

Innovation Delivered — Wind Power (Denmark)

Synfuels (USA)

Fast Failure

Missed Opportunity — Loudcloud (USA/Private)

Increasing Social Value

Hydrogen Cars (USA, 1990s-00s) – Back to Research
Support programme first announced in 2003
Very limited take up or infrastructure development and programme was halted in 2009

Synfuels (USA, 1980s) – Programme Scrapped
Formal tender for synfuel projects in 1980
Funding halted in 1986 following oil price crash - $4.5 billion was spent

Wind Power (Denmark, 1979-2015) – Innovation Delivered
Wind capacity represents 36% of total capacity and generates 41% of total power

LoudCloud (USA/Private, 1999-2001) – Missed Opportunity
Founded in 1999, it was one of the first firms to offer cloud computing as a service
Changed focus to data centre operations after the dot-com bubble due to security concerns around cloud systems

Fig. 4 Innovations can be assessed on their final outcome

generally create a new or improved service or have significant positive externalities (Fig. 4).

From this viewpoint, projects that ended with deployment of a high-value innovation or in fast failure should be seen as desirable. The complexities of the innovation system mean that it is not possible to identify high social value projects with certainty at the start of the innovation process. Once it is realised that a project will not be beneficial enough to justify further support, it should be stopped to minimise wasted resources. The final deployment of a technology is only one stage of what is often a lengthy process that is characterised by high uncertainty. Even if a particular innovation is eventually defined as a slow failure, the pathway that led to that outcome should be analysed to avoid unfairly characterising it as the result of poor public interventions, when it is possible that exogenous shocks rather than poor policy are to blame.

Second, the dynamic process of supporting innovation. Successful public policy is more than achieving a final result of fast failure or technology deployment—the actions taken throughout the whole innovation process should be evaluated. How governments support changes in response to unexpected circumstances is a key part of successful intervention and can easily be overlooked if the dynamic nature of intervention is not fully appreciated.

Sensible responses to shocks are required—support should be withdrawn or further action taken whenever long-term changes in circumstances demand it. When analysing interventions dynamically, the actions taken in response to changing circumstances and shocks carry more weight than the final outcome. Hence, it needs to be understood that not only is failure a feature of innovation, but also that failure of advanced innovations is acceptable under certain circumstances. Wind power was developed in several countries during the 1970s and provides a comparison of how different types of policies at different stages result in diverging outcomes (Fig. 5).

Dynamic intervention along the innovation pathway should ensure that the appropriate market signals at each stage are not suppressed by excessively favourable policies. It is important that an appropriate level of exposure to competition is maintained at each stage of innovation to make full use of market signals in guiding intervention decisions. Providing overly generous budgets or limited exposure to competing technologies late on in the innovation pathway prevents the viability of a technology from being tested, making fair appraisal more difficult (Fig. 6).

It is also important that these signals are interpreted in the correct fashion and that

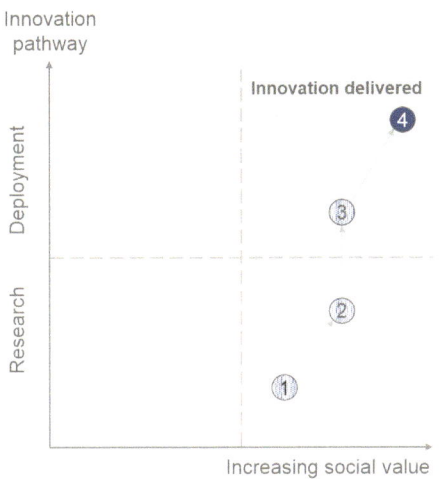

1) **Wind generation has been researched since the 19th century**
   - Given greater importance following oil crises in 1970s.

2) **1970s: Establishment of research and formal standards**
   - Riso National Laboratory – researched and certified turbines.
   - Allowed for effective knowledge-sharing between users, manufacturers and researchers.
   - Focus on small, reliable designs with upscaling later.

3) **1980s: Niche market creation and support**
   - Subsidy system, favourable taxes and supporting policies (wind mapping, grid connection regulations etc.).
   - "Wind cooperatives" encouraged where communities shared a wind turbine – increased dissemination.

4) **1990s-2000s: Scaling up**
   - Increase in wind's % of total generation from 3% in 1995 to 18% in 2005 and then to 41% in 2014.
   - Consistent development of larger capacity turbines.

Fig. 5 Denmark successfully delivered innovation in wind power

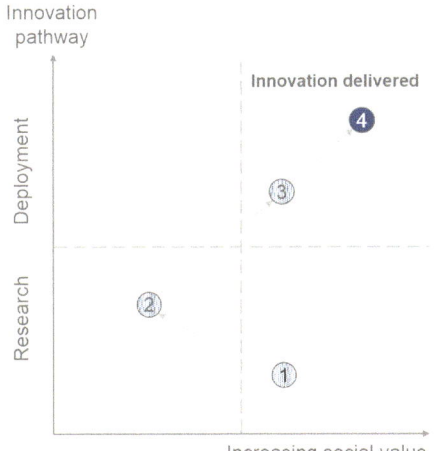

1) **Wind generation has been researched since the 19th century**
   - Given greater importance following oil crises in 1970s.

2) **1970s-1980: Research and initial use of large-scale turbines**
   - R&D focus on large turbines without strong market support.
   - Low demand/incentives for unproven, unreliable technology.
   - 1983 – Growian turbine: 3MW, largest in the world but only 420 hours operation over 5 years.

3) **1990s: Switch to Danish model**
   - Manufactured turbines based on the Danish designs – benefited from technology transfer and knowledge spill-overs.
   - 1990 – feed-in tariffs and priority dispatch introduced for renewables.

4) **2000s: Renewable subsidy programme bolstered**
   - Adoption of more comprehensive subsidy programme – wind now accounts for 20% of total capacity.
   - Gradual lowering of subsidies and switch to tender system for new installations.

Fig. 6 Germany eventually successfully deployed wind capacity

hypotheses are tested and updated rationally over time. Even when signals are obvious, biases or misaligned incentives can skew expectations and lead to irrational beliefs. A part of the solution is to ensure that successful innovation intervention is not defined solely by the output, such as delivering technologies to the deployment stage (Fig. 7).

The deployment of synfuel capacity in the USA illustrates the issues that can arise when targets are rigid and not updated with new information. The US synfuel project committed itself to developing capacity that would no longer be viable in the intended time frame, rather than shifting the focus to developing alternative technologies that would be more valuable in the new context (Table 7).

(2) Monetary interventions

One of the avenues of innovation policy is to provide direct monetary support. This action can be in the form of funding assistance, where the government enters the innovation market as a funder of primary R&D, or as an investor; or it can be non-direct assistance to help improve the investment ecosystem, create long-term incentives for investment and develop mid-stream

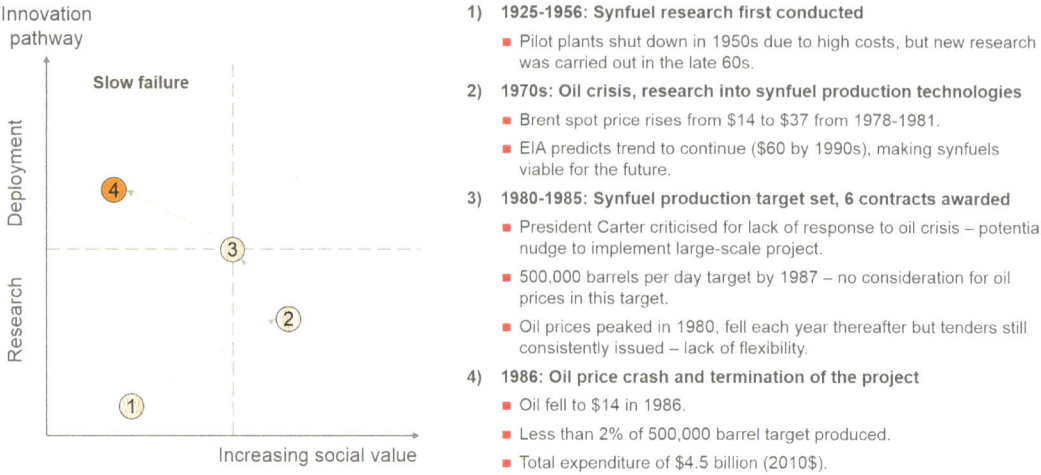

Innovation pathway
Slow failure
Deployment
Research
Increasing social value
1) 1925-1956: Synfuel research first conducted
Pilot plants shut down in 1950s due to high costs, but new research was carried out in the late 60s.
2) 1970s: Oil crisis, research into synfuel production technologies
Brent spot price rises from $14 to $37 from 1978-1981.
EIA predicts trend to continue ($60 by 1990s), making synfuels viable for the future.
3) 1980-1985: Synfuel production target set, 6 contracts awarded
President Carter criticised for lack of response to oil crisis – potential nudge to implement large-scale project.
500,000 barrels per day target by 1987 – no consideration for oil prices in this target.
Oil prices peaked in 1980, fell each year thereafter but tenders still consistently issued – lack of flexibility.
4) 1986: Oil price crash and termination of the project
Oil fell to $14 in 1986.
Less than 2% of 500,000 barrel target produced.
Total expenditure of $4.5 billion (2010$).

Fig. 7 Synfuels in the USA ended as a slow failure. *Note* EIA = U.S. Energy Information Administration

Table 7 A different approach may have led to the programme being better received

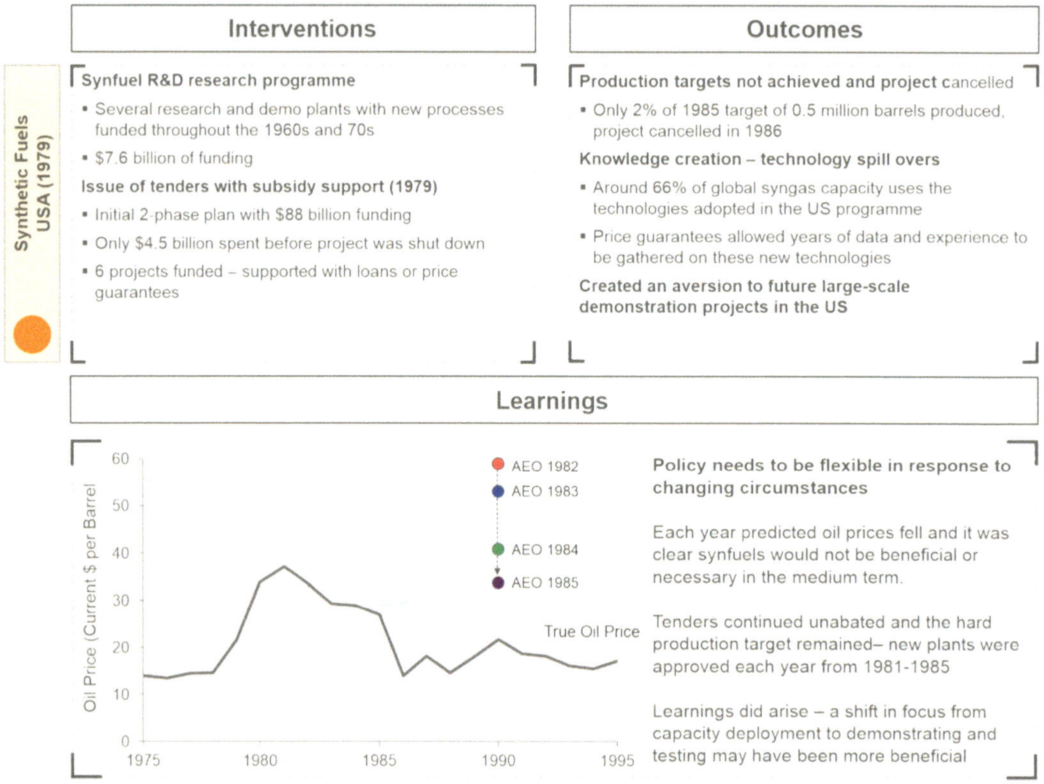
Interventions
Outcomes
Synthetic Fuels USA (1979)
Synfuel R&D research programme
Several research and demo plants with new processes funded throughout the 1960s and 70s
$7.6 billion of funding
Issue of tenders with subsidy support (1979)
Initial 2-phase plan with $88 billion funding
Only $4.5 billion spent before project was shut down
6 projects funded – supported with loans or price guarantees
Production targets not achieved and project cancelled
Only 2% of 1985 target of 0.5 million barrels produced, project cancelled in 1986
Knowledge creation – technology spill overs
Around 66% of global syngas capacity uses the technologies adopted in the US programme
Price guarantees allowed years of data and experience to be gathered on these new technologies
Created an aversion to future large-scale demonstration projects in the US
Learnings
Oil Price (Current $ per Barrel)
60
50
40
30
20
10
0
1975
1980
1985
1990
1995
AEO 1982
AEO 1983
AEO 1984
AEO 1985
True Oil Price
Policy needs to be flexible in response to changing circumstances
Each year predicted oil prices fell and it was clear synfuels would not be beneficial or necessary in the medium term.
Tenders continued unabated and the hard production target remained– new plants were approved each year from 1981-1985
Learnings did arise – a shift in focus from capacity deployment to demonstrating and testing may have been more beneficial

Note AEO = U.S. Energy Information Administration, Annual Energy Outlook.

areas. Doing so can ensure that socially beneficial innovations receive funding when externalities skew the incentives of investment. Governments should use the fact that they have far longer time-horizons than any private investor, which makes them uniquely suited to support projects where success is highly uncertain, benefits will be widespread, adoption rates will be slow and the payback period long. Energy generation technologies generally fit into this category and can be an obvious area for government intervention to have a beneficial impact.

Government can grant private innovators exclusive access to revenue streams as an alternative to direct government financing. Private investment in innovation can be motivated if there is sufficient profit to be made. For example, patents grant a time-limited monopoly for an innovation, which has been used very effectively in pharmaceutical development. In sectors with regulated prices, such as electricity networks, price formulas can be adjusted to incentivise innovation. In highly competitive sectors, government can support innovation by offering to buy a new product at a higher than normal price if the new product meets innovative specifications. The use of public procurement to reward innovation has been effective in, for example, energy efficiency (Table 8).

The government should also intervene when ecosystems of investors are too limited to support good innovations through all stages. Governments can intervene and prevent a good innovation from entering the Valley of Death. However, innovations can also fail to be funded along the pathway because the innovation is simply not good. There is an asymmetric information issue facing governments, who should be pragmatic and not overestimate their ability to discern which innovations are valuable and which are not. However, trade-offs exist for government between picking winners and allowing market forces to determine innovation.

The efficiency of ethanol fuel production in Brazil increased sharply once subsidies were reduced, demonstrating the benefits of market forces and the potential consequences of picking winners. It was never intended to be a cheaper

Table 8 The Human Genome Project also has opportunities to succeed

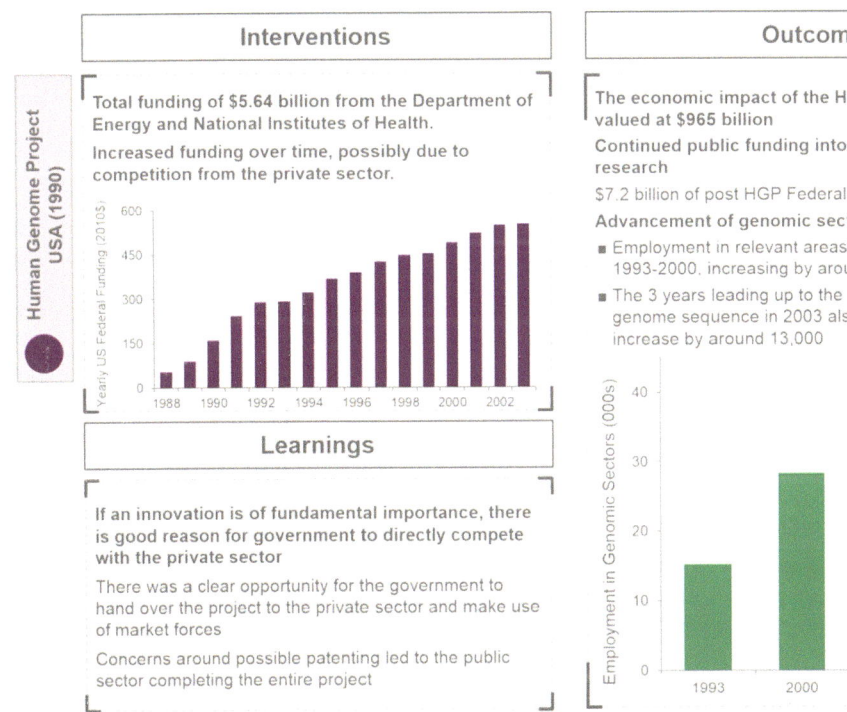

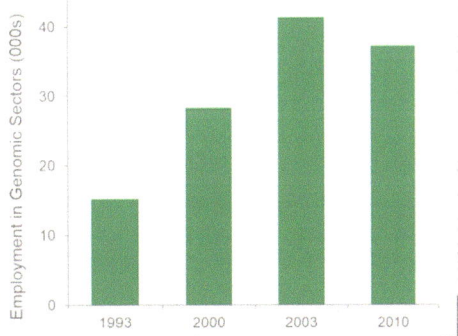

alternative to petrol when the project first started in 1975, but predictions of future high oil prices led to sustained generous subsidies and a belief that ethanol would become competitive over time. After economic difficulties in 1985 and the oil price crash in 1986 guaranteed prices for ethanol were reduced, causing ethanol production to fall for the first time since the programme began. It was only following this reduction in guaranteed prices that ethanol production efficiency began to increase rapidly and consistently. From 1985–95, production costs fell by 45%, with average costs over that period nearly 40% lower than in 1975–85. Efficiency gains were obtained throughout the entire supply chain, from improving agricultural

yields to using larger distilling units and turning waste by-products into heat and energy. This potential for improvement was likely present in the 1970s but did not arise due to overly-protective government policies (Table 9).

(3) **Non-monetary interventions**

A more diverse range of private investors can mitigate the issues arising from a transition along the innovation pathway, as well as the inherent uncertainty of innovation. Cultivating a culture of innovation investment and encouraging more agents to participate facilitates the matching of projects with investors. This helps to resolve the

Table 9 A reduction in subsidies led to increases in the efficiency of ethanol production in Brazil

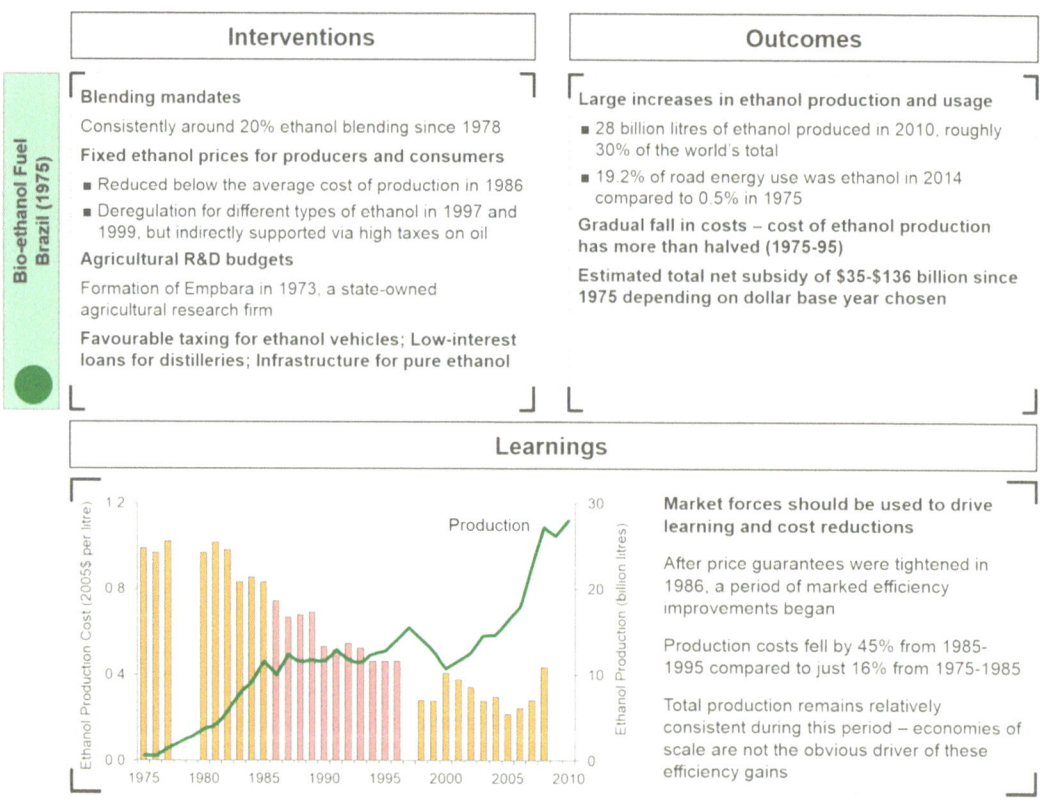

friction of changing capital requirements and risk across stages of the innovation pathway.

Ensuring strong links between different agents and investors eases the transfer of projects between stages and potentially generates benefits from technology clustering and spillovers. Better links between agents help mitigate issues of information asymmetry between investors and make it easier for an innovation to be passed to a more suitable party as it progresses along the pathway. Strong links between agents can also encourage collaboration across sectors, leading to spillovers where a technology is used for a role it was not originally intended for. Both clustering and spillovers have the potential to increase the avenues of use for an innovation, leading to larger impacts from the technology and increased incentives for innovation. A way in which this can be done is by designating institutions to set standards and collect research to encourage knowledge-sharing and collaboration between agents. Reliability is a vital factor for new technologies. Establishing non-profit institutions to develop standards would be a prudent step to avoid similar losses of confidence.

The contrast between the Danish and German experience with wind turbine research and deployment shows the importance of holistic support and the impact of non-monetary interventions. The Riso National Laboratory for Sustainable Energy in Denmark was tasked with developing a certification process for wind turbines and with providing large-scale turbine testing facilities and performing R&D activities. Riso's role allowed it to coordinate interactions between agents in industry, policy and research and to provide technical assistance to manufacturers when required. Denmark also provided strong market support policies, encouraging the adoption of wind power by a wide range of agents. Germany had a heavy R&D focus in the 1980s and a lack of proper consideration for reliability and other measures of support for the initial deployment phase of wind power. Finally, Denmark's adoption and use of wind power has far outpaced Germany's, despite a far smaller R&D spend (Table 10).

3 Major Factors Influencing Global Energy Technology Development and Their Trends

History shows that there are many factors that can influence energy technology development. Some are constant throughout time, such as striving to raise the quality of life. Others are specific to a certain period, such as the discovery of new geological resources, awareness of the need to protect the environment, the spreading influence of other technologies, and accidental events.

3.1 Major Factors Influencing Global Energy Technology Development

The major factors that influence energy technology development in our time are:

(1) **Deep integration of digital and intelligent technologies in the energy sector**

The global financial crisis undoubtedly brought tremendous shocks to the international economy. As World Bank data indicates, global GDP in 2009 reported negative growth for the first time since the 1960s.[5] From a positive perspective, however, people realised that the development and wide deployment of digital and intelligent technologies are very likely to generate game-changing impacts on production and living.

In recent years, new concepts identifying a new industrial revolution have emerged. The most important include:

First, Industry 4.0. This concept explains that human society has passed through three industrial revolutions, characterised by the steam engine, electricity, and electronics and information technology (IT) respectively. The world is now entering the fourth industrial revolution,

---

[5]This impact remains. As the World Bank data indicates, from 2009 to 2015, global GDP grew from $63.12 trillion to $75.24 trillion (calculated in 2010 $), up 2.97% annually. In comparison, from 2002–08, global GDP grew from $51.95 trillion to $64.22 trillion, up 3.6% annually.

Table 10 Comparison between Demark and Germany on wind turbine development

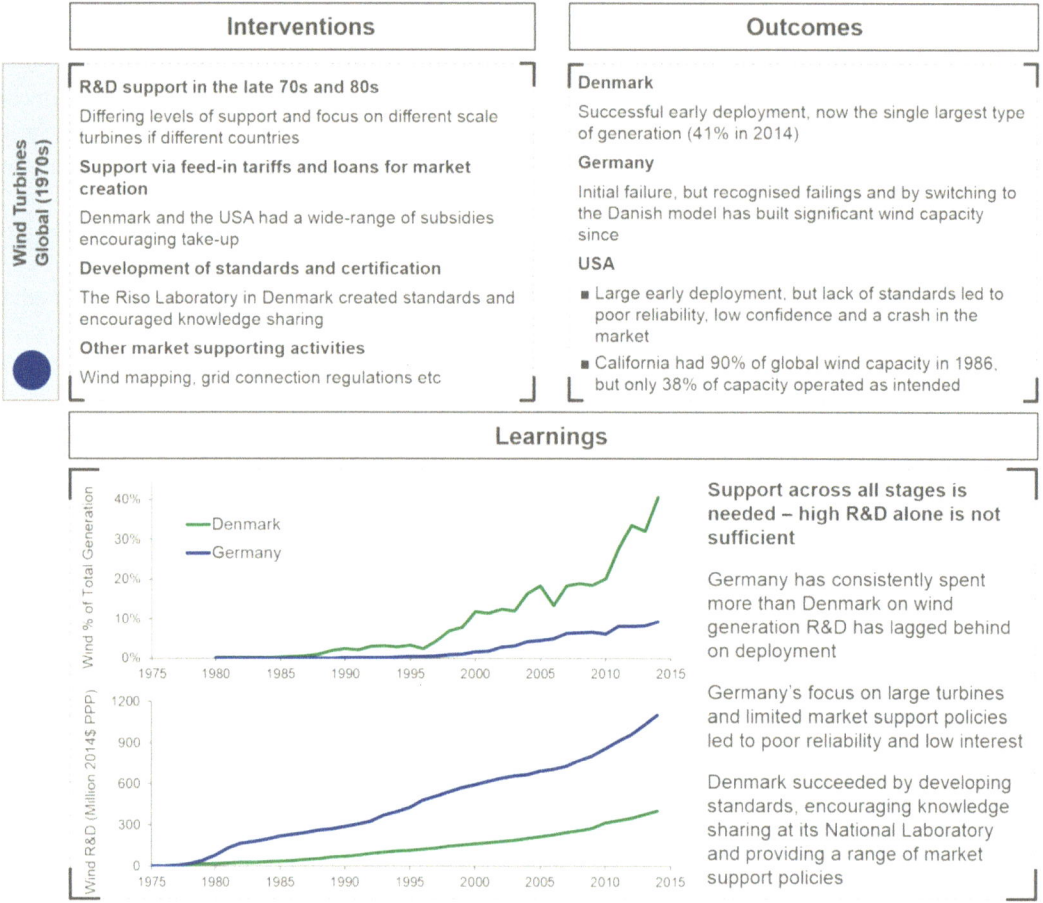

Note PPP = purchasing power parity.

which comprises several components, including cyber-physical systems like smart grids and autonomous vehicles, the Internet of things and cloud computing. Industry 4.0 will have a huge impact on global manufacturing and society.

Second, the industrial Internet of things. In this concept, the countless machines, equipment, facilities and system networks in manufacturing and industry are integrated with computing, information and communications technologies to create a new industrial Internet revolution. The essence of the industrial Internet of things (IIoT) comprises three elements—intelligent machines, advanced analytics and people. The impacts of IIoT are not limited to industry, but also span

transport, healthcare and government. It is projected that by 2025, the IIoT can affect 50% of global economic volume, equivalent to about $82 trillion. According to GE, the IIoT is a direct response to the scarcity of energy resources.

Third, the New Industrial Revolution (NIR). At the 2016 G20 Summit in Hangzhou, China, the G20 New Industrial Revolution Action Plan was released. It says, among other things, that "the ongoing industrial revolution characterised by the intelligent interconnectedness of people, machine and resources driven by the convergence of Next Generation Information Technology and advanced manufacturing, is increasingly blurring the boundary between the physical and the digital world and between industry and services, and presents countless opportunities to harness modern technologies for pursuing enhanced economic growth with the potential for more efficient and environmentally friendly processes". And, "The NIR has the potential to improve productivity and competitiveness, reduce energy and resource consumption, and hence to protect the environment and increase resource efficiency and effectiveness".

Deep integration of digital and intelligent technologies in the energy sector generates significant impacts on energy technology development. First, it drives the development and deployment of advanced technologies like energy efficiency management and smart grids. The International Telecommunication Union estimates that information and communications technology (ICT) could reduce global carbon emissions by 15–40% and cut energy use in industry to a fifth of what it is today. Second, it improves the life cycle efficiency of most existing energy technologies, making them greener. For instance, by using virtual drilling digital technology Shell reduced its drilling costs in Argentina from $15 million to $5.4 million.

3.2 Global Energy Consumption Will Continue to Grow

According to BP, during 1965–2016, global energy consumption grew by a factor of 2.56, from 3,730 Mtoe to 13,276 Mtoe. However, this growth was quite unbalanced—the total energy consumption of OECD countries more than doubled from 2,641 Mtoe to 5,529 Mtoe, while that of non-OECD countries grew from 1,089 Mtoe to 7,747 Mtoe, increasing by a factor of 6.11. In 2007, the total energy consumption of non-OECD countries exceeded that of OECD countries for the first time, and the gap between them continued to widen. There are several reasons for this. For OECD countries, it is mainly due to their entering post-industrialism and achieving significant improvements in energy efficiency; for non-OECD countries, it mainly results from their rapid industrialisation (Fig. 8).

Given that many emerging countries will start to industrialise in the near future, and developing economies will intensify their industrialisation, global energy consumption is projected to continue to grow. According to the United Nations, there were 56 industrialised economies,[6] 31 emerging industrial economies and 78 developing economies in 2015. The industrialisation process in these emerging and developing countries is characterised by energy intensity.

As some research indicates, global energy consumption is projected to increase from 575 quadrillion Btu in 2015 to 663 quadrillion Btu in 2030 and 736 quadrillion Btu in 2040. Energy consumption in OECD countries will remain stable, but due to rapid economic growth, significant population increase and greater access to energy markets, non-OECD countries will contribute most of the growth in global energy consumption. This trend is relatively stable and consistent (Fig. 9).

By region, among non-OECD countries, Asia shows the most obvious energy consumption growth. During 2015–40, the energy consumption in non-OECD Asian countries will increase by 51%. Driven by rapid population growth and abundant domestic energy resources, the African and Middle East countries will also see growth in energy consumption. In comparison, thanks to the improved energy efficiency of new technologies, OECD countries will see modest growth in total energy consumption (Fig. 10).

By industry, the total energy consumption in energy-intensive manufacturing and non-energy-intensive manufacturing will maintain fairly fast growth.

---

[6]As defined by the United Nations Industrial Development Organization (UNIDO), economies can be divided into "industrialized economies", "emerging industrial economies" and "developing economies" by the adjusted manufacturing added value per capita. If an economy's adjusted manufacturing added value per capita is more than $2,500 or GDP per capita more than $20,000 (measured at purchasing power parity), it is an "industrialized economy". If an economy's adjusted manufacturing added value per capita is $1,000–2,500 or its share in global manufacturing added value higher than 0.5%, it is an "emerging industrial economy". The rest are considered "developing economies".

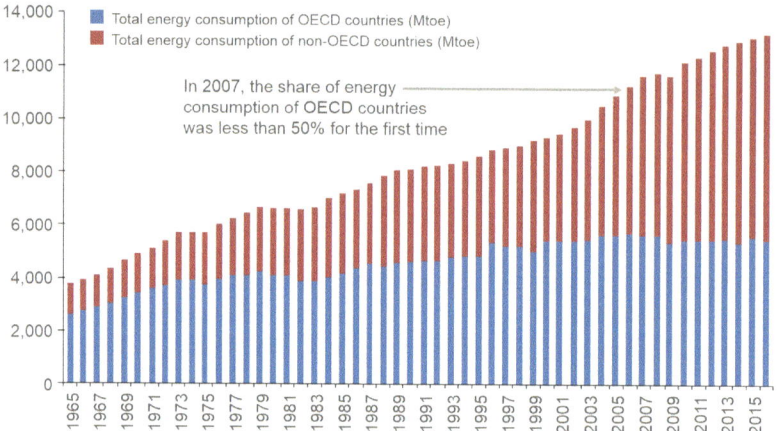

Fig. 8 Changes in total energy consumption of OECD and non-OECD countries (1965–2016). *Source* BP

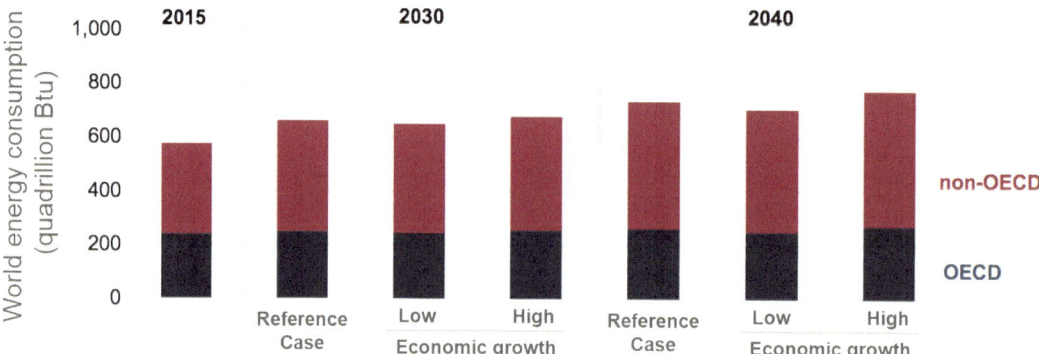

Fig. 9 World energy consumption in three economic growth cases (2015, 2030 and 2040). *Source* U.S. Energy Information Administration

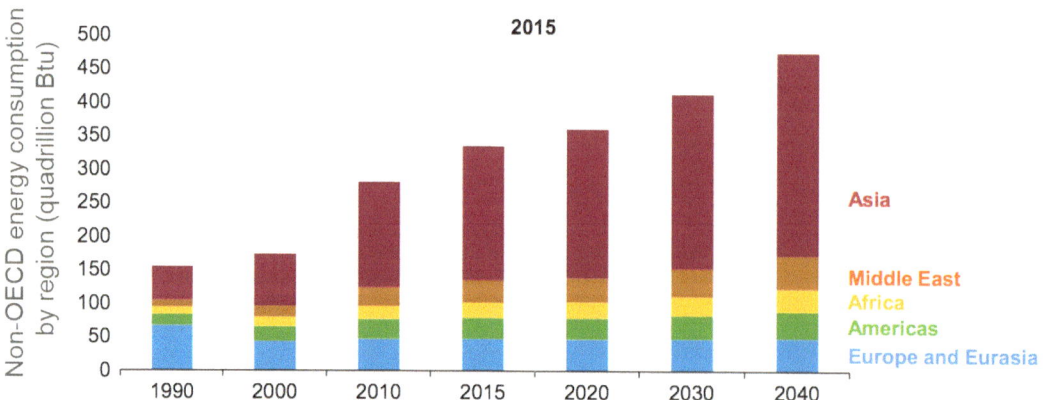

Fig. 10 Energy consumption by region. *Source* U.S. Energy Information Administration

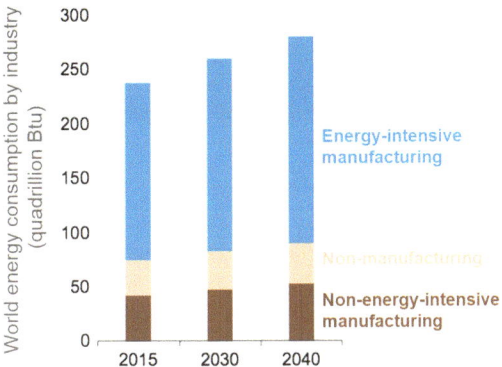

Fig. 11 World energy consumption by industry. *Source* U.S. Energy Information Administration

Such growth trends have varied impacts on global energy transition and technology development. On the one hand, this provides market space for development and deployment of many new energy technologies, and positively drives energy technology innovations. On the other hand, conventional energy forms like coal and oil will continue to play a significant role for a long time, so the energy transition cannot be achieved overnight and shifts between different energy forms will remain for a long time (Fig. 11).

3.3 Great Uncertainty in Global Collaborations to Combat Climate Change

Climate change represents a unique challenge in the present age. Currently, there is plenty of evidence that human activities have significant impacts on the climate; failure to make change may result in catastrophic consequences. As related data show, the number of published papers that reject the conclusion that climate change is caused by human factors is insignificant.

In recent years, major countries have taken action to combat climate change. In December 2015, nearly 200 ratifying parties of the United Nations Framework Convention on Climate Change (UNFCCC) agreed on the Paris Agreement at the 2015 Paris Climate Conference (COP 21). This was the second legally binding climate deal following the Kyoto Protocol of 1997. Taking effect in November 2016, the Paris Agreement is a milestone in the shared commitment to deliver the long-term goal of keeping the increase in global average temperature to well below 2°C above pre-industrial levels, and even limit the increase to 1.5°C. The 2030 Agenda for Sustainable Development, effective as of January 2016, put forward 17 sustainable development goals (SDGs) and 169 targets. Several of the SDGs are closely connected with energy production, transmission and consumption, as well as with addressing climate change. They include "ensure access to affordable, reliable, sustainable and modern energy for all"; "build a resilient infrastructure, promote inclusive and sustainable industrialisation and foster innovation"; "make cities and human settlements inclusive, safe, resilient and sustainable"; "ensure sustainable consumption and production patterns"; and "take urgent action to combat climate change and its impacts". In addition, some countries are proactively promoting carbon taxes and carbon trading systems to help combat climate change at national level.

There are several misperceptions about climate change, as pointed out by the World Bank: (i) climate change is a slow process, but individuals' position on climate change is based on recent experience and observations; (ii) ideology and social loyalty affect how people accept information on climate change; (iii) people tend to ignore or fail to fully understand information expressed as a probability number; (iv) people focus more on current issues than on future ones, but the worst impacts of climate change may not be evident for years; (v) some risks (like climate change) are unclear, and people tend to avoid taking action when facing unknown situations; and (vi) when deciding how to share responsibilities to combat climate change, individuals and organisations tend to follow their own interests.

The intensity and pace of global actions to address climate change impact significantly the future direction of energy technology innovation. Many actions aiming to combat climate change will eventually result in the development and deployment of new and low-carbon energy technologies. Whether the world can reach a more binding and powerful action plan to address climate change will directly affect progress of the shift to new energy technologies.

3.4 International Competition Based on Resource Endowments

The energy technology revolution will eventually generate new and important sectors, which will impact international competition. As a result, all major countries value highly the energy technology revolution. In recent years, the major energy powers have introduced an array of regulations, policies and actions to accelerate energy technology innovation, striving to occupy a vantage point and sharpen national competitiveness. This competition between nations spurs innovation across a diverse range of energy technologies.

The USA under the Obama presidency introduced the All-of-the-Above Energy Strategy, which identified science and energy as a top priority. The strategy aimed to create a complete energy technology innovation value chain, from basic research to final market solutions. It focused on accelerating low-carbon and clean energy technology development—especially solar power and fourth generation and modular nuclear power—and energy efficiency. The USA has also established many new energy R&D and innovation platforms, including the Advanced Research Projects Agency—Energy (ARPA-E) and the Energy Innovation Center to support game-changing energy technology development and effectively integrate resources from companies, universities and research institutions.

Japan has unveiled strategic plans, including the Energy and Environment Innovation Strategy for 2030, which sets guidelines for the country's energy security, economic efficiency, environmental protection and safety. The strategy supports the development of nuclear energy, energy efficiency, renewables, new energy storage technologies, and advanced coal utilisation

technologies like integrated gasification combined cycle (IGCC) and integrated gasification fuel cell cycle. After the Fukushima disaster in 2011, Japan updated its Basic Energy Plan and adjusted its priorities for energy technology development, including speeding up the development of renewables and reducing the share of nuclear energy. The Energy and Environment Technology Innovation Strategy targets for 2030 and 2050 include increased R&D investment to ensure Japan's global leadership in new energy technologies and applications.

The EU has developed strategic plans, including the Energy Roadmap 2050, to highlight the principal role of renewable energy in energy supply and develop ideas for smart grids, carbon capture and storage (CCS), nuclear fusion and energy efficiency. In September 2015, the EU announced the Integrated Strategic Energy Technology Plan (ISET-Plan) to drive the transition to a low-carbon energy system. Similarly, after announcing its nuclear phase-out plan, Germany has prioritised renewable energy and energy efficiency improvement technologies, and generally adjusted its policies for energy technology development and deployment.

Competition between countries can also be reflected in many specific fields. For example, major car manufacturing countries are accelerating the development of new energy vehicles. The USA adopted the EV Everywhere Grand Challenge Blueprint and the Intelligent Transportation System Strategic Plan 2015–19 to promote the development of new energy vehicles and an intelligent transport system. The EU introduced the Strategic Innovation Plan 2020 and the Intelligent Transport System Development Plan to promote a low-carbon and intelligent transport system. Japan's Next-Generation Vehicle Strategy 2010 and Automobile Industry Strategy 2014 prioritise the development of new energy and fuel-efficient vehicles. In China, the Action Plan for Innovation in the Energy Technology Revolution (2016–30) prioritises electric vehicle energy storage and wireless charging technologies in the energy technology revolution.

3.5 Uncertainty Due to Accidents and Changes in Government

Natural disasters and changes in government can have a significant impact on energy technology development.

A case in point was the release of radioactive materials at the Fukushima nuclear power plant in Japan, one of the world's largest nuclear power stations, after it was severely damaged by an earthquake and tsunami in 2011. This drew the world's attention to nuclear power generation, with some countries deciding to phase out nuclear power in favour of other energy technologies. Such policies directly affect the technology structure of the energy revolution.

A change of government can bring uncertainty to energy technology cooperation within and between countries. For instance, after taking office President Trump announced the "America First" Energy Plan, which promotes the development of domestic conventional energy resources—including shale gas, oil, natural gas and coal. He also withdrew the USA from the Paris Agreement. Such policies not only affect energy technology development domestically, they also bring uncertainty to global energy technology collaboration.

3.6 The Potential of Promotional Regulation Cannot Be Ignored

As understanding of people's thinking and behaviour deepens, governments and non-governmental organisations (NGOs) can play an important role in changing energy consumption patterns and driving business model innovation. This also impacts the direction in which energy technology innovation proceeds.

Take Opower, a US-based company, that was acquired by Oracle in 2016. The company sends a home energy report to its residential customers, showing the difference in energy consumption between their home and those of neighbours. This comparison reduces home energy consumption by

2 percentage points, a result equivalent to raising the electricity tariff by 11–20%. The main reason for this success is that people value the constraint that social norms have on their behaviour. When energy saving is considered a social norm, people tend to follow the norm, once they understand the context (Fig. 12).

In addition to the effects of social norms, interventions like information disclosure and setting proper default options can deliver very good results.

4 New Trends in Global Energy Technology

There are two main trends in global energy technology development.

4.1 Decoupling of Economic Growth from Energy Consumption

Annual growth in global energy consumption has been lower than economic growth in recent decades. This is due to improvements in energy technologies and energy efficiency and the impact of digitalisation. During 1965–2015, the average growth rate of global GDP (in 2010 $) was 3.33%, while that of global primary energy use was 2.54%. Some countries like Denmark and states like California achieved economic growth without increasing energy consumption (Fig. 13).

Looking at future trends, energy intensity in developed economies will continue to decline. Rapidly growing developing countries like the BRICS nations (Brazil, Russia, India, China and South Africa) will shift to less energy-intensive industrial structures, resulting in significant decreases in energy intensity. Although the Middle East and Africa will register population and economic growth, their declining energy intensity will help reduce their energy consumption intensity (energy use per square foot per year) (Fig. 14).

As future economic growth will be driven mainly by productivity increases rather than by energy investment, the decoupling of economic

UtilityCo

Home Energy Report
Account number: 1234567890
Report period: 11/00/00 – 12/00/00

We are pleased to provide this personalized report to help you save energy.

The purpose of the report is to:
- **Provide information**
- **Help you track your progress**
- **Share energy efficiency tips**

BOB SMITH
414 NICOLLET MALL GO 6
MINNEAPOLIS, MN 55401

 This information and more is available at UtilityCo.com/reports

Last Month Neighbor Comparison | You used **19% MORE** energy than your efficient neighbors.

How you're doing:

Great ☺ ☺

▶ **GOOD** ☺

More than average

* This energy index combines electricity (kWh) and natural gas (therms) into a single measurement.

Who are your Neighbors?

■ **All Neighbors**
Approximately 100 occupied, nearby homes that are similar in size to yours (avg 2,109 sq ft) and have both electricity and natural gas service

■ **Efficient Neighbors**
The most efficient 20 percent from the "All Neighbors" group

Last Winter Heating Comparison

WINTER ❄

⚠ Last winter, you used **37%**
more energy on heating
than your neighbors.

Best ways to save this winter:

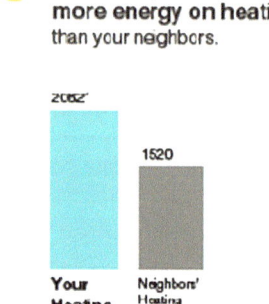

 Lower your thermostat before leaving home
Adjusting by 10° can save 10% on heating costs.

 Seal air leaks around windows and doors
Use caulk or weatherstripping to eliminate cold drafts.

 Let sunshine in for warmth
Open blinds during the day to capture free heat.

For more tips, visit UtilityCo.com/reports

Fig. 12 UtilityCo's home energy report provides energy-efficiency comparisons between households

growth and energy consumption will become more evident. According to the International Energy Agency's World Energy Outlook 2017, in the next two decades the world economy will maintain an annual growth rate of 3.4%, with 75% of that growth driven by productivity improvements, reducing energy intensity even more (Fig. 15).

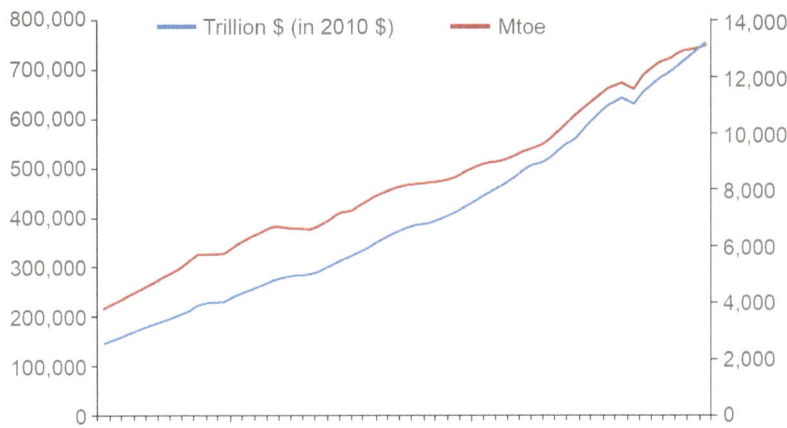

Fig. 13 Global economic growth versus energy consumption (1965–2015). *Source* BP and World Bank

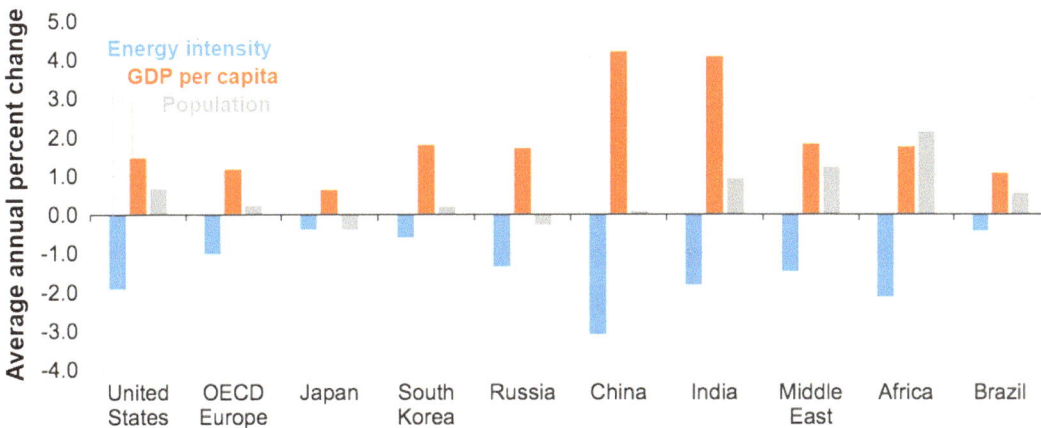

Fig. 14 Energy intensity, per capita GDP and population growth in selected regions (2015–40). *Source* U.S. Energy Information Administration

4.2 Breakthroughs in Clean Energy Technologies

Breakthroughs in energy technologies are an important part of the technology revolution and industrial change. Some clean energy technologies are already at the demonstration stage or in wide deployment and are gradually changing the global energy landscape. The International Energy Agency has projected that by 2030 renewable energy will exceed coal-fired power generation to become the largest power source globally; and by 2040, renewable power will account for more than half of all new generating capacity. In the fuel sector, biofuels and electricity have already replaced oil to some extent.

Clean and low-carbon energy technologies will make a big difference in driving the global energy transition. The 760°C ultra-supercritical power generation technology can improve the net energy efficiency of a coal-fired power plant by 14 percentage points and reduce CO_2 emissions by 30%. Technologies like integrated gasification combined cycle (IGCC), carbon capture and storage (CCS) and pressurised oxygen-enriched combustion are advancing rapidly. H-class heavy-duty gas turbines have been commercialised, and the net efficiency of combined cycle

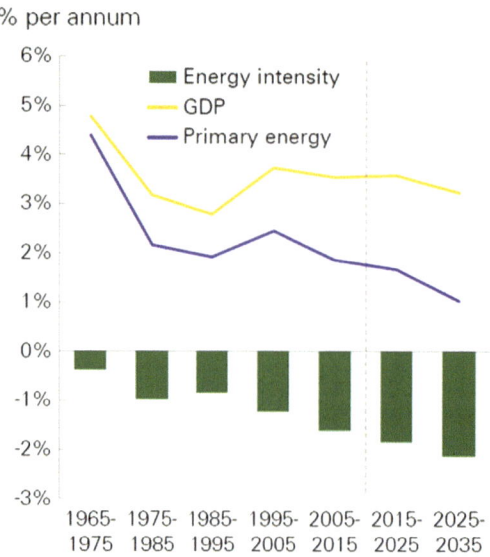

Fig. 15 Growth in GDP and primary energy (2015–35). *Source* BP

power plants improved. Breakthroughs in unconventional oil and gas exploration and development technologies have been made, primarily in North America. As a result, shale gas and tight oil have become new sources of growth in oil and gas. Offshore oil and gas exploration and development is continuously prospecting ever greater depths. China has made big breakthroughs in shale gas exploration and development and in coalbed methane exploration, capacity building and R&D, which are reflected in the rapid growth of reserves and production.

Third-generation nuclear power technology has become mainstream in China's newly built generating units. Fourth-generation nuclear technology has reached the commercial-scale demonstration stage in a new plant featuring secure, modular high-temperature gas-cooled reactors. Renewable energy is gradually becoming an important source of newly added power capacity. The efficiency of solar photovoltaic (PV) systems continuously improves, and the average amortised cost of utility-scale and residential PV systems continues to fall. Concentrated solar power is under demonstration at scale. Wind power technology is maturing and the cost of producing biomass energy is falling.

Pumped storage power generation and lead-acid battery technologies have matured; and thermal storage, compressed-air energy storage (CAES), capacitor and super-capacitor technologies have either matured or are being commercialised. At the end of 2016, ownership of new energy vehicles exceeded 2 million units, and the demonstration of hydrogen-powered vehicles was progressing.

5 International Experience

The purpose of this section is to draw on international experience to identify the conditions for innovative technology to be successfully applied at a scale that revolutionises an energy system. Technologies that achieve high levels of deployment benefit from a supporting set of factors in addition to their technological development. We reviewed the different patterns of innovation and deployment across 12 technologies. This review suggests that four conditions are often collectively sufficient for successful deployment: technology innovation to a level such that deployment is feasible, supply of the inputs the technology requires, demand for the services the technology provides, and markets that incentivise deployment.

Most of the largest G20 energy revolutions were triggered by economic growth, energy security concerns, new market incentives or shocks, rather than by technology. Energy revolutions since the 1970s have been primarily triggered by the following factors. First, supply factors, including local energy resources (the greatest revolutions occur at the extremes, either when resources are abundant or when they are extremely scarce); and connectivity to energy trade (this is often a case of making the necessary investment in import or export capacity). Second, demand factors including rapid economic growth (this is a major driver of revolutions because investment is available and required during periods of growth, and energy networks are built, which, once built, lock-in energy choices), consumer demand for energy services and cleaner and more flexible fuels (this can trigger rapid change, especially when accompanied by new low-cost supplies of energy). Third, market factors including liberalised energy markets (this is especially important when the fundamental cost structures of new or incumbent energy sources change, but these changes cannot feed through to technology choices due to a regulated energy industry). In addition, energy systems are prone to shocks, and these can trigger energy revolutions.

The technologies involved in major energy revolutions can be considered on two dimensions. First, capital intensity. A technology with a high capital intensity, such as nuclear power, requires a large-scale player to deploy it, while a technology with low capital intensity, such as biofuels or compressed natural gas (CNG) vehicles, can be deployed by individuals. Second, network intensity. A technology with high network intensity, such as offshore oil and gas, requires significant investment in a network, which is often delivered by players other than the technology developer. A technology with low network intensity, again such as power generation technologies, can plug and play into an existing network, which means that the technology is not reliant on actions elsewhere in the energy system.

In international experience, high capital intensity and low network intensity technologies have played a major role in energy revolutions. These technologies are often deployed to meet rapid changes in energy demand, such as demand for more energy, secure energy or cleaner energy. This is because governments often have responsibility to meet these needs, and they have tended to favour deployment of large, established, single-fuel technologies—such as nuclear or coal —that can plug into existing networks.

5.1 Basics

History demonstrates that new technology can trigger revolutionary changes in energy systems. Over the timescale of centuries, technology has clearly transformed the energy system. The history of the UK, the first country to industrialise,

shows this very clearly. The invention of the steam engine in 1763 started the Industrial Revolution, leading to increases in coal demand. The invention of the Ford Model T car in 1908 started a period of declining transport costs, leading to increases in oil demand. The opening in 1956 of Calder Hall, the world's first commercial nuclear power plant, led to increases in primary electricity supply. These examples show that the energy system is fundamentally based on technology, and so major changes in technology will be closely related to major changes in energy systems.

Recent decades have provided plenty of reasons for technological innovation in energy. Air pollution has become a major concern. In the 1950s countries such as the UK and the USA passed clean air laws, and by the late 1970s international agreements were formed, such as the Convention on Long-Range Transboundary Air Pollution. Energy security became a prominent concern with the oil crises of the 1970s. Climate change reached international levels of concern with the adoption of the Kyoto Protocol in 1997. These imperatives, coupled with the increasing technological sophistication of the wider global economy, have led to high levels of innovation activity. However, this has not resulted in significant change in the energy system. Since the 1980s, shares of primary energy have been relatively constant between biomass, coal, oil, and modern energy carriers (Fig. 16).

Other factors appear to have been the trigger for revolutionary change. These changes, while often using new technologies, were not always triggered by the development of these technologies. The UK provides examples of this. In 1967, demand for clean air and warmer homes led to a centralised decision to switch more than 40 million appliances from town gas to natural gas. In 1984, the miners' strike severely disrupted coal supply chains. In the 1990s the power sector experienced a dash for gas as market liberalisation enabled recently invented combined-cycle gas turbine (CCGT) technology to compete for the first time. Each of these examples revolutionised the UK energy system, but technology had a supporting, rather than a leading, role.

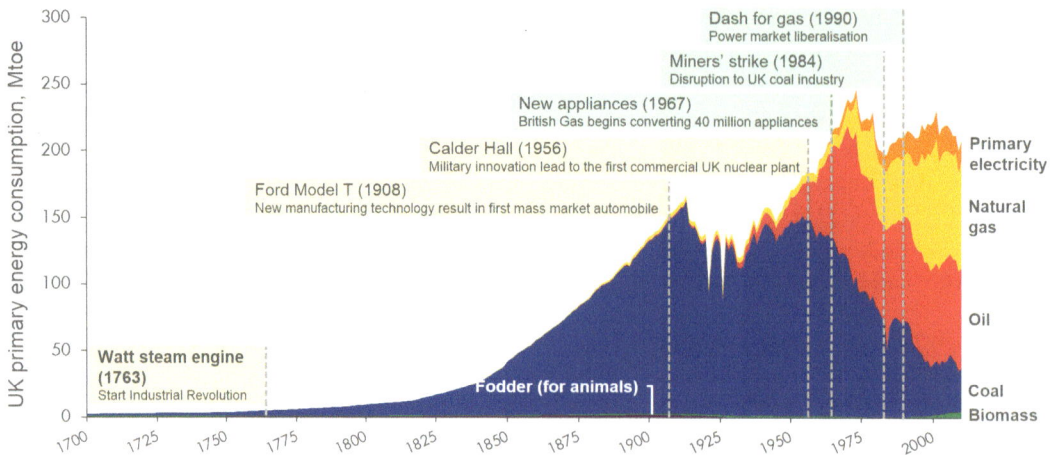

Fig. 16 UK history demonstrates that new technology can trigger revolutionary energy system change

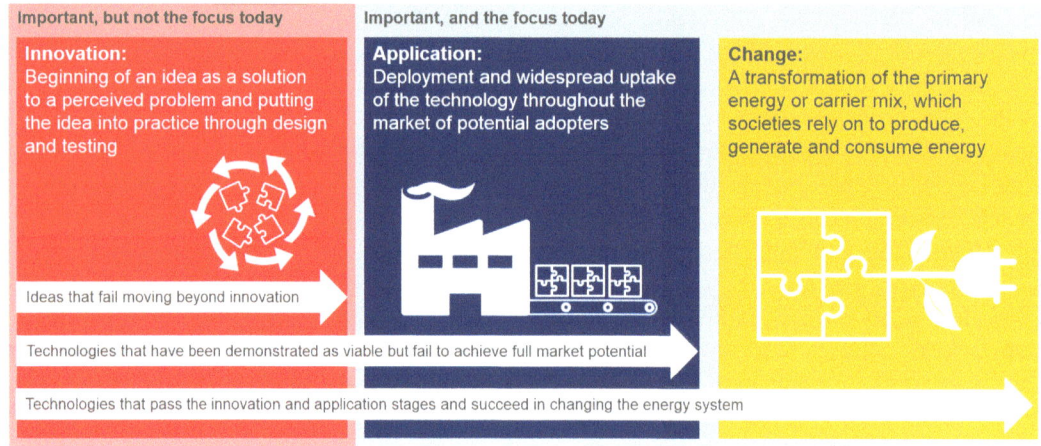

Fig. 17 Major energy technologies must develop through the three stages. *Source* Vivid Economics

The experience of recent decades shows that the contribution of technology to an energy revolution has different stages. First, innovation—the beginning of an idea as a solution to a perceived problem and putting the idea into practice through design and testing. Second, application—the deployment and widespread uptake of a technology throughout the market of potential adopters. Third, change—a transformation of the primary energy or secondary energy carrier mix, which societies rely on to produce, generate and consume energy. For a technology to contribute to an energy revolution it must develop through all three stages (Fig. 17).

An understanding of how innovative technologies can be successfully applied to generate revolutionary system change will be essential for China to deliver its energy revolution. Along with the rest of the world, China faces a major challenge to deliver low-carbon, low-pollution, affordable and secure energy. This challenge will be met by radically changing the energy mix, from fossil fuels to renewable energy. Technological innovation is often seen as the trigger for this change. The recent, rapid cost decreases in wind and solar photovoltaic (PV) technology support this view. However, the rapid improvements in technology have not yet changed the system and, in those countries where the most change has occurred, there has been significant effort to get the conditions right for large-scale application. This experience demonstrates that policymakers must consider factors other than technology that trigger revolutionary energy system change.

Successful innovation is the key first stage that benefits from policy support. Innovation is a complex process, and is undertaken by a network of stakeholders, with at least four distinct stages. First, research and development (R&D)—where original ideas are developed and combined. Second, demonstration—where successful combinations of ideas are tested to see if they can deliver a viable product. Third, niche markets—where technology is deployed at a small scale, often in areas where the performance of the new product is valued over the low costs of existing options, facilitating the route-to-market. Fourth, wide deployment—where technology is mass-produced and costs can compete with existing options. The risk of failure at each stage is high, and technology often falls back to an earlier stage, only to be combined with a new idea or champion that enables the technology to progress again.

International experience of best practice for accelerating innovation is a rich and important topic that may be worthy of specific focus for learnings, as China has significant innovation capabilities that should be put to best use. However, this study focuses on how technology can be successfully applied once innovation has occurred. This is because without the conditions for successful application, innovation efforts are wasted, and these conditions for successful application are interconnected with the goals of China's energy revolution.

5.2 Drawing on International Experience

This study uses quantitative analysis to evaluate the role of technology in delivering changes to the energy system and to summarise the related international experience and the implications for China.

5.2.1 Quantitative Analysis

(1) Definition of revolutionary energy technology

As technology and energy revolutions are complex, so too is the methodology for testing the role of technology in past energy revolutions. A revolutionary energy technology is the application of scientific knowledge that transforms the primary or secondary energy carrier mix relied on by societies to produce, generate and consume energy. The figure below shows how our work focuses on energy technologies, which are a subset of general technologies. Within energy technologies, our work focuses on revolutionary energy technologies that transform the fuel mix. This study does not consider energy technologies that increase the scale of energy production, generation or consumption. Such technologies are important, but the primary challenge for China and other countries in the coming decades is changing the fuel mix, rather than increasing scale. Therefore, our focus is on technologies that change the fuel mix (Fig. 18).

(2) Innovation level

We quantify innovation levels using metrics of hype and R&D. On the one hand, we quantify hype using the frequency of citations of a technology in English language sources since 1970. These data are from Google Ngrams, a dataset of all words cited in written sources digitalised by the Google Books programme up to 2012. On the other hand, we quantify R&D activity using International Energy Agency (IEA) data on public R&D expenditure in OECD countries since 1970. Innovation levels are quantified for hype and R&D for 12 technologies: biofuels, carbon capture and storage (CCS), energy storage, fuel cells, geothermal power, hydropower, hydrogen, industrial energy efficiency, power from nuclear fission, power from nuclear fusion,

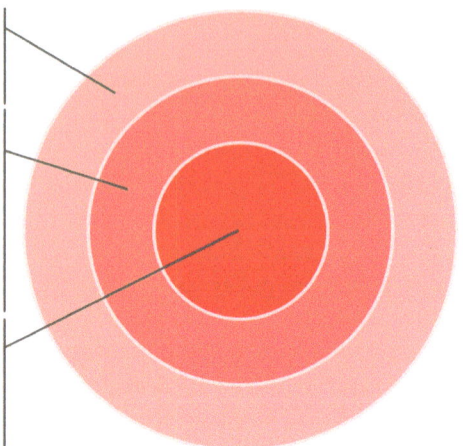

Technology is the application of scientific knowledge for practical purposes

Energy technologies are those applications of scientific knowledge that influence the energy system, ranging from radical new innovations to marginal performance improvements. These include material changes in technological hardware but also changes in social behaviour

Revolutionary energy technologies are applications of scientific knowledge that transform the fuel or carrier mix that societies rely on to produce, generate and consume energy

Fig. 18 Definition of revolutionary energy technology. *Source* Vivid Economics

renewable power, and transport energy efficiency. For each technology we analysed the profile of hype, R&D spend and, where possible, deployment of the technology by country for G20 countries. There are lags between innovation levels and deployment across countries, and we analysed these lags to test when and where innovation was followed rapidly by application. Long lags suggest that other factors must be in place to trigger deployment of a technology.

It needs to be noted that not all technologies have suitable deployment data. For example, technologies such as nuclear fusion have been deployed only in small pilots. Other technologies, such as industrial energy efficiency, do not have comprehensive datasets for deployment. Some technologies are described as revolutionary by industry commentators even though deployment levels are too low to change the energy mix. For example, renewable power, such as wind and solar, is often described as driving a revolutionary change in the energy system, and high growth rates in capacity support this characterisation. However, the share of renewable power in the energy mix is very small, and changes in the mix towards renewable power are not yet on the same scale as the historical changes to gas or nuclear. Our view is that a revolution should be judged on its effect on the energy mix. This perspective is at the heart of our analysis, as we investigate which actions are

needed to transform commentary and hype that a technology is revolutionary into a revolutionary change of the energy system (Fig. 19).

We analyse major groups of technology. A technology, such as nuclear, can include a family of technology subclasses and generations. For example, nuclear reactors can be Magnox reactors, pressurised water reactors (PWRs), boiling water reactors (BWRs), advanced gas-cooled reactors (AGRs) or fast breeder reactors (FBRs). Each of these reactor types is in turn composed of a family of technologies. Developments within major technology groups can have a significant effect on deployment. Figure 20 shows that deployment of nuclear power lagged between countries, which was in part due to the varying development paths of the different technologies they employed. The UK initially built its nuclear power plants with Magnox reactors, the Japanese with BWRs and PWRs, and the French with PWRs. So, the subclasses of technologies can be important to understanding deployment. However, this study focuses on aggregated technology groups for two reasons. First, from an energy revolution perspective, the technology subclass that succeeds is not important, only that some succeed rapidly and at scale. Second, while historical analysis can provide many lessons for innovation at this micro level, it adds greatly to complexity, and suffers from survivor bias.

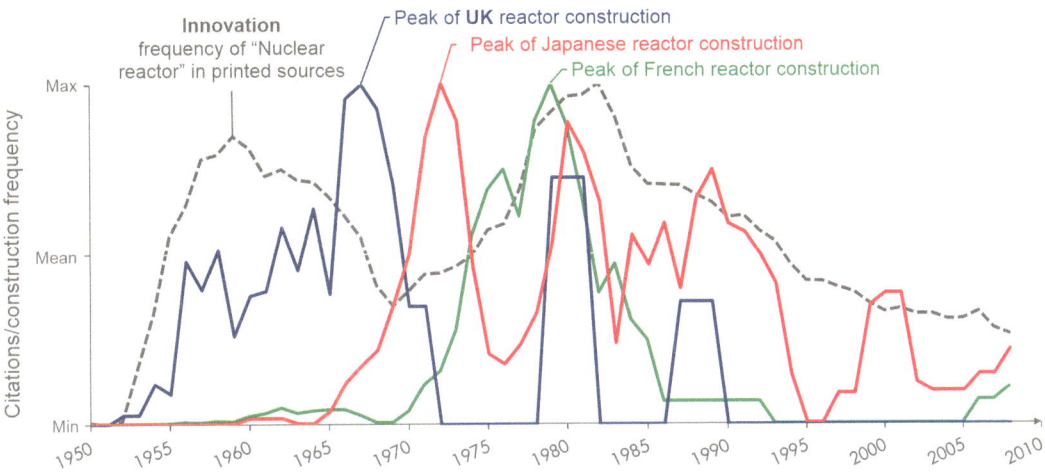

Fig. 19 The lag from citations to deployment of a technology. *Source* Vivid Economics

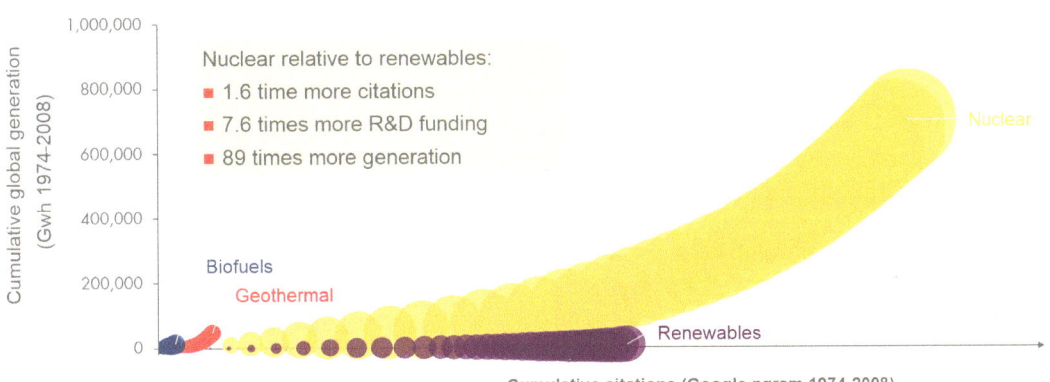

Fig. 20 Hype does not translate into R&D and deployment of new technologies. *Note* Bubble size represents cumulative R&D expenditure. *Source* Vivid Economics

(3) Definition of energy system change

We developed a change metric to quantify energy system change. The energy system is the entire flow of energy from primary sources, through conversion to secondary fuel carriers (primarily power and refined oil products) to final consumption by end-use sectors (primarily buildings, industry and transport). Each component of this energy system, such as the flow of primary energy into power generation, has a particular fuel mix. We consider a revolutionary change to have occurred within a component of the energy system if the fuel mix has changed significantly within a decade. We calculate a single number (a change metric) to identify revolutionary changes in components of the energy system and the overall energy system, for all countries in the world. The change metric is calculated using annual data between 1971 and 2014 for most countries, and between 1960 and 2014 for a smaller set of countries where data are available. The data source is the International Energy Agency's World Energy Balances 2016. There are two crucial steps in calculating the change metric: first, significant changes in the fuel mix must be identified; and second, the magnitude of significant changes must be calculated in a single metric, comparable across countries.

We use a wild binary segmentation algorithm to identify significant changes in the fuel mix. This algorithm identifies out of the ordinary changes in the variance of the rate of change of the fuel mix. That is to say, the algorithm identifies when there is a statistically significant change in the rate at which the fuel mix changes relative to its normal rate of change. Such a test is necessary. The magnitude of significant change is identified by summation of the rates of change in fuels' share of the energy flow between the years identified as significant by the algorithm. For each fuel in a flow, such as primary energy to power, the absolute rate of change in its share of the fuel mix is calculated between the years identified as significant by the algorithm. The change metric for the flow is then the summation of these absolute rates of change. The change metric for the entire energy system is the sum of weighted change metrics for each component in the energy system, where weights are the ratio of absolute energy in that flow to the absolute energy in final consumption.

We identified the common characteristics of cases where technology played a major role in an energy system change by analysing the largest energy system changes in the G20 since 1970. We used the change metric dataset to identify the quantifiably largest energy system changes across countries and over time. We then investigated these periods of greatest change to understand which of the factors identified in our analysis of innovation and deployment (technology, demand, supply or markets) triggered the change. We then combined this analysis of trigger factors with an assessment of the characteristics of technology that played a major role in the set of greatest changes, even when technology was not a trigger factor. Our findings from this assessment were developed into a framework that describes the conditions that, in international experience, have enabled innovative technology to be successfully applied at a scale that revolutionises an energy system.

5.2.2 Findings from International Experience

We analysed the changes in G20 countries across five components of the energy system—primary energy, power, industry, transport and refining, and buildings to determine patterns of change. Additional results for the G7 subset are presented to demonstrate how patterns of change have altered as industrial economies mature into service-led economies over this time period.

(1) Factors driving the successful application of energy technology

The gap between hype, research effort and deployment varies by technology, which suggests that some technologies have supporting factors that others do not. First, hype tends to peak before R&D spending, which peaks before deployment. This suggests that technology eventually has its impact on the energy system many years after its potential is perceived, and that changes flowing from actual deployment do not gather as much notice as the initial innovation. Second, there is significant variation in the gaps between hype, R&D and deployment across technologies. Renewable power is often perceived as driving an energy revolution. However, since 1974, nuclear power has received 1.6 times more citations, 7.6 times more R&D funding and generated 89 times more power.

Technologies that achieve high levels of deployment benefit from a supporting set of factors in addition to their technological development. These include technology innovation to a level such that deployment is feasible, supply of the inputs the technology requires, demand for the services the technology provides, and markets that incentivise deployment. All the factors must be present and aligned if a technology is to move from innovation, through application to changing the energy system.

The case of US tight gas illustrates how the sequencing of supply, demand, market and technology factors can determine when revolutionary change is triggered. US tight gas revolutionised US and, to an extent, global gas markets. This occurred with apparent speed: the first large-scale application of hydraulic fracturing technology was in 2000, with production increasing rapidly from 2006. However, this rapid revolution was supported by factors that had long been in place. The supply of tight gas below ground has been present for millions of years, while above ground the exploration and production industry was mature, and the demand for gas in the USA had been established for decades, with an extensive pipeline network serving the widest possible set of consumers. The USA also has a mature liberalised gas market that provides incentives to any technology that can supply demand. It can be argued that it is this market factor that triggered the shale gas revolution, because it is when gas prices increased due to the prospect of a shortage that hydraulic fracturing technology was applied at a large enough scale to change the energy system. The change metric shows that the fuel mix of US energy production has changed significantly in the 1970s–1980s. It was only when gas prices increased that the change metric rose above historic highs (Figs. 21 and 22).

It is important to note that to reduce technology deployment to four factors is a useful simplification of a complex process. Energy systems are complex systems. This means that there is a large number of factors that induce change to the system, and there are few direct, clear relationships. The simplification is a useful one as differences in the factors of technology, supply, demand and markets describe many of the differences between outcomes for technologies.

(2) **The role of technological development in driving energy revolution**

From 1960 to 1985 energy systems underwent significant change in the fuel mix, the change metric and to the scale of energy consumption. However, since 1985 fuel mixes in G7 countries have been relatively stable, as has energy demand. This is despite increased energy R&D in the 1970s and accelerating rates of innovation in the broader economy (Fig. 23).

The stability of energy systems over the past 30 years is in stark contrast to the revolutionary changes required in the next 30 years due to decarbonisation. Countries tend to change their energy system most when energy demand is increasing, because this is when new energy transmission networks are created, which, once built, lock in fuel choices. Recent energy revolutions have mainly occurred in the upstream and midstream sectors, with the fuel mix that supplies energy use remaining relatively constant after significant change before 1985. In the period

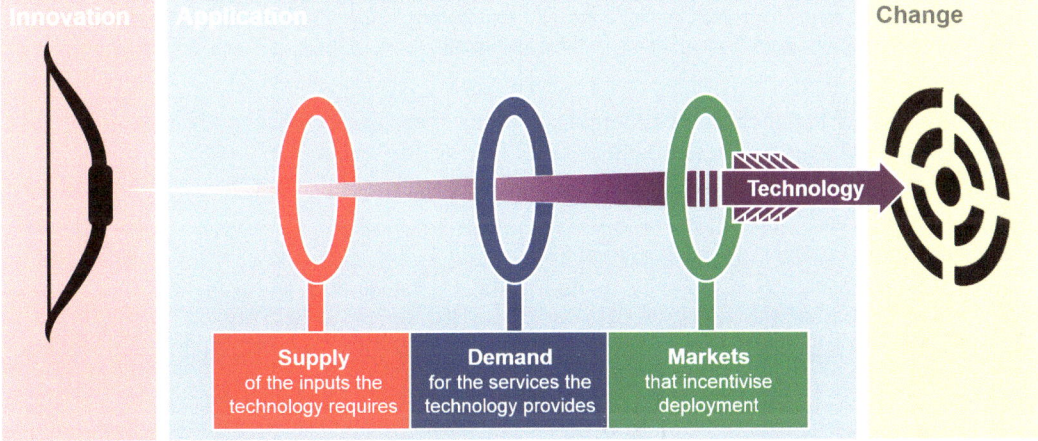

Fig. 21 Technology requires alignment of supply, demand and markets if it is to change energy systems. *Source* Vivid Economics

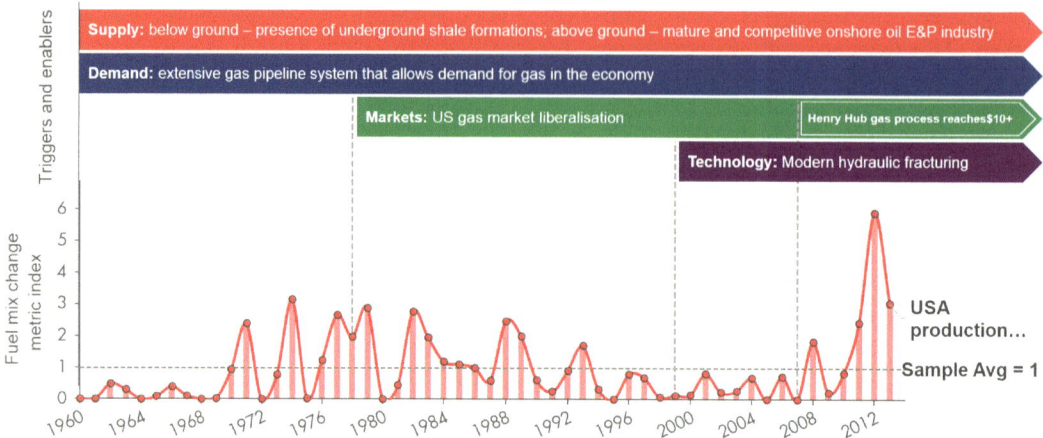

Fig. 22 US tight gas illustrates how the sequencing of supply, demand, markets and technology determines when revolutionary change is triggered. *Source* Vivid Economics

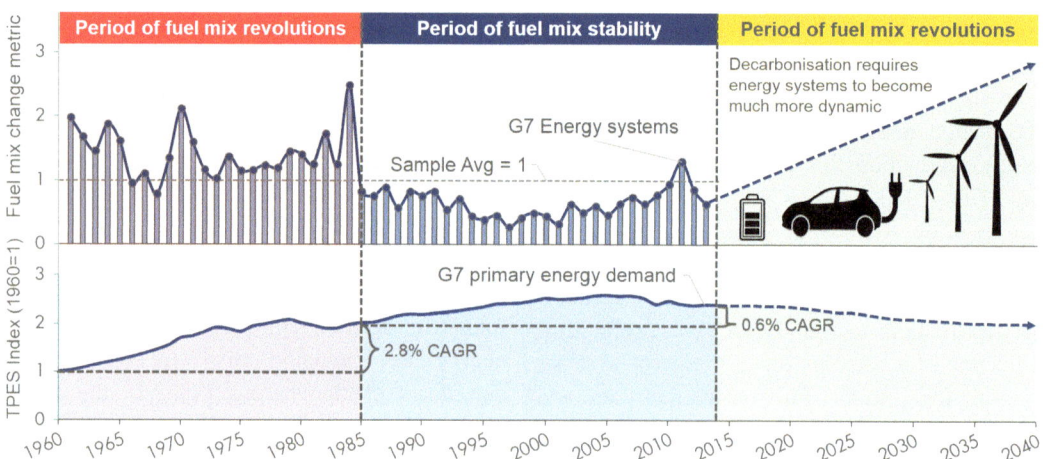

Fig. 23 G7 energy systems have been stable in the past four decades. *Note* TPES = total primary energy supply; CAGR = compound annual growth rate. *Source* Vivid Economics

before 1985, changes in fuel mix for buildings and industry contributed to a significant proportion of the overall change metric. However, in the decades since 1985, change has largely been driven by shifts in production and imports. Transport has remained oil-dominated throughout the period. This suggests that downstream fuel mixes are locked in by their distribution networks, which is a challenge for the future, given that fuels currently used in energy use—such as oil for transport, coal in industrial processes and gas for heating—will

have to change if climate mitigation targets are to be met (Fig. 24).

Across the largest G20 energy revolutions, most are triggered by economic growth, energy security concerns, new market incentives or shocks, rather than by technology. Revolutions are colour-coded: red for a supply-triggered revolution; blue for a demand-triggered revolution; green for a market-triggered revolution; and purple for a technology-triggered revolution, of which there are none (Fig. 25).

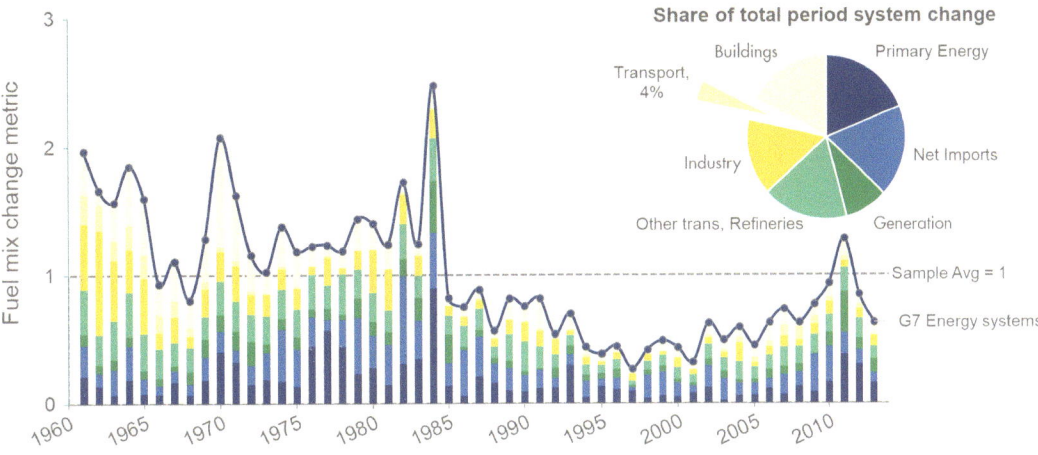

Fig. 24 Recent G7 energy revolutions have mainly occurred upstream. *Source* Vivid Economics

Rank	Primary energy	Power	Industry	Transport/Refining	Buildings
1st	UK, 1970-1980 ↑ Oil and Gas (North Sea)	IDN, 1999-2010, ↑ Coal and Geothermal (Economic growth)	IDN, 1999-2010, ↑ Coal and gas (Economic growth)	ZAF, 1980-1982 ↑ Coal (CTL Apartheid)	KOR, 1990-1994 ↑ Oil (Economic growth and strong currency)
2nd	KOR, 1981-1991 ↑ Nuclear (Energy security)	KOR, 1981-1991 ↑ Nuclear (Energy security)	TUR, 2000-2010, ↑ Gas (Pipeline and LNG)	IND, 1970-1988 ↑ Oil (Shift from rail to road)	DEU, 1990-1991 ↓ Coal (Collapse of East German coal mining)
3rd	FRA, 1974-1991 ↑ Nuclear (Energy security)	FRA, 1974-1991 ↑ Nuclear (Energy security)	MEX, 1994-2010 ↑ Electricity and Coal (NAFTA)	CHN, 2008-present ↑ Gas and Electricity (Market making)	TUR, 2000-2010, ↑ Coal and Electricity (Low cost alternatives to biofuels and oil)
4th	ITA, 2000-2010 ↑ Biofuels and Renewables (Decarbonisation)	GBR, 1990-1998 ↑ Gas and Nuclear (Dash for gas)	UK, 1970-1976 ↑ Gas (North Sea)	BRA, 1982-1986 ↑ Biofuels (Energy security)	UK, 1970-1976 ↑ Gas (North Sea)
5th	IDN, 1999-2010 ↑ Coal (Economic growth)	JPN, 2011-2014 ↓ Nuclear (Fukushima)	FRA, 1980-1984 ↓ Oil (Oil shock)	ARG, 1992-1995 ↑ Gas (CNG market making)	FRA, 1980-1984 ↑ Electricity (Nuclear programme)

Fig. 25 Most G20 energy revolutions are not driven by technology. *Source* Vivid Economics

(3) Characteristics of revolutionary energy technologies

Analysis suggests that, in addition to a technology having supporting supply, demand and market factors, the following characteristics of a technology can influence its successful application. First, capital intensity. A technology with high capital intensity, such as nuclear power, requires a large-scale player to deploy it, whereas a technology with low capital intensity, such as biofuels or compressed natural gas (CNG) vehicles, can be deployed by individuals. Second,

network intensity. A technology with high network intensity, such as offshore oil and gas, requires significant investment in a network, in addition to the capital of the initial technology, which is often delivered by players other than the technology developer. A technology with low network intensity, such as power generation technologies, can plug and play into an existing network.

In international experience, technologies with high capital intensity and low network intensity, have played a major role in energy revolutions. These technologies are often deployed to meet

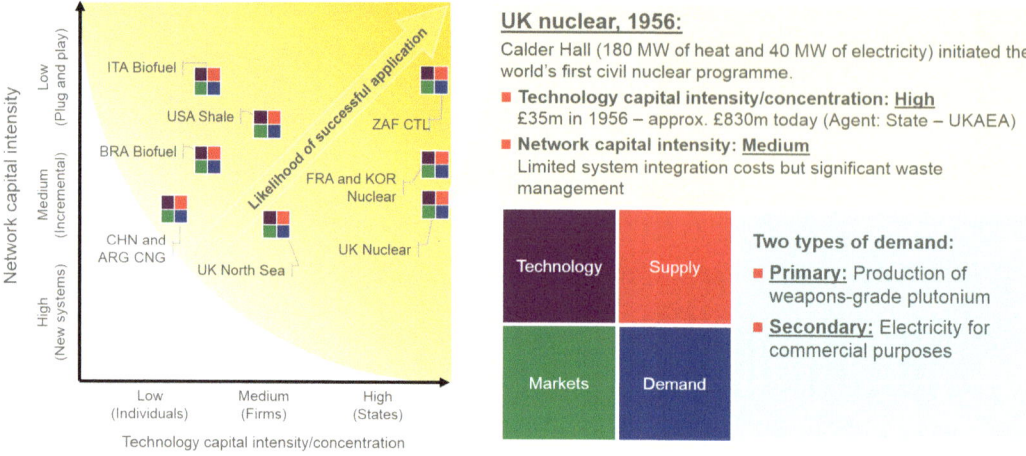

Fig. 26 Revolutionary technologies tend to rely on large state investments and/or require incremental network investments

rapid changes in energy demand, such as demand for more energy, secure energy or cleaner energy. This is because governments often have responsibility to meet these needs, and they have tended to favour deployment of large, established, single-fuel technologies—such as nuclear or coal —that can plug into existing networks (Fig. 26).

6   Current Developments and Potential Impacts of Some Major Energy Technologies

This section analyses the development trends of some energy technologies of significance to China, and their implications for China to deliver energy technology revolution.

6.1   Smart Grids

6.1.1   Current Status of, and Outlook for, Smart Grids

A new round of energy revolution is around the corner. Large-scale use of clean and renewable energy to build a green, smart and sustainable energy system has become an unstoppable trend. The smart grid—integrating the new generation of energy, IT, control and materials technologies —is the key to green, smart and sustainable

development of the energy system. Its significance has been widely recognised.

(1)   China

To accelerate the development of the smart grid, the Chinese government has introduced incentive policies and included smart grids in its strategic plans. In November 2014, the State Council issued the Strategic Action Plan for Energy Development (2014–20),[7] which defines the smart grid as a priority area in energy technology innovation. In July 2015, the National Development and Reform Commission (NDRC) and the National Energy Administration (NEA) released their Guiding Opinions on Boosting Smart Grid Development.[8] This states that the smart grid is an important means to realise the energy production, consumption, technology and system revolutions and the Energy Internet (Internet+). In February 2016, the NDRC, the NEA and the Ministry of Industry and Information Technology published their Guiding Opinions on Promoting the Development of Internet + Smart

---

[7]General Office of the State Council, Strategic Action Plan for Energy Development (2014–20), 2014, pp. 16–17.

[8]National Development and Reform Commission and National Energy Administration, Guiding Opinions on Boosting Smart Grid Development, 2015, p. 1.

Energy,[9] according to which an integrated energy network based on the smart grid will be developed. In April 2016, the NDRC and the NEA released the Action Plan for Innovation in the Energy Technology Revolution (2016–30)[10] and the Roadmap for Major Innovation Activities in Energy Technology Revolution,[11] which describe the plan to develop smart grid power transmission and transform end-user equipment.

China's smart grid focuses on the integrated and coordinated development of power generation, transmission, conversion, distribution, consumption and scheduling. State Grid Corporation of China proposed in 2009 to develop a smart grid based on robust grid architecture, supported by a communications and information platform and incorporating intelligent control, all voltage levels and all parts of the power system, including "power, information and business flows."[12] China Southern Power Grid researches such fields as new energy, flexible DC power transmission, intelligent substations, power distribution systems, integration of distributed energy resources, microgrids, and power use and information and communications technologies. The company is also developing green, reliable, smart and efficient 3C (computer, communications, control) power grids to make them efficient, resource-saving and environmentally friendly.[13]

(2) **USA**

The US government designed a strategic framework for smart grid development in the Energy Independence and Security Act of 2007.[14] The U.S. Department of Energy issued the Smart Grid System Report[15] in July 2009, which defines the scope, characteristics and indicators of smart grids. The new version of the Estimating the Costs and Benefits of the Smart Grid,[16] published by the Electric Power Research Institute in April 2011, devised a method to calculate the cost and benefits of smart grids. The All-of-the-Above Energy Strategy as a Path to Sustainable Economic Growth,[17] issued by the Obama administration in May 2014, proposed developing solar, wind, geothermal and other renewable types of energy to bolster economic growth and protect the environment. In January 2016, the U.S. Department of Energy made public its new blueprint for the modern power grid, which aims to integrate conventional energy, renewable energy, energy storage and energy efficiency to ensure grid reliability and protect it from cyberattack and climate change.[18] Smart grids will, therefore, play a significant role in driving the transition to low-carbon energy.

The USA's smart grid underscores resilience, reliability, affordability, flexibility and sustainability. Currently, the grid transmits electricity from large remote power plants via high-voltage transmission lines to local distribution networks, which deliver the power to industrial, commercial and residential users, primarily with one-way electricity flow. In the future, the smart grid will still need large power plants for energy, but it

[9]National Development and Reform Commission, National Energy Administration, and Ministry of Industry and Information Technology, Guiding Opinions on Promoting the Development of Internet + Smart Energy, 2016, pp. 5–7.

[10]National Development and Reform Commission and National Energy Administration, Action Plan for Innovation in the Energy Technology Revolution (2016–30), 2016, p. 6.

[11]National Development and Reform Commission and National Energy Administration, Roadmap for Major Innovation Activities in Energy Technology Revolution, 2016, pp. 67–69.

[12]Strong Smart Grid https://baike.baidu.com/item/%E5%9D%9A%E5%BC%BA%E6%99%BA%E8%83%BD%E7%94%B5%E7%BD%91/9399809?fr=aladdin.

[13]China Southern Power Grid, Corporate Social Responsibility Report 2016, pp. 19–27.

[14]U.S. Congress, Energy Independence and Security Act of 2007, pp. 293–304.

[15]DOE, Smart Grid System Report, 2009.

[16]EPRI, Estimating the Costs and Benefits of the Smart Grid, 2011.

[17]Executive Office of the President of the United States, The All-of-the-Above Energy Strategy as a Path to Sustainable Economic Growth, 2014, pp. 31–39.

[18]State Grid Energy Research Institute Co., Ltd., Analysis Report on Grid Development and Application of New Technologies in and outside China, Beijing: China Electric Power Press, 2016, pp. 22–23.

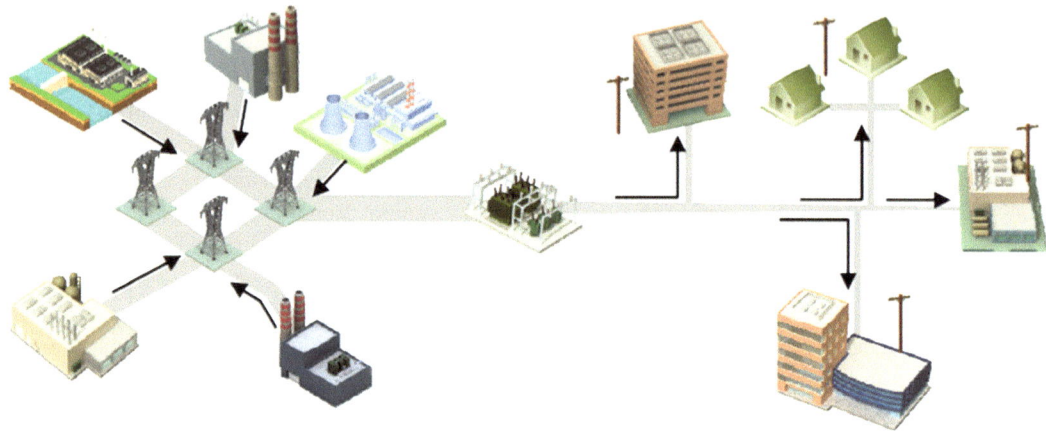

Fig. 27 Conventional power grid

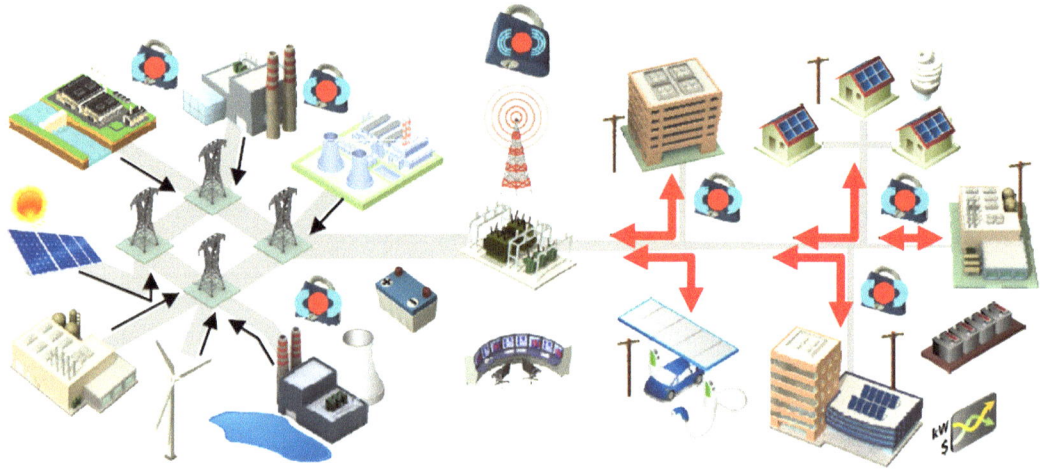

Fig. 28 Future smart grid

will integrate and coordinate different types of load, including distributed power, electric vehicles and smart homes through a communications and control platform. One-way electricity flow will no longer be predominant (Fig. 27 and 28).

(3) **Europe**

Development of the smart grid became an important driver for the EU to deliver its 20-20-20 climate and energy targets for 2020.[19]

The European Commission published Smart Grids: From Innovation to Deployment[20] in 2011, in which it sets out its policy for the future development of the European grid. The Pan-European Ten-Year Network Development Plan,[21] released in July 2014, defines 2030 as an important milestone and describes the overall development path for the European grid. The European Network of Transmission System Operators for Electricity (ENTSO-E) released the

[19]European Commission, Energy 2020: A strategy for Competitive, Sustainable and Secure energy, 2010, pp. 18–20.

[20]European Commission, Smart Grids: From Innovation to Deployment, 2011.

[21]ENTSO-E, Ten-Year Network Development Plan 2014, 2014.

fourth version of its 10-year network development plan[22] in 2016, committing to continue developing the European smart grid.

The Roadmap to Implement the EU's Power Grid Vision,[23] initiated by the European Commission and developed by two major grid operators in Europe, outlines the future European power system. On the one hand, the European smart grid would comprise interconnections to enable power transfer across borders and grid integration of large volumes of renewable energy. On the other hand, it highlights the importance of distributed energy resources and the combination of advanced measurement and control with effective market mechanisms, thus ensuring real-time balance and redundancy in the grid (Fig. 29).

(4) **Japan**

For the Japanese government, the smart grid is a critical tool to develop renewable energy, improve power infrastructure, boost economic growth and hedge against various risks. The Energy Innovation Strategies[24] unveiled by the Ministry of Economy, Trade and Industry in 2016, target complete energy mix optimisation by 2030 by expanding energy investment, improving energy efficiency, increasing the share of renewable energy and reducing greenhouse gas emissions. This would help deliver the national GDP target of JPY 600 trillion for 2030.

Japan holds the view that the smart grid should make power supply efficient, high quality and reliable by integrating large-scale distributed power systems, high-speed communications technologies, distributed energy resources, energy storage devices and other demand-side resources. Japan's smart grids are divided into national, regional and household (building)

levels. Their characteristics differ from level to level. The national level comprises the transmission and distribution networks. Regions include renewable power generation and, given the reliance of renewables on weather conditions, regional demand-supply balance through an energy management system is essential (Fig. 30). Households and buildings focus on the collection of energy consumption data and the optimal control of electric power.[25]

6.1.2 Generic Technologies in the Smart Grid

Smart grid technologies are an important driver for the development of the smart grid. Power distribution and retail are the priority fields for innovation in smart grid technologies. Advanced metering infrastructure, advanced distribution automation, microgrids and the intelligent use of electricity are often seen in the smart grid development roadmaps of major countries.

(1) **Overview**

First, advanced metering infrastructure (AMI). AMI integrates smart meters, communications networks and data management. It allows two-way communication between the grid and end users. It also provides users with time-of-use or real-time measurement data—such as power consumption, voltage, current and electricity prices—to facilitate efficient power consumption by users and support coordinated grid operation.[26] A typical AMI architecture is shown in Fig. 31.

AMI could provide power utilities with a communications network that connects with end-user terminals and improves grid control and visibility with the data uploaded by AMI. It is a very important foundation for the smart grid.

---

[22]State Grid Energy Research Institute Co. Ltd., Analysis Report on Development of Smart Grid in and outside China, Beijing: China Electric Power Press, 2015.

[23]State Grid Energy Research Institute Co. Ltd., Analysis Report on Development of Smart Grid in and outside China, Beijing: China Electric Power Press, 2013, p. 39.

[24]Ministry of Economy, Trade and Industry, Energy Innovation Strategies, 2016, pp. 1–2.

[25]State Grid Energy Research Institute Co. Ltd., Analysis Report on the Development of Smart Grids in and outside China. Beijing: China Electric Power Press, 2012, pp. 40–46.

[26]Advanced Metering Infrastructure and Customer Systems, https://www.smartgrid.gov/recovery_act/deployment_status/ami_and_customer_systems.html (2015).

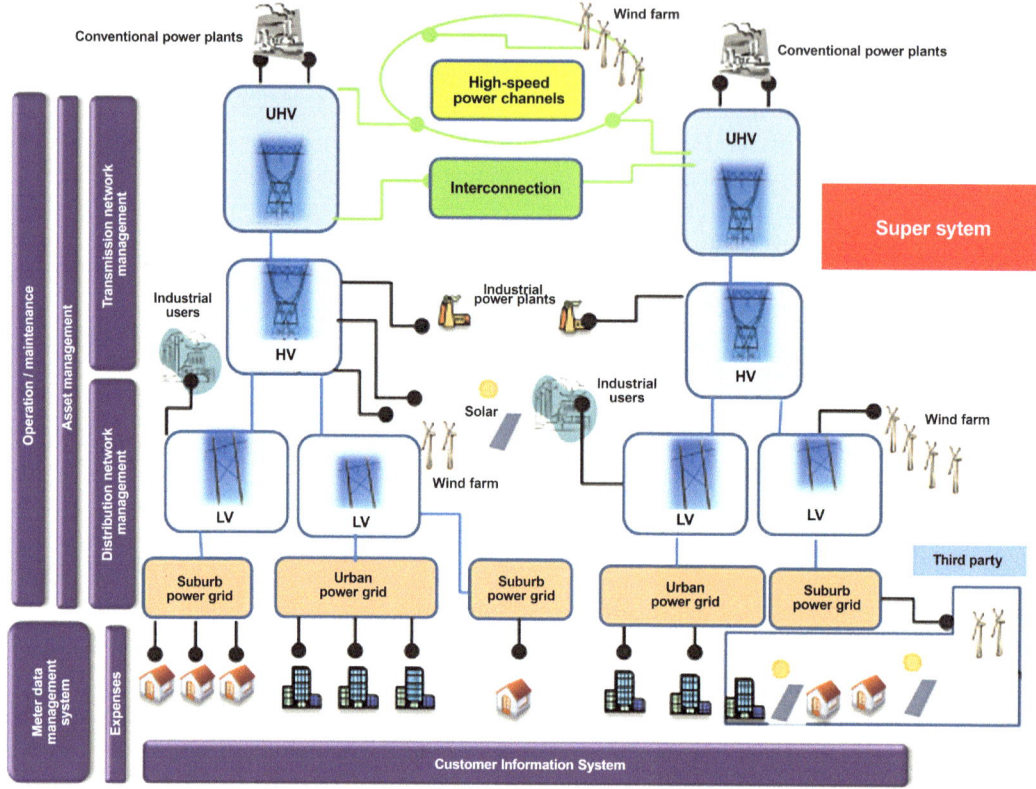

Fig. 29 Pan-European power system 2050. *Source* State Grid Energy Research Institute Co. Ltd., Analysis Report on Development of Smart Grid in and outside China, Beijing: China Electric Power Press, 2013, p. 39

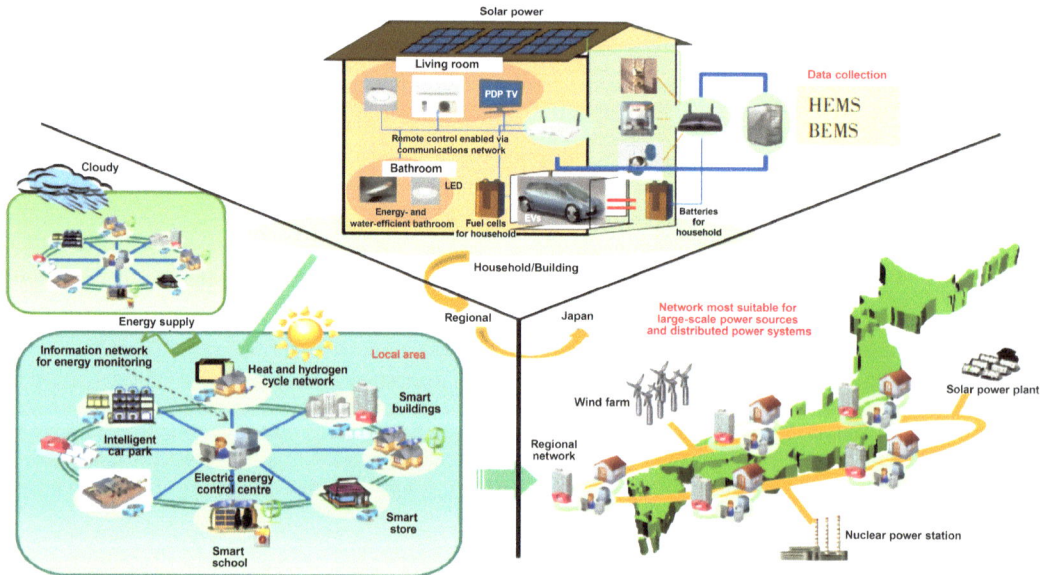

Fig. 30 Japan's smart grid. *Source* State Grid Energy Research Institute Co. Ltd., Analysis Report on Development of Smart Grid in and outside China. Beijing: China Electric Power Press, 2012, pp. 40–46

AMI could provide several crucial measurements —including voltage, current and power—to address user requirements and support remote connection or disconnection, two-way metering, and regular or random reading of measurement data. AMI can also act as a gateway to the user's indoor network, providing them with real-time electricity prices and power consumption data, control over the indoor power load, and laying the foundation for demand response.

AMI provides significant benefits in energy use reduction, peak shaving and blackout recovery. Prior to the installation of AMI, US utility PG&E made 48,000 repair sessions for outage calls each year, which was reduced substantially after AMI was deployed, saving $43 million annually.

Developed countries conduct extensive R&D of AMI in areas like automatic collection of power consumption data, measurement abnormity monitoring, power quality monitoring, and electricity consumption analysis and management. In 2009, China started to develop power consumption data collection systems and deploy smart meters. It introduced 24 technical standards on power consumption data collection and 12 technical standards on smart metering. By the end of 2015, State Grid Corporation of China had installed more than 300 million[27] smart meters.

Second, advanced distribution automation (ADA) uses power electronics, communications and network technologies to integrate topology information, operational and historical data, geographic information and user data with monitoring, protection, control and management of the power distribution network. The configuration and functions of a typical ADA are shown in Fig. 32.

ADA is an integral part of the smart power distribution network. It provides intelligent control of distributed power and energy storage systems, electric vehicle charging and discharging facilities and demand response. It improves power supply reliability, shortens outage recovery times and reduces power cuts and demand rationing. When integrated with other parts of the smart grid, ADA can enhance system monitoring, improve reactive power and voltage management, reduce transmission losses, improve asset utilisation, and optimise operation, scheduling and maintenance activities.[28] According to American Electric Power (AEP), which owns the largest power transmission system in the USA and is one of the country's biggest power generators, ADA reduced the number and duration of outages by 45% and 51% respectively.[29] By minimising use of line patrol vehicles after power failures, ADA can reduce greenhouse gas emissions.

ADA is being developed in countries like the USA, UK, France, Singapore and Japan. Japan stands at the forefront of ADA. In 1999, ADA helped reduce the average outage in Japan to 3 min per household per year.[30] China is vigorously promoting ADA across the country. State Grid Corporation of China had deployed ADA across a third of its operating area in 2016.[31]

Third, microgrids. A microgrid is a small, self-contained power generation and distribution system that can operate independently or be integrated into a larger grid. Microgrids can supply one user, like a campus or military base, or multiple users across an island or geographic area.

The main functions of a microgrid are: first, autonomous operation. A microgrid operates autonomously and stably and meets the demand for power by itself. Second, it reduces power supply volatility by stabilising fluctuations in generation and consumption and by maintaining power and voltage stability at the point of connection with other grids. Third, it provides

---

[27]Analysis on the Development and Market Prospects of China's Smart Meter Industry, 2016, http://www.chyxx.com/industry/201606/426731.html (2016).

[28]Yu Yixin and Luan Wenpeng, Smart Grid, Power System and Clean Energy, vol. 25, 2009, pp. 10–11.

[29]AEP Ohio, Final Technical Report, 2014, pp. 169–216.

[30]Liu Yong and Han Wen, Development of China's Power Distribution Network vs. Construction of Japan's Power Distribution Network, https://wenku.baidu.com/view/e6b524cbfc4ffe473268abb6.html (2016).

[31]State Grid Corporation of China, Social Responsibility Report, 2016, pp. 36–37.

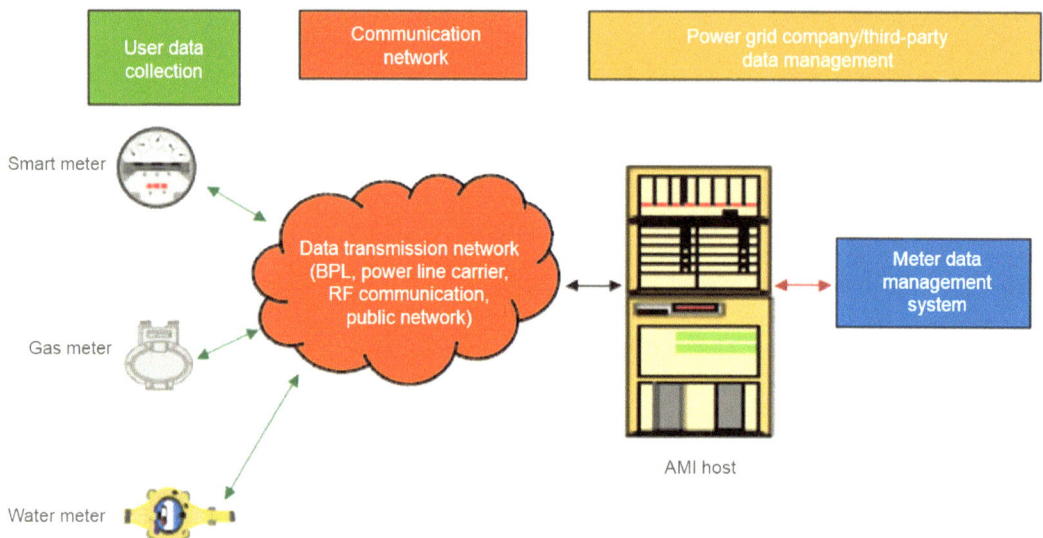

Fig. 31 Typical AMI architecture. *Source* EPRI, Advanced Metering Infrastructure, 2007, p. 1

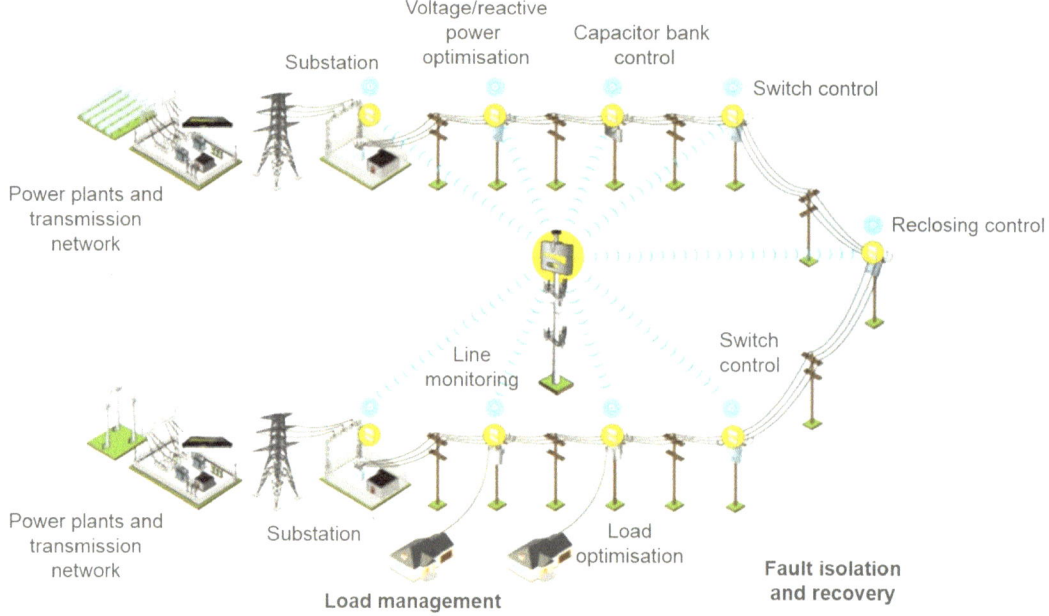

Fig. 32 Configuration and functions of advanced distribution automation. *Source* Distribution Automation, http://ruggedcom.net.ua/applications/electric-utilities/da.html

ancillary services by increasing power output or reducing power loads.

Increasingly mature microgrid technologies enable the integration of renewable energy, which reduces the use of fossil fuel generation and lowers greenhouse gas emissions. Flexible and parallel operation of the microgrid with the public grid enables peak shaving and better use of grid equipment. Developed countries have carried out in-depth research on microgrid

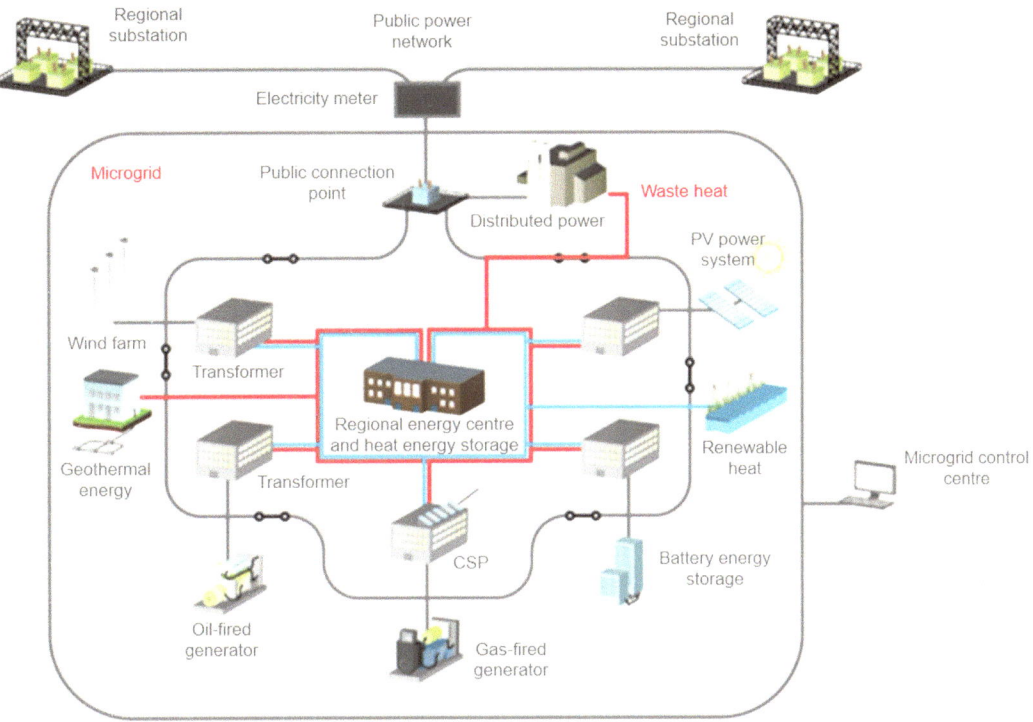

Fig. 33 Microgrid. *Source* NYPrize: Building Community Microgrids in New York. http://earthdesk.blogs.pace.edu/2014/02/03/nyprize-building-community-microgrids-in-new-york/, (2014)

technologies and constructed several pilot projects, such as Continuon's (now Liander) facility in the Netherlands, EDP's microgeneration facility in Portugal, the Mad River Park microgrid in Vermont, USA, and the Aichi microgrid in Japan. China also has several pilot microgrid projects, including those at Sino-Singapore Tianjin Eco-City; Henan University of Economics and Law; the Guangdong Foshan combined cooling, heat and power (CCHP) microgrid system; and the integrated solar, wind, diesel, energy storage and seawater desalination system on Zhejiang Dongfushan Island (Fig. 33).

Fourth, the intelligent use of electricity is an important area that demonstrates the advantages of the smart grid. It optimises resource allocation, enables peak shaving and reduces the cost of electricity by guiding users to manage their power consumption efficiently. As smart homes, electric vehicles and other power-driven devices

evolve and become more flexible, technologies related to demand response, battery charging and discharging will make two-way intelligent electricity use services possible.

Demand response enables end users to change their power consumption behaviour in response to market price signals or incentive mechanisms. Demand response is mainly incentive- or time-based electricity pricing. Incentive-based demand response rewards users who consume less electricity during peak periods, thus lowering power loads. Time-based pricing strategies guide users to change their power demand by providing them with electricity pricing data (Fig. 34).

The rapid development of electric vehicles (EV) could have a major impact on the future power distribution system. EV-related technologies are divided into vehicle technologies (including battery management, engine, power control, safety, etc.) and charging/discharging technologies (such as battery swapping, two-way grid interaction and battery cascading).

The development of intelligent electricity use technologies improves equipment utilisation, reduces the cost of operation and maintenance and lowers energy consumption. Demand response can smooth out the load curve and reduce power supply costs in short-term power markets. If there is a power shortage or wholesale prices are high, demand response can adjust prices to level out price fluctuations. In long-term power markets, demand response reduces peak power demand to avoid or slow down the need for new investment; and it improves safety and power system stability by taking advantage of users' response to electricity prices.[32] The USA has abundant experience of demand response and has standards and an industry alliance for automated demand response known as OpenADR. China started to research demand response and launch pilot projects in 1998, making great progress. In 2016, the Action Plan for Innovation in the Energy Technology Revolution (2016–30)[33] proposed research on demand response-based technologies to make China's power consumption more intelligent.

Electric vehicles and hybrid electric vehicles (HEVs) have a significant impact on power distribution and use. As EV ownership increases and battery performance improves, EV batteries can be used as mobile energy storage units, charging during non-peak hours and supplying electricity to the grid during peak hours, thereby reducing valley-peak fluctuations in demand and improving grid efficiency. In microgrids with a high proportion of renewable power, electric vehicles can be used to store energy during periods of high renewable output and low load, and discharge the energy into the grid when

renewable power output is low and demand high (vehicle-to-grid, V2G), which strengthens the grid's capacity to absorb renewable power (Fig. 35).

EVs and HEVs are more energy efficient than fossil fuel vehicles. In the USA,[34] the deployment of smart charging facilities could increase the share of EV mileage by light vehicles by 9 percentage points (from 64% to 73% of the total). Compared with fossil fuel vehicles, EV and HEV light vehicles use 2–5% less energy. Currently, China has built a proprietary standard system of EV charging/battery swap facilities and is constructing a network of rapid-charging stations along urban roads and motorways. A rapid charging network has been built from Beijing–Harbin, Beijing-Hong Kong–Macao, Beijing–Shanghai, Shanghai–Chengdu, Shanghai–Chongqing, on Beijing ring roads and the Hangzhou Bay ring expressway, covering 95 cities and 14,000 km of expressway.

(2) Demonstration projects

1. AEP GridSMART Demonstration Project, USA

The AEP GridSMART Demonstration Project comprises nine technical demonstration domains, including advanced metering, home area networks and redistribution management. Its advanced metering infrastructure (AMI) and demand response capability have made remarkable achievements in reducing carbon and PM2.5 emissions and improving grid efficiency (Table 11).

After the deployment of AMI, the average CO_2 reduction was 16.91 tonnes per month, amounting to 406 tonnes per year. AMI saved AEP from reading meters on-site, avoiding 5,694 miles (9,163 km) of travel per month, and about 68,326 miles (109,960 km) per year. Assuming that driving one mile generates 423 g of CO_2 on average, this amounts to reductions in CO_2

[32]Zhao Xin and Gao Shan, Demand Response and Advanced Metering in the US Electricity Market, in Power Demand Side Management, vol. 9, 2007, pp. 68–69.

[33]National Development and Reform Commission and National Energy Administration, The Action Plan for Innovation in the Energy Technology Revolution (2016–30), 2016, pp. 8–10.

[34]DOE, The Smart Grid: An Estimation of the Energy and CO2 Benefits, 2010, pp. 3.25–3.27.

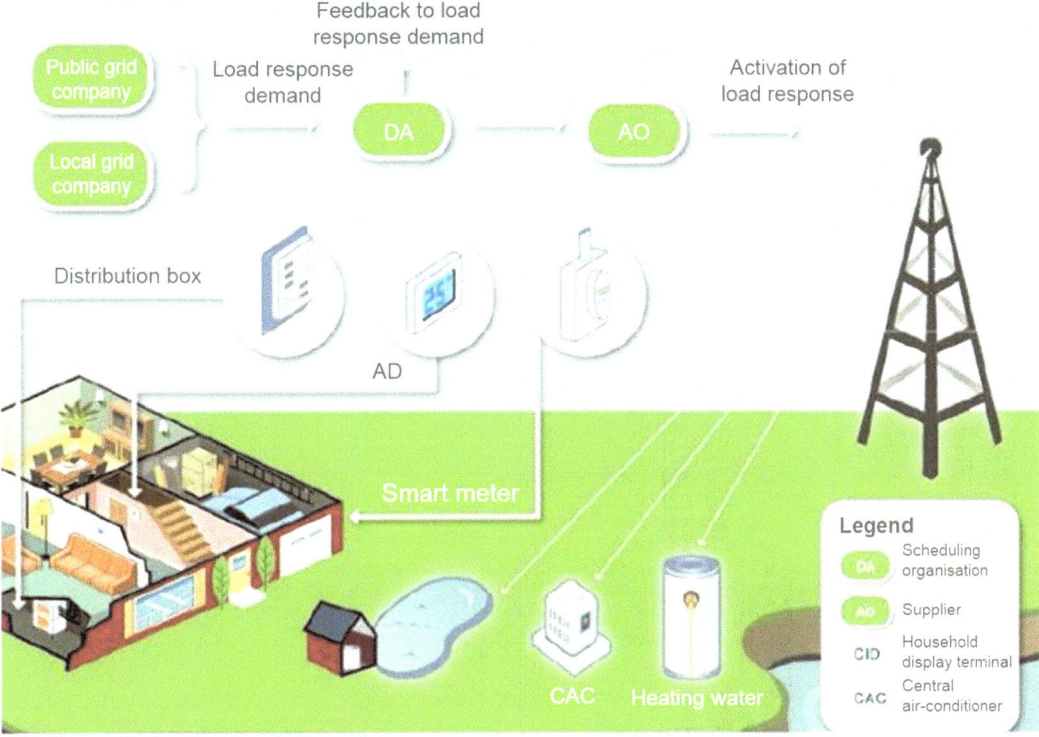

Fig. 34 Demand response. *Source* Rodan Energy, Demand Response and the Smart Grid in Ontario, 2012, p. 11

emissions from driving of 2,408 tonnes per month or 28,903 tonnes per year (Figs. 36, 37).

Moreover, the project saved 0.956 kg of NO_x on average per month, amounting to 22.9 kg over the two years of its duration, as well as 0.220 kg of SO_x per month on average (5.3 kg over the two years) and 0.191 kg of PM2.5 per month on average (4.6 kg over the two years).

The project's demand response products include SMART Shift, SMART Shift Plus and SMART Choice. SMART Shift and SMART Shift Plus provide users with electricity price information for different time periods in power supply contracts. SMART Choice provides users with quasi real-time electricity price information that is updated every 5 min to guide users' power consumption behaviour. Demand response plays an active role in reducing energy consumption and fossil fuel emissions and in peak load shaving.

As shown in Fig. 38, SMART Shift and SMART Shift Plus users consume less power and emit less CO_2. Calculation[35] results show that under these three modes, nearly 196 tonnes of CO_2 could be reduced.

Figure 39 shows that SMART Shift and SMART Shift Plus users consume less power and reduce their emissions of SO_x, NO_x and PM2.5 by about 749 kg, 335 kg and 284 kg respectively.

2. Sino-Singapore Tianjin Eco-City Smart Grid demonstration project

The project comprises distributed power generation, microgrids and energy storage systems, intelligent substations, power distribution automation, equipment status and power quality monitoring systems, visualisation platform, power

---

[35] AEP Ohio, Final Technical Report, 2014, pp. 115–116.

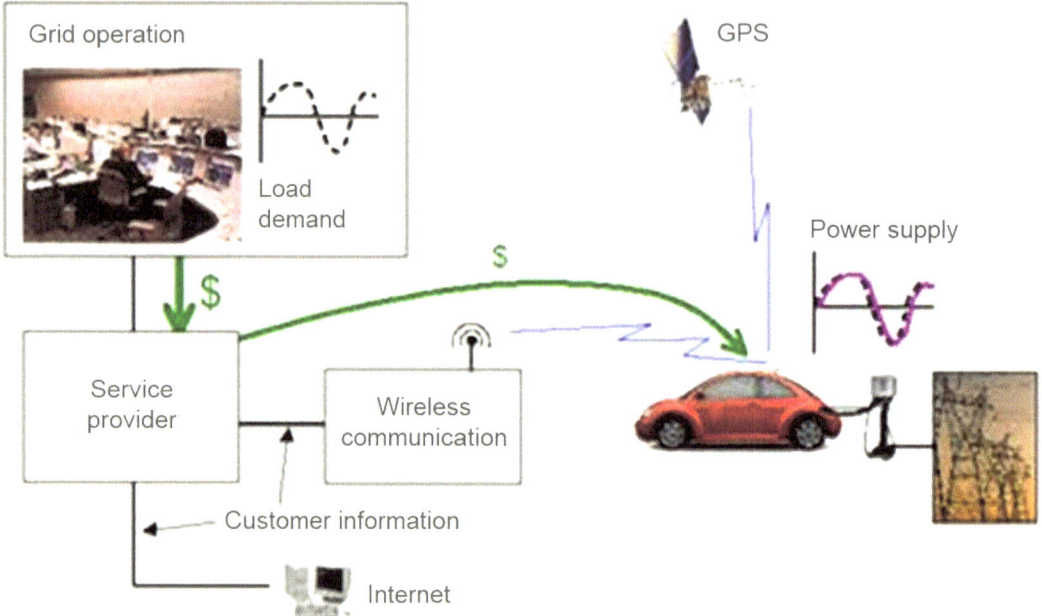

Fig. 35 Vehicle-to-grid. *Source* A market study on hybrid vehicles and the concept of V2G. https://www.dolcera.com/wiki/index.php?title=A_market_study_on_Hybrid_vehicles_and_the_concept_of_V2G

Table 11 Profile of the AEP GridSMART Demonstration Project

Item	Number
Residents	100,000
Employees in commercial and industrial fields	10,000
Peak load:	
Summer	800 MW
Winter	650 MW
Total electricity sold:	3.5 million MWh
To residential users	1.2 million MWh
To industrial and commercial users	1 million MWh
Total number of substations	16
Total number of power distribution lines	80
Total length of power distribution lines	3,000 miles
Total length of power transmission lines	0

AEP Ohio, Final Technical Report, 2014, p. 6

consumption data acquisition system, intelligent community/building, EV charging facilities, intelligent service, and a communications and information network (Fig. 40).

Different types of distributed energy are connected to the grid, including 40 MW of solar power, 10 MW of biomass power, 125 MW of wind power and 1.5 MW of gas-fired combined

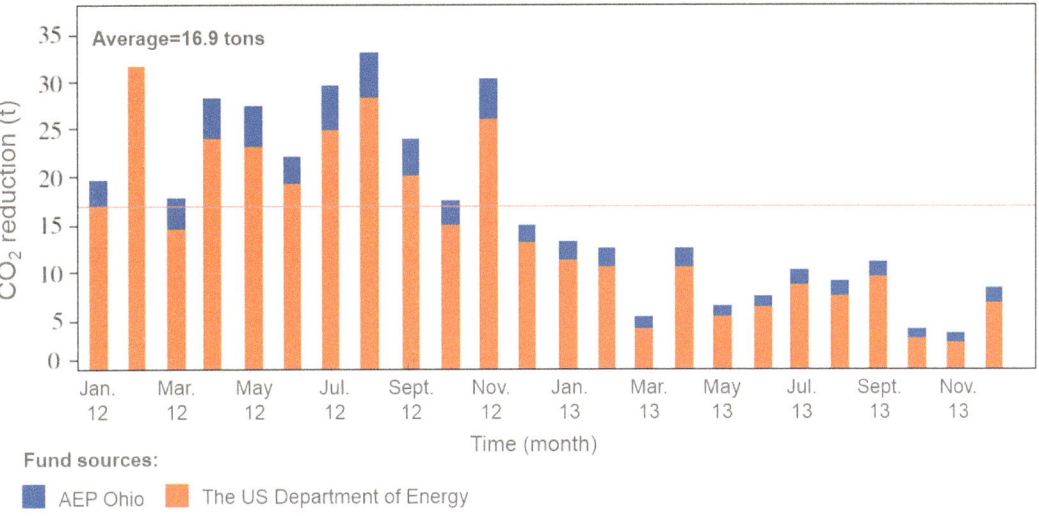

Fig. 36 CO_2 reduction after deployment of AMI. *Source* AEP Ohio, Final Technical Report, 2014, p. 35

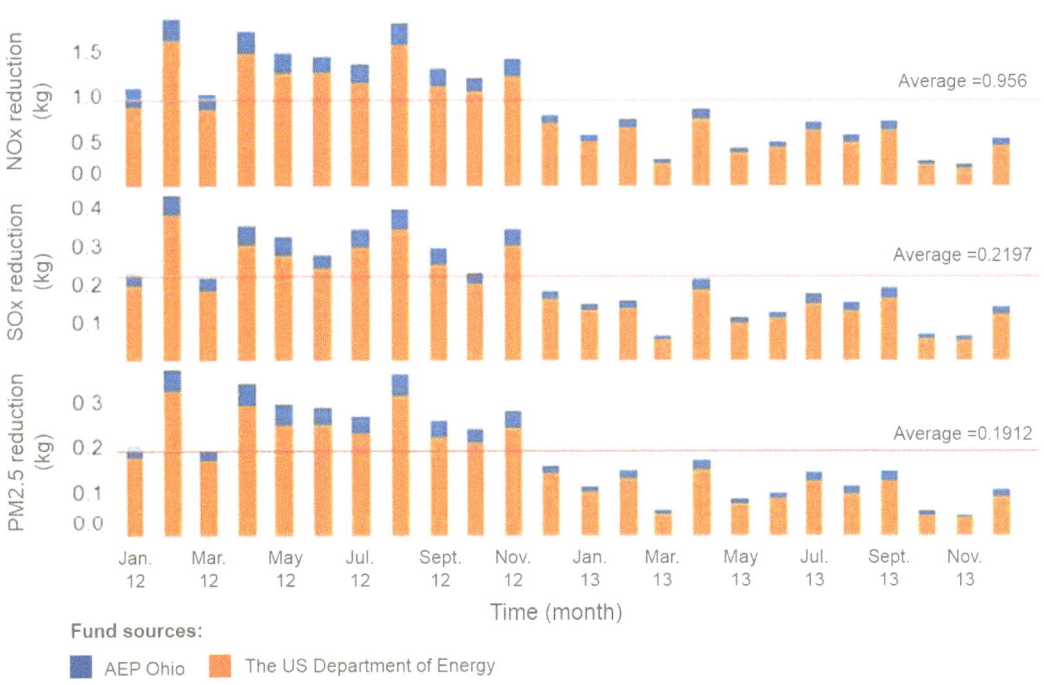

Fig. 37 Reduction of air pollutants after deployment of AMI. *Source* AEP Ohio, Final Technical Report, 2014, p. 39

cooling, heat and power. The microgrid is supplied by distributed power from a 30 kWp solar power system and 6 kW wind turbines. The energy storage system is a 15 kW × 4 h lithium-ion battery. Microgrid loads include 10 kW of lighting and 5 kW of EV charging piles. Intelligent control of the grid is realised through a microgrid energy management system. Thanks to the deployment of technologies like power distribution automation, equipment monitoring system, intelligent

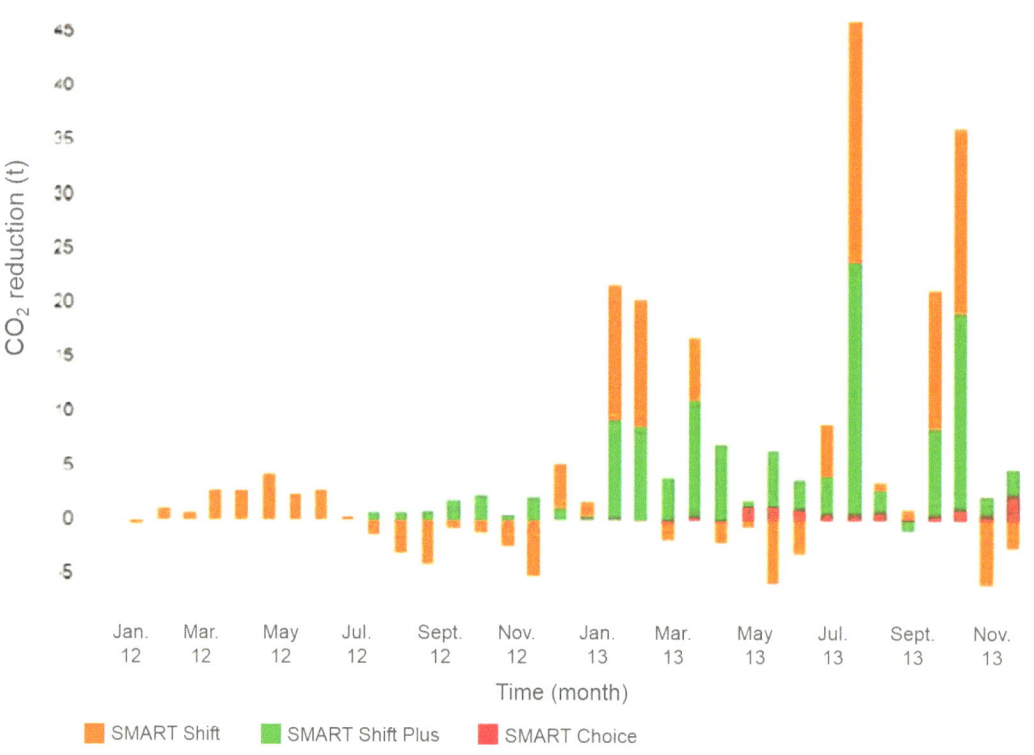

Fig. 38 Monthly increase/reduction of CO_2 emissions with SMART Shift, SMART Shift Plus and SMART Choice. *Source* AEP Ohio, "Final Technical Report, 2014, p. 115

scheduling and smart substations, the eco-city's power supply reliability, voltage qualified rate and N-1 pass rate stand at 99.999%, 100% and 100% respectively, and the overall line loss is reduced by 1.18%, improving energy supply reliability.[36]

The Sino-Singapore Tianjin Eco-City Smart Grid demonstration project aims to verify new smart grid technologies, specifications, equipment performance and to test smart grid technologies in a comprehensive way. Its economic benefits include lower investment costs, reduced line losses, improved power supply reliability, lower operation and maintenance costs and enhanced operating efficiency (Fig. 41). The project saves 1,074 tonnes of fuel oil and 5,929 tonnes of standard coal equivalent (SCE) per

year and reduces CO_2 emissions by 18,488 tonnes annually.[37]

3. Henan University of Economics and Law microgrid project

The project is located at the university campus. The microgrid comprises one 380 kW solar photovoltaic (PV) system and a 2×100 kW/100 kWh energy storage system. It supplies seven dormitory buildings. The microgrid controls the power distribution system for the buildings and canteens in Power Distribution Zone IV at the university, including two energy storage systems and 32 low voltage power distribution lines. It also communicates with the power scheduling system. During operation, the

---

[36]Full Record of Smart Eco-city Part of deployment: Overview of the Sino-Singapore Tianjin Eco-City Smart Grid Demonstration Project. http://www.sgcc.com.cn/ztzl/newzndw/sdsf/09/254912.shtml (2011).

[37]State Grid Energy Research Institute Co. Ltd., Analysis Report on the Development of Smart Grids in and outside China. Beijing: China Electric Power Press, 2013, pp. 103–106.

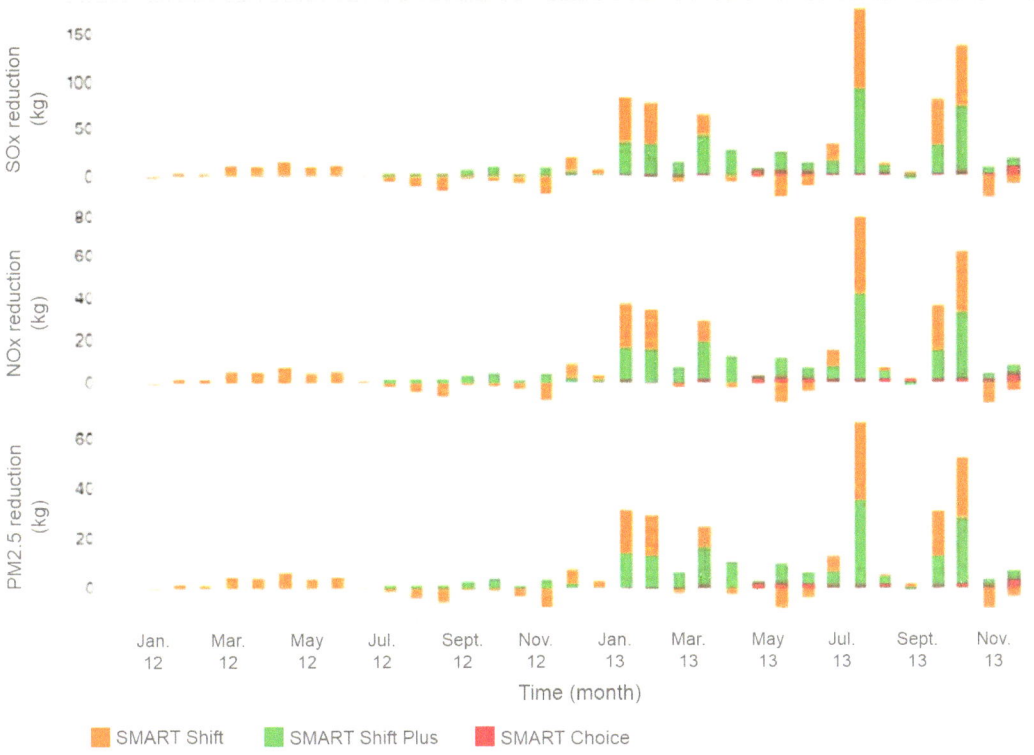

Fig. 39 Monthly increase/reduction of SO_x, NO_x and PM2.5 with SMART Shift, SMART Shift Plus and SMART Choice. *Source* AEP Ohio, Final Technical Report, 2014, pp. 118–119

microgrid minimises power consumption from the public grid by storing excess energy from the PV system when output is greater than demand, and by releasing that stored power into the microgrid during peak hours.[38]

Taking into account annual sunshine hours at Henan University, PV power generation from the microgrid could supply about 880,000 kWh of electricity per year, helping the university save RMB 492,800 annually. If its service life is 20 years,[39] the cumulative power output will be 20 GWh, which means a saving of RMB 11.2 million in the university's energy spend. Calculations

based on the demonstration results show that the project will generate 2.336 GWh of electricity per year, amounting to 58.40 GWh in 25 years. This means direct economic benefits of RMB 32.704 million. Compared to thermal power, the microgrid could save about 21,030 tonnes of SCE and avoid 39,960 tonnes of CO_2, 684 tonnes of SO_2, 615 tonnes of NO_x, 975 tonnes of dust and 19,565 tonnes of ash.[40]

6.1.3 Outlook

The smart grid is an important enabler of the energy revolution. Statistics from the International Energy Agency[41] show that energy conservation and energy efficiency improvements enabled by the smart grid can reduce CO_2

---

[38]Successful joint debugging of the first distributed PV power generation and grid operation and control pilot project in Henan, http://news.163.com/11/0214/09/6SRI3GBM00014AED.html.

[39]Successful joint debugging of the first distributed PV power generation and grid operation and control pilot project in Henan, http://news.163.com/11/0214/09/6SRI3GBM00014AED.html.

[40]Successful joint debugging of the first distributed PV power generation and grid operation and control pilot project in Henan, http://news.163.com/11/0214/09/6SRI3GBM00014AED.html.

[41]IEA, Technology Roadmaps: Smart Grids, 2011, p. 28.

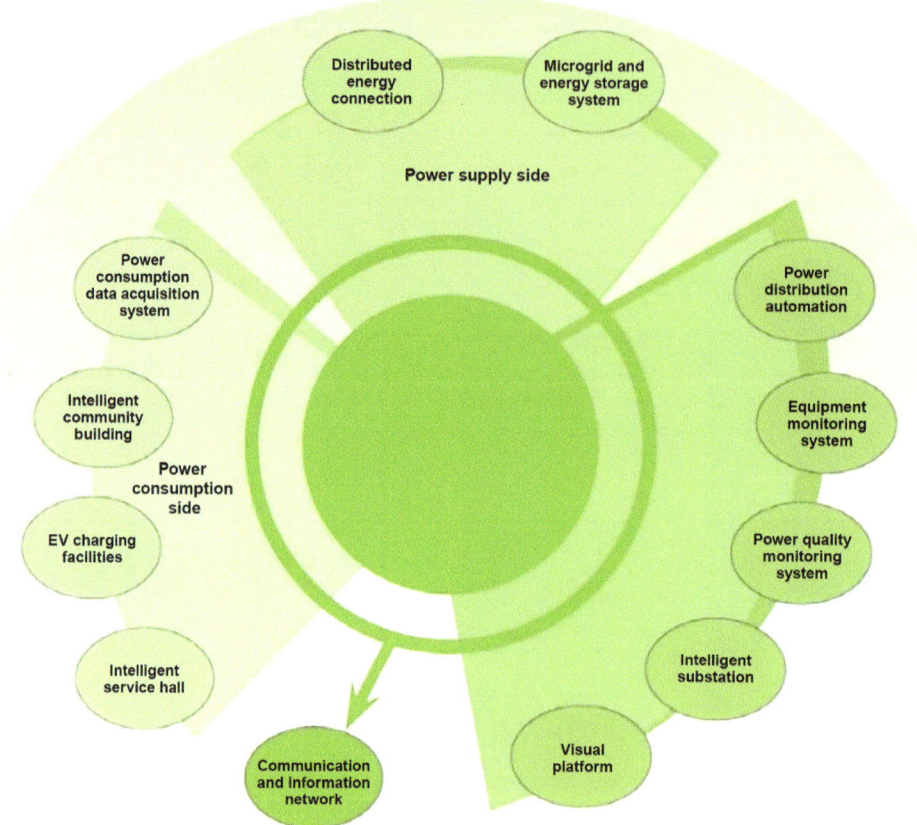

Fig. 40 Architecture of Sino-Singapore Tianjin Eco-City Smart Grid demonstration project. *Source* State Grid Energy Research Institute Co. Ltd., Analysis Report on the Development of Smart Grids in and outside China. Beijing: China Electric Power Press, 2013, pp. 103–106

emissions by more than 840 million tonnes by 2030, and renewable energy and EV charging can reduce CO_2 emissions by more than 3 million tonnes. As exploration deepens, technology innovation will expand the scope and forms of the smart grid.

First, the smart grid will continue to grow at high speed. Global power demand in 2030 will be double that of 2000.[42] To meet increasing demand, more investment will be needed in smart grids. According to the International Energy Agency,[43] China will invest at least $96 billion in the smart grid by 2020, and the world will invest $2 trillion by 2030.[44]

Second, the share of new energy in the energy mix will be higher. Climate change is a global concern. Most countries base their energy strategies on greenhouse gas emission targets, using those targets to calculate the share of new energy needed. The Institute of Electrical and Electronics Engineers (IEEE) predicts that by 2030, wind and solar will account for more than a third of global power supply.[45] The EU has set the following targets for 2030: the share of renewable energy will be at least 32%, the

[42]IEA, Technology Roadmaps: Smart Grids, 2011, p. 26.
[43]IEA, Technology Roadmaps: Smart Grids, 2011, p. 21.

[44]$2 Trillion will be Invested in the Global Smart Grid Market by 2030, http://smartgrids.ofweek.com/2012-01/ART-290010-8470-28595501.html.
[45]IEEE, IEEE Vision for SG 2030, 2013, pp. 80–81.

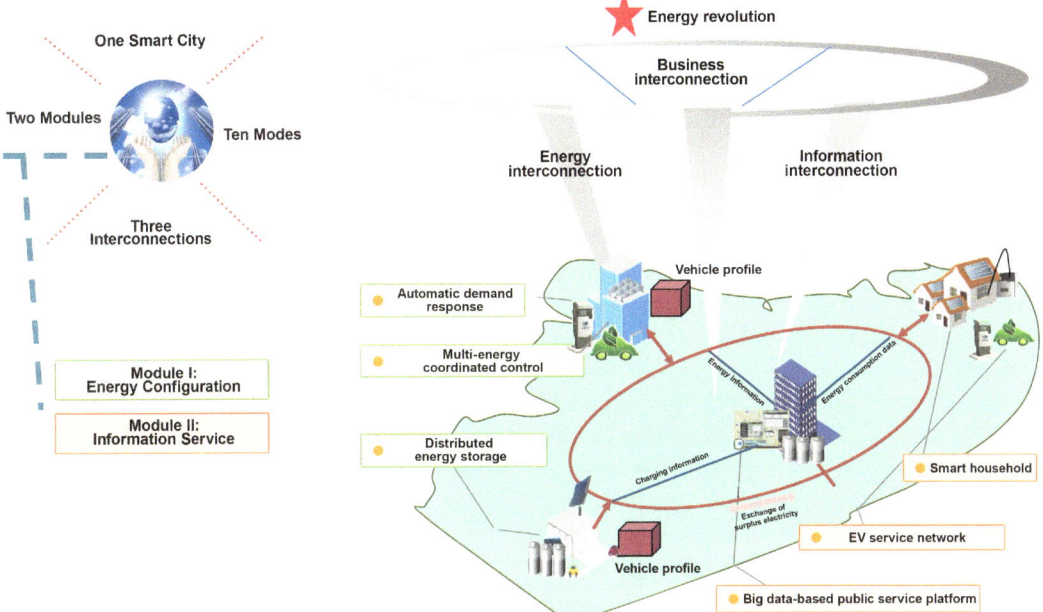

Fig. 41 Sino-Singapore Tianjin Eco-City Smart Grid demonstration project. *Source* Full Record of Smart Eco-City Part of deployment: Overview of the Sino-Singapore Tianjin Eco-City Smart Grid Demonstration Project, http://www.sgcc.com.cn/ztzl/newzndw/sdsf/09/254912. shtml, (2011)

improvement in energy efficiency will be at least 32.5%, and greenhouse gas emissions will be at least 40% lower than in 1990. The Energy Innovation Strategies unveiled by the Ministry of Economy, Trade and Industry of Japan aim to increase the share of renewable energy to 22–24% by 2030 and reduce greenhouse gas emissions by 26% (compared with 2013). Thanks to continuous innovation, China expects to make breakthroughs in wind, solar, biomass, geothermal and marine energy technologies in terms of efficiency, costs and flexibility. Distributed energy systems, comprising multiple types of energy—wind, solar, energy storage and small gas turbines—will be widely deployed and efficiently integrated. This will help China achieve its goal of increasing the share of non-fossil energy in primary energy consumption to around 20% by 2030.

Third, large-scale grid connection of intermittent energy makes customer requirements increasingly demanding and grid operation more complex. To improve grid visibility and control, IT and automation technologies will be widely applied, making grid operation and control more intelligent. IEEE holds the view that[46] full integration of grid and information and communications technologies (ICT) will improve grid monitoring speed by about 1,000 times. ICT-related equipment will become a major component of modern grid investment. Global grid modernisation needs around $6.9 trillion invested by 2030, of which about $1.7 trillion (one-fourth of the total) will be invested in ICT-related equipment. By 2030, China's smart grid will achieve deep integration of information and power flows and become a highly integrated information and physical network. Power system coordination and control of energy source, grid and load will become more intelligent and efficient, and operation safer and more stable. Low-cost and secure quantum communications technologies will become an R&D priority to ensure information security in the smart grid.

Fourth, two-way interaction between supply and demand will improve. Deep integration

---

[46]IEEE, IEEE Vision for SG 2030, 2013, p. 58.

between the grid, big data, cloud computing, the Internet of things, mobile Internet and Internet + (the Energy Internet) offers huge opportunities to enrich power services. By innovating new services and creating win-win business ecosystems, power utilities can smooth load volatility and reduce grid investment. According to Power Perspectives 2030,[47] demand response will account for 10% of daily load in the EU by 2030, reducing demand for grid and backup capacity by 10% and 35% respectively, thereby saving EUR 7 billion and EUR 25 billion. The International Energy Agency[48] forecasts that more than 20 million battery electric vehicles and hybrid EVs will be sold worldwide by 2030. The Japanese government, through its Japan Revitalisation Strategy, plans to increase the proportion of new energy vehicles in Japan to 50–70% by 2030. EV upstream and downstream industry chains in China see continuous growth and breakthroughs in battery manufacturing and battery charging and swapping in the coming years. EVs will be faster and easier to charge. High-density and low-cost energy storage technologies will be innovated, driving commercial operation of large-capacity distributed energy storage systems and contributing to peak shaving and grid efficiency.

Fifth, the smart grid can conserve energy and make energy use more efficient, delivering significant economic and environmental benefits. According to the All-of-the-Above Energy Strategy as a Path to Sustainable Economic Growth,[49] if the USA achieves its energy efficiency improvement target of more than 70% it will reduce CO_2 emissions by 3 billion tonnes in 2013–30 and save hundreds of billions of dollars in investment. Research by the EU[50] shows that

energy efficiency improvements will reduce its load demand by 14% by 2030. This will lower demand for grid and backup capacity by 55% and 31% respectively, saving about EUR 299 billion in investment. China needs to improve energy use efficiency. Energy efficiency technologies will become more integrated and intelligent. Industry, buildings and transport will be priority fields for innovation. Integrated energy use technologies, including cascading, will evolve in the future. Energy monitoring and measurement will be more accurate, substantially improving energy system use efficiency.

6.1.4 Problems and Suggestions

(1) Problems

First, creating the smart grid is a huge, wide-ranging project. It involves social, economic, policy, regulation and science and technology aspects, as well as numerous stakeholders —government, power system users, equipment and service suppliers, financial institutions, research institutions and consultancies. The interests and requirements of these stakeholders vary greatly. Existing management, pricing, investment and financing mechanisms need to be improved, and the various stakeholder interests taken into consideration and balanced.

Second, breakthroughs need to be made in core technologies. China has mastered and locally manufactured some key smart grid technologies and equipment, but it still relies on imports, especially in the fields of direct current, power electronics and renewable energy. There is much room for performance improvement and cost reductions in key technologies and equipment. Moreover, industry and national standards for the smart grid have yet to be developed, putting China behind in terms of international smart grid standards.

Third, business models need to be improved. Successful business models can move smart grid development forward and demonstrate the benefits. Many new and value-added businesses are starting up in intelligent EV charging, coordinated operation of renewable power and energy

[47]European Climate Foundation, Power Perspectives 2030: On the Road to a Decarbonized Power Sector, 2015, p. 11.

[48]IEA, Technology Roadmaps: Smart grids, 2011, p. 12.

[49]Executive Office of the President of the United Stated, The All-of-the-Above Energy Strategy as a Path to Sustainable Economic Growth, 2014, p. 8.

[50]European Climate Foundation, Power Perspectives 2030—On the Road to a Decarbonized Power Sector, 2015, p. 53.

storage systems, one-stop services for intelligent electricity use, and demand response. However, they are not yet mature, and their profitability is weak, slowing down their commercialisation.

(2) **Suggestions**

First, government should lead smart grid innovation. Plans should be made to motivate power utilities, equipment manufacturers, users and other market players to collaborate on smart grid development and achieve win-win results. Finance and taxation, science and technology, and the support policies necessary to create the smart grid should be researched and implemented. A scientific electricity pricing system that reflects key aspects of the power market should be established, including ancillary services like frequency regulation and peak shaving. International communication and cooperation should be strengthened to help smart grid technologies, equipment and standards go global.

Second, R&D of key smart grid technologies should be improved. Attention should be paid to big data, cloud computing, the Internet of things, mobile Internet, artificial intelligence and other new technologies. Investment in scientific research should be made at an early stage to encourage innovation. New theories, methods and technologies should be explored. A complete and open system of domestic smart grid technical standards should be drawn up. International cooperation on smart grid standards should be bolstered, and businesses and research institutions encouraged to participate in defining a body of international smart grid standards.

Third, smart grid business models should be created. The decisive role of the market in resource allocation should be brought into full play, and industry alliances should be established to facilitate development of consistent technology and product standards. Smart grid development and operation should benefit businesses and users to drive innovation and change in the power services industry. Technologies and approaches that combine the Internet and energy should be explored to facilitate innovation in smart grid business models.

6.2 New Energy Technologies

6.2.1 Developments in, and Outlook for, New Energy Technologies

(1) **New energy technologies**

There is no single definition of new energy at present. Generally, new energy refers to the new energy sources and technologies that differentiate it from conventional (old) energy. Broadly speaking, new energy means new sources and technologies, including those for energy development, conversion, use and support. In a narrow sense, new energy refers to sources only, including non-hydro renewables, unconventional and future energy. For the purpose of this report, new energy refers to the latter, especially non-hydro renewable energy.

Non-hydro renewable energy includes solar, wind, biomass, geothermal and marine energy. Power generation is the principal way to use non-hydro renewable energy. Non-hydro renewables can be used by consumers after conversion into power. Global installed capacity of non-hydro renewable energy is increasing in step with technical progress.

Wind power: Wind can be divided into onshore and offshore generation. Onshore wind turbine manufacturing technologies have already matured. As onshore wind has reached saturation point in some European countries, wind power generation has moved gradually from onshore to offshore, and from offshore to deep water. Wind energy is now the most mature (technically) and promising (in terms of large-scale development and commercialisation) new energy.

Solar power: Solar can be divided into solar heating and solar power generation. Solar water heating and solar photovoltaic (PV) power generation are the most mature technologies at present. Solar PV converts sunlight directly into power through photovoltaic panels. Concentrated solar power (CSP) technology produces electricity by focusing solar radiation onto heat absorbers with light concentrating technology. The heat is then used to convert water into high-pressure steam to power steam turbines.

Biomass energy: As a renewable energy source that contains carbon, biomass can be used to generate power, or heat (gas) or produce liquid or gaseous fuels or used directly or indirectly for materials. Biomass use spans hybrid systems, where it is used to partially substitute fossil fuels in existing assets, for example in power station cofiring; blended biofuels; and dedicated uses such as biomass-fuelled combined heat and power.

Geothermal energy: Geothermal energy is used to generate heat and power. Shallow geothermal and hydrothermal energy heating and cooling technologies are basically mature. Shallow geothermal energy is used mainly with heat pumps. In geothermal power generation, high-temperature dry steam power is the most mature and cheapest technology, followed by high temperature wet steam. The cost efficiency of low- and medium-temperature geothermal power generation technologies needs to be improved.

Marine energy: Marine energy is used mainly to generate power in tidal, wave, current, temperature difference and salinity gradient applications. The development cost is generally high. Tidal power generation is the most mature and competitive.

(2) **Global deployment of new energy technologies**

New energy technologies (NET) can be classified in four stages according to their technological maturity. NET at the commercialisation stage refers to those technologies with great potential that have been widely accepted, such as wind and solar photovoltaic. NET at the demonstration stage refers to those that are proven in industrial-scale demonstration projects and are about to be commercialised, but which are faced by uncertainties in terms of go-to-market, such as concentrated solar power. NET at the quasi-demonstration stage refers to those that have reached, or are about to reach, industrial demonstration, like second-generation biofuels. NET that have yet to be verified refers to those whose potential is recognised but that still have a long way to go to be commercialised, such as, nuclear fusion and combustible ice.

Wind and solar photovoltaic (PV) are deployed at scale and therefore promising. Wind and PV power are characterised by: (i) predictable cost—with technical progress and scale-up, their cost will be comparable with conventional power in the future; (ii) clear and great resource potential; and (iii) the ability to be scaled up because of the vast size of China. Wind and PV power generation require little water and have very limited impact on the environment.

Considering the maturity, resource potential and development cost of different new energy technologies, this study focuses on wind, solar PV and concentrated solar power (CSP). In wind power generation, large-capacity, low-speed wind turbines will be the future trend. Breakthroughs continue to be made in solar power technologies and efficiency.

Wind turbine capacity, height and output continue to increase across the globe. Developed countries made great breakthroughs in wind turbines in the 1980s, with capacity reaching 75 kW and the hub height 20 m. In the 1990s, turbine capacity rose to 300–750 kW and hub height to about 30–60 m. These wind turbines dominated medium and large wind farms. In the 21st century, to generate more power and use land more effectively, turbine capacity increased to several megawatts and hub height to 70–100 m.

Lightweight high-tower low-velocity wind turbines predominate. High tower is the key to low wind speeds and high-shear wind farms. Vestas is now the leader in the field of lightweight high-tower wind turbines. Its all-steel 120 m towers weigh only 226 tons, 30% lighter than other all-steel or steel-concrete towers. Goldwind, GE and Siemens' wind turbines with 120 m towers are mostly prototypes.[51]

Solar PV technologies are divided into crystalline silicon, thin-film and new cell technologies according to the cell material and manufacturing process used.

1. *Crystalline silicon cell technology.* This is the mainstream technology at present. It includes monocrystalline and polycrystalline silicon. Crystalline silicon cell manufacturing is becoming more and more diversified. Aluminium back surface field monocrystalline and polycrystalline silicon cells are manufactured at scale, and their average conversion efficiency has reached 19.8% and 18.5% respectively. The conversion efficiency of the monocrystalline and polycrystalline silicon cells using passivated emitter rear cell (PERC) technology is about 0.5 percentage points higher. With rapid technological progress, N-type crystalline silicon cells have entered small-scale production. The conversion efficiency of monocrystalline and polycrystalline silicon cells, with a new structure and using new technologies, could be improved substantially in the future. N-type crystalline silicon cells using passivated emitter rear totally diffused (PERT) technology, heterojunction with intrinsic thin layer cells (HIT), interdigitated back contact (IBC) solar cells and other back contact cells will be the future trends.

2. *Thin-film cell technology.* Industrial-scale thin-film cell technologies have gradually matured and have bright prospects. Thin-film cells include silicon-based thin film, copper indium gallium selenide (CIGS), cadmium telluride (CdTe) and gallium arsenide (GaAs). The innovation space of silicon-based thin film cell technology is limited, with market

share decreasing in recent years. Currently, CdTe and CIGS thin-film cells are mainstream, with the highest conversion efficiency in laboratory testing exceeding 22%. Mass production of GaAs cells has not been achieved due to high costs.

CSP technologies are divided into trough, tower, dish-Stirling and linear Fresnel.

Trough CSP. Trough is the first CSP technology to be commercialised and has the largest share of commercial CSP plants globally. The current status of this technology is: (i) trough CSP has a simple design and is low cost; (ii) multiple concentrating heat collectors (troughs) can be connected in series or in parallel to form a large-capacity CSP system; (iii) its concentration ratio is low, typically 50–80 suns, and it is difficult to increase the temperature of its heat transfer medium, which is usually around 400 °C; and (iv) due to the long heat transfer loop, trough CSP loses large amounts of heat and has a low system efficiency of about 11–15%.

Tower CSP. The current status of this technology is: (i) a tower CSP system has a high concentration ratio of 300–1,000 suns and a high system operating temperature of 500–1,400°C; (ii) thanks to its short heat transfer loop, tower CSP systems have small heat loss and a high system efficiency of about 14%; (iii) tower CSP is suitable for large-scale and large-capacity commercial applications; and (iv) the system is costly and requires heavy capital investment, and its design and control system are complicated.

Dish Stirling CSP. The current status of this technology is: (i) it has a high concentration ratio of around 1,000–3,000 suns and an operating temperature close to 1,000°C. Its peak conversion net efficiency can reach 30%; (ii) the capacity of a dish Stirling system is usually 5–50 kW, with unit costs high; and (iii) the cost of generating power does not depend on the size of the project, as the plant can be used as a distributed power system or a megawatt-level power station connected to the grid.

Linear Fresnel CSP. Linear Fresnel is a simplified version of trough CSP. The current status of this technology is (i) it uses flat tracking

---

[51]Stage Grid Energy Research Institute Co. Ltd., Analysis Report on Power Generation with New Energy in China, 2017, pp. 36–39.

mirrors instead of the parabolic-shaped collectors in trough CSP. The mirrors are close to the ground, have a low wind load, a compact simple structure, and high land use efficiency; (ii) the heat transfer tubes do not need to be a vacuum, which simplifies the design and lowers the overall cost of the system; and (iii) the system's concentration ratio, operating temperature and system efficiency are low.

Currently, trough CSP technology is mature; tower CSP technology is approaching an early stage of maturity and has great potential; Fresnel and dish technologies are still at the demonstration stage. There is still much to do in CSP technologies.

(3) Trends in related technologies

First, wind turbines will use longer and lighter blades of modular design, thereby improving power generation efficiency. The number of direct drive (gearless) wind turbines is expected to increase.

1. Wind turbines will use longer and lighter blades with modular design. As wind resource-rich regions with low wind velocity become hot spots for development, the swept area per kilowatt generated needs to be increased to capture more energy. Compared with the past decade, the rotor diameter and rated power of wind turbines have increased by 70% and 50–100% respectively. Currently, the average rotor diameter of low-velocity turbines is 116 m and is forecast to reach 160 m in the next decade, with swept area and annual availability expected to double. However, a new problem will then arise—ultra-long blades make transport and hoisting on challenging terrain difficult. Road construction, piling and hoisting costs will get higher and safety risks could increase. Modular blade technology simplifies production, ensures product quality and facilitates transport and erection.

2. The proportion of direct drive (gearless) wind turbines is expected to rise. Direct drive (including excited and permanent magnet direct drive) wind turbines is a hot area of research. Germany-based Enercon and other vendors using excited direct drive systems have around an 8% share of the global market. These turbines feature stable performance and mature technology. Permanent magnet direct drive wind turbines have no gearbox and avoid some of the mechanical faults associated with gearboxes. The magnet remains stable, withstanding vibrations and temperature variations.

Second, improved conversion efficiency and lower cell production costs characterise solar PV technologies.

1. PV cell conversion efficiency is improving. The conversion efficiency of monocrystalline silicon cells is currently about 19%, while that of third-generation polycrystalline silicon cells is around 18%. New technologies will improve crystalline silicon cell conversion efficiency over time. Passivated emitter rear cell (PERC) technology improves cell conversion efficiency by adding a dielectric passivation layer to the back of the cells.

2. Cell production costs are falling. Fierce competition is forcing cell manufacturers to lower production costs in several ways. The first method is to improve conversion efficiency through technical progress, primarily through metal wrap through (MWT) and interdigitated back contact (IBC) technologies. Experience shows that when cell conversion efficiency is improved by 1%, production costs are lowered by 7%. The second method is to reduce material consumption and thus costs. The cost of cell processing comes mainly from the slurry. It is difficult to lower the cost of slurry because it contains silver and other commodities. Therefore, manufacturers choose to reduce

slurry consumption per unit with various techniques. The third way is to decrease the thickness of the silicon wafers. Over the past three decades, silicon wafer thickness has been reduced from 450–500 μm in the1970s to 180–200 μm today, lowering cell production costs by more than half.

Third, integrated solar PV power plants will become larger. Building-integrated PV systems are already widely deployed and the number of off-grid PV systems will grow. PV power plants with a capacity of more than 1 GW are already under construction. Building-integrated PV systems provide numerous benefits, including small footprint, lower investment costs, low transmission losses, low aesthetic impact and high energy efficiency. They are often located in load centres and easy to integrate with the local grid. Off-grid PV systems will be more widely deployed in remote regions without access to electricity.

Fourth, large-scale CSP technology will shift gradually from trough to tower and other technologies with a high concentration ratio and high conversion efficiency.

There is much scope for the future development of CSP technologies. Better efficiency and improved cost effectiveness are two examples of areas of focus. Tower CSP technology, with its high concentration ratio, large system capacity and high efficiency, is a hot area of R&D at present. Tower CSP is expected to become the main CSP technology, enabling the large-scale development of CSP across the globe. Dish Stirling technology also has a high efficiency level and will probably be used in distributed power systems in the future.

Large capacity and low heat storage costs is the way to improve CSP system efficiency. R&D currently focuses on: (i) improving power generation efficiency by increasing the system operating temperature or expanding plant capacity or reducing heat loss from the heat absorbers; (ii) reducing solar island costs by lowering equipment expenditure and optimising design; (iii) using high-capacity heat storage to ensure 24/7 electricity supply and meet grid

requirements; and (iv) reducing plant energy and water consumption.

6.2.2 Current Developments and Difficulties in China's New Energy Sector

(1) **Development environment for China's new energy technologies**

First, China's installed capacity of new energy increased substantially to 237,720 MW in 2016, accounting for 14% of the country's total. In 16 provinces, new energy has become the second largest energy form. The grid-connected capacity of wind, solar and biomass reached 148,640 MW, 77,420 MW and 11,660 MW respectively, accounting for 62%, 33% and 5% of the total installed capacity of integrated new energy.[52] Newly added installed capacity of solar PV exceeded that of wind power for the first time and contributed half of new PV installed capacity worldwide. China passed the USA to become the leader in wind power output for the first time. China has ranked first as the country that adds the most wind and solar PV installed capacity annually for many years.

Second, the technical standards for new energy are gradually being developed. The National Energy Administration founded the Wind Power Technical Committee for Standardisation in the Energy Industry in 2011 and issued the Framework of Wind Power Standards System,[53] which covers the following aspects: wind farm planning, design, construction, installation, operation, maintenance and management; wind power connection management technologies; wind machinery and equipment, and wind power electrical equipment. In 2014, the Standardisation Administration of China

---

[52]White Paper of State Grid Corporation of China on New Energy Development 2017, pp. 4–11.

[53]Notice from NEA on Printing and Distributing the Rules on Development of Wind Power Standards, Articles of Association of Wind Power Technical Committee for Standardisation in the Energy Industry and Framework of Wind Power Standards System (Guo Neng Ke Ji, 2010, No. 162).

issued seven national standards on solar power generation, which cover quality and performance testing of solar cells.

(2) **Current development of China's new energy technologies**

First, wind power technologies. China has localised megawatt-level wind turbine design and manufacturing technology in place for the whole wind power value chain. Several Chinese wind power businesses are among the world's top 10. The technical level and reliability of Chinese wind power equipment is world-class. Up to now, China has focused on the design and manufacture of wind turbines and turbine components below 3.6 MW but is now developing technologies for turbines of 5 and 6 MW.

The rotor diameter of low-velocity wind turbines continues to increase. The regions with low wind velocity in central and east China have become new hotspots for wind power development. Large-scale development of wind parks in those regions needs low-velocity wind turbine technology. A range of ultralow-velocity wind turbines was launched in 2013. Low-velocity wind areas account for more than 60% of China's exploitable wind resources.

Second, solar PV technologies. A world-leading complete PV value chain covering polycrystalline silicon purification, silicon rods, ingots and wafers, as well as cells, panels and system integration, has taken shape in China and is expanding rapidly. PV power generation has become a strategic, emerging and internationally competitive industry for China. In 2015, China had 16 polycrystalline silicon companies, with a total production capacity of 190,000 tonnes (excluding metallurgy)—they produced 165,000 tonnes of polycrystalline silicon, which was 47.8% of global output. Four of them are in the global top 10 in terms of production capacity. China's total silicon wafer production capacity in 2015 was 64.3 GW, with output reaching 48 GW, 26.3% higher than the previous year and 79.6% of the global total. Of the 10 largest silicon wafer manufacturers in the world, nine are in mainland China. China's crystalline silicon cell production capacity in 2015 was 49 GW, with output 41 GW, 24.2% higher than the previous year and about 66% of the global total. Among the top 10 cell manufacturers (in terms of output) in the world, seven are in mainland China. Total solar panel production capacity in China in 2015 was more than 71 GW, and output 45.8 GW. Among the top 10 panel manufacturers (in terms of output) in the world, six are in mainland China.[54]

Third, CSP technologies. CSP plants generally consist of concentration, heat absorption, a thermodynamic power cycle, power generation and heat storage systems. Key equipment includes concentrating mirrors, heat collectors and a heat storage system. Technology development in concentrating mirrors is focused on improving the durability and precision of the reflective materials. Heat collectors influence heat absorption efficiency and include evacuated collector tubes and chamber absorbers. The heat storage system—which comprises heat transfer fluid, molten salt and metal materials—is the key to large-scale and uninterrupted operation of a CSP system.

Generally, more than 90% of the equipment and materials used by CSP plants are manufactured in China and at a technical level very close to world-class. China lacks only long-term operational experience. However, core components and materials like trough ball joints, tower CSP tubes, high-temperature molten salt and molten salt pumps still need to be imported.

(3) **Problems with China's new energy technologies**

First, basic research on wind power is insufficient in China. Key equipment and materials are still imported. China's route to progress in wind power technologies is through introduction, digestion, absorption, integration and innovation.

---

[54]China Photovoltaic Industry Association, Roadmap for China's PV Industry, 2016, pp. 2–4.

China still lags behind developed countries in terms of calculating aerodynamics and load, flow field analysis and other basic R&D fields, as well as large wind turbine design, bearings, the main control and pitch control systems, and other high-end technologies. For example, leading overseas wind turbine manufacturers have commercialised 4–7 MW wind turbines. Prototypes of 8 MW wind turbines are at the installation and testing stage. Wind turbine companies in Europe and North America are already designing 10 MW turbines and are exploring and researching 20 MW models. There is a gap between China and those countries.

Second, solar PV technologies are facing severe challenges to their leadership in the latest cell technologies. Interdigitated back contact (IBC), heterojunction with intrinsic thin layer (HIT), passivated emitter and rear cell (PERC), metal wrap through (MWT), bifacial and other efficient cell technologies are developing rapidly across the world and about to be commercialised. But there is still a gap between China and the developed world, especially in IBC and HIT cells.

Third, the cost effectiveness of CSP technologies needs to be improved. Practical experience is insufficient and technical standards are incomplete. China's CSP technologies are now at the stage of pilot demonstration. Equipment manufacturing and project development technologies are not mature. There is much to be done in technological development. The investment cost per unit of capacity is still high. CSP-related equipment manufacturing, design, construction, operation and maintenance standards are relatively undeveloped. Experience needs to be accumulated through project construction and operation.[55] Even through China

has introduced support policies, only a few trough and tower CSP power plants are in operation or under construction.

(4) **Outlook for China's new energy technologies**

First, the outlook for China's wind power technologies. The cost of wind power will further decrease to a level below that of conventional power by 2030. China's wind power industry will maintain strong momentum, but its growth will slow down. Wind power installed capacity will exceed 450 GW by 2030. Lighter and longer blades, integrated drive chains and taller, lighter and easier-to-install towers will be developed. The average cost per kWh of wind power will decrease by more than 35% to RMB 0.35/kWh by 2030.

Second, the outlook for China's solar PV power technologies. The cost of solar PV will fall to a level lower than conventional power by 2030, and installed capacity will increase. China's solar PV industry will maintain strong momentum. PV installed capacity will exceed 400 GW by 2030. The conversion efficiency of monocrystalline and polycrystalline silicon cells in commercial applications will reach 25% and 21% respectively, and that of thin film cells will reach around 18%. The average cost per kWh of PV power will fall by more than 50% to RMB 0.31/kwh by 2030, making it more competitive than wind power.

Third, outlook for China's CSP technologies. Tower technology will mature by 2030 to lead the global CSP sector into the stage of large-scale development. China's CSP industry will see rapid development, with installed capacity expected to reach 30,000 MW by 2030. Air solid-state particle heat absorbers will be deployed gradually. Ceramic, solid-state concrete heat storage technologies will be developed, but molten salt heat storage will still be the mainstream technology for large CSP plants. The average cost per kWh of tower CSP power will decrease by more than 40% to RMB 0.42/kWh by 2030, getting closer to PV power and thus more competitive.

[55]Huang Qili, Zhang Zhengling, Zhang Ke, Li Qionghui and Huang Bibin, The wind power and solar power parts of the Report on Strategic and Emerging Industries of China, 2016.

6.2.3 The Role of New Energy in China's Energy Technology Revolution

First, new energy technologies have become an important means for China to ensure energy security and sustainable supply and overcome the challenge of climate change. New energy provides great support for the energy transition. Given the complexity of geopolitics, transnational energy flow will see great change. Energy security is at the heart of national energy strategies. Global climate change is a huge challenge. Onshore wind and solar energy resources are abundant, more than enough to meet global energy demand. With technological progress and the use of new materials, the efficiency of wind, solar, marine and other new energy sources will be improved. Their cost effectiveness and market competitiveness will be enhanced, making them the dominant energy types of the world.

Second, new energy technologies have become a strategic, emerging and internationally competitive industry in China. The energy technology revolution will shift energy development from resource-dependence to technology-dependence. Innovation in energy and power technologies has become a priority for all countries to sharpen their competitiveness and strive to become technology leader. New energy is an emerging industry with great potential, but it is also a strategic and internationally competitive sector for China.

Third, replacing fossil energy with new energy has become a strong driving force behind the fourth industrial revolution. Innovation in energy technologies, integration of information technologies, the Energy Internet and smart grids, energy storage, power system interconnections, plug-in and fuel cell electric vehicles, will all be pillars of the fourth industrial revolution.

6.2.4 Policies for Promoting New Energy Technologies

First, more effort should be invested in scientific research. This would enable breakthroughs in R&D and the design and manufacture of software and equipment, and it would enhance China's proprietary technologies and manufacturing level. National R&D centres for wind power technologies should be set up. Basic research in key wind power technologies should be carried out to help businesses tackle common technical difficulties. These should include: (i) R&D of advanced and large wind turbines, low-velocity turbines and other key wind power technologies, supported by renewable energy development funds and national scientific and technological research projects; (ii) national R&D and testing platforms for blades, drive systems and other wind power components should be built to improve the performance of wind turbines, wind farms and wind power equipment.

Second, innovation in PV technologies should be driven forward and R&D in key CSP technologies strengthened. China should: (i) increase innovation in PV technologies and continue to focus on improving the conversion efficiency and production cost of cells and panels. Solar power technologies should be digitalised to make plants more intelligent and harness the benefits of big data and cloud computing; (ii) strengthen R&D in key CSP technologies. More investment should be channelled into R&D of solar tracking controllers, plant control, manufacturing and system integration technologies. Research on mirrors and heat collector technologies, heat storage, heat transfer materials and thermal-to-power conversion systems suitable for west China should be carried out. Materials, technologies and processes should all be improved. Generating efficiency should be enhanced and costs lowered to enable China to become the largest producer and biggest user of CSP technologies.

Third, demonstration projects for CSP, marine and other new energy technologies with great potential in the short and medium terms should be built to accelerate technological progress and accumulate planning, design, construction, operation and management experience. This would also lay a solid foundation for policy studies, commercialisation, international competitiveness and cost reduction. These projects could explore different technical routes to find the best solutions for China's geographical and meteorological conditions.

Fourth, the development of new energy technologies should be closely coordinated with smart grid, energy storage and Internet + (Energy Internet) technologies. China should: (i) regard the smart grid as an important tool to promote renewable energy development and use and, therefore, a major component of the future power grid. The smart grid would facilitate the integration of renewable power, including distributed power and large-scale wind farms; (ii) use the Internet to revolutionise energy production and consumption, improve energy efficiency and reduce emissions. Information sharing between upstream and downstream businesses should be strengthened to coordinate operations between power plants and grids. Both non-fossil and fossil energy should be used to generate power; and (iii) energy storage is a key technology and an important means to integrate new energy in the power system, because it can mitigate voltage fluctuations and stabilise the grid. Research and projects in and outside China show that large-scale energy storage is an important solution to connect new energy to the grid and facilitate its use.

6.3 Energy Storage

6.3.1 Current Developments and Trends in Global Energy Storage Technologies

(1) **Current technology developments**

After more than a decade of development, energy storage technologies are beginning to take off. Statistics up to 2017 show that the cumulative installed capacity of global energy storage projects was 169.2 GW. Pumped storage and electrochemical energy storage accounted for the largest share of installed capacity at 97% and 1.3% (ranking No. 3) respectively. The installed capacity of electrochemical energy storage

projects worldwide was 94.4 MW, up 551% year-on-year and 50% month-on-month.

The UK, Australia, the USA and China markets see rapid growth. In 2017, the UK, China and Japan were the top three countries in terms of installed capacity. Nearly all these projects were deployed in grid-connected centralised renewable energy and ancillary services. Australia, the USA and the UK were the top three in terms of the installed capacity of projects planned or under construction. These projects would likewise mainly be deployed in grid-connected centralised renewable energy and ancillary services, accounting for 91% of the total capacity of such projects.

By application, up to 2017, the installed capacity for ancillary services was 31.5 MW, accounting for 33% of the total installed capacity of energy storage projects worldwide. These projects were located mainly in the UK, Germany and Belgium, for example, those in Bristol and Darlington in England. They became part of the European balancing market in the form of independent energy producers or in joint operation with gas power plants to provide primary frequency modulation.

In terms of current energy storage market capacity, pumped storage remains dominant (98% market share), although electrochemical energy storage is rapidly gaining market share. In the global market, the top three electrochemical energy storage technologies are lithium-ion, sodium-sulphur and lead-acid batteries, accounting for 53%, 29% and 9% of the total market respectively. In China, the top three electrochemical energy storage technologies are lithium-ion, lead-acid and flow batteries, which account for 57%, 28% and 10% of the total market respectively.

Mainstream energy storage technologies comprise four categories: physical, chemical, electromagnetic and others. Physical energy storage refers mainly to pumped storage, compressed air energy storage and flywheel energy storage. Chemical energy storage technologies are developing fast and attracting the most attention, mainly in lead-acid, lithium-ion, flow, and sodium-sulphur batteries. Electromagnetic energy storage includes super capacitors and superconducting magnetic energy storage. Others includes fuel cells and metal-air batteries. In addition, there are many technologies at the frontier stage of development. These can be divided into two categories—the first is improving or optimising conventional technologies like lithium-sulphur batteries and liquefied air energy storage; the second is designing and developing new technologies, such as lithium-air and aluminium ion batteries.

(2) Global development trends

Energy storage is about to become an important driver of change in the energy sector. The International Energy Agency (IEA) forecasts that the USA, Europe, China and India will increase their energy storage capacity for grid-connected electricity by 310 GW by 2050, at a cost of at least $380 billion. A study by McKinsey & Company says energy storage will play a game-changing role and have significant impact on the global economy by 2025. Its predicted market value is $0.1–0.6 trillion.

The USA, Japan and Europe have made national R&D plans for energy storage technologies. As a result, their technology development and demonstration activities are making rapid progress. Utilities like grid operators, large energy equipment manufacturers and some small- and medium-sized technology companies see a great future for, and are making inroads into, the energy storage market.

According to the IEA's Technology Roadmap: Energy Storage (2017), energy storage is of value for most energy systems, but the technologies vary greatly in terms of maturity. Currently, some small-scale energy storage systems are cost-competitive in remote communities and off-grid applications. Large-scale heat storage technologies are cost-competitive in heating and cooling applications. But more public support for research and development of energy storage technologies is needed.

The European Commission has introduced the EU Strategic Energy Technology (SET) Plan and materials roadmaps for energy applications and low-carbon energy technologies. They describe the research and innovation activities for materials critical to the development of 11 energy technologies (wind, photovoltaic, concentrated solar power, geothermal, electricity storage, power grids, bioenergy, novel materials for fossil fuels—including carbon capture and storage, hydrogen and fuel cells, nuclear fission and energy-efficient materials for buildings) over the next 10 years.

The Materials Roadmap Enabling Low Carbon Energy Technologies considers energy storage an important technology that can improve the controllability and flexibility of the European electricity system. Currently, most energy storage technologies are too costly and technically inadequate for system-level deployment and integration. Materials often restrict performance improvement, and are a decisive factor in the cost-effectiveness, efficiency and reliability of energy storage deployment in the grid. The commercialisation of large-scale energy storage technologies is a priority task.

The Electrical Energy Storage Roadmap describes an overall R&D plan for energy storage systems and technologies. For instance, low-cost, safe and sustainable electrochemical and electrolyte materials have super-electrochemical, thermal and mechanical properties, as well as a long life and the ability to withstand extreme conditions. They can be used to innovate design and manufacturing processes in lithium-ion and redox-flow batteries, pumped storage, compressed air energy storage (CAES), electrolytic capacitors, superconducting magnetic energy storage and flywheel energy storage.

The roadmap focuses on the development of new electrochemical pathways and on verifying emerging technologies like metal-air and solid-state batteries and liquid metal systems. It describes four pilot projects that demonstrate industry-scale, high-speed and low-cost deployment of electrical double-layer capacitors, lithium-ion batteries, flywheel rotors and motors, and compressor and dielectric materials resistant to high heat and pressure that are used in CAES heat storage containers. It also describes another five pilot projects for testing and verifying whether these advanced storage technologies are durable and can be reused in different market environments. The roadmap and pilot projects complement the formation of a pan-European network that pools industrial and scientific resources for research and innovation activities. In addition, the Roadmap also suggests establishing an education and training centre for electrochemical and energy storage.

6.3.2　Current Developments in Energy Storage in China

(1)　Policies related to energy storage in China

Since 2005, China's policies on energy storage have gradually evolved, from early-stage technical exploration to a technology roadmap, industry guidelines and support mechanisms. China's early polices focused on funding scientific research in battery materials and technologies and proof of concept. As technologies and applications progressed, the government introduced policies to reform the electricity system, incentivise energy storage and create support mechanisms.

Energy storage is part of the national energy strategy and industry development plan. The government has issued a series of 13th Five-Year Plans (2016–20) for Strategic Emerging Industries Development, Renewable Energy Development, Energy Development, and Energy Technology Innovation, all of which develop the positioning and direction of energy storage.

The government has also introduced several support policies for energy storage development. Examples are polices like the National Energy Administration's for promoting energy storage in electricity ancillary services market mechanisms in north, north-west and north-east China; and guidelines for developing energy storage technologies and an energy storage industry.

The former of the two policies was issued in 2016. It stipulated that each province should select no more than five electric energy storage facilities to be part of a pilot project on peak load and frequency regulation ancillary services. It

encouraged power generators and end users to take part in the market. And, it set out the principles of results-based compensation.

The second of the two policies was introduced in 2017. It defines the pathway and describes application scenarios for China's energy storage development. In the future, compensation policies for the application scenarios will be defined and the standards for those advanced energy storage technologies eligible for compensation will be introduced. In combination with electricity system reform, the policies for energy storage pricing will be researched and a timetable for their introduction drawn up. In Stage 1 of the policy, during the 13th Five-Year Plan (2016–20), the following focus areas will drive energy storage to early commercialisation: start up pilot and demonstration projects, research and develop key technologies and equipment, draw up a preliminary system of standards for energy storage technologies, create new business models and foster competitive market players. By Stage 2, during the 14th Five-Year Plan (2021–25), energy storage projects will be widely deployed to enable the transition from early commercialisation to scale development, thus becoming a new growth point in the energy sector.

The energy storage industry is increasingly attracting attention and its importance is growing. Many energy planning policies introduced by the Chinese government—including the Outline for the 13th Five-Year Plan (2016–20), the Strategic Action Plan on Energy Development (2014–20), and the Action Plan for Innovation in the Energy Technology Revolution (2016–30)— identify energy storage as a major research and development field for the future. Energy storage will play an important role in connecting renewable energy to the grid and in electricity ancillary services and retailing.

(2)　Current developments in China's energy storage technologies

In the 12th Five-Year Plan (2011–15), China's energy storage technologies and industry development had three characteristics.

First, different technologies were developed in a coordinated way and were characterised by continuously and rapidly decreasing costs. As costs fell, new technologies and materials were developed.

Second, pilot applications gradually evolved into commercial demonstration and the most promising energy storage applications were identified. There were two stages to this process. In 2011–13, demonstration projects were designed mainly to verify the technology and application—they lacked financial considerations. By 2013–15, the projects began to explore business models and payback periods, gradually transiting to commercial demonstration. At the same time, it became clear which applications held the most promise for development: power generation, ancillary services, power transmission and distribution, renewable energy and user consumption, electric vehicles and the Energy Internet.

Third, the energy storage industry explored pathways to large-scale commercial development. In 2015, China's installed electrochemical energy storage capacity was 141.1 MW, compared to only 2.4 MW in 2010. As installed capacity grows, the energy storage value chain improves, and manufacturers are motivated to invest.

The energy storage industry is now using three mainstream commercial application models, each of which offers several market opportunities. For example, wind farms equipped with energy storage can reduce wind curtailment and increase revenues by providing ancillary services. And consumers can use energy storage to save money by not using electricity—or they can earn money by releasing stored energy into the grid—at peak periods.

Energy storage is growing rapidly. In 2017, the cumulative installed capacity of energy storage projects in China was 27.7 GW. Of this, pumped storage had by far the largest share at 99%, followed by electrochemical energy storage at 318.1 MW or 1% of the total, an increase of 18% on the previous year. The installed capacity of China's newly opened energy storage projects was 22.8 MW, up 114% year-on-year.

By region, east China has the largest installed capacity of newly opened energy storage projects at 12.2 MW, which is 53% of the total. All these projects were deployed by users, mainly by industrial parks, helping companies reduce their electricity costs by adapting to peak and off-peak prices.

By application, the newly opened projects were deployed in grid-connected centralised renewable energy plants and end-user applications. End-user installed capacity was the largest at 17.8 MW, 78% of the total. This equates to 67% growth year-on-year and 287% month-on-month (Table 12).

6.3.3 The Role of Energy Storage Technologies in China's Energy Revolution

First, energy storage may be the best solution to address the grid connection challenge facing large-scale clean energy. Energy storage is an important factor in the energy transition, because when large-scale renewable energy enters the transmission system, it impacts the entire power system due to its unstable supply. China has severe wind, solar, hydro and nuclear curtailment problems, with more than 100 GWh of electricity wasted annually. Energy storage can help coordinate unstable renewable energy supply with energy consumption and smooth over fluctuations in production and demand. It shares this role with IT technologies like the smart grid, Internet of things, and big data analysis and management.

Second, energy storage is a key enabler for the deep integration of IT and energy technologies. Unlike other types of energy—oil, gas and coal—electricity is used as soon as it is produced. Historically, it could not be stored cost effectively, with the exception of pumped storage, which is costly and is used in hilly and mountainous regions. Energy storage has the potential to break this constraint, allowing electricity to be stored until needed or when prices are lowest. When integrated with the Energy Internet, it enables generation, distribution and consumption to be optimised in a way not previously possible.

Third, energy storage will be a key enabler to start the smart electricity revolution. As electricity is a carrier of both energy and information,

Table 12 China supports energy storage related development policies

Date	Department	Document	Main content
2005	NDRC	Guidelines for the Development of the Renewable Energy Industry	Identifies two battery projects as priorities and promotes the pilot deployment of energy storage technologies
2009	NPC Standing Committee	The Amendment to the Renewable Energy Law of the People's Republic of China	Policies to support the new energy and energy storage industries
2010	State Grid Corporation of China	The 12th Five-Year Plan for the Smart Power Grid	Development of smart grids
2010	China Southern Power Grid	China Southern Power Grid on Supporting New Energy Development	Deployment of new energy and energy storage technologies
2011	National Energy Administration (NEA)	The 12th Five-Year Plan for National Energy Technologies	Lists the critical technologies for developing energy storage and multi-energy systems
2012	The Ministry of Finance, the Ministry of Science and Technology, and NEA	Circular on Implementing the Golden Sun Solar Demonstration	Considers introducing subsidies for energy storage facilities
2014	General Office of the State Council	Strategic Action Plan on Energy Development (2014–20)	Names energy storage as one of nine major innovation fields for the first time; proposes energy storage support capabilities to effectively address wind, hydro and solar curtailment
2014	NEA	Circular of the NEA General Office on the 13th Five-Year Plan for Solar Energy Development	On creating a new energy system that combines distributed solar power, solar thermal, geothermal, energy storage and distributed natural gas
2015	CPC Central Committee and State Council	On Deepening Reform of the Electricity System	Encourages the deployment of energy storage and IT technologies to make energy use more efficient
2015	NEA	Guidelines for Developing New Energy Microgrid Demonstration Projects	Energy storage as a critical microgrid technology
2016	CPC Central Committee and State Council	Outline for the 13th Five-Year Plan	Large-scale energy storage is listed as a critical technology. It is needed to support the development of strategic emerging industries, including distributed energy. China should develop distributed new energy technologies for scale deployment
2016	NEA	Circular of the National Energy Administration on promoting energy storage in electricity ancillary services market mechanisms in north, north-west and north-east China	Energy storage is clearly defined as part of the electricity market. Power generation, retail companies and users are encouraged to invest in energy storage facilities. Provinces are

(continued)

Table 12 (continued)

Date	Department	Document	Main content
			encouraged to include energy storage facilities when planning new centralised power generation bases and ensure their operation is coordinated and optimised with the grid
2016	National Development Reform Commission (NDRC), NEA	Action Plan for Innovation in the Energy Technology Revolution (2016–30)	Create an integrated and intelligent energy technology system by: (i) improving the flexibility and cost-effectiveness of energy storage and peak shaving; (ii) promoting deep integration between energy technologies and IT; (iii) optimising the entire energy system to ensure energy resources are used effectively; and (iv), prioritising distributed energy, energy storage, smart grids, the Energy Internet and energy efficiency in industry, buildings and transport
2016	NDRC, the Ministry of Industry and Information Technology, and NEA	Made in China 2025—Energy Equipment Implementation Plan	Identifies 15 fields for energy equipment development, including nuclear power, oil and gas exploration and development, gas turbines, smart grids and energy storage
2016	NDRC and NEA	The 13th Five-Year Plan for Electricity Development	Advanced grid and energy storage technologies are listed as one of 18 priorities. The Plan says China will develop many energy storage demonstration projects, including large-scale mechanical and electrical energy storage, molten salt and chemical energy storage. The goal is to reduce significantly the construction costs per kilowatt of capacity and accelerate development and deployment
2016	NEA	The Administrative Regulation for National Electricity Demonstration Projects	It states that electricity demonstration projects will be included in the national electricity development plan and defines clear requirements for their evaluation, selection and approval. It also makes clear that demonstration projects are covered by the support policies in the Circular of the National Energy Administration on the Regulations for Major National Energy Technology Demonstration Projects

(continued)

Table 12 (continued)

Date	Department	Document	Main content
2016	NDRC and NEA	The 13th Five-Year Plan for Energy Development	The plan states that China will channel resources into renewable energy, especially into connecting new energy, energy storage and microgrids to the grid; build the Energy Internet (Internet +); improve grid regulations; develop advanced and energy-efficient technologies; and strive to become the leader in energy technology
2016	NDRC	The 13th Five-Year Plan for Renewable Energy Development	Identifies eight priority tasks, one of which is to develop energy storage technology demonstration projects in renewable energy and improve the cost-effectiveness of energy storage
2016	State Council	The 13th Five-Year Development Plan for Strategic Emerging Industries	Develop a new energy industry; accelerate the development of advanced nuclear power, solar PV and CSP, large-scale wind power, energy storage and distributed energy; improve the cost-effectiveness of new energy and build and adapt the electricity system to support a large share of new energy; facilitate multi-energy systems and their coordination and optimisation; and lead the energy production and consumption revolutions
2017	NEA	Guidelines for Developing Energy Storage Technologies and an Energy Storage Industry	The strategic importance of energy storage in China's energy industry is made clear for the first time. It sets out the development goals for energy storage over the next 10 years and underscores five priority projects related to energy storage
2017	NEA	Guidelines for Energy Work 2017	Identifies several energy storage projects throughout the country that were to be completed or started in 2017
2017	The Special Committee of Energy Storage under the China Strategic Alliance of Smart Energy Industrial Technology Innovation, the Special Committee of Energy and Water Supply Prices under the Price Association of China	Energy Storage Subsidy Policy Consulting Meetings (1st, 2nd and 3rd rounds)	Energy storage subsidy proposals for flow cell and all-vanadium redox flow batteries, and from Chinese battery manufacturers CATL and Lishen In the first meeting, companies differed greatly in their proposals on how to subsidise energy storage. In the third meeting in 2017, the focus was on reaching agreement on the best way to proceed

the integration of power and electronics provides a powerful platform for the smart grid. As energy storage technologies and electricity market reform progress, large-scale renewable energy can be used to power industrial production for the whole of society; and data storage, information processing and decision analysis can be integrated into the industrial system and become the engine that starts the smart electricity revolution, after which it will then undertake the historical mission of driving the energy revolution and providing industrial civilization with green energy.

6.3.4   Strategy and Policies for Energy Storage Development

First, the strategic positioning of energy storage in the energy revolution should be clarified. The energy revolution is an important engine for China to develop a modern energy system and accomplish the energy transition. Energy storage will play an important role in this. Physical and chemical energy storage, hydrogen and fuel cells, and heat storage technologies will be developed and compete against each other. Eventually, one or two of them will win out. Energy storage will be widely deployed and change the structure and business model of the power sector, thus helping to build a modern energy system.

Second, technology and the energy system should evolve in a coordinated fashion. The biggest obstacle facing energy storage is high cost. The most practical applications are likely to be with users. Other applications like generation and transmission have limited growth potential. Energy storage should be coordinated with other technologies like heat storage and control to deliver synergies and avoid rushing into the market alone. It should be used as a support service for the power distribution network, industrial parks and similar applications.

Third, effort should be made to tailor a market mechanism conducive to the value of energy storage. China should: (i) identify cost-effective niche markets and support the deployment of energy storage technologies in these markets. It should incentivise the development of existing energy storage technologies to improve their

efficiency and flexibility; (ii) create a strong market and regulatory environment by eliminating price distortions and vested interests; and support R&D and demonstration projects for early-stage energy storage technologies, including high-temperature heat storage and scalable battery and hybrid energy storage systems; (iii) use the market mechanism to tap the potential of pumped storage power stations; (iv) follow the market economy principle of "who benefits and who shares costs" to adjust the two-part tariff mechanism; and (v) in combination with the electricity market reform process, create a competition-based pricing system and bidding rules for electricity ancillary services to encourage pumped storage power stations to participate in the market.

Fourth, a standards system should be developed for the energy storage industry. China should: (i) benchmark its industry against international leaders, make incremental improvements in energy storage technologies, assess existing energy storage facilities, and evaluate the potential for energy storage in regions and energy markets; (ii) develop international and national collaborations on data to speed up research, monitor progress and unblock R&D bottlenecks; and (iii) based on the experience from pilot projects, improve the related industry standards.

6.4   Long-Distance Power Transmission

6.4.1   Drivers and Characteristics of Long-Distance Power Transmission

(1) **Drivers**

Misalignment between energy sources and load centres is common in many countries and regions. Load centres generally do not have abundant energy sources. They enjoy strong economic growth and have high population concentrations, resulting in a large and increasing demand for power. In contrast, regions with abundant energy sources are often characterised by low power consumption due to an underdeveloped economy and small population. For example, more than 80% of Russia's power generation resources are distributed in the east (Siberia has abundant hydropower and coal resources), but load centres are located in the west. Historically, China's load centres are in the central and eastern regions, but its energy sources are mainly in the west and north. For instance, coal, wind and solar power and other new energy sources are largely in west and north China, and hydropower resources in south-west China.

Such misalignment is determined by geography, but shaped and intensified by social and economic development. In countries like China and Russia, large-scale, long-distance and efficient power transmission is needed to connect energy resources with load centres.

(2) Characteristics

Long-distance power transmission technologies are mainly divided into: extra-high voltage and ultra-high voltage (EHV/UHV) AC, EHV/UHV DC, flexible AC/DC power transmission, multi-terminal direct current (MTDC) power transmission, and new technologies like fractional frequency transmission, superconducting transmission, half-wavelength transmission, etc.

10–220 kV AC transmission is the most common technology in large power systems. AC transmission at low voltage over hundreds of kilometres results in substantial electrical losses. One solution is to increase the voltage level to EHV. Another is to convert AC to DC for long-distance transmission (DC has lower power losses). EHV generally refers to 330–765 kV AC and ± 500 to ± 660 kV DC transmission technologies.

UHV means a voltage level of at least 1,000 kV for AC transmission and ± 800 kV or more for DC transmission. UHV transmission systems can carry more power at a higher voltage over longer distances and with lower losses than EHV.

6.4.2 Current Trends in Global Long-Distance Power Transmission

(1) Application analysis

First, technological progress.

Several major economies have developed technologies and equipment critical to long-distance power transmission since the 1960s. The former Soviet Union, Japan, the USA and Italy initiated UHV power transmission.

The former Soviet Union was one of the earliest countries to investigate UHV transmission, and the only country (except China) to have UHV AC transmission projects in operation up to now. The former Soviet Union started to construct the Siberia-Kazakhstan-Ural 1,150 kV UHV AC transmission project in 1980 to transmit electricity from Siberia to load centres in the European part of the country. The project started to operate at its intended rated voltage in 1985 but was later reduced to 500 kV due to technical difficulties.

Research and experiments in UHV technology have been carried out in the USA but have not been deployed. In 1974, American Electric Power and General Electric conducted audible noise, radio jamming and other tests at the UHV test station at Pittsfield, Massachusetts. A 1,000–1,500 kV three-phase test line was constructed by the Electrical Power Research Institute in 1974. It provided experience in electromagnetic operating environments, tower installation and transformer design.

Japan started to construct 1,000 kV transmission and substation projects in 1988. By 1999 it had built two 1,000 kV power transmission lines with a total length of 430 km and one 1,000 kV substation. One of the two lines, called the North-South Line, is 190 km long and starts from the nuclear power plants along the north-west coast to the Tokyo region further south. The second line, the East-West Line, is 240 km long and connects power plants along the east coast of the Pacific Ocean This line has been operating at reduced voltage because of lower load requirements.

Italy started to build a 1,050 kV UHV experimental line of 3 km in 1984 at the Suvereto experiment station. It was completed in October 1995 and operated until December 1997 at a rated voltage of 1,050 kV, providing some operational experience.

Second, practical deployment in projects.
Countries in Europe, Africa, Asia and the Americas have constructed long-distance power transmission projects and developed long-term plans for them in recent years. These plans include the European Ten-Year Network Development Plan, US Grid 2030, and EHV/UHV power transmission projects in India, Brazil and countries within China's Belt and Road Initiative for global development.

The European Network of Transmission System Operators for Electricity (ENTSO-E) published the draft of a new pan-European Ten-Year Network Development Plan (TYNDP) in July 2014. TYNDP sets 2030 as the milestone for Europe to achieve its energy targets and it describes the coming investment programmes for the European power transmission network. Its guidance is mandatory and ensured by legislation. According to TYNDP 2014, Europe needs to invest EUR 150 billion in the construction and upgrading of about 48,000 km of high voltage power transmission lines for the pan-European grid. This investment is needed to transfer renewable energy from around the North Sea to other parts of Europe. The large-scale development of renewables will require the construction of more long-distance and large-capacity transmission projects to interconnect countries in a pan-European power grid. TYNDP 2014 expects the average transmission capacity of the European grid to be doubled by 2030.

In Brazil, the Belo Monte UHVDC transmission project[56] connects vast hydropower resources along the Amazon River and branches in the north with the Rio de Janeiro and Sao Paulo load centres in the south-east, up to 2,500 km away. Long-distance, large-capacity UHV power transmission is the only way to harness these hydropower resources and transport the energy to where it is needed.

The joint venture between the State Grid Corporation of China and the Brazilian utility Electrobras won the order for the ±800 kV UHVDC power link in 2014. The line, 2,092 km long, starts from the Belo Monte hydropower plant on the Xingu River and ends at the Estreito converter station in Iberaci. It is the first ±800 kV UHVDC power transmission line in the Americas. It was inaugurated in 2017.

In 2015, State Grid Corporation of China won a 30-year concession to build and operate a second ±800 kV UHVDC transmission link from the Belo Monte hydropower station. Belo Monte has an installed capacity of 11.2 GW, the second largest in Brazil. The transmission project includes one ±800 kV UHVDC power transmission line of 2,518 km, a converter station at each end and auxiliary projects. The link has a power transmission capacity of 4 GW and is scheduled to start commercial operation in 2020 (Fig. 42).

India (UHVDC)[57]: to transmit hydropower from the north-east and thermal power from central regions, India has built two ±800 kV UHVDC links: Biswanath Chariali–Agra (1,728 km) and Champa–Kurukshetra (1,365 km). The two links transmit surplus power from eight states in the north-east to the north, where there is a shortage of electricity. This not only powers industrial development in the north, but delivers economic benefits to the north-east, where the economy needs development.

---

[56]Official website of Ministry of Commerce of China: http://www.mofcom.gov.cn/article/i/jyjl/l/201507/20150701052029.shtml.

[57]Electric Power Construction in India and its UHV DC/AC Power Transmission Plans, He Dayu, Electric Power, Issue 2, 2008.

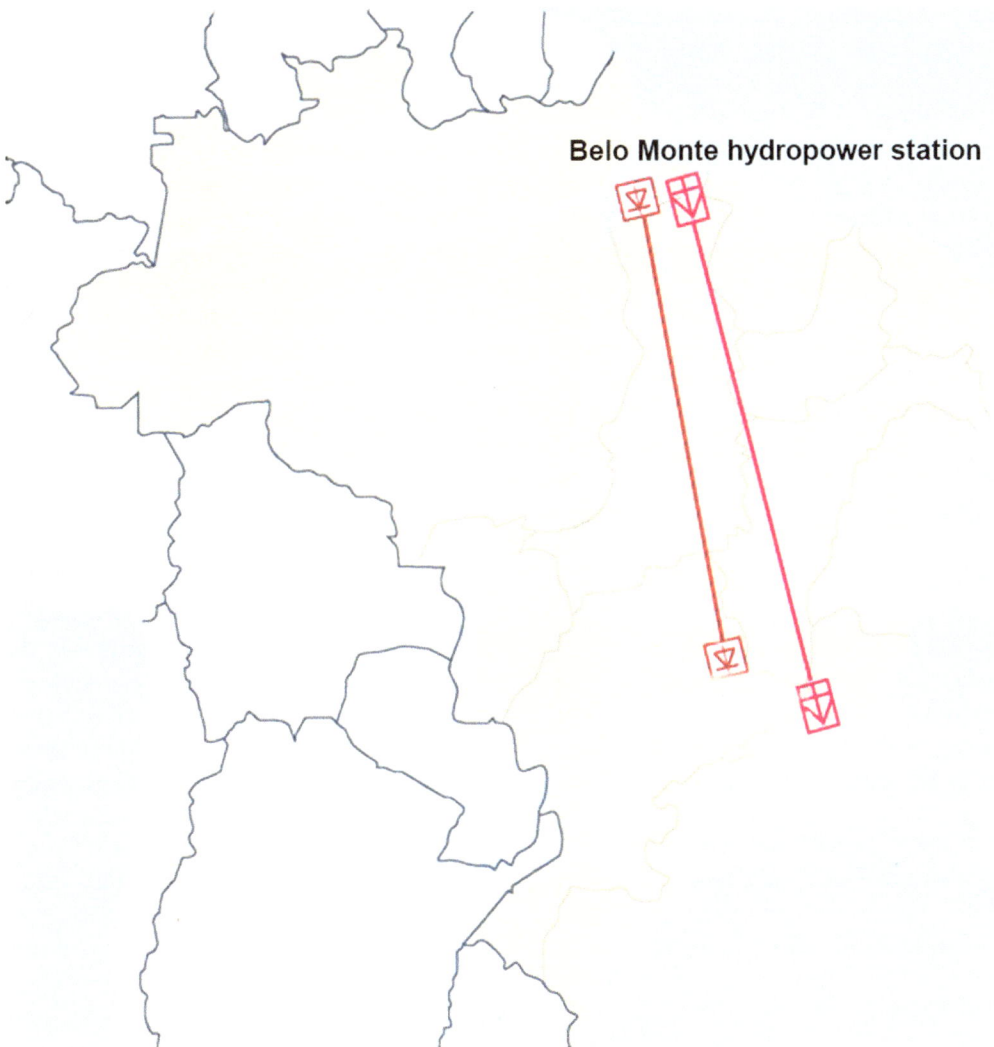

Belo Monte hydropower station

Fig. 42 Belo Monte UHV DC project, Brazil. *Source* White Paper on Global Energy Internet Development Strategy, Global Energy Interconnection Development and Cooperation Organization, Feb. 22, 2017, Beijing, http://www.chinasmartgrid.com.cn/news/20170223/622201.shtml

In addition, India is researching and testing 1,200 kV UHVAC power transmission links. It plans to construct six 1,200 kV UHVAC lines to interconnect its power networks by 2020 and transfer hydropower from north-east India to the rest of the country.

In Egypt, the Egyptian Electricity Transmission Company's (EETC) 500 kV power transmission project[58]—about 1,210 km long and costing around $760 million—is the first deal concluded under Sino-Egyptian capacity cooperation in the BRI framework. It is the largest power transmission line with the highest voltage and widest coverage in Egypt. The project was executed on an engineering, procurement and construction (EPC) basis by China Electric Power Equipment and Technology Co., Ltd., a wholly owned subsidiary of State Grid Corporation of China. It has increased transmission capacity from the gas-fired power plants in the

---

[58]http://www.ceec.net.cn/art/2017/6/12/art_11096_1397882.html.

Nile Delta, strengthened the national grid, and improved economic development and the efficient use of electricity. In addition, the project boosted upstream and downstream industries including power supply, electrical equipment and raw materials in Egypt and the Middle East. It created about 7,000 jobs for Egypt and achieved win-win results for Sino-Egyptian cooperation.

In Pakistan, the Matiari-Lahore HVDC power transmission project is a priority of the China-Pakistan Economic Corridor of infrastructure projects. The link will be about 900 km long and have a capacity of 4 GW. State Grid Corporation of China is developing the project on a BOOT (build, own, operate and transfer) basis. When completed in 2021 the link will transport much-needed power from the south, where it is generated, to load centres in central Pakistan.

The EuroAsia Interconnector is an HVDC project that will connect the Greek mainland with Crete, Cyprus and Israel using multi-terminal flexible DC technology. The interconnector is part of the European network of interconnected electricity grids that provide bi-directional electricity supply through national transmission operators. It will enable the three partner countries (Israel, Cyprus and Greece) to better utilise their recently found offshore natural gas by converting it into electricity for their own use and for transmission to the rest of Europe. When completed in 2023, the interconnector will be the longest HVDC cable link in the world at 1,518 km and the deepest subsea cable at 3,000 m. The link will have a power capacity of 2 GW.

In the USA, Grid 2030—A National Vision for Electricity's Second 100 Years[59] is a report published in 2003 by the U.S. Department of Energy. The report clearly defines the vision for the future US grid and emphasises the importance of nationwide interconnection and interconnection with Canada and Mexico. Grid 2030 consists of three major elements: (i) a national electricity framework of high-capacity

transmission corridors that link the east and west coasts and connect with Canada and Mexico; (ii) interconnected regional grids—power from the corridors is distributed to regional networks by long-distance AC transmission or, in some cases, by expanded DC links; and high-capacity DC interties are used to link adjacent, asynchronous grids; (iii) local, mini-grids and microgrids. The nation's local distribution systems are connected to the regional networks, and through them to the national corridors. Real-time monitoring and information exchange enable markets to process transactions instantaneously and on a national basis. Customers have the ability to tailor their electricity supply.

(2) **Related technologies**

1. Long-distance power transmission technologies Factors such as growing demand for power, the challenges of integrating new energy with the grid, intelligent power consumption and ever stricter environmental requirements, are making it increasingly difficult for conventional transmission links to meet future requirements. Many countries have carried out research on EHV and UHV and other new power technologies.

UHV power transmission delivers many benefits, including large capacity, long distance, low energy losses, small footprint and cost effectiveness. Other new power transmission technologies have some obstacles to overcome and have yet to be widely deployed, but they offer superior benefits over conventional methods in certain applications. With development and progress, those technologies are expected to be deployed in projects.

First, UHV power transmission. AC and DC hybrid power networks represent the future for UHV power transmission. UHVAC and UHVDC have their own characteristics, but they are also complementary. UHVAC is economically competitive for distances from 1,000–2,000 km. UHVDC power transmission is more cost effective for distances of more than 1,500 km. UHVDC is, therefore, used to transmit large volumes of electricity from energy bases over long or ultra-long distances. UHVAC is used for

---

[59]https://energy.gov/oe/downloads/grid-2030-national-vision-electricity-s-second-100-years.

grids with higher voltage levels and for cross-regional interconnections.

Second, flexible DC power transmission. Flexible DC is now used to connect wind farms to the grid, transfer high-quality power to city centres, and create DC networks for remote regions and islands—these are the main applications. Flexible power transmission improves the quality of unstable renewable energy in the grid. For some remote regions characterised by low and fluctuating loads and long transmission distances, HVAC and conventional HVDC power transmission technologies are not technically and economically feasible. Flexible DC power transmission transports power through DC lines to remote regions and load islands, supporting local economic development with emission-free electricity.

Third, multi-terminal DC (MTDC) power transmission transports electricity from multiple power sources to multiple load centres. It is more cost effective than conventional HVDC power transmission, which links one converter station to another via HVDC transmission lines or cables. For example, Tibet has great hydropower potential and will be an important energy source for China in the future. However, the existing power transmission corridor is inadequate, and the single hydropower station that generates power from the Jinsha, Lancang and Nujiang rivers in east Tibet is small. An MTDC system could be used to transfer power from multiple plants on the three rivers to multiple receiving terminals near load centres. Several MTDC power transmission systems are already in operation in China, Africa, Europe, the USA and shortly in India. MTDC power transmission projects will play an important role in future long-distance and large-capacity power transmission in many countries.

Fourth, superconducting power transmission has already been demonstrated in power distribution because of its low voltage level and short power transmission distance. Many countries are researching its application in long-distance power transmission. For example, the Netherlands is researching superconducting transmission at a rated voltage of 50 kV, and South Korea is examining high-temperature applications. The USA is researching three-phase resistance and a saturated core high-temperature superconducting fault current limiter using the second generation of high-temperature superconducting materials. Superconducting DC power transmission could become a future technology trend. It is more efficient than superconducting AC because it loses less power in transmission. It is also more cost effective than AC at the same transmission capacity. China and Japan have both carried out experiments in superconducting DC power transmission. The Institute of Electrical Engineering at the Chinese Academy of Sciences is constructing a demonstration project to supply power to an electrolytic aluminium plant.

Fifth, fractional frequency power transmission has become a research priority for transporting large-scale offshore wind power, as it is potentially more cost effective and reliable than alternative technologies. However, research is still at the theoretical and simulation stage. Although it does merit deeper study and deployment in real projects. The technology is a combination of fractional frequency power transmission and power electronics, which together have the potential to produce benefits in terms of system design, control, capacity, reactive power and harmonics. Another application for this highly promising technology is hydropower.

Sixth, half-wavelength AC power transmission means that the electrical distance of power transmission is close to one power frequency half-wavelength, i.e. 3,000 km or 2,600 km ultra-long distance three-phase AC power transmission. A lossless half-wavelength AC line is like an ideal transformer with a transformation ratio of −1. Sending-end and receiving-end voltage is at the same level but opposite in phase. It is suitable for ultra-long distance and ultra-large capacity power transmission. A.A. Wolf and his colleagues in the former Soviet Union came up with the idea of half-wavelength AC power transmission as early as 1940. Potentially, half-wavelength AC power transmission has several advantages over conventional AC power transmission over long distances—such as no need for reactive compensation equipment and switching stations; high transmission capacity; and it is less costly than UHVDC. Although there is no half-wavelength power transmission project at present, the potential cost-effectiveness of UHV half-wavelength power transmission makes it promising for the future.

2. Selection criteria

The research and development of EHV and UHV technologies and equipment provides a strong foundation for the future application of long-distance power transmission. Demand for large-capacity, long-distance power transmission is the main driver. Technology selection is influenced by national conditions and trends.

Future growth in global power demand will come mainly from non-OECD countries. India, Brazil and other developing countries with escalating demand need to build large power source bases to supply electricity to load centres, which often lie far away. UHV AC and DC is the technology of choice.

North America and Europe are characterised by low growth in power demand and a transition to low-carbon and green energy. They have plans to develop large wind power, solar energy and clean energy bases. Advanced technologies including MTDC, flexible AC/DC and UHVAC could be used to construct power transmission networks with higher voltage levels and more flexibility, enabling renewable energy to be connected to the grid and consumed in more areas.

6.4.3 Current Developments in Long-Distance Power Transmission in China

(1) The policy environment

China's energy sources and load centres are separated, often thousands of kilometres apart, making long-distance power transmission important. China's energy sources are abundant in the west and north, and scarce in the east and south. Large energy bases are located mainly in the north, north-east, north-west and south-west where the power load is low. The distance between these bases and load centres in central and east China is 1,000–3,000 km, longer than the cost-effective distance of conventional EHV power transmission. The size and distance of energy flows will continue to grow, imposing challenges on power transmission in terms of capacity, distance and efficiency.

Long-distance power transmission is also a means to address the severe ecological and environmental problems caused by pollution and haze in east and central China. It can help improve the environment both in energy bases and load centres, enabling high-efficiency coal-fired power plants to be built in coal-producing regions in the west and north. This power can be transmitted in large volumes over long distances to central and east China, sharing the environmental burden more equally across regions and easing environmental pressure in load centres.

At the policy level, the central government attaches great importance to the energy transition. It has introduced policies and plans to develop long-distance power transmission and strengthen energy security. From 2005, UHV was included in national strategic plans, including the Outline of the National Medium- and Long-term Science and Technology Development Program (2006–20),[60] published in 2006, and the 12th Five-Year Plan for the National Economic and Social Development of China,[61] issued in 2011.

The Strategic Action Plan on Energy Development (2014–20)[62] released in 2014 proposed the clean and efficient development of coal power, the construction of large coal power bases and power transmission corridors and the development of long-distance and large-capacity power

---

[60]http://www.most.gov.cn/mostinfo/xinxifenlei/gjkjgh/200811/t20081129_65774.htm.

[61]http://www.gov.cn/2011lh/content_1825838.htm.

[62]http://www.mlr.gov.cn/xwdt/jrxw/201411/t20141119_1335668.htm.

transmission technologies. It also proposed expanding the capacity of the West-East power transmission project and implementation of the North-South transmission corridor. The National Energy Administration issued the Circular on Accelerating and Promoting the Construction of 12 Major Power Transmission Corridors under the Air Pollution Prevention and Control Action Plan,[63] also in 2014. These corridors include the Xilingol League-Shandong, Huainan-Nanjing-Shanghai, West Inner Mongolia-South Tianjin and Yuheng-Shandong UHVAC projects; the East Ningxia-Zhejiang, Xilingol League-Taizhou, Shanghaimiao-Shandong, North Shanxi-Jiangsu and North-west Yunnan-Guangdong UHVDC projects; and three ±500 kV power transmission lines.

In 2016, the National Energy Administration released the 13th Five-Year Plan for Electricity Development.[64] Among other things the plan proposes to add 130 GW of capacity to the West-East transmission project by 2020. This will increase the project's capacity to around 270 GW, reducing annual standard coal equivalent (SCE) consumption by more than 100 million tonnes in east and central China, thereby reducing emissions and air pollution in the receiving regions.

(2) Current developments

China began to study UHV technologies in the 1980s. Large-scale research and demonstration projects began in 2004. China has made breakthroughs in UHV in recent years, developing key technologies and the ability to manufacture UHV transformers, reactors, 6 in. thyristors, high-capacity converter valves and other key equipment. China has built the South-east Shanxi-Nanyang-Jingmen 1,000 kV UHVAC pilot and demonstration project, Yunnan-Guangdong (5,000 MW) and Xiangjiaba-Shanghai (7,000 MW) ±800 kV UHVDC demonstration

projects, which have been operating safely and stably for a long time, proving that China's UHV power transmission technologies are mature and can be deployed widely.

Up to June 2017, China had built 18 UHV projects, with two more under construction. Among the 20 UHV projects, seven are AC and 13 are DC, with a total length of nearly 30,000 km. The East Junggar-South Anhui ±1,100 kV UHVDC power transmission project is the world-record holder in terms of length (3,324 km), voltage (1,100 kV DC) and capacity (12,000 MW) (Tables 13 and 14).

The longest UHVDC power transmission line in service is Jiuquan-Hunan ±800 kV. It opened in June 2015 and runs through Gansu, Shaanxi, Chongqing, Hubei and Hunan. The two converter stations, one at either end, have a combined converter capacity of 16,000 MW. At 2,383 km, Jiuquan-Hunan is the longest UHVDC line in the world at present. It is a trans-regional power transmission corridor mainly for wind, solar and other renewables. In time it will become a west-east power transmission super-highway, enabling the bulk transmission of wind and coal power from Gansu to meet the current power shortage in central China.

The first UHVAC link in China is the South-east Shanxi-Nanyang-Jingmen 1,000 kV UHVAC power transmission pilot and demonstration project: The line is 640 km long. Phase 1 was energised in January 2009 and was the world's first UHVAC line to enter commercial operation. Phase 2 opened in December 2011. It has a transmission capacity of 5,000 MW and can transport 25,000 GWh of power per year. The project connects the north China and central China power grids and is an important north-south transmission corridor. It delivers coal-fired power from north China to the south in winter and surplus hydropower from central China to the north in summer. It also provides backup power in case of outages, delivering substantial economic and social benefits.

The longest UHVAC power transmission line in service is Yuheng-Weifang. The 1,000 kV UHVAC project opened in May 2015. It runs through Shaanxi, Shanxi, Hebei and Shandong.

---

[63]http://www.zjdpc.gov.cn/art/2014/6/2/art_981_653664.html.

[64]http://www.nea.gov.cn/2016-11/07/c_135811086.htm.

Table 13 UHV transmission in China (up to June 2017)

DC	Project	Starting point	Destination	Capacity (MW)	Voltage level (kV)	Construction commencement date	Operation date	Length (km)
1	Jinping—South Jiangsu	Sichuan	Jiangsu	7,200	±800	2009	2012	2,100
2	Xiluodu—West Zhejiang	Sichuan	Zhejiang	8,000	±800	2012	2014	1,700
3	Xiangjiaba—Shanghai	Sichuan	Shanghai	6,400	±800	2007	2010	1,907
4	Kumul—Zhengzhou	Xinjiang	Henan	8,000	±800	2012	2014	2,210
5	East Ningxia—Shaoxing	Ningxia	Zhejiang	8,000	±800	2015	2016	1,720
6	Jiuquan—Hunan	Gansu	Hunan	8,000	±800	2015	2017	2,383
7	North Shanxi—Jiangsu	Shanxi	Jiangsu	8,000	±800	2015	2017	1,119
8	Xilin Gol League—Taizhou	West Inner Mongolia	Jiangsu	10,000	±800	2015	2017	1,628
9	Shanghaimiao—Shandong	West Inner Mongolia	Shandong	10,000	±800	2015	2017	1,240
10	Chuxiong—Huidong	Yunnan	Guangdong	5,000	±800	2006	2010	1,412
11	Pu'er—Jiangmen	Yunnan	Guangdong	5,000	±800	2011	2015	1,400
AC	Project	Starting point	Destination	Capacity (MW)	Voltage level (kV)	Construction commencement date	Operation date	Length (km)
1	Southeast Shanxi—Nanyang—Jingmen	Shanxi	Hubei	5,000	1,000	2006	2009	1 × 654
2	Anhui—East China	Anhui	East China	6,000	1,000	2011	2013	2 × 656
3	North Zhejiang—Fuzhou	Zhejiang	Fujian	3,000	1,000	2013	2014	2 × 603
4	Xilingol League—Shandong	West Inner Mongolia	Shandong	7,000	1,000	2014	2016	2 × 730
5	West Inner Mongolia—South Tianjin	West Inner Mongolia	Tianjin	6,000	1,000	2015	2016	2 × 608
6	Huainan—Nanjing—Shanghai	East China Ring Grid		6,000	1,000	2014	2016	2 × 738
7	Yuheng—Weifang	Shaanxi	Shandong	6,000	1,000	2015	2017	2 × 1,048

The 2 × 1,049 km double-circuit line is the longest UHVAC line in the world. Part of the north China UHV AC/DC grid, the link has helped Shaanxi and Shanxi develop their energy bases to supply power to other regions. It has also increased the capacity of the north China power grid and effectively reduced power shortages in central and east China.

Table 14 UHV projects under construction in China (up to June 2017)

AC	Project	Starting point	Destination	Capacity (MW)	Voltage level (kV)	Construction commencement date	Operation date	Length (km)
1	East Junggar —south Anhui	Xinjiang	Anhui	12,000	±1,100	2016	2018	3,324
2	Jarud— Qingzhou	East Inner Mongolia	Shandong	10,000	±800	2016	2018	1,234

(3) **Economic and social benefits**

First, the South-east Shanxi-Nanyang-Jingmen UHVAC power transmission project.

As the first 1,000 kV UHVAC link in the world, this pilot and demonstration project has played an instrumental role in optimising grid resources in the 10 years since it opened. In dry winter seasons, it transfers thermal power from the north to Hubei. In wet summers, it delivers surplus hydropower from south-west China, including Sichuan, to the north China power grid, relieving power shortages in Shandong. The complementarity of hydropower and thermal power not only avoids hydro curtailment in the wet season, it reduces coal consumption in north China and increases the use of clean energy.

Fossil energy sources are scarce in Hubei. Its coal reserves account for less than 1% of China's total. 98% of the coal used for Hubei's power generation needs to be procured from other provinces. Its hydropower has been fully developed and is transmitted to other regions. 60% of Hubei's power consumption is supplied by thermal power, exposing it to increasingly serious energy constraints. The South-east Shanxi-Nanyang-Jingmen UHVAC project supplies 5,000 MW of thermal power from north China to Hubei. This avoids the transport of more than 7 million tonnes of thermal coal per year, which if converted into electricity would be equal to the output of the 2,700 MW Gezhouba hydropower plant in Hubei. Up to the end of October 2016, the link, which has a capacity of 5,720 MW, had transmitted 55,600 GWh of electricity, effectively eliminating the power shortage in Hubei. Calculations show that the price of electricity transmitted by the link to Hubei is RMB 0.02–0.07/kWh lower than the current benchmark thermal power price in Hubei. According to a report published by Greenpeace, the transmission of every 100 GWh through UHV reduces particulate matter (PM2.5 and PM10) and sulphur dioxide/nitrogen oxide (SO_2/NO_x) from load centres by about 7 tonnes, 17 tonnes and 450 tonnes respectively.[65]

Second, the Anhui-East China UHVAC power transmission project.

The Anhui-East China 1,000 kV UHVAC power transmission line, which starts at Huainan and ends at Shanghai, has been operating since 2013, delivering multiple benefits to Huainan and east China. Huainan's power generation companies transmit around 24,000 GWh of electricity to east China through the UHV project every year, generating more than RMB 17 billion in sales revenue and RMB 3 billion in taxes annually, which boost the local economy. In addition, coal is used locally in Anhui at RMB 30/tonne, rather than being transported over long distances to Shanghai at RMB 70/tonne. This reduces transport-related pollution. The dust, exhaust gas and lines of trucks on the roads of Huainan are no longer to be seen. In addition to supplying Shanghai and Zhejiang with 6,000 MW of capacity and 24 billion kWh of electricity, the link enables Shanghai and Zhejiang to reduce coal consumption by 10.8 million tonnes per year, equivalent to six 1,000 MW thermal generating units. This has cut emissions of CO_2, SO_2 and NO_x in Shanghai by 21 million

---

[65]http://www.cet.com.cn/, "Power Expressway" Empowered by UHV Technology Brings Enormous Economic Benefits, http://power.in-en.com/html/power-2269268.shtml.

tonnes, 50,000 tonnes and 56,000 tonnes per year, significantly improving the air quality in China's second biggest city.[66]

(4) **Main problems**

Generally, China's UHV technology is world-class in terms of technology and capacity. However, there are still some issues to be resolved.

First, UHV increases the complexity of the power system and puts it at greater risk. Effective precautions need to be taken to ensure system security. AC-DC interconnections and increasing volumes of new energy change the system profoundly. Once interconnected, regional grids are interdependent and impact each other mutually. This increases overall complexity, risk and uncertainty. If a commutation failure occurs on a ±800 kV DC line, it generates an 8,000 MW power surge, which is 3–4 times that of AC. If a DC commutation failure occurs, requiring multiple restarts, an instantaneous 21,000 MW power surge and 8,000 MW repeat surges will be generated, which could cause an outage.[67] Higher transmission voltage levels and larger amounts of new energy increase the risk to grid security.

Second, there is still a quality gap between locally manufactured UHV equipment and advanced international technologies. Some key components still need to be imported. Supply channels for drawings, materials, user manuals, software, and some other spare parts are simplex. This increases the cost and complexity of operating and maintaining UHV projects.

6.4.4 The Role of Long-Distance Power Transmission in China's Energy Revolution

(1) **Role**

Long-distance power transmission provides strategic support for the development of clean energy in China. It is an important tool to shape a national market and optimise resource allocation, and an integral part of the modern power system.

First, long-distance power transmission enables electricity to be transferred in bulk across the country. As China's energy sources and load centres are in different regions, trans-regional electricity flows with multiple sending and receiving ends is the future. The capacity, distance and types of electricity flow will only increase. As a result, China needs to speed up development of long-distance power transmission and plan in a centralised manner the links between sending and receiving regions.

Second, the development of long-distance power transmission delivers comprehensive benefits, including regional interconnections, inter-basin hydropower transfer, trans-regional allocation, peak shaving, the complementary and coordinated use of hydropower and thermal power, and less need for backup capacity. Trans-regional allocation reduces the need for receiving regions in east and central China to build coal-fired power plants. New power demand is supplied mainly with local nuclear power and with electricity from other regions. Peak shaving is used to optimise the power mix and better integrate clean renewable energy.

Third, long-distance power transmission can help connect and consume clean energy. As clean energy is intermittent and far from load centres, grid connection is faced with two challenges: (i) clean energy poses more demanding requirements than conventional power on the grid's peak shaving capacity; and (ii) the ability of clean energy-abundant areas to use all the energy locally is limited. China can make the most of its national peak-shaving resources and new energy capacity only when grid connection and consumption of clean energy in large areas is

---

[66]http://www.xinhuanet.com/, Calculation of Economic and Environmental Benefits of UHV Technology from the Perspective of Anhui-East China UHV Power Transmission Project, http://news.xinhuanet.com/2016-11/01/c_1119830177.htm.

[67]Tang Yong, speaker at the 5th China Power Development and Technology Innovation Forum, April 27, 2017, http://shupeidian.bjx.com.cn/news/20170428/822880.shtml.

possible. To allocate its hydro, wind and solar resources optimally, China must develop long-distance, large-capacity and low-loss power transmission technologies.

(2) Medium- and long-term development of China's grid

The separation of energy resources from load centres determines the development pathway of the grid. China now has six AC synchronous power grids: North China-Central China, East, North-east, North-west, South China, and Tibet, which operate according to jurisdiction and grid requirements. Connection between these grids and the ability to allocate energy resources are weak.

State Grid Corporation of China has issued a national grid development plan in which China will optimise its six regional power grids into two UHV synchronous grids characterised by clear sending-receiving terminals and AC-DC coordinated development in the medium and long terms. The two grids are: West (North-west and Sichuan, Chongqing and Tibet), East (North, East and Central China; Heilongjiang; Jilin and Liaoning; and Inner Mongolia).[68]

The two grids will enable the coordinated generation and bulk UHV transmission between regions of hydropower and thermal and of wind and solar power. This will allow clean energy to be deployed and used at scale in west China and to be consumed efficiently and sustainably in east China.

6.4.5 Strategy and Policies for Long-Distance Power Transmission

(1) Strategic positioning

Long-distance power transmission is the key to a modern electricity-centred energy system. China is faced with energy security challenges, including shortage of fossil fuels, rising demand

for energy and import dependency. In addition, pollution is becoming more and more serious. The way to overcome these challenges is by following the electricity-centred energy strategy. This will give full play to electricity's role as an energy conversion hub, allowing multiple types of energy to be converted into power and used efficiently. In an electricity-centric China, power transmission will reduce the need to transport primary energy and allow renewables to be integrated and transmitted at scale, ultimately replacing fossil energy.

Long-distance power transmission is a sector in which China can exercise its Go Out strategy. China has the largest number of UHV projects in the world. It has UHV engineering experience and expertise, a strong UHV product R&D programme, and experience in delivering huge transmission projects. It has world-class UHV technologies and production capacity. Thanks to the smooth delivery and operation of Belo Monte in Brazil and other overseas projects, China's UHVDC technology is yet another national business card alongside high-speed rail.

China also dominates the development of international UHV standards. Three Institute of Electrical and Electronics Engineers (IEEE) UHV standards, developed under the leadership of China, have been published in recent years: the Guide for Voltage Regulation and Reactive Power Compensation for Systems of 1,000 kV AC and Above (IEEE P1860), the Guide for On-Site Acceptance Tests of Electrical Equipment and System Commissioning of 1,000 kV AC and Above (IEEE P1861), and the Recommended Practice for Overvoltage and Insulation Coordination of Transmission Systems at 1,000 kV AC and Above (IEEE P1862). The IEEE UHVAC Power Transmission System Technical Committee, founded in 2013 and chaired by China, significantly strengthens China's influence in the field of global grid standards. The fact that China's UHV standards have become international ones means that China's UHV technologies are internationally advanced.

Long-distance power transmission is an important facilitator of China's Belt and Road Initiative (BRI), enabling power interconnection

---

[68]http://www.sgcc.com.cn/xwzx/gsyw/2015/12/330711. shtml.

with surrounding countries. Interconnected infrastructure is a priority of the BRI. Countries taking part in the BRI, with their sovereignty and security concerns respected, can build an infrastructure network connecting Asia, Europe and Africa. By planning infrastructure construction and aligning technical standards, BRI strengthens cooperation between countries and promotes the construction of cross-border power transmission corridors. With large-capacity, low-loss long-distance power transmission technologies, China can build a corridor that interconnects grid infrastructure across Asia-Pacific. Given the resource endowment differences and strong energy complementarity of the countries of the Silk Road Economic Belt (a component of BRI that applies to countries along the original Silk Road between China and Europe), China can import surplus power from neighbours through long-distance power transmission, stimulating economic development in those countries and securing power for China at lower cost.

(2) **Policy proposals**

First, the intensive development of large-scale renewable energy bases should drive research on UHV and other new long-distance power transmission technologies. These wind, hydro and solar power bases require grid connection and super-large capacity and ultra-long distance transmission systems, both within China and across neighbouring national borders. These technologies require further research and development, especially in UHV AC and DC and in new transmission technologies like half-wavelength, superconducting, wireless and pipeline.

Second, China should expand research on UHV grid control and simulation to minimise the risk of cascading power outages. Currently UHV power transmission technologies are faced with three major challenges to grid reliability: (i) the wider application of power electronics increases the possibility of cascading outages and makes grid stabilisation more complex and simulation analysis more difficult; (ii) an outage can have a greater impact on interconnected grids, and

power fluctuations can increase the risk of a grid-wide outage; and (iii) poor voltage stability at the receiving end can increase if large changes are made to the power source mix. China therefore needs to increase its research on UHV power grids in areas such as simulation analysis, grid analysis, voltage stability and grid control.

Third, China should consolidate UHV's Go Out advantage. China needs to maintain its international leadership in DC transmission technologies, strengthen UHV's going-global advantage, build a world-class brand characterised by Made in China 2025 and Lead by China, increase the influence of Chinese businesses in target markets, thus laying a solid foundation for the export of China's UHVAC power transmission technologies and equipment.

Fourth, China should evaluate the economic and social benefits of long-distance power transmission in a comprehensive, scientific and systematic way. UHV power networks are characterised by high safety risks, heavy investment and large scale. They should be assessed according to their input-output ratio and in relation to their economic and social benefits and safety.

Fifth, China should strengthen early-stage planning and demonstration of UHV projects. UHV has helped China become a world leader in power transmission. However, UHV projects are a long process requiring precise and scientific preliminary studies. Project execution usually takes 5–8 years or even longer. The projects often run through many provinces and over long distances, resulting in much cross-provincial coordination work and high safety risks. Research institutes need to be engaged at each stage of the planning process. Targets and schedules should be adjusted dynamically as conditions change during the planning period.

6.5   Nuclear Power

The next 10 years will be a watershed for nuclear power development. Between the first oil crisis in 1973 and the Three Mile Island nuclear accident in the USA in 1979, around 170 units were built

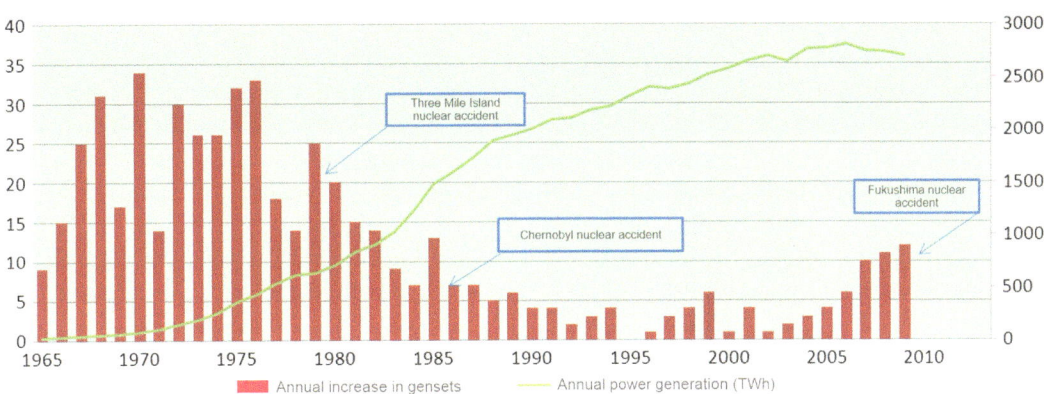

Fig. 43 Annual installed capacity and output of nuclear power

and put into operation across the globe. Many of those plants have a 40-year lifespan that will expire by 2020. Nearly a third of them will be decommissioned in the following 10 years if their operating life is not extended. Although new nuclear power capacity can replace decommissioned plants, it remains to be seen if this new capacity can smoothly deliver sustained development of the nuclear power industry.

6.5.1 Global Nuclear Power History

(1) Cyclical fluctuations

Global nuclear power history is cyclical. It is a history of fast expansion that is halted by a nuclear accident. Driven by global economic growth and an energy shortage, nuclear power entered the first rapid expansion cycle in the 1960s and 1970s. At times more than 30 nuclear power units were put into operation annually. After the Three Mile Island nuclear accident, the expansion of nuclear power slowed down. The second growth cycle occurred in the late 1990s when climate change and energy security made nuclear power attractive again. Three Mile Island and the Chernobyl nuclear disaster in the former Soviet Union in 1986 brought safety technologies and safety supervision to the fore. The Fukushima nuclear accident in Japan in 2011 brought nuclear power into hard times again. After Fukushima, safety and supervision again

became an issue. Nuclear power is now in a third cycle (Fig. 43).

(2) The impact of accidents on nuclear power development

The three nuclear power plant accidents—Three Mile Island, Chernobyl and Fukushima—had far-reaching effects on the development of nuclear energy. First, the annual increase in installed capacity shrank from 25 units to fewer than five after Chernobyl. Second, nuclear power technologies were upgraded to increase safety and procedures were improved to minimise human error. Multiple layers of independent and redundant protection were added (defence in depth) to strengthen response to failures, accidents or fires. Generation 2 technologies were improved, and Generation 3 and 4 reactors were developed, significantly enhancing safety and security. Third, supervision of nuclear safety became more and more sophisticated. After Three Mile Island, the USA improved its safety culture. The Institute of Nuclear Power Operations (INPO) was formed to set industry-wide performance objectives, criteria and guidelines for nuclear power operations. After Chernobyl, the International Atomic Energy Agency (IAEA) improved its nuclear safety standards, guidelines and procedures. In 1986 it published two treaties: the Convention on Notification of Nuclear Accident, by which states agree to notify others if a nuclear accident occurs within their jurisdiction; and the Convention on Assistance in Case of Nuclear Accident or Radiological Emergency, by which states that have signed the convention agree to help other states in which a nuclear accident has occurred. Fourth, more importance was attached to risk communication and public engagement. The complexity of nuclear power technologies makes it difficult for the public to understand, which creates antipathy. Nuclear accidents made people concerned about nuclear radiation. Governments started to focus on risk communication and public engagement to win the public over to nuclear energy.

(3) Occurrence and resolution of safety issues

The main line of nuclear power technology development is characterised by initial attention to engineering equipment, later to human operation, then to response to natural disasters. Increasingly reliable engineering equipment and standard operating procedures gradually replace human actions.

(4) Safety challenges

Nuclear power has characteristics that no other power generation technology has. It is safe, clean and sustainable, but if an accident occurs it has the potential to destroy people and the environment. International governance and oversight are therefore essential to ensure that safety is prioritised and risk is minimal.

The special characteristics of nuclear power pose numerous challenges on nuclear power safety. Nuclear power is highly sensitive clean energy with far-reaching effects if a major accident occurs. It needs the support of modern national governance. Its special characteristics pose strict requirements for scientific decision-making, public engagement and safety supervision. The rapidly decreasing cost of renewable energy and ever-stricter regulations on

nuclear power safety supervision challenge the cost-effectiveness of nuclear power.

The US Nuclear Regulatory Commission's (NRC) safety standards have played an important role in minimising risk. After Three Mile Island, the NRC defined two safety thresholds. First, the risk of instantaneous death to the individuals around a nuclear power plant, due to a nuclear accident, should not exceed one thousandth of the sum of the risks of instantaneous death to the whole of society from other accidents. Second, the risk of death by cancer to individuals around an operating nuclear power plant should not exceed one thousandth of the sum of the risk of death by cancer to the whole of society from other causes. Over the past 30 years and within the framework of the two thresholds, safety standards were improved, facilitating the safe and long-term development of nuclear power.

Experience shows that global nuclear power plants meet the two thresholds. However, the public does not accept nuclear power. Even though nuclear power does not increase the risk of death and cancer, the public's concern about nuclear power remains strong. Why? The two millesimal safety thresholds set more than 30 years ago do not meet current safety requirements. Fukushima exposed their shortcomings. Part of the public's concern after Fukushima is about the impact of radiation on marine life. The same goes for China's inland nuclear power plants, where the public is concerned about the impact on the Yangtze River's ecology. Environmental safety is not included in the two NRC safety thresholds.

The absence of environmental safety standards causes many problems. It is difficult to measure and compare the environmental impact of different industrial sectors effectively. The public's faith in nuclear power is therefore ambivalent. The lack of quantitative environmental safety thresholds means there is insufficient scientific evidence to justify improvements to safety standards. But more and more safety features are added, increasing the construction and operating costs of nuclear power plants. But the real effect of these features is hard to evaluate. These problems accumulate and increase the complexity of decision-making regarding nuclear power.

6.5.2 Global Experience of Nuclear Energy

(1) Long-term, clearly defined strategies are essential for the development of nuclear power

France defines nuclear power as an important strategic choice for energy development. After the first oil crisis in 1973, the French government introduced a policy to generate all its electricity with nuclear energy. The government established standards and a development programme, then gradually replaced fossil fuel-fired power plants with nuclear power stations. Nuclear power currently supplies three-quarters of the country's electricity. France has maintained continuity in nuclear power since the 1970s, winning first-mover advantages and blazing an independent trail to success.

In Japan, the strategic positioning of nuclear power is clear. The Long-term Plan for the Development and Use of Atomic Energy, issued in 1956 and revised every five years, provides a guideline for Japan's nuclear power industry. In 1978, nuclear power was defined by Japan as its base energy for the 21st century, helping the public and business community clearly understand national policies and facilitating the creation of a nuclear power industry, including R&D and nuclear fuel systems. In the 20 years following the Chernobyl disaster, Japan developed from an importer of nuclear power technologies to an exporter. Nuclear power generates a third of Japan's electricity.

South Korea's nuclear power strategy and programme enabled it to develop nuclear power independently. To realise its long-term nuclear energy objectives, South Korea established its Comprehensive Nuclear Energy Promotion Plan (CNEPP) in 1997, which it updates every five years. CNEPP includes long-term nuclear policies and objectives, targets, budget and investment plans. Clear planning and a firm strategy have enabled South Korea to use its limited human and financial resources in the best possible way. As a result, it has an independent nuclear power industry, innovative national brands and a sharp competitive edge in the nuclear world.

(2) Government policies support nuclear energy development

The USA has introduced preferential policies to encourage nuclear power investment and construction since the 1950s. In the initial phase of nuclear power (1955–62), the United States Atomic Energy Commission (AEC) clearly defined government and business responsibilities. It introduced four rounds of preferential policies to attract investment in nuclear power. The AEC became the US Nuclear Regulatory Commission (NRC) in 1975. The new organisation significantly enhanced nuclear power development efficiency by reforming the licence issuing system and simplifying the approval process. In 2005, to encourage business to invest in and construct nuclear power plants, Congress passed the Energy Policy Act, which provided tax incentives, loan guarantees and other investment support policies.

Japan granted different tax benefits for each stage of a nuclear power plant project to reduce the cost of development and deployment. Technologically, the government cooperated with power utilities to construct the first generating unit of each series, developed new technologies to test pilot plants and then transferred proven technologies to businesses for use. In terms of finance and taxation, investment funds were raised through fiscal means. For example, the Power Development Promotion Tax Law stipulates that tax benefits should be granted to nuclear power and other emerging energy industries every year through special accounts.

India reduced business risk by granting government subsidies and financial support. The financial budget allocated by India's Department of Atomic Energy to support the development of nuclear energy in 2004–05 reached $963 million, of which $540 million was awarded to atomic energy projects and $424 million to nuclear power projects.

(3) Sustained, stable and large-scale

To create a nuclear power industry, major nuclear power countries adjusted their industrial systems, gradually forming an industry suitable for their national conditions. Adjustment is characterised by government steering, relatively centralised control of the nuclear power industry, marketisation, and capacity building of the main suppliers, plant owners and advanced engineering firms.

In the USA, a market-oriented industry, dominated by large businesses and involving many project owners, was shaped in the industry's early years. In the middle of the last century, Westinghouse and GE were the first to master civil nuclear power technologies and become suppliers of nuclear power equipment, after taking part in the development of military nuclear power technologies. In the initial stages of nuclear power development, key suppliers provided turnkey services or advanced engineering to the project owners. During later stages, the market players became stronger and more diversified. When nuclear power grew in the 1970s, powerful and professional project management companies entered the field to provide project services to plant owners. They soon became major players in nuclear power plant construction. Some nuclear power plant owners chose to construct and manage projects themselves, combining project ownership with advanced engineering. Such a hybrid system is possible because of the USA's size, maturity and large number of small and medium-sized power utilities.

France formed a system with a single project owner, and with advanced engineering separated from the main suppliers. Having introduced technologies and management methods from the USA, France reformed its original nuclear industry to the system it has today: the Atomic

Energy Commission oversees research and development of military nuclear technologies and infrastructure, while nuclear fuel recycling is operated by professional companies. The Atomic Energy Commission and Électricité de France (EDF) formed Framatome, which later became the nuclear power business of Areva, before changing its name back to Framatome in 2018. Framatome, which is 75% owned by EDF, designs, manufactures and installs components, fuel, and instrumentation and control systems for nuclear power plants. EDF manages all nuclear power plants in France.

Japan's solid industrial base accelerated the formation of an approach using multiple project owners, suppliers and advanced engineering companies. After starting with technologies from Westinghouse and GE, Japan switched to its own local suppliers—Mitsubishi, Toshiba and Hitachi —to provide turnkey nuclear power plants to project owners. Japan's power transmission system is divided into 10 zones, each operated by an electric utility. After accumulating experience in how to build and operate nuclear power plants, the 10 utilities started to develop and invest in nuclear power plants independently, with the support of the main suppliers.

In South Korea, the government gradually deregulated the nuclear power industry. As the only nuclear power project owner, Korea Electric Power Corporation plays multiple roles as an investor, constructor, owner and operator. It held 41% of the shares of Korea Heavy Industries, through the assistance of the government, which enabled it to develop into a key nuclear island equipment supplier. Once this was achieved, Korea Electric Power Corporation divested its holding in Korea Heavy Industries, enabling the company to operate independently. Korea Heavy Industries was acquired by Doosan in 2001.

(4) Re-innovating older technologies is a pathway to success for late-mover countries

Some late-mover nuclear power countries— Japan and South Korea in particular—gained much valuable experience as they progressed through the nuclear power process, from technology introduction to independent operation.

First, a strong organisation and huge investment by government strengthen the independence of national nuclear power technologies. Proven nuclear technologies were researched under the auspices of the government, or jointly with businesses funded by the government, until they reached the pilot stage or commercialisation. To develop a nuclear power capability independently, Japan and South Korea invested in R&D and developed new technologies. The ratio of investment to independent R&D and technology development is 1:4 in South Korea and 1:8 in Japan.

Second, human and material resources are allocated in a centralised manner by government to achieve nuclear power independence. This process of resource allocation includes adapting the country's industrial system; creating a network of main suppliers; developing a professional advanced engineering capability; and establishing an entire integrated supply chain, from design to equipment supply, construction and operation.

(5) The public has the right to know and participate in decision-making on major issues of nuclear power

Nuclear power developers in the West regard the public's right to know and participate in decision-making as crucial to the smooth development of nuclear energy. Nuclear power countries have established a public intervention mechanism for nuclear power development and safety management.

Two rounds of public consultations are carried out before a nuclear power plant is approved in the USA. The Atomic Energy Act stipulates that a public hearing must be held, which covers a range of issues, including licence approval, suspension and abolition; construction permits; and licence-related regulations.

In France, nuclear power plants have to go through a national debate procedure that is open to the public at various levels, from special community sessions to local government.

France's nuclear power safety authorities emphasise that nuclear power decisions must be public and transparent. Government authorities allow residents living near nuclear power plants to track the status of the plants and provide information and even risk evaluation.

Japan adopted the right to public veto in 1996. Construction of a nuclear power plant must be approved by the public first. In Japan, nuclear power safety authorities disclose nuclear facility accidents and radioactivity control data to the public on dedicated websites.

(6) Laws and regulations ensure the safe and orderly development of nuclear energy

The legal system is the basis for policymaking. Major nuclear power countries passed their version of an atomic energy act early. These acts typically included regulations for each subsector of nuclear power development. This resulted in a sophisticated legal system that ensures the peaceful use of nuclear power and a strong foundation for policymaking.

The USA passed the Atomic Energy Act in 1946 and amended it in 1954. The act covers civil and military nuclear activity, the use and management of nuclear materials, the research and development of nuclear facilities, the organisational and management system, international activities, and the rights and responsibilities of stakeholders.

Japan developed the Atomic Energy Basic Law in 1955, which regulates management organisations and R&D institutions, the development and procurement of nuclear materials, nuclear fuel and reactor management, patents, radiation prevention, compensation, and so on.

India's Atomic Energy Act was passed in 1948 and amended in 1962. The act stipulates that all nuclear power and nuclear fuel recycling activities are controlled by the central government.

South Korea, France and other nuclear power countries have also established comprehensive legal systems to govern their atomic energy industry.

(7) **Strong, independent supervision and administration ensure the safety of nuclear power facilities**

Developed nuclear power countries value nuclear radiation safety supervision. They set up relatively independent nuclear safety regulators and provide human, material and financial resources for them. This ensures that the safety supervision laws are implemented effectively, and that the quality and safety of nuclear facilities during construction and operation are guaranteed.

The US National Regulatory Commission (NRC) is an independent nuclear safety supervision and management agency tasked with protecting public health and safety related to civil nuclear facilities and materials. It has two independent advisory committees and one nuclear safety and licence deliberation panel. The panel is responsible for providing advice and holding hearings on major nuclear safety supervision issues. The government ensures the NRC can perform its tasks effectively in terms of funds and people.

The Direction générale de la sûreté nucléaire et de la radioprotection (DGSNR), the main nuclear safety regulator in France, ensures nuclear safety and radiation protection in France on behalf of the government. DGSNR develops and implements safety policies and measures for civil nuclear facilities and discloses nuclear safety information to the public. The Institut de radioprotection et de sûreté nucléaire (IRSN) is the main technical backup force under DGSNR; it researches, examines and appraises nuclear facilities and activities. In addition, France has cross-ministry agencies and an atomic energy commission to coordinate its nuclear power sector on issues such as policy, military and civil nuclear power, and to prevent nuclear energy regulators from making one-sided decisions that prejudice the development of France's nuclear power industry.

Japan's nuclear safety regulators include the Cabinet Office; the Ministry of Education, Culture, Sports, Science and Technology; and the Ministry of Economy, Trade and Industry, each of which undertakes different nuclear safety supervision duties. The Atomic Energy Committee and Nuclear Safety Commission under the Cabinet Office review nuclear research, development and use policies, assess the safety review of the administrative departments and hold public hearings. The Ministry of Education, Culture, Sports, Science and Technology supervises safety at nuclear power facilities except for the plants themselves. The Nuclear and Industrial Safety Agency (NISA), set up by the Ministry of Economy, Trade and Industry, supervised nuclear and industrial safety. It was replaced by the Nuclear Regulation Authority in 2012, following the Fukushima nuclear disaster.

6.5.3 The Development of China's Nuclear Power Safety Strategies

(1) **Scientific scope of safe nuclear power development**

Human beings have been exposed to safety issues since ancient times. Safety threats can be divided into two categories. The first is survival threats in nature like wild animals, natural disasters and hunger. The second is safety threats from human factors, such as wars, the struggle for land and global warming.

Safety is a measurement of risk affordability. It is a relative concept, rather than an absolute one. Safety is also dynamic not static, which means the objective is to strike a balance between interests and cost. Safety is measured against several criteria. When evaluating a safety risk, its probability and consequences should be taken into consideration. The scientific community usually measures safety with risk: risk = the frequency of an event multiplied by its consequences.

In nature, safety means that risks are controlled at an acceptable level. Complete elimination of safety risks is not possible. Most of the factors affecting nuclear safety are technical ones. First, thanks to the massive investment in research and development over the past 50 years, most safety problems in nuclear power plants are technically solvable and a plant's technical safety level can be assessed. Second, effective technical protection measures are taken against radiation hazards during the whole nuclear energy project cycle—site selection, design, manufacture, construction, commissioning, operation and decommissioning. The same applies to the fuel cycle—exploration, mining, milling, conversion, enrichment, fabrication, power generation and waste disposal.

(2) Nuclear power safety strategies in the new era

The objectives of China's nuclear energy programme are national safety, energy security and environmental protection. To achieve these goals China should develop an advanced and world-class nuclear power industry that operates at the highest levels of safety, technological innovation, closed nuclear fuel cycle systems, and environmental protection.

The first priority is energy development. China will fight two battles during the energy transition for a long time to come. The first battle is to meet energy demand while completing the processes of industrialisation and urbanisation. The second battle is to cope with the increasing pressure to safeguard the environment. This means that China desperately needs clean energy. Nuclear power is indispensable to China's development.

The second priority is environmental protection. The policy of safety first is the key to public and environmental safety. Nuclear accidents are sudden, extremely impactful and hard to recover from. Safety requirements are far stricter in nuclear power than in other types of energy. A nuclear accident would jeopardise not only the momentum of the nuclear power industry in China, but also social and political stability. China has promised the international community that it will uphold its national responsibility for nuclear safety.

Third, an industry-wide nuclear power safety system needs to be established. Safety risks could be present during the whole process of nuclear energy development, generation and use,

involving all participants in the value chain. Independent safety supervision is an important measure to ensure safety. But ultimately, safety comes from great site selection, design, manufacture, construction, commissioning, operation and decommissioning. The same factors ensure a high level of plant availability and power supply and a sharper competitive edge for China's nuclear power industry.

The objectives of independence should be realised by relying on industry projects to enhance China's nuclear power technologies and equipment manufacture. Nuclear power regulations need to be developed and improved, and the nuclear power safety management mechanism put in place and optimised. The nuclear power safety supervision system needs to be improved for plants under construction and in service. A national nuclear accident response mechanism should be established to enhance China's emergency response capabilities. A national nuclear energy technology innovation programme consisting of key technologies, equipment, demonstration projects and technical innovation platforms needs to be established. Advanced pressurised water reactors in the million-kilowatt category should be mainstream, and efforts should be made to develop new technologies like high temperature gas-cooled reactors, fast reactors and small modular reactors. Nuclear fuel supply and spent fuel treatment systems should be improved to secure nuclear power's long-term development.

Fourth, environmental safety standards for nuclear power need to be explored. Exploring environmental safety standards in nuclear power is an historical mission for China to be among the world's best. China is expected to reach this top level in third-generation technology. The countries at the peak of nuclear power technology are being replaced—Russia and South Korea's nuclear power technologies are still highly rated in the global market, and China has started to enter the UK's nuclear power market.

Environmental safety standards are an important step from industrial to ecological civilisation. The development of environmental safety standards will be an important phase in the transition and reform of industrial civilisation. The emergence of any civilisation is the outcome of a successful response to tough environments. The Fukushima nuclear accident highlighted the shortcomings of global industrial civilisation. In this sense, the nuclear power industry has been pushed to the front of the transition of industrial civilisation.

6.5.4 Pathways and Measures to Develop Nuclear Power Safely

(1) **Develop safe, authoritative, stable and efficient nuclear energy development strategies**

Guided by the principles of "strategic necessity, environmental safety, guiding standards, coordination and stability", the strategy and policy of "efficiently developing nuclear energy while ensuring safety" should be enriched and improved. A mandatory, authoritative and strategic development plan should be drawn up, guided by the Scientific Outlook on Development (one of the guiding socioeconomic principles of the Communist Party of China), which emphasises the comprehensive, harmonious and sustainable development of China for the Chinese people. After adequate and scientific demonstration, the plan should cover strategic positioning, development guidelines and objectives, the technical route to achieve those objectives, R&D, the nuclear fuel cycle, resource guarantees, independent equipment manufacture, training people in nuclear power, and make the plan into an integral part of national energy programmes.

First, the Scientific Outlook on Development should be followed to promote the development of nuclear energy, improve nuclear energy's position in China's energy development strategies, and include nuclear energy in the national energy development programmes.

Second, China should strengthen research and development of nuclear energy strategies and plans, put in place the relevant organisations and funds, draw up nuclear energy development strategies and plans, define the targeted installed capacity of nuclear power and priority projects, optimise project schedules, and prepare an authoritative medium- and long-term national nuclear energy development plan.

Third, the actions and procedures to implement strategic plans should be consistent, and the duties and rights of government agencies, businesses and scientific research institutions should be defined, thus truly implementing the strategic plans and generating good results.

Fourth, when developing strategy studies, policy discussions and plans around "independent design, independent R&D, independent manufacturing and independent operation", some important strategic decisions, technology pathways and policies should be included in national policy papers and implemented in the projects.

(2) **Establish a nuclear energy law and regulatory system based on the Atomic Energy Law**

Atomic energy legislation is urgently needed to provide a reference point for important decisions on nuclear activities. More than 20 years have passed since China proposed the Atomic Energy Law. Today, the principles of the Atomic Energy Law are still being discussed. Requisite laws for the development of nuclear energy are absent. Due to the severe delay in passing legislation, conflicts and problems that arise in nuclear energy development are regulated with temporary policies. This is not in line with China's nuclear power development requirements.

First, the Atomic Energy Law should be drafted as soon as possible to provide legal protection for the development of nuclear energy. The legislation should be effective, practical and focused, and it should define in a comprehensive manner the timelines and protocol for disclosing information on nuclear accidents.

Second, legislation for specific aspects of nuclear energy should be drawn up in an orderly manner and by priority. The regulations for nuclear damage compensation, spent fuel management and radioactive waste management should be introduced as soon as possible.

A relatively complete Atomic Energy Law should be introduced by 2020.

Third, safe and smooth operation of nuclear power plants should be ensured by finalising nuclear power safety regulations and procedures, tightening safety law enforcement and supervision, adopting strict labour discipline and standard operating procedures, and strengthening the nuclear accident emergency system.

(3) **Establish a separate, independent, authoritative and specialised nuclear power safety supervision system**

In light of the nuclear supervision reform trends in other countries and the actual problems in China, four nuclear power safety supervision reform principles—"separate, independent, authoritative and specialised"—should be followed to strengthen nuclear safety supervision and enable China to develop nuclear power at large scale in the future.

First, the correct separation of functions. In China, management and safety supervision of nuclear power are already separated. The major task at present is to clearly define departmental functions.

Second, focus on specialised legislation and functions. Legislation on nuclear energy should contain provisions on not only promoting nuclear energy development, but also ensuring nuclear power safety. Absence of either will have negative impacts on the healthy development of the nuclear power industry. In addition to basic safety supervision for nuclear facilities and materials, it is vital that liabilities and the emergency response of third parties are included in the nuclear power safety assurance system.

(4) **Explore and develop environmental safety standards for nuclear power within the framework of the three millesimal safety targets**

Environmental safety standards for nuclear power should be developed within the framework of the three-millesimal safety targets in a three-step approach.

Step 1: China should focus on the key concerns of the people by assessing the risk to public health and the potential impacts of radiation and thermal pollution on the environment. It should select typical evaluation objects, launch pilot and demonstration projects, and compare the impacts of other industrial activities. A pilot risk evaluation project on environmental safety should be launched using existing data and facts. And an environmental safety standard development mechanism—based on data and evidence that can be communicated, questioned and optimised —should be established.

Step 2: The scope of environmental safety targets should be enriched with the experience gained from the pilot and demonstration projects to expand risk evaluation and evaluation objects. Evaluation capacity building should be strengthened by standardising safety evaluation guidelines. A mechanism to upgrade nuclear power environmental standards should be established through information disclosure and public engagement.

Step 3: As data and facts are accumulated and the environmental risk evaluation system improved, environment insurance and reinsurance systems need to be introduced. These in turn will substantially improve China's environmental risk evaluation, prevention and mitigation ability, allowing it to be an integral part of the state's governance capability. Nuclear power environmental safety standards should be improved by comprehensively comparing risks and through a mature risk evaluation system. A development mechanism featuring interaction between, and checks and balances for, environmental safety standards, nuclear power technologies and environmental safety insurance should be formed.

(5) **Invest more in scientific research to enhance nuclear technologies**

Investment in nuclear science capacity building should be centralised and improved. In particular, investment in major nuclear science projects and research and testing facilities should be secured to ensure stable financial support for basic nuclear research.

First, investment in key disciplines should be extended to gain approval for national projects of large advanced pressurised water reactors. More investment should be made in R&D of third-generation nuclear technologies, especially on technology demonstration projects of large advanced pressurised water reactors. Production facilities for these Chinese reactors should be in place by 2020, laying a solid foundation for the future large-scale development of nuclear power.

Second, the research and development of prototype fast reactor nuclear power plants and their fuel cycle technologies should be strengthened. More money should be invested in basic R&D to support independent innovation in nuclear energy.

Third, a high-quality, competitive and open nuclear power education system should be established by taking full advantage of China's nuclear energy education resources at universities and research institutions. A programme should be initiated to improve the development of professional technicians, operations and management personnel and skilled workers.

Fourth, talent development channels should be expanded. High-level talent should be given a leading role through talent development programmes and by promoting international cooperation and exchange. Scientific and technological innovation team building should be strengthened. Motivation should be enhanced by increasing remuneration and incentives for nuclear researchers, thus improving innovation and creativity.

(6) **Establish a scientific decision-making and interaction mechanism to convince the public of nuclear energy's strategic importance**

The public's right to know and participate in the decision-making process on nuclear power development should be protected. Information on risk and benefits evaluation should be provided, and public participation in the debate about nuclear power should be allowed. The national strategic requirement for the harmonious development of energy, society and the environment should be met through interactive decision-making with the public. Public opinion should be eased scientifically through the main issues, using various channels and approached from different perspectives to eliminate the fear of nuclear power. Third parties, including independent associations, should be encouraged to help supervise the process. Energy education facilities should be built to popularise nuclear knowledge free of charge to increase the public's faith in safety.

First, industrial management and decision-making authorities should cherish and value the public's trust in nuclear power and protect their right to know and take part in the making of nuclear power development decisions. Public participation in the discussion about nuclear power plans should be allowed, and the national strategic requirement for the harmonious development of energy, society and the environment should be met through interaction with stakeholders.

Second, safety regulators, including the National Nuclear Safety Administration and the Ministry of Ecology and Environment, should strengthen their supervisory role and establish transparent report channels to eliminate public concern. Nuclear power monitoring facilities should be set up to make the disclosure of radiation information more transparent and alleviate the public's concern about nuclear power. The risk and benefits evaluation results should be made public to help people better understand the risks and benefits of nuclear power.

Third, nuclear knowledge should be popularised among sensitive groups to eradicate irrational perceptions. In particular, basic information on radiation and nuclear accidents like Chernobyl should be provided scientifically and objectively to reduce rumour-spreading.

Fourth, the government should do its best to involve local communities and leverage their public influence through diverse media channels, thus boosting trust and winning the public's support.

Fifth, residents living close to nuclear power plants should be rewarded. Local government should, together with environmental authorities and nuclear power companies, bear people's interests in mind and create long-term health records to eliminate unnecessary fears. In addition, thermal pollution and electricity tariffs should be researched, and benefits and compensation awarded.

6.6 Unconventional Gas

6.6.1 Current Developments and Trends in Global Unconventional Gas

(1) Definition and classification of unconventional gas

Unconventional gas refers to gas resources that previously could not be explored and exploited economically by conventional methods and techniques. It is characterised by its abundance, poor physical properties of reservoir (porosity: less than 10%; permeability: lower than 1×10^{-3} μm^2).

Unconventional gas usually includes tight gas (short for tight sandstone gas), shale gas, coalbed methane and natural gas hydrate. Tight gas means that the natural gas deposits in sandstone (carbonatite, volcanic) strata are tighter than in conventional reservoirs. In China, tight gas is called low, very low or ultra-low permeability gas. Coalbed methane refers to the gas absorbed on the surface of coal and in micro-fissures in the coal bed. Shale gas means that natural gas exists in a free state in minute pores or fissures inside the shale and adsorbed on the surface of minerals and organics. Natural gas hydrate is a solid crystalline compound composed of water and natural gas formed at low temperature and high pressure, also called combustible ice.

(2) Current developments in unconventional gas in major countries

First, tight gas. Tight gas resources are abundant (about $210 \times 1,012$ m^3) across the globe and distributed widely in Asia-Pacific, North America, Latin America, the former Soviet Union, the Middle East and North Africa. Survey results from the U.S. Geological Survey show that around 70 basins with tight gas have been discovered in the world.

A dozen or so countries and regions, including the USA, Canada, Australia, Mexico, Venezuela, Argentina, Indonesia, China, Russia and Egypt are exploring and exploiting tight gas deposits. The USA and Canada are leaders in this regard. Tight gas exploration and exploitation in the USA started in the late 1970s when national gas output decreased substantially, aggravating the imbalance between supply and demand. So, the US government introduced a series of tax incentives and subsidy policies to encourage the development of unconventional gas and low-permeability gas deposits, leading to significant breakthroughs. As a result of the support policies, tight gas became a major part of US gas production, hitting 60 billion cubic meters (bcm) and 100 bcm in 1990 and 1998 respectively and amounting to 175.4 bcm in 2010 (accounting for about 29% of total US gas output). Currently, in the USA, tight gas is being developed in the Greater Green River, Denver, San Juan, Piceance, Powder River, Uintah, Appalachian and Anadarko basins in the Rocky Mountains region. Canada's tight gas is mainly distributed in Alberta in west Canada. Canada drilled its first industrial tight gas well in 1976. The Hoadley and the Milk River gas fields discovered later are promising prospects for tight gas. Some 6,400 km^2 of land have tight gas deposits in Canada, with a geological reserve of 42.5×10^{12} m^3.

Breakthroughs in key technologies are an important driver behind the rapid development of tight gas in North America. The comprehensive application of horizontal well drilling, underbalanced drilling, well completion and gas reservoir protection technologies in the 1990s boosted gas yield sharply, resulting in rapid increases in tight gas production in North America.

Second, shale gas. Shale gas resources have not been evaluated worldwide yet. Most of the volume data is estimated from existing geological information. Incomplete estimates by Rogner

Fig. 44 Gas output in the USA by product, 2007–15. *Source* U.S. Energy Information Administration

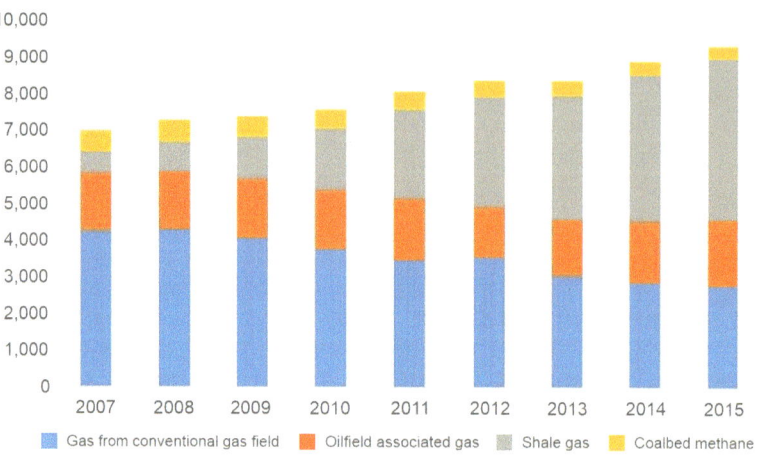

in 1997[69] show there are around 922 trillion cubic metres of unconventional gas in the world, nearly half of which is shale gas (456 trillion cubic metres, mainly in North America, Asia, Latin America, the Mediterranean and Australia). Recent data released by the U.S. Energy Information Administration (EIA) show that geological reserves of shale gas in 137 sets of shale formation in 95 shale gas basins in 42 countries (including the USA) in 10 geographical regions of the world are around 1,013 trillion cubic metres. 220.69 trillion cubic metres of shale gas are technically recoverable, up 33.67 trillion cubic metres from the reserves announced in 2011.

Pre-project evaluation and fundamental research of shale gas resources have been carried out in more than 30 countries. Large-scale commercial exploitation has been achieved in the USA and Canada. China has entered the stage of massive development. Other countries are still at the initial stage of shale gas development. As the earliest pioneer in terms of shale gas exploration and exploitation in the world, the USA possesses the most mature technologies. The country produced 438.2 bcm of shale gas in 2015, accounting for 47% of its total gas production. Shale gas became the largest contributor in its gas mix (Fig. 44).

Technologies and policies empowered the rapid development of shale gas in the USA. The success of the US shale gas revolution has driven research and attempts at emulation across the world. It is widely believed that technology, policy, market, regulation and infrastructure are the five drivers behind shale gas evolution in the USA.

Technological progress is the principal driver. Thanks to continuous technological innovation, the cost of exploiting US shale gas has been reduced to only half of that of conventional gas. The cost of drilling and constructing a shale gas well at Eagle Ford in Texas, with a horizontal segment of around 1,500 m, was $0.86 per cubic foot, which was drilled and completed within 20 days.

Second, preferential policies provided great support for shale gas development. The US government invested significantly or set up specialised research funds to subsidise preliminary R&D and exploration studies on shale gas. It was estimated in 2012 that the US government had invested more than $6 billion in unconventional gas exploration and production since the early 1980s, almost $2 billion of which was used for training and research. In addition, the US government transferred the preferential tax policies granted to the upstream development of conventional oil and gas to shale gas development. The government also introduced five tax incentives for the oil and gas industry: intangible

[69]H-H. Rogner, An assessment of world hydrocarbon resources, in Annual Review of Energy Environment, 1997, 22, pp. 217–262.

drilling costs tax deduction, tangible drilling costs tax deduction, rental deduction, allowing a working interest to be classified as active income, and extending the depletion allowance to small producers. These incentives strongly encouraged drilling and development investment by medium- and small-sized enterprises, and accelerated shale gas exploration and production.

Third, the open market environment is a powerful impulse. The shale gas exploration and development market in the USA is mature and features diversified participants and sound competition. There are thousands of shale gas companies at present, with more than 2,000 drilling rigs. 85% of shale gas is produced by small and medium-sized companies, which take the lead in making technological and industrial breakthroughs. Large companies enter and participate in the market by acquiring and merging small and medium-sized ones. This creates a market environment characterised by a mix of large. medium and small companies, an optimal combination of specialisation and collaboration, and the efficient flow of capital throughout the value chain.

Fourth, reasonable regulation acts as an important guarantee for the shale gas industry. The US government values regulation in shale gas exploration and production and reasonably delegates regulatory power to state governments. The power to regulate interstate energy business activities is shared by federal and state governments. In case of conflict between federal and state regulations, the former prevails. Should federal standards be lower than state standards, both standards apply. The federal government intervenes finitely in shale gas through supervision of environmental regulations and interstate pipeline access. Specific regulatory power covering the time and place to exploit, as well as gas well standards, is delegated to states.

Fifth, complete infrastructure and third-party access supports the development of shale gas. Gas pipeline networks and urban utility gas facilities in the USA are well developed, greatly reducing the need for early development investment and market risk. Statistics from the U.S. Energy Information Administration show that the lower 48 states now have 490,000 km of pipelines, of which 349,000 are interstate and 141,000 intrastate. The complete separation of gas exploitation from transport, the regulation of pipeline transmission prices and the deregulation of gas prices powerfully support the commercialisation of shale gas.

Third, coalbed methane (CBM). According to the data from the International Energy Agency, global coalbed methane resources are abundantly and widely distributed in Russia, Canada, China, the USA, Australia and other countries (Table 15).

The USA, Canada and Australia are leaders in CBM development. The USA produced and used coalbed methane on a large scale in the 1990s. In 2007–09, the USA produced more than 56 billion cubic metres (bcm) of CBM, a record equivalent to 7% of US gas output. CBM output decreased year by year thereafter to 33.5 bcm in 2015 (accounting for 3.6% of gas output). Canada and Australia made breakthroughs in CBM development and commercialisation, but on a far smaller scale than the USA. Canada's CBM output increased rapidly after 2004 to 7.5 bcm in 2010 (accounting for 6% of Canada's gas production). Australia's CBM yield in 2005 was only 1.8 bcm, which increased sharply by 40% to 7.4 bcm in 2010 (equal to more than 13% of Australia's gas production).

Breakthroughs in key technologies and cost effectiveness are the two most important factors for CBM industrialisation. The US government conducted groundbreaking research in the 1970s by investing around $400 million in the San Juan and Black Warrior basins. This generated the theory of desorption-diffusion-seepage and the process of drainage-depressurisation-gas

Table 15 CBM reserves in major countries (trillion cubic metres)

Russia	Canada	China
17–113	18–76	37
USA	Australia	Germany
22	8–14	3

Source IEA

recovery, which established a basic theoretical system for exploiting and processing CBM. In the following decade, the US government invested more in research and completed an assessment of the nation's CBM resources. It also spent $6 billion on drilling experiments and on proving the technical feasibility of CBM recovery, both of which are the pre-conditions for CBM industrialisation.

Canada's experience also proves that breakthroughs in exploration and production technologies are a prerequisite for the development of a CBM industry. Canada increased output per well substantially and then achieved mass production after making significant advances in multi-branch horizontal wells, coiled tubing fracturing and nitrogen foam fracturing. These successes were the result of a technology R&D programme tailored to Canada's own CBM conditions.

Around 30 years of subsidies lay behind the growth of a healthy CBM industry. The US government supported coalbed methane development projects with tax subsidy policies and a special-purpose fund started in accordance with the Crude Oil Windfall Profit Tax Act of 1980 for financing the development of unconventional energy with windfall profit taxes on conventional energy. The tax subsidy policies were implemented in two stages. During stage 1 (from 1980 to 2002), CBM was subsidised for much of the period. Stage 2 began with the new Energy Policy Act of 2003, which set the subsidy threshold at single well production capacity of no more than 56,700 m^3 per day. At their highest, the subsidies accounted for half the market price of CBM. During the 30 years, the US government spent billions of dollars subsidising CBM.

Fourth, natural gas hydrate. The carbon content in gas hydrate is estimated to be more than double that in other known fossil fuels, making it a next-generation strategic energy source. Many countries and regions have surveyed hydrate deposits, discovering deposits at more than 130 sites. The first country to recover natural gas hydrates is the former Soviet Union, which pilot-produced gas hydrates in the 1970s at Messoyakha, Siberia, with depressurisation and inhibitor injection methods. After that, Canada,

the USA and Japan launched pilot gas hydrate drilling and production projects and made significant progress. The pilot gas hydrate production project at Mallik, Canada in 2008 proved the feasibility of the depressurisation method.

6.6.2 Current Developments and Challenges in China's Unconventional Gas

(1) **Current developments**

First, tight gas. China has huge tight gas potential. Preliminary estimates using the analogy method show that China has 10 trillion cubic metres of recoverable tight gas. The cumulative proven reserve rate is only 18% at present. The huge potential could be tapped by speeding up exploration and production. The biggest deposits of tight gas in China are in the Ordos and Sichuan basins, followed by the Tarim, Junggar and Songliao basins, which together hold 90% of China's total tight gas resources.

The key technologies for recovering tight gas are basically mature. China has made great progress in recent years by learning from other countries about the main technologies of tight gas exploitation, including vertical, cluster and horizontal well-staged fracturing. Fracturing reformation increases output per well to 10,000–20,000 m^3 per day. At the central Sulige gas field in Inner Mongolia, for example, average daily single-well output has been a steady and cost-effective 10,000 m^3 for four years.

Reserves and output are increasing rapidly. In recent years, geological reserves and output of tight gas have increased by 300 bcm and 5 bcm respectively year-on-year. Cumulative proven geological reserves of tight gas were 3.3 trillion cubic metres at the end of 2011, accounting for 40% of China's total geological gas reserves. 1.8 trillion cubic metres of tight gas were recoverable, accounting for about a third of recoverable gas reserves in China. Tight gas output in 2011 amounted to 25.6 bcm, accounting for about a quarter of the country's total gas production.

Second, shale gas. China has huge shale gas resource potential. Although China's shale gas

Fig. 45 Comparison of prediction data of China's recoverable shale gas reserves

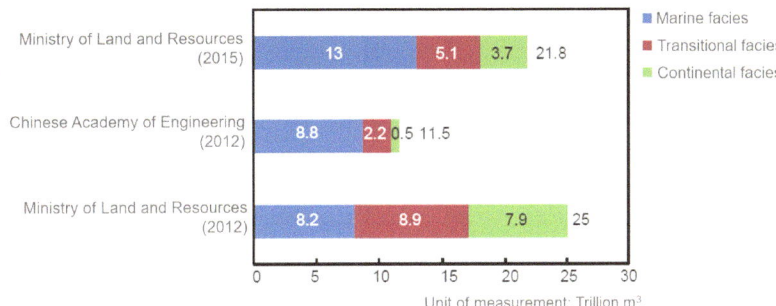

reserves have not been ascertained yet, preliminary estimates by Chinese and foreign researchers show that shale gas resources in China are abundant, diversified, widely distributed and highly promising. The resource base needed for fast exploration and production is already in place. According to the report titled Results of the Survey and Assessment of Shale Gas Resource Potential of China and Favourable Zone Optimisation published by the Ministry of Land and Resources of the People's Republic of China in March 2012, China's geological reserves of onshore shale gas are 134.42 trillion cubic metres, of which 25.08 trillion cubic metres (excluding Qinghai-Tibet) are recoverable. In 2015, the resource assessment results from the Ministry of Land and Resources of the PRC showed that China had 21.8 trillion cubic metres of technically recoverable shale gas, including 13 trillion cubic metres in marine facies, 5.1 trillion cubic metres in transitional facies and 3.7 trillion cubic metres in continental facies (Fig. 45).

Breakthroughs have been made in exploration and exploitation. China had granted 44 exploration permits covering 144,000 km^2 by 2015, proving geological reserves of 544.1 bcm. Many shale gas wells in and on the periphery of the Sichuan Basin generate industrial gas flow during exploration of the marine facies shale stratum of the Silurian Longmaxi Formation, proving great resource and development potential. Exploration of the continental facies shale stratum of the Triassic system in the Ordos Basin has also generated gas.

Four national shale gas demonstration blocks have been set up in China, i.e. Sinopec's Fuling demonstration block in Chongqing, CNPC's Changning-Weiyuan demonstration block in Sichuan, CNPC's North Yunnan and Guizhou-Zhaotong demonstration block, and Yanchang Petroleum's Yan'an demonstration block in Shaanxi. China's annual shale gas production in 2016 reached 7.88 bcm, third after the USA and Canada. Commercialised and large-scale development has been achieved at the Jiaoshiba (Fuling), Changning-Weiyuan and Zhaotong blocks. Up to the end of 2016, China had invested RMB 8.8 billion in exploration and production of shale gas, leading to 50 drilled exploratory wells and 92 development wells.

Support policies and mechanisms have been strengthened. The Ministry of Finance and the National Energy Administration unveiled subsidy policies for the development and use of shale gas in 2012. From 2012 to 2015, shale gas exploitation companies received subsidies from the central budget at a rate of RMB 0.4 per cubic metre. In 2015, the two authorities announced that the subsidy would continue to apply to shale gas development and use during the 13th Five-Year Plan (2016-20), albeit at the lower rate of RMB 0.3 per cubic metre for the first three years and RMB 0.2 per cubic metre for the last two years.

The Shale Gas Industry Policies of the National Energy Administration, published in 2013, provides rules and guidance on industrial regulation, demonstration block construction, technical policies, markets and transport, economical use and environmental protection for the healthy development of the shale gas industry. The shale gas joint development mechanism for joint ventures was explored and developed during the 12th Five-Year Plan (2011–15). Joint ventures between Sinopec/CNPC and local enterprises have been formed at Fulin (Chongqing), Changning (Sichuan) and other shale gas blocks.

Third, coalbed methane (CBM). China has abundant CBM resources. According to the 2015 evaluation by the Ministry of Land and Resources of the PRC, China's geological reserves of CBM in the five major blocks were 30.05 trillion cubic metres, of which 12.5 trillion cubic metres were recoverable. The North China block boasts the biggest deposits, with geological reserves accounting for 46% and recoverable deposits 41% of the total. Most of China's CBM resources are in the Erdos and Qinshui basins. In 2016, another 57.612 bcm of geological reserves were proven, a huge increase on the previous year.

The CBM industry enjoys rapid development. Driven by coal mine gas governance and a broadening supply of gas, China began to research, explore and exploit CBM in the 1990s. Small-scale industrialised production of CBM was achieved during the 11th Five-Year Plan (2006–10). By 2016 China had invested RMB 1.591 billion in CBM exploration and production, with 87 drilled wells and 97 development wells. 4.495 bcm of CBM was produced on land in 2016.

The industrial policy system for CBM is basically in place. During the 12th Five-Year Plan (2011–15), the State Council issued Opinions on Further Strengthening Coal-mine Gas Prevention and Control, followed by Opinions on Further Accelerating Coalbed Methane (Coal-mine Gas) Extraction and Exploitation. Government authorities followed up on these policies by issuing the Coalbed Methane Industry Policies and Action Plans for Coalbed Methane Exploration and Exploitation. They also introduced preferential policies, such as refunding VAT on electricity generated with coal mine gas, which improved the CBM industrial policy system. Major coal provinces developed and implemented support policies. In addition to subsidies from central government, Shanxi and Shaanxi provinces subsidised the extraction and use of CBM at RMB 0.1 per cubic metre. In

Hunan province, each new gas-fired power plant receives RMB 800,000–1 million as an incentive. Anhui, Henan and Guizhou provinces have established special-purpose funds to support the extraction and use of coal mine gas.

Fourth, natural gas hydrate. Natural gas hydrate deposits are abundant in polar tundra sandstone and seabed sandstone. Estimated reserves are currently 83.7 trillion cubic metres, mainly in the South China Sea and Qinghai-Tibet Plateau tundra. China has made breakthroughs in natural gas hydrate development. The Blue Whale 1 rig is in production and 20 key technologies have been innovated. China is the first country to produce combustible ice in a safe and controllable pilot project in an argillaceous silt reservoir. Natural gas hydrate was successfully produced in a trial project in May 2017 in the South China Sea for 60 consecutive days. 309,000 m^3 of gas were produced in total. The project set a world record in terms of duration and volume.

(2) **Problems and challenges**

China's unconventional gas has entered the stage of rapid development, but obstacles still exist in technology, systems, policy and cost-effectiveness.

First, China does not completely master the technologies. CBM deposits in China are complex and demanding in terms of development technologies. Fundamental breakthroughs in basic theories and technical processes have yet to be made for the development of CBM from low-permeability, high-stress tectonic coal. The economical use of low-concentration coal gas and surface hole extraction in mining areas also needs to be improved. As for shale gas, China has not mastered deep-stratum development technologies yet, making the extraction of shale gas resources at depths greater than 3,500 m in south Sichuan a great challenge.

Second, there is insufficient market competition. Currently, the exploration and production of tight gas, CBM and shale gas are dominated by several state-owned enterprises that are supervised and regulated by the central government.

Access channels for private capital are not smooth, making effective output difficult to achieve. Mining rights overlap is a longstanding issue. Effective constraints through the policy of "coalbed methane before coal mining" are not in place. Mining rights in favourable shale gas zones often overlap with conventional oil and gas mining rights. In addition, the technical services market is underdeveloped and averse to improving exploration and production technologies and equipment through market competition, making cost reductions and production increases difficult.

Third, effects of support policies and incentives have been weakened. Support policies, including refunded coalbed methane VAT, are not well implemented in some regions. The subsidy for shale gas is lower than during the 12th Five-Year Plan (2011–15). There are no subsidies for tight gas. Furthermore, the Chinese government has repeatedly lowered natural gas prices, which has weakened the effects of subsidies and tax reduction and exemption measures.

Fourth, exploration and production are not always cost-effective. Shale gas wells are characterised by high single-well investment costs, long implementation cycles and rapidly decreasing output over time. This brings heavy capital pressure and investment risk to shale gas development businesses. Gas content and pressure increase, and coalbed gas permeability decreases, with depth, making CBM extraction more difficult. In some regions, tight gas deposits are in difficult geological conditions. The cost of exploiting the deposits is high and the output-input ratio is low, leading to poor economic returns or even losses.

6.6.3 The Role of Unconventional Gas in the Energy Revolution

Unconventional gas is the key to the sustainable development of the oil and gas industry and an enabler for the global oil and gas production revolution. It can safeguard China's energy supply security, revolutionise energy production and consumption, and help sectors of the economy develop and expand.

Unconventional gas can boost China's gas production and improve gas supply security. China's gas output is forecast to increase from 135 bcm in 2015 to 350 bcm in 2030, of which more than 80% will come from unconventional sources. However, gas import volumes will increase and reliance on overseas gas will intensify. Between 2020 and 2030, more than 40% of gas will be imported, that figure will exceed 50% in 2050. China, due to natural constraints, has limited potential to increase conventional gas production. Unconventional gas is, therefore, the key to keeping gas imports below 50% by 2030. Increasing the development of unconventional gas is essential for China's gas supply security (Table 16).

Tight gas and shale gas are the two most important growth points for China in unconventional gas, as they were for the USA as well. China's tight gas reserves are relatively known and have been explored with mature technologies —the two preconditions for fast development and use. It is expected that China will build a

Table 16 China's future gas supply capacity

	Conventional gas	Tight gas	Shale gas	Coalbed methane (CBM)	Total volume of domestic gas	Imported gas	Proportion of imported gas (%)
2015	126		4.6	4.4	135	61.4	31
2020	130	37	30	10	207	150	42
2030	130	100	80	40	350	250	42
2050	130	100	100	40	370	430	54

Note Unit of measurement: billion cubic metres (bcm)
Source The 13th five-year plan for natural gas, Research Institute of Petroleum Exploration and Development (2016)

complete, efficient and low-cost tight gas technology system by 2020, and that the Erdos, Sichuan and Tarim basins will reach peak production of 37 bcm by 2020. Tight gas exploration and production at the Dzungaria, Songliao, Turpan-Hami, Bohai Bay and Qaidam basins will achieve breakthroughs by 2030, with output rising to 100 bcm. China has abundant shale gas resources and has made preliminary advances in their exploration and production. Development of Sichuan and Yunan's marine facies shale gas is expected to take off around 2020. Production at the Fuling, Changning, Weiyuan and Zhaotong shale gas demonstration blocks will reach 30 bcm by 2020. Exploration and production of marine facies in the organic-rich shale formation of the Palaeozoic Erathem era in south China and of lacustrine and marine-terrigenous facies in organic-rich mudstone strata in north China could make great progress by 2030, achieving a production capacity of 80–100 bcm of shale gas.

Natural gas hydrate is expected to become a key contributor to the global oil and gas production revolution. In addition to boasting world-leading rigs, China has made breakthroughs in natural gas hydrate development and key technologies. It has successfully produced gas, safely and controllably in a pilot project, from combustible ice in an argillaceous silt reservoir ahead of other countries, meaning that China is a world leader in terms of natural gas hydrate development. In the past two decades, North America has led the global shale oil and gas revolution through the broad application of horizontal drilling and hydraulic fracturing. With its continuously evolving technologies and equipment, China strives to become the first country to commercially recover natural gas hydrate by around 2030. This will enable China to become internationally competitive in natural gas hydrate exploration and production and guide the revolution in global oil and gas production.

The exploration and production of unconventional gas should be strongly promoted in an orderly manner. China's unconventional gas resources are diversified, abundant and the key to the sustainable development of its gas industry. However, quality, technical maturity and extraction processes differ from one type of unconventional gas to another.

First, the development of tight gas should be regarded as a priority and implemented to effectively replace conventional energy sources. Breakthroughs should be made in multi-layer fracturing of thin reservoirs in vertical and horizontal wells and low-cost well drilling and completion technologies, so as to build a complete, efficient and low-cost tight gas technology system. Support policies should be adopted, such as tax incentives for imported equipment, tax deductions for exploration costs, and tax deductions and exemptions for businesses.

Second, more investment should be made in shale gas and coalbed methane to make them the main means to increase gas production. Establishing mature and proprietary exploration and production science and technologies, suitable for China's shale and coalbed reserves, should be accelerated through research. The survey and assessment of national shale gas and coalbed methane resources should also be accelerated. Advances should be made in key technologies and equipment, including seismic interpretation, measurement while drilling, rotary steering and fracturing monitoring.

Third, the commercialisation of natural gas hydrate as a strategic substitute should be achieved as soon as possible. Resource surveys and development studies should be completed within the next 10 years and commercial exploitation by around 2030. This will enable China to develop strong international competitiveness in natural gas hydrate exploration and production.

6.6.4 Strategic Positioning and Policy Suggestions

First, the positioning of unconventional gas as a major driver of gas output growth should be clearly defined. The Chinese government's Energy Production   and Consumption Revolution Strategy (2016–30) clearly describes China's gas development direction and targets. The strategies and plans developed by central government should be well implemented by now. However, clear national gas development strategies for 2050 should be developed. The positioning of unconventional gas as a key driver of gas output growth, and the strategic goals and means of delivery, should be clearly defined. Generating huge monopoly profits by virtue of large-scale investment and the relative scarcity of resources is the traditional pathway for the international oil and gas sector. Faced with increasingly fierce competition from renewable energy sources, the oil and gas industry must be transformed and revolutionised. Low-cost, green development should be the main strategy of the oil and gas revolution.

Second, the oil and gas industry requires deep reform. The Ministry of Land and Resources of China listed shale gas as an independent mineral energy resource and invited bids for the exploration rights. Allowing non-oil and gas companies and private enterprises to bid for rights breaks the upstream monopoly of the major state-owned oil companies, which is a big step towards oil and gas reform. However, more than 70% of unconventional oil and gas resources overlap with conventional deposits. The blocks with the highest-quality resources are still being explored by the three major oil companies and Yanchang Petroleum, because bidding has not yet been introduced. The next step should be to reform the administration of mining rights for existing oil and gas blocks. It is suggested that exploration and mining rights be awarded by tender, auction or listing. In addition, a mining rights evaluation system, competitive pricing system and mining rights transfer system should be introduced to create an orderly mining rights market. This would attract more businesses to invest in and exploit oil and gas resources, thereby increasing China's oil and gas production.

Third, technology and equipment innovation should be improved. Strong technology innovation and advanced equipment are prerequisites for the oil and gas production revolution. It is suggested that exploration and production technologies suitable for the characteristics of China's unconventional oil and gas resources should be improved by importing, absorbing and innovating advanced technologies through joint R&D and foreign cooperation. A key technology and equipment system with Chinese characteristics should be developed and deployed at large scale. Local equipment manufacturing should be improved to generate independent intellectual property rights. Specifically, in terms of exploration and production technologies, breakthroughs should be made in drilling and completion, reservoir stimulation, micro-seismic monitoring and other key unconventional gas development technologies. Focus should be directed on research on exploration and production technologies adapted for deep and ultra-deep quasi-continuous tight sandstone and viscous oil in the East China Sea and deep-water oil and gas in the South China Sea. In terms of equipment, priority should be given to the development of unconventional and marine oil and gas resources. R&D of large acidification and fracturing equipment, semi-submersible rigs and production platforms, drilling ships, jack-up drilling platforms, floating production, storage and offloading vessels, geophysical prospecting ships, underwater production systems and other key systems and equipment should be supported.

Fourth, the gas pricing mechanism should be reformed to speed up the development of natural gas in China. Those parts of the value chain where price can be shaped by market competition should be deregulated to unlock market vitality. Pipeline transmission is a natural monopoly that should be priced under government regulation to ensure market fairness. Wellhead and end-user (excluding residential users) gas prices should be deregulated steadily and determined by the market. The gas transmission tariff and distribution fee should be regulated by the government, and tiered residential gas prices established to adjust the residential gas price reasonably. In addition, interruptible gas pricing and peak and off-peak prices should be introduced, and peak-shaving pricing applied to gas-fired power.

Fifth, fiscal and taxation policies should be standardised and improved as soon as possible. Because of resource constraints, the best way to reflect resource value and scarcity is by implementing a system whereby mineral resources are procured, and mining rights awarded, through the market. The market will link mining rights to resource reserves, quality and other objectives that reflect the value of the deposits. The businesses that win the mining rights will do their best to improve the rate of recovery, thus creating a good interest-driven mechanism.

Natural gas is urgently needed to adjust the energy mix and protect the environment. To promote the exploration and production of gas, especially shale gas, CBM and other unconventional resources with great potential, it is necessary to introduce tax incentives. In particular, resource tax reductions or exemptions, preferential policies for corporate tax and VAT for businesses that use gas as their main energy source.

Sixth, the environmental supervision of the exploration and production of unconventional gas should be improved. China suffers from frequent environmental pollution events, and the public's environmental sensitivity is rising. In the future, environmental risks will intensify as unconventional gas output increases. China must pay attention to relevant environmental issues and prevent pollution and incidents from occurring, as this would impact the industry's development. The government must therefore create an environmental regulatory regime, while deregulating market access. Specifically, China should: (i) develop an overall plan for setting up a regulatory organisation to ensure environmental regulation is efficient; (ii) accelerate the development and implementation of the laws, regulations, technical standards and specifications for environmental regulation during the development of the oil and gas sector; (iii) build fundamental environmental regulation capabilities and improve the research and development of environmental technologies to provide technical support for regulation; (iv) improve the information disclosure mechanism and broaden channels for public engagement; and (v) intensify the penalty criteria for environmental pollution and improve the reward and penalty mechanism.

7 Strategies and Policies

7.1 Main Characteristics of the Evolution of China's Energy Technology Policies

7.1.1 Energy Technologies Pass Through Four Development Stages During Macro Energy Trends

Energy is fundamental to the economy. It provides important material support to social and economic activities and is influenced by macroeconomic trends. The energy industry has passed through four historical stages since its initial reform and opening up. Each stage has different characteristics that match the different stages of China's economic development, its strategic orientation and support policies.

First, China's initial energy development strategies (1978–93). After the reform and opening up, and to increase energy supply as fast as possible, the Chinese government boosted investment to accelerate the development of the energy industry. However, due to the long construction cycle of energy infrastructure, the severe energy shortage became a bottleneck to China's economic growth. During this period, the guiding idea for China's energy strategies was to increase energy supply, and the energy technology strategies focused more on introducing production technologies to address the supply shortage of coal, electricity and other energy sources.

Second, adjusting and stabilising China's energy development strategies (1994–2003). The energy bottleneck in China's national economic development was more or less eliminated in the late 1990s. During this period, China's economic growth underwent a transition from overheating to soft landing, and a balance was achieved between energy supply and demand that had not been possible for years. The relative oversupply of energy changed China's strict control over consumption.

China's energy technology strategies focused more on improving energy supply quality and the technologies themselves. The Chinese government introduced technology strategies and policies for optimising energy consumption and encouraging the use of clean energy.

Third, the rapid expansion of China's energy development strategies (2004–12). In the 21st century, China's energy-intensive industries developed rapidly thanks to industrialisation and urbanisation. This resulted in soaring energy demand and consumption. This huge change in the energy supply-demand relationship forced the Chinese government to rethink its energy strategies. Besides conventional exploration and development technologies, China's energy technology strategies now focused on industry, transport and buildings, especially on innovating and promoting energy-efficient technologies.

Fourth, optimising and improving energy development comprehensively (2012–present). As China's economy enters its new normal of slower growth, the country's energy demand growth is also slowing down. The principal objective of energy development has shifted from meeting basic energy supply requirements to satisfying people's growing need for energy to support a better life. The energy industry is also undergoing a shift from energy quantity growth to energy quality improvement. It is imperative to improve energy development quality through advanced technological innovation and widespread implementation of the green development philosophy.

Table 17 clearly shows that China has successfully explored an energy development path with Chinese characteristics that is comprehensive, coordinated and sustainable. Each energy technology strategy and policy system is created within these macro-strategies for energy development.

7.1.2 China's Energy Industry and Energy Technology Strategies and Policies Are in Parallel and Mutually Beneficial

China's energy technology strategies and policies have transitioned from a traditional planned economy approach to a market economy system. Policy evolution is influenced by many factors, including national economic development strategy, the development of natural resources and reserves, society's demand for energy, as well as institutional factors like China's classification of industries. China's energy technology strategies and policies run along two lines.

The first is the energy industry. Previously, China's energy technology strategies and policies were mainly initiated and promoted by former government departments like the Ministry of Coal Industry, the Ministry of Nuclear Industry, the Ministry of Electric Power, the Ministry of Petroleum, the Planning Commission, the Information Commission, and the Economic and Trade Commission. With China's institutional restructuring, energy technology strategies are now developed by ministries and agencies like the National Development and Reform Commission, the National Energy Administration, the Ministry of Industry and Information Technology, the Ministry of Housing and Urban-Rural Development, and the Ministry of Transport.

For example, the State Council published Key Points of National Energy Technology Policies in 1986, which gave the technology policy requirements for coal, oil, natural gas and hydropower. These requirements included speeding up coal development; increasing the economic benefits of oil field development; prioritising hydropower; using oil and gas resources reasonably and improving the way that petroleum refining and oil products are allocated; making coal processing, combustion and conversion technologies better and improving product allocation. In the 1990s, the government issued Outline of China's Energy Efficiency Technology Policies, which focused on energy demand. Specifically, China prioritises energy efficiency when developing energy technologies, focusing on making research breakthroughs in critical energy efficiency technologies for energy-intensive industries, and guiding private capital into energy efficiency applications.

The second is energy technology. Government departments like the Ministry of Science and Technology (MOST) and the Ministry of

Table 17 China's energy strategy orientation in different historical periods

Period	Orientation of energy strategies
6th Five-Year Plan (1981–85)	Strengthen energy development and conservation to meet the demand for national economic growth; and follow the guidelines of "adapting energy development to local conditions, developing multiple and complementary sources of energy, integrating energy use and achieving real results" to ensure reasonable energy use and conservation in rural areas
7th FYP (1986–90)	Put equal focus on energy development and conservation; take measures on price, tax and credit to accelerate energy production and reduce energy consumption, and thus gradually diminish the energy shortage; continue implementing the guidelines of "adapting energy development to local conditions, developing multiple and complementary sources of energy, integrating energy use and achieving real results" to ensure reasonable energy use and conservation in rural areas
8th FYP (1991–95)	Put equal focus on energy development and conservation, highlighting the importance of conservation; implement the guidelines for adapting energy development to local conditions, coordinating the development of hydro and thermal power and nuclear energy; speed up the construction of coal mines, integrating coal exploitation, processing, transport, sales and use, and implement the guidelines for maintaining stable development in east China and developing west China; and improve energy infrastructure in rural areas
9th FYP (1996–2000)	Put equal focus on energy development and conservation, but prioritise conservation; adjust the energy production and consumption system, and improve energy production efficiency; simultaneously carry out energy development and environmental governance, and resolve the pricing of energy products; improve the exploration and development of oil and gas resources and develop new energy with electricity at the centre and coal as the basis; speed up the commercialisation of rural energy and develop it into an industry, and improve its service system; and follow the principle of adapting energy development to local conditions to advance small-scale hydro, wind, solar, geothermal and biomass energy systems
10th FYP (2001–05)	Proactively develop coalbed methane resources and make more effort to research and develop clean coal technologies; put equal focus on oil and gas, channel more effort into the development of offshore oil, increase the share of natural gas in total energy consumption, and build overseas oil and gas supply bases to diversify imported oil; establish a national strategic oil reserve to safeguard national energy security; construct and improve urban and rural power grids and increase grid interconnection across the country; deepen the reform of the electricity system to gradually separate power plants from grids and implement the policy of awarding grid connection contracts through bidding; develop new and renewable energy (hydro, wind, solar and geothermal) and nuclear power
11th FYP (2006–10)	Follow the guidelines of prioritising conservation, rely primarily on domestic resources with coal as the basis, diversify energy development and optimise the energy production and consumption system; build a stable, affordable, clean and secure energy supply system to develop coal in an orderly manner and electricity and renewable energy strongly, and speed up the development of oil and gas; reinforce the policy orientation of energy conservation and energy efficiency to maximise the benefits of saving energy
12th FYP (2011–15)	Drive changes in how energy is produced and used, and build a modern energy system that is secure, stable, affordable and clean; promote the clean and efficient use of conventional energy and accelerate the development of new energy—hydropower (on condition that the environment is safeguarded) and nuclear power (providing safety is assured); and develop smart grids and increase the strategic oil and gas reserves

Education have developed a series of technology innovation strategies and policies around basic science, technological innovation and industry development that provide a solid basis for innovation in energy technologies. For instance, MOST was lead for the Outline of the National Medium- and Long-Term Development Plan for Science and Technology in 2005, which prioritised energy technologies. The plan drew up guidelines on independent innovation, major technological breakthroughs and development support to achieve progress in energy technology and sustainable energy development. The research focused on the efficient development and clean use of coal, addressing the shortage of liquid fuels, renewable energy, and energy-efficient technologies.

7.1.3 China's Energy Technology Innovation Has Made Some Achievements, But Gaps Remain

In recent years, China has significantly improved its ability to innovate and independently manufacture energy and equipment technologies. It has also built an impressive array of world-class energy technology demonstration projects. Specifically, China has: (i) mastered in part the critical equipment and technologies needed for exploring and developing shale gas, tight oil and other unconventional energy sources; made coalbed methane exploration and development possible at scale; independently developed and manufactured a deep-water semi-submersible drilling vessel with the ability to drill at depths of 3,000 m, and equipment packages like a large gas liquefaction system and electric motor-driven compressor unit for long-distance pipeline transmission; deployed internationally advanced oil and gas exploration and development technologies in complex terrain; and developed oil refining technologies in the 10 million tonne per year category; (ii) developed green and safe coal mining technologies, industrialised coal-intensive processing technologies, including gasification, liquefaction and pyrolysis, and carried out demonstration projects that use low-grade coal in higher-quality applications; (iii) deployed more

ultra-supercritical coal-fired power generation units than any other country; made breakthroughs in large integrated gasification combined cycle (IGCC) power generation, carbon storage demonstration projects and 700 °C ultra-supercritical coal-fired power generation technologies; developed large 1,000 kV ultra-high voltage AC and ±800 kV ultra-high voltage DC technologies and complete equipment packages at a world-leading level; and rapidly developed smart grid and energy storage technologies; (iv) developed the AP1000 nuclear reactor and critical equipment and material manufacturing technologies; started construction of the Hualong 1 nuclear power demonstration project that uses Generation 3 technologies independently developed in China; and started construction of the first commercial nuclear power demonstration project to use high-temperature gas-cooled reactor technology and an independently designed and manufactured nuclear-level digital instrumentation and control system; and (v) taken onshore wind power technologies to an internationally advanced level, made breakthroughs in and successfully demonstrated offshore wind power technologies; and developed solar photovoltaic power at scale, demonstrated concentrated solar power technologies, and made advances in critical cellulosic ethanol.

Even though China has made great progress in energy technologies, there is still a big gap compared with the world's energy technology powers and the qualities needed to lead the energy revolution. First, China lacks core technologies—key equipment and materials need to be imported. For example, the key technologies in fields like Generation 3 nuclear reactors, new energy and shale gas are often introduced from overseas then slowly developed in China over a long time. In other technologies—like gas turbines and high-temperature materials, offshore oil and gas exploration and development—China is behind. Second, relations between industry, universities and research institutions are not close enough and the position of companies as innovation driver is not clear. Insufficient use is made of the valuable innovation opportunities that major energy projects and energy technology

R&D offer; the disconnect between innovation and industry demand still exists. Third, improvements need to be made in the innovation system and mechanisms, market allocation of innovation resources, intellectual property protection and management, and talent development, management and motivation. Fourth, long-term strategic plans are absent. The current energy policy system does not put technology innovation at the centre, and national technology innovation strategies and a development roadmap for energy are absent.

China's strategies and policies are transitioning from a planned economy to a market economy in energy and other industries alike. In the traditional planned economy, China's energy policies had a direct effect on the structure, organisations and planning of the energy industry. Although planning did not focus on energy technology polices, there were technology requirements even though the policies did not exist.

7.2 There Are Opportunities to Lead Global Energy Development During the Global Energy Transition

7.2.1 China Has Seized the Opportunity to Reshape Production

In the past four decades, China has seized the opportunity offered by globalisation to reform its labour market. It has evolved from a family-centred system and used the booming manufacturing industry to grow the entire economy. As a result, China has made the historic transition from an agricultural country to an industrial one. This enabled labour to leave the land and work in industry. The tax sharing system reform triggered competition between counties and created a collective atmosphere of "competition, learning and catching up". Globalisation allowed China's industry to become the workshop of the world and the Chinese people to become rich. However, extensive energy development and use resulted in severe environmental problems, reflecting the contradiction between the people's need for a better life and unbalanced development.

7.2.2 The New Industrial Revolution Is Driven by the Transition to Smart Power

The global energy landscape is undergoing major change thanks to the deep integration of energy and information and technology innovation. In the next four decades, the optimal distribution of energy and information will become a key driver in the modernisation of industrial civilisation. Electricity will play an historical role in driving China to leap forward from being a rich country to a strong power. Every technology revolution is a time for substituting new for old social development drivers and is a window for replacing world powers. In the past 300 years, the two previous technology revolutions were driven by energy changes. The steam engine and the electric motor replaced human strength and created world powers like the UK and the USA respectively.

China is embracing the technology revolution driven by information and human intelligence. Electricity is a carrier of both energy and information. A pan-power network, integrating electricity and electronics, is an important platform for the deep integration of electric power and intelligence. The intelligent electricity revolution will play an increasingly important role in history. Electricity market reform is an important engine for the smart electricity revolution; it has the historical mission of integrating smart decision-making with industry and modernising industrial civilisation with green energy.

Major economies consider energy technology a breakthrough for the new technological and industrial revolution and have developed various policies and measures to achieve leadership and sharpen their national competitiveness. For example, the USA has published the All-of-the-Above energy strategy, Japan has issued the Innovative Energy Strategy for 2030, and the EU has released the Energy Roadmap 2050.

7.2.3 The Historical Opportunity to Deeply Integrate Energy and Information Will Help China Achieve Its World Power Strategy

Science and technology determine the future of energy and create the energy of the future. Technology innovation plays a decisive role in the energy revolution, so it must be placed at the centre of energy development. The priorities, timetable and roadmap for innovation in the energy technology revolution should be made clear.

Energy technology policies are part of national technology policies. They are the code of conduct for energy technology development by which a country or party, under given historical and actual conditions, delivers political, economic and social goals. As part of the overall policies of a country, energy technology policies set the direction and guide the strategies of that country's energy development. They are a key to building an innovative country and one of eight major industry technology innovation pathways for the future. Energy technology innovation is an important means to drive industry forward, effectively address international competition, and make China an innovative country.

7.3 Strategies and Policies to Achieve an Energy Technology Revolution

7.3.1 Overall Goals

By 2020, China's independent energy innovation capacity will significantly improve and a batch of critical technologies will achieve major breakthroughs. China's dependence on imported energy technologies and equipment, components and materials will significantly decline, and the international competitiveness of China's energy industry will substantially increase. The energy technology innovation system will take shape and support China in building a prosperous society in a comprehensive way.

By 2035, China will have developed a relatively complete energy technology innovation system and improved its capacity to innovate energy technologies independently. Moreover, China's energy technologies will reach an internationally advanced level. They will help China to develop its energy industry in a sustainable and environmentally sound way and place China among the world's top energy technology powers.

7.3.2 Development Ideas

China should: (i) put independent innovation at the centre of energy technology innovation, strengthen basic research in energy, improve originality, and aim for game-changing technology innovation; (ii) give businesses the leading role and facilitate the efficient and reasonable distribution of innovation resources; (iii) accelerate the shift in government function from R&D administration to innovation service; (iv) make breakthroughs in key and frontier technologies that constrain energy development and drive revolutionary progress, carry out pilot and demonstration activities in major energy projects; and (v) improve collaborative innovation between government, companies, universities, research institutions and users; encourage innovation in major technologies, equipment, manufacturing, demonstration projects and technology platforms; and combine international and domestic resources to deliver innovation in energy technologies.

7.3.3 Support Policies

The energy technology revolution needs interaction between, and the coordinated evolution of, systems, policies and markets. The government needs to develop national strategies for energy technology innovation and energy equipment development; companies need to follow the guidelines of "business is the main actor in a field of market-oriented collaboration between government, companies, universities, research institutions and users". The development of energy talent needs to be improved and the nurturing of exceptional talent encouraged. Specifically, the following support mechanisms need to be enhanced:

First, improve the energy technology innovation environment. China should develop and improve as quickly as possible the laws and regulations on energy, the support policies and legislation on the commercialisation of research findings, intellectual property protection, and standardisation. China should also improve the life cycle closed-loop evaluation system for energy technology projects, strengthen in-process and post-project supervision and services, and focus on evaluating innovation performance.

Second, make business more dynamic in technology innovation. China should develop a business-led energy technology innovation system to boost the motivation of business in innovation and promote "widespread entrepreneurship and innovation".

Third, consolidate the foundations of energy technology innovation. China should deepen its reform of scientific research institutions in energy and the scientific research system in universities. It should develop a team of talented inter-disciplinary managers who have knowledge and experience in markets and strategy.

Fourth, improve the investment and financing mechanism for technological innovation. China should leverage the advantages of policy and development-oriented finance and increase its support for key applications in energy technology.

Fifth, innovate the support mechanism for tax, pricing and insurance. China should implement tax policies beneficial to energy technology innovation, improve the method for calculating R&D investment by energy companies, reduce the tax burden of energy companies, and implement preferential policies on assets like resources, energy and land.

Sixth, deepen international cooperation and exchange on energy technology. China should develop international strategies for energy technology innovation and develop comprehensive, multi-level international cooperation on energy technology. China should also leverage major energy projects under the Belt and Road Initiative to evolve China's advanced energy technologies, equipment and standards into benchmarks for global energy development.

Special Report 4: China's Energy System Revolution

Shi Yaodong and Angus Gillespie

Innovation plays a crucial role in China's energy system transition and revolution. At the 6th meeting of the Central Leading Group for Financial and Economic Affairs in June 2014, General Secretary Xi announced that China would: (i) carry out an energy system revolution to put China's energy development on the fast track; (ii) reform the energy industry into an effective and competitive market system with a market-oriented pricing mechanism; and (iii) transform the government's regulation of the energy industry and improve the energy legislation system.

There have been many system constraints on the sustainable development of China's energy industry for a long time. For example, state monopoly in the oil, gas and power sectors has restricted market competition and kept private capital investment low. Government price control of those three sectors has prevented the market from allocating resources efficiently and distorted the prices of some products. To address these constraints, the government has introduced a series of guidelines and reform roadmaps over the past few years. These include the Opinions of the CPC Central Committee and the State Council on Further Deepening the Reform of the Electric Power System (March 2015), the Opinions of the CPC Central Committee and the State Council on Advancing the Reform of the Pricing Mechanism (October 2015), the Opinions on Deepening the Reform of the Oil and Gas System (May 2017), and the Action Plan for Energy System Revolution (July 2017). These guidelines and roadmaps clearly identify the objectives, pathways and mechanisms for deep energy system reform.

With the introduction of these guidelines, action plans and related support measures,[1] the

DRC Team Lead of Special Report 4: Shi Yaodong from the Research Department of Industrial Economy, DRC of the State Council of China.

Shell Team Lead of Special Report 4: Angus Gillespie, Former Vice President Group CO2, Shell Global Solutions.

Contributors: Philip Gradwell and Cameron Hepburn from Vivid Economics; Li Weiming, Chen Jianpeng and Zhou Jianqi from DRC of the State Council of China; Liu Xiaoli from the Energy Research Institute, NDRC; and Fan Jingli and Wu Lin from the China University of Mining and Technology.

S. Yaodong (✉)
Research Department of Industrial Economy, DRC of the State Council of China, Beijing, China

A. Gillespie
Former Vice President Group CO2, Shell Global Solutions, The Hague, the Netherlands

[1]For example, the supporting documents for power system reform include the Implementation Opinions on Promoting the Reform of Power Transmission and Distribution Pricing Mechanisms, the Implementation Opinions on Promoting the Development of the Power Market, the Implementation Opinions on Building and Consistent Operation of Power Trading Institutions, the Implementation Opinions on the Orderly Deregulation of Power Generation and Consumption Planning, the Implementation Opinions on Promoting Power Retail Reform, and the Guidelines on Strengthening and Regulating the Supervision and Management of Backup Coal-Fired Power Plants.

© The Author(s) 2020
Shell International B.V. and the Development Research Center (DRC) of the State Council of the People's Republic of China (eds.), *China's Energy Revolution in the Context of the Global Energy Transition*, Advances in Oil and Gas Exploration & Production, https://doi.org/10.1007/978-3-030-40154-2_5

reform of China's energy institutions and system has made positive progress.

Reforming China's energy system is a formidable task and cannot be achieved overnight. Li Wei, Director of the DRC of China's State Council and Chinese lead of this DRC-Shell collaborative research project, said in a recent speech that in the context of the global energy revolution, there are still some deeply rooted contradictions and problems for China to address before it can deliver a clean, low-carbon, secure and efficient modern energy system.[2] In Special report 4, we discuss how China can develop and deepen its energy system revolution and institutional innovations, and we put forward a series of constructive policy proposals.

1 Factors and Trends in Energy System Reform

1.1 Energy Supply and Demand: Global Energy Oversupply and Strong Energy Demand Growth in Asia's Emerging Economies

Historical experience suggests that it is difficult to implement energy system reform at a time of rapidly growing energy demand, as the priority is to ensure energy supply security and meet demand. In recent years, affected by such factors as slowing world economic growth and industrial restructuring, global energy demand has shown little movement. On the other hand, the shale gas revolution and large-scale investment and development in response to high energy prices have resulted in global energy oversupply and lower energy prices. As China's economy enters the new normal of slower growth, demand for energy will decline and lead to overcapacity in the country's energy sectors, including coal, electricity and oil. Thus, it is unlikely that there will be sharp fluctuations in energy demand and

large-scale energy supply shortages. To drive energy system reform forward, improvements in energy use efficiency can help strengthen relatively stable internal and external environments.

The annual growth rate in China's energy demand is forecast to fall below 2% by 2035, from 8% in 2000. This is attributed to China's slowing economic growth, improved energy efficiency and changing patterns of consumption. China's energy demand is increasingly less dependent on energy-intensive industries like steel and cement. Instead, future energy demand will be closely linked to economic restructuring—more structural adjustments mean less energy demand, and vice versa. For example, if China's economic structure shifts closer to that of the USA, energy demand will decrease. As the global economy grows, energy demand will also grow. Almost all new energy supply in 2014–35 will be consumed by rapidly developing economies. According to BP Energy Outlook 2035 (2016), the average annual growth rate in world primary energy demand in 2014–35 will be 1.4% and world total energy demand will increase by 34% in the same period. In the new normal economy, China's energy demand will grow slowly but sustainably. In 2025–35, China will account for less than 30% of the increase in global energy demand, compared to 60% in the past decade (BP 2016).

The share of oil, natural gas and coal in global energy demand has been stable over the past decade (Fig. 1). In 2016, the share of oil, natural gas and coal production was 38.95%, 28.55% and 32.5% respectively.

Oil and natural gas production shows slight but steady growth over the past decade, compared to coal, which has declined. As coal-dominated energy producers begin to seek alternative energy sources, the energy system gradually evolves (Fig. 2).

Global primary energy demand did not change significantly in 2006–16 (Fig. 3). Oil had the largest share of global primary energy demand, reaching 33% in 2016. Although many countries are now reducing their consumption of fossil fuels to lower their CO_2 emissions, this did not have much impact on energy demand. Renewable energy is expanding and its

---

[2]Speech made by Director Li Wei at the China Grand Energy Transition Forum 2017, August 19, 2017.

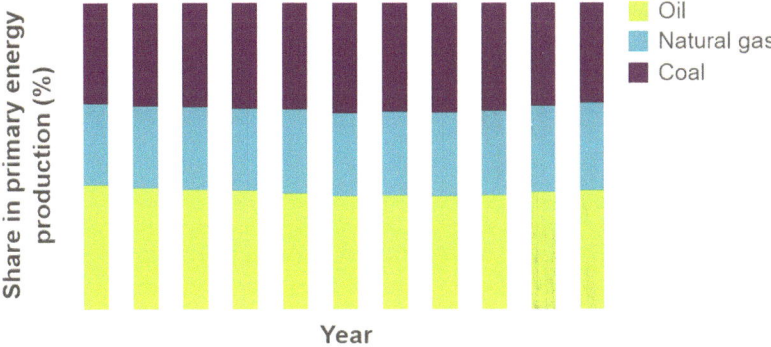

Fig. 1 Share of oil, natural gas and coal in global energy production. *Source* BP Statistical Review of World Energy 2017

Fig. 2 Global primary energy production. *Source* BP Statistical Review of World Energy 2017

production landscape is changing. In 2016, demand for renewable energy grew by 12% (including wind, geothermal, solar, biomass, waste-to-energy and biofuels, excluding hydro-power). Although this is lower than the average growth rate of 15.7% in 2006–16, it still represents the biggest annual increase ever (a rise of 55 Mtoe, which is more than the decline in coal demand). In the same year, China became the world's largest renewable energy producer ahead of the USA. Although renewable energy only accounts for 4% of global primary energy demand and the average growth rate of some renewable energy sources, including nuclear and hydro, is only 2.64%, new energy is expected to remain the major driver of the energy revolution and will play a major role in driving economic growth in the future.

Figure 4 shows that growth in energy demand to 2035 comes entirely from emerging economies, with China and India accounting for more than half of the increase. In contrast, oil demand in OECD countries will continue to fall steadily. In terms of oil supply, non-OPEC countries are the main source of increased supply, producing

11 million barrels per day (MMbbl/d) compared to 7 MMbbl/d in OPEC countries. The increase in non-OPEC oil supply comes entirely from the Americas: shale oil from the USA, deep-sea oil from Brazil, and the oil sands of Canada (BP 2016) (Fig. 5).

BP estimates (BP 2016) that the average annual growth rate of global natural gas demand will reach 1.8% in 2014–35, making natural gas the fastest growing fossil energy source. This robust growth is the result of focused supply and environmental policy support. The increase in natural gas demand comes mainly from emerging economies, with about 30% from China and India and more than 20% from the Middle East. Industry and power generation are behind the increase, whereas in OECD countries the increase is primarily from power generation. World shale gas production is rising. In 2014–35, the average annual growth rate of shale gas production is expected to reach 5.6%, and its share of total natural gas production will be almost 25%. In 2014–25, nearly all the increase in shale gas output will be contributed by the USA. By 2035, China is expected to become the

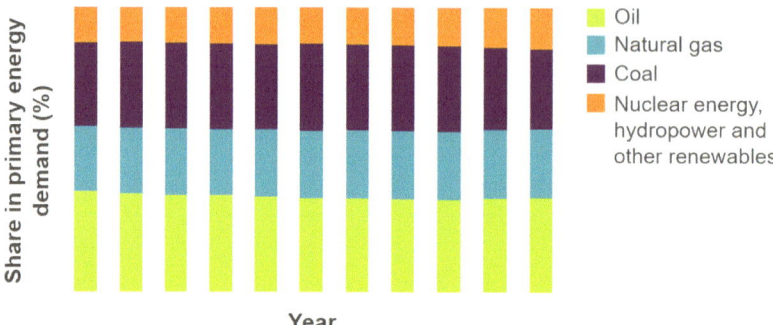

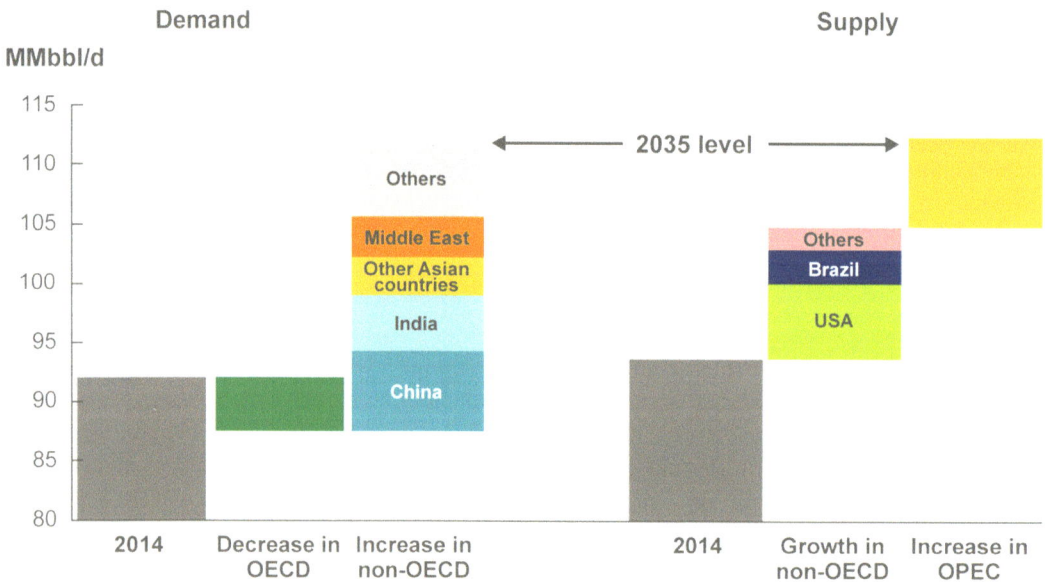

largest contributor to increased shale gas production. Global coal demand is forecast to drop sharply to an average annual growth rate of only 0.5% in 2014–35. This is largely due to slowing growth in coal demand and rebalancing of the economy. Even so, China will remain the world's

World energy consumption

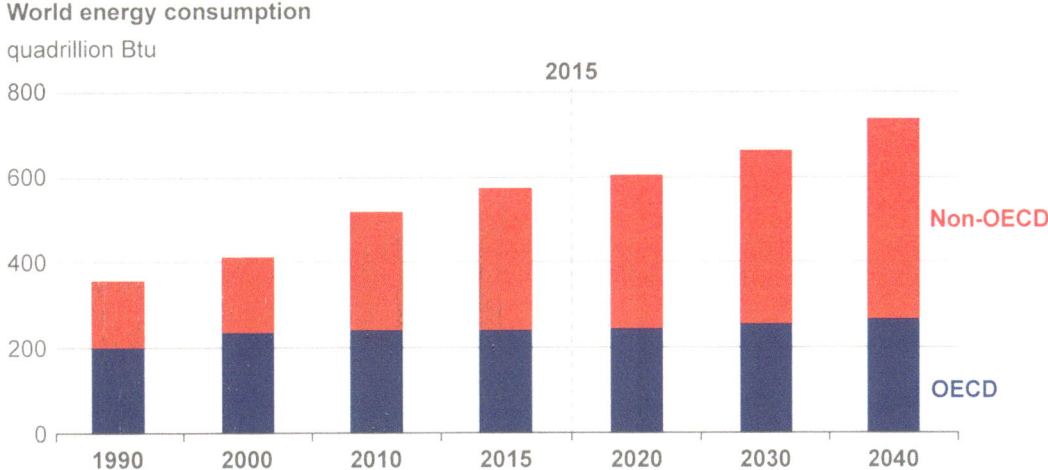

Fig. 6 Energy consumption in OECD and non-OECD countries. *Note* Btu = British thermal units. *Source* EIA World Energy Outlook (2017)

largest coal market, consuming almost half of global output in 2035. In the same period, India will register the highest growth in coal demand (435 Mtoe) and become the world's second largest coal consumer ahead of the USA. Although strong growth in coal demand in India and South East Asia can offset lower demand in the USA and the EU, it is unlikely that India and South East Asia will drive the global coal market as China did previously.

Global demand for hydropower and nuclear power will grow steadily at an average rate of 1.8% and 1.9% respectively, due mainly to growing demand for energy in Asia. China's unprecedented development of hydropower will come to an end, settling at an average annual growth rate of 1.7% in 2014–25. Brazil will have the second highest growth rate (after China) and will replace Canada as the world's largest hydropower producer. China's nuclear power demand will grow rapidly, climbing to an average annual growth rate of 11.2%, which is higher than China's growth rate in hydropower over the past two decades. China's demand for nuclear power is forecast to double by 2020 and increase ninefold over the current level by 2035.

According to the U.S. Energy Information Administration's (EIA) World Energy Outlook 2017, global energy consumption will increase

by 28% between 2015 and 2040, with more than half of the increase coming from non-OECD countries in Asia, including China and India (Fig. 6). Growth in energy demand from Asia's emerging economies is the result of their strong economic performance. Rising economic growth and growing energy demand will intensify competition for energy supply. It is therefore important that these countries seize the strategic opportunities of the Belt and Road Initiative to drive international energy cooperation and deepen energy system reform.

1.2 Major Adjustments to the World Energy Landscape: Diversified Energy Supply and Increased Regional Energy Collaboration

After 2005, the share of shale gas in total US natural gas production increased rapidly—from 5.4% (1 trillion cubic feet) in 2006 to 56% (15.2 trillion cubic feet) in 2015.[3] In 2009, US natural gas production exceeded that of Russia, making it the world's largest natural gas producer. US

---

[3]https://www.eia.gov/dnav/ng/ng_prod_shalegas_s1_a.htm.

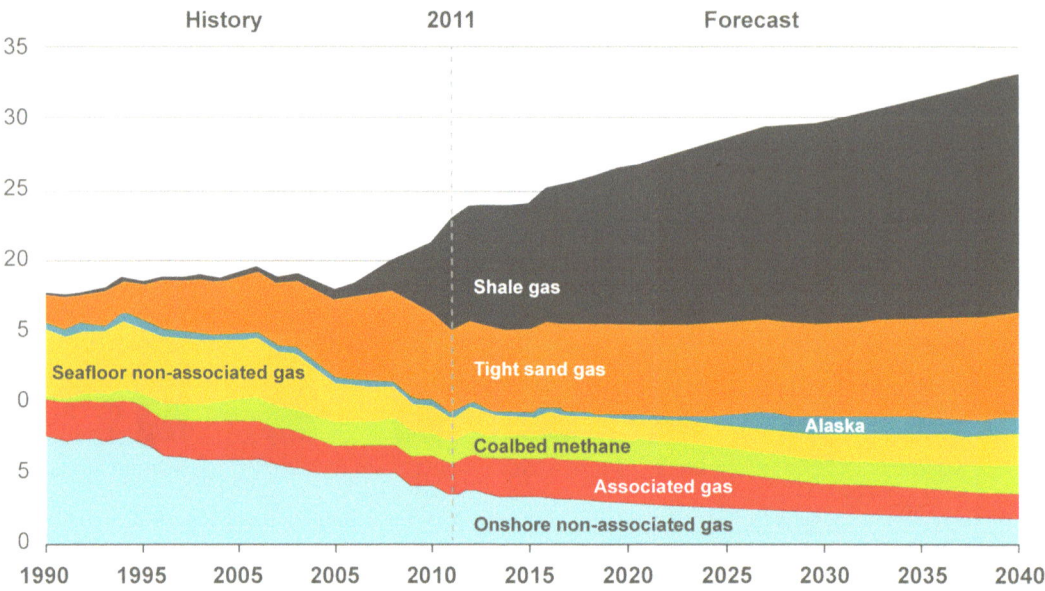

Fig. 7 Natural gas production in the USA (trillion cubic feet). *Source* IEA (2017)

shale gas production is forecast to exceed 20 trillion cubic feet by 2040, accounting for more than half of its total natural gas production.[4] New proven natural gas reserves in the USA continue to increase and are currently estimated to be 2,300 trillion cubic feet. At the present level of consumption, the reserves are sufficient to last for nearly 100 years (Fig. 7).

With its substantially increased oil and gas production, the USA is likely to turn its long-advocated slogan of energy independence into reality. The USA has reduced its dependence on oil imports from 60% in 2005 to 25% in 2016.[5] The National Intelligence Council predicts that the USA will achieve energy independence by 2030 and become an oil self-sufficient country and a major natural gas exporter. Canada —with proven oil reserves of 172.19 billion barrels in 2015—has 10.1% of the world's total proven oil reserves, the largest after Venezuela and Saudi Arabia. It also has about 95% of the world's proven oil sand resources. With abundant conventional and unconventional oil and gas resources, Canada has become a new energy

superpower. According to the International Energy Agency (IEA 2017), Canada's oil production will reach 30–60 MMbbl/d by 2030. The rise of North American energy intensifies the diversification of energy supply.

The US shale gas revolution sparked a shale gas investment boom across the world. In 2011, the USA calculated its domestic shale gas resources and those of 32 other countries. In 2013, it expanded its assessment to include 137 shale formations in 41 countries, including shale oil resources in addition to shale gas. The assessment found that there are highly abundant shale oil and gas resources in the world, corresponding to 10% of the world's recoverable crude oil reserves and 32% of the world's recoverable natural gas reserves. Russia has the most abundant shale oil reserves, followed by the USA, China, Argentina and Libya. China has the highest shale gas reserves, with Argentina, Algeria, the USA, Canada, Mexico and Australia also richly endowed.[6] Currently, Europe and Australia have increased their efforts to explore and exploit shale gas resources, and China is

---

[4]EIA, Short-Term Energy Outlook 2017.

[5]https://www.eia.gov/tools/faqs/faq.php?id=32&t=6.

[6]https://www.eia.gov/analysis/studies/worldshalegas/pdf/fullreport.pdf.

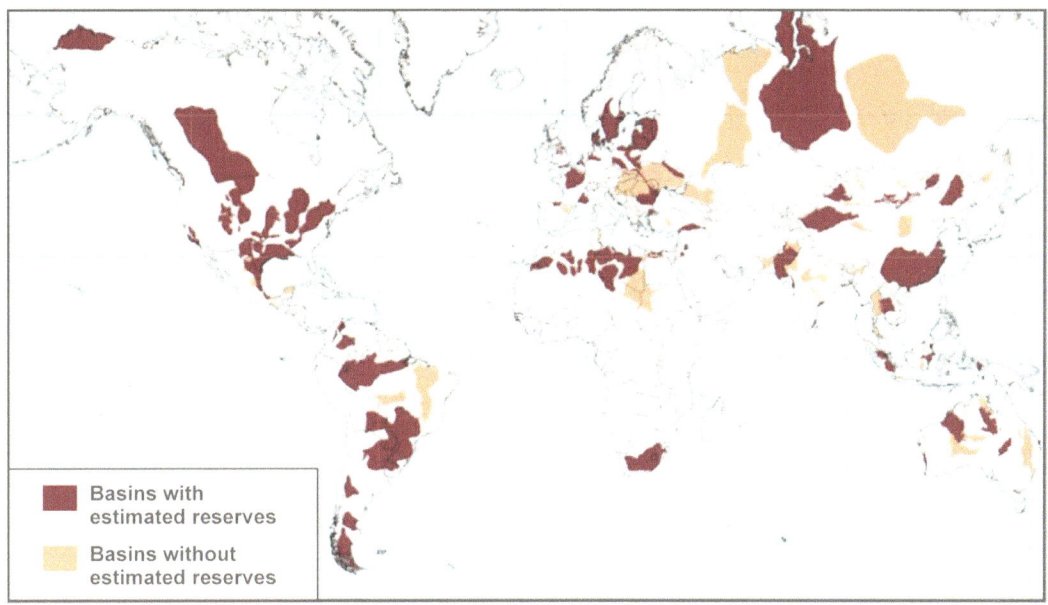

Fig. 8 Distribution of basins with shale oil and gas resources across the world. *Source* https://www.eia.gov/analysis/studies/worldshalegas/

starting large-scale commercial development of its shale gas resources. The IEA forecasts that unconventional natural gas will account for about half of world total natural gas production in 2035, with most produced in China, the USA and Australia. In 2035, China's shale gas production will stand at 13 billion cubic feet per day (Fig. 8 and Tables 1 and 2).

The shale gas revolution in the USA has caused a ripple effect in energy and related fields across the world. The impacts of this revolution were first reflected in natural gas prices, which in the USA fell from $10 per million British thermal units (MMBtu) in 2005 to around $3/MMBtu today. In time, as costs rise, the price is expected to increase by about 2.4% per year, reaching $7.8/MMBtu in 2040. Compared with prices in Asia and Europe, the USA will have a clear advantage in prices over the longer term. The IEA estimates that the cost of liquefied

Table 1 Top 10 countries in terms of shale oil reserves in 2015[a]

Ranking	Country	Reserves billion (bbl)
1	USA	78
2	Russia	75
3	China	32
4	Argentina	27
5	Libya	26
6	UAE	23
7	Chad	16
8	Australia	16
9	Venezuela	13
10	Mexico	13
	Countries' total	**419**

[a]https://www.eia.gov/analysis/studies/worldshalegas/

Table 2 Top 10 countries in terms of shale gas reserves in 2015[a]

Ranking	Country	Reserves (trillion cubic feet)
1	China	1115
2	Argentina	802
3	Algeria	707
4	USA	623
5	Canada	573
6	Mexico	545
7	Australia	429
8	South Africa	390
9	Russia	285
10	Brazil	245
	Total	**7,577**

[a]https://www.eia.gov/analysis/studies/worldshalegas/

natural gas (LNG) exports from the USA (including the cost of liquefaction, transport and gasification) can be held below $10/MMBtu, which is still competitive when compared with the current level of $16/MMBtu and $12/MMBtu in Asia and Europe respectively. As the USA gradually deregulates natural gas exports, its impact on the world's natural gas market will be increasingly prominent (IEA 2017) (Fig. 9).

As a result of lower natural gas prices, many coal-fired power plants in the USA have switched their fuel from coal to natural gas, resulting in falling coal and electricity prices. This brings a great opportunity to regenerate the manufacturing and chemical industries. Chemical companies that shut down plants in the USA several years ago, due to high natural gas prices, are planning to reopen production facilities and are again using low-price natural gas as a raw material to produce ethylene, synthetic ammonia, chemical fertilisers and diesel fuels.

In short, the large-scale extraction and use of unconventional oil and gas resources make the Americas one of the most important resource exporters after the Middle East, Russia and North Africa. US oil and gas exports will impact energy trade between Europe and Russia for several decades—reducing Europe's dependence on its neighbour and diversifying its energy import structure. To safeguard energy exports, Russia will strengthen energy cooperation with countries in East and South Asia. The reduction in Europe's energy dependence on Russia will further consolidate the political, economic and military alliance between Europe and the USA and Canada. Russia's political and economic cooperation with China, and with emerging economies in Asia, will also grow. China will strengthen its energy cooperation with other countries through the Belt and Road Initiative (BRI). BRI covers two high-quality energy-rich regions: Russia-Central Asia; and the Gulf Region-Western Europe, the former endowed with oil and gas, the latter with advanced energy technologies and widely deployed renewables. Stronger international energy cooperation through BRI will connect Central Asia, North-east Asia, South East Asia, Europe and the Americas, creating regional energy communities to deliver win-win cooperation.

1.3 Protecting the Environment and Combating Climate Change Are Global Concerns

The large-scale development and use of fossil energy severely impacts the air, water and environment. Fossil fuel use emits a large amount of

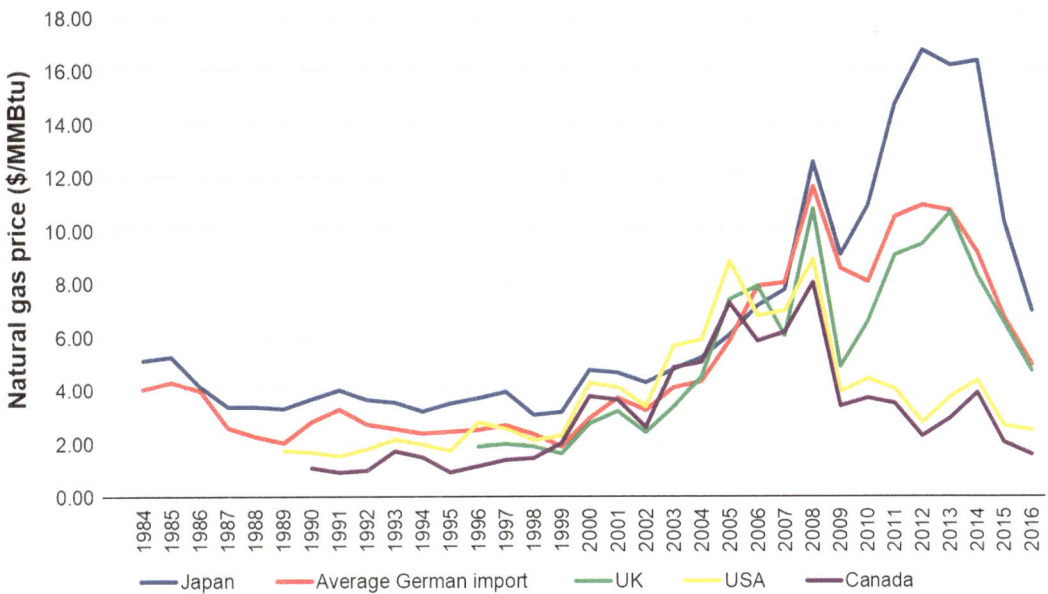

Fig. 9 Price of imported natural gas ($/MMBtu). *Source* BP (2017)

pollutants into the air, including carbon dioxide (CO_2), sulphur dioxide (SO_2), nitrogen oxides (NO_x) and soot. Global SO_2 emissions are about 90 million tonnes per year (IEA 2017). They acidify soil and rivers and erode buildings and historical sites. Around 30% of China's soil surface contains sulphur deposits above the critical level. NO_x emissions from fossil energy use impact land, rivers, marine ecosystems and the ozone layer. NO_x emissions from urban traffic and thermal power plants are the main source of PM2.5 atmospheric particulate matter. Increasing amounts of particulate matter from thermal power plants, transport and other industrial sectors cause widespread and severe haze, threatening human health. In addition, fossil energy extraction and use consume water resources and cause severe pollution.

According to the International Energy Agency's World Energy Outlook 2017, 20% of the world's population live in areas with water shortage. Global water consumption in energy production was 600 billion tonnes, about 15% of total global water consumption. Its impacts include wastewater discharge from coal

production and marine and groundwater pollution from oil and gas exploitation. Large-scale development of conventional energy resources can damage vegetation and landforms. Renewables and other types of new energy also impact the environment, such as visual pollution from wind turbines and the disposal of nuclear waste.

Most CO_2 emissions from human activities are from burning fossil fuels. As energy consumption grows, CO_2 emissions also increase. According to the Fifth Assessment Report (AR5) of the Intergovernmental Panel on Climate Change (IPCC), the concentration of CO_2 in the atmosphere reached 392 parts per million (ppm) in 2012, and the global average temperature rise over the last century was 0.74 °C. The report suggests that global climate warming has become an indisputable fact, which can be evidenced by the rise in atmospheric and sea temperature, large areas of melting snow and ice, and rising sea levels. According to the data monitored by the US National Oceanic and Atmospheric Administration (Fig. 10), global average CO_2 concentration exceeded 400 ppm in July 2017, which is about 40% higher than 100 years ago.

Some climate experts believe that the CO_2 concentration of 400 ppm represents a critical value that cannot be reversed. When the critical value is exceeded, the global temperature rise will reach 2°C.

Global warming has negative impacts on the ecosystem. The combined average temperature over global land and ocean surfaces for April 2016 was 1.10 °C above the 20th century average of 13.7°C—the highest temperature departure for April since global records began in 1880. Sixteen of the 17 warmest years on record have occurred since 2000. Global warming does not just make winters warmer and summers hotter, it also causes major environmental crises, such as glacial retreat in the Qinghai-Tibet Plateau and melting ice and snow in the Antarctic and Arctic. Research indicates that if the global average temperature rise exceeds 2°C compared to pre-industrial levels, many species with poor adaptability will die out; if the global average temperature rise exceeds 4°C, grain yields will fall sharply and fishing productivity will be significantly reduced, exposing global food supply to high risk.[7] Global warming increases the frequency and intensity of extreme weather conditions, and accelerates the melting and shrinking of glaciers, the largest freshwater reserves on Earth, causing sea levels to rise. Millions of people will be threatened by environmental disasters like floods, droughts, typhoons, water scarcity and the submergence of coastal areas, islands and low-lying littoral cities. According to the World Bank, the losses inflicted by extreme weather conditions are increasing—the average annual loss in the 1980s was valued at $5 million. This has risen to $200 billion in the past decade.[8] Some scientists predict that the average global temperature will rise by at least 3°C by the end of this century, which will result in the extinction of a large number of species (Fig. 11).

From the United Nations Framework Convention on Climate Change (UNFCCC) in 1992 to the Kyoto Protocol in 1997 and the Paris Agreement in 2015, the international community has been increasingly concerned about climate change and made vigorous efforts to reduce greenhouse gas emissions. The Paris Agreement sets ambitious goals for reducing greenhouse gas emissions. It established an approach based on the nationally determined contributions (NDCs) of each signatory country to reduce national emissions and adapt to the effects of climate change; and it reiterates the UNFCCC's principle of the common but differentiated responsibilities of individual countries. The Paris Agreement represents the first consensus reached by the international community to combat climate change together, and the first proportionate response by the world's political system to environmental threats. It also brought politicians and academics together, in agreement and on the same side. However, the process is full of twists and turns— the Trump administration announced withdrawal from the Paris Agreement in June 2017 and cancelled the Clean Power Plan introduced by the Obama administration in October 2017, both of which obstruct efforts to combat climate change. The Trump administration is also widely criticised by the international community for its regression and non-action about global climate change. To deliver the goal of emissions reduction, all countries must gradually reduce their use of fossil energy. It is, however, extremely difficult to arrange international negotiations on climate change and align developed and developing countries on climate protection goals, emissions reduction responsibilities and financial investments. As the country with the most advanced industry and the highest cumulative carbon emissions per capita, the USA should accept its responsibilities. Its decision to withdraw from the Paris Agreement is disappointing. Despite the intricate political and economic games behind the negotiations on climate change, the trend of international collaboration on combating climate change cannot be reversed (Fig. 12).

The sense of urgency felt by much of the world about controlling environmental pollution and combating climate change provides operational space for energy system change. Pollution like haze and environmental pollution constraints

---

[7]http://www.ccchina.gov.cn/Detail.aspx?newsId=68179.

[8]http://news.china.com.cn/world/2013-11/20/content_30647909.html.

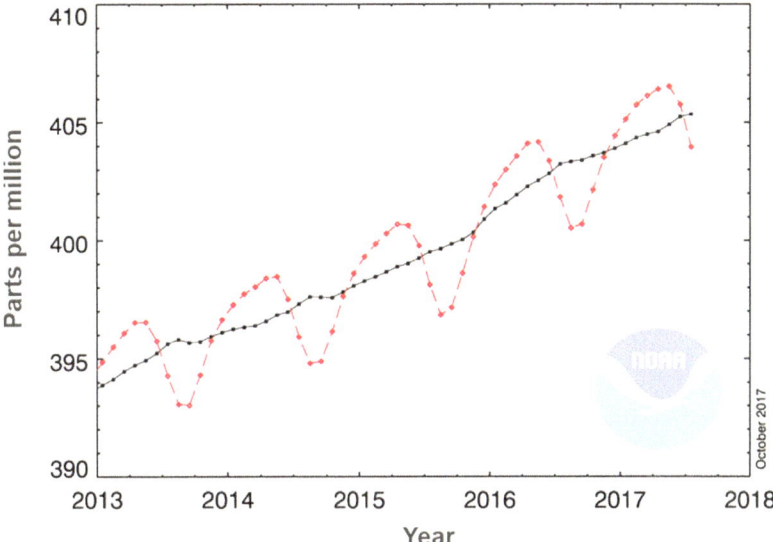

Fig. 10 Global monthly mean CO_2 concentration. *Source* US National Oceanic and Atmospheric Administration, NOAA (2017) (https://www.esrl.noaa.gov/gmd/ccgg/trends/global.html)

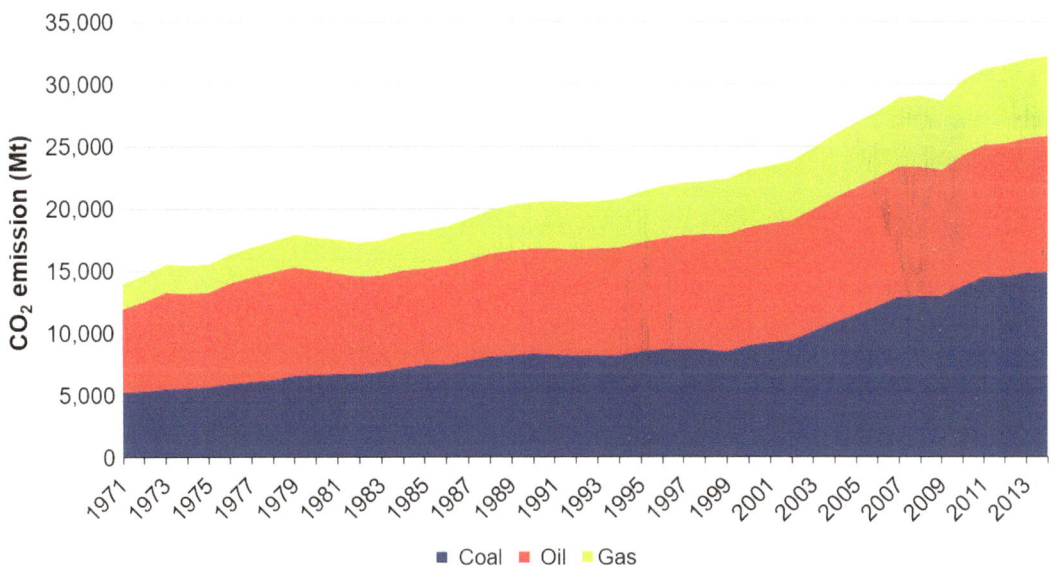

Fig. 11 Global CO_2 emissions from fossil fuels. *Source* IEA (2016)

have made the public aware of the effects of energy consumption on the environment. As a result, the public generally supports energy system reform to control haze and alleviate energy constraints. People are willing to choose cleaner energy consumption and share the cost of environmental protection. As past international experience shows, the public's recognition of clean development can force the government and business to focus on clean development—the Montreal Protocol on substances that deplete the ozone layer is a case in point (Fig. 13).

The green energy era began at the start of the 21st century. The principal trend in the global

Fig. 12 GDP, energy consumption and CO_2 emissions.

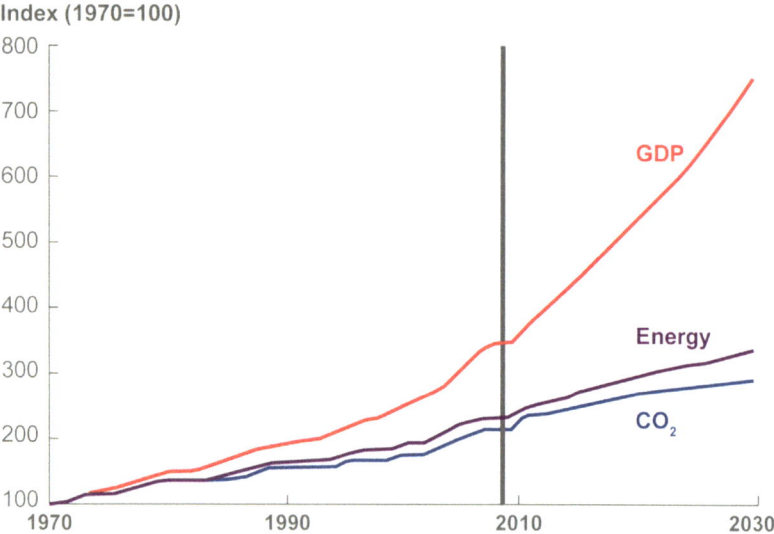

Index (1970=100)

energy transition is the shift from fossil fuels to a sustainable, clean and low-carbon energy system. In developed economies, the share of low-carbon energy in the supply system is increasing. In 1974, Japan issued a new energy development plan to raise investment in developing and using solar, geothermal and hydrogen energy and synthetic natural gas, and identified the development of solar energy as a national strategy. In 2004, Japan introduced its strategic vision of developing new energy technologies, including solar and wind, into a pillar industry valued at JPY 3 trillion. This would include reducing oil use from 50% to 40% of its total energy consumption and increasing the share of new energy to 20%. In recent years, European and American countries have adopted a goal-oriented and systematic approach to achieving an energy transition by 2050. For example, as mentioned in the EU Energy Roadmap 2050, the share of renewables in total energy consumption will be more than 55% in 2050. A study by the U.S. Department of Energy said that renewables can meet 80% of power demand by 2050.

Clean energy is becoming a megatrend in global energy development. As the world's largest energy consumer, China is already targeting an energy revolution. In energy technology, the Action Plan for Innovation in the Energy Technology Revolution (2016–30) and the 13th Five-Year Plan (2016–20) for Energy Technology Innovation have defined 15 innovation pathways for China's energy technologies in the medium-to-long term. According to the Strategy of Energy Production and Consumption Revolution (2016–30), by 2020, the year before the 100th anniversary of the Chinese Communist Party, China will fundamentally change its model of extensive growth in energy consumption. Total energy consumption will be held within a 5 Btce limit and the share of coal consumption will be decreased. The share of non-fossil energy will be 15%, and energy consumption per unit of GDP will be 15% lower than in 2015. By 2030, the share of non-fossil energy and natural gas in total energy consumption will be around 20% and 15% respectively. Increases in energy demand will be met mainly by clean energy.

1.4 A Sound Legal System Will Secure the Energy Market and Guide the Energy Revolution

Energy in developed economies is based on a relatively mature market system. A sound legal system across the entire energy value chain gives developed economies a head start in energy development. For instance, Japan introduced a

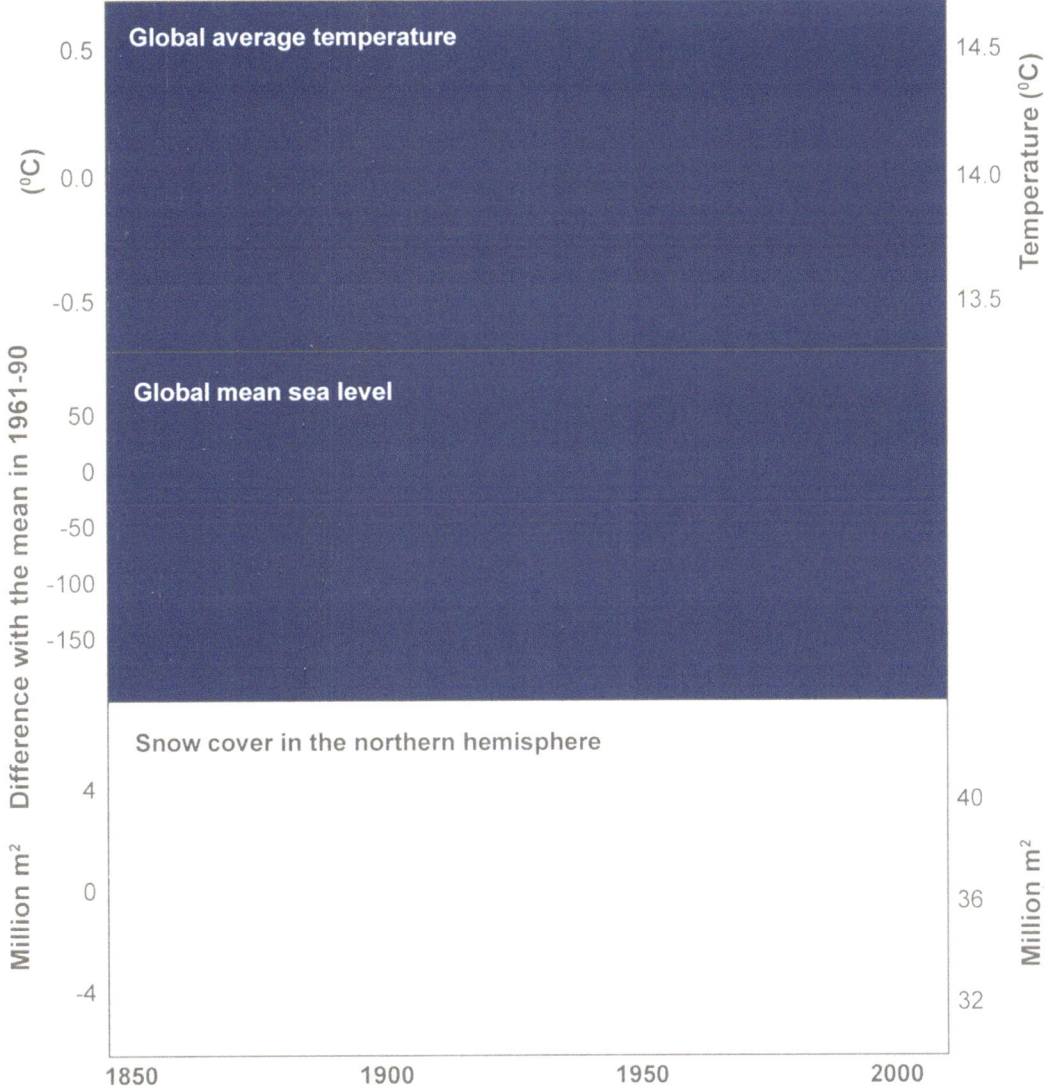

Fig. 13 Change in global average temperature, sea level and snow-cover in the northern hemisphere. *Source* Intergovernmental Panel on Climate Change (IPCC), Fourth Assessment Report (2007)

series of laws to implement emissions reduction measures and control energy demand growth in all sectors. To lower energy consumption, the USA issued a national energy efficiency policy and regulations and the National Appliance Energy Conservation Act. To reduce emissions, the USA passed the Clean Water Act, the Clean Air Act and the Solid Waste Disposal Act, and others.

Developed economies are at the forefront of new energy legislation. The UK government

passed the Climate Change Act 2008 and the Energy Act 2011, which cover green energy options, energy efficiency and low-carbon technologies. The House of Representatives passed the American Clean Energy and Security Act of 2009 (ACES) to drive development of clean energy and energy security in the USA. In Japan in 2011, the House of Councillors and the House of Representatives passed the Act on Special Measures for Renewable Energy to promote new energy technology innovation and reduce dependence on nuclear power. Germany introduced the Renewable Energy Sources Act in 2014, which requires the share of new energy consumption to eventually exceed 50%. Earlier, in March 2009, Germany passed the Renewable Energies Heat Act to increase the share of renewable energy in heat production.

Renewable energy is an irreversible trend. According to the International Renewable Energy Agency, by the beginning of 2016 173 countries had set renewable energy development goals and 146 countries had introduced related support policies. For instance, by 2050 Denmark will be free from fossil fuels and in Germany renewable energy will account for 60% of energy consumption.

China has yet to introduce an energy law to support energy development. In May 2016, the National Energy Administration of China issued the Energy Legislation Plan 2016–20, identifying "five laws and four regulations" as priority projects. These include the Energy Law of the People's Republic of China and laws on electric power, the coal industry, oil and gas, oil and gas pipelines, nuclear power administration, offshore oil and gas pipelines, national petroleum reserves, and energy supervision and administration. As the energy industry develops, the gap between safeguarding energy security and regulating and managing energy is increasingly apparent. To reform the supervision and administrative system of the energy industry in China, it is important to implement the Energy Law of the PRC.

In recent years, China's new energy development has accelerated. Installed wind and solar power capacity is at a world-class level. However, there are still some challenges for China to address, including wind and solar curtailment, destructive exploitation of geothermal energy resources, and management and operational mechanisms for new energy development. Solving these problems will largely depend on adjusting relevant policies and laws and regulations, and by drawing on those of other countries.

1.5 The Energy Technology Revolution: IT, Smart Technologies and the Energy Internet

The energy landscape is undergoing major change. This is an age where technological breakthroughs are continually made in energy technologies and the energy system undergoes game-changing evolution. It is the age of systemic change, where the electricity market is increasingly deregulated and existing business and profitability models in the energy industry are shaken. This is the age of the Energy Internet (Internet+), where the Internet and energy are integrated and disruptors spring up to accelerate the game-changing process.

According to Jeremy Rifkin, a renowned American trend expert, the revolutionary combination of communications technologies and energy is breeding the third industrial revolution. Advanced information and communications technologies like the Internet of things (IoT), big data and cloud computing create the Energy Internet by reshaping energy production, transmission, marketing and use. The Energy Internet, with the smart grid as a carrier, is the inevitable result of the extension of Internet development into the energy and other industries. New technologies like cloud computing, IoT, big data and E-commerce connect people with things and enable dispersed components like information and the grid to be consistently managed. In this way, grid production and management can gradually emerge from their decentralised silos into a culture of centralised collaboration, thus improving business, management and innovation. As the integrator of energy and information, the Energy Internet will drive technological and industrial revolution and have a broad and profound impact.

Many governments and businesses are exploring Energy Internet projects. The US FREEDM project is building the Internet of energy: a network of distributed energy resources that intelligently manages power using secure communications and advanced power electronics. The German eTelligence project used Internet technologies to build a real-time electricity balancing and trading system, which manages intermittent, fluctuating levels of energy output using load adjustment to integrate new energy sources with the grid. It also provides a real-life example of how energy allocation can be adjusted using a real-time electricity trading system. And in China, the State Grid Corporation has initiated the concept of the "global energy Internet", a globally connected smart grid that is designed to allow clean power transmission through ultra-high voltage transmission systems.

There are many other pilot projects that are focused on achieving optimal integration of distributed energy and microgrids. For instance, GCL's distributed micro-energy grid and ENN's Ubiquitous Energy Network (UEN) are both examples of a complementary regional multi-energy Internet that works across an extended industry chain. GCL has mainly focused on solar photovoltaic and combined heat and power, although it also operates in natural gas and smart energy; while ENN is mainly focused on fuel gas and the processing of fuel gas for power generation and cooling and heating services.

For traditional power grid businesses, the Energy Internet brings significant challenges. In particular, the deregulation of the distribution network and the openness and sharing that are inherent in the Energy Internet will significantly weaken these companies' control of the grid. The availability of more competitive products and services will result in high-value customers leaving traditional power grid companies; this can already be seen in the continual customer loss that major grid companies in other countries have suffered after electricity market reform. Traditional power grid enterprises therefore need to: (i) change their mindset to one of active competition, maintain their strength on the demand side of the distribution network, and proactively develop integrated energy services like combined cooling, heat and power; (ii) develop clean energy on the supply side and in

alignment with the relevant national policies; and (iii) focus on and incubate promising and competitive industries to rapidly connect the value chain. In short, traditional power grid companies need to transform themselves and become more competitive in the market.

For traditional power generators and other types of energy company, the era of the Energy Internet provides both opportunities and challenges. Power generation companies can gradually move from being behind the scenes to take centre stage and interact directly with customers. They need, however, to diversify their business portfolio. As demand growth for power slows down, power generation companies will need to adjust their business rapidly, because they are no longer competing solely with other power generation businesses, but with companies along the entire power value chain. To stay ahead in the Energy Internet, traditional power generation companies will need to focus on clean and distributed energy, and proactively develop integrated energy services that provide decision support for the demand-response system. In brief, they should be customer-oriented and target the end-user market.

For new energy companies, the Energy Internet will bring substantial financial benefits; many of these companies have already achieved fame and wealth from it. The conservative approach of the traditional energy companies, which often translates into a wait and see strategy, means that they have yet to show up fully as competitors. New energy companies should, therefore, take advantage of this opportunity and capitalise on their strengths by developing benchmark projects and industry standards, in readiness for a future counter-attack by the energy majors. In short, new energy companies should develop a flexible approach in order to proactively guide industry trends.

For power equipment companies, the Energy Internet undoubtedly poses higher demands on those implementing the Made in China 2025 strategy. For these companies, their top priorities are to address the weak integration of energy and the Internet, and to support and drive improvement in power equipment manufacturing. They should also consider collaborating with other companies to jointly develop solutions in the Internet of things and artificial intelligence. In brief, power equipment companies need to modernise and exercise their strengths in intelligent manufacturing.

For Internet and IT companies, big data is an important cornerstone of the Energy Internet. In April 2016, the National Development and Reform Commission (NDRC) issued its Guidelines on Promoting the Development of Internet+ Smart Energy. In these guidelines, the NRDC proposed developing energy big data service applications and outlined the requirements for IT companies and for the integration and secure sharing of big data, business service systems, and industry management and supervision systems. However, Internet and IT companies need to explore how to capture energy big data effectively and integrate it with other big data to maximise data value. In short, Internet and IT companies need to unlock the value of energy data.

1.6 New Energy Development and Storage Technologies: Cheaper and Better Renewable Energy Accelerates the Growth of a Low-Carbon Power System

According to Bloomberg New Energy Finance's New Energy Outlook 2017, the cost of solar power and onshore wind power will drop by 66% and 47% respectively by 2040, and the operating cost of renewable energy will be lower than that of most fossil fuel power plants by 2030. As the report indicates, the transition of global power systems towards low-carbon energy will be faster than in previous predictions—the total global carbon emissions from power systems will peak in 2026, and by 2040 will be 4% lower than in 2016. One of the report's authors says: "Thanks to rapidly decreasing solar and wind power costs and the increasingly important role of various battery technologies, including electric vehicle

batteries, in balancing power supply and demand, the green power system represents an irreversible trend across the world". The report also points out that solar and wind power will dominate the future power system. It estimates that by 2040, 72% of total global new investment in power generation will be in renewable energy ($7.4 trillion of the $10.2 trillion total). Solar power investment will be $2.8 trillion and its installed capacity will increase by a factor of 14, and wind power investment will be $3.3 trillion and its installed capacity will rise by a factor of 4. By 2040, wind and solar power will account for 48% of global installed capacity and 34% of global power output, significantly higher than the current levels of 12% and 5% respectively.

Solar energy will pose more challenges for coal-fired power generation. Currently, the levelised cost of energy of solar photovoltaic power is only a quarter of the 2009 level and is expected to fall by a further 66% by 2040, i.e. the solar power that can be purchased with $1 will be 2.3 times that of the current level. In Germany, Australia, the USA, Spain and Italy, the price of solar power is at the same level as coal power.

The cost of offshore wind power is expected to decrease faster than onshore wind power. Thanks to greater experience, intense competition, less risk and the substantial effects of economies of scale in large wind farms, the cost of offshore wind power will fall sharply by 71% by 2040. The cost of onshore wind power will drop by 47%, from a level that has declined by 30% over the past eight years. This is due to lower wind turbine costs, improved efficiency and more streamlined operations and maintenance.

New and flexible energy capacity, such as battery storage, will also facilitate the development of renewable energy. The market for lithium-ion batteries for energy storage systems is projected to grow to at least $239 billion by 2040. As a result, there will be increasingly intense competition between utility-scale battery storage and natural gas-fired power generation. As a report from the International Renewable Energy Agency shows, the cost of battery storage for some applications may fall by up to 66% by 2030. Declining battery prices could also increase the installed capacity of battery storage by a factor of 17, creating new business opportunities.

The increasing use of electric vehicles (EVs) will raise demand for electricity as well as help balance the power grid. By 2040, EVs will consume 13 and 12% of the power output in Europe and the USA respectively. Since EVs can be charged at peak times of renewable energy generation or when the wholesale electricity price is low, the power system will be able to accommodate intermittent energy sources such as solar and wind power better. The development of EVs will drive a reduction of 73% in the cost of lithium-ion batteries by 2030. At the same time, the use of fixed power storage technologies like EV batteries helps decarbonise energy end-use applications.

Household photovoltaic (PV) systems will become increasingly popular. By 2040, the output from rooftop PV systems will account for 24, 20, 15, 12, 5 and 5% of the total power output in Australia, Brazil, Germany, Japan, the USA and India respectively. The development of large renewable energy systems will squeeze the demand for power generated by coal- and natural gas-fired power plants. Even with the increased power demand from EVs, large fossil fuel-fired power plants will be under pressure to maintain their profitability.

1.7 Accelerating Change in Global Energy Governance: From OPEC and the IEA to Win-Win Cooperation

Global energy governance in the second half of the 20th century was epitomised by the ongoing struggle between the Organization of Petroleum Exporting Countries (OPEC), representing the interests of oil producers, and the International Energy Agency (IEA) under the Organization for Economic Co-operation and Development (OECD), which represents the interests of Western oil consumers.

In the 21st century, countries are increasingly concerned about energy security, as a result of which global energy governance has gradually improved. As the most important international energy organisation, the IEA has shifted from its initial goal of "preventing oil supply disruption" to "maintaining and improving system response to oil supply disruptions, promoting reasonable energy policies, strengthening cooperation with non-IEA countries, and industrial and international organisations across the world, operating a permanent international oil market information system, improving the global energy supply and demand structure, driving forward international collaborations on energy technologies through development of alternative energy and improvement of energy utilisation efficiency, and helping integrate environmental and energy policies". To this end, it has taken a series of measures: strengthening exchange and negotiations with OPEC to promote stability in the international oil market; developing cooperation with Russia and emerging economies such as China and India through the G7 platform; and improving energy efficiency and promoting clean energy to address the huge challenges of global energy development and climate change. In 2012, the IEA put forward six basic principles to build an "efficient energy world". These include making the economic benefits of energy efficiency more visible, so that regulation can encourage the promotion of energy-efficient technologies.

Global energy governance has developed as the energy industry has evolved. Over 150 years, the energy industry has become a major driver of progress in the world and a core industry in most major countries. However, as the number of participants in the world energy market gradually increases, global energy governance has become ever-more complicated. Despite the current mainstream message of win-win cooperation, the diverse economic and political features of energy make global governance a difficult task in the long term. It will take several decades for the world's energy consumption to shift from fossil fuels to new energy alternatives, and the global game that currently surrounds fossil energy production, transport and consumption will remain and even intensify while the transition takes place. This can significantly affect international energy cooperation. In particular, the inherent structural conflict between energy producers and consumers makes dialogue and collaboration between them difficult. However, sovereign countries will continue to play a dominant and irreplaceable role in global energy governance, and their input and influence will be essential when it comes to building a new global energy governance system and navigating the energy technology revolution.

1.8 Global Energy Majors Accelerate Their Transformation into Integrated Energy Companies

In recent years, as environmental problems have become increasingly severe, a global consensus has formed around the shift to low-carbon and clean energy. In response to the need for a change in the overall structure of energy consumption, all countries are increasing their policy support for large-scale development of the new energy sector. As a result, the installed capacity of new energy is continuously increasing, the technical costs are rapidly decreasing and the return on investment is steadily improving.

International oil majors, including BP, Shell and Total, began investing in new energy in the 1990s. However, the profits that were available from continuously rising international oil prices after 2009 meant that they reduced their new energy investments or dropped parts of their new energy businesses altogether. After June 2014, the international oil price plunged and then settled at a sustained low price. At the same time, the oil majors recognised the existential role of climate change and the need for oil and gas companies to respond to societal concerns on climate change and play their part in the transition to a lower carbon world.

BP is the international oil giant with the deepest and widest footprint in new energy. the company has identified biofuels as a priority in its new energy business. According to BP, by 2035, the number of vehicles worldwide will grow to 1.8 billion units, double the current level.

By then, despite the relatively sufficient availability of oil, the pressure from climate change and carbon emissions will increase, which BP suggests can be mitigated by biofuels. With more than 40 years of experience in hydrogen production and more than 10 years of experience in operating hydrogen refuelling stations, BP has been named the energy partner of the world's two largest hydrogen demonstration projects in Europe and the USA. BP and the Ministry of Science and Technology of the People's Republic of China have cooperated successfully on hydrogen energy projects, including China's first hydrogen refuelling station. Wind power is one of the largest components in BP's renewable energy business. BP currently operates 14 wind farms in seven states of the USA, with 2,200 MW of capacity.

Turning to Shell, new energy technology now accounts for a fifth of Shell's annual R&D budget and is expected to become an important growth business. Raízen, a biofuels joint venture between Shell and Cosan, has evolved into the third largest biofuels company in Brazil, producing more than 2 billion litres of bioethanol and more than 20 billion litres of other industrial and transport fuels annually. In hydrogen, Shell plans to build 390 hydrogen retail sites by 2023, including 230 sites using Shell products. Shell has several wind farms in the USA and the Netherlands, with annual wind power output exceeding 500 MW. In 2017, Shell acquired NewMotion, which operates more than 30,000 eV charging stations in western Europe. The acquisition highlights how Shell has strong expectations that this represents the future trend in vehicle energy.

In the context of lower oil prices, Total plans to invest $500 million annually in new energy and expects to increase its share of the new energy market to 15–20% by 2035. Total's solar energy has been listed among the world's top three in terms of business size. A leading biofuels producer in Europe, Total began its development of biofuels in 1992, including the first generation of ethyl tert-butyl ether from ethanol and vegetable oil methyl ester. Currently, Total is developing the second generation of biofuels. In 2011, Total paid $1.4 billion for a 66%

shareholding in SunPower, the world's second largest solar panel manufacturer. SunPower made net profits of $246 million in 2014 and became a key pillar of Total's business performance. In 2016, Total acquired Saft Groupe, a France-based battery manufacturer, for €950 million. This company ranks 15th in the world in fields like nickel-cadmium batteries, high-performance disposable lithium batteries and lithium-ion satellite batteries.

It is clear that these traditional energy titans have had a profound influence on technological progress, especially in new energy, while carrying out their own strategic readjustments.

2   Current Developments in China's Energy Industry

In 2016, China's energy supply and demand situation was generally improving. Structural reform of the supply side was under rapid implementation, and there was steady progress in realigning the energy system. However, the traditional sectors—coal, coal-fired power generation, refining and chemicals—were still running at overcapacity, and the development of clean energy was facing major challenges.

2.1   Slight Growth in Energy Demand and Significant Progress in Changing the Energy System

In 2016, China's total energy demand was 4,360 million tonnes of coal equivalent (Mtce), up 1.4% (60 Mtce) on 2015. This was 2.2% lower than the average annual growth rate during the 12th Five-Year Plan (2011–15) and 5.3% lower than that for the 13th Five-Year Plan (2016–20).

The first reason for this change was a sharp decrease in China's coal consumption, especially scattered coal. In 2016, China's total coal consumption was 3,780 Mt, down 4.7% (185 Mt) from 2015. This was the third consecutive decrease since 2014, and it slowed down growth in total energy consumption. Between 2011 and

2015, China's total energy consumption increased by 7.4%, an annual growth rate of less than 1.5%. This was in sharp contrast to the average annual growth rate of 7.9% between 2000 and 2011. China's energy consumption per unit of GDP fell by 18.2% between 2011 and 2015, exceeding the goal of the 12th Five-Year Plan (2011–15); it then declined by a further 5% in 2016. This shows that China is supporting economic and social development, while reducing energy consumption and using less resources. Individual sectors tell a similar story: between January and November 2016, coal consumption in the power and steel industries decreased by 0.4% and 0.6% respectively, compared to the same period in 2015; coal consumption in building materials remained at the same level as 2015; coal consumption in chemicals increased by 7.2%; and coal use in other sectors decreased by 10.7%. Coal consumption in power, steel, building materials and chemicals represented 47.9%, 16.3%, 13.8% and 7.2% respectively of total coal use during the same period, up 0.6, 0.1, 0.3 and 0.6% from 2015, while coal consumption in other sectors accounted for 14.9% of the total, down 1.5% from 2015. The reduction in coal consumption in other sectors, especially of scattered coal, was driven mainly by the policies on air pollution control introduced in the previous two years. From 2016, major projects designed to reduce coal consumption, such as modernising coal-fired boilers and substituting waste heat and shallow geothermal energy for coal in household heating, were developed in the Beijing-Tianjin-Hebei region to reduce coal consumption. The scope of this pilot coal reduction and substitution programme was gradually expanded to include the Yangtze River and Pearl River deltas, as well as Liaoning, Shandong and Henan. The scope was also extended from power projects to non-power projects.

The second influential factor was that China's power consumption grew steadily and the demand for power from service industries and households increased significantly. The National Energy Administration's data show that China's total power consumption was 5,920,000 gigawatt-hours (GWh) in 2016, up 5% from

2015 (an increase of 282,500 GWh). Specifically, the growth rate of power consumption by the service industries and urban and rural households was 11.2% and 10.8% respectively, far higher than the growth rate of 2.9% in manufacturing. The service industries and urban and rural households accounted for 56% of the country's total power consumption.

The third factor was that consumption of oil products kept growing, though there was a clear difference between petrol and diesel. In 2016, as China restructured its economy and modernised consumption patterns, use of oil products continued to rise. According to the Bureau of Economic Operations Adjustment of the National Development Reform Commission (NDRC), China's total consumption of oil products was 289 Mt in 2016, up 5% on 2015. Specifically, petrol consumption grew by 12.3%, due to the rising number of vehicles on the road. At the same time, macroeconomic trends and industrial restructuring led to a decline in diesel consumption, down 1.2% from 2015, although there was a gradual increase in demand for industrial diesel. Aviation fuel continued its high growth trend, up 10.4% on the preceding year but 7% points lower than previously due to the popularity of high-speed rail. High consumption levels of oil products and significant imports of crude oil meant China's apparent oil consumption reached 556 Mt in 2016, up 5.5% on the previous year.

The fourth factor was that China's natural gas consumption picked up, though its share of total energy use remained low. Data from the National Bureau of Statistics (NBS) show that China's total apparent natural gas consumption reached 208.6 billion cubic metres in 2016, up 8 or 4.6% points on 2015 (an increase of 15.5 billion cubic metres). However, natural gas's share of primary energy consumption remained at a low 6.3, only 4% higher than in the early 21st century. Driven by a market-oriented natural gas pricing system and new controls on air pollution, natural gas has begun to replace coal projects in many provinces. And on the residential side, a tiered gas pricing mechanism has been launched to encourage the shift to gas. In addition, natural gas consumption

in several major sectors has significantly increased. According to the NBS data, the power sector used the most natural gas in 2015; gas consumption in the coal, gas and water production and supply sectors reached 35.27 billion cubic metres, up 30.87% on 2014 (an increase of 8.106 billion cubic metres).

The fifth factor is a significant increase in the use and share of non-fossil energy. In 2016, use of commercial non-fossil energy was around 541 Mtce, accounting for 12.4% of China's total energy consumption. Adding non-commercial new energy brings the share to 13.3%, a rise of 1.3% on 2015. Initial estimates show that consumption of primary electricity reached about 1,701,000 GWh, up 11.2% (171,000 GWh) on 2015. Non-fossil energy use rose by about 60 Mtce, making it the largest contributor to the increase in consumption.[9]

So, by taking a broad range of measures, China continued to optimise its energy system in 2016. Coal's share of energy consumption decreased for the third consecutive year, and non-fossil energy was the main source of new energy use. Coal's share of primary energy consumption was 62%, down 1.7% points on 2015, whereas oil's share was 18.3%, similar to 2015. The share of natural gas and fossil fuels in primary energy consumption was 6.4% and 13.3% respectively, up 0.5% and 1.25% on 2015. Changes in China's patterns of energy production and consumption over the past decade have impacted the relationship between supply and demand, driving reform (Fig. 14).

2.2 Sharp Decrease in Energy Production and Significant Rise in Clean Energy Supply

In 2016, China's total primary energy production was about 3.46 Btce, down 4.2% from 2015. Coal and crude oil production sharply decreased,

---

[9]Xiao Jianxin et al., Analysis of China's Energy Development in 2016 and the 2017 Energy Outlook for China, in Energy of China, Issue 3, 2017.

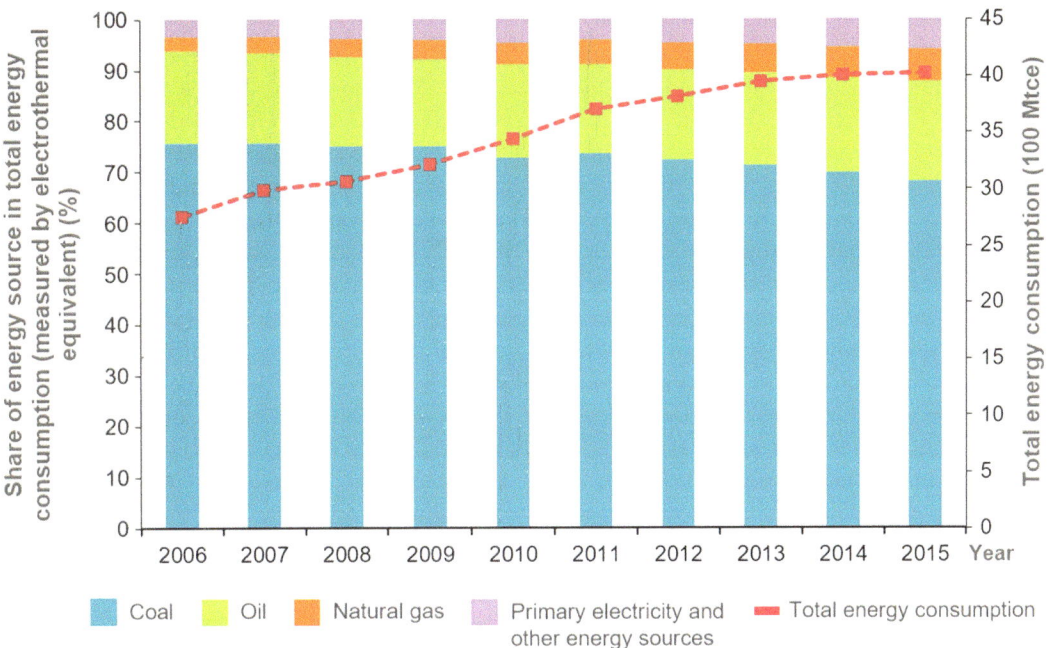

Fig. 14 Total energy consumption and share of various energy sources. *Source* National Bureau of Statistics of China (2017)

and natural gas and primary electricity slightly increased.

Coal production declined sharply in 2016, with raw coal production at 3.41 Bt, down 340 Mt (9%) on the previous year. This was consistent with the trend of declining coal consumption over the previous three years. There was a slight increase in the second half of the year, as some high-quality, safe and efficient capacity was released in August and September. The State Council of China announced in its Opinions on Alleviating Overcapacity in the Coal Sector to Secure Sound Development that the approval of new coal mines and mine modernisations to create additional capacity would be stopped from March 2016. For any new coal mines that were genuinely essential, the policy was to reduce outdated capacity and replace it with high-quality capacity. Thanks to the efforts to eliminate overcapacity, excess coal capacity was reduced by 290 Mt, which also helped stabilise the coal price.

There was also a clear drop in crude oil production, though refined crude oil output maintained its growth rate. In 2016, spurred by lower

international oil prices, China's crude oil producers further reduced inefficient production. Data from China's NBS show that crude oil production was 199 Mt in 2016, down 6.9% on the previous year. This was also the first year since 2010 that crude oil production was less than 300 Mt, with the annual decrease exceeding 10 Mt for the first time. At the same time, regulations around the import and use of crude oil were relaxed. As a result, driven by high demand from the vehicle and aviation sectors, China's crude oil processing capacity grew rapidly, reaching 541 Mt (up 3.6% on the previous year). In particular, crude oil processing capacity in Shandong, a province with several local oil refining companies, was 101 Mt, which made Shandong the first province to exceed 100 Mt in crude oil processing capacity. In 2016, China's production of oil products stood at 345 Mt, an increase of 2.4%.

Natural gas production, on the other hand, rose slightly, and shale gas output grew rapidly. Coalbed methane (CBM) production also increased. In 2016, due to slower growth in domestic natural gas demand and a significant increase in natural gas imports, China's production of conventional natural gas was 136.9 billion cubic metres, up slightly (1.7%) on 2015. Shale gas production maintained strong momentum, with production in 2016 reaching 7 billion cubic metres, up 52.2% on 2015. CBM developed slowly—CBM ground extraction reached 4.5 billion cubic metres, a slight increase of 1.7%, and CBM use was 4.2 billion cubic metres. Production of coal gas was 1.6 billion cubic metres, up 14.3% on the previous year. In 2016, the production and use of coalmine gas were 17.9 billion cubic metres and 8.8 billion cubic metres respectively, an increase of 96 and 148% compared to 2010.

Installed power capacity continued to grow, and non-fossil energy's share of the power mix increased. According to data from the China Electricity Council, by the end of 2016, China's installed power capacity reached 1,650 gigawatts (GW). Specifically, installed capacity and output of hydropower were 330 GW (up 3.92% on 2015) and 1,174,800 gigawatt-hours (GWh) (up 5.58% from 2015). The installed capacity and output of nuclear power were 33.64 GW (up 23.83%) and 213,100 GWh (up 24.39%). The installed capacity and output of grid-connected wind power were 147.47 GW (up 12.79%) and 240,900 GWh (up 29.78%). The installed capacity and output of grid-connected solar power were 76.31 GW (up 80.91%) and 66,500 GWh (up 68.51%). The installed capacity of non-fossil energy, including hydro, nuclear, wind and solar power accounted for 36.6% of China's total installed power generation, while output from non-fossil energy generating units amounted to 29.14% of the total. New installed capacity in 2016 reached 120 GW, including 48.36 GW of thermal power, 19.3 GW of wind power, 34.54 GW of solar power, 11.7 GW of hydropower and 7.2 GW of nuclear power. Non-fossil energy accounted for 60% of China's total new installed power capacity, the fourth consecutive year it exceeded 50%. This is clear evidence that China is continuing to optimise its power system. In 2016, China's total power output was 5,990,000 GWh and the share of non-fossil energy was 28.4%, up 1.5% points on 2015. The new power

output is equivalent to the 2016 output of two Three Gorges hydropower stations.[10]

The breakdown of China's primary energy production in 2006–15 is shown in Fig. 15. Raw coal production gradually declined, but is still the main component in energy production. In 2015, raw coal's share of total primary energy production reached 78.2%, while the share of other energy sources exceeded 10%. Natural gas production increased at an average annual growth rate of 5.27%, reaching 5.3% of total primary energy production in 2015. The average annual growth rate of primary electricity and other energy sources was 6.94%. The share of other energy sources also increased, the result of China's changing energy system. In the future, the market share of natural gas and renewable energy sources will increase, and their production share could also continue to increase.

China should: (i) develop an energy supply system based on solar, wind, hydro, nuclear and biomass as well as other clean energy sources, complemented by fossil energy such as coal and oil; (ii) encourage energy producers to invest more in clean energy technologies and production, give clean energy priority access to the market, and increase the share of clean energy in the energy supply system; and (iii) implement appropriate taxes and financial policies on fossil fuel energy production and supply, and reduce the development of existing conventional fossil energy resources.

2.3 Energy Efficiency Policies Gradually Take Effect and Controls on Energy Consumption and Energy Intensity Produce Results

Energy efficiency is a key component in building a greener civilisation. Energy efficiency is not just about reducing energy use, but also about improving energy productivity, sharpening

---

[10]Xiao Jianxin et al., Analysis of China's Energy Development in 2016 and the 2017 Energy Outlook for China, in Energy of China, Issue 3, 2017.

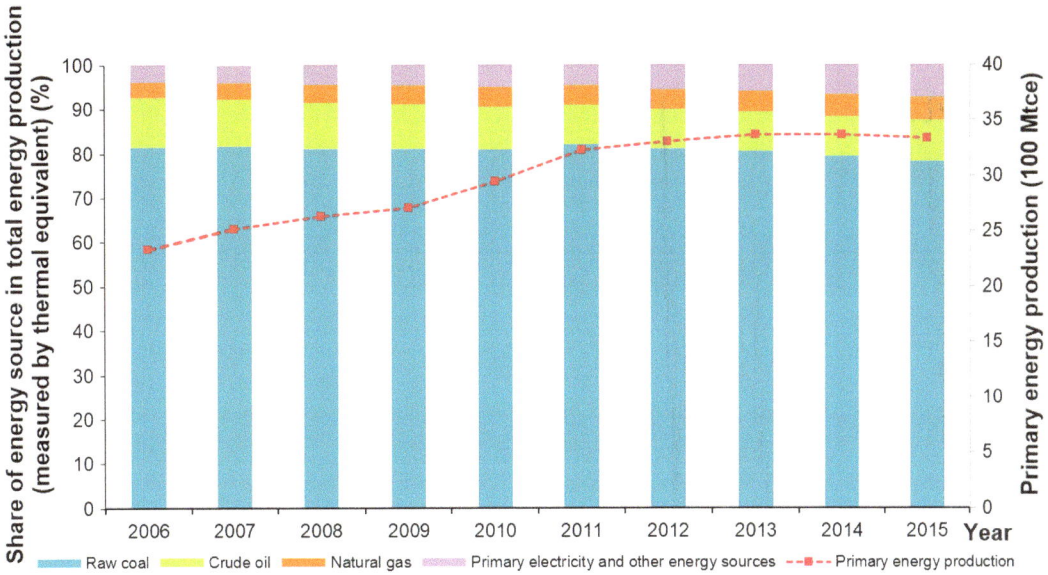

Fig. 15 China's primary energy production and share of various energy sources. *Source* National Bureau of Statistics of China (2017)

industrial competitiveness and creating an efficient and modern energy system. The measures to achieve this include reshaping patterns of production and living, restructuring the economy, optimising the industrial system, increasing the share of clean energy, and making energy use more efficient. Energy efficiency can pave the way for a green and low-carbon society and help realise the centenary goals to mark the founding of the Communist Party of China in 2021 and the People's Republic of China in 2049. Those two goals are building a moderately prosperous society in all respects by 2021 and transforming China into a modern socialist country that is prosperous, strong, democratic, culturally advanced and harmonious by 2049.

In 2016 the State Council of China issued the Overall Work Plan for Energy Conservation and Emissions Reduction for the 13th Five-Year Plan (2016–20), outlining targets for energy efficiency: by 2020, the energy consumption per unit of GDP should be RMB 10,000, 15% lower than in 2015, and total energy consumption should remain less than 5 Btce. Moreover, the goal of controlling total energy consumption and energy intensity would be delegated to the provinces. As part of the 2016 plan, the National Development and Reform Commission (NDRC) and other authorities were entrusted by the State Council to evaluate provincial government success in delivering the energy efficiency goals of the 12th Five-Year Plan (2011–15). The provinces also had to define their 2016 and 2017 targets for controlling total energy consumption and energy intensity. In 2017, NDRC and its partner authorities assessed the provincial governments' progress in fulfilling those targets. Twelve ministries and commissions, including the NDRC, the Ministry of Science and Technology, the Ministry of Industry and Information Technology and the Ministry of Finance, jointly issued the Action Plan for National Energy Saving during the 13th Five-Year Plan (2016–20). The plan defined the top 10 energy-saving measures to achieve the targets, including deploying energy-efficient products and promoting energy saving. The introduction of regulations on energy saving—including the Measures for Energy Conservation Supervision, the Measures for the Energy Conservation Review of Fixed-Asset Investment Projects, the Plan for Developing the Energy Saving Standard System and the

Measures for the Administration of Energy Efficiency Labels—helped drive energy saving.

With policy support, China began to promote energy efficiency in key sectors. In 2016, the Energy Efficiency Forerunner Programme, designed to make China a leader rather than a follower in energy efficiency, was extended. In the same year, the NDRC and the General Administration of Quality Supervision, Inspection and Quarantine of the People's Republic of China published the Energy Efficiency Forerunner Product Catalogue for household refrigerators, flat-panel TVs and adjustable-speed air conditioners. The NDRC provided investment support to improve energy efficiency in energy-intensive sectors, promote energy conservation, and improve energy use in urban street lighting, airports, stations and ports. China issued the Measures for the Administration of Energy Conservation in Major Energy-Consuming Enterprises, launched the "100-1,000-10,000" energy conservation programme,[11] and drove the development of an online energy consumption monitoring system of major energy-consuming enterprises. The Ministry of Housing and Urban-Rural Development made great progress in implementing the green building initiative. The Ministry of Transport promoted the development of a modern, integrated transport system and established a green transport network.

Thanks to the efforts made by these stakeholders, China's total energy consumption was effectively controlled in 2016, and the energy consumption per unit of GDP reduced by 5% compared to the previous year, thus exceeding the targets for 2016. Overall, China reduced energy consumption by 230 Mtce, equivalent to about 500 Mt of CO_2 emissions.

2.4 Fossil Fuel Energy Prices Rebound and the Electricity Price Continues to Decline

In 2016, oversupply in the international market slackened and China's energy supply reforms were strengthened. As a result, the price of major energy sources rebounded after levelling out, though there was much variation among them.

The price of coal increased slightly in the first half of 2016, rose significantly in the third quarter (Q3), then stabilised at the end of 2016. The price rise in those six months could be seen in the Bohai-Rim Steam-Coal Price Index (BSPI) for 5,500 kilocalories per kg (kcal/kg) thermal coal, which reached RMB 401 per tonne at the end of June, RMB 30/t higher than earlier in the year. Between July and October the price rose quickly, and by the end of October exceeded RMB 600/t, RMB 236/t higher than at the start of the year. At the beginning of Q3, several advanced capacity projects came into operation and dealers released their stockpiles, substantially increasing the amount of coal on the market. Between November and December, the coal price slowly declined, RMB 14/t lower at the end of 2016 than its high point for the year. The coal market then stabilised.

The international oil price rose irregularly after hitting the bottom, and prices of oil products in China increased correspondingly. At the beginning of 2016, Brent and West Texas Intermediate (WTI) prices had plunged to $27.88 per barrel (bbl) and $26.21/bb respectively, the lowest point of the year, after which they gradually picked up. By the end of 2016, Brent and WTI prices had doubled—both exceeding $50/bbl. However, due to an easing in both supply and demand, rebalancing was slower than expected. In addition, the lowering costs and improving efficiency of US shale oil also restricted a rise in oil price. In 2016, of the 25 adjustment cycles for oil products in China, there were five downward adjustments, 10 upward adjustments and 10 non-adjustments. The annual cumulative price rises of petrol and diesel were RMB 1,015/t and RMB 975/t.

The international natural gas price remained at a low level, while that of imported natural gas

---

[11]The "i00, 1,000, 10,000" energy conservation initiative places the top 100 energy-consuming enterprises in China under national regulation, the top 1,000 energy-consuming enterprises under the regulation of their respective provincial governments, and other high energy consuming enterprises under the regulation of lower-level governments.

significantly declined. In 2016, the gas price in international markets dropped—the average annual Port Henry price, UK National Balancing Price (NBP) and Japan Liquefied Natural Gas Import price were 2.49/$ per million British thermal units (Mbtu), $4.64/Mbtu and $6.8/Mbtu respectively, down 5%, 30% and 36% respectively on the previous year, and the price differences between the three indexes narrowed. According to statistics from the General Administration of Customs, China's average natural gas import price was $305/t in 2016, down 27% on 2015. In particular, the average price of imported pipeline gas and imported LNG were $270/t and $343/t, 31 and 24% lower than in 2015.

The electricity price continued to decline, which helped lower the cost to the real economy. In January 2016, China introduced the coal-electricity price linkage mechanism, and lowered the feed-in tariff for coal-fired generating units and the retail price for general industrial and commercial use by RMB 0.03/kWh, helping businesses to reduce their electricity expenditure by about RMB 22.5 billion. The feed-in tariff for renewable energy decreased: the benchmark feed-in tariff for onshore wind power in Class 1, 2 and 3 resource-rich areas was decreased by RMB 0.03/kWh, and in Class 4 resource-rich areas by RMB 0.01/kWh. The benchmark feed-in tariff for solar photovoltaic power in Class 1 and 2 resource-rich areas was decreased by RMB 0.1/kWh, and in Class 3 resource-rich areas by RMB 0.02/kWh. The pilot reform of power transmission and distribution pricing passed the strict cost supervision and review test, reducing power transmission and distribution prices by 16.3%.

2.5 Imports of Major Energy Sources Grow Rapidly and Oil and Gas Imports Hit a Record High

In the context of lower international energy prices, China's imports of coal, crude oil and natural gas grew rapidly in 2016, with growth rates exceeding 10%. As a result, China's dependence on oil and gas imports rapidly increased. Exports of oil products also grew significantly.

The strong rebound in China's coal price stimulated a significant increase in coal imports. International coal prices, affected by trends in bulk energy commodities like oil, rose only slightly. Meanwhile, lower shipping prices made imported coal more price-competitive in China's south-east coastal areas. All these factors led to coal imports increasing from May 2016 onwards, with total coal imports in 2016 reaching 256 Mt, up 25.2% from 2015, while coal exports were 8.78 Mt. Net coal imports were 247 Mt, 48 Mt (24.2%) higher than in 2015.

Due to a build-up in oil reserves and changes in the regulations regarding the right to import crude oil, China's crude oil imports grew sharply to 381 Mt in 2016, equal to that of the USA. This figure was 45 Mt higher than in 2015, reflecting an annual growth rate of 13.6%. Driven by this significant growth, China's dependence on oil imports hit a record high of 64.4, 3.9% points higher than in 2015.

Despite the limited growth in domestic demand, China's annual net exports of oil products reached 32.55 Mt, a substantial increase of 11.2 Mt compared to 2015. Exports of oil products accounted for 10.7% of China's total crude oil processing capacity, and 17.9% of the total net oil product exports of the Asia-Pacific region in 2016, 5.2% points higher than in 2015. China became the third largest exporter of oil products in Asia-Pacific, after Japan and South Korea.

In 2016, due to lower international natural gas prices, rising natural gas demand in China and many new long-term contracts, China's natural gas imports resumed their high growth, though this came at the cost of lower production at domestic gas fields. Data issued by the National Bureau of Statistics of China show that natural gas imports were about 74.5 billion cubic metres in 2016, an increase of 21.9% on the previous year. LNG imports increased particularly sharply, by 11 billion cubic metres. In addition, China's dependence on natural gas imports reached 34.2%, up 3.1% points.

2.6 Variations in Energy Company Profitability and Demands that the Petrochemical and Power Sectors Develop Sustainably

In 2016, the diverging price trends of major energy sources caused business performance to vary in the different energy sectors. In particular, coal companies significantly improved their operating performance, while upstream oil and gas companies made losses and midstream oil and gas companies enjoyed steady profits. The profitability of coal-fired power plants declined and non-fossil energy companies registered losses.

In the first half of 2016, a slight rise in the coal price helped to improve the performance of coal companies. Between January and April, the coal mining and coal washing and preparation sector reported total profits of RMB 960 million, down 92% on the same period in 2015. From May onwards, however, coal company profits increased month by month. In 2016, the cumulative profits of the coal mining and coal washing and preparation sector reached RMB 109.09 billion, up 223.6% on 2015. In contrast, the total profits of the mining industry decreased by 27.5% against 2015. Total profits of the coal mining and preparation sector were 60% those of the mining industry, significantly improving the performance of coal companies.

The performance of oil and gas companies varied greatly—oil and gas exploration and development suffered losses but refining and chemicals made huge profits. According to the CNPC Economics & Technology Research Institute, due to lower international oil prices, the profits of China National Petroleum Corporation (CNPC), which has a large share of the upstream oil and gas exploration and production business, decreased sharply by 94.34% in 2016, while China National Offshore Oil Corporation (CNOOC), which is also in the upstream business, suffered a huge loss of RMB 7.735 billion in the first half of 2016. Benefiting from national legislation on oil products, China Petroleum & Chemical Corporation (or Sinopec), which has a higher share of the refining and chemical

businesses, reported profit growth of 11.2%. In the context of lower oil prices, the performance of international and Chinese oil companies was weaker. As a result, reducing investments and costs and improving operational efficiency was a common remedy for most oil companies.

In 2016, the profit margins of coal power companies were squeezed significantly due to lower feed-in tariffs, rising coal prices, a sharply decreasing electricity price and lower output. According to statistics from the China Electricity Council, between January and November 2016, the total profits made by the top five Chinese power generation groups amounted to RMB 54.2 billion, down 45% on the same period in 2015. In particular, the profits of coal power businesses dropped by 67.4%. Preliminary estimates show that the reduction in profits in China's coal power sector resulting from lower feed-in tariffs, rising coal prices and lower output were RMB 110 billion, RMB 7 billion and RMB 7.4 billion respectively.

Due to difficulties connecting renewable energy to the grid, non-fossil energy companies suffered large losses. In 2016, the restrictions on connecting non-fossil energy to the grid became a severe problem—about 150,000 GWh of clean power was not used effectively, equivalent to a reduction in turnover of RMB 60–80 billion. This undoubtedly affected the profitability of non-fossil energy companies. In addition, lower renewable energy subsidies also resulted in higher financial losses for these businesses. Initial estimates show that the gap that renewable energy subsidies failed to cover in 2016 exceeded RMB 60 billion.[12]

2.7 Coalbed Methane Is Developing Well, but Challenges Need to Be Addressed Urgently

On November 24, 2016, the National Energy Administration of China released the 13th

---

[12]Xiao Jianxin et al., Analysis of China's Energy Development in 2016 and the 2017 Energy Outlook for China, in Energy of China, Issue 3, 2017.

Five-Year Plan (2016–20) for Coalbed Methane (CBM) Development and Use. The 3rd Five-Year Plan (1966–70) for the CBM sector is the guideline for China's exploration and use of CBM during the 13th Five-Year Plan. The 13th Five-Year Plan is the decisive stage in building a moderately prosperous society in all respects and a critical period for adjusting the energy system towards diversified energy supply and clean energy. The CBM sector faces both opportunities and challenges, but the opportunities tend to outweigh the challenges.

On the one hand, the external environment is generally favourable. China has implemented the necessary energy supply structural reforms to: (i) increase the share of non-fossil energy and natural gas in total energy production and demand; (ii) raise the share of natural gas in total primary energy demand to 10%; and (iii) encourage the development of CBM. China has also implemented an innovation strategy and accelerated the local development of critical technologies and equipment. As a result, the technological bottlenecks constraining CBM development are expected to disappear. As China introduces increasingly strict requirements for safety in coal mines, the safe extraction of coal-mine gas is a fundamental means to prevent gas explosions. China has made a solemn commitment to the international community to decrease CO_2 emissions per unit of GDP by 40–45% by 2020, and that CO_2 emissions will peak around 2030. These factors bring a hard-won opportunity for the rapid development of the CBM sector.

On the other hand, the sector still has some challenges that need to be addressed urgently. CBM is still in its infancy. The size of the market is small and competition is weak. In recent years, the price of CBM has dropped significantly due to the adjustment of natural gas prices, which offsets the positive effects of subsidies. Moreover, gradually rising production costs and slowing investment in CBM exploration and development have created a large gap between the production and use of CBM. In some regions, CBM does not have open and fair access to natural gas pipelines, and the transmission and

distribution facilities of some CBM projects are inadequate. The large number of potentially dangerous mines in China increases the difficulty of extracting gas. In addition, the gas extracted from coal mines is of low concentration and very difficult to use. China therefore needs to make vigorous efforts to address these challenges during the 13th Five-Year Plan (2016–20).

2.8 Steadily Promote Energy Development in the Belt and Road Initiative

In September and October 2013, during his visits to Kazakhstan and Indonesia, President Xi Jinping proposed building the Silk Road Economic Belt and the 21st Century Maritime Silk Road, which together form the Belt and Road Initiative (BRI).

Energy cooperation is an important part of the BRI, which connects energy consumption markets in Europe and Asia with major energy exporters in the Middle East, Central Asia and Russia. The BRI covers two high-quality fossil energy-rich regions—Russia-Central Asia and the Persian or Arabian Gulf—as well as Western Europe, which has advanced energy technologies and widely deployed green energy. Building the BRI with other countries can promote stability in the energy market in Eurasia and drive the transition to green energy.

In March 2016, the Outline of the 13th Five-Year Plan (2016–20) for the National Economic and Social Development of the People's Republic of China was published. The BRI was included in the Five-Year Plan because it is an important means to coordinate regional development in China and it opens up and actively involves China in global governance.

The 13th Five-Year Plan (2016–20) for the Development of the Energy Industry states that China will: (i) take the domestic and the international markets into consideration and use all available resources to implement its energy cooperation strategy; (ii) seize the major opportunities of the BRI to interconnect energy infrastructure, expand international cooperation on capacity sharing and actively participate in global energy governance; (iii) speed up the development of energy cooperation projects to interconnect energy infrastructure in BRI countries and regions; and (iv) conduct research on cross-border transmission systems and collaborate on power grid modernisation.

The 13th Five-Year Plan (2016–20) for the Development of the Power Sector states that China will: (i) use all available domestic and international resources to promote international cooperation on power system equipment, technologies, standards and engineering services; and (ii) drive the building of interconnected power grids and encourage Chinese power companies to participate in the construction and operation of overseas power projects.

The 13th Five-Year Plan (2016–20) for the Development of the Coal Sector states that China will: (i) strengthen international cooperation on coal to sharpen the competitiveness of China's coal sector; (ii) steadily develop international coal trading to promote the development and use of overseas coal resources and expand overseas engineering, procurement and construction projects and technical services; and (iii) promote international win-win cooperation on coal capacity to develop overseas coal resources, build support infrastructure and invest in upstream and downstream coal businesses.

The 13th Five-Year Plan (2016–20) for the Development of the Oil Sector states that China will: (i) improve the quality and benefits of international cooperation on oil; (ii) optimise the pace of investment and asset portfolio; diversify overseas investments, investors and modes of cooperation; and integrate energy and finance—thus boosting the Going Global strategy of Chinese oil enterprises; (iii) optimise and promote oil and gas cooperation with Russia, the Middle East, Africa, the Americas and Asia-Pacific; and (iv) collaborate on building interconnected infrastructure in BRI countries.

The 13th Five-Year Plan (2016–20) for the Development of the Natural Gas Sector states that China will: (i) implement the BRI and strengthen cooperation with natural gas producers to develop a diversified supply system and

safeguard natural gas supply security; (ii) establish a multi-level coordination regime to ensure supply security and safety with countries with cross-border natural gas pipelines; and (iii) cooperate with natural gas consumers in North-east Asia to create a regional natural gas market and improve China's right to speak on natural gas pricing.

The 13th Five-Year Plan (2016–20) for the Development of the Renewable Energy Sector states that China will: (i) develop the renewable energy value chain to sharpen the international competitiveness of China's renewable energy sector and play an active role in driving the global energy transition; and (ii) combine the development plans and construction demands of BRI countries to deliver signature collaborative projects in a timely fashion, thus encouraging a joint Going Global strategy for renewable energy consulting, design, contracting, equipment supply and operating companies.

2.9 Strongly Promote Energy Reform

2.9.1 Supply-Side Structural Reform of the Coal Sector Will Cut Overcapacity

China has introduced policies to identify the measures necessary to cut overcapacity in the coal sector. In early February 2016, the State Council of China issued Document No. 7, which set the goal of shutting down surplus coal production facilities. Between March and July the authorities introduced eight policies to identify the support measures needed to reposition surplus workers, administer funds for bonuses and subsidies, manage land and mines for new capacity, and set environmental constraints. Despite the sharp slowdown in demand for coal, China's coal sector still has strong market potential in the medium and long terms, according to the Research Report on Overcapacity Cutting and Development Strategies in China's Coal Sector, 2017–21. The coal sector should, therefore, grasp the opportunities to implement structural reform of coal supply, seek policy support, shut down outdated capacity and cut overcapacity, thereby transforming and modernising the industry.

China has made major progress in shutting down coal production facilities. With the help of support policies, China has placed strict controls on unsafe production, overcapacity and the illegal construction of mines, and has introduced reward and subsidy measures. As a result, 290 Mt of cumulative coal capacity was cut in 2016, which is almost 60% of the overcapacity scheduled for removal during the 13th Five-Year Plan (2016–20). As stated in the Report on the Work of the Government published in March 2017, China will also shut down at least 150 Mt of coal production facilities and reduce steel production capacity by around 50 Mt. At the same time, China will suspend or cancel 50 GW of coal-fired power generation capacity.

China made limited progress in 2016 in restructuring the coal industry and reducing production capacity. The main reasons for this are: (i) the companies to be merged had great difficulty guaranteeing re-employment for workers; and (ii) companies to be restructured lacked funds in the first half of 2016; when those funds were made available in the second half of the year they were considerably lower than expected, which slowed down the restructuring process.

Reducing overproduction is an important part of the goal of "cutting overcapacity, reducing excess inventory, deleveraging, lowering costs and strengthening areas of weakness" in order to structurally reform the coal industry. In accordance with the decisions of the Central Committee of the Communist Party of China and the State Council, China's coal sector made further efforts to reform the industry and cut overcapacity. Various measures were taken, including restricting new capacity, shutting down outdated production facilities and excess capacity in an orderly manner, controlling illegal mine construction and expansion, and reducing production. As a result, coal production declined, reducing oversupply to some extent.

From July 2016, as the excess coal inventory lessened and coal demand grew, the coal price rose rapidly and some regions faced shortages.

The reasons for the shortages were several. They include rising thermal power output, increased storage of coal for winter, better control of over-the-limit road transport and adjustments to the railway network, reduced overcapacity, production controls and curtailing illegal production. However, it should be noted that coal demand did not increase sharply: coal consumption in the first 11 months of the year declined by 1.6% against the previous year. There are still many coal mines that are a safety risk or close to resource depletion, use outdated technologies and equipment, do not conform with safe production conditions and coal industry policies, or have a production capacity of less than 300,000 tonnes per year, all of which make cutting coal overcapacity a challenge.

2.9.2 Electricity System Reform Measures Are in Place and Making Significant Progress

A multi-level pilot project was rolled out across China in 2016 (excluding Xizang). Twenty-one provinces implemented a pilot scheme for comprehensive electricity system reform, nine provinces and the Xinjiang Production and Construction Corps launched a pilot system to reform power retailing, and three provinces piloted grid connections for renewable energy.

The power transmission and distribution price reforms covered provincial power grids. In 2016, China introduced the Measures on Supervision and Review of Power Transmission and Distribution Prices with the aim of identifying and reducing unneeded assets and unreasonable costs in grid companies and establishing an incentive and restraint mechanism. In addition, China disclosed the power transmission and distribution prices of 12 provincial grids; the average historical grid cost reduction ratio stood at 16.3%. Power generation and power consumption underwent orderly deregulation. The formation of electricity trading institutions was basically completed in 2016. It involved creating a medium-to-long-term electricity trading mechanism in 28 provinces to pilot cross-provincial trading, along with arrangements for direct trading between power users and power generation companies. In 2016, the amount of electricity traded exceeded 1 million GWh, accounting for about 19% of China's total power consumption.

In addition, China established an access and exiting mechanism for market participants and a new regulatory system, developed a pilot reform scheme for power distribution services, and included coal-fired backup power plants within the scope of the regulations.

2.9.3 Oil Sector Reform Features "Comprehensive Promotion, Major Breakthroughs and Priority Trials"

First, a pilot programme was launched to reform the management of oil and gas mining rights. In 2015, Xinjiang piloted open bidding for the exploration of conventional oil and gas blocks. And in October 2016, Xinjiang was identified as the pilot province for comprehensive energy reform, with the focus on deregulating market access to the oil and gas sector.

Second, the reforms deregulating the right to use imported crude oil and the right to import crude oil took effect. In 2016, an increasing number of non-state enterprises were awarded the necessary permits to import and use crude oil, and the quota for imported crude oil was increased significantly. To address issues like quota reselling, China implemented new allocation principles, along with various other adjustments and a strict assessment regime. This effectively guaranteed the fair allocation of imported crude oil and maintained sound market development.

Third, overcapacity cutting policies were extended to the refining and chemical sectors to improve oil product quality. In 2016, China introduced several policies—including the Guidelines on Adjusting the Structure, Promoting the Transformation and Increasing the Profits of the Petrochemical Industry; and the Development Plan for the Petrochemical and Chemical Industries, 2016–20—to solve the problem of production overcapacity in refining and chemicals. Tax incentives were also introduced to increase oil product exports and reduce domestic output of oil products. In addition, oil product quality was further improved—in 2016, 11 provinces in east China supplied China V petrol and diesel; this was extended across China in 2017.

Fourth, the oil product pricing system was improved to reflect market developments. In 2016, in the context of sharply falling international crude oil prices and various difficulties affecting domestic oil and gas production, China adjusted the oil product pricing system and lowered the limit for oil product price regulation. This will mitigate risks that may arise from any future increase in oil product prices.

Fifth, China accelerated the reform of state-owned oil enterprises and encouraged the participation of private capital. In 2016, CNPC restructured its engineering and construction businesses, Sinopec restructured its regional oil engineering services, and CNOOC integrated its refining and chemical companies. In addition, state-owned oil enterprises allowed private capital to become stakeholders, thereby creating mixed ownership businesses. A case in point is Sinopec, which sold a 50% stake in Sichuan-to-East China Gas Pipeline Co., Ltd.

2.9.4 The Natural Gas Sector Is Reformed and Achieves Remarkable Success

First, there was progress in reforming natural gas pricing into a market-oriented system. In 2015, China delivered the goals of the three-step transition in natural gas pricing reform. In 2016, China took natural gas pricing reform even further by identifying pricing policies for gas storage facilities, promoting market-oriented natural gas pricing for chemical fertilisers, and by piloting reform of the city-gate gas price in Fujian. The price of non-residential gas, which accounts for more than 80% of China's total natural gas consumption, was almost all negotiated between the supply and demand parties, apart from a small amount of residential gas necessary to maintain people's living standards. The market-oriented pricing system drove down the price of natural gas, reducing the cost of gas use for companies.

Second, regulation of the natural gas transmission tariff and distribution fee has been steadily strengthened. Following the principle of "deregulating gas sources and retailing, and controlling gas transmission and distribution", China introduced several policies to regulate natural gas transmission and distribution. These included measures for administering the gas pipeline transmission tariff and for supervising and reviewing pricing costs, which required the cross-provincial gas pipeline operating companies to establish independent accounting processes before June 1, 2017. China also effected measures for the supervision and administration of local gas transmission tariffs and distribution fees, requiring all provinces to introduce the necessary support policies to reduce end-use gas prices.

2.9.5 Policies to Promote the Development of Renewable Energy Are Introduced and an Assessment System Is Developed

First, in 2016, a guiding framework for renewable energy development and use was introduced. China quantified the share of non-fossil energy for the first time and clearly defined the requirements to develop, connect and measure renewable energy in all provinces.

Second, a system was developed to guarantee the purchase of wind and solar power. In 2016, China issued the Measures to Guarantee the Purchase of Electricity Generated by Renewable Energy and made public the volume of wind and solar power purchased in all provinces. This was an important institutional change to help reform the legal and power systems, and a significant starting point to guarantee grid connection of

renewable energy and ensure market access for new energy in the future.

Third, the green certificate and trading system for renewable energy was formed. The green certificate is an evidence-based means for evaluating how well power supply companies are delivering their planned share of non-hydro renewable energy. As electricity system reform continues and a carbon trading system is developed, market conditions for the green certificate and trading system will gradually improve, and the green certificate will become a channel to secure new funding for renewable energy projects.

2.9.6 Reform of State-Owned Energy Enterprises Accelerates

First, China has developed the "1+N" policy. "1" refers to the Guidelines on Deepening Reform of State-owned Enterprises (2016). Specifically, 10 pilot reform projects were implemented in accordance with the Guidelines and led by the State-owned Enterprise Reform Steering Group under the State Council. "N" stands for the support pillars, which include the development of a mixed ownership economy, improvement of the state-owned asset management system, supervision of state-owned assets to prevent their erosion, development of a robust legal system, introduction of equity and dividend incentives, separation of social responsibilities previously undertaken by state-owned enterprises (SOE), supervision and management of state-owned asset transactions, SOE restructuring, illegal operations, investment accountability, employee stock ownership pilot projects, and so on.

Second, China is improving corporate governance by forming pilot boards of directors. The scope of the pilot includes China Baowu Steel Group, State Development & Investment Corporation (SDIC) and China General Nuclear Power Corporation (CGN). A system for recruiting managers and professional executives will be piloted at centrally run enterprises and their subsidiaries, including SDIC and China Railway Signal & Communication (CRSC).

These pilots will be extended to include eight other SOEs—China Shenhua, Baowu, China Minmetals Corporation (CMC), China Merchants Group (CMG), China Communications Construction Company Limited (CCCC), China Poly Group, China Cheng Tong and China Reform Holdings Corporation Limited.

Third, China is using mergers and acquisitions to integrate enterprises. In May 2015, with approval from the State Council, China Power Investment Corporation (CPIC) and State Nuclear Power Technology Corporation merged to form State Power Investment Corporation (SPIC), with assets of more than RMB 700 billion and annual operating revenues of around RMB 200 billion. SPIC broke the monopoly in nuclear power of China National Nuclear Corporation (CNNC) and China General Nuclear Power Group (CGN) and made SPIC one of the country's top-three nuclear power groups, integrating upstream and downstream businesses. In the future, SPIC plans an initial public offering, with the ambition to become China's largest nuclear power group.

3 The Current Energy System and Existing Problems

3.1 Energy Legislation System

3.1.1 Current Developments

China's legal framework for the energy system has started to take shape, providing the legal basis and guarantees for energy development and government administration. Energy law enforcement is an important part of China's energy administration. However, legal supervision and compliance remain in their early stages.

(1) **The energy law framework is taking shape**

China's energy legislation is complex in structure and scattered in content. It is a mixture of old and new laws and regulations.

The elements of China's energy legislation system comprise laws, administrative regulations, departmental regulations, mandatory standards and specifications, local laws and regulations, and international conventions. The legislative hierarchy consists of national institutions and the State Council, below which are departments, local institutions with legislative powers, and local government.

The participants involved in the energy system include producers, traders, consumers, government at all levels and other energy market entities. The provisions regarding the rights and responsibilities of energy market participants are scattered across different energy laws and regulations.

China's energy legislation system covers exploration and production, processing and conversion, distribution, transport and use. The system is an integral part of the much larger ambition to build an ecological civilisation and robust legal system. Provisions relating to downstream energy are linked to environmental protection and combating climate change.

(2) **Energy law enforcement has become an important part of energy administration**

Energy law enforcement is an important part of China's energy administration.

The main energy law enforcement entities include national and local authorities for development and reform (pricing), land and resources, environmental protection and safety. Departments and administrative bodies relating to commerce, customs and quality supervision also play a role.

Enforcement mainly takes the form of traditional law enforcement and administrative methods like supervision. Central government is generally better than local government at enforcement. The number of enforcement agents necessary to cover a sector as large as energy makes it difficult to find good-quality personnel who satisfy professional requirements.

Results are typically related to capability. Because the basis for enforcement is sometimes unclear, it can be difficult to enforce legislation effectively over national energy infrastructure. Enforcement by local government often meets intervention from other parties.

(3) Legal supervision and compliance in the energy sector is still at an early stage

The development of a judicial system for energy law is just beginning. Due to the long-standing pattern of administrative enforcement, energy cases resolved by the judicial system are rare, and the final decision tends to be one of administrative action.

Energy law compliance currently relies more on self-discipline. The development of a system requiring compliance with energy law is still at an early stage and the existing self-governing industry organisations continue to play an important role.

Legal supervision is still in its infancy. The activities of non-governmental organisations play a role, but the development of a consistent legal supervisory system still has a long way to go.

3.1.2 Analysis of Existing Problems

In general, the inconsistency and lack of coordination in China's energy law and regulatory system need to be addressed. Problems such as dependency on administrative enforcement, a lack of judicial practices, inadequate punishment and slack law enforcement exist. In addition, poor coordination between the energy legislation system and energy reform is set to become a major barrier to China's energy revolution.

(1) Current problems in China's energy law and regulatory system

The fact that a fundamental energy law and specific laws for key energy sectors have been lacking for a long time has impeded the sustainable development of China's energy industry. The Energy Law of the People's Republic of China (PRC) should be the foundation of the energy industry and the bedrock of the entire energy legislative system. Currently, however, the Energy Law of the PRC is in draft form only.

At the same time, the long absence of laws in key energy sectors has resulted in ineffective non-law-based regulation of construction, management and operations. Currently, there are no specific laws for the oil and gas sectors, or for

energy utilities and energy product marketing and services.

Some regulations still bear the hallmarks of a planned economy, and others are poorly adapted to the needs of a market economy. For example, the Electric Power Law of the PRC lacks rules for electricity trading, power generation mechanisms and power system development. The Coal Industry Law of the PRC has many provisions that are ill-adapted to the current state of the coal sector, and which need to be amended and improved as soon as possible. In addition, China's energy laws and regulations are excessively based on principles and can be impractical. Many energy laws and regulations do not have the necessary support regulations for implementation. Their impracticality is due mainly to lack of clarity on rights and responsibilities. It is, therefore, necessary to define the rights and responsibilities of government and market participants in each part of the energy value chain. In addition, there are departmental regulations that are clearly intended to suit the interests of the department or authority concerned.

The provisions regarding specific services in China's energy industry are scattered among different laws and administrative, local and departmental regulations. As such, their effectiveness varies. In the absence of consistent guidelines, differences between the laws and regulations of different jurisdictions create inconsistencies in regulating the same energy services.

(2) Energy law enforcement is decentralised and relies on administrative processes

Constrained by the energy legislation and administrative system, China's energy law enforcement relies strongly on administrative processes. Even in those areas of energy law instituted since China's reform and opening-up, only rarely are energy issues resolved through judicial channels. Energy issues with a clear law enforcement basis and defined rights and responsibilities are mainly resolved through energy regulation (see below). Some policy goals are realised through administrative enforcement, for example by shutting down overcapacity, and these actions tend to be taken by local government. Technically speaking, these measures do not have a legal basis; their main purpose is to implement government policy.

China's energy law enforcement is decentralised and spread among different levels of government and departments. Enforcement is mainly the responsibility of the authority for energy policymaking, regulation and administration. The right to enforce energy law relating to the economy is decentralised to the National Development and Reform Commission and the National Energy Administration; and the right to enforce energy law with regard to society is with the Ministry of Environmental Protection, as well as those authorities responsible for quality and safe production. The enforcement of energy law also involves the relevant agencies of local government. China's energy and environmental law enforcement are separate, especially in the upstream and midstream energy sectors.

China lacks personnel and expertise in its energy law enforcement force, especially at grassroots-level energy administration, where enforcement teams are often too small and lack professionalism and independence. This results in incomplete or slack enforcement, no deterrence for severe violations or only minor penalties for infringements, rules and regulations that are ignored, and an absence of regulation of enforcement institutions.

Energy law enforcement is not carried out on a regular basis, and there is a severe lack of enforcement while activities are taking place. Instead, China's energy law enforcement is limited to unannounced or scheduled inspections and the post hoc results of energy activities.

At the same time, punishments for activities that break the energy laws are inadequate. China's energy law enforcement focuses more on correcting behaviour, and the punishments for violations tend to be light. In the economic enforcement field, the penalty for violating fair competition is far less than the illegal profits that

may have been made. In the social enforcement field, the excessively light punishments for non-compliant emissions are far lower than the cost of controlling those emissions.

(3) **Poor coordination between the energy legislation system and energy reform**

The slow development of energy legislation has hindered energy system reform. Some of China's key energy laws were developed in the early years and not amended for a long time, so they are poorly adapted to the requirements of modern energy development. Held back by the constraints of these energy laws, reform at different levels of the energy industry has achieved only limited improvement of the energy system. Despite local successes, the ideal situation, in which energy legislation is aligned with energy reform, has not been realised, and so the reforms have had no effective legal basis.

The values underlying some energy laws are not aligned with the direction of energy system reform. From the market's perspective, some of China's existing energy laws were based on a guiding principle of "meeting demand with sufficient supply", and this clear supply-oriented bias makes the laws increasingly ill-suited for market-oriented energy supply and demand. From the government's perspective, the existing energy laws and regulations are unclear regarding the roles, rights and responsibilities of the government, and contain only very limited descriptions of the relationship between the government's functions (including planning, supervision and regulation) and the market.

The lack of clear definitions of rights and responsibilities in the energy laws results in energy system reforms lacking a legal basis. Specifically, government bodies in the energy sector do not have clear powers that they can use to act, and energy market participants do not have clear descriptions of their rights and responsibilities. These gaps stifle effective planning, operation and regulation in these natural monopolies.

3.2 The Energy Administration System

3.2.1 Current Developments

In general, China's energy administration system has become relatively centralised in the upstream and midstream sectors, with coordinated administration now a regular practice. Energy regulation and policies are becoming increasingly important, although it is still difficult to distinguish between regulation and intervention. Energy planning is gradually transitioning to a market-oriented system from a planned economy approach. Energy strategies are increasingly valued and the energy reserve system is undergoing improvement.

(1) **Centralisation of upstream and midstream energy administration**

Gradually, China's energy administration has shifted from a decentralised to a relatively centralised format.

The National Energy Administration (NEA) is China's most important energy administration body. As a national vice-ministerial organisation under the National Development and Reform Commission (NDRC), the NEA oversees such key energy sectors as power, coal, oil, natural gas, nuclear power and renewable energy. Its main responsibilities include: (i) preparing drafts of energy-related laws and regulations, formulating and implementing energy development strategies, plans and policies, promoting and drafting energy system reform, and coordinating major issues in energy development and reform; (ii) engaging the relevant personnel to develop energy industry policies and related standards, and approving fixed-asset investment projects in the energy industry; (iii) guiding and promoting technological progress in the energy sector; (iv) managing nuclear power development, including the development and implementation of plans and standards, carrying out nuclear power-related R&D and organising emergency management of nuclear power plants; (v) taking charge of energy conservation and ensuring

comprehensive resource use in the energy industry; (vi) conducting energy forecasting, releasing information and participating in operational coordination and emergency preparations; (vii) supervising and regulating the power market; (viii) licencing, supervising and managing power production safety, reliability and emergency response; (ix) taking the lead in promoting international energy cooperation, and reviewing or approving major overseas energy investment projects; and (x) participating in energy policy-making in areas such as resources, finance and taxation, environmental protection, climate change, and making recommendations on energy pricing and imports and exports.

The NEA has 12 departments and employs 240 executive staff, including 42 leaders. The NEA departments and their main responsibilities are listed in Table 3. The NEA's internal organisational integration involves two major changes.

First, Market Supervision and Electric Power Safety have become two new departments. Electric Power Safety covers the responsibilities of the former State Electricity Regulatory Commission (SERC), and Market Supervision expands the responsibilities of Electric Power Safety to include new energy, coal and oil and gas, as well as promoting market-oriented energy development. Thanks to the responsibilities added by the new Market Supervision department, the NEA will gradually play a dominant role in energy system reform and the formation of an energy market.

Second, the former Policies and Regulations functions of the NEA and SERC have been combined to form Legal and Institutional Reform, which is mainly responsible for research on major issues in energy law, regulations, supervision and institutional reform.

China's local energy administration system comprises two levels—six regional energy bureaus and 12 local energy offices. However, as the local energy administration system was formed by merging the electricity regulatory agencies under the former SERC into the NEA, the main responsibility of China's regional

Table 3 NEA internal organisation

Name	Main responsibilities
General Office	Administrative affairs; disclosure of government affairs; security and confidentiality; petitions; energy statistics; forecasts and early warning; coordination of the National Energy Commission Office
Legal and Institutional Reform	Research on major energy issues; drafting laws and regulations on the development, supervision and management of energy system reform; reviewing the legality of documents; supervising enforcement, reviews and litigation; work related to energy system reform
Development Planning	Carrying out research and making proposals on energy development strategy; creating energy development plans, annual plans and industry policies; participating in research on controlling China's total energy consumption; guiding and supervising work related to controlling China's total energy consumption; and comprehensive energy services
Conservation and Scientific Equipment	Providing guidance on energy conservation and the use of resources by the energy industry; work related to technological progress and equipment; developing energy industry standards (excluding coal)
Electric Power	Developing and implementing plans and policies on thermal generation and power grids; electricity system reform; coordinating and balancing power supply and demand.
Nuclear Power	Preparing and implementing nuclear power development plans and policies; emergency management of nuclear power plants
Coal	Developing and implementing plans and policies on coal development, coalbed methane (CBM) and the conversion of coal into clean energy products; work related to the reform of the coal sector; managing stakeholders in CBM development; shutting down outdated coal production facilities; the management and use of CBM
Oil and Gas	Developing and implementing plans and policies on oil and gas development and refining; work related to the reform of the oil and gas sector; management of national oil and gas reserves; supervision and management of commercial oil and gas reserves
New and Renewable Energy	Providing guidance for and coordination of new, renewable and rural energy development; developing and implementing plans and policies for new energy, hydropower, biomass energy and other renewable energy sources
Market Supervision	Preparing plans for power market development and regional power markets; regulating the power market and power distribution, supply and non-competitive power generation services; handling disputes in the power market; conducting research and providing proposals on electricity tariffs; supervising and reviewing the electricity tariff criteria for ancillary services; recommending policies on universal power services and supervising their implementation; and regulating fair access to oil and gas pipeline facilities
Electric Power Safety	Developing policies on the safety of non-nuclear power plants, the construction of power system projects and supervising project quality and safety; supervising and managing power safety and production, reliability and emergency response; supervising and managing the safety of hydropower dams; investigating and handling accidents in power production and engaging the participation of relevant stakeholders
International Cooperation	Promoting international energy exchange and cooperation; conducting negotiations and signing agreements with foreign energy authorities and the International Energy Agency; developing energy opening-up strategies, plans and policies; coordinating overseas energy development and use
Party Committee (HR)	Personnel and organisational management, team building, discipline and supervisory work in the NEA and regional energy administrations directly under it; party work in the NEA and the Beijing-based organisation directly under it

energy bureaus and local energy offices is still regulating the power sector.

The NDRC and the State Council are responsible for the approval of major energy projects (including infrastructure); and the NDRC, under the control of the central government, is responsible for energy pricing (mainly in the upstream and midstream sectors) and its supervision and management. The development and reform department under local government is responsible for the approval of local energy projects, and the local development and reform (pricing) department is responsible for local energy pricing (mainly with regard to downstream use).

Some government administration functions related to energy have been assigned to other departments. For example, the Ministry of Land and Resources administers the mining rights for oil, natural gas and natural uranium; the central and local land and resources authorities administer the coal mining rights under their jurisdiction; the State Administration of Coal Mine Safety is responsible for safety in coal exploration and development, and the State Administration of Work Safety is responsible for safety in conventional energy production; and customs authorities administer energy imports and exports. In addition, the State-owned Assets Supervision and Administration Commission (SASAC) invests in large state-owned energy enterprises.

Coordinated administration has become normal practice. Some major energy policies are jointly released by the NDRC and NEA. Basically, the NEA can coordinate the administration of all energy sources, except nuclear power and coal, which have special requirements for safety. To undertake overall administration of China's energy system, the NEA must coordinate with the NDRC and local government, and effectively perform its governmental functions of planning, regulation and supervision (Table 4).

Table 4 China's energy administration institutions and their responsibilities (2014)

Institution	Responsibilities
National Energy Commission (NEC)	National energy development strategies and major issues and activities
National Development and Reform Commission (NDRC)	Overall plans, addressing climate change, and statistics; energy conservation, clean energy development mechanism, development and use of new and renewable energy, and demonstration and promotion of energy conservation and emissions reduction
National Energy Administration (NEA)	Information, strategies, plans, strategic reserves, emergency management of nuclear accidents, institutional reform and international cooperation; energy equipment and technology, energy industry events
The Ministry of Land and Resources	Resource management (mining rights)
The Ministry of Water Resources	Rural hydropower strategy, plans, institutional reform, international cooperation and information; small hydropower projects, rural power grids and technologies
The Ministry of Industry and Information Technology	Emergency management of nuclear accidents, nuclear industrial planning and international cooperation; nuclear research, major technologies and equipment, energy conservation and emissions reduction; fuel efficiency of motor vehicles and development of new energy vehicles
The Ministry of Science and Technology	Technical research, research on basic science; development and implementation of science and technology plans, and promotion of oil and gas pipeline technology advances
The Ministry of Finance	Fuel tax, and financial and tax policies related to renewable energy development, gas use and energy-efficient products
The Ministry of Construction	Energy-efficient buildings and urban sewage and biogas
The Ministry of Agriculture	Promotion of biogas, biomass energy and solar energy
The Ministry of Transport	Planning and implementation of green transport
The Ministry of Environmental Protection	Control of emissions from energy-consuming machines and tools, and pollution

(2) **Energy regulation and policies play an increasingly important role**

Government regulation and policies are playing an increasingly important role. The government has promoted a shift in energy production by: (i) introducing an assessment system for energy conservation and emissions reduction; (ii) restraining total energy consumption by shutting down outdated production facilities and cutting overcapacity; and (iii) stimulating the development of new and clean energy with tax incentives. Although these are all government-led mandatory policies, they have delivered genuinely strong results to date. The ability of government regulations and policies to incentivise or disincentivise the energy industry is critical to the efforts to create an ecological civilisation and promote sustainable energy development.

However, government is still intervening in market access. The government has imposed selective control of the energy sector and set many access thresholds in terms of investment scale, site choice, equipment level and product sourcing for energy development projects. Compared with large enterprises and foreign-owned companies, small and medium-sized enterprises and domestic companies are faced with high restrictions on access.

(3) **Energy planning promotes balance and adapts to economic and social development**

Planning is an important government function in the energy system and is increasingly adapted to economic and social development.

Energy planning has been shifting from "meeting demand with sufficient supply" to balancing supply and demand. China's energy market has entered an era of diversified competition, where energy substitution occurs at multiple levels and supply and demand are changing rapidly. The previous energy market, driven by supply, has changed, with the demand side now playing an increasingly important role.

Energy planning functions have also changed, from mandatory to directive. Previously, energy planning started from the supply side and focused on state-owned energy enterprises in key sectors. With the reform of the energy market and enterprises, the mandatory planning function is gradually weakening and energy plans are rapidly assuming a more directive role.

Energy planning has gradually become scientific, consistent and democratic. The practices in the 12th and 13th Five-Year Plans (2011–15 and 2016–20 respectively) prove that the government values research during the early stages of energy planning and seeks opinions from all walks of life, striving to ensure that energy plans are in the common interests of all stakeholders.

(4) **Energy strategies are valued factors in decision-making**

China values energy strategies and considers them an important basis for decision-making. Both central and local government develop strategies for each energy source and each part of the energy value chain, as well as medium- and long-term strategies for some key sectors. China's Strategic Action Plan on Energy Development (2014–20) of the General Office of the State Council describes the major actions and safeguards required to implement energy strategies.

(5) **Energy reserves are valued and emergency response stocks have improved**

Energy reserves are the responsibility of the energy administration under the central government. Reserves of fossil fuels are mainly physical (energy products) and cannot respond quickly to ongoing events. In terms of energy source, China's reserves are mainly oil and coal. The oil reserve is of strategic significance. Due to China's high self-sufficiency in coal, coal reserves are not kept at strategic levels. In addition, natural gas reserves are mainly used for balancing gas supply with demand, so natural gas storage facilities are small and used for peak shaving.

National strategic oil reserves are the responsibility of the National Energy Administration (NEA). China has built a national oil reserve centre. The NEA is responsible for building and managing national oil reserve bases, collection and storage, rotation and use of strategic oil reserves, and monitoring supply and demand in international and domestic markets.

Natural gas reserves are used mainly to adjust gas supply. China's gas reserves have difficulty meeting normal demand. Due to the small scale of gas reserve storage facilities and absence of a peak-shaving gas pricing mechanism, the peak-shaving capability of gas reserves is below international levels. At the end of 2014, the increased production of residential gas and weak downstream gas demand resulted in higher inventories at several liquefied natural gas terminals, but there is still not enough space to store more spot gas.

Coal reserves are not as crucial as oil reserves. China has begun building an emergency coal reserve system. Many provinces are building emergency coal reserves, and large coal enterprises, utilities and ports already have emergency coal reserves. Currently, China's thermal power plants consider a coal stock of 15 days to be normal and of seven days to be a warning. The thermal coal stock of a power generation company should not be less than 20 days. In terms of administration, China's emergency coal reserve system is mainly the responsibility of the National Development and Reform Commission (NDRC), the Ministry of Finance and government departments for transport, railway and energy. Organisations like the China National Coal Association (CNCA) and State Grid provide coal market monitoring, early warning services and information support. The enterprises responsible for emergency coal reserves build dedicated storage facilities. Overall planning and regulation of emergency coal reserves are carried out by national government and specific task fulfilment is the responsibility of enterprises and ports.

In the event of a severe natural disaster or emergency, the NDRC and the Ministry of Finance make decisions on using reserves based on applications from provincial governments or suggestions from organisations like the CNCA or State Grid. NDRC and the Ministry of Finance then give the order to use the emergency reserves

to the relevant enterprises that hold the reserves. Since normal production and operation are ongoing, reserves are rotated so that used stocks are replaced with new, and at least one rotation per quarter takes place.

Enterprises that undertake coal emergency reserve tasks can apply for bank loans, and the national government can grant subsidies for newly built reserve facilities and reconstruction or expansion projects. For enterprises that have fulfilled their emergency reserve tasks, the central finance department will pay interest on their loans or on the capital they have invested in the emergency reserves, as well as subsidise other costs such as site occupancy and storage charges.

China's energy emergency reserve system is designed to respond to energy emergencies, ensure secure and stable energy supply, and make short-term decisions on energy security. At the centre of emergency management is the "inter-ministerial coordination mechanism for ensuring coal, electricity, oil and transport security". China's energy emergency reserve system is decentralised and short term, with the relevant central enterprises performing the emergency management function. First, the NDRC has the largest say in energy emergency management and also has administrative responsibility for energy regulations. Second, the responsibilities of the NEA include ensuring energy emergency supply, managing oil reserves, guiding implementation of administrative measures by national oil companies, and emergency management of the power sector, which was previously the main responsibility of the former SERC. Third, the State Administration of Work Safety is responsible for leading production safety and emergency rescue work in key energy sectors, such as coal and petrochemicals. In addition, the State Administration of Science, Technology and Industry for National Defence is responsible for emergencies in nuclear power. The general administration office and the energy authority of local governments cooperate with the central government in carrying out energy emergency management. China's energy emergency management has played an important role in addressing energy emergencies, such as the 2008 snow disaster in south China, conflicts between the coal and power sectors, and regional oil and gas shortages.

3.2.2 Analysis of Existing Problems

China's existing administrative system lacks consistency and the level is low. Problems such as decentralised administration, unclear rights and liabilities, and overlapping functions remain. Administrative coordination between various departments regarding strategy, planning and policy is poor. The culture of putting approval ahead of public service has not fundamentally changed. In addition, the management of energy reserves and emergencies lacks public engagement.

(1) **Inconsistent administrative oversight and no independent high-level institution**

There are many agencies that have some responsibility for energy administration, but they lack a centralised organisation. In many places they suffer from overlapping remits, where different departments have joint administrative responsibility for specific areas, which can be highly problematic. The National Energy Commission was established to improve energy strategy and overall coordination, but its role has been restricted by the way its organisational structure was set up, and it does not coordinate the administrative activities of the various departments effectively. The National Energy Administration (NEA) is responsible for overall energy planning and development, as well as for supervising the energy industry as a whole, but the administrative functions for specific energy sectors are decentralised to the Ministry of Industry and Information Technology, the Ministry of Water Resources, the Ministry of Agriculture, the State Administration of Work Safety, the Ministry of Environmental Protection and the Ministry of Commerce. This administrative approach of "divided policies from various sources" not only increases the cost of coordination, but also results in unclear responsibilities, conflicting authorities and inefficient administration.

Definitions of rights and liabilities in the energy sector are unclear, both horizontally (between departments of the central government) and vertically (between central and local government). The first problem is that China's energy-related administrative institutions perform their functions in accordance with the Plan on Defining Functions, Internal Organisation and Staffing, and therefore the power granted to them is not based on law. The absence of a fundamental energy law means that the rights and liabilities of energy administrative institutions also lack a legal basis. The next difficulty is that the body framing energy regulation is a subsidiary of the institution that administers those same regulations, and this non-independent approach, which bundles administration and regulation together, hinders effective energy law enforcement. Another problem is that local energy regulation is vulnerable to pressure from local government. Finally, the administrative systems in various energy sectors have their own specific features (except for oil and gas). In those energy sectors that adopt a hierarchical administrative approach, there is a clear lack of coordination between central and local administrative institutions.

The NEA is a vice-ministerial department. It therefore ranks below other resource administration departments that are ministries, such as the Ministry of Water Resources and the Ministry of Land and Resources. For some major energy decisions, the NEA still needs to seek approval from the National Development and Reform Commission (NDRC). All this makes it difficult for the NEA to play its coordinating role effectively. A typical example where overall oversight is needed is regulating energy volume and energy prices. The NEA is mainly responsible for energy volume and the NDRC controls the administration of energy pricing, and this separation does not sit well with the strong link between the two areas. Inevitably, conflicts arise between the measures for adjusting volume and those for adjusting price, and as a result the policies introduced by both departments tend not to deliver good results.

(2) Outdated approach to administration and overcomplicated approval processes

Due to the typically large scale of investment required in the upstream and midstream sectors of the energy industry, and the need for productivity allocation and cross-provincial coordination, the requirement that the administrative function be vertically hierarchical and horizontally decentralised creates an overcomplicated approval process. Moreover, there are some specific requirements that make it difficult to access non-state-owned capital. Central enterprises entering this field also need to handle their relationship with local government, for whom the conflict between administration and interests is difficult to manage.

There is also a lack of public services in the energy sector, which can be clearly seen in the provisions for information disclosure. First, there is insufficient disclosure of government information, including missing or inadequate updates regarding administrative licencing and laws and regulations, as well as too little disclosure of documents and notices that have a direct impact on the development of the energy industry. Second, the mandatory information disclosure system for natural monopolies like oil and gas pipelines has not been established, and information disclosure platforms are absent. Third, there is no administrative system for making data public, especially in basic research. As a case in point, the system for collecting and submitting data on oil and gas resources was introduced in China several years ago, but due to slack enforcement the relevant administration has had difficulty obtaining and managing these data. The delay in building databases and disclosing information has severely affected the integration of mining rights administration processes.

(3) Poor and uncoordinated energy plans make it difficult to balance supply and demand

The coordination between energy plans at different levels and in different sectors is poor. Some energy plans are conditional on each other.

Other energy plans (especially local plans) do not take energy supply and demand sufficiently into account. The energy plans for some sectors have not been effectively implemented or even completed for some time. Coordination between central and local energy plans is poor. Typically, energy plans are more often completed on the supply side and poorly implemented on the demand side. The evaluation of energy plans and their implementation is far behind schedule.

Planning for supply and demand balancing tends to consider volume ahead of how price affects energy demand. Although changes in the scale and structure of energy demand are assessed within the context of socioeconomic development, the current (government-led) pricing mechanisms for oil, gas and electricity make it hard for price changes to reflect market supply and demand. When energy plans are implemented, price tends to severely restrict the market's role. Moreover, the administration of energy volume and price are separate responsibilities of the NEA and NDRC respectively, which also contributes to the tension between volume and price when energy supply and demand are considered.

The lack of coordination between plans for power consumption and generation has become increasingly acute, which can be clearly seen in generating equipment availability. In 2014, the average availability of several types of generating equipment in China hit a record low. In addition to weak power demand due to the sluggish economy, the distribution of power generating units also caused problems. For example, the start-up of Unit 2 of the Hongyan River nuclear power plant was postponed even though it was ready, because the power supply capacity of the north-eastern power grid had not been properly considered at the planning stage. The lead time for nuclear power plant construction increases the likelihood of mismatches between installed power capacity and power demand.

(4) Energy strategies lack national support and an implementation mechanism

China lacks national medium- and long-term energy strategies. Despite the introduction of the

Strategic Action Plan on Energy Development (2014–20), national energy strategies are unclear. Local and national energy strategies are not coordinated, and the guidance provided by energy strategies lacks relevance for energy plans. In addition, there is no strategy implementation mechanism or policy support.

The energy revolution needs clear goals and a roadmap, coordinated safety measures and relatively stable policies. The energy revolution and medium- and long-term energy strategies are carried out by means of dynamic economic and social development and synchronised reforms, including those of government, finance, science and technology. Some of these dynamically developing reforms are interdependent.

(5) **The energy reserve and emergency response systems need to be updated**

China's energy reserves commonly suffer from several problems, exemplified by oil and coal reserves. These problems are: (i) a powerful legislative guarantee for strategic reserves is lacking—there is no clear legal definition of the boundary between government and market, or of the rights and liabilities of the various energy industry participants, regarding levels of reserves of different energy sources; (ii) an incomplete administration system—the storage and release conditions for energy reserves need to be improved: price is the key, but the necessary supervisory system is absent; (iii) the method for holding reserves needs to be improved—static energy reserves result in excessively high costs and relatively limited reserve size; (iv) poor levels of IT functionality in energy reserve administration, leading to insufficient disclosure of information, especially on reserve capacity and volume; (v) inefficient reserve storage facilities—the logistics of energy reserves still need to be improved; and (vi) physical reserves are concentrated in large central energy enterprises, while commercial oil and gas reserves are growing slowly.

China's energy emergency management is designed mainly for domestic, local and short-lived, unexpected energy incidents; it would have great difficulty responding effectively to global energy events that had wider or deeper implications. First, China's energy emergency management is limited to guaranteeing supply in the case of an emergency. Second, China's energy emergency management is decentralised and lacks consistent national management and coordination. It is very much set up for emergencies involving a specific energy source or a particular part of the energy value chain and is inadequate for responding to global energy events that involve several energy sources. Moreover, the geographical scope of energy emergency response is limited to China, and no international energy emergency management has been set up. Third, the development of an energy emergency material guarantee system is inadequate. China's oil reserves still have some way to go to meet the International Energy Agency's 90-day oil stock security criteria, and the size and capacity of the power grids and the natural gas pipeline network do not meet emergency demand levels. Fourth, energy reserve and emergency-related legislation is still lacking, and related systems need to be improved, including the use and storage of energy reserves, supervision, evaluation, IT-based administration and funding systems.

China's energy reserve and emergency management suffer from poor information disclosure and public engagement. As there is no information disclosure system, participants in the energy sector have difficulty obtaining information in a timely fashion. Since commercial energy reserves are not mandatory and a reasonable pricing mechanism for energy reserves has yet to be established, private capital is not motivated to enter the field.

3.3 Energy Market and Circulation System

3.3.1 Current Developments

China's energy market and circulation system are still very much a monopoly. This is rooted in the energy system and institutions and is reflected in network facilities. China's energy market system is still in its early stages. The level of marketisation varies for different energy sources—the coal sector has completed its market-oriented reform; the downstream oil and gas sector is highly deregulated, but upstream still suffers from severe monopoly problems; and the market-oriented reform of power transmission, distribution and retail is urgently needed. In addition, the institutional barrier in natural monopolies has become a major hurdle in the reform of the energy circulation system.

(1) **Slightly more competition in the oil and gas market, but resource monopoly remains**

As a result of the 1998 reform and restructuring of the oil and gas sector, and the related reform and policy measures that followed, an energy production and circulation system has taken shape, covering upstream and downstream, domestic and overseas trading, and production, transmission and sales. The market consists of the top four oil groups—China National Petroleum Corporation (CNPC), China Petroleum & Chemical Corporation (Sinopec), China National Offshore Oil Corporation (CNOOC) and Yanchang Petroleum—alongside various other oil and gas enterprises of different ownership types and sizes, with the top four dominating and forming in effect an oligopolistic or monopolistic competition landscape.

Competition in refining and chemicals, wholesale and retail, has increased. In recent years, state-owned oil enterprises such as CNOOC and Sinochem, which have crude oil resources, have rapidly enlarged their refining presence, intensifying the competition. Because CNOOC and Sinochem have limited downstream distribution resources, the construction and operation of refineries backed by CNOOC and Sinochem investment will increase competition in the energy wholesale and retail sectors.

At the same time, easier entry terms into wholesale oil products have helped diversify market participants. CNOOC, Sinochem and some private oil and gas companies have all been awarded an oil product wholesale licence. Compared with the early 21st century, the competition landscape of the wholesale oil product market has evolved, and competition has gradually intensified.

CNPC, Sinopec and CNOOC have retained their resource monopoly, restricting the development of other market players. In terms of market structure, domestic crude oil and gas resources are concentrated in state-owned oil enterprises, especially CNPC and CNOOC. Wholesale is dominated by CNPC, Sinopec and CNOOC, but in recent years competition in refining, oil production and natural gas wholesale markets has intensified somewhat. There are numerous players in the energy retail market, but because the resource market is dominated by CNPC, Sinopec and CNOOC, private companies dealing in oil products and natural gas other than the top four oil groups do not have a complete vertical presence throughout the market, and they rely on the top four oil groups for resources and even marketing.

(2) **Electricity market reform has started and needs to be deepened**

China's electricity market has more than a decade of experience, during which time small reforms in different parts of the electricity value chain have been continuously carried out and have delivered good results. In 2002, China separated generation from the power grid, and gradually developed pricing mechanisms in power generation, transmission, distribution and sales. In 2004, China introduced the benchmark feed-in tariff policy and began to set and announce regularly the feed-in tariff for new generating units in each province.

The main problems in China's electricity sector are reflected in its market structure. In power generation a pattern of diversified competition has taken shape. The vast majority of power generation enterprises are state-owned and state-controlled, including the top five large power generation corporations (China Huaneng, China Datang, China Huadian, China Guodian and State Power Investment Corporation, SPIC) and the top four small power generation groups (State Development & Investment Corporation, SDIC; Guohua Electric Power; CR Power; and China General Nuclear Power Group, CGN). The installed capacity of these companies accounts for 60% of the country's total. There are also some power generation enterprises that are wholly owned or controlled by local government. At the same time, power transmission, distribution and sales have not yet been deregulated, and power supplies in all provinces come mainly from local monopoly power grid enterprises. Electricity sales from central grid enterprises account for more than 80% of the country's total electricity demand. Independent market-oriented trading has not yet been established in China's electricity market—competition has been introduced in power generation, but monopolies remain in transmission, distribution and sales. Grid companies with power transmission and distribution networks have a monopoly in both wholesale and retail. Finally, power generation plans are developed in coordination with government at all levels and with power plants and grids. This has created an administrative system influenced by multiple factors, such as regulation policies, energy conservation and emissions reduction, and the requirements of administrators.

The new electricity market reform plan has been released in the Opinions on Further Deepening the Reform of the Electric Power System. The goal of "establishing and improving the market mechanism in the power sector that features 'legal basis, separation of government and enterprises, compliant market participants, fair trading, reasonable prices and effective supervision" stated in the Opinions echoes actual requirements. The priorities and pathway of electricity market reform can be summarised as "deregulating a new power distribution and sales market, deregulating electricity prices other than transmission and distribution tariffs, and deregulating power generation other than that for safe and efficient power operation and supply reliability, and others. The reform framework is defined as "deregulating power generation, sales and end use, and control of power transmission and distribution", which is basically what would

be expected. Generally, this round of electricity market reform has been carried out with a common-sense approach and has taken both the requirements of reform and the principle of practicality into consideration. Compared with "Document No. 5", released in 2002, the Opinions is of more practical significance.

(3) **Coal sector market reforms are complete, though the long-term contract price for thermal coal needs to be tested in the market**

The coal sector has ended its dual-track price system and completed its market-oriented reform. Apart from thermal coal, China's coal market was already market-based. In 2013, the cancellation of priority contracts and the ending of the dual-track price system for thermal coal completed the market reform of the coal sector. Although the contradiction between market-oriented coal and planned electricity appears to have been resolved, the long-term contract price for thermal coal, reached through difficult negotiations and compromise between all parties involved, has yet to be tested in the market.

3.3.2 Analysis of Existing Problems

China's energy circulation system is largely monopolistic. Administrative monopoly is prominent in investment access and resource acquisition in upstream primary energy. The fact that there is only a limited number of market participants in the oil and gas sector (especially in upstream and midstream) affects the overall efficiency of the sector.

(1) **The upstream and downstream oil and gas monopoly affects the overall efficiency of the value chain**

The monopoly in upstream oil and gas persists. Despite mutual penetration of each other's dominant market areas by CNPC, Sinopec and CNOOC, the "separation of upstream and downstream, separate administration of domestic and overseas trading, and separation of offshore and onshore oil and gas" in the oil and gas sector, which was formed at the end of the 20th century, has remained relatively stable, and China's current regulations and administration reinforce this situation.

First, the unequal status of different market participants makes access difficult for non-state-owned investors, so it is hard to create a fair market environment. This is not conducive to improving the overall efficiency of the oil and gas sector or sharpening the competitiveness of oil and gas enterprises, nor does it help increase the exploration and development of domestic oil and gas resources.

Second, local monopoly markets form between upstream, midstream and downstream, and within each part of the oil and gas value chain, which leads to market separation and makes it difficult to establish a consistent and open market system.

Third, various market structures—a national oil and gas exploration and mining rights trading market, a pipeline capacity trading market, a reserves market and a futures trading market—are all yet to be established. As a result, the market has very little scope to play a role in allocating resources; resource use efficiency could therefore be improved. As a large oil and gas consumer, China does not have enough say in international oil and gas pricing.

(2) Many interrelated problems hinder the development of the electricity market

The Opinions is just a guideline. If the requirements stated in the Opinions are truly to be implemented and the reform goals achieved, then support policies and regulations, along with implementation documentation, are needed. Problems to be solved in the power sector include: (i) the relationship between power grid enterprises and relatively independent electricity trading institutions urgently needs to be defined; (ii) power transmission and distribution tariffs need to be set independently; (iii) market-oriented power generation and sales prices, other than those for public goods, need to be phased in; (iv) cross-subsidies for electricity prices need to be gradually decreased; (v) the

bilateral trading market needs to be kick-started by increased market participation and a cross-provincial market mechanism; (vi) power sales reform needs to be steadily promoted and power distribution and sale services need to be deregulated to admit private capital in an orderly manner; and (vii) the results of reform need to be safeguarded by legislation.

(3) Barriers in the coal market and severe decentralised administration remain

Of all the energy sectors, coal has the most relaxed administration. However, despite the substantially reduced level of government intervention in the coal market, there are still many problems in the sector. First, the involvement of multiple departments in its administration and the complexity of coal circulation results in high costs. Second, there are acute transport bottlenecks, such as poor allocation of transport capacity. Third, coal logistics management is inadequate, and there are still no emerging coal circulation companies and logistics centres. Fourth, supervision of the coal circulation process is poor and there is a serious problem with pollution. In addition, poor quality control is holding back the sustainable development of the coal market.

There are still many obstacles in coal circulation. First, coal logistics generally divides into west-to-east and north-to-east coal transport, and any imbalance in economic transformation may reinforce such supply and demand differences. Second, rail transport plays a more important role in the coal logistics system, which is of greater significance for coal transport in Xinjiang and eastern Inner Mongolia. Third, water transport is an important method of north-to-south coal transport. The coal volume transported by water increases year by year, and the established directions of coal flow are undergoing change.

The coal sector features the loosest administration of all primary energy sources. Central government, local government and coal associations all play a role in the administration of coal. Although administrative streamlining has allowed some tasks to be erased or delegated to local government, there has been no obvious improvement in decentralised administration.

3.4   Energy Pricing System

3.4.1   Current Developments

The government continues to intervene in energy pricing. In each part of the energy value chain the pricing guidance reflects market dynamics poorly. Overall, there are many problems in China's energy pricing system: in price formation, composition and regulation, and in the coordinated administration of production and pricing. Pricing in upstream primary energy has gradually been deregulated, but government intervention remains influential in the pricing of energy products for major end users. The ground has been prepared for market-oriented pricing reform in the oil and coal sectors, but gas pricing reform still needs more work. The energy tax and finance systems and policies play an important role at some stages of reform and have a positive impact on end-user energy prices.

(1) **Problems with China's energy pricing systems severely constrain energy system reform**

China has been steadily promoting energy pricing reform for years, but deeply rooted contradictions have not been resolved. Distortions in pricing cause resources to be miscalculated, structural imbalances and extensive overdevelopment. Although the policy of "replacing pricing reform with adjustment" has had partial success, prices and price adjustments that work in the short term have not developed into a long-term pricing mechanism and a workable pricing settlement for the entire energy system.

Energy pricing reflects the various drawbacks in China's energy system and is a core component of energy system reform. Unreasonable energy product prices and slow-moving pricing reform severely constrain the smooth transition to a new development process and a change in overall economic structure. Administratively, energy planning and balancing supply and demand are mainly the responsibilities of the NEA, while energy pricing and administration are the responsibilities of the NDRC. This administrative separation of production volumes and pricing is the major cause of China's energy pricing problems, many of which are outlined in the following paragraphs.

First, due to the unfair pricing mechanism and the government's forceful intervention in pricing, prices do not reflect the true value of energy products. Market-oriented pricing in competitive energy fields and related areas is not common practice. For historical reasons and due to the specific characteristics of energy resources, the prices of China's energy products have been set or strictly controlled by the government for a long time. As a result, energy prices cannot effectively reflect supply and demand and the scarcity of energy resources, which restricts the role of price in resource allocation.

Second, pricing calculations and taxation are unreasonable and full-cost accounting is not well implemented. In the current price calculation of energy resource products, cost items are entered at a low level that does not provide reasonable compensation for the energy resource production process. Neither do they reflect the cost of controlling environmental pollution in the resource development process nor the cost of safe production. Moreover, external costs are not internalised. Inconsistent taxation and energy costs do not reflect the price relationships of energy products, resulting in users lacking any cost constraints for some energy products. This is detrimental to the correct alignment of the energy system.

Third, the incomplete price regulation system, poor regulatory expertise and inadequate information disclosure hinder effective price regulation. When the price regulatory body regulates prices in natural monopolies it tends to lack real-time cost information and feedback; its price and cost accounting and supervisory capability are, therefore, weak. Moreover, there are insufficient levels of personnel, finance, materials and IT in the price regulatory body, and inadequate transparency of price regulation and public engagement. At the same time, lack of transparency and cross-subsidies due to monopolistic pricing are the main reasons why effective supervision is difficult to deliver. Some energy sources and links in the energy value chain completely lack cost and price regulation, resulting in a hardened price monopoly in natural monopolies like power grids and oil and gas pipeline networks.

Fourth, the disconnect in energy pricing results in intermittent price transmission and obvious contradictions between pricing in different links of the energy value chain, which affects the ability of the market to allocate energy resources effectively. This phenomenon is especially evident in the power sector, for example in the disconnect between the sales price and cost of generation, and between the power transmission and distribution tariffs and the feed-in tariff. In addition, the links with international and domestic market pricing are weak. Government pricing does not respond promptly to market dynamics, and China also lacks a voice in international energy pricing.

Fifth, separating production volumes and price amplifies the shortcomings in China's energy pricing system. Energy planning and balancing supply and demand are mainly the responsibility of the NEA, and energy pricing and administration are the responsibility of the NDRC. This administrative system, which separates production volume and price, is the major cause of China's energy pricing problems. When supply is tight, the effects of the separation are not obvious, but in the context of diversified energy supply and demand, especially during the transition to market-oriented energy pricing, the administrative separation is increasingly ill-adapted to the requirements of energy development.

(2) **Deregulation of upstream pricing is gradual, but government intervention continues**

Pricing in China's upstream energy sector has gradually been deregulated. The deregulation of pricing in the coal and oil industries has been implemented in phases: upstream coal pricing

has been deregulated for years, and the crude oil price is independently determined and based on the international oil price. The implementation of netback pricing has made the upstream natural gas wellhead price redundant. Power generation pricing has evolved from a price based on each generating unit to a benchmark price, and government interventions are gradually decreasing.

Government intervention still occurs in downstream energy, especially in the pricing of major end-use energy products. Natural gas and electricity end-user prices in particular are still controlled by local government, due to the reliance of these two energy sources on natural monopoly networks and facilities and the public service attribute associated with them.

(3) **Pricing reform in the oil and coal sectors is completed, but hurdles remain in electricity and gas**

Oil pricing has become market oriented. China's crude oil prices are now negotiated or independently determined by oil-producing enterprises, based on the international oil price. The price of crude oil supplied by national oil companies is negotiated by the buyer and seller on the principle that the cost of delivering domestic onshore crude oil to refineries should be equivalent to the cost of delivering imported crude oil from international markets to refineries. The price of crude oil supplied to local refineries from national oil companies is set by referring to the price of mutually supplied crude oil between two oil groups. The price of crude oil produced by CNOOC and other companies is independently determined by the corresponding producer with reference to the international oil price.

The pricing of oil products is linked to international oil prices, but a degree of government intervention remains. Having connected the government-guided moving-average price with the international oil price, the government can influence end-user oil prices more through tax policies. In March 2013, the government announced an improved oil product pricing mechanism, whereby the price adjustment cycle of oil products would be shortened from 22 to 10 working days, and the price adjustment restriction of ±4% of the energy source's average price in the international market would be cancelled. In general, China's current oil product pricing mechanism follows the basic principle of following the international oil price, but to ensure a relatively stable domestic oil price it implements the government-guided pricing approach. The existing oil product pricing mechanism covers the following five aspects: (i) differentiated oil product prices—either government-guided price or government-controlled price; (ii) adjustment of petrol and diesel prices based on changes in the international crude oil market price every 10 working days; (iii) adjustment of oil processing profit margins and thus oil product prices, based on changes in the international crude oil price; (iv) the NDRC sets the ceiling retail price of petrol and diesel for all provinces and major cities; and (v) defines the difference between wholesale and retail prices.

A dynamic price adjustment mechanism that reflects market supply and demand and the scarcity of resources has been preliminarily established in the natural gas sector. China has gradually implemented the netback pricing approach for pipeline gas since 2011, but the price of pipeline gas has been restricted by gas source price and limits on pipeline transmission and distribution. The implementation of the netback pricing approach represents a major step in rationalising the relationship between the natural gas price and the alternative energy price. China's natural gas market is still supplier-dominated and non-competitive, and the gas end-user price is determined by such factors as gas source prices, government-controlled long-distance gas transmission tariffs and local government-controlled gas distribution fees and sale prices. Hence, compared to alternative energy sources, the competitiveness of the natural gas price is uncertain. Liquefied natural gas (LNG) pricing has been reformed and is market-oriented, but the price of LNG entering the pipeline network as peak-shaving gas remains unclear.

Due to the complicated network of interests associated with the price of electricity, the market-oriented reform of electricity pricing has been much discussed, but without reaching a conclusion. China's electricity price is currently considered both from the power generation side and from the demand side. China's current electricity price system covers mainly feed-in tariffs, transmission and distribution tariffs and the power sale price. On the generation side, different power generation companies set different feed-in tariffs in light of local conditions, and the NDRC sets the benchmark feed-in tariff. On the supply side, local government sets different prices according to the type of end user—a dual pricing system (including basic price and meter price) is generally applied to large industrial users, and a tiered electricity pricing system is applied to residential users.

Coal pricing has been market-oriented for a while. The double-track price system used for thermal coal pricing was ended along with priority contracts in 2013. The publication of the Guideline of the General Office of the State Council on Deepening the Market-Oriented Reform of Thermal Coal marked the official ending of the dual-track price system, which had been in place for more than 20 years. The NDRC no longer sets the annual framework for allocating rail transport capacity for cross-provincial coal to local governments.

3.4.2 Analysis of Problems

In general, the government still intervenes heavily in energy pricing in China, and there is insufficient guidance on end-user energy pricing. Energy prices reflect poorly market supply and demand and the attributes of each energy commodity. The current energy pricing system struggles to reflect the interactive relationship between production volume and price, and there is insufficient transparency in pricing. In addition, the energy tax and subsidy systems fail to provide useful guidance on energy production and consumption.

(1) Transparency and timeliness in the oil product pricing system

Despite the quasi-market-oriented nature of oil product pricing, the contradiction between operating a transparent pricing system and allowing for delayed price adjustments has not been solved. The risk-free arbitrage behaviour that this gap allows clearly generates instability in market supply, which in turn has an impact on the stability of the oil product market, especially when oil prices fluctuate wildly. A market-based oil product pricing mechanism will address these problems and allow the government to still intervene through tax and other means.

(2) There is no market-oriented gas pricing mechanism; the price of gas does not reflect its actual value

There is no market-oriented gas pricing mechanism, and the difference in gas price between regions is unjustified. In general, the price of gas does not properly reflect the attributes of natural gas as a commodity. Gas pricing reform can be completed independently, but it should be coordinated with pipeline network reform.

The price determined by netback pricing is unreasonable—it is obviously too high, and the ceiling city-gate gas price defined by current policies has become the de facto gas source price in each province. As research indicates, this price level is about 30% higher than the total cost of gas, so there is a great amount of profit built in. However, the high price squeezes the profit margin of downstream industries and discourages any significant increase in gas's share of total energy consumption.

Gas price adjustment policies are incomplete and the residential gas price is low. The gas price adjustment policies introduced in Beijing, Hebei and Shanghai in 2011 do not affect the residential gas price, allowing it to stay at a low level. This results in an underlying expectation among residents that low-cost gas will be available. In contrast, the price of industrial gas is higher than that of residential gas.

The gas pricing system is incomplete and distorted. There is no clear pricing mechanism and no administrative system that differentiates between peak and off-peak consumption, gas prices that allow supply cut-off at any time and tiered gas prices. There are no specific requirements on how to calculate cost and price or to determine who is responsible for supervision, which makes it difficult to set prices.

(3) Electricity pricing is a major problem in energy pricing reform

Electricity pricing reform is dependent on electricity system reform. Two main problems need to be addressed.

The first problem is the arbitrary pricing mechanism. On the grid connection side, the feed-in tariff is decided by the government. Without market competition, it is difficult to motivate power enterprises. On the power transmission and distribution side, power grid operators do not decide transmission and distribution tariffs; their revenue is generated by the difference between the power wholesale price and the feed-in tariff. On the sales side, wholesale prices do not reflect user demands and result in limited options for consumers. Cross-provincial power supply lacks a mechanism that can motivate both the supplier and the buyer. In general, electricity prices in China are decided by administrative bodies and do not reflect the real supply and demand relationship and production costs. A scientifically based pricing mechanism has yet to be established.

The second problem is the large cross-subsidies. On the grid connection side, the hydropower feed-in tariff is significantly higher than that for other types of power generation. On the wholesale side, the subsidies for different types of use overlap. For instance, urban areas subsidise rural areas, industrial and commercial users subsidise residential and agricultural users, users of high voltage subsidise users of low voltage, and consumers of large loads subsidise those of low loads. As a result, the electricity retail price does not reflect the true cost of power supply and fails to guide consumers to use electricity efficiently. The existence of cross-subsidies also hinders market-oriented reform in power generation and sales.

In addition, although coal pricing has become market-oriented, the price does not include externalities, so it needs to be adjusted through institutional reforms such as an environmental tax and resource tax.

3.5 Energy Regulation System

3.5.1 Current Developments

China's energy regulation system is not independent or sufficiently professional, and its supervision capabilities are inadequate to safeguard fair market operation. The organisational structure of China's energy regulation system is mostly situated at lower administrative levels, and the rights and liabilities it sets out are unclearly defined and decentralised. The system prioritises economic regulation. It bundles administration and regulation together, resulting in weak supervision and enforcement that require improvement.

(1) **Regulatory functions are decentralised and multi-level**

Generally, China's energy regulation system is: (i) decentralised across several authorities, with both a horizontal division of labour (regulating market access, investment, cost and price, etc.) and a vertical division of labour (regulating the value chain of the different energy sources); (ii) multi-level (including the State Council, central ministries and commissions, and local government); and (iii) features separate economic regulation and societal supervision.

Economic regulation of the energy sector is at a high administrative level, being mainly the responsibility of the central government. China's economic and social supervisory functions for oil and gas, coal, electricity and nuclear power are decentralised across various departments and authorities, whereas the economic regulatory functions for the energy value chain are mainly

concentrated in central government. For example, the regulatory functions for market access, cost and pricing in the power sector are in the hands of different departments, including the NDRC, the Ministry of Finance and the NEA. A single economic regulatory function may involve different departments. For example, the supervision of electricity investment involves the NDRC, the NEA and local government bodies.

Significant regulatory functions are mainly undertaken by the NDRC and the NEA. As a macroeconomic management agency, the NDRC also coordinates China's economic and social development and the regulatory means to enact it. The establishment of the NEA in 2013 highlights how energy regulation has been strengthened. Its supervisory activities focus on "improving the energy regulation and administrative system, strengthening energy regulation and administration, controlling total energy consumption, developing the energy market and maintaining energy market order" (see Table 5).

(2) **Economic regulation has been prioritised and social supervision is gradually being improved**

The energy sector is a priority for government regulation, but China's energy regulation is still in the early stages of transition away from planned supervision and administration. Energy regulation prioritises economic regulation, while the social supervision of energy use remains relatively weak. Currently, economic and social supervision are separate.

In general, energy regulation in China means economic regulation. Economic regulation consists mainly of supervising market access, pricing, investment, costs and market trading conduct. These functions are mainly decentralised to departments like the NDRC and NEA. China has strict supervision of market access and investment, but the supervision of trading conduct is not very effective. Economic regulation takes place at a higher level, and (with the exception of coal) is typically implemented by central government.

Table 5 China's energy regulation institutions and their responsibilities (2014)

Institution	Responsibilities
NDRC	Supervision of pricing and investment (access)
NEA	Technical standards, supervision of investment (access), market (order), electricity safety and universal service
The Ministry of Water Resources	Technical standards for hydropower in rural areas
The Ministry of Finance	Energy-related financial norms and accounting regulations
The State Administration of Work Safety (the State Administration of Coal Mine Safety)	Supervision of energy-related (production) safety
The State Administration of Science, Technology and Industry for National Defence under the Ministry of Industry and Information Technology	Review and management of the import and export of nuclear materials, regulation of nuclear materials (circulation)
The National Nuclear Safety Administration under the Ministry of Environmental Protection	Supervision of environmental and nuclear safety
The General Administration of Quality Supervision, Inspection and Quarantine	Quality standards and product standards
The Ministry of Commerce	Criteria for oil product market access and supervision of the oil product market

Source Based on the Plan on Defining Functions, Internal Organisation and Staffing, and official websites

China is attaching increasing importance to social supervision in the energy sector. The supervisory functions for the environment, safety and health involve externalities and public goods, and tend to be centralised in departments or administrative institutions such as those for environmental protection and safety at work. Social supervision, however, tends to follow the guiding principles of local administration.

(3) Administration and regulation have been bundled together, but regulation is weak

China's institutional organisation of energy regulation generally bundles administration and regulation together. This reflects the relationship between energy regulation and other energy administrative functions. For example, supervision, planning, policy, regulation and public services are centralised in one department, like the NDRC or NEA. These departments are responsible not only for developing investment, operation and administrative policies in energy sectors like power, natural gas and coal, but also for supervising the implementation of these policies. Despite the relative independence of these organisations and the formation of some supervisory departments and bureaus, regulation is still weaker than other functions, such as planning and policy.

In the long term, however, especially as the energy pricing mechanism gradually improves, the separation of administration and regulation will continue. Regardless of which approach to administration and regulation is taken, bundled or separated, the following tasks need to be undertaken: increase energy regulation; gradually raise the access threshold to the energy market; improve technical standards, environmental requirements and safety standards; and improve the management of energy projects and the safety supervision of energy enterprises. In addition, the scope of energy regulation should be enlarged and the supervision of costs in sectors with a natural monopoly should be increased.

(4) The energy regulation system has taken shape

China's energy law and regulation system has taken shape. Energy-related laws, administrative regulations, departmental regulations, normative

documents and mandatory standards and specifications form the legal basis for energy regulation, and they all play their part in modifying and
regulating energy development.

The legal foundation of China's energy regulation system is spread across various energy
laws and regulations, with some parts of the
energy sector subject to several different energy
laws and regulations. The laws that can be used
as the basis for energy regulation include the
Mineral Resources Law of the PRC, the Electric
Power Law of the PRC, the Coal Industry Law of
the PRC and the Energy Conservation Law of the
PRC. There are also some provisions on energy
regulation in other related laws, such as the
Water Law of the PRC, the Environmental Protection Law of the PRC, the Law of the PRC on
Prevention and Control of Radioactive Pollution
and the Production Safety Law of the PRC.
Beyond this, the departmental regulations passed
by the State Council and by ministries and
commissions serve as the basis for supervision of
various energy sectors, for example the Regulation on Electric Power Supervision (2005) and
the Regulation on the Emergency Response to
and Investigation and Handling of Electric Power
Safety Accidents (2011). The former State
Electricity Regulatory Commission (SERC)
introduced more than 60 regulations on supervision and more than 160 normative documents.
Finally, the mandatory standards and specifications set the limits and procedures for key energy
sectors, so they provide the most direct basis for
supervision.

3.5.2 Analysis of Problems

In general, China's energy regulation lacks
independence, is positioned at a low administrative level, needs higher standards of professionalism, and has difficulty carrying out the
supervision needed of a modern energy market.
As a result, there is not enough consistent and
effective supervision in key energy sectors,
especially in natural monopolies.

**(1) Supervisory agencies are positioned at a
lower administrative level and they are
not independent**

First, there is currently no independent high-level
energy regulatory agency. An approach that
bundles administration and regulation together
has been in use for years. During this time there
has been no independent, dedicated agency
responsible for energy regulation. This lack of
independence has severely restricted how energy
regulation is carried out.

Second, the existing energy regulation agencies are at a lower administrative level. Among
the major agencies that perform energy regulation, the NEA is at the department/bureau level
and the NDRC is at the division level. Only the
National Nuclear Safety Administration is relatively independent; it is a national agency
administered by the Ministry of Environmental
Protection on behalf of the central government.
However, the special nature of nuclear safety
supervision makes it different to most other types
of energy regulation.

Third, there is no national agency that consistently carries out energy regulation. The
decentralised nature of supervision has prevented
the synergies from the energy regulation system
as a whole being realised, so that often a situation
arises where many parties have responsibility for
something, but none of them delivers anything.
When the NEA was formed in 2013 it was given
the energy regulation function, but there are more
than 10 national government departments
involved in energy sector supervision for oil and
gas, power, coal, nuclear power, renewable
energy and others.

Fourth, when it comes to supervisory efficiency, the excessively decentralised regulatory
function increases the difficulty of coordinating
supervisory activities, which compromises efficiency to some extent. The contradictions and
conflicts between various departments, and
between central and local government, are

unavoidable. To give one example, if the supervision of pricing, cost and quality was well coordinated it would help to deliver optimal results; but in reality pricing supervision is the responsibility of the NDRC, cost supervision is the responsibility of the Ministry of Finance and NEA, and the NEA is also responsible for supervising service quality and standards. This division of functions runs counter to a logical solution, which is to supervise pricing, cost and service quality together. The current administrative silos and the non-sharing of information undoubtedly affect the efficiency of supervision and the effective implementation of China's energy policies.

(2) Consistent and effective supervision is absent in key energy sectors, especially in upstream and the midstream

First, consistent and effective supervision is absent in key energy sectors. Thanks to independent, professional and consistent supervision, the former State Electricity Regulatory Commission (SERC) achieved good results in the supervision of electricity. In other key energy sectors, including oil, gas and coal, the decentralised, fragmented and inconsistent supervision available does not deliver a professional service, with knock-on effects in the various energy sectors.

Second, supervision is particularly inadequate in some key links of the energy value chain, such as in natural monopolies. For example, there is no effective supervision of dominant enterprises that might abuse their market rights in the oil, gas and power sectors.

Third, there is a large rift between energy regulation and environmental supervision. In the renewable energy sector, supervision of the resource and environmental problems caused by renewable energy development has yet to be introduced. And when it comes to social supervision of oil and gas exploration and development, the energy and environment supervision authorities do not play a significant role, with the result that action on these issues depends on the self-discipline of oil and gas companies.

(3) An incomplete basis for supervision makes it difficult to meet the requirements of modern energy regulation

There are many reasons for China's incomplete energy regulation system. Large gaps in legislation and delays in ratifying laws and regulations in key energy sectors make it difficult to meet modern energy regulation requirements.

First, the absence of significant energy laws and regulations has resulted in an incomplete regulatory system. The most important energy law is still awaited—the Energy Law of the PRC was prepared at the end of 2007, but the date of its release is still uncertain. In addition, there are no energy regulatory laws, so there is no high-level legal basis for defining energy regulation rights and liabilities. In addition, laws for some key energy sectors are missing. For example, the long-term absence of the Oil and Gas Law of the PRC and the Atomic Energy Law of the PRC has left gaps in energy consumption legislation.

Second, obsolete laws and delayed amendments have failed to meet the requirements of energy development and supervision. Based on obsolete legislative concepts, some energy laws, such as the Coal Industry Law of the PRC and the Electric Power Law of the PRC, are essentially traditional administrative regulations. Legislation in the oil and gas sector is emergency response-oriented, temporary and out of date. Moreover, the amendment of some energy-related legislation is slow. A case in point is the amendment of the Mineral Resources Law of the PRC, which, due to theoretical disagreements and the involvement of other non-energy sectors, has still not been finalised after years of discussions.

Third, the absence of essential implementation and support regulations leads to poor operability. For example, oil and gas legislation, such as the Coal Industry Law of the PRC, suffers from unclear regulatory boundaries, unclear definition of rights and liabilities, and uncertainty regarding its scope and conditions. The lack of implementation and support regulations results in legislation that is ineffective and unsuited to practical implementation.

(4) **Inadequate and unprofessional government supervision**

The NEA has built a dedicated supervisory department that aims to extend the successful experience gained in electricity supervision to other energy sectors, especially fossil fuels. The oil, gas and power sectors are strikingly different in terms of technical and economic characteristics and supervisory demands. China's oil and gas supervision is still in its infancy and the gap between it and the advanced supervision found elsewhere in the world is wider than in electricity supervision. Coal supervision, however, has Chinese characteristics, which make it hard to apply the experience and practices of electricity supervision to fossil energy.

In addition to the two dedicated supervisory departments of the NEA, the agency under the former SERC—that was merged into the NEA and positioned as a dedicated supervisory department—has found it difficult to address the supervisory requirements of multiple energy sources. Furthermore, China's energy regulation organisations suffer from inadequate funding and low-level equipment, technology and IT capability.

Finally, given non-existent social supervision, non-governmental organisations that try to fill the breach have no legal status within energy regulation, and it is difficult for them to obtain useful information and access the necessary communication channels.

4 China's Energy System Transformation: Progress, Conflict and System Design

4.1 The Urgent Need for Energy System Transformation

Energy supply pressure is mounting. Energy consumption in China has been increasing steadily. As a result of its continued economic momentum and the declining energy consumption of the major mature economies, China has overtaken the USA to become the largest energy consumer in the world. Energy supply in China, rich in coal but short of oil and gas, is limited, unable to meet growing demand. This has led to an expanding gap between supply and demand, which shows no sign of improvement. While striving to address its energy security, China is becoming more and more dependent on oil and gas imports. Meanwhile, the sources of these imports are relatively concentrated geographically and susceptible to global turbulence, increasing the risk to energy security. In the future, unsustainable energy supply will give rise to uncertainty in energy development, bringing about an energy security challenge that cannot be ignored.

China's energy product pricing mechanism is ineffective. Market signals do not accurately reflect the supply-demand relationship and resource scarcity, as well as externality costs such as environmental pollution. Energy pricing that is not market-led, including lower energy prices imposed by regulation, leads to excessive energy consumption and impedes the effective allocation of scarce resources. For a long time, China's economy has been dominated by energy-intensive secondary industries, which account for a large amount of the total energy consumed by industry. China's energy intensity is 1.55 times the global average (2016), significantly impairing the efficiency of the country's economic growth.

Massive energy consumption brings severe environmental pollution and environmental damage. The direct effects can be seen in the greenhouse gas emissions from unnecessary energy use that pollute the environment, damage biodiversity and impact the climate, as evidenced by the frequent occurrence of extreme weather, such as acid rain and haze. In a time of global decarbonisation, high pollutant emissions seriously impair China's image in the world and restrict its potential for exports, employment, revenues, investment and economic growth.

China's energy system needs to adapt to a new productivity trend. China has become the world's largest energy producer and consumer, with an energy system comprising coal, electricity, oil, gas, new energy and renewables, as well as significantly improved technical sophistication and energy use for living and production. However, a few factors, including inadequate market competition and pricing mechanisms, an outdated energy legislation system and inefficient regulation, have all restricted the growth of China's energy industry. It is imperative that the Chinese economy shifts from being resource/energy/pollution-intensive to a healthier, innovation-driven mode of growth, and that it moves away from quantity and speed to quality.

Transformation of the energy system, as an instrumental part of the energy revolution, means pursuing goals in five different areas: energy availability guarantees, environmental pollution controls, economic restructuring, price shock absorption, and energy security.

First, energy system transformation can lead to improved energy availability. Energy underpins the development and security of a nation. Given an energy portfolio that features an abundance of coal and very little oil and gas, as well as insufficient use of nuclear, wind and solar power, a more streamlined energy system is vital to increase the use of new energy and provide long-term energy supply stability in China.

Second, energy system transformation alleviates environmental pressure on the energy industry. With energy consumption increasing, pollutant emissions from China's energy sector are becoming a critical challenge. Especially in recent years, climate change and the environmental impact of the energy sector have drawn much attention. In China, extreme weather, such as acid rain and haze, has become a major threat to people's health. Energy system transformation will, therefore, improve efficiency in the sector and promote clean and green initiatives, which will reduce the resource and environmental pressure on the sector.

Third, energy system transformation facilitates economic restructuring. China's extensive economic growth over the past three decades has resulted in a rigid energy consumption structure as well as waste and the inefficient use of energy sources. Given this context, energy transformation provides an effective path to optimise the economy: the efficient use of energy sources will promote economic restructuring and ensure sustained efficient operation of the economy.

Fourth, energy system transformation can alleviate the impact of energy prices on businesses and people. An efficient energy pricing system accurately reflects the supply-demand relationship and guides sensible production and consumption and promotes the efficient use of resources. If it is to succeed, energy system transformation must take into account resource scarcity, affordability, social fairness and sustainability.

Fifth, energy system transformation helps improve energy security. Against the backdrop of an increasingly complex global political and economic landscape, energy security is essential for the smooth operation of the Chinese economy, as well as for the sustained progress of Chinese society. Using market mechanisms, energy system transformation leverages pricing to balance energy supply and demand, ultimately contributing to national energy security.

4.2 Status Quo and Major Conflicts in China's Energy System

Since opening up, China has implemented a series of transformative initiatives in energy development and exploitation, market access, pricing, investment and financing, international trade, and administration. The time when only government invested in the energy sector has gradually changed. The old system of planning and administration, where demand was dependent on availability, has also changed, to cover supply and demand as well. There has been solid progress in transforming the energy pricing mechanism, and an energy market has taken initial shape. A few energy sectors are no longer under central control, and the institutional climate for energy development has improved, which plays an instrumental role in guaranteeing supply-demand balance and driving social and economic development. Specifically, reforms like price control removal, administration-enterprise unbundling and empowering enterprises to become market participants have revitalised the coal market, leading to a golden decade (2002–12) for coal. Initiatives such as restructuring the three national oil and gas companies in 1998, and loosening market access and price control have improved the oil and gas sectors and made China one of the 10 largest oil producers. The initiatives to set up two grid companies, five power generation groups and the China Power Regulatory Commission in 2002 have essentially broken the hold of the monopolies. As a result, issues that were common in the planning era (such as administration-enterprise bundling and grid-generation conglomerates) have been resolved, allowing competition between multiple participants in the power generation market. All these initiatives have effectively released the potential for productivity improvements and given impetus to the development of China's energy industry.

Despite all this, it should be noted that, in the context of the global energy industry and in the drive to deepen reforms, the current energy regime does not satisfy the demands on energy production. Neither does it support the consumption revolution nor drive the development of the socialist market economy. Some deep conflicts and issues still need to be addressed urgently, as summarised below.

4.2.1 A Modern Energy Market System Has yet to Take Shape

Since opening up, China has shifted from a planned market economy to a socialist market economy and has carried out a series of institutional reforms in investment, taxation and pricing. Despite the reform initiatives in the energy industry, there are still several challenges to be addressed, including slow progress in some sectors, the unclear boundary between administration and business, government intervention in economic activities, unfair competition between market participants of different legal status, and the absence of consistent market access criteria. Other concerns include the government's control of coal and oil exploration and mining rights, imperfect administration, and poorly regulated market flow. An open, transparent and well-regulated energy market access system has yet to take shape. The coal, oil and power

markets are highly fragmented: the system for assessing and approving new entrants is more like an administrative procedure, with approval more likely to be given to large state-owned enterprises, which poses higher barriers for privately owned and foreign companies. Energy production is still under government control, by means of power output and oil and gas production scheduling. Even now, more than three decades after opening up, it is thought-provoking to realise that sometimes a market participant can decide neither price nor production volumes. There are other administrative and regulatory challenges that need to be addressed: the imperfect energy market mechanism and taxation system; the financial and taxation policies that are poorly suited to encouraging the development of new and renewable energy; and the restricted range of incentives to support development of the energy industry, including subsidies, financial discounts and tax incentives. Also, in the context of global economic integration, China still needs to integrate itself properly with the global energy market and use its international influence to shape regional energy markets.

The organisational structure of the energy industry is still less than optimal, constraining effective competition in some critical areas. In general, there are major differences in industrial organisation between different types of energy (especially in energy supply), including market barriers to private capital, a lack of competition in some sectors and excessive competition in others, and even surplus capacity. In the oil and gas sector, the three leading national oil companies maintain obvious first-mover advantages in onshore oil and gas production and pipeline operation and management (CNPC), petrochemicals and oil products (Sinopec), and offshore oil and gas exploration and production (CNOOC), while each of them still demonstrates shortcomings in value chain integration. Effective competition is still required in some major areas to improve the overall efficiency of the industry. Another major concern is the increasingly obvious overcapacity in the refining segment.

In short, the energy system challenges that need to be addressed are: imperfect market system, monopoly in some segments, pipeline-operation bundling, dispatch-operation bundling and competition restrictions. Most energy companies in China are large state-owned enterprises, with few private investors, giving an unbalanced mix of market participants. The government still controls the licencing of coal, oil and gas exploration and mining rights, and the rules necessary to ensure consistent market access and fair market competition are still not in place. So, depending on their legal status, companies are not competing on the same level playing field. Grid operators still have holdings along the entire value chain of power transmission, distribution and sales, and a fair and well-regulated market competition mechanism has still not been put in place. Integrated operations encompassing exploration, development, refining, transmission, imports and sales have more or less been realised in the oil and gas industry, but the industry still does not allow diverse participants to compete in individual sections of the value chain. A spot and futures market for oil and gas products has not been established, and regional international energy markets are less and less influential in the wider world.

4.2.2 The Energy Pricing Mechanism Is Still Imperfect

The deregulation initiative has made slower progress and there is still no fair pricing mechanism—the price of oil products, gas and electricity is still decided by the government. The price structure is unjustified. The cost of infrastructure like pipelines is not systematically checked, and externality costs related to the environment have not been internalised. Price distortion remains: electricity and gas prices for residential use have been below cost for some time, and cross-subsidies abound. The energy taxation system is still imperfect, and the structure and levels of resource taxation are unreasonable. Other concerns to be addressed include: poorly aligned fiscal and taxation policies for coal, oil, gas and renewable energy; a restricted range of incentives to support development of the new energy industry, including subsidies,

financial discounts and tax incentives; and incomplete pricing, fiscal and taxation systems that fail to reflect the supply-demand relationship in the energy products market, the scarcity of energy products, and their environmental impact.

The government still intervenes in the energy market and it is still impossible to deploy energy resources by means of price. In a market system, the most efficient information is the price signal—precise and flexible price signals can successfully adjust supply and demand, provide guidance for investment and optimise resource allocation. With the exception of coal, there is still no market-based pricing mechanism for other types of energy—grid and sales tariffs are still decided by the government, and the price for oil products is still not decided by market competition, though it has been aligned with the international market. Current energy prices in China do not reflect supply and demand, resource scarcity and environmental impact, or provide guidance for consumption, investment and resource allocation. Over the course of a decade of reform, the monopoly in the coal sector has been broken up to enable competition. However, state-owned enterprises still maintain their dominance in the oil and gas and power transmission and distribution sectors, and a diversified range of market participants has still not materialised. The enterprises active in these sectors are less motivated to improve productivity, and their resource allocation is inefficient.

4.2.3 Government Administration Still Needs to Be Improved

The relationship between government and the market still needs to be improved and the government needs to accelerate its transition to a different role. On the one hand, the government intervenes too much in the market and its administration is too specific, especially when it comes to project approval. The procedure for project approval is relatively complex and some of the initiatives to shift approval from upper- to lower-level government have failed to achieve the desired results. On the other hand, there has not been enough research on energy strategy, which is not clearly aligned with the goals of the

Two Centenaries.[13] Energy planning is less sci-entific, authoritative and actionable than it should be, and the alignment between different plans is poor. Project approval tends not to be aligned with implementation.

Even though reforms have been carried out in the energy industry, progress in some areas is slow. The boundary between administration and business is less clear than it should be, and the government intervenes too much in energy-related economic activities. Other con-cerns to be addressed include the impossibility of achieving fair competition between market par-ticipants of different legal status, the absence of consistent market access criteria, the over-representation of state-owned capital, and the lack of efficient market-led pricing mecha-nisms. In 2002 China replaced the examination and approval system with the verification system, with the aim of clarifying the obligations of the government and business, streamlining the pro-cess, and regulating investment. However, in the real world, the verification system looks very much like administrative approval, which, com-pared with practices in other developed countries still has several flaws. In addition, project veri-fication is usually more favourable to large state-owned enterprises, while setting higher entry barriers for privately run businesses.

4.2.4 Imperfect Regulation by Government

The regulatory system for energy is still not independent, and the regulatory responsibilities for market entry, pricing, investment, costing and transaction management are spread across dif-ferent authorities, including the development and reform commissions and the energy administra-tors. The regulatory function is still weak, and the regulation of oil, gas, coal, new energy and the Energy Internet is clearly inadequate. There

is no effective rules-based regulation of the market. Industry regulation of oil and gas pipe-line safety, the transfer and assignment of mining rights, fair third-party access, scientific resource development, clean production and reasonable energy consumption is incomplete. In addition, the competence of professional regulators is poor. In particular, regulation of technical mat-ters and of compliance with standards and spec-ifications barely meets the needs of conventional and new energy, nor of the energy transition.

There is a lack of independent and profes-sional regulators. The existing decentralised regulatory system, characterised by the integra-tion of administration and regulation, fails to achieve the results expected, especially in market regulation.

The level of coordination and alignment between different regulatory departments is poor —regulatory goals, effectiveness and speed across different departments and between central and local government are inconsistent. Decen-tralisation results in some regulatory functions being absent or inadequate. Another concern is the focus on market access approval without attaching sufficient importance to post-access regulation. In other words, the government focuses its regulatory efforts on project identifi-cation and approval, not on regulating project execution and post-project delivery. In addition, the government addresses economic aspects, such as investment approval, product and service price, and production scale, but fails to regulate externalities like resource conservation, safety, the environment and quality. Other challenges include an inadequate legal foundation and a lack of strict and scientific energy regulatory standards.

Monopolies in midstream transmission are not clearly identified, and their regulation is inade-quate, especially of power grids and gas pipe-lines. Grid operators monopolise power purchase and sale. The building of a regional power mar-ket has progressed slowly, power tariff reform is sluggish, and the separation of transmission and distribution and the bidding mechanism for grid connections have yet to be implemented. Gas transmission still awaits effective regulation and

---

[13]The Two Centenaries mark the founding of the Communist Party of China (in 2021) and the People's Republic of China (in 2049). The goals are to build a moderately prosperous society in all respects by 2021 and transform China into a modern socialist country that is prosperous, strong, democratic, culturally advanced and harmonious by 2049.

fair access is difficult. National oil companies still monopolise investment and construction of the pipeline network, while bundling and intra-company transactions continue in gas transmission, distribution and sales. In some regions there are still pipeline monopolies.

4.2.5 Incomplete Legislative System

China's energy legislation system is incomplete, as evidenced by the lack of a fundamental energy law, the delayed adoption and revision of other laws, and the incomplete scope and poor operability of the system. Provisions are spread across laws, administrative regulations, regional regulations and departmental rules, with varying degrees of enforcement. Due to the lack of consistent legislative guidelines and fundamental principles, the links between specific regulations at different administrative levels are absent. Some departments legislate regulations for the benefit of the department itself. Other notable concerns include an excessive reliance on administrative enforcement and inadequate punishment for infringements.

China is now confronted with severe resource and environmental constraints, pressure to reduce greenhouse gas emissions, and the challenges of maintaining energy security and driving the energy technology revolution. China's initiative to combat climate change requires that its carbon emissions peak around 2030. While energy consumption will still rise in the future, the abundant clean energy resources of hydro, wind and solar have not been exploited effectively. This suggests there is huge scope for adjustment in the energy system. The dependence of oil and gas on imports is 70% and 40% respectively, and China is the largest importer of oil and gas from the Middle East, making energy security a challenge. The development of emerging technologies like the Energy Internet will radically change the industry. This will force the government to change its regulatory and administrative practices, and drive innovation in industrial organisation and government administration,

which in turn will set new demands on the energy revolution.

In brief, to solve the above issues and address emerging situations and challenges, it is necessary to transform China's energy system urgently. In this critical window, when international energy prices are low, China will waste no time in initiating its energy system revolution, based on the consensus of all stakeholders.

4.3 System Design for China's Energy System Revolution (2030)

4.3.1 Guiding Ideas

In the spirit of the 18th and 16th National Congress of the Communist Party of China in 2012 and 2002 respectively, China should embark on its energy system revolution with the goal of building an ecological civilisation. This should focus on accelerating energy transition in line with the development concepts of "Innovation, Coordination, Green, Openness and Sharing". The market should play a decisive role in resource allocation, while the government should carry out its administrative and regulatory roles better in order to break down the barriers to energy sustainability and shape a modern, open, complete and competitive energy system. Such a system will prioritise renewables and gas, ensure the well-coordinated development of centralised and distributed energy, soundly balance supply and demand, and improve energy efficiency, all for the benefit of the national economy, people's livelihoods and the environment.

4.3.2 Fundamental Principles

(1) **Market orientation**

Taking into account industry characteristics and laws, China will distinguish between areas of natural monopoly and areas of competition and give participants equal access to the market. Efforts will be channelled towards creating an "effective market + efficient administration", with the focus on building the market, enhancing market regulation, maintaining market order, ensuring fair competition, mobilising and motivating market participants and shaping a modern energy market system.

(2) **Access criteria**

While loosening the restrictions on market access, China will introduce substantive standards for safety, the environment and energy efficiency. China will also transform and modernise the energy mix to alleviate the harmful effects of energy production and consumption on the environment and release the benefits of reform as much as possible.

(3) **Energy security**

Given the fact that it is a large energy consumer and importer, China will maintain its energy security by integrating international and local markets, sharing international resources, competing globally and playing an active role in global energy governance.

(4) **Public benefits**

China will carry out its transformation of the energy system with a well-coordinated package of goals and with the aim of providing clear economic and social benefits—including providing the public with clean, affordable and high-quality energy and ensuring a reliable energy supply.

4.3.3   Strategic Goals

(1) **Market system perfection**

China will create an open and modern energy market system, underpinned by comprehensive legislation and orderly competition. A market competition landscape—where huge energy companies act as the backbone and a diversified mix of energy production, transmission and sales companies of different legal statuses and sizes

coexist—will effectively address concerns about inequality among market participants, market segmentation and disorderly competition.

(2) **Pricing mechanism perfection**

Prices in competitive segments will be decided by the market. Prices in natural monopoly segments will largely be supervised and regulated by the government. Together, these will create a pricing mechanism and fiscal and taxation system that accurately reflect the supply-demand relationship, resource scarcity and environmental impact, and address concerns about unfair price policies and pricing mechanisms.

(3) **Government administration improvement**

The boundary between government and market will be clarified and an industry strategy, overall plan, regulations and standards, and high-level energy administration bodies for energy reserves and emergency response will be created. Following the tenet that "Things prohibited by laws cannot be done, things not prohibited by laws can be done, and duties enforced by laws must be performed" will address the challenges caused by the absence of unified and independent high-level energy administration bodies.

(4) **Effective market regulation**

Unified, independent and professional market regulators will be set up to form a modern energy regulatory system. The new system will have a clearly defined responsibility matrix, and will pursue fairness and equality, transparency and efficiency, and effective regulation, thereby addressing problems such as the bundling of administration and regulation, decentralised or missing regulatory functions and poorly enforced regulations.

(5) **Complete legislative system**

A complete, well-structured, well-aligned and unified system of energy law will be created, with associated regulations, standards and laws for power, coal, oil and gas. This will support national energy security and sustainability, and address existing problems, such as the lack of unified fundamental principles that underpin legislation, and the lack of coordination, consistency and cohesion in the current energy legal system.

4.3.4 Strategic Priorities

(1) **Creating a modern energy market system**

Natural monopoly segments will be separated from competition-based segments, and the market access mechanism will be improved to encourage diversified participants to invest in different segments of the energy industry in an orderly manner. Trading regimes for the energy market will be introduced or improved, and a modern hierarchical energy market system, including links between the national market and multiple regional markets, will be formed. Power system operators capable of performing dispatching and transactions independently will be established to separate the power transmission and distribution businesses. Oil and gas and coal exploration and mining rights will be acquirable only through bidding and market competition. Oil and gas pipeline operators will be encouraged to gain exclusive rights, pipeline transport services will be completely separated from sales, and fair third-party access to pipelines and other infrastructure will be allowed. Deployment of a global Energy Internet that connects everything and comprehensive energy service markets will be accelerated. An energy system that facilitates neutral peer-to-peer interconnection between centralised and distributed energy sources, as well as between energy storage and load devices, will be deployed.

(2) **Reshaping the energy market pricing mechanism**

A scientific pricing mechanism for monopoly segments will be created, combining constraints and incentives, and based on cost supervision and audits. By introducing scientific, transparent

and consistent regulation, monopoly segments will be encouraged to undertake healthy and sustainable development and to reduce their costs. Prices in competition-based segments will be deregulated and shift to a market-led pricing mechanism. Performance-based, complete, independent and incentive-guided power transmission and distribution pricing systems will be created. Pricing reform in inter-provincial and local power transmission will be introduced. Studies will be conducted to determine the correct tariffs for regional distribution grids, and the formation of a complete power transmission and distribution regulatory system will be accelerated. Studies will be conducted on methods that can be used to publish power transmission and distribution fees and price information, as well as methods to determine and distribute voltage-based costs. The price of oil products and gas will be deregulated and determined by market competition. Apart from the gas distribution network, charging for the use of other oil and gas pipeline infrastructure will be decided by the market. Subsidy and assistance mechanisms will target people in need and industries serving the public good. Cross-subsidies will be eliminated and an energy price regulation regime will be perfected to establish a reasonable price ratio between different types of energy.

(3) **Creating an efficient energy administration regime**

A central regulatory body for state-owned natural resources and the environment will be formed to oversee land use, environmental conservation and recovery, pollutant emissions and enforcement. Planning, policy, standards and incentives will be used to administer energy industry development in a holistic manner. The boundary between government and market will be clarified, and complete lists defining powers and duties will be created. Approved plans will be carried out effectively, approval criteria clearly defined, approval procedures streamlined and administrative approval powers revoked or delegated, all of which will reduce government intervention on specific issues. Renewable energy integration and curtailment will be thoroughly addressed, and inter-provincial transmission will be incorporated into the national strategy for long-term power transmission and use. Power generation will be completely deregulated and new energy generation in all regions will be guaranteed the sale of a minimum number of hours per year. Peak shaving and standby auxiliary service market mechanisms will be completed, motivating improvements to fossil power flexibility and the deployment of new peak shaving and energy storage facilities. Spot energy markets will be formed to benefit from the lower marginal cost of renewable energy.

(4) **Creating an effective energy regulation system**

China will separate administration from regulation, establishing independent and professional regulators, and building on the dual-level (central and regional) regulatory system. The responsibilities of regulators will be clearly defined, focusing on economic aspects, while regulation of the societal dimension will be enhanced to ensure fair competition in natural monopoly segments like pipeline infrastructure. Regulatory capacity and efficiency will be improved to maintain fairness in market competition.

(5) **Creating a modern energy legislation system**

China will adopt its Energy Law to provide a legal basis for the creation and revision of other energy-related laws and regulations. The Electric Power Law will be revised, the Oil and Gas Law will be researched and drafted, and the Coal Law will be improved, all of which will provide a legal basis for the creation, implementation, assessment, supervision and adjustment of power, coal, oil and gas strategies. The Energy Conservation Law and Renewable Energy Law will be implemented, while consistent regulatory, coordination, holistic decision-making and public consultation mechanisms will be formed or improved. Studies will be carried out on

formulating the Energy Regulation Provisions, and energy regulatory rules, methods and procedures will be created.

5 Reforming the Oil and Gas Sector

5.1 Progress in the Reform of China's Oil and Gas Sector

The pace of market-oriented reform of China's oil and gas sector accelerated after the 18th National Congress of the Communist Party of China in 2012. A series of reform policies and measures covering upstream access, market-based pricing, pipeline network reform, market regulation, and management of crude oil imports and exports was introduced. The Central Committee of the Communist Party of China and the State Council unveiled the Opinions on Deepening the Reform of the Oil and Gas Industry in May 2017, which defines the direction, targets, methods and tasks of reform. The move to reform the oil and gas industries energised the market, enabling it to allocate oil and gas resources fairly and create a favourable policy environment for the sustainable and steady development of China's oil and gas industries.

5.1.1 Pilot Mining Rights Reform of Upstream Resources Has Made Great Progress

Due to the long-term monopoly in oil and gas exploration and production in China, mining rights are highly concentrated in several large state-owned oil and gas corporations, making it difficult for private capital to gain access to upstream resources. Because the mining rights transfer and exit system is defective, the mining rights market cannot operate efficiently, and barriers to trade are common, leading to low oil and gas exploration, poor exploitation efficiency and insufficient investment. To resolve these problems, the Chinese government has launched pilot oil and gas mining rights management reform programmes.

Public tender was applied to shale gas blocks. Following the first round of public tenders to attract private investment to shale gas in 2011, the Ministry of Land and Resources initiated the second round in September 2012. The two rounds of public tendering involved 24 shale gas blocks, covering a total area of 20,002 square kilometres in Chongqing, Guizhou, Hubei, Hunan, Jiangxi, Zhejiang, Anhui and Henan. There were successful bids for 21 blocks. The bid winners include: large state-owned enterprises like China Huadian Corporation and China Shenhua Energy Company; energy groups invested in by local government, such as Chongqing Energy Investment Group and Tongren Energy Investment Group; and two private enterprises (Huaying Shanxi Energy Investment Company and Beijing Titan Source Natural Gas Resources Technology).

There were breakthroughs in mining rights reform in conventional oil and gas blocks. The Ministry of Land and Resources initiated public tenders for six conventional oil and gas blocks in Xinjiang in 2015. Thirteen enterprises submitted bids, including state-owned oil companies, local energy corporations and publicly traded petrochemical companies. At the end of 2017, the Ministry of Land and Resources commissioned the Xinjiang Land and Resources Trading Centre to list the mining rights for five oil and gas blocks in Xinjiang for public transfer. The transfer term was set at five years, two years longer than that specified in the 2015 tender. Private companies were also allowed to take part in this pilot reform of upstream oil and gas. The objective was to expand investment in oil and gas exploration and exploitation and diversify upstream investors. Awarding the blocks by public tender broke the state-owned oil companies' monopoly over upstream exploration and exploitation, helping the reform process to make substantial progress.

Competitive bidding was also used to transfer shale gas and coalbed methane exploration blocks. Guizhou and Shanxi provinces signed agreements with the Ministry of Land and Resources to develop shale gas and coalbed methane resources. On the instructions of the Ministry of Land and Resources, the Guizhou Provincial Government auctioned the Zheng'an block, where Anye Well 1 is located, in August 2017 to accelerate the exploration and production of shale gas. Besides Guizhou, other provinces and regions also selected shale gas exploration blocks for transfer through competitive bidding. The Department of Land and Resources of Shanxi Province transferred the mining rights for 10 coalbed methane blocks (covering a total area of 2,043 square kilometres) through public bidding in November 2017 to seven local enterprises, including Shanxi Blue Flame Coalbed Methane Group. This was the first batch of coalbed methane transferred in China after the reform of the mining rights system. It motivated local governments and energised the market.

Many support policies have been introduced to deepen mining rights reform. The Reform Scheme of the Transfer System of Mining Rights and Plan for the Reform of the Mineral Resource Royalty System was adopted in December 2016. Its purpose was to promote the transfer of mining rights through competitive bidding, restrict the transfer of mining rights through private agreement and build a new mineral resources royalty system with Chinese characteristics. As stated in the 13th Five-Year Plan (2016–20) for Natural Gas, the transfer of exploration blocks through competitive bidding to eligible market participants in a fair and open manner would help create an exploration and production system dominated by large state-owned oil and gas enterprises, but which would also include private companies.

5.1.2 Market-Oriented Pricing Reform is Progressing Rapidly

In China, the natural gas price is administered at different levels. The city-gate price is administered by the price authorities under the National Development and Reform Commission (NDRC), and the sales price after the city-gate station is administered by local government pricing authorities. Gas price is composed of ex-plant price, city-gate price[14] and end-user price.[15] Ex-plant price, trunk pipeline transmission tariff and city-gate benchmark price are set by the NDRC. The local pipeline gas distribution fee is set by the provincial government's pricing authorities, as is the urban end-user price.

Gas price deregulation in China has been making steady progress since the 18th National Congress of the Communist Party of China in 2012. Following the principle of "maintaining the administration of pipeline transmission tariffs in a natural monopoly and deregulating gas sources and prices", a complete price regulatory system covering the natural gas value chain from trans-provincial long-distance pipelines and provincial short-distance pipelines to city/town gas distribution pipelines has been established.

The non-residential gas price was fully reformed, step by step, through pilot implementation in selected regions. The three steps of gas price deregulation were completed during the 12th Five-Year Plan (2011–15). Pilots for gas pricing mechanism reform were first carried out in 2011 in Guangdong and Guangxi. Measures included shifting gas price management from factory to city-gate, capping the maximum price, changing the pricing method from cost-plus to netback, and building a dynamic adjustment mechanism linking gas price to the price of fuel oil, liquefied petroleum gas and other alternative energy sources. The new gas pricing mechanism, based on the knowledge gained from the two pilots in Guangdong and Guangxi, was introduced nationwide in 2013. The price of stock gas and incremental gas was adjusted separately. The price of incremental gas was first linked to the price of alternative energy. Then the price of stock gas was linked to the price of alternative energy in three steps. At the beginning of 2015, the price of stock gas and incremental gas was loosely tied to conditions in domestic and

overseas markets. In this way, the non-residential gas price was fully reformed.

The residential gas price has been gradually adjusted. In order to take people's well-being into account, China used a dual pricing system for residential and non-residential city-gate gas prices. The residential city-gate price had not been adjusted since 2010, which meant that the average residential city-gate gas price was about 20% lower than the non-residential price. With changing domestic and international markets, deepening reform of non-residential gas prices and inconsistent pricing mechanisms for residential and non-residential gas, the gas price became increasingly difficult to manage and was a constraint on gas supply security. To address the residential and non-residential city-gate gas dual pricing system, the NDRC released the Circular on Adjusting Residential City-Gate Gas Price in May 2018. This changed the ceiling price administration system into a benchmark price administration system. It stated that the residential gas price should be linked to the non-residential city-gate gas benchmark price, and that the supplier and buyer should be allowed to negotiate a city-gate price that is no more than 120% of the base level (with no lower limit). Given the difference between residential and non-residential gas prices in some provinces, the national pricing authority allows them to reform residential gas price stepwise. The price increase in residential gas price may not exceed RMB 0.35 per cubic metre in 2018, and the remaining price difference should be adjusted a year later. At this point, residential and non-residential gas are linked by both the pricing mechanism and price level, and the price of gas in competitive parts of the value chain is decided by the market.

The price of shale gas, coalbed methane, coal-to-gas and other types of unconventional gas were deregulated and allowed to fluctuate in 2013, as was the price of liquefied natural gas in September 2014 and the price of gas supplied directly to users (except fertiliser companies) in April 2015. The price of gas used for chemical fertilisers, and the gas storage purchase, sale and service price were deregulated one by one in

---

[14]City-gate price = ex-plant price + pipeline transmission tariff.

[15]End-user price = city-gate price + local pipeline gas distribution fee.

2016. A pilot reform of the city-gate price was also carried out in Fujian province.

After several years of reform, the marketisation of the residential gas price had clearly improved. Before the reform, the residential gas price was managed by the government. After the reform, the price of non-residential gas (which accounts for more than 80% of China's total gas consumption) was determined by the market, of which 50% was completely determined by the market. For the remaining 30%, a flexible mechanism, whereby the price may be 20% higher than the maximum gate station price, was applied.

A price regulation system covering all nodes in the gas transmission and distribution networks was created by reforming the pipeline transmission pricing mechanism. To resolve the long-term absence in China of clear and sophisticated gas pipeline transmission pricing and supervisory and review mechanisms, the NDRC issued the Measures for the Administration of Natural Gas Pipeline Transport Prices (tentative) and the Measures for the Supervision and Review of Natural Gas Pipeline Transport Pricing Costs (tentative) in October 2016. These measures specify the scope and objectives of price regulation, the methods and procedures of price management, as well as some key indicators based on the principle of "allowable cost + reasonable profit". As a result, a new gas pipeline transmission pricing method has been established. The price no longer differs from one pipeline to another. Rather, it is assessed and then fixed, based on the cost of transmission (made public by the government) and the pricing formula. Pipeline transmission costs that are not part of the tariff are clear to all parties. This standardises pricing and provides clear and verifiable charging criteria for third parties wanting access to the transmission network.

Because the price regulation rules for gas distribution are not yet complete, the gas distribution price varies between regions, with some regions charging a higher price than others. The NDRC issued the Guiding Opinions on Strengthening Regulation of Gas Distribution in June 2017, which defines the methods for setting the gas distribution price. It also defines the key indicators and parameters and the requirements needed to improve regulation, build cost constraints and create motivation mechanisms, as well as requiring disclosure of business information. The Guiding Opinions established the price regulation framework for downstream urban gas distribution, thereby building a complete price regulation system covering all parts of the gas system, from transmission to distribution. In addition to alleviating the cost burden on users by lowering the high distribution price in some regions, the Guiding Opinions lays a solid foundation for the future separation of distribution and sales and the opening of the gas distribution network to third parties.

In the first half of 2017, the NDRC reviewed the pricing of 13 trans-provincial gas pipeline transmission companies. The pipeline transmission price had decreased by around 15% on average, saving businesses using downstream gas about RMB 10 billion. Shaanxi, Jiangsu, Zhejiang, Hebei, Yunnan and Jiangxi developed their own transmission and distribution price regulations, as required by central government, and reduced their pipeline transmission and gas distribution prices, cutting the cost for businesses by more than RMB 4 billion. Supervision and review of transmission and distribution prices enabled pipeline transmission companies to reduce costs and increase efficiency, benefitting gas users. Most importantly, the process opened the pipeline network and marketised gas trading. After the long-distance pipeline transmission price had been reviewed and fixed, two pipeline transmission companies immediately opened their pipelines to third parties.

5.1.3   Access to Infrastructure Was Further Deregulated

Systemic shortcomings severely impede the efficient use of infrastructure and fair competition among market participants. To resolve these problems, the NDRC and the NEA issued the Measures for the Administration of Natural Gas Infrastructure Construction and Operation and the Measures for the Supervision and Administration of the Fair Opening of Oil and Gas

Pipeline Network Facilities (tentative) in April 2014. The measures state that: "the state shall encourage and support various kinds of capital to participate in the investment and construction of natural gas infrastructure included in integrated planning"; "third parties shall be allowed to use gas infrastructure, including liquefied natural gas (LNG) terminals"; "oil and gas pipeline network operators shall, in the case of spare network capacity, share their pipeline network facilities equally with third-party market participants and provide transport, storage, gasification, liquefaction, compression and other services"; and "oil and gas pipeline network facility operators shall use facility capacity to the full, ensure services are maintained for existing users, and provide access to the facilities for new users in a fair and non-discriminatory manner". After these measures were declared, some oil companies provided third parties with access to their LNG terminals. For example, CNPC provided Beijing Gas Group with access to its Dalian and Tangshan LNG terminals and the Yongqing-Tangshan-Qinhuangdao pipeline, thereby providing unloading, storage, gasification and transmission services for imported LNG (450 million cubic metres of gas was transmitted in 2016). CNPC's Caofeidian terminal received 15,000 tonnes of LNG imported by China Gas from Nigeria. Sinopec transmitted 210 million cubic metres of natural gas for Kunlun Gas, China Gas, China Resources Gas and Shanxi Guohua Energy. CNOOC transmitted 54 million cubic metres of natural gas in its Guangdong gas pipelines for China Resources Jiangmen and ENN Dongguan.

To improve the supervision and regulation of fair access to oil and gas pipeline facilities and to provide information that could form the basis for fair access, the NEA issued the Notice on Information Disclosure of Access to Oil and Gas Pipeline Facilities in September 2016. The notice specifies the parties required to disclose information, the information to be disclosed, the methods of disclosure and the arrangements for supervision and regulation. CNOOC, Sinopec and CNPC published on their websites the information required on all their infrastructure,

including long-distance pipelines and LNG terminals. Some provinces, including Shanxi, also disclosed information on their pipelines to the public. After three years, gas infrastructure information was transparent, providing the basis for fair access to, and supervision and regulation of, the facilities.

5.1.4 Reform of the Right to Import Crude Oil and the Right to Use It

Crude oil is imported to China by state-owned and non-state-owned trading companies. Permits for state-owned trading companies are issued by the State Council. Those companies with permits include Sinopec, CNPC, CNOOC and Sinochem. Import volumes are not limited. Non-state-owned trading companies are managed by quota. Their import volumes are capped. Permits and import volumes are managed by the Ministry of Commerce. Crude oil imported by non-state-owned trading companies can enter approved refineries only. There were more than 20 non-state-operated trading companies before 2012, importing around 10% of China's total imported crude oil. China began to deregulate crude oil import rights in 2012, granting them to state-owned enterprises like China National Chemical Corporation, and in 2014 to private companies like Xingjiang Guanghui Petroleum.

China began to speed up reform in 2015, by extending the right to import and use crude oil to approved Chinese refineries. The NDRC issued the Notice on Issues Concerning Use Management of Imported Crude Oil in February 2015. This allows crude oil processing companies that meet energy consumption, quality, environmental and safety conditions to use imported crude oil, on condition that they shut down outdated capacity or construct gas storage facilities of a certain size. The notice marks the point when crude oil import rights were officially opened up and the eradication of oil and gas monopolies began to gain pace. In 2018, 32 local refineries (excluding those operated by Sinochem) were granted permits to import more than 90 million tonnes of crude oil. The Ministry of Commerce issued the Circular on the Application by Crude

Oil Processing Enterprises for Non-State-Owned Trading and Import Qualification in July 2015. The circular allows non-state companies approved to import, export and refine oil products—and which meet energy consumption, quality, environmental, safety and storage conditions—to apply for approval to import crude oil. Up to August 2016, 16 companies had obtained the right to import crude oil, importing 62.57 million tonnes in total.

Further deregulation adjusted the original oil import-export management system and laid an institutional foundation for the formation of a transparent and open oil refining market featuring orderly competition and diversified market participants. The move brought new opportunities to small and medium-sized oil companies in trade, financing and logistics. The utilisation rate of private refineries was improved significantly. Statistics show that average capacity utilisation of Shandong's refineries increased from 41.2% in June 2015 to 60% in the first half of 2017.

5.1.5 Oil and Gas Exchanges Were Steadily Established

The introduction of natural gas exchanges is an important means to reform the natural gas pricing mechanism and strengthen China's voice in international natural gas pricing. The role of natural gas exchanges is repeatedly mentioned in government documents, such as the 13th Five-Year Plan (2016–20) for Natural Gas, the Notice on Clarifying the Price Policies for Gas Storage Facilities, the Notice on the Policies Regarding City-Gate Station Price in Fujian Province and the Notice on Promoting Deregulation of the Price of Gas Used for Chemical Fertiliser Production. Gas suppliers and users are encouraged to participate in oil and gas exchanges and other platforms to establish a transparent gas price through trading.

The Chinese government accelerated the introduction of natural gas markets, successfully building two trading platforms, one in Shanghai and the other in Chongqing. The Shanghai Petroleum & Gas Exchange, founded in December 2014, was put into trial operation in July 2015 and official operation in November 2016.

Unilateral gas trading volume exceeded 6.5 billion cubic metres in 2015 and 15 billion cubic metres in 2016, accounting for around 8% of total gas consumption in China. The volume was expected to exceed 50 billion cubic metres in 2017. LNG bidding for 12 coastal provinces and cities is now conducted through the Shanghai Petroleum & Gas Exchange, and 174,800 tonnes of LNG have been traded in total. The price reached by bidding fully reflects local LNG supply and demand. The Shanghai Petroleum & Gas Exchange carried out the first round of bidding for pipeline gas in September 2017. The exchange plays an important role in aligning supply with demand and establishing a reasonable price. With its influence growing, the Shanghai Petroleum & Gas Exchange attracts wide attention in and outside China.

Chongqing Petroleum and Gas Exchange was unveiled in January 2017 and put into trial operation in April 2018. Some provinces and cities, including Xinjiang, Zhongwei, Shenzhen and Hubei, are actively researching the formation of regional natural gas exchanges.

Natural gas exchanges are important outcomes of oil and gas price deregulation. They provide influential support for the deepening of reform. Though their size is not yet large, their role of allocating resources in a market-oriented and optimal fashion and in providing equal competition opportunities to downstream users is now widely recognised.

China has created an oil futures market. Despite China's position as the world's largest crude oil importer, the second largest oil consumer and the fourth largest oil producer, China has inadequate influence in the crude oil market, and suffers from the Asian premium of paying more for crude oil from the Middle East than European or US refiners. The development of oil futures provides a method of price discovery and drives the reform of China's oil products pricing system. It can also help eliminate the Asian premium, hedge against oil finance risks and promote RMB internationalisation. Crude oil futures were officially listed on the Shanghai International Energy Exchange (INE), a unit of the Shanghai Futures Exchange, in March 2018.

It is the first futures market on the Chinese mainland open to outside investors.

5.2 Major Problems in the Existing Oil and Gas System and Mechanisms

Following the intensive introduction and implementation of reform policies and measures after the 18th National Congress of the Communist Party of China in 2012, China's oil and gas market-oriented reform made great progress. Measures include improving the oil and gas pricing mechanism, marketising the allocation of oil and gas resources, enhancing oil and gas supply capacity, and improving oil and gas resource use. However, there are still some underlying institutional problems in the oil and gas sector that urgently need to be addressed.

5.2.1 Unsound Market System and Inadequate Market Competition

First, the administrative monopoly is strong in exploration and production, restricting the extent to which oil and gas production capacity can be improved. Oil and gas exploration and production is a monopoly in China. According to the Administrative Measures for Registration of Mineral Resources Prospecting Blocks and the Administrative Measures for Registration of Mineral Resources Exploitation issued by the State Council in 1998, the companies engaged in exploration and exploitation of oil and gas resources are subject to mandatory approval by the State Council. So far CNPC, Sinopec, CNOOC and Yanchang Petroleum have been approved. Oil and gas mining rights need to be registered, which has been a long-term policy. The mining rights of the four companies cover all favourable blocks. At present, China's registered oil and gas exploration rights cover 4 million square kilometres of land, of which 3.9 million square kilometres belong to CNPC, Sinopec and CNOOC. Registered oil and gas exploitation rights cover about 118,000 square kilometres of land, of which 117,000 square kilometres (99%

of all registered exploitation rights) are owned by CNPC, Sinopec and CNOOC. In recent years, China began to explore the transfer of mining rights through competitive bidding for shale gas and coalbed methane exploration blocks and carried out pilot public bidding for conventional oil and gas blocks in Xinjiang. However, oil and gas mining rights are still highly concentrated in a few large state-owned oil and gas companies, leading to limited market competition, a low degree of marketisation and inefficient operation. Since the procedures for transfer and exit from oil and gas mining rights are not mature, and since supervision and follow-up measures are absent, it is difficult for private capital to enter the market. A diversified upstream market based on orderly competition has yet to be formed. As a result, there is insufficient investment in oil and gas resource exploration and exploitation and the potential for increased output and lower costs is restricted. Furthermore, enclosed gas blocks are not exploited. This prevents blocks of potential resources from being fully exploited.

Second, the lack of diversity in infrastructure investors and insufficiently deregulated investment access result in infrastructure construction lagging behind market development. In China, trunk gas pipelines and branch pipelines are mainly invested in and built by CNPC, Sinopec and CNOOC. Regional branch pipelines are mainly invested in and built by CNPC, Sinopec, CNOOC and local capital. CNPC's gas pipelines account for more than 70% of total gas pipeline investment. Because of this lack of diversity in pipeline investors, China's gas pipeline construction does not match demand. During the 12th Five-Year Plan (2011–15), China constructed about 25,800 kilometres of trunk and feeder gas pipelines, accounting for only 58.5% of the planned 44,000 km target. Although pipeline investment and construction are not so far behind that they hinder gas supply, it is still difficult for private natural gas companies to enter the market, due to constraints like insufficiently deregulated investment access and long project approval processes. As a result, it is hard to unlock the huge potential in gas pipeline investment and construction—the quantity and speed of pipeline construction does not meet demand, and the efficiency and economic benefits of pipeline investment and construction need to be improved. Some provinces with their own pipeline companies want gas pipelines to be built and operated by these companies in a centralised manner, excluding other companies from investing in and constructing gas pipelines in their province. In other provinces that do not have such policies, the monopoly policy is exercised during project approval. Gas storage facilities also lack diversity. The country's 12 gas storage facilities are owned by CNPC and Sinopec. Six private gas companies have jointly constructed one town gas storage facility, but it is yet to be provided with working gas.

Third, gas infrastructure is highly integrated, which hinders downstream market development. China's gas sector has not yet separated transmission from distribution and sales. As a result, the highly integrated upstream, midstream and downstream monopoly business model dominates. For example, about 98% of upstream gas is supplied by CNPC, Sinopec and CNOOC. These three companies have built about 95% of trunk oil and gas pipelines and account for 90% of the receiving capacity at China's 17 liquefied natural gas (LNG) terminals. They are also the principal actors in gas production, purchase and sales. As pipeline construction and investment needs to have gas sources and a market, the upstream monopoly creates a barrier for pipeline investment that prevents other participants from entering the field. In addition, the upstream monopoly restricts direct gas purchase transactions with large users and hinders growth in imported gas, which has a negative impact on the development of the downstream natural gas market.

Fourth, most of China's trunk gas pipelines are owned by CNPC, Sinopec and CNOOC, but they are poorly interconnected. And most provincial pipelines were built jointly by provincial state-owned enterprises and the three oil and gas titans. Typically controlled by provincial state-owned companies, the provincial pipelines are not interconnected with national trunk pipelines. Access to pipelines and fair

services is difficult for newcomers, even for state-owned enterprises supervised and managed by central government. Why? Under the vertically integrated model of operation, pipelines help large oil and gas companies integrate exploration with imports and sales. Large oil and gas companies with upstream resources do not want to lose the lucrative midstream pipeline transmission market and be restricted in their sales activities. There is not enough pressure on them to open pipeline infrastructure to others. Also, large oil and gas companies view LNG terminals as gas supply points, rather than public infrastructure. There is, therefore, a conflict of interest between maintaining the price of gas high over the long term, providing third-party access to infrastructure, and performing the duty of ensuring supply. The existing oil and gas system can meet the demands of market-oriented reform, but it has become a bottleneck impeding the rapid, healthy and sustainable development of the natural gas industry.

5.2.2 The Market's Ability to Determine Price Is Insufficient

First, market-oriented pricing has not been implemented in competition-based sectors. The price of most oil products and natural gas is currently set by the government, rather than by the market. Since the conflict between the transparency of the oil products pricing mechanism and the hysteresis of price adjustments has not been resolved, the oil product price mechanism cannot fully reflect supply and demand in China, or the scarcity of resources or environmental impact. As a result, it cannot effectively stimulate or restrain the efficient development and reasonable consumption of oil and gas resources. As for the natural gas pricing mechanism, the current city-gate price is a benchmark price set by government, including ex-plant price (imported gas price) and the pipeline transmission tariff. This pricing model, with its two links, impedes third-party access to pipeline facilities. Although the Circular on Adjusting the Residential City-Gate Gas Price of May 2018 stated that China would gradually unify the residential

and non-residential city-gate gas prices to eliminate cross-subsidies, the residential gas price is
still lower than the cost of gas supply. Furthermore, cross-subsidies have still not been completely eradicated. In some regions, price
adjustment lags behind the city-gate price, so
changes in the upstream market are not reflected
in price in a timely fashion.

Second, the gas price system needs to be
improved. On the supply side, the construction of
gas storage in China has lagged far behind
market development for some time. This is due to
the high investment cost of underground gas
storage, a shortage of options for recovering
investment in cushion gas, the absence of a
peak-shaving gas price, and the failure to establish a market-oriented pricing mechanism for gas
storage services and peak-shaving gas capacity.
As indicated in the Report on the Oil and Gas
Industry in and Outside China 2017, up to the
end of 2017, China had constructed 12 gas
storage facilities (clusters), the peak-shaving
capacity of which stood at 10 billion cubic
metres. The working gas capacity of 8 billion
cubic metres accounted for 3.4% of China's gas
consumption that year, much lower than the
global average of 12%. In developed countries,
the working capacity of gas storage accounts for
20–30% of their total gas consumption. The
severe shortage of gas storage peak-shaving
capacity was an important cause of the national
gas shortage at the end of 2017. On the demand
side, the absence of interruptible and peak and
off-peak gas prices dampens the motivation to
expand the interruptible user base.

Third, natural gas exchanges help push market
reform of the gas pricing mechanism and restore
the commodity attributes of gas. Although the
Shanghai Petroleum & Gas Exchange has started
operations and the Chongqing Oil and Gas
Exchange will start soon, China's natural gas
exchanges are still in their infancy. Spot trading
has begun with small quantities, but futures
trading has yet to start. The gas trading volume
of the Shanghai Petroleum & Gas Exchange
exceeded 15 billion cubic metres in 2017, about
6.3% of China's total gas consumption. There is
a large gap between China and the USA and

Europe. Only a small volume of gas is traded via
competitive bidding. At the Shanghai Petroleum
& Gas Exchange, pipeline gas is traded mainly
by reference to the provincial city-gate price set
by the NDRC. Competitive bidding is limited to
low-volume LNG trading. Deals are mainly
concluded by negotiation, with trading models
still being explored. The services offered by the
exchanges need to be diversified and the trading
system needs to be improved. The following
challenges remain: gas pricing policy factors that
restrict trading, the absence of a competitive
upstream market, and pipeline networks and
infrastructure that do not provide fair access to
third parties. This means that the natural gas
exchanges can only trade in volume; they cannot
determine benchmark prices. As such, it is difficult for them to replace government-set benchmark prices in the short term.

5.2.3 Poor Government Administration and Supervision

China has made great progress in reforming oil
and gas management, but administration and
supervision still need to be improved.

As regards resource development and use, oil
and gas supply security and environmental protection, too much attention is paid to reviewing
and managing market access qualifications, and
not enough to overseeing project implementation
and effects. Public services for information
management and modernisation technologies are
not in place. Important information is sometimes
not disclosed in advance or on time. The government intervenes too much in the market.
Government management is too low-level and
specific, such as focusing on investment
approval, price setting and output control. The
power, duties and interests of central and local
government are not balanced. The central government has strong power and limited duties, and
local government has little power and unlimited
duties, leading to inefficient administration and
poor outcomes.

Oil and gas strategies have been missing for a
long time. Industry plans and arrangements
(major projects), policies, standards and specifications, and laws and regulations are not aligned

and harmonious, so synergies cannot be created. For example, the positioning of natural gas development strategies is not clear, and the scepticism about the cleanness of natural gas needs to be addressed. Plans for the oil and gas sector, limited to seeking advice, are not harmonious with other plans, such as those for land, the sea and natural reserves. Key content is not well aligned. In addition, national oil and gas pipeline construction plans do not incorporate local government plans for land and urban and rural development, making coordination during implementation difficult. And industry associations do not play their roles to the required extent.

The regulatory system is incomplete. First, the functions and duties of oil and gas authorities are decentralised. The policy goals of central and local government departments are different and their work rate is not aligned, making coordination difficult. Second, regulatory activities do not have a clear interface. Their efficiency is poor and their effects are weak. The absence of relevant laws and regulations means that there is insufficient basis for effective problem solving. For instance, China has not yet developed an energy law, an oil and gas law or natural gas law, and the policies related to opening pipeline networks and gas infrastructure to third parties are issued by ministries or commissions in the form of circulars or measures. Third, there is not enough diversification in the regulators and the means of regulation, making it difficult to implement reforms that call for "administrative streamlining, delegation of powers and service optimisation". Government regulators rely mainly on standard administrative methods and peer-to-peer supervision and review.

5.3 Objectives of Oil and Gas Industry Reform

5.3.1 The Idea Behind the Reform

Reform of China's oil and gas industry is guided by Xi Jinping's Thoughts on Socialism with Chinese Characteristics for a New Era, the 19th National Congress of the Communist Party of China in 2017, the 2nd and 3rd Plenary Sessions of the 19th Central Committee of the Communist Party of China in 2017, and the development concepts of "Innovation, Coordination, Green, Openness and Sharing". The reforms should be implemented across the whole oil and gas value chain in an orderly manner, with the focus on supply-side structural reform, including exploration and production, pipeline transmission, circulation, refining and chemicals, enterprise reform, government regulation, and the enactment, amendment and abolition of oil and gas regulations. The relationship between the market and government should be clarified to enable the market to play a decisive role in resource allocation and the government to play a more efficient role.

5.3.2 Basic Principles

The market's decisive role in resource allocation should be given full play. Industry characteristics and the laws of market development should be followed. Barriers to entry and monopolies in the oil and gas industry should be eliminated to enable eligible market participants to enter competition-based sectors. This will improve competition and efficiency, energise the market and raise oil and gas supply capacity. In addition, government should play a more efficient role in macroeconomic regulation, market supervision and regulation, and service provision.

Oil and gas is an integral part of the nation's economy. Reform of the oil and gas sector has a direct impact on economic development, resource security and environmental safety, people's livelihoods and social stability. Its positive and adverse impacts on the social economy should be fully considered. Industry characteristics and the laws of the market economy should be followed to reform upstream, midstream and downstream in an orderly and problem-oriented manner. An overall reform plan that clearly defines the direction and goals of reform should be developed, targeting breakthroughs at key points of the plan and implementing pilot projects to gain experience and test solutions.

China is a large oil and gas producer and consumer. Ensuring national energy security under open conditions should be the first objective. Domestic and international markets and resources should be planned in a holistic manner. The survey, assessment and exploration of domestic oil and gas resources should be strengthened, and the resources developed in an orderly and cost-effective way to improve China's domestic supply capacity. China should also actively develop overseas resources to diversify imported oil and gas supply, thus lowering risk and enhancing supply security.

Different targets and interests should be acknowledged. Economic returns and social benefits should be combined organically. Public interest should be better protected. Oil and gas supply capacity and quality should be improved to make high-quality, clean and affordable oil and gas resources widely available to the public.

5.3.3 Objectives of Reform

Market-oriented reform of the oil and gas value chains should be accelerated, as required by the Central Committee of the Communist Party of China and the State Council in the Opinions on Deepening the Reform of the Oil and Gas System. This will be achieved by capitalising on the strategic opportunity to deepen reform of the oil and gas sector, thanks to the abundant supply resources available in the international oil and gas market and low oil and gas prices. The goal is to build a modern oil and gas market that has been opened up in an orderly fashion, featuring fair competition, effective regulation and governed by laws. By 2030 the market will play a decisive role in allocating oil and gas resources and government will play a more efficient role. Oil and gas supply capacity and efficiency of use should be improved rapidly, and the substitution of energy types should be realised as soon as possible. In this way, economical oil and gas resources will be available to people, and the strategic objectives of energy production and consumption will have been achieved.

More effort should be channelled into restoring the commodity attributes of oil and gas to allow prices to be determined by supply and demand. Government focus will shift, from intervening in operations and setting prices for oil and gas producers to developing strategies, plans, policies and rules to create a fair market. Government should strengthen its regulations on market access, transactions, monopolised segments, pricing and costs. It should also improve the standards, rules and procedures of the regulatory process to create a standardised, orderly, open and transparent supervisory and regulatory system.

Oil and gas supply capacity and energy efficiency will be improved. Reform of the whole value chain should be deepened to make the competition-based segments of oil and gas companies more dynamic. Technical innovation should be pushed to reduce production costs and improve supply capacity. Other targets include: using domestic as well as overseas resources; mitigating market risk; improving transmission and service; improving oil and gas use efficiency; ensuring strategic oil and gas security; and operating safely and cleanly along the whole value chain.

The substitution of energy types should be achieved as soon as possible. Petroleum consumption should be guided. The proportion of natural gas should be increased significantly to develop natural gas into one of the main energy sources as soon as possible, thereby achieving the strategic objectives of energy production and consumption.

5.3.4 Pathways to Reform in the Oil and Gas Sector

To achieve the above objectives, oil and gas sector reform should be as follows.

(1) **Reform the oil and gas mining rights management system and establish a mining rights market**

Reform of the oil and gas mining rights management system should "be market-oriented, diversify investors in an orderly manner, strengthen supervision and regulation, and improve the tax and fee system".

To ensure orderly, stable, efficient and green exploration and production and the efficient allocation and use of oil and gas resources, management of mining rights should be performed at a high administrative level. China can use the lessons learned from the trial implementation of the competitive transfer model for shale gas and coalbed methane (CBM) exploration blocks in provinces like Guizhou and Shanxi. These were jointly sponsored by the provincial government concerned and the relevant ministry. Provinces should be selected to delegate shale gas and CBM mining rights to local government, thus motivating local government to drive the exploration and development of shale gas and CBM resources throughout the country. To eliminate the uneven distribution of tax revenues between central and local government, the mechanism by which oil and gas benefits are shared by government needs to be improved to ensure that resource development benefits local people.

The Mining Rights Transfer System Reform Plan issued by the State Council in 2017 clearly states that China's mining rights (including oil and gas exploration and exploitation) will shift from agreements to public auction. The plan requires that all mining rights transfers are made through listing and that agreement-based transfers will be restricted. The price of mining rights transfers will be determined by the market. This competitive transfer approach will be fully implemented in China's oil and gas exploration blocks. It is suggested that the relevant authorities develop support policies for the national mineral resources royalty system as soon as possible, and that differentiated benchmarks be set for onshore and offshore oil and gas exploration blocks and unconventional oil and gas resources.

The experience gleaned from the pilot auction of exploration rights for the first and second batches of conventional oil and gas blocks in Xinjiang should be extended to other provinces. Companies registered in the People's Republic of China that have a Chinese entity as their controlling shareholder, have sufficient financial resources and a sound financial and accounting system, and can independently bear legal liabilities, are allowed to hold oil and gas mining rights, thus breaking the upstream monopoly.

However, bid winners should be selected on their financial and technical strength, rather than on the level of investment committed, to ensure that blocks are transferred to companies that are truly interested in oil and gas and to avoid the need for passive supervision later. The lessons learned from the auction of shale gas and oil and gas exploration rights in Xinjiang, as well as the pilot reform of CBM mining rights in Shanxi, should be studied. Small and medium-sized enterprises with different ownership systems should be allowed to exploit the abundant oil and gas reserves owned by CNPC, Sinopec and CNOOC that have not been used and for which there is no development plan in the short term. Non-public enterprises and capital, as well as companies with superior technical strength, are encouraged to form joint ventures and bid for oil and gas mining rights. Finally, an exploration and exploitation system involving diverse participants should be created through mining rights reform, enabling the market to determine resource allocation.

The lowest limit of investment for oil and gas exploration blocks should be set at different levels in different regions for different mineral varieties and for different phases of exploration. Given the low cost of ownership of exploration rights, and to encourage more companies to exit from idle blocks and attract other investors into exploration, the upper limit of exploration investment should be raised for the first and second years and lifted completely for the third year. However, the upper limit of exploration investment and exploration royalties for difficult and high-risk blocks could be reduced. For example, exploration and exploitation of unconventional oil and gas resources are far more difficult than for conventional resources. A lower exploration investment threshold and exploration royalty should be applied to unconventional oil and gas blocks based on geological conditions and/or difficulty and the time required for exploration. The cost of owning oil and gas block mining rights and the amount to be paid to central government in the form of royalties should be set at a suitable level. Mining rights should be rationalised and the tax and fee system reformed.

Block circulation includes the transfer of exited and existing blocks and circulation between businesses. First, measures to restrict block exit should be developed to improve the exit mechanism, including: (i) economic means to raise the charges for holding exploration rights; and (ii) administrative means to force mandatory exit of a proportion of the same basin whenever an exploration right is extended. It is suggested that differentiation be applied according to the difficulty of exploring in different regions. For example, the lower limit of exploration rights investment for the Tarim and Sichuan Basins, which have complex geological conditions and a long drilling period, should be different from other basins or regions. The years in which such blocks are under assessment should be extended appropriately. Second, the measures to administer transfer of existing blocks should be improved. Such measures exist already, but there are no relevant assessment agencies or detailed rules of implementation and no precedent for oil and gas block transfer. It is suggested that rules on the paid use of reserves and valuation methods that are fair, reasonable and consistent with international practice should be studied and developed, and that world-class reserve assessment agencies should be cultivated to create conditions for mining rights circulation and reserves trading. Third, circulation of existing blocks should be accelerated. After analysing successful circulation cases, such as that of Daqing Oil Field entering the Tadong Block, the four major oil and gas companies (Sinopec, CNPC, CNOOC and Sinochem) should be allowed to transfer blocks and trade reserves among themselves using market-oriented approaches to revitalise mining rights resources.

To encourage businesses to exit idle blocks and make room for other investors, the initial investment of those exiting businesses should be compensated. If a licence holder exits and the exploration right is taken by another company, that company should provide compensation for the previous investment of the licence holder. Compensation criteria could be investment in seismic exploration, research or drilling.

China will cancel the franchise of the three major state-owned oil and gas enterprises for onshore cooperation with foreign businesses and allow other Chinese companies to cooperate with foreign businesses under government regulations. Companies should be motivated—and capital, advanced technologies and management experience leveraged—to improve China's ability to supply oil and gas, especially unconventional resources. The franchise for offshore cooperation should also be steadily deregulated. Pilot reforms could be carried out, initially with CNPC and Sinopec. After which the criteria for awarding offshore oil and gas mining rights should be researched by examining these pilot reforms. Subsequent laws and regulations should be introduced step by step.

While granting access to upstream resources, China should improve management and share geological data. It is suggested that the government should form a national geological data organisation, giving it the right to set up an oil and gas geological data submission system and empowering it to require businesses (including national oil companies) to submit geological data. Non-confidential geological data should be made available to all companies to facilitate data and information sharing.

(2) **Accelerate reform of the oil and gas pipeline network and build an independent and diversified system of oil and gas infrastructure**

The guiding principle behind the reform of the oil and gas pipeline network is: "drive the independence of trunk pipelines of large state-owned oil and gas companies step by step to unbundle pipeline transmission and sales; improve the mechanism for fair access to the oil and gas pipeline network to make trunk pipelines and intra- and inter-provincial pipeline networks open to third-party market participants". China will focus on addressing the main factors constraining the development of its natural gas industry: infrastructure construction, undiversified construction companies and operators, low level of pipeline interconnection, bundled transmission and sales, and ineffective implementation of third-party fair access. China is striving to achieve independence for oil and gas pipelines and create a nationwide pipeline network by 2020 to support an oil and gas production, supply, storage and sales system, and deliver the ultimate goal of a coordinated oil and gas value chain by 2030.

Pipelines and other infrastructure link upstream to downstream in the oil and gas sector. The reform of the pipeline network, and its effect on upstream-downstream reform, has attracted widespread attention. The experience of countries with a mature natural gas market suggests that natural monopolies like pipelines should be separated from upstream and downstream businesses, and that pipeline companies should divest their bundled sales services. Pipeline companies should operate independently and only provide a transmission service, thus helping to create a competitive and efficient oil and gas market. The independence of oil and gas pipelines not only provides the strongest guarantee for fair access to infrastructure, it also eases supervision of pipeline transmission costs, accelerates the formation of an efficient pipeline service market and helps marketise the whole industry.

In recent years, factors such as low oil and gas prices, slower growth in gas consumption, a significant decline in the operating revenues of CNPC, Sinopec and CNOOC, and the upcoming pipeline reform have reduced pipeline infrastructure construction and made investment in upstream exploration and production less attractive. If reform is not accelerated, it will restrict the construction of gas pipelines and have an adverse effect on gas supply security. On the other hand, as oil and gas system reform deepens, private enterprise will enter the upstream market to explore and exploit resources, import oil and gas, invest in infrastructure and develop the natural gas trading market. As a result, third-party demands for fair access to infrastructure, especially gas pipelines, will increase. Independent operation of pipelines will undoubtedly facilitate fair access and promote sector-wide market-oriented reform. The DRC-Shell work stream believes that China urgently needs to speed up reform of infrastructure, including the trunk pipelines of large state-owned oil and gas enterprises, and introduce the plan to reform China's oil and gas pipeline network as early as possible. This will create pipeline independence and develop pipeline services, thereby ensuring rapid development of the natural gas market.

The reform of local pipeline networks should be vigorously encouraged. Laws and regulations should be improved to ensure fair access. Judging from international experience, a third-party fair access policy can balance the financial interests of midstream pipeline owners, upstream producers and downstream consumers, and attract capital to invest in infrastructure construction. To encourage private capital to invest in the construction of China's oil and gas infrastructure and to improve the use of existing infrastructure and thus the efficiency of the entire sector, it is suggested that infrastructure operators (including trunk oil and gas pipelines, intra- and inter-provincial pipeline networks, LNG terminals and underground gas storage facilities) should provide transmission services to third parties on a non-discriminatory basis. To ensure third-party access, the national energy administration should improve the Measures for Regulation of Fair and Open Access to Oil and Gas Pipeline Networks and the Measures for Administration of Natural Gas Infrastructure Construction and Operation, and develop the Rules for Implementation of Fair Access to Natural Gas Infrastructure. In so doing it should define: how infrastructure operators can provide independent services like pipeline transmission to all users in a fair and just manner; how they should disclose information; and define the legal liability for breach of fairness and openness, thus laying an institutional foundation for fair access to pipelines. An efficient system should be created in which regulators can review and approve the qualifications of prospective upstream and downstream users applying to operate infrastructure or gain access. Companies failing to grant fair access or discriminating against third parties should be penalised and publicised.

Restrictions should be gradually eliminated to diversify investment. According to the Medium and Long-term Plan for Oil and Gas Pipeline Networks, issued by the NDRC and NEA in 2017, by 2025 China's total length of oil and gas pipelines will be 240,000 kilometres. This includes 77,000 km of crude oil and oil product pipelines, which increases by 29,000 km the 27,000 km of crude oil pipelines and 21,000 km of oil product pipelines at the end of the 12th Five-Year Plan (2011–15), and 163,000 km of gas pipelines (an annual increase of 10,000 km on the 64,000 km of gas pipelines at the end of the 12th Five-Year Plan). The next 10–15 years will see oil and gas pipeline construction peak. Because the construction of oil and gas pipelines is capital-intensive, and China needs to build many pipelines, a policy that restricts investment to CNPC, Sinopec and CNOOC will restrain China's pipeline development and slow the growth of its natural gas market. Restrictions on private capital funding pipeline construction should, therefore, be progressively eliminated, motivating the private sector to invest in pipeline construction and operation. To this end, the following aspects should be reformed. First, the infrastructure investment project approval system should be reformed by softening the review terms and criteria and by streamlining the approval procedure, thereby attracting private capital to invest in and construct gas infrastructure included in the national and provincial plans. Second, new pipeline investment that is open to third parties should, if reasonable, be prioritised for approval. Third, low-interest or interest-free loans or subsidies should be offered to those companies building key interconnecting pipelines with low return on investment, to increase their returns without raising users' costs. Fourth, investors should be allowed to invest in, construct, operate and manage LNG terminals and gas storage facilities as independent legal entities. This will diversify ownership and operation of LNG terminals and gas storage facilities and speed up their construction.

Government should—based on the pipeline needs for crude oil, natural gas and refined oil products—plan pipeline deployment and create standards for investment, construction, operation and pricing. In particular, provincial and inter-provincial pipelines should be interconnected to support regulations. Unified standards should be developed for natural gas from different sources feeding into the pipeline network to ensure stable quality, usability and safety. The pressure grade of newly built pipelines should be carefully determined by referring to proven experience from other countries. Unified design and manufacturing parameters for gas usage facilities and gas volume measurement criteria should be defined to ensure technical interconnectivity.

Supervision and approval of transmission and distribution tariffs should be improved to reduce transmission and distribution costs. First, in accordance with the principle of China "maintaining administration over pipeline transmission tariffs in natural monopolies and deregulating gas source and sale prices", a pricing system that is based on "allowable cost + reasonable profit" and which combines constraints and incentives should be established. Pricing administration, cost supervision and review, and price regulation of gas pipeline transmission and urban gas distribution should be improved This will allow independent gas transmission tariffs and distribution fees to be set accurately, remove unreasonable costs and charges at pipeline nodes, and lower the costs of long-distance pipelines, intra- and inter-provincial pipelines and gas distribution pipelines. Second, financial independence in the gas distribution, gas sales and engineering services of urban gas utilities should be promoted. Third, large users that meet the conditions for direct supply should be allowed to purchase gas from a provincial pipeline network, local branch pipeline or city gas franchise. Given insufficient pipeline capacity, the absence of third-party access and the excessive pipeline transmission tariff, large users that meet the conditions should be allowed to construct pipelines for direct supply, thus forcing provincial pipelines to lower costs, be competitive and improve efficiency. Fourth, under the guidance of the national gas market regulator, independent gas market regulations should be established at the provincial level to implement independent supervision of fair access to gas pipelines, regulate prices and quality of products and services, and maintain an orderly gas market.

Local pipeline reform is a key part of China's market-oriented reform of the oil and gas pipeline network. Local reform should be carried out alongside long-distance pipeline reform, or the latter will be one step short of success. Reform measures for local pipeline networks include establishing independent gas pipeline companies, unbundling pipeline transmission and gas sales, and allowing fair third-party access to infrastructure. First, the creation of independent provincial pipeline companies and the unbundling of pipeline transmission and gas sales should be accelerated. Provincial pipeline companies should not operate gas sources or purchase or sell gas. They should only provide a gas transmission service. Second, provincial pipeline companies must allow fair and non-discriminatory access to pipelines to third parties, and provide all users with pipeline transmission, gasification, liquefaction and compression services. Third, local pipeline construction should be deregulated to encourage private capital to construct gas infrastructure included in the provincial plan. The rate of construction of intra-provincial pipelines should be planned, taking into account oil and gas resources and demand. The principle of "adapting pipelines and storage tanks to local conditions" should be followed to allow storage to play a dominant role in cities with lower gas demand, thus avoiding unnecessary pipeline construction. Fourth, intra-provincial pipeline interconnection should be promoted by developing consistent technical pipeline standards and by setting a reasonable lowest load indicator when estimating the transmission tariff for connecting pipelines, thereby ensuring gas supply security. Fifth, in provinces with excessive transmission and distribution, reform should reduce the number of nodes to lower gas transmission and distribution costs. Users should be encouraged to negotiate prices direct with gas source companies to encourage fair competition on the supply side.

(3) **Promote oil and gas pricing reform and create natural gas exchange centres**

In accordance with the principle of "maintaining administration over pipeline transmission tariffs in natural monopolies and deregulating gas source and sale prices", China's oil and gas pricing reform should: (i) improve the oil product pricing mechanism and allow the market to play the decisive role; and (ii) gradually deregulate gas source and sale prices, improve the price protection mechanism for low-income groups and establish a market-oriented oil and gas pricing mechanism. The government should only regulate those pipeline transmission and distribution tariffs under natural monopoly, and restrict price intervention to abnormal oil and gas price fluctuations and to emergencies. This would enable the government to transition its role from pricing to regulation and ultimately to marketise oil and gas prices.

The oil product price mechanism should be further improved. After several rounds of price mechanism reform, China's pricing system for refined oil products has become increasingly market-oriented. In the current and future international energy market, with abundant oil supply and rapid development of alternative energy sources, the external conditions for market-oriented oil product pricing are good. China's oil product market is characterised by three features. First, market participants are diversified. Private refining and chemical companies, oil product businesses and service stations have developed rapidly, and their market penetration matches that of the three state-owned enterprises: CNPC, Sinopec and CNOOC. As the crude oil import quota is gradually deregulated, the quota granted to local refineries, including private ones, has exceeded 100 million tonnes, allowing oil product sources to diversify. Second, there is regular market competition. Excess oil refining capacity has intensified competition in the oil product market. This means that government-guided crude oil and oil product prices have declined in significance. Third, there are regular price adjustments. As the market-oriented pricing mechanism for crude oil and oil products continuously improves and China's economy grows, affordability for people and businesses has significantly improved, and oil product price adjustments are no longer a sensitive issue. The correct conditions for market-oriented oil product pricing now exist. It is suggested that domestic oil product pricing should be fully deregulated by 2020, and that supply and demand should play the decisive role in pricing. In addition, the regulation of price, oil quality and tax revenue should be strengthened, and the anti-monopoly law enforcement system improved, thus preventing monopoly pricing and maintaining a fair and orderly market.

China has adopted the netback pricing approach to administer city-gate benchmark prices across the country, reformed gas transmission and distribution pricing, and unified residential and non-residential gas prices, laying a good foundation for a gas pricing mechanism where price is determined by market competition. It is suggested that market-oriented reform should progress steadily in stages and that the principles of "pilot implementation followed by extensive deployment", "incremental capacity increases" and "deregulating while rationalising the industry" should be followed, in order to deliver the goal of open, fair, competitive and orderly market-oriented gas pricing reform.

In the short term: first, the pricing system should be improved. After unifying the residential and non-residential benchmark city-gate prices, all provincial pricing authorities should improve the measures for administering residential tiered gas pricing. This will allow them to define the consumption volume and price level of each tier, and establish an upstream and downstream linked price mechanism. Second, pricing policies like interruptible gas prices, peak and off-peak prices and peak-shaving prices should be introduced as soon as possible. To ensure stable gas supply, especially in winter in north China, an interruptible gas price should be set to increase the number of interruptible users and offset the lack of gas storage facilities. China should guide construction of gas storage facilities by defining peak and off-peak prices for different

gas-consuming areas and times and encourage users to adopt peak-shaving. The requirements for establishing a gas storage peak-shaving ancillary service market, as stated in the Opinions on Accelerating Gas Storage Facility Construction and Improving the Gas Storage Peak-Shaving Ancillary Service Market, which were issued in April 2018, should be implemented. Purchase and sale prices by gas storage facilities should be determined by market competition, and the cost of constructing gas storage facilities owned by city gas companies should be lowered. Third, modification of metering and billing methods should be accelerated. Volume-based billing should be replaced by calorific value. This would help solve the difficulty of feeding natural gas of different quality from different sources into pipelines and of opening infrastructure to third-party access, and it would align with the international gas trading system. Measures to prevent illegal billing behaviour should be improved in terms of technology and management. Fourth, China should explore how to set up pilot projects to deregulate end-user gas prices. After summarising the lessons learned from Fujian's city-gate price deregulation pilot, additional pilots to deregulate end-user gas prices should be started in provinces like Sichuan, Chongqing and Jiangsu to facilitate later national deregulation of end-user gas prices.

In the medium to long term, once upstream operators are diversified, third-party access to infrastructure has been achieved, and prices at gas exchanges reflect supply and demand, China should stop controlling city-gate prices and deregulate gas source and end-user prices, allowing them to be determined by supply and demand. The pricing authority should only regulate the transmission costs and tariffs of inter-provincial long-distance pipelines, branch pipelines, intra-provincial pipelines and distribution pipelines under natural monopoly, thus achieving the goal of fair, competitive and orderly gas price reform.

The introduction of gas trading markets should be accelerated. In the short and medium terms, the Shanghai Petroleum & Gas Exchange, Chongqing Petroleum & Gas Exchange and other regional gas exchanges should form China's gas spot price index. First, unconventional gas (coalbed methane and shale gas), whose pricing is completely market-oriented, should compete with one another through platform trading. Companies should be encouraged to trade surplus capacity in pipelines, LNG terminals and gas storage facilities at the exchanges to increase the economic benefits for infrastructure operators and reflect the value of peak-shaving. Second, market trading participants—including gas suppliers, independent traders, large users and local gas companies—should be fostered to drive trading growth stably. Third, non-residential gas should be traded at the exchanges to ensure openness and transparency within two to three years. Fourth, regional exchanges should be established in an orderly manner, while accelerating the development of the Shanghai and Chongqing petroleum and gas exchanges. These regional gas exchanges should be developed in accordance with the principle of "different focus, mutual support, moderate competition and coordinated development". They should be located in the main consumption centres and infrastructure hubs, including Xinjiang, Shenzhen, Beijing, Tianjin, Hebei, Zhongwei and Hubei. Fifth, spot and medium- and long-term trading should be encouraged through listing and competitive bidding. This will promote competition between different types of gas and allow price to reflect fully the following: market demand and supply; price fluctuations in alternative energy sources; seasonal, peak and off-peak price differences; and other market factors—enabling China to create a gas spot price index as soon as possible. In the long term, the Shanghai Petroleum & Gas Exchange should become an international gas exchange open to the Asia-Pacific region and even the world. Gas futures, gas options and other financial products should be launched once the spot gas market is large enough and market participants in upstream and downstream oil and gas are diversified and compete, and infrastructure is interconnected and open. Later, an integrated natural gas exchange—where financial transactions and spot trading complement each other and develop hand in hand—should be built to shape Chinese and even Asia-Pacific gas spot and futures price indexes and to influence the global gas market.

(4) **Fiscal and taxation policies for the oil and gas sector need to be improved**

The aim of fiscal and taxation reform includes clarifying the interests of different stakeholders, supporting and guiding the development of unconventional and deep-water oil and gas resources, adjusting the distribution of tax revenues between central and local government, promoting the development and use of green resources, and building a fiscal and taxation system that balances the interests of different stakeholders.

Tight gas subsidy policies should be introduced as soon as possible. The geological reserve of China's tight gas resources is 25–30 trillion cubic metres, equivalent to 45–50% of China's conventional gas reserves. Moreover, 50% of these tight gas resources can be exploited, so it is the most practical available natural gas resource and has great development potential. However, tight gas development relies on hydraulic fracturing technology to deliver industrial capacity, and its development costs are two to four times that of conventional gas. Moreover, its single well production is low and steady production is for a short time only. The government has suggested that it grant a subsidy of RMB 0.2 per cubic metre for newly opened tight gas wells.

Shale gas will remain a key contributor to China's gas output growth in the future. Given this, subsidies for shale gas exploitation are important for China's development of shale gas. It is suggested that differentiated subsidies for shale gas resources should be researched. For instance, subsidies for shale gas resources in marine facies in south China should differ, depending on whether they are efficiently or inefficiently extractable at a depth above or below 3,500 m.

To reduce China's increasing reliance on imported oil and gas and enhance domestic supply security, the government should introduce differentiated tax policies for unconventional, deep-water, low and enhanced recovery of oil and gas to encourage companies to invest in the development of these resources.

The mechanism for sharing oil and gas tax revenues and royalties between central and local government should be adjusted. Due to decentralisation and the unreliability of local taxation, funds for local oil and gas development are limited. As China's economy develops, local governments are subject to increasingly strong demands to support local economic growth and raise people's living standards with the tax revenues from oil resource development. It is therefore suggested that the tax mechanism for oil and gas revenues and royalties be reformed to ensure that resource development benefits local people. Environmental compensation mechanisms should be established to levy a carbon or environment tax on resource development and use to increase local revenues.

A risk exploration fund should be established to develop strategic oil and gas fields. The government could use RMB 3–5 billion from oil tax revenues to establish a risk fund, which could be used for high-risk exploration led by the government. The government can engage oil companies to carry out initial exploration and drilling in high-risk fields by commissioning their services. If recoverable resources are discovered the field can be transferred through competitive bidding, and the earnings added to the risk fund to create a long-term mechanism.

Fiscal and taxation policies to support the construction of gas storage facilities should be developed. The delays in gas storage construction have become one of the major constraints to China's gas supply security. It is suggested that in provinces that lack the capacity to fulfil gas storage peak-shaving tasks, the government can invest in gas storage construction in several ways, such as by allocating funds from the central government budget. The VAT rate for gas storage should be lower, and a tax refund could be given for the first five years to support early-stage construction of gas storage facilities.

(5) Standardise government administration and create an effective oil and gas regulatory system

For the government to exercise its role in market supervision, regulation and services to the full and achieve the reform objectives for the oil and gas sector, government functions should be transformed and the methods of regulation overhauled.

First, government administration should be standardised. Government should intervene less in the market and in business activities. Administration should focus on the development of strategies, plans, policies and market rules and the creation of a fair and impartial industry. Second, plans for the oil and gas industry should be developed in a scientific way. They should be linked to economic and financial development, land use, infrastructure construction, environmental protection, safety, transport and technological innovation. A plan development mechanism should be established to make planning more scientific and authoritative. Plans should have a dynamic adjustment mechanism, and environmental impact assessments should be improved in accordance with relevant laws. Third, the development of industry standards should be accelerated. Macro-management based on planning, policies and standards should be applied to the oil and gas sector. Fourth, government services should be optimised. Energy information and geological data sharing platforms should be built. The government's role of driving technological innovation should be emphasised.

An effective oil and gas regulatory system should be created. China should: (i) separate administration and regulation, and form independent, unified and specialised regulatory authorities, and improve the regulatory system at the national and provincial levels; (ii) consider setting up an independent centralised energy regulatory system within the National People's Congress to define regulatory responsibilities (mainly economic regulation), improve social regulation and ensure fair competition in natural monopolies like pipelines; and (iii) improve the efficiency of government regulation, prioritising regulations for processes and activities like fair market access, trading behaviour, natural monopolies, taxes, pricing, safety and environmental protection, and combining legal tools, administrative means, specifications and public engagement to enable collaborative regulation.

To enable the market to play the decisive role in resource allocation and the government to play a more efficient role, those parts of existing laws that are outdated or conflict with the oil and gas sector or hinder its reform should be repealed or modified to facilitate oil and gas mining rights reform. For example, those parts not suitable for a market system in the Mineral Resources Law of the People's Republic of China, the Rules for the Implementation of the Mineral Resources Law of the People's Republic of China, the Regulations on Registering Mineral Resource Exploitation, the Administrative Measures for the Registration of Mineral Resource Exploration Blocks, the Measures for the Registration Administration of Mineral Resource Exploration, the Environmental Protection Law of the People's Republic of China and other laws and regulations should be amended as soon as possible. In addition, it is suggested that the Oil Law and Gas Law be studied and used as soon as possible to: (i) standardise the business activities of the companies and public bodies active in the oil and gas sector; (ii) standardise the supervisory and administrative activities of government and the actions of other stakeholders; (iii) improve the development, use and conservation of oil and gas resources; and (iv) enable the healthy development of the oil and gas sector and other related sectors. The development of energy regulations should be completed as soon as possible to define regulators' roles, functions and responsibilities, regulatory procedures and the decision-making mechanism; standardise information disclosure and the dispute settlement mechanism; and define the penalties for violations. All these actions will give legal support to supervisory and regulatory work and help them become authoritative.

6 Reforming the Coal Sector

6.1 Progress and Trends in Global Coal Industry Reform

6.1.1 More and More Countries Have Adopted a Coal Exit Strategy

Reducing emissions from coal combustion has become an important aspect of cooperative global governance. The international community has maintained a high degree of unanimity regarding the global fight against environmental pollution and climate change. The Paris Agreement was adopted with the unanimous agreement of 195 countries. The USA officially announced its withdrawal from the Paris Agreement in June 2017 and submitted withdrawal documents to the United Nations in August 2017, receiving widespread criticism. Reducing pollutant emissions is an irreversible trend and an important issue in global governance. The USA's withdrawal has not yet had a material impact on global emissions reduction and will not shake the worldwide determination for global climate governance or change the development course of clean energy. Coal is a fossil fuel energy with a high carbon content. Reducing coal consumption has therefore become a focus of global emissions-reduction cooperation.

Most European countries have begun to phase out coal-fired power plants. Eurelectric—the sector association for the European electricity industry, which has around 3,500 member companies—and the European power industry will not "invest in or construct new coal-fired power plants after 2020". This plan is supported by 26 EU member states. Leaders of EU countries reached consensus in October 2014 regarding the objective of "reducing greenhouse gas emissions by 40% over 1990" and agreed to disinvest gradually in the coal industry and end subsidies for fossil energy. For example, France stopped providing financial support for overseas-owned coal-fired power plants without carbon capture and sequestration in May 2015. Major financial institutions no longer invest in coal mining projects. The Norway Sovereign Fund began to disinvest in coal companies in June 2015. Scotland shut down its last coal-fired power plant in March 2016. England, Finland, Portugal and Austria plan to phase out coal-fired power plants by 2025, 2029, 2030 and 2025 respectively. In addition, EU countries plan to cancel coal-related subsidies by 2018. Austria, England, Belgium and Finland have all taken power industry decarbonisation measures.

All OECD member states strictly control subsidies for coal-fired power plants. Thirty-four OECD member states reached agreement in November 2015 to restrict export credits for coal-fired power plants using older technology, granting allowances only to "coal-fired power plants that meet the strictest environmental standards", and tightening financial support for the coal industry.

The USA, Canada and other countries have gradually weakened policies supporting the coal industry. The USA began to restrict the export of coal technologies before the OECD. Major banks in the USA have cancelled financial support for coal projects, as have the Export-Import Bank of the United States, the World Bank and the European Investment Bank. Canada backed carbon capture and sequestration technologies for coal-fired power plants in August 2011. It is possible that it will completely phase out coal-fired power plants by 2050. Many regions in Canada have introduced decarbonisation measures for the power industry.

Many countries levy resource taxes. Denmark, Finland, Sweden and Norway are pioneers in these taxes. In Canada, the Alberta government began to levy CAD 20 ($15) on every tonne of carbon emitted from January 1, 2017. France started to levy a carbon tax on coal power in 2017.

In summary, the policies for exiting coal can be divided into two categories: (i) restrictive policies on the supply side to reduce coal production; and (ii) policies on the user side that set targets for the transition to new energy as a means to reduce coal demand.

6.1.2 The Strategic Positioning of Coal Differs from Country[*] to Country

Many countries are heavily dependent on coal. For example, in Germany, coal accounts for more than 20% of the energy mix and coal-fired power for more than 40% of the power mix. In South Korea, the figures are around 30% and almost 40% respectively; and in India, they are higher—about 50% and 60% respectively. China is one of a minority of coal-based countries. Coal accounts for around 63% of China's energy mix, and coal-fired power for more than 70% of its power mix. A few countries may become more dependent on coal, instead of less. Japan is an example. Coal accounts for about 25% now, in both its energy mix and power mix. By 2030, the share of coal in its energy mix and power mix is likely to increase to 30% and 26% respectively.

To ensure energy supply security, some countries increase their use of coal. For example, Japan made a new energy plan after the Fukushima nuclear accident, which involves building 41 coal-fired power plants over the next decade to ensure energy security. To meet increasing energy demand and improve economic development, India plans to increase coal output year by year, and has introduced support policies on the environment, land approval and on granting exploitation rights to private capital.

The positioning of the coal industry differs from country to country. Most countries are actively reducing emissions from coal combustion and optimising their energy and power mix in many ways. But their attitudes on whether to exit or continue to use coal are different, mainly influenced by their resource endowment, industrial structure and phase of development. Significantly decreasing coal's share of the energy mix is impossible for many countries in the short and medium terms. For example, in Germany and the USA, the coal-exiting policy has been ignored by coal and other traditional industries. Japan has increased its coal power to ensure energy security. In keeping with its phase of development, India's coal consumption is increasing rapidly, and its coal policies are aimed at stimulating coal supply.

6.1.3 Clean Energy Is Rapidly Replacing Coal, Thanks to Incentives for Users

Countries aiming to abandon coal, even though they may be highly dependent on it in the short and medium terms or increasing their use of it, might yet actively take part in global emissions reduction governance and issue the necessary policies to encourage power generation with clean energy. North America is expected to increase the share of clean energy consumption to 50% by 2025. Argentina plans to invest $2.1 billion in wind, solar, biomass power and small hydropower stations. Sweden plans to make 100% of its energy green by 2040. Germany encourages distributed energy through a high level of energy efficiency tax refunds and through financial support for energy-efficiency equipment. It is estimated that the proportion of renewable power in total electricity consumption will increase to 40–45% by 2025, and to 55–60% by 2035. England supports distributed power projects through subsidies, credit loans and intelligent metering. Italy supports distributed power projects through an energy tax, credit loans, high prices, consumption subsidies and other preferential policies—the feed-in tariff for distributed power is 50% higher than the domestic retail price. In Chile, the proportion of new energy will reach 50% by 2035 and 70% by 2050.

The cost of generating power with clean energy decreases gradually as technologies improve. Meanwhile, the environmental cost of traditional fossil energy is increasing. For that reason, and to promote the development of the clean energy industry, some developed countries have begun to end subsidies for power generation with clean energy. For example, Germany ended its policy for a preferential feed-in tariff for renewable power in June 2016. The European Union began to limit subsidies from member states to the renewables industry in 2017. In the future, governments will increasingly use market mechanisms as the clean energy industry matures. The policies facilitating the replacement of coal by clean energy will gradually change into incentives to make power generation efficient.

6.1.4 Government Support for the Coal Industry Still Exists

Many countries once dominated by coal still have a large coal industry, with many coal workers. Support policies of different types are, therefore, still needed from government. The USA, for instance, launched in 2015 the Partnership for Opportunity and Workforce and Economic Revitalization (POWER) initiative, which helps coal communities adapt to the new energy landscape and reposition their economies. The key points in the Federal Government's America First Energy Plan include ensuring self-sufficiency in energy, revoking the Climate Action Plan of the previous administration, and supporting and promoting the clean development of coal. These policies will help revitalise the coal industry. In addition to building more coal-fired power plants, Japan introduced policies to increase investment in scientific research on energy saving and emissions reduction for coal combustion and support the modernisation of coal-fired power plants.

Government support focuses on technical innovation, energy saving and emissions reduction. After analysing different countries' coal support policies, we found that governments support innovation in coal combustion technologies, rather than simply supporting expansion of coal's share of the energy mix, to encourage energy saving and emissions reduction. Replacing coal with clean energy will take a long time, and coal power is characterised by low-cost and proven technologies. That is why many countries strongly support energy-saving and emissions-reduction technologies in the coal sector, while optimising their energy and power mix. They hope to achieve low-carbon use of high-carbon energy by minimising the emissions from coal combustion.

6.1.5 Coal Companies Are Folding, Merging or Being Sold

Coal companies are going bankrupt. Others are merging or being sold. The rapid replacement of coal by clean energy, the global economic recession after the global financial crisis of 2008, and China's declining coal demand since 2014 have led to oversupply in the international coal market. In the USA, Patriot Coal, Walter Energy, Alpha Natural and Arch Coal filed for bankruptcy in 2012, 2015, 2015 and 2016 respectively. Peabody Energy, the largest coal company in the world and engaged in coal exploitation for 100 years, joined the list of insolvent coal companies in April 2015. For coal companies that had not filed for bankruptcy, the situation is frustrating: most of them have been brought to the verge of bankruptcy due to long-term losses. For example, Murray Energy in the USA announced a mass layoff programme in 2016 to reduce its losses, and the net loss of New World Resources, the largest private coal company in eastern Europe, was around €233.6 million in 2015. In 2015, more than 80% of Chinese coal companies made a loss. The Chinese economy improved in 2017. Output of domestic raw coal picked up. Output from January to September increased by 5.7% over the previous year, which is 16.2% points higher than in 2016. But low efficiency, debt, state-owned enterprise reform and other structural problems in China's coal industries remain. Merging and restructuring large coal companies is on the agenda again, and a new trend of vertical integration has emerged. Shenhua Group, the largest coal company in China, and China Guodian Corporation merged to form China Energy Investment Corporation Limited in 2017. Some local state-owned coal companies are also exploring horizontal and vertical integration, aiming to sharpen their competitive edge through resource optimisation and reorganisation.

International consortia have divested their coal businesses. Total of France, the fifth largest energy company in the world, ended its coal production and sales activities in 2015. The total value of the coal assets divested by the Anglo-Australian coal mining giant Rio Tinto since 2013 is more than $4.7 billion, and it sold 40% of its holding in Australia's Bengalla coal mine for more than $600 million. In addition to divesting 60% of its assets in 2016, global mining company Anglo American sold 51% of its holding in Dawson coal mine and 70% of its interest in Foxleigh coal mine, both in Queensland, Australia. Vattenfall, wholly owned by the Swedish government and the third largest energy supplier in Germany, announced in 2016 that it would gradually retreat from the German coal market. In the same year, the largest sovereign wealth fund in the world, the Norwegian Oil Fund, sold its shares in 52 coal-related companies.

Some coal companies have seized the opportunity to expand their business. Exxaro Resources of South Africa acquired coal assets from Total of France for $472 million in 2014, $45 million lower than their value the previous year. Indonesia's state-owned coal company Bukit Asam entered Australia's coal market in 2016 by acquiring one of Rio Tinto's largest coal mines in Australia.

Bankruptcy and merger and acquisition are commercial activities, as well as means of government gaining control. Asset restructuring of international coal companies, an unavoidable result of the industry's descent, is changing the structure of the global coal industry.

6.2 Progress, Status and Assessment of China's Energy System Reform

6.2.1 Reform Has Progressed Despite the Industry's Downturn

The years 2010–16 were a transitional period for China's coal industry, with industrial growth dropping from high to medium and institutional reform deepening. The golden decade for China's coal industry was 2002–12, when annual coal output increased on average by almost 10%. The growth rate of the Chinese economy slowed down in 2012, leading to overcapacity in the coal industry. Shortage was rapidly replaced by oversupply. The growth rate of coal output dropped substantially to 0.74% in 2013, then declined into negative growth for three years from 2014, before re-entering positive growth in 2017. In the context of the downturn, coal industry reform has made progress in many aspects.

China has been using dual coal pricing for a long time. Coal traded by sellers and buyers using a market-based pricing mechanism, through which the price fluctuates in line with market conditions, is called market coal. Previously, coal producers and thermal power plants on the state's list of key enterprises traded coal at an annual national coal fair price under a planning mechanism called planned coal. During the golden decade, the price of planned coal was RMB 200 lower than that of market coal, resulting in years of coal power combat. China unified the two prices in 2013, which marketised the coal market.

An important component of China's administrative system reform is "administrative streamlining and decentralisation, the combination of deregulation and regulation, and service optimisation". Coal authorities ended the Coal Production Permit and the Approval Needed to Found Coal Businesses in 2013. Many local governments also introduced the reform policies, delegating approval and improving administrative supervision and management. Administrative supervision and management is becoming increasingly market-oriented and IT-based. Government functions focus more and more on industrial services.

Overcapacity in China's coal industry emerged during the 12th Five-Year Plan (2011–15), reaching a peak in 2015. From November 2011 to January 2016, the Bohai-Rim Steam-Coal Price Index (BSPI) dropped by 56.51% from RMB 853/tonne to RMB 371/tonne. According to data from the National Bureau of Statistics of China, the profit margin of the coal mining and washing sectors decreased from 14.06% in May 2011 to 1.76% in December 2015, and to its lowest point of 1.55% in August 2015. In 2015, more than 90% of large and medium-sized coal companies made a loss. China's coal industry, characterised by its long history, large size and

huge number of employees, was an important pillar of the national economy. A fast and substantial downturn in the industry brings financial and social risks and threatens the overall stability of the Chinese economy. The coal industry therefore became a key sector for China's supply-side structural reform under the new normal of slower economic growth. The structural reform of the coal industry is focused on reducing capacity, optimising supply and achieving the reform objectives of reducing inefficient supply and increasing efficient supply. The State Council issued the Opinions on Reducing Overcapacity in the Coal Industry to Achieve Development by Solving the Difficulties in February 2016, which set out the objective of reducing capacity in the coal industry. China cut coal capacity by 290 million tonnes in 2016, exceeding the target of 250 million tonnes. China will strive to meet the capacity reduction target of 150 million tonnes in 2017.

Coal mine safety management has always been an important area of coal industry reform. Since 2010, reform has focused on shutting down or reorganising small coal mines, reforming safety management and increasing safety investment. First, small coal mines that did not meet safety requirements were shut down. More than 3,000 small coal mines, each with an annual output of less than 300,000 tonnes, were closed by the end of 2015. The second measure was to merge small coal mines that met safety requirements with large companies. Shanxi Province implemented the plan between 2009 and 2012, shutting down 1,545 mines. Other major coal producing provinces also merged large companies with small coal mines. Henan merged and restructured 466 small coal mines in 2010. Hebei began to regulate small coal mines in 2011, shutting down 237 during the 12th Five-Year Plan (2011–15) through merger, restructuring and integration of resources. Shandong launched a programme to merge and restructure small coal mines in 2012; by 2015 there were 52 fewer coal mines than in 2010. The third measure was to increase investment in safety. This became part of management procedures and was well promoted and well implemented. Reform was based on management and supervision at each level and on clearly defining responsibilities, including those of regulators, government and coal companies. In particular, the responsibilities of company management were clarified.

Emissions of air pollutants from coal use are strictly controlled. The Chinese government has improved its air pollution prevention work in recent years. Coal combustion is one of the main sources of air pollutants, so coal-fired power plants and steel companies are two important areas of government supervision.

The government began to build a national monitoring system of pollution sources in 2010 and has been improving it ever since. Once the system was completed, the government passed the Environmental Protection Law (which provides strict and comprehensive regulation) in January 2015, and the Atmospheric Pollution Prevention and Control Law in January 2016, which codifies emissions regulation. Also, in 2012 and 2015, the government substantially raised emission standards for the steel industry. Hebei province, a major steel producer, imposed limits and substantially raised emission standards in 2014 by incentivising coal-fired power plants to implement super-clean emission standards. Finally, there has been a push to reduce coal consumption. Air contamination in Beijing, Tianjin and Hebei is a serious problem. Reducing regional coal consumption is an important method of governance. Beijing is implementing the coal-to-gas programme. Beijing, Tianjin, Langfang and Baoding plan to introduce coal-free zones by October 2017. Shijiazhuang, Tangshan, Handan and Anyang have adopted temporary policies to limit steel production capacity to 50% in the cold season.

Tax reform of the coal industry since 2010 is centred on two measures. First, reform of the coal resource tax, which aims to move from a fee system to a tax system, as well as encouraging more efficient use of resources. Second, changing business tax into value-added tax. Unlike the reform of the coal resource tax, the business tax to VAT reform is a national reform, not just a coal industry reform. It involves paying VAT instead of business tax on taxable items. The

business tax to VAT programme, launched in May 2016, has been fully implemented in the coal industry, just as it has in other sectors. The tax reform programme, which aims to relieve the tax burden on small and medium-sized companies and eliminate double taxation on large and medium-sized companies, is another example of tax reform in China following the tax sharing reform of 1994.

The reform of state-owned coal companies has been accelerated. The extensive losses of China's coal industry were caused by oversupply, the heavy debt burden of state-owned coal companies and the weak profitability of most non-coal businesses. The Chinese government therefore regards speeding up the reform of state-owned coal companies as an important way to help the industry become profitable. In addition, accelerating the reform of state-owned enterprises is an important aspect of the coal industry's supply-side structural reform.

The first stage was to restructure state-owned capital investment corporations through pilot programmes. Central and local state-owned coal companies have become pilot reform targets for state-owned capital investment corporations. The plan is now being developed and implemented. The state-owned capital management system and mechanism are being transformed, as is the organisational structure.

The second stage was to integrate, merge and restructure resources. Some major coal suppliers in Shanxi, Shaanxi, Heilongjiang and Hebei formed large local state-owned coal enterprises before 2010 by integrating state-owned coal resources. Henan further integrated its coal resources after 2010. To improve the safety and efficiency of coal mines, state-owned coal companies were encouraged to integrate private coal resources. The coal companies owned by central government began to integrate resources during the 13th Five-Year Plan (2016–20). The State-Owned Assets Supervision and Administration Commission under the State Council introduced new policies in 2016. These require companies owned by central government that are marginally involved in the coal industry to exit the industry and transfer their resources to coal enterprises owned by central government. For example, Shenhua Group and China Guodian Corporation merged into China Energy Investment in 2017.

The third stage was to remove state-owned coal companies of their social responsibilities, which were a legacy from the previous round of reform of state-owned enterprises. To relieve the burden on coal companies during the economic downturn, government at all levels began to remove these responsibilities from coal companies and made great progress in doing so.

The fourth stage was to reduce coal capacity. Local governments and state-owned coal companies actively shut down old mines and zombie capacity.

The final stage was to promote mixed ownership. As planned by state-owned enterprise reform, state-owned coal companies are exploring ways to carry out mixed ownership reform and are expected to make progress during the 13th Five-Year Plan (2016–20).

6.2.2 Reforms Are Characterised by Marketisation and More Administration

Deregulating the coal price, abolishing coal production permits and requiring companies to obtain approval to start a coal business are significant steps towards market-based reform. The coal market's competitive attributes have been strengthened. Intermediary coal service organisations are more mature. To meet the need for market competition, the authorities made regulations more market-oriented. First, laws, regulations, rules, policies and standards were customised for the coal market. Second, a market information system was developed to monitor and analyse the coal market in real time. Third, plans, guidance and support policies for coal trading, logistics and reserves were introduced to marketise the coal industry.

As circulation becomes more market-oriented, the government is improving its administrative means to regulate the coal industry, especially regarding capacity reduction and price stabilisation. At first, in 2016, China put more effort into capacity reduction. The State Council announced

in January 2016 that capacity reduction should "first take place in the steel and coal industries", and that "coal capacity should be reduced significantly" through "the rule of law and marketisation". To implement the capacity reduction plan, the State Council published the Opinions on Reducing Overcapacity in the Coal Industry to Achieve Development by Solving the Difficulties in February 2016. After which, support documents from government departments for human resources, finance, safety supervision, land, quality inspection, taxes, and environmental protection were issued. The provinces, cities and autonomous regions that are major coal and steel producers also set targets, plans and measures for capacity reduction. Strongly driven by policies, the coal industry achieved its capacity reduction objectives ahead of schedule in 2016.

Administration played a leading role in capacity reduction during the market recovery in 2016. This was enabled by the following five factors: (i) the macroeconomy reported slower but stable growth; (ii) market-driven capacity closures and output reduction during the earlier downturn reduced supply; (iii) legal measures forced some outdated capacity to be shut down; (iv) the government strictly controlled the addition of new capacity; and (v) some production restrictions adopted in 2016, such as limiting mine operations to 276 working days a year, controlled output during the rapid price recovery. Factor (i) was the foundation. Factors (ii) and (iii) were essential, but the policies of tightening supply were more crucial and accelerated the recovery of the coal market. China's output of raw coal in 2016 was 9.4% lower than in the previous year. The government paid close attention to price fluctuations and took the necessary measures to stabilise the coal price. Once coal capacity reduction had delivered results, the coal price went up. To stabilise the price, the government actively influenced long-term agreements with key accounts, under which coal was traded at a price lower than the market price, restraining price rises. Finally, the government strengthened the administrative regulatory regime for safety, making it a priority to regulate the people in charge of coal companies.

State-owned coal companies have started to reform themselves in accordance with China's state-owned capital management system. Measures include reforming state-owned capital investment corporations, introducing mixed ownership, organisational restructuring and developing market mechanisms. Centrally run state-owned coal companies have made breakthroughs in vertical integration along the value chain. Local state-owned coal companies are exiting from inefficient mines, transforming state-owned capital investment corporations and restructuring.

After the economic downturn, China's coal companies developed expertise in energy industry trends and began to transform and optimise their business structure in two aspects: (i) exiting inefficient coal, coal-fired power generation, coal chemicals and coal machinery, and reorganising internal resources to allow them to exit those fields; (ii) planning for new emerging business, developing modern services and intelligent manufacturing, and using R&D and other resources to make the necessary breakthroughs.

The tax burden on the coal industry is still heavy. After the business tax to VAT reform was carried out, the taxes on some business areas of coal companies decreased, such as in equipment manufacturing. But for other business areas, the tax burden increased because there were fewer allowable deductions. Overall, businesses are paying more taxes since the business tax to VAT reform. After the coal resource tax was reformed, the tax rate on price and resource-related charges differed from region to region. It is not unusual to find coal companies for whom tax liabilities have increased under both reforms.

6.2.3 China's Coal Industry Reform Still Faces Many Difficulties

Market-oriented reform conflicts with administrative measures, as revealed in the following four cases.

First, there is a conflict between the government's regulatory function and business governance. Government regulates macro-aspects of the economy and should not intervene at the micro level of business operation and management. Government and business are on two different levels of administrative, social and economic governance. They should cooperate. The introduction of market-based reform requires the boundary between government and the market to be defined. Only by keeping government within bounds and allowing businesses to perform can the market's decisive role be unleashed. If the administrative process is not well controlled, the boundary between government regulation and business governance can become blurred, with the government crossing the line and intervening in business operations. This could be seen during the drive to stabilise the coal price and improve safety. Government authorities were encouraging large coal companies to trade with key accounts at a contractual price, which meddles in the coal companies' business and works against the market-oriented coal pricing mechanism. The process of improving coal companies' safety implicitly regards them as targets for administrative management, which conflicts with their responsibilities as market players and hinders the reform of the state-owned capital management system.

Second, there is a conflict between the target to reduce production capacity in the medium to long term and the current supply shortage. Coal industry losses changed into industry-wide profitability in 2017, making it more difficult to make further capacity reductions. One cause of the supply shortage is that the government's compulsory measures make it harder for the market to meet demand. Capacity reduction should therefore continue. However, improvements in the business environment and the universality of the government's capacity reduction policies still allow inefficient and uncompetitive production capacity.

Third, there is a conflict between reducing and exiting inefficient supply and increasing efficient supply. Staff relocation and debt repayment are the inevitable challenges of capacity reduction. Local government and financial institutions have introduced some protective measures to mitigate social and financial risks. Despite being uncompetitive, some inefficient mines have been operating for a long time. In addition, the government's administrative regulatory system has played an important role in capacity reduction since 2016, though its inflexibility has restricted the development of efficient capacity to some extent.

Fourth, there is a conflict between market-oriented regulations and government administration. The government has emphasised the market's decisive role and deepened reform along the lines of "administrative streamlining and decentralisation, the combination of deregulation and regulation, and service optimisation". However, macro-control of the coal market is still dominated by administrative processes. To achieve their environmental and safety objectives, some local governments were inflexible in their administrative guidance, which was not based on law and resulted in unnecessary financial losses for compliant enterprises. There are also problems with the coal contract price encouraged by government. The government abolished dual coal prices with the aim of deregulating coal pricing and forming a market-oriented coal pricing mechanism. Trading coal with key accounts at a contractual price lower than the market price is consistent with the laws of economics. However, if the contractual price does not follow the price trend, or differs greatly from the market price, it could be concluded that the contractual price has been affected by the government's macro-control, rather than being determined by supply and demand. If the contractual price is similar to the old planned price, then clearly the government has not created an effective market-oriented mechanism for coal pricing.

China's coal circulation regulatory system under the new normal of slower economic growth faces new challenges. The market is now the vector for transactions, which has a profound influence on the coal industry.

First, a modern coal circulation regulatory system is not yet in place. The parties engaged in coal circulation are numerous, decentralised and small. These factors make government regulation more difficult. The current regulatory system cannot cover all parties for the following reasons: (i) policy constraint is not effective for each party involved in circulation. Regulations on trading and coal quality cannot fully supervise and restrain complex markets; (ii) coal circulation regulations are not clearly distributed between central and local government, making the regulatory system messier and more complicated; and (iii) development of the auxiliary coal market has lagged behind. This dates back to the start of the reform regime and is characterised by poor contractual performance, scattered market information, an unsound credit rating mechanism and the absence of financial regulation.

Second, there is no market-based carrier optimising supply and demand through the coal circulation regulatory system. Coal circulation regulation needs to allow an existing market-oriented coal pricing mechanism to play its role and stabilise market price fluctuations. However, there is currently no effective carrier. The way to stabilise price fluctuations is by improving supply and demand in a market-based way, rather than intervening in businesses' production and operations, which disturbs the market pricing mechanism. The current coal circulation regulatory system enables regulators to keep track of coal market dynamics but does not allow them to optimise supply and demand.

Third, the regulatory regime transforms coal circulation. Supply-side optimisation has a profound impact on coal circulation. Future coal circulation will not be limited to coal trading and the transport of coal from one place to another but will expand into higher levels of circulation. For instance, the value added to raw coal during circulation will increase, logistics will be modernised and capital and information flow services improved. A coal circulation regulatory regime should not only standardise and manage circulation, it should guide the transformation of circulation. However, the current coal circulation regime does not cover higher levels of circulation services. Such a regime has not yet been designed.

Fourth, China needs a coal circulation regulatory system that meets the demand for green and low-carbon energy. To meet this new requirement, the coal circulation regulatory system prioritised three things: the standardisation of coal quality, restricting the use of bulk coal, and the introduction of policies that support the use of clean coal. Clean coal is a highly beneficial way to transform China's coal production and consumption and create new industry trends. However, the clean coal regulatory regime is not systematic enough in its approach and needs to be aligned with China's coal circulation system. Restrictions and incentives are needed: (i) the industrial heating sector needs to be reformed and aligned with the use of clean coal; (ii) restrictive regulations are not strictly enforced, giving some businesses and consumers opportunities to break them; and (iii) the incentives to drive technical innovation in clean coal and modernise industrial and public sector equipment are not in place, nor are consumption subsidies.

It is difficult to reform state-owned coal enterprises due to the large number of people employed in the industry and the heavy debt burden that many companies carry, both of which restrict innovation and transformation. The reform of China's coal industry is strongly related to the reform of state-owned coal companies, which play an important role in China's coal industry. The top 10 coal companies in China, in terms of coal output, are all owned by the state. They produced around 1 billion tonnes of coal between January and September 2017, accounting for more than 60% of raw coal output by large national coal companies. Large state-owned coal companies—characterised by their large size, diverse business activities and long history—have inefficient production capacity that they need to exit. For many companies, inefficient capacity has become a serious problem, leading to surplus workers and a high debt-to-asset ratio. These companies enter a vicious cycle where the heavier the personnel and debt burdens, the more difficult it is for them to exit inefficient capacity. The recovery of the Chinese economy in 2017 played an unusual role by acting as a buffer, allowing state-owned coal companies the time and space to transform and

innovate. A strong internal organisation is a precondition for such transformation and innovation. Internal weaknesses need to be eliminated first. Only then can business creativity be stimulated through institutional and mechanism innovation within the larger reform of the coal industry. The reform of state-owned coal companies poses many questions, such as: What will happen to the workers? How can the company repay its debts? Where will the money come from? There is currently no answer to these questions, which is slowing down the process of industry-wide reform.

6.2.4 Evaluating China's Coal Industry Reform

The coal industry is a pioneer in market-oriented reform. Reform started in the 1980s, and has never stopped, but has had different characteristics during different periods. As the Chinese economy moved from a high growth rate to the new normal of lower growth, and environmental governance became more intense, a supply shortage of coal was replaced by oversupply. At present, supply is tight because of the initiatives on capacity and emissions reduction. Changes in the market have accelerated the reform of the coal industry. However, difficulties in business operation and the pressure to prevent air pollution have pushed the government to strengthen its macro-control.

The reform of China's coal industry has the following characteristics: it is an advanced market, with better regulations and deepener reform.

At first, the marketisation of the coal industry made steady progress. Coal authorities abolished various policies, including dual coal prices, the coal production permit and the need for approval to start a coal business. At the same time, coal companies actively explored transitioning from coal producer to energy supplier. The logistics, information flow and capital services for the coal market were improved. Driven by government and businesses, the coal market became more and more sophisticated.

Next, the government strengthened its macro-control. The government's plans to reduce production capacity achieved remarkable results in a short time. It not only reduced overcapacity but improved the business environment of the coal industry as well. The government also, by combining capacity reduction with emissions reduction and focusing on restricting new capacity and even exiting coal, promoted supply-side reform. To suppress the recovery of the coal price, the government took other measures as well, such as encouraging coal companies to agree low-price contracts with key accounts.

Finally, the reform of state-owned enterprises deepened. To help state-owned enterprises escape the conflicts and issues that urgently needed to be addressed, the government eliminated deep-rooted institutional and mechanism problems. The papers regarding reform of state-owned enterprises were adopted after 2015, gradually forming the 1+N policy. The State-Owned Enterprise Reform Leading Group of the State Council carried out the Ten Pilot Reform Programmes in 2016, which summarise the present round of state-owned enterprise reform. The reform of state-owned coal companies has focused on state-owned capital investment corporations, introducing mixed ownership, divesting coal operations from centrally administered state-owned enterprises whose core business is not coal, and on restructuring and merging companies. The reform of centrally administered coal companies is now making rapid progress, whereas that of local state-owned coal companies is slower.

The coal market experienced a significant downturn after 2010 and was faced with unprecedented environmental pressure. On the one hand, a supply shortage turned into oversupply, which benefitted market-oriented reform. Dual coal prices were abolished. On the other hand, the downturn resulted in financial losses for the industry. Stabilising growth and mitigating risk were especially urgent. Given market and environmental pressure, administrative regulation became essential. As a result, the reform of China's coal industry after 2010 progressed in marketisation and government administration. Administration played a more important role after 2016, not only in reducing production capacity and improving environmental protection, but also in stabilising the coal price once supply and demand had been optimised. The government achieved remarkable results in the short term.

The coal industry regained profitability in 2016. However, the drawbacks of administrative regulation reform were becoming increasingly evident. Coal supply and demand depends on several factors. Reducing production capacity tends to lead to a supply shortage in some regions and in certain types of coal, and exiting mines affects people's livelihoods, even though it helps reduce emissions. Trading with key accounts at a contractual price should be encouraged. Once the price difference between contractual price and market price reaches a certain threshold, a dual price system could reappear. The goal of reform is to build a sophisticated coal market, rather than reduce capacity and stabilise prices. The government's reform measures were indispensable, but an administrative mechanism cannot replace the market mechanism or prevent it from fulfilling its basic functions of optimising the system and determining price. The administrative mechanism should be combined organically with the market mechanism. Therefore, in the long term, China needs to build a coal market system that is aligned with the long-term objectives of marketisation and that manages the relationship between the market and government.

6.3 Guidelines and Pathways for Future Reform

6.3.1 Guidelines

The revolution in the coal industry needs to resolve the principal contradictions constraining its further development. Reform is the most important path. The journey begins by gaining a clear understanding of the coal industry's role in China's economic and social development, and of how China is striving to achieve its long-term objectives of building a market-oriented, efficient and sophisticated coal industry management system by facilitating revolution through reform and being problem-oriented. There are six aspects to consider: (i) handle the relationship between the long-term target of creating a sophisticated management system and the short- and medium-term policies of the coal industry, while protecting the transformation process from disturbance by an emergency; (ii) the government's basic regulatory functions in the coal industry are to avoid price fluctuations and mitigate risk, as the coal sector is of fundamental importance for jobs and economic growth; (iii) once a sophisticated coal market system has been created, control the degree of government administrative regulation and ensure it does not cross the boundary between government and market; (iv) allow the domino effect of reform to take effect at carefully defined breakthrough points of the coal market system; (v) redefine the powers and responsibilities of coal mine safety management, and improve and standardise coal company governance; and (vi) use state-owned capital to optimise the coal industry and improve the industry's operations.

6.3.2 Handle the Relationship Between the Long-Term Management System and the Short- and Medium-Term Policies of the Coal Industry

Coal industry reform should have long-term objectives, but the market deals in short- and medium-term cycles. This makes it difficult to combine long-term system reform with short- and medium-term regulation. Supply shortages will not persist in China's coal market in the long run. Moderate oversupply is the target, which is also conducive to improving the coal market system. Shifting from supply shortage to moderate oversupply is a long process, during which labour pains are inevitable. China's coal market entered a downward trend in 2012, reaching the bottom of the cycle in 2015. Markets have the ability to self-regulate. From 2012 to 2015, with an imbalance between supply and demand, many small- and medium-sized coal companies that were not competitive left the market. In 2015, the coal industry made large financial losses. If no other measure other than self-regulation had been used, more coal companies would have shut down. To reduce coal capacity, the government implemented many policies and measures, which were successful after about a year. Then the coal market was short of supply. Compared with the historical price, the coal price in 2016–17 did not rise sharply, reaching only the historical mid-range, though this had a great impact on the bulk raw materials market. Stabilising the coal price was on the government's agenda again. The goal of ensuring safety also involved a conflict between long-term objectives and short- and medium-term performance. In the long term, coal mine safety needs to be managed by law. Only with a sophisticated legal system could a long-acting safety management mechanism be built. But in the short and medium terms, the government officials in charge could strengthen the rulebook to improve coal mine safety management and shift more responsibility onto businesses. During the reform of the coal industry, the central government set long-term reform objectives. However, when the administrative authorities introduced the policies and measures, they did so in response to short- and medium-term problems. It is not unusual for short- and medium-term policies and measures, even emergency policies and measures, to conflict with long-term reform objectives. Therefore, to achieve the long-term reform of the coal industry, the relationship between the long-term coal management system and short- and medium-term emergency policies should be properly managed. Specifically, institutional reform of the coal industry needs to achieve long-term objectives. Short and medium-term government regulation is essential, but it should not be used at the expense of long-term institutional reform. Although short-and medium-term regulation that runs counter to long-term reform can deliver results quickly, it will generate new bottlenecks that hinder the long-term reform process.

6.3.3 The Government's Regulatory Function Needs to Be Defined Clearly

Coal will remain China's main energy source for a long time to come, although its share of energy consumption will decline. Forecasting of future consumption trends suggests that the coal industry's role in the Chinese economy will change. Its function of supporting energy supply, the economy and society will remain, but its status as a pillar industry will weaken. Coal, which is abundant in China, is a low-cost energy that will continue to account for a high proportion of the country's energy consumption. The coal industry has many workers and assets and high levels of debt. What happens in the coal industry has a strong chain effect on midstream and downstream industrial enterprises and the service sectors. Stability in the coal industry is important.

As China's main energy source, coal stabilises and supports the economy and society. The government therefore needs to impose macro-regulation on short- and medium-term fluctuations in the coal market to minimise economic, social and safety risks. The two functions are related. If the coal price is stable, economic and social risks are low. However, even if the price is steady, coal mine closures, unemployment, insolvency and other economic and social risks still exist. In addition, safety should be prioritised.

6.3.4 Controlling Administrative Regulation by Market and Legal Means

Administrative regulation focuses on short- and medium-term problem-solving and tends to pursue short-term performance at the cost of long-term objectives. The root cause of this is that administrative regulation is not well controlled. Government directly intervenes in the production and operation of coal companies, crossing the line between government and the market. But the status of the coal industry has changed. It has opened up. Government regulation of the coal market should also change in alignment with ongoing coal industry reform, replacing administrative regulation with market and law-based regulation.

To control administrative regulation successfully requires a sophisticated coal industry system to be in place. Only with complete standards and laws can the government's use of market and legal regulation be fully developed. So, to make the most of market and law-based regulation, the coal market system must be improved continuously.

6.3.5 The Coal Market System Should Be Developed at Carefully Defined Points

China's coal market made great progress after 2010. Once dual coal prices were abolished, the pricing mechanism became more market-oriented. Some regional and local coal markets are becoming increasingly mature. Coal transport, trading, finance and information services are becoming more and more professional. The shortage of rail capacity has improved radically. But the coal market system still has some shortcomings, such as incomplete inventory standards, noncompliant inventory management, multiple information systems and inaccurate delivery of finished products. These are problems related to coal circulation, which clearly indicate that China's coal circulation system is unsophisticated and that the necessary means to regulate circulation are not in place. A sophisticated coal market system is the precondition for government macro-regulation using market and legal means. So, the coal market system needs to be further improved to lay a foundation for institutional reform of the coal industry. The key is to select accurate breakthrough points. Inventory standards, inventory management, unified information systems (the key) and accurate delivery of finished products are four areas to be improved. Coal market information, as an important enabler of government regulation of coal circulation, includes coal trading, inventory, flow and demand, and could engage the other three areas to improve the current coal market system.

6.3.6 Adjusting the Powers and Responsibilities of Coal Mine Safety Management

Minimising safety risks is a key part of the government's macro-control. Safety management of coal mines is special. The coal mines of many large and medium-sized coal companies and groups are owned by subsidiaries. The state-owned Assets Supervision and Administration Commission administers the rights of investors in state-owned coal companies. Coal mine safety management responsibilities in the Production Safety Law conflict with those in the Company Law. According to the latter, the main responsibility of shareholders is not safety management. The direction of coal industry reform is towards marketisation and legalisation. Another priority for coal industry reform is to align the Production Safety Law with market law. Current legislation on the powers and responsibilities of coal mine safety management are not consistent with China's ongoing state-owned capital management system reform. In the long term, the powers and responsibilities of coal mine safety management need to be redefined. They could be adjusted moderately to alleviate investors' administrative management burden.

6.3.7 The Role of State-Owned Capital Is to Optimise the Coal Industry

Improving the state-owned capital management system is a priority for China's reform of state-owned enterprises. Many state-owned coal companies are now defined as state-owned capital investment corporations. This redefines asset management as capital management, and industrial asset operation as industrial capital investment. These changes will have a significant impact on the future structure of the coal industry, leading to two major reform tasks: standardisation of corporate governance and the introduction of mixed ownership. Adjusting the powers and responsibilities of coal mine safety management is one way to standardise corporate governance. Mixed ownership reform and the flow of state-owned capital are complementary. In the future, the state-owned capital of state-owned coal companies will optimise and improve coal industry operations.

6.4 Policy Suggestions for Deepening Coal Industry Reform

6.4.1 Optimise Macro-Regulatory Institutions

China's reform of the coal industry relies on effecting change in government functions. Only through administrative streamlining and decentralisation, a combination of deregulation and regulation, and service optimisation, can government authorities achieve the institutional objectives of reform. Government functions should focus on guidance and services, which are closely related to the basic functions of the coal market.

Guidance can be used in two ways: to guide government regulators on the strategic frontline of ensuring economic growth; and to help optimise the coal market by constraining and stimulating the regulatory system. The market, which is a bridge between upstream and downstream, should guide the transformation of coal production and consumption.

Service has several purposes: to help government regulators integrate decision-making and public services into a macro-control platform for the coal market; to facilitate macro-control by the central government; and to ensure that production and consumption develop harmoniously and that circulated resources are shared.

Related policy suggestions include: (i) harnessing the promising opportunities unearthed by the reform of party and state institutions to reorganise government's macro-regulatory authority and optimise its functions; (ii) enhancing the government's macro-strategies for the energy industry (including coal), and the functions of monitoring, forecasting, early warning and law-related regulation of the coal market; and (iii) reducing administrative micro-management.

6.4.2 Improve the Regulatory System

The first challenge for China's coal industry reform is that the regulatory system is not sufficiently sophisticated. Building a coal circulation regulation system in line with China's national characteristics is an important area for institutional innovation.

Coal circulation regulations should be aligned with a regulatory system that combines centralisation and decentralisation, encourages central-local collaboration, and strengthens self-regulation. Multiple authorities, at different levels and covering different types of participant—including regulatory and other government authorities, central and local government, administrations and associations—should take part.

The regulatory regime should cover various functions of the coal market, including coal quality, commercial fields, logistics, capital, and information flow. Multi-functional supervision is limited rather than universal. This means that it should regulate key processes such as quality standards, environmental protection, contractual performance, credit supervision, risk supervision, and coal mine safety supervision.

Coal mine safety regulations need to be aligned with the reform of the state-owned capital management system and should take advantage of IT and intelligent technology. Regulators should consider redrawing the boundary of corporate governance to relieve shareholders and investors of the burden of administrative management.

6.4.3 Build a Modern Market Information Network

Institutional reform of the coal industry is impossible without the support of IT and communications technologies. The purpose of building a coal market information network is very clear: it is to create a coal market macro-control platform to provide data for government regulation. Suggestions are outlined below.

First, a data system that integrates production data and capacity regulation information should be built. It should pool production data from different sectors and form a 3D real-time capacity data network, providing technical support and scientific references for capacity governance by government. The transition from a data system in which designated companies upload their data at intervals, to one in which business and regional data is transferred in real time, will enable the data network to spot companies forging data or information in a timely fashion. Moreover, the common practice in developed economies of penalising companies that forge information should be followed. Social groups and individuals should be incentivised to engage in capacity supervision, and safe whistle-blowing channels should be provided to supplement governance.

Second, coal circulation should become increasingly intelligent to ensure logistics and processing are visible, important coal market information is collected, and key performance indicators of the regulatory system are measured.

Third, intelligent development should be aligned with the introduction of integrated coal logistics parks, and centralised real-time data acquisition points should be set up at different nodes in the coal circulation system.

Fourth, vertical and horizontal barriers in administration should be eliminated, and the data interfaces of the regulatory platform should enable data to be gathered from different administrations and local government authorities.

Fifth, coal industry associations and big data companies should collaborate to create an intelligent coal market information network that uses big data to make coal circulation regulation more market-oriented.

6.4.4 Accelerate Coal Circulation Reform

Reform of China's coal circulation lasted more than 30 years, achieving great outcomes. But coal circulation reform was not carried out systematically, except in some areas, such as coal pricing mechanism reform, coal market system building and coal regulatory reform.

Under the new normal of slower economic growth, the coal market has experienced fundamental change. Coal circulation reform was faced with external and internal pressures. One of the objectives of regulatory innovation is to speed up reform, including the reform of coal circulation. The regulatory system acts on the external part of the industry. The internal impetus for coal industry reform should be developed. Given the fundamental role of the coal industry in the economy and society, central government should focus on coal circulation reform. Suggestions are outlined below.

First, the optimisation and transformation of coal supply in China should be coordinated with air pollution prevention, the energy demand revolution and the development of traditional industries like steel in order to study coal circulation reform systematically. Details of the reform should reflect the global nature of coal circulation, including logistics and the flow of capital and information. The objective of coal circulation regulation reform can only be turned into actions through design excellence and systematic implementation.

Second, the parties involved in circulation should be encouraged to innovate. Reform of the coal industry needs to stimulate the industry's development. This should be in the form of technical and methodological innovation, including clean coal processing technologies, logistics, process visualisation, information integration and new circulation methods. The launch of new pilot projects should be considered.

6.4.5 Deepen the Reform of State-Owned Coal Companies

China should seize the policy opportunities of supply-side reform and state-owned capital management reform in order to speed up the optimisation of capital stock and eliminate inefficient capacity. At the same time, efforts should be made to promote the two-way flow of capital, coordinate exit and entry into the market, optimise the internal structure of the coal industry, improve the operations of coal companies, and accelerate innovation and transition through reform.

Staff and debt are the two main barriers to exiting inefficient capital stock in traditional enterprises. The challenge of repositioning staff can be addressed, but the debt barrier is more

difficult to surmount. Debt is the priority when it comes to exiting inefficient capital stock. The policies to reduce debt and make short-term debt repayments focus on debt-for-equity swaps. This shows that exiting inefficient capital stock can be combined with swapping debt for equity.

The current challenges include: (i) Inefficient capital that needs to be exited cannot rely on debt-for-equity swaps. In the past, assets undergoing a debt-for-equity swap could rapidly enter a new round of the profit cycle, but inefficient assets with high debt now have little market value. Financial institutions typically issue a debt-for-equity swap based on profit expectations for assets with potential profitability, not for inefficient assets without any prospect of profitable development; (ii) the entity that issues the debt-for-equity swap is the parent company of a group, not the subsidiary operating the inefficient asset. In the early 21st century, a group would have relatively undiversified businesses, and the group would be the principal debtor. The case is different now: a large group typically has a diversified business portfolio, and the debtor of an inefficient asset is usually the subsidiary that operates the asset. If the debt-for-equity swap is carried out within the group, it does not solve the challenge of exiting the debts of inefficient assets. Suggestions are outlined below.

First, the purpose of debt-for-equity swaps should be made clear—it is to spin off inefficient assets and lower the asset to debt ratio of companies, and, more importantly, to improve the operational efficiency of the real economy.

Second, an inefficient asset disposal platform should be established. Inefficient assets can be packaged and optimised or reorganised by business group or region to facilitate consistent implementation of the debt-for-equity swap policy.

Third, exit channels in financial institutions should be innovated. The financial policies for cutting overcapacity can be leveraged, and the opportunities to reform state-owned capital investment and introduce mixed ownership can be used to innovate financial capital exit channels in a market-oriented manner.

7 Market-Oriented Reform of the Power Industry

7.1 Review of International Experience

Vivid Economics, a UK think tank, and a Shell International team of experts summarised the global market-oriented reform of the power industry and made suggestions for the future reform of China's power industry.

7.1.1 The Evolution and Effects of Power Market Deregulation

China and the World: Economic Transformation through Structural Reform, prepared by Dr. Mallika Ishwaran, Senior Economist, Shell International, for the 2017 China Development Forum, summarises the reform of the global power market. The paper points out that power market reform has evolved over several decades. Liberalised power markets, such as those of the USA and UK, were initially designed to meet the dual objectives of energy affordability and energy security (the energy dilemma). Starting with the UK in the 1990s, many countries have liberalised their power markets, based on standard market design criteria involving privatisation, competitive wholesale power generation and retail power supply markets, fair third-party access to transmission and distribution networks, and unbundling the ownership of transmission and distribution networks from power generation and retail supply. Liberalisation has been premised on encouraging competition—and competitive pricing—as a way to reduce electricity prices and develop diversity in supply sources to ensure supply security (Fig. 16).[16]

The first phase of power market liberalisation started in 1990 with the UK. This entailed unbundling power generation and retailing activities from the natural monopoly of

transmission and distribution. In parallel, a process of privatisation also occurred, particularly in the competitive segments of the supply chain. This form of liberalisation spread to Norway, Chile, Argentina, New Zealand and Australia in 1991, and began to spread across the USA from California in 1994. The European Commission published directives in 1996 that encouraged more countries across Europe to liberalise.

Following the 1990s liberalisation, notions of a standard market design began to coalesce around maximising the role of competition, rather than around government involvement, in order to optimise dispatch and investment in power generation and retail. Whereas the costs of production for power generation and the supply of electricity to the consumer could be minimised through competition, grid infrastructure, as a natural monopoly, was best delivered through the regulation of a single company.

Phase 2 of power market liberalisation came in the 2000s. This second phase of reform began with increasing focus on regional integration, harmonisation and environmental objectives. The European Commission and US Federal Energy Regulatory Commission began to promote standard market designs by attempting to prescribe best practice and facilitate trade across borders. The challenge of addressing climate change has meant that the energy dilemma facing policymakers has evolved into an energy trilemma—delivering affordable, secure and decarbonised electricity. Current liberalised power markets, which were designed to deliver energy dilemma objectives, have struggled to cope. The novelty and high initial cost of renewables has required generous government subsidies (undermining affordability) and their contrasting economics with fossil fuel generation has reduced the ability of the wholesale power market to incentivise investment and balance demand and supply (undermining security of supply).

Consequently, power markets across the world are in a state of flux, as countries have started decarbonising their power supply and the levels of renewables penetration have increased. Germany, which has led the way in renewables adoption, has nonetheless struggled to decarbonise energy

---

[16]Structural Power Market Reform to Support a Decarbonised Economy, March 2017. Prepared by Dr. Mallika Ishwaran, Senior Economist, Shell International, for the 2017 China Development Forum on China and the World: Economic Transformation through Structural Reform.

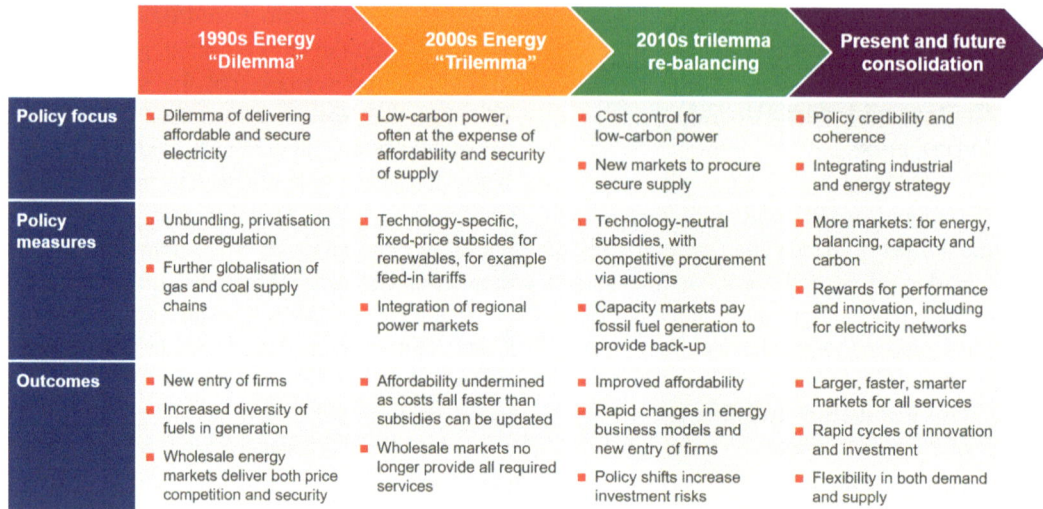

Fig. 16 The evolution of power market policy focus, measures and outcomes. *Source* Structural Power Market Reform to Support a Decarbonised Economy, Mallika Ishwaran, Shell International, March 2017

supply, because coal plants remain a low-cost source of secure supply.

The UK implemented phase 1 of its power market reform in the mid-2000s but has had to rapidly update policy to bring costs under control and maintain sufficient supply. This has created uncertainty for the up to £100 billion of planned investment in power generation from 2015 to 2020 (60% of UK infrastructure spending in the period).

In recent years, innovation in market reform has begun to address the challenges of balancing the energy trilemma. These have focused on auctioning renewable subsidies and capacity payments to control costs and on incentivising new solutions to provide flexibility and improve network management.

Further reform will be needed to deliver a net zero emissions power system. The supply of flexible power is not yet sufficient, and demand for flexibility will increase as more energy end-uses are electrified, such as transport, which will likely increase peak loads. Currently, options such as batteries require multiple revenue streams to be economically viable, requiring changes to how value is distributed across the electricity supply chain. At the same time, the manufacture of renewable power equipment and

the addition of a technology layer on top of energy assets, such as with smart meters and the Internet of things, links the transition to a net zero emissions energy system with broader industrial strategy aims.

The efficient coordination of an increasingly complex net zero emissions power system will be most efficiently delivered through greater use of markets. However, due to the challenges of low-carbon power, the appropriate market structures will be different from the historical standard market design.[17]

7.1.2 Overcome the Market Challenges of Renewable Power

Another important task during global energy system reform is to overcome the market challenges of renewable power. Variable renewables, such as solar and wind, are a crucial part of a net zero emissions power system, but they have characteristics that undermine the efficient operation of liberalised power markets as currently configured.

---

[17]Structural Power Market Reform to Support A Decarbonised Economy, March 2017. Prepared by Dr. Mallika Ishwaran, Senior Economist, Shell International, for the 2017 China Development Forum on China and the World: Economic Transformation through Structural Reform.

Variability. A key feature of many low-carbon power technologies is that they are variable generators. This means that they generate on an as-available basis rather than on demand, i.e. they require the sun to be shining or the wind to be blowing in order to generate power. Flexible power generation and/or storage capacity is required to make up for this variability and to accommodate renewables into the power grid in a way that maintains security of supply. Furthermore, variable renewables, such as wind and solar, may frequently deviate from forecast levels even when the sun is shining and the wind is blowing, requiring greater short-term balancing between electricity demand and supply to correct for forecast errors. Power systems are already designed to manage variability. However, with greater renewables penetration, the demand for more frequent and greater levels of balancing are also likely to increase.

Low operating costs. Variable renewables, such as wind and solar, have zero or a very low marginal cost of operation once they have been built. This, combined with their variable nature (or intermittency), is already impacting the efficiency of existing liberalised wholesale power markets and the ability of these markets to deliver long-term investment in capacity. Specifically, renewable energy generators, bidding in the wholesale power market at their operating (or marginal) costs, have lowered average power prices. At the lower average price, conventional plants have found it difficult to recover their costs, and additional incentives have been required to invest in long-term capacity necessary to manage intermittency. This has reduced the ability of wholesale power markets to deliver energy security alongside the other two trilemma objectives.

Locational constraints. Renewables tend to be less flexible in terms of their location compared to conventional (fossil-based) power plants, requiring investment in grid expansion to accommodate and integrate them into the power system. They can be located far from demand centres, with optimal sites for wind and solar only coincidentally located close to urban areas and with network infrastructure required to connect supply with demand. Small-scale renewables, such as roof-top solar, may sell power into local distribution networks, but network investment is required to allow these new two-way flows of power without congestion.

Table 6 summarises the evolution of liberalised power markets and the new market signals necessary to accommodate a significant increase in renewables' share of the power mix.

7.1.3 Build Efficient Net Zero Emissions Power Markets

Most countries are likely to iterate towards efficient power market structures and policies based on the cost reductions and efficiency gains they are likely to provide, both in terms of cheaper investment and electricity. These benefits are only likely to increase with greater electrification

Table 6 Additional power market incentives required in a net zero emissions world

Characteristics of an NZE electricity system	Required signals
Low operating costs – renewables bid low in energy market, reducing prices	Revenue stream to invest in capacity
Variability – output changes and is not necessarily available on demand	Strong scarcity signals with high temporal resolution to incentivise flexibility
Locational constraints – generation may not be near demand	Strong locational signals; sufficient grid investment

Note NZE = net zero emissions
Source Structural Power Market Reform to Support a Decarbonised Economy

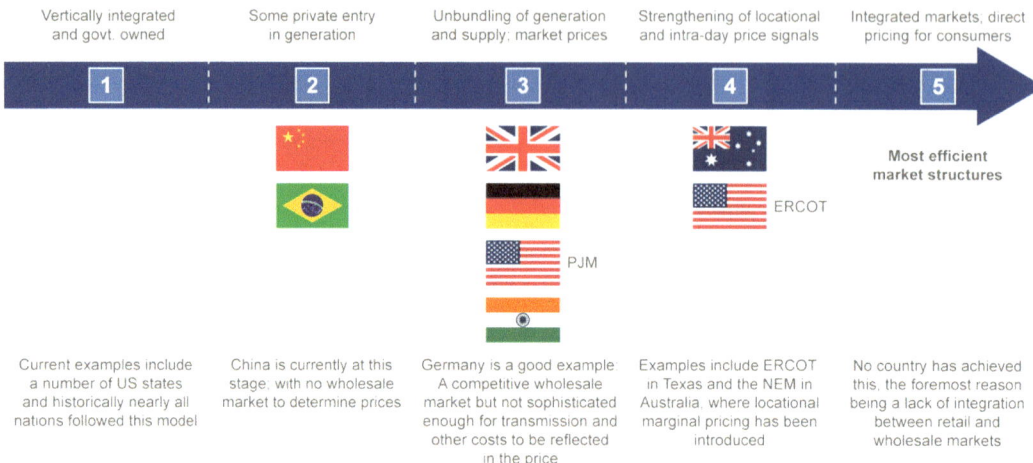

Fig. 17 Progress towards efficient power market structures. *Source* Vivid Economics

of energy use and decarbonisation of electricity supply.

Progress towards efficient net zero emissions power markets will depend on a country's circumstances, such as its level of economic development and its domestic energy resource endowments. Looking at a range of seven countries across this spectrum (Australia, Brazil, China, Germany, India, England and the USA), most are making progress. For example, unbundling along the supply chain to increase competition and using market prices (Fig. 17). However, currently these price signals are not very sophisticated. Only in Australia and the Electric Reliability Council of Texas (ERCOT) in the USA do prices provide strong locational and temporal signals, and in no market are consumers exposed to real-time prices. China and Brazil are the furthest behind, with limited competition between state-owned enterprises.

In terms of progress towards decarbonisation, government action, rather than market competition between fuels, currently drives decarbonisation (Fig. 18). Governments tend to decide the level of renewables by fixing the level of available subsidies, but then look to allocate these subsidies competitively. The most efficient pathway for decarbonising the power sector is for the level of renewables to be determined through

a strong carbon price and ensuing market competition. Only the UK gets close, with its contracts for difference and carbon price floor. Germany and China, while delivering high levels of decarbonisation, are doing so through strong government involvement and at a high cost.

Efficient management of the power grid is currently delivered according to regulatory directives, rather than incentives (Fig. 19). Arrangements for grid management vary considerably across countries, with none basing investment on economic incentives. Instead, all seven countries regulate grid provision, with varying degrees of oversight. The UK and Germany are closest to being efficient, with performance-based regulation that rewards efficiency. At the other end of the spectrum, Brazil, India, and China reward grid provision on a cost-plus basis, which gives little incentive for reducing costs. Australia and PJM and ERCOT in the USA are in the middle of this spectrum, with independent oversight and benchmarking of cost estimates, which provide indirect pressure for efficient grid management.

Finally, policy credibility depends on the broader political, institutional and public environment. Even with efficient power market structures and policies, the ability to attract investment depends on policy risk and whether

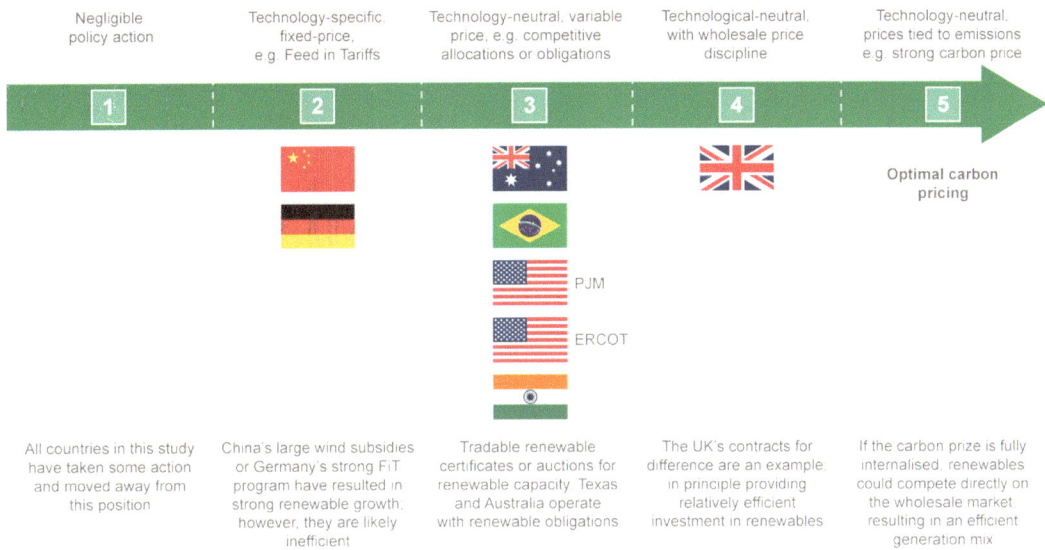

Fig. 18 Progress towards carbon pricing and market competition between fuel sources. *Source* Vivid Economics

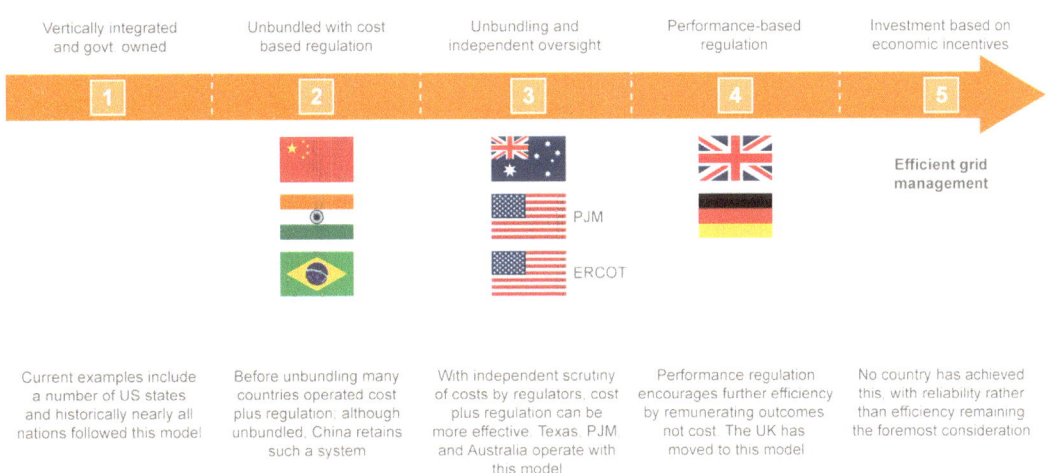

Fig. 19 Progress towards efficient management of the power grid. *Source* Vivid Economics

there is an expectation that particular arrangements will persist, or at least change predictably. If policy is seen as credible, this will minimise the perceived risk of investment. Credibility is a quality of the policy environment, which depends on institutional capacity, political predictability, public support and the coordination of objectives across power market services. Governments can also take additional measures to increase credibility and reduce investment

risk, such as passing primary legislation (as the UK did with the 2008 Climate Change Act), writing legal contracts, providing guarantees and releasing information on the performance of first-of-a-kind investments.

The pressure on policymakers to ensure that power markets function efficiently in delivering affordable, reliable and low-carbon electricity is only going to increase with higher levels of renewables penetration and greater reliance on electricity as an energy carrier. From a policy-maker's perspective, there are two key characteristics of well-functioning and efficient power markets in a net zero emissions world[18]:

- market structures that provide efficient price signals, both for near-term electricity dispatch and for long-term investment in the sector; and
- energy and environmental policies that correct for market failures and support the efficient operation of power markets.

These two factors are necessary for the efficient operation of net zero emissions power markets, without which meeting the energy trilemma objectives becomes significantly more challenging and expensive. For example, in an environment of rapidly declining technology costs, markets are better placed than policymakers to competitively and efficiently procure the necessary investments. However, policymakers have a fundamental role in providing the necessary regulatory and policy frameworks to underpin the efficient operation of these markets.

7.1.4 Market Structures for Providing Efficient Price Signals

The price signal from wholesale and retail power markets for conventional generation covers the range of energy provision services: supply of electricity, investment in generation capacity, and balance between the demand and supply of electricity. Variable renewables only provide energy and do so intermittently. This increases the importance of capacity and balancing, with markets potentially being required for these services as they are no longer provided as part of the bundle that has traditionally been provided by a single generation asset and paid for through wholesale and retail electricity prices.

The three requirements for efficient price signals are discussed below (see also the left-hand side of Fig. 20).

Efficient electricity markets. Through continued liberalisation of wholesale and retail power markets (including spatially larger and integrated markets that maximise the diversity of supply) to increase competition and security of supply, and through greater time-of-use and locational pricing with shorter settlement periods to increase the strength and specificity of the price signal for efficient electricity dispatch.

Investment in capacity. Through a technology-neutral capacity market (including both supply and demand measures and sufficient penalties to ensure delivery of contracted capacity), where efficient price signals for new investment in capacity are generated through competition among providers of this capacity. By themselves, short-term price signals from wholesale and retail markets are not likely to be enough to incentivise investment in long-term capacity at high levels of renewable energy penetration, thus requiring capacity markets.

Competitive balancing services. Short-term, grid-specific balancing and flexibility services may also be required to accommodate renewables. These would be required to supplement the price signal from wholesale and retail markets, which may be constrained by political economy considerations to protect end users from unexpected price spikes or due to forecast errors. Procuring these services competitively will provide them at the least possible cost and encourage the innovative use of new technologies, such as batteries and technologies that enable demand-side response. The need for such a market could decline as the time required to dispatch from a power plant decreases, and wholesale and retail power markets become more instantaneous.

7.1.5 Energy and Environmental Policies that Correct for Market Failures

Efficient policy approaches and frameworks to correct for market failures are also needed to support the efficient operation of power markets. Three key characteristics of the required policy framework are described below (see also the right-hand side of Fig. 20).

---

[18]Structural Power Market Reform to Support A Decarbonised Economy.

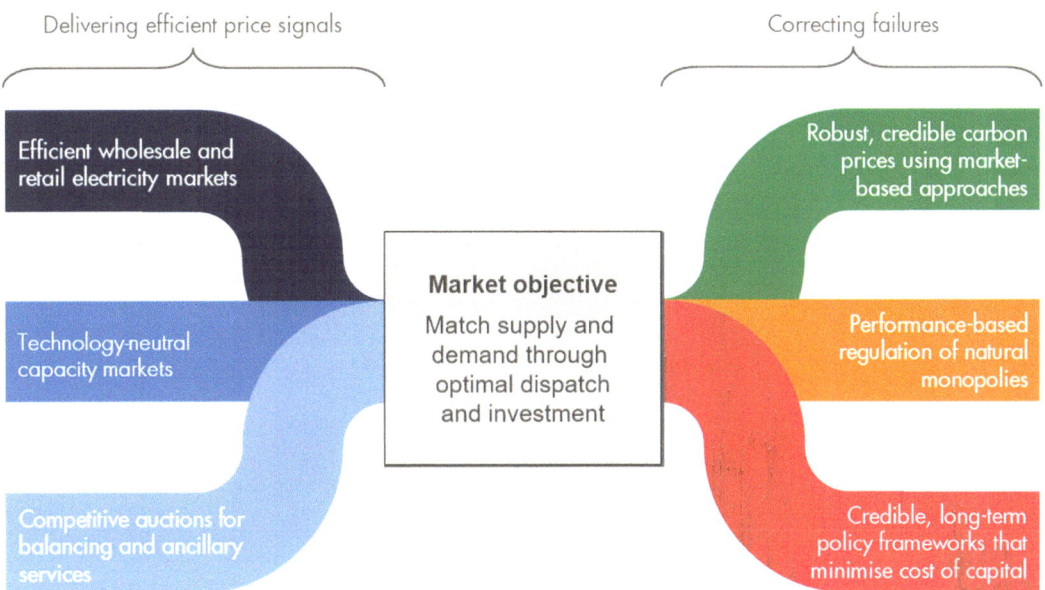

Delivering efficient price signals

Efficient wholesale and retail electricity markets

Technology-neutral capacity markets

Competitive auctions for balancing and ancillary services

Market objective

Match supply and demand through optimal dispatch and investment

Correcting failures

Robust, credible carbon prices using market-based approaches

Performance-based regulation of natural monopolies

Credible, long-term policy frameworks that minimise cost of capital

Fig. 20 Efficient power market structures and policies for a net zero emissions world

- The optimal way to incentivise low-carbon generation or encourage energy efficiency is through a robust and credible carbon price, preferably through market-based approaches such as a carbon tax or emissions trading. In its absence, there is likely to be a proliferation of a patchwork of more expensive and less effective policies, such as renewable energy subsidies and mandated performance standards.

- The natural monopoly of power transmission and distribution networks is most efficiently dealt with through independent ownership by a single company whose performance is regulated by an independent regulator. While this is a well-developed area of regulatory economics and policy, it will need to evolve to accommodate consumers that both demand and supply electricity; create flexible infrastructure with natural monopoly characteristics, such as large-scale storage and interconnectors; provide the returns needed for investment in grid extension and

expansion to integrate renewables; and grant incentives for innovation in smarter grids.

- Maintaining a stable, predictable and credible policy framework across the range of energy and environmental policies that affect power markets is essential in order to minimise the cost of capital and incentivise the necessary long-term investments in the sector. Minimising policy risk is even more critical on the path to a net zero emissions world, given the long-term, large-scale and capital-intensive nature of the required investments. For example, the recent cycle of iterative power sector reform and learning has unsettled investors, even as the technology has matured to the point of mass deployment. A balance should be struck between the need for policymakers to respond to new circumstances and the need for stable revenues over long asset lifetimes, for example by making policies predictable, with changes announced well in advance and conditional on specified changes in circumstance.

Capacity market
- Supplies capacity to produce electrons, priced per MW
- Drives investment decisions
- Determines the mix of technologies providing power in the long term (years)

Dispatch market
- Supplies electrons, priced per MW-hour
- Drives consumption decisions
- Determines the mix of technologies providing power in the short term (day to day)

Balancing market
- Supplies extra supply and demand at short notice so that supply and demand always match
- Also supplies ancillary services to maintain grid frequency

Fig. 21 The power system comprises three markets, which can be managed through central control or market mechanisms. *Source* Low Carbon Power Markets: Lessons from International Experience, prepared by Vivid Economics for the DRC-Shell Markets Work Stream, April 2017

7.2   Use of Central Control or Market Mechanisms in Seven Countries[19]

Vivid Economics prepared a report summarising international experience in power market reform for the DRC-Shell Markets Work Stream. The report describes the choices made in seven countries on managing their power systems by central control or market mechanisms. Management of power systems in Australia, Brazil, China, Germany, India, the UK and the USA (PJM, covering 13 states in the north-east, and ERCOT in Texas) are compared, based on the framework outlined in Fig. 21 and the major market features listed in Fig. 22.

Power market structure. Most countries are slowly moving away from central control to market mechanisms, as shown in Fig. 22. This trend is expected to accelerate with the increasing penetration of renewables. While no country has achieved completely liberalised and integrated markets, most countries have been evolving in this direction over the past three decades. We

observe three main archetypes of market structure:

- centrally controlled power markets with limited entry in generation (China, Brazil). China currently belongs to this group with no competitive wholesale market;
- unbundling of generation and supply; wholesale market prices (the UK, PJM in the USA, Germany and India). Germany is a good example of this: the country has a competitive wholesale market, but it is not sophisticated enough to reflect adequately transmission and other costs in market prices. Therefore, some government intervention is necessary; and
- increased market liberalisation and strengthening locational and price signals. In rare cases, for example ERCOT in Texas and NEM in Australia, locational marginal pricing has been introduced.

Market power. In order to ensure efficient investment and operation of grid infrastructure (natural monopolies), most countries currently use direct regulation, rather than incentive-based instruments, as shown in Fig. 23. For grids, options for regulation can be plotted along a spectrum, from government-owned and operated

---

[19]This part is prepared by Vivid Economics. See Low Carbon Power Markets: Lessons from International Experience prepared by Vivid Economics for DRC-Shell Markets Work Stream, April 2017.

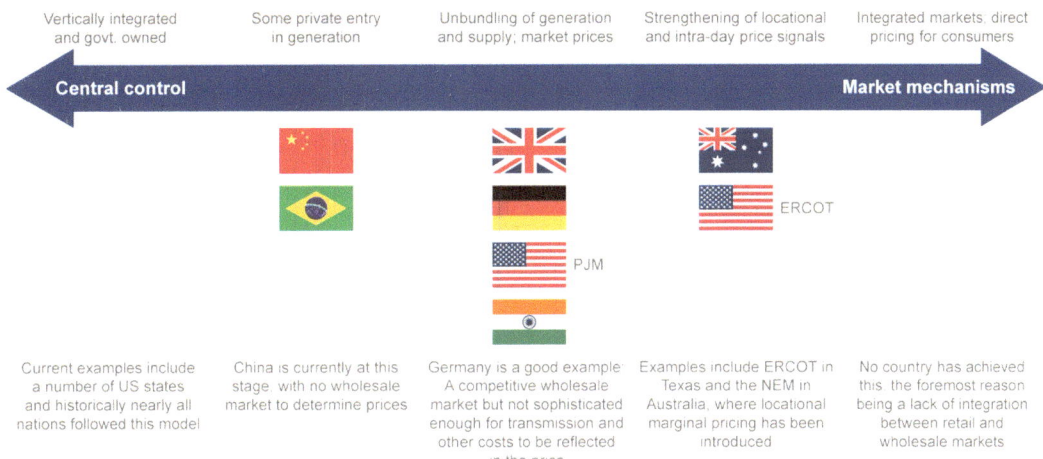

Fig. 22 Most countries are slowly moving from central to market mechanisms; this is expected to accelerate with decarbonisation. *Source* Low Carbon Power Markets: Lessons from International Experience, prepared by Vivid Economics for the DRC-Shell Markets Work Stream, April 2017

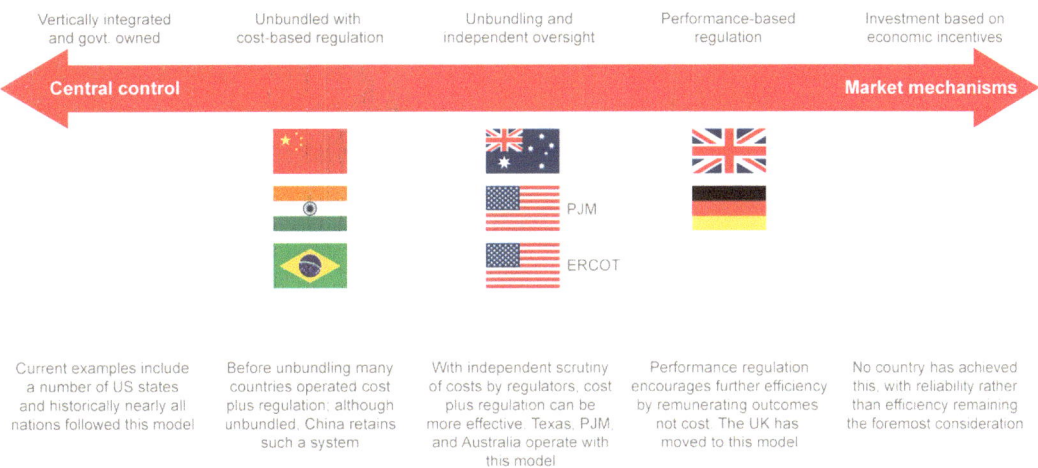

Fig. 23 Market power is mainly managed through regulation, rather than incentives, especially the power grid. *Source* Low Carbon Power Markets: Lessons from International Experience, prepared by Vivid Economics for the DRC-Shell Markets Work Stream, April 2017

to an unbundled sector with an independent service operator, where the government regulates the cost of service provision. To avoid moral hazard, performance- or incentive-based regulation can be introduced. However, as Fig. 23 suggests, the most advanced countries on the liberalisation spectrum (Germany and the UK) have opted for performance-based regulation, such as the UK price caps. China is at the other end of the spectrum—although the Chinese

sector is unbundled, the country still operates cost-plus regulation.

Decarbonisation. The move to lower emissions is currently driven by central control, rather than by market mechanisms, as shown in Fig. 24. Given political constraints to the roll-out of carbon pricing, other options that preserve competition while directing investment to cleaner generation have been introduced. On the more liberalised part of the spectrum, those options

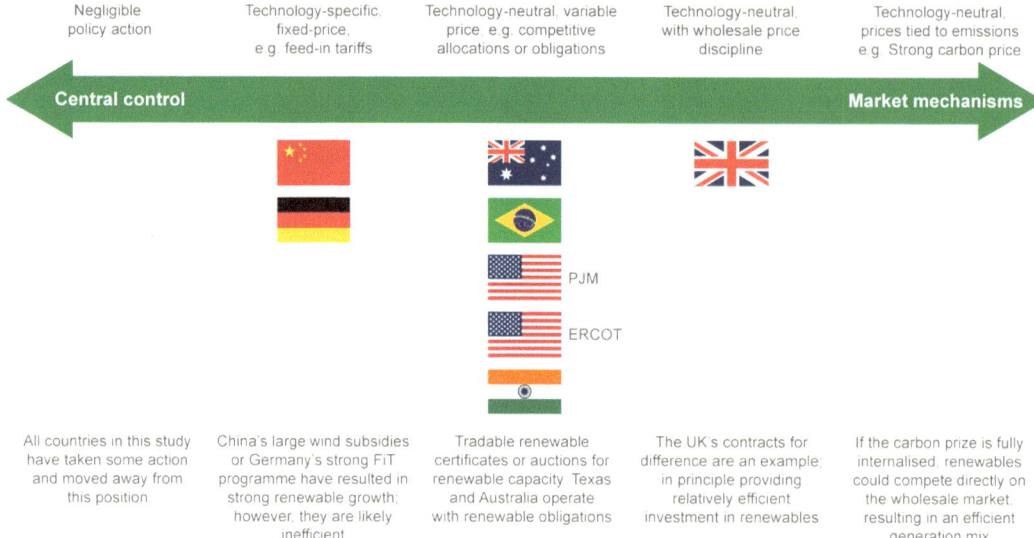

Fig. 24 Decarbonisation in most countries is driven by central control, rather than by market mechanisms. *Source* Low Carbon Power Markets: Lessons from International Experience, prepared by Vivid Economics for the DRC-Shell Markets Work Stream, April 2017

include auctions for low-carbon electricity generation contracts, such as those currently in operation in Brazil and the UK. At the other end of the spectrum, China has adopted the more traditional way, which involves incentivising renewable investment through targeted fixed price measures, such as subsidies.

Policy credibility depends on the broader political, institutional and public environment. Even in a liberalised market, policy is important as it sets the market arrangements. The ability to attract investment depends on policy risk and whether there is an expectation that particular arrangements will persist, or at least change predictably. If policy is seen as credible then the perceived risk of investment will be minimised. Credibility does not have a scale from central control to market mechanisms but is a general quality of the policy environment. It depends on institutional capacity, political predictability, public support and the coordination of objectives across electricity market services. Governments can also take additional measures to increase credibility and reduce investment risk, such as passing primary legislation, writing legal contracts, providing guarantees and releasing

information on the performance of first-of-a-kind investments.

7.3 Global Trends of Power System Transformation

More and more countries have shifted the focus of energy development to clean energy in recent years, accelerating the global development of renewables and further optimising the energy consumption mix. Green, low-carbon, clean and efficient energy development has become a major trend and many countries have taken new measures regarding the market system, market mechanisms and regulation policies.

7.3.1 Adapt to the Energy Transition and Innovate a New Market System and Mechanisms

Increasing the share of renewables in the energy mix and deepening the energy transition imposes new challenges on the conventional power market system. In many countries, the power market system is already optimised and the types of energy traded is comprehensive. Those systems are better able to achieve the objectives of allocating international resources efficiently, making the power market competitive, channelling investment into new and low-carbon power generation, meeting consumer demand, and facilitating the use of new technologies.

The European Union published the New Energy Market Design report in 2015, according to which a new power market mechanism to create the European Energy Union will be built. It includes developing an international short- and long-term power market capable of attracting investment, perfecting a market mechanism suitable for renewables, improving the transnational capacity mechanism and strengthening the coordination between wholesale and retail markets. Regional cooperation will be strengthened, an integrated power market will be built and the energy supply security of the whole European Union will be achieved.

To achieve more flexible and efficient power supply and ensure power security, Germany passed the Electricity Market Act in 2016 to help build Electricity Market 2.0. The act focuses on renewables. It strengthens the procedure for adjusting the price signal, expands the balancing market, encourages supply-side management, and builds a fair power transmission tariff mechanism and capacity standby mechanism.

Great Britain launched a new round of power market-oriented reform focusing on low-carbon development in 2014 by holding the first contract for difference and capacity market auctions and amending the relevant rules and trading mechanism. California improved its real-time market, replacing the hour-ahead market with the 15-min market, and launched the energy imbalance market to adapt to the output of renewables.

7.3.2 Build Large Resource Application Platforms by Expanding the Market

To promote the development of renewables, diversify energy supply and ensure energy security, transregional and transnational resource allocation is urgently needed. The scope of the power market has expanded again and again, as with the Southwest Power Pool (SPP) in the USA.

Western Area Power Administration, Basin Electric Power Cooperative and Heartland joined the SPP in 2015, and power transaction volumes between the USA and Canada increased. EU member states have continued to strengthen regional cooperation on energy. Latvia, Estonia and Lithuania signed the Baltic Energy Market Interconnection Plan in 2015. The five major power exchanges in Europe signed an agreement to develop a unified European transnational intra-day trading platform, the purpose of which is to optimise the intra-day transnational and transregional trading plan and promote an integrated European power market. Sixteen grid operators in Europe announced an integrated day-ahead market for grids in central and western Europe, and in central and eastern Europe. Six countries in the western Balkans signed a memorandum of understanding on regional power market cooperation. A power market trading platform has started to facilitate the integration of the power markets and grids of countries in the Union for the Mediterranean. The European Union and Turkey strengthened their cooperation on energy market integration and renewables development.

7.3.3 Pay Close Attention to the Construction of Energy Infrastructure that Helps Create Large Power Markets

With the expansion of the power market and the large-scale development and use of clean energy, many countries are investing in the development or modernisation of transnational grids to support wide power transactions and optimise resource allocation:

- Grain Belt Express, a cross-state high-voltage transmission project in the USA, was approved by several states. The project, costing $2 billion and 1,255 km in length, aims to integrate and transmit the wind power resources of Kansas;
- the European Union announced that transnational transmission capacity of each member state will account for at least 10% of its installed capacity by 2020, and 15% by 2030;

- transmission lines connecting the Iberian Peninsula, Baltic Sea region, Ireland and Great Britain will be built. To this end, the EU set up the South-west Europe High-level Working Party to encourage Spain, France and Portugal to construct energy infrastructure through technical support and regular monitoring, and to strengthen interconnections between the peninsula and other parts of the EU;
- two new transmission lines connecting Lithuania, Poland and Sweden are under construction to interconnect the Baltic states with Sweden and Poland for the first time;
- the EuroAsia Interconnector, a high-voltage transmission project that will connect the power grids of Israel, Greece and Cyprus with continental Europe, is under construction;
- to adapt to the rapid development of natural gas and wind power generation, Australia continues to invest heavily in cross-regional interconnections. As estimated by the Australian Energy Market Operator, Australia needs to invest AUD 24 billion in the transmission network by 2030 and AUD 120 billion in the power distribution network; and
- Russia has promoted the construction of an energy bridge with North Korea, while accelerating connection to the grids of surrounding countries like Azerbaijan, Belarus, Estonia, Georgia, Kazakhstan, Latvia, Lithuania, Mongolia and Ukraine.

7.3.4 Build a Strong Regulatory Mechanism to Ensure Effective Competition in the Power Market

As power market reform moved forward, countries gradually realised that strong regulation is an important condition for effective competition. Increasing the powers of the regulators and improving the content and methods of regulation has been an important trend in power market reform in recent years.

- The US Federal Energy Regulatory Commission (FERC) issued policies and regulations to improve wholesale power, capacity and auxiliary service market pricing, and regulates the demand-response resources entering the market in accordance with the latest measures for demand-side management.
- As suggested by Germany's Electricity Market 2.0, the German Federal Network Agency (Bundesnetzagentur) should issue regulatory measures and publish reports on the wholesale market at least once every two years, making the regulation of market abuse and monopoly actions more open and transparent.
- An amendment to Japan's Electricity Business Act was adopted at the plenary session of the House of Councillors. The act defines the plan for phase 3 of reform and states that an independent electricity regulatory commission will be formed.
- Australia released its national energy market reform roadmap, according to which regulating investment by grid operators and determining a reasonable rate of return for them is an important part of the reform.
- FERC issued rules on market supervision, regulation and analysis and strengthened power market information disclosure. This requires regional power wholesale markets to make power sales and transmission information public within a specified time, make power market prices more transparent and ensure users pay a reasonable price.
- To promote the development of low-carbon power and provide users with affordable electricity, the UK announced a plan for a new round of reform measures that includes price support for low-carbon energy through contracts for difference and a capacity market for procuring flexibility to balance the intermittency of variable renewables like wind and solar. This should improve investment in low-carbon energy and increase capacity. The government is also considering adjusting the transmission tariff mechanism to reflect the tariff for different types of generation, including prosumers supplying distributed renewable generation to the grid and consuming electricity from the grid during times of low or no renewables generation.

7.4 An Evaluation of Progress in China's Power System Reform

7.4.1 The Problems Facing Electricity System Reform

As far as policy background is concerned, the Decisions of the Central Committee of the Communist Party of China on Major Issues Concerning Comprehensively Deepening Reforms was adopted in November 2013 at the third plenary session of the 18th Central Committee. The Decisions clearly state that "economic restructuring is still the focus of deepening reform comprehensively. Appropriate handling of the relationship between the government and the market is still the core issue of economic reform, which will enable the market to play a decisive role in allocating resources and the government to play its role better... In natural monopoly industries, in which state-owned capital continues to be the controlling shareholder, we will carry out reform focusing on the separation of government administration from enterprise management; the separation of government administration from state assets management, franchise operation and government oversight; the separation of infrastructure networks from operations; and relax control of competitive businesses based on the characteristics of different industries; and make resource allocation more market-oriented... and improve the mechanism whereby prices are determined mainly by the market. Any price that can be determined by the market must be left to the market, and the government should not improperly intervene. We will push ahead with pricing reforms of water, oil, natural gas, electricity, transport, telecommunications and other sectors, while relaxing price control in competitive areas".

At the sixth meeting of the Leading Group for Financial and Economic Affairs held in June 2014, the Four Energy Revolutions were mentioned for the first time, i.e. the revolutions in energy demand, energy supply, energy technologies and of the energy system. Regarding the revolution of the energy system, it is stated explicitly that "the commodity attributes of energy should be restored. A market structure and market system featuring effective competition, and a mechanism whereby energy price is determined mainly by the market, should be built".

Since the electricity system reform of 2002, which separated generation from transmission, the absence of many power market mechanisms has continued to hinder progress. These missing mechanisms include: a trading mechanism, the absence of which reduces resource use efficiency; a market-oriented pricing mechanism, which is missing because price relationships are not rational; government functions have not been fully defined, which prevents the regulatory system from regulating the power market effectively; nationwide and sector-wide planning and coordination mechanisms are not in place; the development and use of new energy and renewables is encountering difficulties due to an incomplete market development mechanism; law-making and amendments are lagging behind, restricting power marketisation and its healthy development; and the market credit system is developing slowly.

After observing and comparing China's power industry management system with that of OECD countries, Vivid Economics pointed out that the institutional framework for China's power industry is very complex. No agency has complete control over the power industry or the authority to coordinate other agencies' actions. Unlike regulators in OECD countries, the power regulatory function in China is exercised by the National Energy Administration (NEA), which is administered in turn by the National Development and Reform Commission. The NEA is not an independent body. A problem arises from such a management system: planning and pricing decisions at the central and local levels may be influenced by political objectives. This top-down planning approach for power grids and the lack of coordination with provincial and regional grid companies may lead to inefficient investment decisions and poor coordination between generation and transmission investments. These investment patterns create pockets of generation where electricity supply is abundant, but other areas where it is scarce, due to limited transmission capacity. As a result, while China has significant renewable generation capacity, large volumes of renewable energy are curtailed, increasing overall electricity costs and carbon emissions.[20]

According to Vivid Economics, the transition of China's energy system to a low-carbon and decentralised system is likely to exacerbate these inefficiencies and create new challenges. As the Chinese economy and energy sector decarbonises, it will be increasingly difficult to maintain a high level of grid reliability at an affordable cost. Moreover, decentralisation of electricity resources requires significant investment in smart capabilities and creates challenges for efficient planning and delivery of infrastructure across different networks.[21]

7.4.2 Differences in the New Round of Electricity System Reform

When different stakeholders reach a consensus on the problems in the power industry and the necessity of reform, differences arise regarding the reform path to be selected, mainly due to the positioning of grid companies in the power market.

One view is that the new round of power system reform should follow the same path as the last round in 2002 to complete unaccomplished reform tasks. These tasks include separating transmission from distribution, making dispatching organisations and trading institutions independent from grid companies, enabling the market to determine transmission and distribution prices, building regional power markets and creating a power spot market.

People who hold this view generally believe that integrated grid companies that operate transmission, distribution and sales hinder competition due to their control of the grid and large size, making effective regulation impossible. In

their opinion, the precondition for a competitive power market is to restructure these companies. The basic approaches to restructuring in a Chinese context are:

- separate grid assets from non-grid assets, and transmission assets from distribution assets;
- make dispatching an independent, neutral organisation regulated by government because dispatching is a public service, not a grid company asset;
- build power exchanges involving various stakeholders and explore integrating power dispatching centres with power exchanges;
- create a power market dominated by regional power markets to avoid trans-provincial barriers and provincial oligopolies; and
- build power spot markets as soon as possible, based on the principle of "no spot, no market".

Another view is that the new round of power system reform should develop a practical and effective solution to the problems facing the Chinese power industry. The people holding this view believe that power reform is a long-term and progressive process, and that radical reforms like separation and restructuring will not solve the problems but will impact supply safety and grid stability. They suggest the following reform approaches:

- maintain control of the power grid but deregulate power generation and sales: gradually introduce competition into power generation and sales to develop diversified market participants and create a fair, competitive environment;
- stick to the principle that the power grid is a natural monopoly: maintain integrated management of transmission, distribution, dispatching and trading; strengthen government regulations and provide fair and efficient network services; and
- create a unified, open, competitive and orderly power market in which resource allocation and macro-control are combined organically: forge a sound legal system,

---

[20]Vivid Economics, Electricity Grids in Transition, Final Report. Prepared for the DRC-Shell Markets Revolution Work Stream, October 2017.
[21]Ibid.

change government functions, improve the regulatory system and build a scientifically based price mechanism.

The path arrived at in this section of the report is based largely on the second approach but incorporates some of the principles of the first approach.

7.4.3 Targets and Tasks of the New Round of Power System Reform

The ZhongFa No. 9 Document of 2015 marks the launch of the new round of power system reform. It defines the targets and tasks of reform in the light of the difficulties faced in building market mechanisms for China's power industry.

According to the document, the targets and tasks of the new round of reform are as follows.

The objective of reform is to make power a commodity by building a power industry market mechanism, developing a rational pricing mechanism, deregulating competitive segments in an orderly manner, diversifying suppliers, controlling energy consumption, enhancing energy

efficiency, improving safety and reliability, ensuring fair competition, and protecting the environment.

The main tasks of reform are "three deregulations, one facilitation and three improvements". These are: deregulate (in an orderly manner) pricing in competition-based fields, except transmission and distribution; open distribution and sales to private capital in an orderly manner; lift control over power generation in an orderly manner, except peak-shaving; make power exchanges relatively independent; strengthen government regulations; improve power planning; and make power supply safe, efficient and reliable (Fig. 25).

Generally, the three deregulations, one facilitation and three improvements are all related to building power market mechanisms. The purpose of deregulating price in competition-based fields is to build the market price mechanism. The purpose of lifting control of sales and power generation is to build a market competition mechanism for generation and sales. The purpose of making power exchanges independent is to equip the market with fair and efficient trading

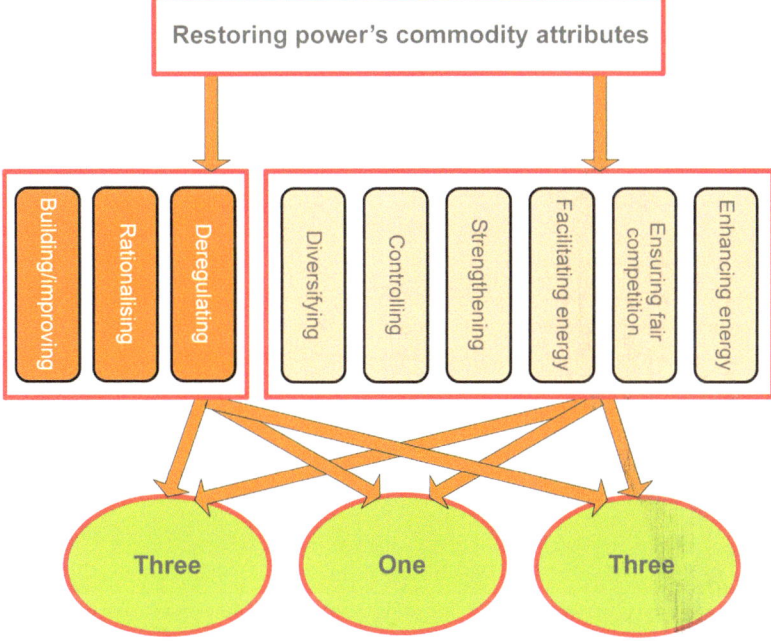

Fig. 25 Targets and tasks of the new round of power system reform

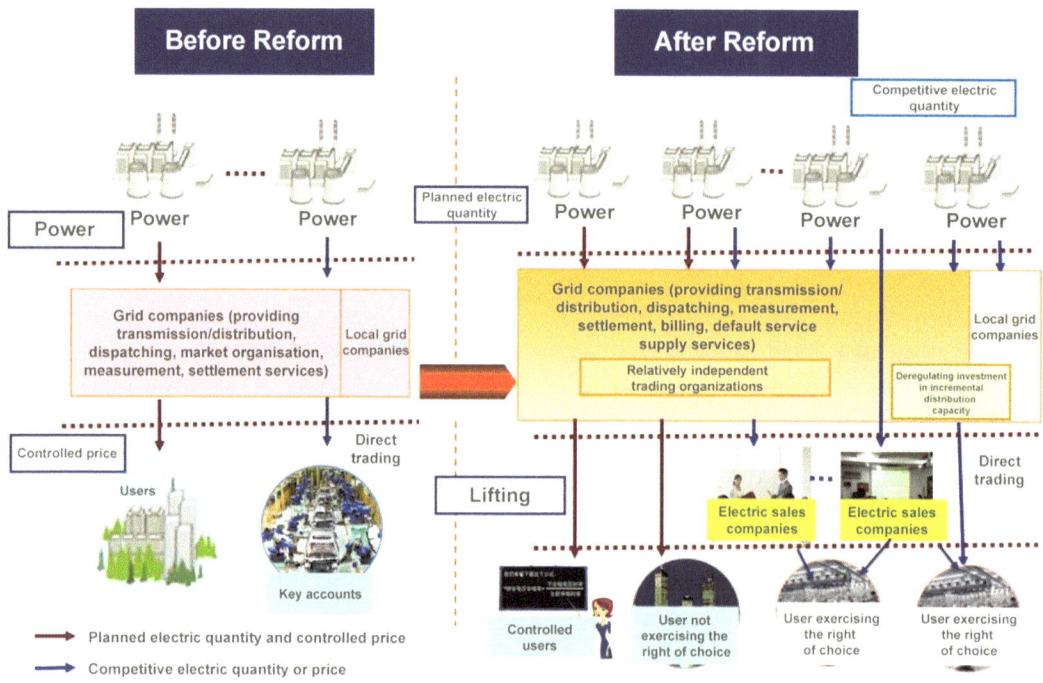

Fig. 26 Competition in the power market before and after reform

services and trading platforms. The purpose of the three improvements is to change government functions and build an effective regulatory system for the power market. The purpose of transmission and distribution price reform and of deregulating investment in distribution is to improve regulation of the power grid—the physical basis of the power market—and facilitate the development of distribution networks (Fig. 26 and Table 7).

Table 7 Measures, objects and purposes of the new round of power system reform

Measure	Object	Purpose
Deregulate price in competitive fields	Price mechanism	Build a market price mechanism
Lift control of sales and power generation	Competition mechanism	Build market competition mechanisms in generation and sales
Make power exchanges relatively independent	Trading platforms	Equip the market with fair and efficient trading services and trading platforms
Strengthen government regulation and power planning; make power supply safe, efficient and reliable	Regulation	Change government functions and build an effective regulatory system for the power market
Reform transmission and distribution pricing, deregulate investment in power distribution	Physical basis of the power market	Strengthen regulation of the power grid—the physical basis of the power market—and facilitate the development of distribution networks

7.4.4 Progress in the New Round of Power System Reform

(1) Reform policies

The ZhongFa No. 9 Document defines the targets and tasks of China's power system reform. The six main support documents published in November 2015 provide a specific working plan for the reforms. The National Development and Reform Commission and National Energy Administration are formulating detailed rules for their implementation. The Administrative Measures for Access and Withdrawal of Electricity Sale Companies and Administrative Measures for Orderly Deregulation of Power Distribution Networks were released in October 2016.

The Opinions on Promoting Transmission and Distribution Tariff Reform sets three objectives: (i) build a reasonable, scientific, transparent and independent transmission and distribution tariff system that is well regulated with clear rules; (ii) form a transmission and distribution pricing mechanism based on the principle of "allowable cost + reasonable profit"; and (iii) define government funds and cross-subsidies. The main content and requirements are shown in Tables 8 and 9.

The Opinions on Establishing Power Trading Institutions and Standardising Their Operation defines the objectives of "building a competitive market structure and market system", as required by the ZhongFa No. 9 Document, and of establishing relatively independent and standardised

Table 8 Opinions on promoting transmission and distribution tariff reform

Main content	Main requirements
Gradually expand the scope of pilot transmission and distribution tariff reform	Include the regions carrying out pilot power system reform in the scope of pilot transmission and distribution tariff reform Allow regional characteristics to be embodied in transmission and distribution tariff calculation parameters, the price adjustment cycle and total income regulation patterns
Carefully calculate the transmission and distribution tariff	For pilot regions, the National Development and Reform Commission performs centralised cost supervision, calculates transmission and distribution tariffs for each voltage class, creates balance accounts for pilot regions, and supervises and regulates income and price levels For non-pilot regions, research and calculate transmission and distribution tariffs for each voltage class
Promote cross-subsidy reform by category	Gradually reduce cross-subsidies for industrial and commercial users, and correctly manage cross-subsidies for domestic and agricultural users Grid companies report the amount of tariff cross-subsidies for different types of user during a transitional period, which will be recovered through transmission and distribution tariffs after being reviewed by price authorities After transmission and distribution tariff reform, the transmission and distribution tariff for each voltage class and the cross-subsidies borne by domestic and agricultural users will be calculated
Policies on transmission and distribution tariffs for power trading during the transitional period	In those regions where transmission and distribution tariffs have been set, power is traded at the approved tariff In those regions where separate transmission and distribution tariffs have not been set, sales-side reform will be promoted

Table 9 Opinions on advancing the creation of power markets

Main content	Main requirements
Market composition	Composed mainly of medium- and long-term markets and spot markets
	Medium- and long-term markets cover power trading on a multi-year, yearly, quarterly, monthly and weekly basis and for interruptible loads, voltage adjustment and other auxiliary services
	The spot market mainly covers day-ahead, intra-day, real-time power trading and trading of standby, frequency modulation and other auxiliary services
	After conditions mature, capacity market, power futures and derivatives trading will be explored
Market mode	Includes decentralised and centralised markets
	Under the decentralised mode based on medium- and long-term physical contracts, power users and producers themselves determine daily output and load curve during the day-ahead period; the deviation is adjusted through day-ahead, real-time balance trading
	In centralised markets, risk is managed through medium-and long-term contracts for difference, and centralised price competition is applied with the help of spot trading
Market system	Divided into regional and provincial (municipal) power markets, which are not graded
	Nationwide optimal allocation of resources is almost achieved, mainly by the Beijing and Guangzhou power exchanges
	Set up spot markets in large and provincial markets that offer optimal allocation of resources
Market participants	Includes power producers, power suppliers (local power grids, bulk sale counties, high-tech industrial parks and economic and technological development zones), power sales companies and power users. The users involved in market trading should be those with large capacity and loads whose access voltage reaches the threshold
Market operation	Detail provisions by trading organisation or implementation, medium- and long-term trade contracts, day-ahead generation plans, intra-day generation plans, pricing of competitive fields, market settlement, safety assessment, congestion management, emergency response, market supervision and regulation
Credit system	Create a market participant credit rating system, annual information disclosure system, loyalty incentives and dishonesty penalty mechanisms

power trading institutions. Its main content and key requirements are shown in Table 10.

The Opinions on Implementing an Orderly Liberalisation of Power Generation and Consumption Planning lists the following objectives: (i) build a power operation mechanism that features orderly competition and is strongly supported, and that shifts the power system gradually from plan-oriented to one in which the market plays a decisive role; (ii) ensure power supply to households and other users who do not have bargaining power by establishing a power generation and purchase priority mechanism, granting preferential grid access for public and peak-shaving purposes. The main content and requirements of the document are shown in Table 11.

The Opinions on Promoting Power Sales Reform has the following objectives: introduce competition into power sales, open power sales

to private capital, expand user choice, develop power sellers in different ways, and form a competition-based environment of diversified buyers and sellers. Its main content and requirements are shown in Table 12.

The main objectives of the Guidance on Strengthening and Standardising the Supervision and Management of Privately Owned Coal-Fired Power Plants are: gradually make privately owned power plants equal with state-owned power plants, promote the orderly development of privately owned power plants, increase the absorption of clean energy, improve energy use efficiency, and ensure fair market competition. Its main content and requirements are shown in Table 13.

The Administrative Measures for Access and Withdrawal of Electricity Sales Companies provides clear stipulations on access, exit, classification and scope of business of power sales

Table 10 Opinions on establishing power trading institutions and standardising their operation

Main content	Main requirements
Functions and positioning	Non-profit trading institutions provide market participants with standardised, open and transparent power trading services under government regulation Construct, operate and manage market trading platforms, organise market trading, provide settlement and relevant services, gather bilateral contracts between power users and producers, register and manage market participants, and disclose and release market information
Organisational form	Separate the trading business of grid companies, establish trading institutions in accordance with the articles of association and rules approved by government Trading institutions could be companies partly controlled by grid companies, their subsidiaries or members of grid companies
Market management committee	Build a market management committee consisting of representatives from grid companies, power producers, power sales companies and power users. The committee is responsible for researching and discussing the articles of association of trading institutions and trading and operational rules, and for coordinating matters related to the power market Market management committees adopt a discussion mechanism, such as voting by category of market participant Decisions of market management committees are executed after being reviewed. The National Energy Administration and competent authorities should have the power of veto
System and framework	Set up relatively independent regional and provincial (municipal) trading institutions in an orderly manner. These include the Beijing and Guangzhou power exchanges and other trading institutions servicing regional power markets. Encourage trading institutions to expand the scope of trading services and promote market integration
Staffing and sources of income	Trading institutions could be staffed with existing grid company employees. Senior managers should be recommended by market management committees and appointed in accordance with applicable laws and procedures. Trading institutions may charge market players reasonably for registration fees, annual fees and service charges for trading
Dispatching	Trading institutions are responsible for market trading. Dispatching organisations are responsible for real-time trading, balancing and system security. For day-ahead trading, functional boundaries should be defined based on circumstances, practical operation and lessons learned

companies, and plays an important guiding role in standardising the power sales market. Its main content and requirements are shown in Table 14.

The Administrative Measures for Deregulating Power Distribution Networks in an Orderly Manner deregulate investment in incremental power distribution network capacity. Its main content and requirements are shown in Table 15.

(2) Reform progress

Power system reform. Up to the end of 2016, the National Development and Reform Commission had designated: Yunnan, Guizhou, Shanxi, Guangxi, Beijing, Hubei, Sichuan, Liaoning, Shaanxi, Shandong, Anhui, Henan, Xinjiang and Ningxia as comprehensive power reform pilot provinces; Guangdong, Chongqing, Xinjiang Production and Construction Corps, Fujian, Heilongjiang and Hebei as power sales deregulation pilot provinces; and Gansu, Hainan and Shanghai as electricity system reform pilot provinces.

Pilot transmission and distribution tariff reform was carried out in six provincial power grids (western Inner Mongolia, Anhui, Hubei,

Table 11 Opinions on implementing an orderly liberalisation of power generation and consumption planning

Main content	Main requirements
Build a purchase priority system	Priority purchasers: primary industry, domestic users, important utilities supplying the service and public service industries The measures to protect priority purchase rights: share generating units, improve demand-side management, implement in an orderly manner, and ensure power supply to remote and poor regions
Build a priority generation system	Priority producers: planned renewables, peak-shaving/frequency modulation and cogeneration, hydropower, nuclear power, power generated with waste heat/pressure/gas; power generated in accordance with trans-provincial and interregional national plans, local government agreements or for historical reasons Measures to protect priority generation: (i) leave space for planning; (ii) improve power output and absorption; (iii) forecast output in a centralised manner; and (iv) organise and implement replacement power and make priority power tradeable
Ensure electricity balancing	Manage priority generation: preferentially arrange renewables generation to fill output gaps; arrange peak-shaving and frequency modulation demand as needed; arrange cogeneration using waste heat, pressure or gas; and manage hydropower and nuclear power Organise direct trading: ensure load characteristics do not deteriorate, avoid an increase in peak-shaving pressure. Direct trading should not affect demand for heating. In regions with a high proportion of hydropower, direct trading should distinguish between the wet and dry seasons Capacity deduction methods: to motivate power producers, the amount of power traded may be converted into generation capacity based on the users' maximum load utilisation hours, local industrial users' average utilisation hours, or the upper limit
Actively promote direct trading	Carry out pilot power market projects in regions with the right conditions. Carry out market-oriented trading in non-pilot regions in accordance with the Opinions on Implementing an Orderly Liberalisation of Power Generation and Consumption Planning. Maintain power load characteristics and avoid irrational competition
Lift control of generation planning in an orderly manner	Gradually increase the amount of power traded: the market for industrial and commercial users of 110 kV and above could be deregulated, followed by that for industrial and commercial users of 35 kV and above. When the conditions have evolved and matured, the market for all users of 10 kV and above could be deregulated Establish a power market system: gradually lower trading access requirements for users, power sales companies and power producers to expand the size of the market Improve the emergency supply mechanism: encourage enterprises and users of priority generation to enter the market voluntarily. Renewable power could take part in market competition with subsidies and be protected by a renewable energy quota system

Ningxia, Yunnan and Guizhou) in 2015. Included in the pilot transmission and distribution tariff reform of April 2016 were the provincial power grids of Beijing, Tianjin, southern Hebei, northern Hebei, Shanxi, Shaanxi, Jiangxi, Hunan, Sichuan, Chongqing, Guangdong and Guangxi; the provincial power grids covered by the national power system comprehensive reform pilots; and the power grids of north China. In August 2016, the National Development and

Table 12 Opinions on promoting power sales reform

Main content	Main requirements
Market participants in power sales	Power sales companies fall into three categories: (i) power sales companies that are grid companies; (ii) power sales companies with the right to operate a power distribution network and use private capital to invest in incremental power distribution networks; and (iii) independent power sales companies without the right to operate a power distribution network and that do not provide minimum supply services Grid companies, eligible power producers, qualified high-tech industrial parks and economic and technological development zones, distributed power and microgrid owners, utilities including water, heat and gas suppliers, energy-saving service providers, private capital and private enterprises may invest in or form power sales companies to carry out power sales activities
Access and exit mechanisms for power sales market participants	Access and exit of market participants, not subject to administrative approval, need to be included in an annual list of provincial government announcements. Such market participants need to make commitments and be registered at trading institutions. Power sales companies need to be independent legal entities registered in accordance with the Company Law of the People's Republic of China
Business and trading activities of market participants	The core business of power sales companies is electricity purchase and sales, which could include energy performance contracting, energy-saving strategies, electricity use consulting and other power-related value-added services Before selling electricity to users in the quantity and at the price agreed with them, power sales companies may trade with power producers freely or through trading platforms. Distributed power or microgrid users may delegate electricity purchase and sales to power sales companies. The power purchase price for users involved in market trading consists of the market transaction price, the transmission and distribution tariff (including line loss and cross-subsidy) and government funds Grid companies offering measurement, meter reading, billing, settlement, installation, repair and other power supply services are responsible for settling electricity bills and ensuring safety of the funds. Grid companies ensure minimum power supply in the regions they serve
Deregulating investment in incremental distribution capacity	Encourage the development of the power distribution business through mixed ownership. Deregulate investment in incremental distribution capacity for qualified market players Private capital investing in and wholly controlling incremental power distribution networks, i.e. with the right to operate the power distribution network, also has the same rights as the grid companies in the same power supply area and should fulfil the same responsibilities and obligations There could be multiple power sales companies in one power supply area. But only one of them has the right to operate the power distribution network and provide minimum supply services
Build credit system and risk prevention mechanisms for the power sales market	Establish a market participant credit rating mechanism and blacklist. Government could intervene in the market in the case of serious abnormalities

Table 13 Guidance on strengthening and standardising the supervision and management of privately owned coal-fired power plants

Main content	Main requirements
Improve planning guidance and standardise construction scientifically	The requirements for planning, selection, approval, construction and grid integration of newly built (or expanded) privately owned coal-fired power plants (except backpressure and waste heat/pressure/gas plant units) should in principle be the same as those for state-owned power plants. State-owned power plants are not allowed to change into private ownership
Improve operational management and take part in auxiliary services	Grid-connected privately owned power plants should follow dispatching disciplines, undertake their responsibilities and obligations to provide safe and stable power supply, and be covered by auxiliary service assessment or compensation in accordance with the Detailed Rules for the Administration of Grid-Connected Power Plants and the Detailed Rules for the Administration of Ancillary Services of Grid-Connected Power Plants
Assume social responsibilities by paying charges and fees	Electricity generated by privately owned power plants and used by the owner is subject to government funds and policy-related cross-subsidies. These monies should not be exempted without permission or collected selectively by local government at any level. Enterprises with grid-connected privately owned power plants should pay a system standby fee to the grid company based on the agreed standby capacity at the rate set by each province
Reduce the use of coal combustion	Replace privately owned coal-fired power plants with renewable energy generation
Define the market participants in market trading	Enterprises with privately owned power plants that are not able to meet their own power demand could purchase electricity after paying a fee set by government. The fee should be in accordance with applicable laws, regulations and policy-related cross-subsidies
Define accountability and improve supervision and management	Privately owned power plants should name the parties accountable, improve organisation and coordination, carry out specialised regulation, improve project management, standardise operation and modernisation, and strengthen supervision and inspection

Reform Commission extended the transmission and distribution tariff reform pilot to all provincial power grids except Tibet. Cost supervision and review were completed in January 2017.

Power sales reform. The first power sales company—Shenzhen Shendianneng Electricity Co., Ltd.—was founded in March 2015. After the Administrative Measures for Access and Withdrawal of Electricity Sales Companies was released in October 2016, existing companies could carry out electricity sales activities after applying to expand their business activities and completing the access procedures. The number of power sales companies increased sharply 3.5-fold in three months. The National Enterprise Credit Information Publicity System shows that there were 5,410 power sales companies in China at the end of 2016.

Reform of trading institutions. Up to the end of 2016, two national power trading institutions, the Beijing Power Exchange and the Guangzhou Power Exchange, had been founded, and 32 provincial power exchanges (except Hainan and Tibet) established, covering almost the whole country.

Reform of incremental power distribution. The National Development and Reform Commission (NDRC) and the National Energy

Table 14 Administrative measures for access and withdrawal of electricity sales companies

Main content	Main requirements
Access mechanism	Define access conditions for power sales companies. Replace administrative licencing with confirmation of registration. Process of market entry: sign the letter of commitment, submit registration information online, announce and register at the National Development and Reform Commission, National Energy Administration and the third-party credit information service provided by government
	Power producers, power-related construction companies, high-tech industrial parks, economic and technological development zones, water/gas/heat utilities and energy-saving service providers in an incorporated capacity, and which meet the access conditions for power sales companies, could carry out power sales activities after applying to industrial and commercial administrations to expand the scope of their business and complete the procedures above
	Existing qualified high-tech industrial parks, economic and technological development zones and companies that build or operate power distribution networks could voluntarily change into power sales companies after completing the procedures above
Scope of business	Power sales companies may provide users with other services, including but not limited to energy performance contracting, comprehensive energy saving, energy use consulting, electric equipment operation and maintenance, and other value-added services
Market-based trading	Power sales companies may purchase or sell electricity through the power market, from or to power producers, or through trading institutions. Power sales companies could select trading institutions for trans-provincial and interregional purchase. There could be multiple power sales companies in one power distribution area. The same power sales company could sell electricity in many power distribution areas in its province
Establish a credit rating system	Build a power sales company credit rating system. Establish a penalty mechanism for breaking power industry law and for immoral behaviour

Administration (NEA) issued an order in August 2016 that requires provinces to start pilot incremental power distribution deregulation programmes and carry out 100 incremental power distribution pilot projects across the country as soon as possible. The order also requires incremental power distribution to be deregulated through mixed ownership. The NDRC and the NEA defined 105 projects, including the Yanqing Smart Power Distribution Network, as the first batch of pilot incremental power distribution reform projects in November 2016.

Market trading. Monthly and annual bidding in Guangdong power markets made great progress. Up to September 2016, 43,960 GWh (RMB 0.033/kWh cheaper on average) had been traded directly in Guangdong, exceeding the planned annual target of 4,200 billion kWh. A total of 15,980 GWh (RMB 0.073/kWh cheaper on average) was traded through monthly bidding, exceeding the annual target of 14 billion kWh.

Up to the end of 2016, Guangdong had conducted seven monthly centralised bidding rounds, trading 15,980 GWh in total. The number of power sales companies increased from 13 to 154. The electricity sold by power sales companies accounted for 70% of the total. The discount granted by the power producers dropped from a high point of RMB 0.148/kWh in April to RMB 0.037/kWh in September (Table 16).

Table 15 Administrative measures for deregulating power distribution networks in an orderly manner

Main content	Main requirements
Planning guidance	Include incremental power distribution network projects with the power distribution network plans developed by local power authorities Incremental power distribution networks need to be finalised in provincial plans to ensure that incremental power distribution networks are consistent with national power development strategies and industrial policies and meet market participant transmission requirements
Competition and openness	Encourage private capital to actively take part in incremental power distribution. Determine investors through market competition
Rights and responsibilities	Private capital investing in an incremental power distribution network is responsible for its operation and management. Investors should follow national technical specifications and standards, fulfil their obligations on safe and reliable electricity supply, minimum supply and social services, while obtaining a reasonable return on investment
Scope of activities in the power distribution network	Invest in, construct and operate incremental power distribution networks that meet transmission demand and planning requirements, and expand power distribution networks through mixed ownership Companies that control existing assets in power distribution networks, including high-tech industrial parks, economic and technological development zones, local power grids and bulk sale counties, could apply to local power authorities for the right to invest in and operate power distribution networks
Operation of power distribution networks	Private capital that invests in and controls the incremental power distribution network has the right to operate the network. Those companies that meet the access conditions for power sales companies may carry out power sales activities Those grid companies with an incremental power distribution network have the right to operate the network. Only power distribution activities can be carried out in power distribution areas. The grid company's competitive power sales business should be gradually shifted to an independent power sales company Power producers are not allowed to invest in or build dedicated lines through which power plants supply power directly to users or that connect the plants to the incremental power distribution network they invest in

Table 16 Power traded through bidding in Guangdong, up to March 9, 2016

	Trading volume (GWh)	Share of power sold by power sales companies (%)	Discount granted by power producers (li (one thousandth of a yuan)/kWh)
1st trade	1,050	64.85	125.55
2nd trade	1,450	68.68	147.93
3rd trade	1,400	82.92	133.28
4th trade	1,870	62.07	93.90
5th trade	2,660	75.50	58.87
6th trade	3,550	75.86	43.38
7th trade	4,000	75.00	37.42

7.5 Principles and Roadmap for Future Reform[22]

The power market comprises multiple markets and multiple market failures. The primary objective of these markets is to match supply with demand through optimal dispatch and investment. This is delivered through electricity, capacity and balancing markets, which face market failures in the form of the carbon pricing externality, regulation of natural monopolies and the need for policy credibility to underpin long-term and long-lived investment decisions. These market structures and failures will exist in a net zero emissions future, but their relative importance and sophistication will change with greater electricity demand and a greater share of supply from renewables.

While the six characteristics of market efficiency (in terms of price signals) and policy efficiency (in terms of correcting for market failures) described above are essential from the design perspective, their importance will vary based on regional, national and local circumstances. For example, countries with large endowments of non-intermittent, low-carbon generation[23] will have less need to implement all six elements, although they are still likely to require some evolution of efficient pricing mechanisms and supporting policy frameworks. These countries will have less need for a short-term balancing and ancillary services market and are likely to require less sophisticated

capacity markets.[24] However, for most countries relying on variable renewables to decarbonise power, there will be significant pressure for reform of the power sector.

For centrally planned economies like China, the challenge will be to achieve the right balance between market forces and policy interventions. On the one hand, central planning through the Five-Year Plan process provides policy certainty, security of supply and the ability to distribute the burden of costs away from consumers to other parts of the system. On the other hand, market forces, through efficient market design and supporting policy frameworks, are likely to be a less expensive and more adaptive approach, especially in the context of dynamic and changing power demand and technologies. The value of such an adaptive approach is even greater, given the likely scale and pace of change required to achieve net zero emissions in the second half of the century.

Progress towards efficient net zero emissions power markets will not be smooth. It is likely to happen at different speeds in different geographies, with countries experimenting with different approaches; for example, the different approaches taken by the UK and Germany, with the former relying on capacity auctions and the latter on creating strategic reserves to deliver long-term capacity and resilience in the system.[25] However, the direction of travel is clear, with many countries already making progress towards more efficient power market structures and policies, driven in large part by the efficiency gains they provide. This trend is only likely to intensify with greater electrification of energy use and with the greater imperative to decarbonise power in a way that keeps costs down and delivers a secure supply of electricity.

---

[22]Following the publication of the Opinions of the Central Committee of the CPC and the State Council on Further Deepening the Reform of the Electric Power System (Zhongfa No. 9, 2015), which provides a roadmap for power system reform in China over the coming years, the National Development and Reform Commission, the National Energy Administration and other ministries and commissions released specialist action plans. In this section, we include the opinions and suggestions of foreign experts, especially Vivid Economics and Shell International, which help us to analyse the future direction and actions of power reform from the perspective of international experts.

[23]For example, hydropower in Norway, bioenergy in Brazil, nuclear power in France and the abundant fossil fuel and huge carbon capture and sequestration potential of North America and other countries.

[24]For example, to deal with seasonal intermittency rather than the more frequent intermittency associated with variable renewables like wind and solar.

[25]Strategic reserves provide capacity and flexibility by keeping mothballed or older (usually thermal) generation plants available, whereas capacity auctions provide a fixed revenue for long-term capacity and flexibility in the system.

7.5.1 Reform Difficulties and Approaches to Resolve Them

A reforming power market needs to solve six problems urgently.

First, how to allocate power resources optimally; how to design an effective market mechanism to transmit abundant power resources in west China to the load centres in east China; how to make the market set a reasonable price for clean energy; and how to design an effective auxiliary services market mechanism to absorb clean energy across provinces and regions.

Solutions:

- market scope: design a nationwide unified electricity market to allow the free flow of resources across the country;
- roll-out: starting with provincial markets, promote deep integration of provincial markets step by step to eliminate the trade barriers between provinces with a market mechanism;
- increase the types of trading in provinces and regions: gradually introduce trans-provincial and interregional spot trading (starting with medium- and long-term trading), and increase the number of participants by engaging users and power sales companies in trans-provincial and interregional trading; and
- build a market mechanism that promotes the inter-provincial and interregional grid connection of clean energy, and gradually introduce auxiliary service market mechanisms, such as peak shaving, to increase trading volumes of clean power between provinces and regions.

Second, how to launch provincial markets; how to design the power market while taking into consideration the differences between provinces in terms of market concept, environment and maturity; how to coordinate the relationship between planning and the market; how to coordinate the relationship between medium- and long-term trading and spot trading; how to coordinate the relationship between intra-provincial power resources and other provinces' power resources; and when to make the spot market open.

Solutions:

- choose a suitable power market model based on local conditions, the market strength of power plants, power supply and demand, lessons learned from established markets, technical conditions and new energy; and
- increase the types of trading categories and expand power trading step by step, taking medium- and long-term trading first, and then spot trading.

Third, how to solve the problem of dominant companies when establishing a provincial power market. Oligopoly is a serious issue in some provincial markets, such as Zhejiang and Qinghai. In the process of supply-side reform, centrally administered state-owned energy enterprises are very likely to be restructured, intensifying their dominance of provincial power markets.

Solutions:

- expand the scope of the power market, through coupling or fusion, to reduce the concentration of power producers in any one market;
- build a strict regulatory system, setting an upper limit for the market share of power generation companies, forcing those with a large share to disclose information; and
- spin off or sell the assets of those power generation companies with a high market share.

Fourth, how to change a provincial market into a national market, in which each provincial market chooses its path of evolution and support measures based on actual conditions.

Solutions:

- build the national market by creating provincial markets, then integrate several provincial markets at their transmission end-points, before finally integrating all provincial markets;
- Integrate provincial markets in order of mutual openness, then relax access conditions for inter-provincial market trading, before unifying market rules; and
- build inter-provincial markets and the national market according to the principle of "medium- and long-term markets first, then the spot market".

Fifth, how to eliminate historical price provincialisation. The tariff in some provinces that are traditionally low-price regions could increase after integration with the national market, which carries the risk of creating social unrest.

Solutions:

- start with provincial markets, expand the scope of the national market and level the electricity tariff in different markets step by step;
- in the early stages of market creation, low-priced power could be partially reserved for industrial users to avoid a drastic impact on their costs after the market is built; and
- set tariff buffers and a protection mechanism for domestic and other public users.

Sixth, how to settle the problem of stranded costs. In the process of creating power markets, some power producers that were previously protected by the planning system may suffer damage to their interests or even go bankrupt, leaving them unable to recover their investment. For example, companies that invested in power plants under the planning system and have out-standing loans might not recover their investment because they are badly positioned for market competition due to their high costs.

Solutions:

- develop a variety of power products, such as contracts for difference and medium- and

long-term contracts and futures, to avoid tariff fluctuations and ensure stable revenues for all market participants;
- use short-term planned power allocation to prioritise power plants that are still recovering their investment, or have recovered their investment but are in difficulty, to give them some stability; and
- restructure the assets of power plants and power generation companies separately, in accordance with the principle of sharing reform costs.

7.5.2 Principles of Reform

To achieve the economic benefits of liberalised electricity markets, international best practice suggests implementing a set of principles for efficient network provision.[26]

First, proceed towards full liberalisation of the wider power system. Efficient investment in and operation of the overall power system is a necessary precondition for the efficient supply of electricity. This requires liberalisation of sectors suitable for competition (fuel production, generation, retail), use of markets to procure key services (capacity, balancing), and the pricing of externalities (such as air pollution and carbon emissions).

Second, align network-provider incentives with public policy objectives. Make the incentives dependent on the provision of a reliable and affordable supply of electricity by controlling monopoly behaviour and ensuring that prices reflect underlying costs:

- *Reform power network institutions.* Power networks are natural monopolies. It is therefore critical that their incentives are aligned with public policy objectives. A monopoly is incentivised to underinvest in new infrastructure and to charge prices that are higher than its costs. A state-owned company may be incentivised to prioritise short-term political objectives, rather than longer-term public

---

[26]This part is provided by Vivid Economics. See Vivid Economics, Low Carbon Power Markets: Lessons from International Experience for the DRC-Shell Markets Work Stream, 2017.

policy objectives. These incentives can be mitigated through institutional reform of the power network. One option is to reform the network company's incentives through performance-based regulation enforced by an independent regulator. Another is to separate network operation and ownership through the creation of an independent system operator (ISO). The UK and most European countries currently use performance-based regulation, while the USA uses the ISO model across its transmission systems, for example in PJM (the transmission system in the north-eastern states), California and New York.

- *Consider the use of locational pricing.* Efficient network investment and operation make use of information on network congestion. If implemented, locational (nodal or zonal) pricing can help reveal the cost of network congestion. Nodal pricing is used in several US states, Argentina, Chile, Ireland, New Zealand, Russia and Singapore, while zonal pricing has been adopted by most European countries and Australia. However, locational pricing has disadvantages as well as advantages. Importantly, locational pricing is most effective once time-of-use pricing is fully implemented across network users.

Third, take action to meet the challenges of a decarbonised system. Rising electrification and improvements in the efficiency of electrical appliances will increase the uncertainty over the future volume and location of demand for transmission capacity. Flexible resources such as electricity storage and demand response can substitute for new network investment, so long as sufficient investment incentives are present.

- *Designate strategic zones for transmission-scale renewable generation to reduce planning and investment uncertainty.* Renewable energy resources may be located far from demand centres and thus require large-scale transmission investment. Uncertainty over the

volume and location of generation can be mitigated through zoning.

- *Ensure there is money available to encourage flexible resources to offer a full range of system services.* The flexible resources needed for decarbonisation provide several system services, such as balancing and frequency response, but there may be underinvestment if the markets do not exist for these services. Several electricity markets in the West run demand curtailment markets, allowing flexible resources to generate revenues.

Fourth, prepare for the development of a decentralised power system and its associated digitalisation by coordinating investment in decentralised resources, their control, balancing, security and data flows.

- *Coordinate investment in decentralised resources.* A coordination problem arises when independent developers that lack information about the plans of other developers make similar investments, creating overinvestment or, if the developers are risk-averse, underinvestment. Either way, the result can be inefficient. Solutions include formal processes for multilateral resource planning and the publication of current and consented resources, as used by the transmission and distribution system operators of Spain and Ireland.

- *Determine how decentralised resources will be controlled.* While distribution networks today are largely passive, an active distribution network is capable of accommodating distributed resources. As the electricity system becomes more active and complex, a single system operator may start to rely on intermediaries, such as virtual power plants and partners such as distribution system operators, to assist with system balance. New systems of control, with new computational requirements, administrative rules and institutional characteristics, may then be employed, reflecting new operational vulnerabilities.

- *Balance data transparency with security.* As information and communications technology infrastructure expands, the amount of data from the electricity system grows. The data systems need their own infrastructure, with public access to facilitate competition and optimise operations. Meanwhile, the distribution of data across resources creates new risks of cyberattack and privacy loss, solved through adequate protocols.

7.5.3 Reform Roadmap

Vivid Economics and Shell's international experts proposed a roadmap for the development of future network arrangements in China. The roadmap is based on the following guiding principles:

- *Strong markets need strong government.* Market-based solutions have the potential to identify and deliver cost-effective investment and operation of electricity systems. However, both markets and natural monopoly networks benefit from a strong government to take an active role in ensuring that institutional incentives are aligned with public policy objectives. Network incentives can be aligned through separation of roles (unbundling) or strong regulation.
- *The institutional framework can be developed progressively.* Wholesale institutional reform is challenging and disruptive. At the outset, small changes in current practice and small-scale pilots may provide proof-of-concept sufficient to build consensus for larger-scale reforms.

The main points of the proposed roadmap are as follows:

(1) **Immediate actions**

- *Continue the market liberalisation programme.* Phase 1 of the DRC-Shell cooperation proposed a programme of power market liberalisation; consistent with this, China's 13th Five-Year Plan (2016–20) aims to improve the systems by which markets play the decisive role in resource allocation. It will be important to continue with the market liberalisation programme to deliver a more advanced and efficient power system.
- *Rationalise investment planning.* Clearly defined metrics for reliability and economic efficiency help network planners to identify efficient investments. Meanwhile, the application of the beneficiary pays principle encourages investment that increases productivity and avoids diverting national resources to stimulate regional output. Together, these approaches, when applied at national and regional level, facilitate greater interconnection and sharing of generation services.
- *Implement a coordinated approach to investment.* The use of a common investment framework enables generation and network investment planning to be coordinated, harnessing strategic decisions and market enterprise. For example, the framework might set out the role of generation zones for large-scale renewables alongside alternatives such as small-scale distributed generation.
- *Implement smart system architecture.* Smart, distributed resources are crucial to affordable decarbonisation. Before building the distributed resources, the system architecture to manage it should be determined. At a minimum, this includes deploying smart meters to introduce time-of-use pricing to consumers on the distribution network; creating upstream information and communications architecture; and carrying out R&D to develop technical solutions for the smart grid.

(2) **Move towards an efficient pricing mechanism**

- *Deregulate prices.* Cost-reflective pricing can signal efficient investments and operations to decision makers. The deregulation of prices can proceed sequentially, with further price reform contingent on the success of previous reforms. Price reform could commence upstream and progress downstream, beginning with input fuels and progressing through generation, network access and ending with retail, with provision to protect retailers if wholesale prices rise above retail prices before deregulation is complete.

- *Create harmonised trading arrangements between transmission systems.* Use prices to determine interconnector flows between provincial and regional transmission systems, signalling which provinces or regions could benefit from new investment. The prices would stimulate lower-cost generators to respond to demand.
- *Implement time-of-use pricing.* Time-of-use pricing allows consumers and flexible resources to respond to variation in generation costs and demand. Again, it can proceed sequentially, starting with larger consumers, such as industrial facilities with flexible production schedules, and ending with smart household appliances.
- *Consider locational pricing.* Similarly, locational pricing can signal efficient investments and operations, but it brings disadvantages as well as advantages. China may consider locational pricing, following the implementation of time-of-use pricing. That is, signal geographical network constraints once demand peaks have been shaved. Zonal pricing is a potential intermediate step between uniform and full nodal pricing.
- *Protect end users.* The deregulation of retail electricity may result in rent-seeking by retailers, raising consumer prices. This can happen if consumers do not switch suppliers readily or if for other reasons competition is not effective. Policies to protect consumers could be developed alongside any deregulation of retail electricity prices.

(3) Begin market trials

- *Create small-scale trials.* Create small-scale trials to procure competitively new transmission investments, non-network alternatives to new transmission assets, ancillary services

and so on. Use competitive tenders or auctions for the trials. These may provide proof-of-concept and experience with innovative, cost-effective solutions. To be successful, procurement must be open-access and transparent.

- *Progressively introduce market procurement.* If the competitive procurement trials are successful, they can be scaled up and wider market procurement progressively introduced, where appropriate, across each transmission system. Market procurement can help to reveal information about the relative costs and benefits of a range of generation technologies.

(4) Make institutional choices

- *Develop transmission network institutions.* Options include the status quo, an enhanced role for market procurement and a regulated transmission system operator (TSO) or independent system operator (ISO). International experience of the regulated TSO or ISO model is yet to reveal the best performer of the two, so it matters more to adopt a good-quality institutional model early than to choose between the options.
- *Select a model of control for decentralised resources.* Initially, when the number of resources is small, the transmission system operator may be able to control them directly. However, as the number of resources increases, and as temporal and locational pricing become more sophisticated, the computational, commercial and contractual capacity of a single operator model may be exceeded, and new models of control may be needed.

8 Build a Unified and Dynamic National Carbon Market

To control total carbon and carbon intensity, China plans to launch a national unified carbon market by 2018, which will cover key industries such as steel, power, chemicals, construction materials, papermaking and nonferrous metals. After national carbon exchanges are created, China will become an integral part of the international carbon trading market. The pressure on China to reduce carbon emissions will be largely alleviated by means of the market. Launch of China's carbon exchanges will drive the transformation and modernisation of traditional industries and sharpen the international competitive edge of China's low-carbon sectors.[27]

The Notice on Carrying out Pilot Carbon Emissions Permit Trading issued by the National Development and Reform Commission (NDRC) in October 2011 approved the launch of pilot carbon emissions permit trading in Beijing, Tianjin, Shanghai, Chongqing, Hubei, Guangdong and Shenzhen. The pilot carbon exchanges in the seven provinces and municipalities were opened between June 2013 and June 2014. The NDRC released the Notice on Key Points for the Launch of a National Carbon Trading Market in January 2016, according to which Phase 1 of the national carbon market should cover petrochemicals, chemicals, construction materials, steel, nonferrous metals, papermaking, power, aviation and other high-emitting industries. However, China's carbon market uses a baseline method to set the carbon quota, which is stricter than the historical method and more demanding in terms of technologies and data.

Given the progress of current pilot carbon markets, when building the carbon market, China should: improve the system of laws and regulations; coordinate allowance allocation; establish a unified trading platform and pricing mechanism; and optimise the design of the national carbon market to include guidance and benchmarks for its creation and improvement.

---

[27]Liu Junyan and Fu Jingyan, Lessons from Guangdong's Carbon Market Practice for Building a National Unified Carbon Market, in Science and Technology Management Research, 2016, 36 (13): pp. 237–242 and 254; Xue Rui, China's Carbon Market Response in the Context of the Paris Agreement, in Ecological Economy, 2017, 33(02): pp. 45–48 and 128; Peng Sizhen, Chang Ying and Zhang Jiutian, Reflections on Major Problems in the Development of China's Carbon Market, in China Population, Resources and Environment, 2014, 24(09): pp. 1–5.

8.1 Summary

Carbon markets are a key instrument for delivering cost-effective mitigation. Mitigation of carbon emissions is in the best interest of all nations, because the cost is less than that of unconstrained climate change. The cost of mitigation can be minimised via carbon markets, especially if carbon markets are the primary driver of mitigation in an economy, if the carbon price is robust to shocks, and if the price provides incentives for long-term investment.

However, competing objectives often require trade-offs that are reflected in carbon policy. Carbon markets, such as an emissions trading system (ETS), can help deliver benefits that are in line with other policy goals, for example improved air quality or resource efficiency. However, carbon markets can also adversely affect some policy goals, like industrial competitiveness, or be undermined by shocks, such as when a recession leads to an oversupply of permits that undermines the ability of the carbon price to incentivise long-term investment. These competing objectives are often legitimate, and shocks should be expected. The challenge to policymakers is, therefore, how to take these into account when designing carbon markets, without reducing the cost-effectiveness of mitigation.

We call carbon markets that accommodate competing objectives and shocks while maintaining cost-effectiveness robust markets, while those that fail to do so are fragile. Unfortunately, most carbon markets in international experience are fragile, having become stuck in a transition trap of managing competing objectives and shocks via countervailing policies that further undermine the markets. As China's ETS will double the quantity of emissions covered by a carbon price globally, when it is fully implemented, it is imperative that it is robust, or else carbon markets may lose support, which would seriously undermine decarbonisation efforts.

International experience suggests that carbon markets can be robust if certain design features are incorporated. Carbon markets have four sets of design options. These are options around: (i) the creation of carbon units, such as targets and cap setting; (ii) the distribution of allowances; (iii) governance of the market, such as coverage, banking, offsets, price controls and trading arrangements; and (iv) policy interactions with other decarbonisation policies. We analysed the experience of the European Union ETS, the Regional Greenhouse Gas Initiative in the USA, the New Zealand ETS and the California ETS, for lessons on which choices in each of these sets of design options lead to a robust carbon market.

China looks likely to use a slow start to emissions trading. As such it should consider adopting price controls and a clear reform path to reduce the risk of becoming stuck in the transition trap. Design trade-offs in the early stages of carbon pricing are common and can result in fragility. This means that an active approach to price management, such as a tight price corridor, can be used to ensure that incentives are maintained while the market develops. This can be supported by signalling the direction of climate policy. For instance, committing to specific changes through legislation can provide clear guidance on the direction of policy. This can help maintain prices and public support for carbon markets in the event of a shock.

While the next stage of China's ETS is being planned, policymakers should consider whether China should adopt an evolutionary or a revolutionary approach to carbon markets. Climate change has aspects that suggest a balance between evolutionary and revolutionary change. The rate and level of the required emissions reduction suggests the need for revolutionary change. However, the difficult politics of such a change, and limitations in institutional capacity, suggest that an evolutionary approach may be more feasible. The choice of approach is for Chinese policymakers to make. We provide a detailed roadmap for both options, across the four sets of design choices, so that the required actions and pace of reform are clear, whether an evolutionary or revolutionary approach is taken. These roadmaps, presented in Sect. 8.4, reflect best practice from international experience, drawn from the examples provided in Sect. 8.3.

Table 17 illustrates what an evolutionary versus revolutionary approach may look like in China. An evolutionary approach would start slow, providing assistance to companies and a low, but stable price; while a revolutionary approach would rapidly make the ETS a central force in economic decisions through strong price support and low levels of assistance.

Regardless of whether China takes an evolutionary or revolutionary approach to carbon markets, international experience suggests some general lessons for China:

Table 17 Evolutionary and revolutionary change can differ in all aspects of scheme design

	Evolutionary	Revolutionary
Creation	■ Target setting based on national circumstance, gradually transitioned to global carbon goal by 2030 ■ Bottom-up cap setting to 2020 ■ Net carbon neutrality after 2050	■ Move to binding, absolute targets and top-down cap setting by 2020 ■ Targets aligned with goal of achieving global net carbon neutrality before 2050
Distribution	■ Move from grandfathering to production-based allocations by 2020 ■ Assistance to emissions intensive industries only with high assistance rates, declining gradually over time ■ Auctioning of remaining units	■ End grandfathering as soon as possible, immediate development of process level baselines ■ Assistance limited to emissions-intensive, trade-exposed industries, and quickly declines ■ 100% auctions by 2030
Governance	■ Start with low ($10-20) price floor or MSR, may also include price ceiling, both growing slowly ■ Phase in coverage starting with electricity generators and major industrial energy users ■ Slow movement to linking with international partners	■ High (>$40) price floor that increases rapidly ■ Immediate coverage of all energy, transport and industrial emissions, with full banking ■ Agriculture and land sectors covered by 2025 ■ Near-term linking with willing international partners
Policy interactions	■ ETS operates as part of a broad policy mix, with other policies playing dominant role before 2030 ■ Carbon market gradually becomes more influential in economic decisions	■ ETS plays immediate major role in investment decision-making, becomes the major driver of decarbonisation investment by 2025 ■ Other policies aligned with decarbonisation objective

Note MSR = market stability reserve
Source Vivid Economics

- developing an ETS requires consideration of three key questions: how many units to create, how to distribute them, and how to govern their trade;
- the supply of emission units should reflect the nature of the climate policy problem in the longer term. This requires movement to caps and targets that reflect the absolute global emission reductions required;
- some auctioning is important for price formation early on, but movement to full auctioning may be gradual;
- allocating free permits may be necessary, but costs can be reduced through benchmarking and limiting allocations to cost pass-through in constrained industries; this requires addressing data constraints as a priority;
- the early learning phase of an ETS brings a risk of price shocks, so that price controls may be needed;
- broad coverage increases market efficiency and stability, and should be sought as rapidly as politically feasible;
- full banking is preferable to support investment from the outset, so long as caps do not create hot air;
- designing an ETS that is robust to shocks is essential for success. ETSs to date have suffered from fragility, with poor design interacting with cyclical, technological and/or supply shocks to undermine their effectiveness; and
- investing in early data collection can help improve policy outcomes and better ETS design.

Overall, China has an important opportunity to learn from international experience and deliver a robust carbon market. While the design of China's forthcoming ETS is largely finalised, revisions are expected by 2020. Therefore the next few years provide an important opportunity to learn from the initial stage of China's ETS and determine which path, evolutionary or revolutionary, China should take in the next phase of its ETS, drawing on the best practice of international experience to ensure a robust design and taking into account the competing objectives and shocks the ETS is likely to face.

8.2   Introduction

8.2.1   Policy Objectives, Market Failures and Trade-Offs

Responding to climate change requires a transformation of energy systems. The world's fossil-fuel-dependent energy mix is not consistent with the Paris Agreement objective to limit global temperature rise to well below 2°C above pre-industrial levels. If this objective is to be achieved, it means that all countries must take action to ensure that their future energy needs are provided by low-emission energy sources.

This transformation requires significant investment in the development and deployment of low-emission technologies and practices. The energy sector is capital-intensive, relying heavily on long-lived assets with high upfront costs. Further, there are multiple technologies that can deliver low-carbon energy. These include mature technologies like nuclear and hydropower, and rapidly developing technologies such as solar, wind and battery storage. This means that policies seeking to drive an energy transformation must both mobilise finance to fund investments and provide a mechanism to choose between competing technologies and providers.

The range of options for low-carbon energy can come with widely varying costs. These costs will be determined by a range of different factors, including, for instance, the development of technology, the geographic location of different energy technologies, and network and learning effects associated with the roll-out of these technologies. This means that while there are many different ways in which an energy transformation can be pursued, these options may come with very different costs.

Achieving climate change mitigation cost-effectively frees resources for alternative uses and enables society to pursue a broader range of opportunities to increase well-being. Climate change mitigation is only one of many outcomes that society seeks to achieve; other outcomes include greater material well-being, higher rates of employment and an engaged and content population. However, these outcomes all have some degree of cost associated with them.

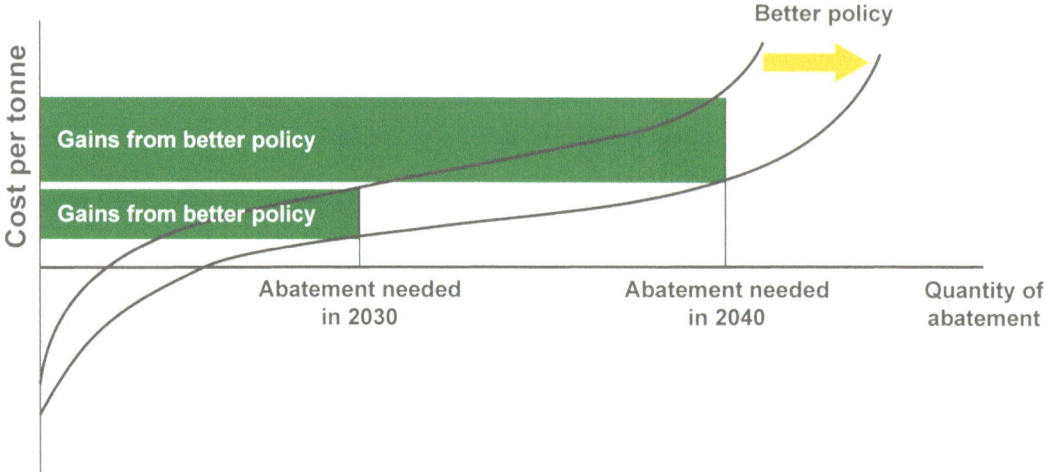

Fig. 27 The gains from better climate policy will grow over time, as low-cost mitigation is exhausted and more mitigation is required. *Source* Vivid Economics

By reducing the costs of achieving climate mitigation, more of a society's resources can be devoted to achieving other objectives.

As the level of climate change mitigation required is likely to increase over time, the cost-effectiveness of a given policy mix becomes increasingly important. If dangerous climate change is to be avoided, then over the next 30 years action on climate change must accelerate to achieve the required balance between sources and sinks of greenhouse gases. This implies a rapid progression in the level of mitigation and the movement to higher-cost forms of mitigation. This means that the difference between an expensive policy mix and a cost-effective policy mix will become greater in absolute terms as time progresses. In order to limit the impact of climate change mitigation on achieving other policy objectives, the focus on cost-effectiveness should increase over time.

The need for a cost-effective mechanism to identify and allocate competing mitigation options has led many jurisdictions to turn to markets. In doing so, individual incentives can be used to identify mitigation options that are suitable to a specific context. Markets enable decision-making processes to be aggregated across individuals, in turn ensuring that the broadest possible set of information and

mitigation options is being considered. This can reduce the incremental and overall cost of moving towards a low-carbon energy mix.

Policy objectives do not exist in isolation, and progress in achieving one objective may have an impact on society's ability to achieve others. This is true for climate change mitigation, where achieving lower emissions cost-effectively can make some policy objectives easier, while making others more difficult. For instance, achieving cost-effective climate mitigation may improve local air pollution by simultaneously reducing emissions of particulate matter. However, it may also make other objectives more difficult; for example, maintaining industrial competitiveness may be harder if a carbon price is applied in one jurisdiction but not another (Fig. 27).

Achieving effective climate policy is made more difficult by the existence of multiple market failures and the interaction of different policies. Carbon pricing is an example of a policy that targets a market failure by seeking to internalise the externality of unpriced climate change. However, other externalities are also relevant to climate policy, such as:

- coordination failures, which may result in inadequate investment in the research,

Fig. 28 Achieving cost-effective abatement makes some policy objectives easier, but may require trade-offs with others. *Source* Vivid Economics

development and commercialisation of low-emission energy technologies;

- network effects, which may create barriers to the expansion of low-emission transport; and
- asymmetric information, which means that opportunities to improve energy efficiency are left untapped.

Different market failures may require different policy interventions, not all of which are compatible. For instance, encouraging investment in low-emission technologies through a feed-in tariff may result in lower emissions from the energy sector and a lower carbon price. This means that there are trade-offs made between objectives, and within the policy mixes that are used to achieve them (Fig. 28).

Balancing and achieving competing societal objectives subject to constraints is a central role of government. There is a range of legitimate objectives that government seeks to achieve at the same time, using imperfect policy

mechanisms. Policy objectives that are commonly affected by climate policy include:

- economic objectives: for instance, ensuring that increased energy costs do not reduce competitiveness in a perverse manner by avoiding carbon leakage;
- social objectives: for instance, ensuring that the costs of transition are fairly distributed;
- environmental objectives: for instance, the specific mitigation options that are taken up will determine the pollution co-benefits that occur at the local level; and
- political objectives: for instance, as different segments of society will be affected in different ways, climate policy choices may have an impact on a government's political support.

The existence of potentially competing objectives means that trade-offs may be necessary, with social preferences determining the relative valuation of different objectives, and the

Fig. 29 The social objectives frontier represents all potentially optimal outcomes based on societies' preferences regarding two objectives. *Source* Vivid Economics

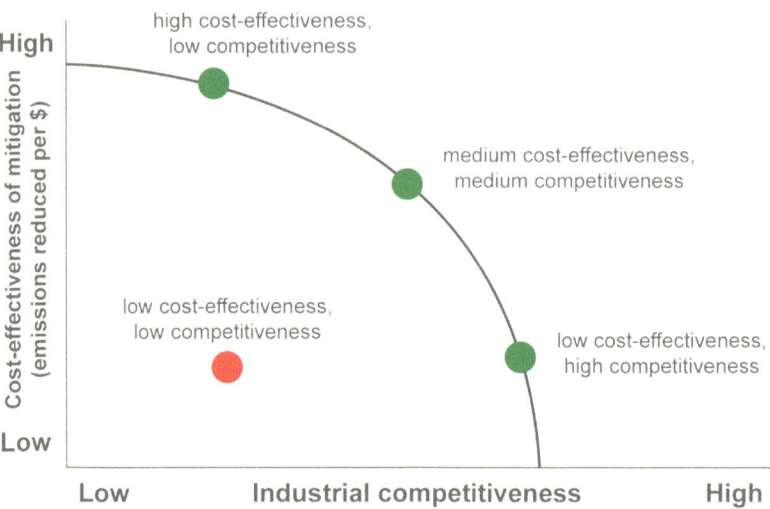

preferred quantity of these objectives that would be obtained with a given constraint.

The necessary trade-offs between two competing objectives can be represented as a frontier. This frontier shows the maximum extent to which different combinations of two competing objectives can be achieved. Any point on this frontier represents a potentially optimal social outcome given a specific set of societal preferences. This in turn will be reflected in different policy designs which may be capable of achieving these objectives to a greater or lesser extent.

Achieving social objectives requires trade-offs. However, poor management of these trade-offs often leads to suboptimal outcomes. When policy results in an outcome inside the social objectives frontier, each objective is achieved to a lesser extent than is possible. For instance, the red dot in Fig. 29 shows low cost-effectiveness and low competitiveness outcomes, where both outcomes can be improved on by moving closer to the frontier.

8.2.2 Fragile Carbon Markets and the Transition Trap

Carbon markets to date have not delivered on the objectives for which they were developed. Across several carbon markets, persistently low prices have undermined incentives to invest in new, low-emission technologies, and have caused policymakers to turn to overlapping policy instruments.

Poor policymaking has resulted in fragile carbon markets incapable of recovering from economic shocks. A carbon price should seek to deliver a price signal that is both efficient and robust. An efficient carbon price is one that is sufficient to deliver the changed production and consumption patterns that are required at least cost. A robust carbon market will also ensure that this price signal is stable and predictable in the long run. This is necessary to reflect the long-run scarcity value of greenhouse gas emissions, and to drive investment in new technologies and capital so that costs may fall in the future. In contrast, a fragile carbon market will fail to deliver the incentives needed to drive the long-term emission reductions required, at the lowest cost possible over time. The fragility of a carbon market may not be obvious until it is hit by a shock. To date, however, almost all carbon markets have been hit by some form of shock that has contributed to their ineffectiveness.

There is a range of shocks that can negatively impact a carbon market. These include:

- demand shocks: for instance, the EU recession reduced industrial demand, production and emissions;
- supply shocks: such as the oversupply of credits in the form of certified emission

Fig. 30 An effective carbon price will be both efficient and robust. *Source* Vivid Economics

reductions and emission reduction units, which depressed prices in the EU ETS; and

- technology shocks: such as the shale gas revolution that changed the structure of the east-coast US energy industry and depressed prices in the Regional Greenhouse Gas Initiative.

Often policymakers will seek to limit opposition to a new policy through a slow start. A slow start may have different elements in different contexts. However, most are characterised by seeking to limit the reallocation of resources between groups in the economy. For instance, a slow start may see the grandparenting of free allocations to existing firms, the limitation of coverage to only a small set of emission sources, or significant access to low-cost emission reductions through the use of international units or offsets (Fig. 30).

A slow start occurs when other objectives are prioritised above the cost-effectiveness of the carbon market in the short term. This may focus particularly on increasing political support for the carbon market, given the political challenges that can occur in establishing a new tax and changing the competitiveness of industries. This limits the

impact of carbon pricing on the attainment of other policy objectives in the short term. However, policymakers may intend to strengthen it in the future as the carbon markets gain acceptability and political support. However, these policies may also result in systematic fragility, which means that when a shock occurs it can result in persistently low prices.

A slow start can make new carbon markets susceptible to the transition trap. The transition trap occurs when policymakers decide to take a slow start to carbon markets, but an economic or technology shock interrupts the reform process, shifting the market to a suboptimal equilibrium before it can be strengthened. When this occurs, it can reduce the credibility of carbon markets and create support for the introduction of overlapping measures, such as feed-in tariffs to support renewable energy investment. The proliferation of these overlapping measures further reduces prices in the carbon market, driving a self-enforcing cycle that keeps carbon prices low. This is shown in Fig. 31.

The transition trap helps to explain the persistently low prices in carbon markets to date. Carbon prices have proved insufficient to drive investment in new low-carbon technologies and

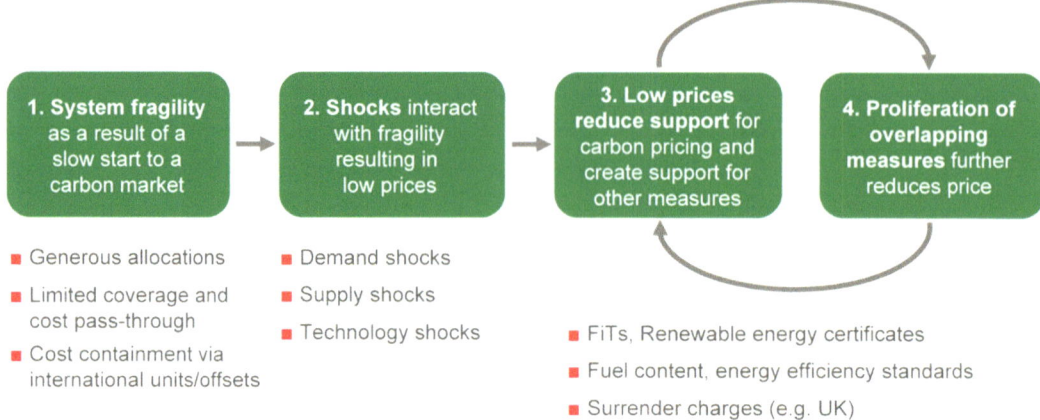

Fig. 31 The transition trap explains how a carbon market may become stuck in an ineffective equilibrium. *Source* Vivid Economics

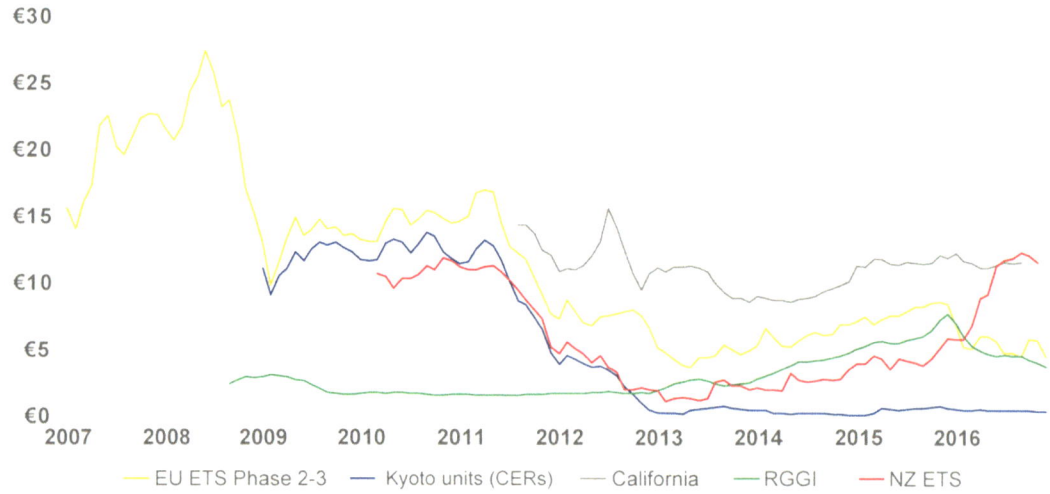

Fig. 32 Carbon markets have experienced persistently low prices. *Note* EU ETS Phase 2 and 3 prices refer to December settlement futures contracts, the most commonly used contract type; RGGI prices are taken from quarterly auction results; and NZ ETS prices were graphically digitalised and are indicative only. RGGI auction prices have been converted to tonnes. All prices have been converted to euro using monthly average exchange rates. CER = certified emission reductions. *Source* Intercontinental Exchange and Quandl (2016), Intercontinental Exchange (2016), Climate Policy Initiative (2016) RGGI (2016) and International Carbon Action Partnership (2016). Exchange rates from OECD (2016)

assets, which have instead largely been incentivised through overlapping policy, such as minimum renewable energy generation requirements, energy efficiency standards and feed-in tariffs.

China can draw on the lessons from international experience to avoid the transition trap and develop a robust carbon market. If well designed,

carbon markets can be a cost-effective mechanism for achieving emission reductions alongside attaining other objectives. International experience provides a range of lessons applicable to China as it considers the role of carbon markets, and how it can improve market design and functioning in the future (Fig. 32).

8.3 International Experience

Creating a carbon market is a process of setting rules for the creation of carbon units, deciding how they should be distributed, establishing rules for market governance and trade, and managing interactions between the carbon market and other policies:

- creating carbon units is the process of determining targets and caps;
- distributing carbon units involves deciding on the level and manner of distribution of free carbon units, and the use of revenue-raising distribution mechanisms, such as auctions;
- governance includes rules regarding coverage and cost pass-through, the banking of units, the use of domestic and international offsets, linking with other emissions trading systems, the use of price controls and the impact of design on the development of the secondary market; and
- managing policy interactions involves considering the appropriate role of carbon price and complementary, overlapping and countervailing policies.

8.3.1 Creating Carbon Units

Targets and coverage decisions will largely determine an ETS's cap—the number of carbon units that can be created. Targets determine the long-term trajectory of emissions in a jurisdiction, and therefore the long-term supply of units in the carbon markets. Caps give the number of emission units that are added to a carbon market in each compliance period. They are generally calculated as the target level of emissions less the expected emissions from sectors that are not covered by the ETS, known as uncovered sectors, as illustrated in Fig. 33.

To ensure that emissions do not exceed a jurisdiction's target, emissions from uncovered sectors must be considered when establishing a cap. This involves estimating the likely trajectory of these emissions based on an assessment of economic conditions and the impact of mitigation policies on uncovered sectors. In the EU, emissions from uncovered sectors are also subject to a target through the effort-sharing decision, providing further guidance on the likely emissions trajectory.

Jurisdictions often overestimate the trajectory of future emissions, as a preference for

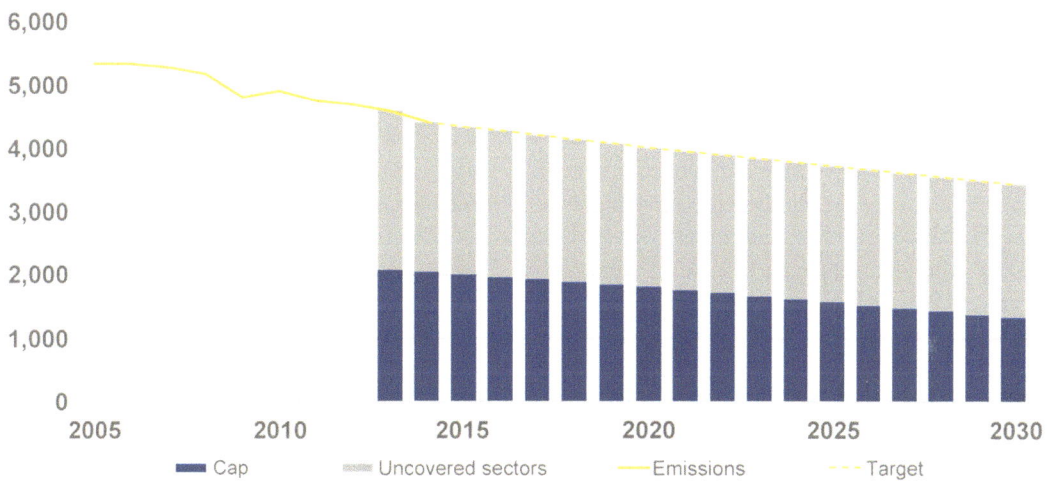

Fig. 33 Targets and coverage decisions largely determine ETS caps. *Source* Emissions from 2005 to 2014 sourced from Eurostat (2016); 2014 emissions cap and linear reduction factor from the European Commission's Directorate-General for Climate Action (2016); other calculations by Vivid Economics

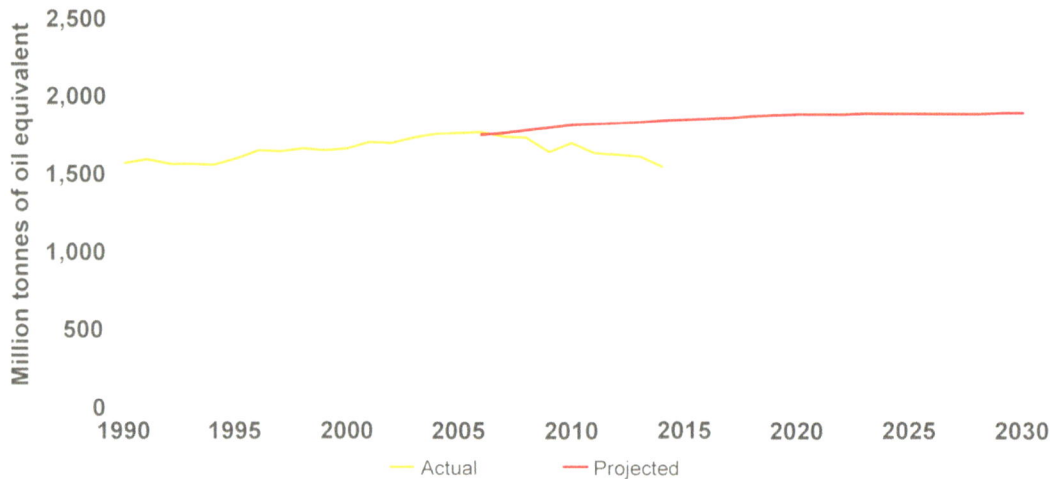

Fig. 34 Projections of continued growth in energy use may have resulted in the adoption of a weaker target in the EU. *Source* European Commission (2006) and Eurostat (2016)

conservative assumptions creates a projection bias towards the continuation of current trends. This can have real effects on the functioning of a carbon market. For instance, in 2006, EU projections suggested continued growth in energy use. However, the recession that followed resulted in a rapid decline in energy use and emissions, as shown in Fig. 34. An assumption of continued energy sector growth would mean that the EU's 2020 emissions target was less ambitious (in terms of deviation from business as usual) than was originally intended. The impact of unexpected outcomes (shocks) can be mitigated by adjusting supply or demand for permits to maintain a relatively stable price trajectory.

The response to short-term market shocks is best dealt with by adjusting caps, while structural oversupply should be rectified by adjusting targets. Cap and target-setting arrangements should provide the market with a clear direction regarding the trajectory of future caps, while maintaining the flexibility to respond to changed conditions. Rule-based supply mechanisms like market stability reserves and contingency reserves are considered in the discussion on price controls below.

Cap-setting mechanisms should provide the market with a clear direction regarding the

trajectory of future caps, while maintaining flexibility to respond when conditions change. Flexible target and cap-setting arrangements can include the use of conditional targets, establishment of institutional review mechanisms and the use of active cap management approaches:

- conditional targets enable governments to respond to changed circumstances. A price-contingent target would increase targets when prices are low;
- institutional mechanisms can be used to review caps and targets, for instance Australia's Climate Change Authority provided the government with advice on the appropriate range of emissions targets; and
- cap management mechanisms could be established within an independent body, such as a carbon central bank, to alter medium-term supply to manage prices.

8.3.2 Distributing Carbon Units

Allocations determine who gets carbon units and what price, if any, they pay for them. Auctions remain the best way to establish the value of units on the primary market. However, in many cases, a jurisdiction will allocate units to businesses with liabilities under the carbon price, free of charge, to offset potential impacts on competitiveness and avoid carbon leakage. Carbon leakage may occur when domestic companies are unable to pass on their cost increases because of competition from companies overseas that are not subject to an equivalent carbon price. This can mean that emission reductions that occur domestically are offset by emission increases overseas, as production can transfer to jurisdictions with a lower—or no—carbon price.

When allocations are done poorly, they can contribute to overallocation, which can increase the fragility of the carbon pricing system. There are two main approaches to free allocations:

- grandparenting, where assistance is tied to historical emissions; and
- production-based benchmarking, where assistance is tied to the level of actual production.

Grandparenting on its own does not cause overallocation. However, it is often part of a broader set of arrangements that cause overallocation and drive low prices. Grandparenting does not of itself prevent carbon leakage, as the provision of free units occurs regardless of the level of production. Grandparenting can be used as a transitional measure while benchmarks are

being developed for other allocation methods (Fig. 35).

The design of a benchmark determines the mitigation options that are used:

- process benchmarks: allocation is based on the emissions intensity of a process used to produce a specific good, which incentivises process efficiency;
- product benchmarks: allocation is based on the average emissions intensity of the production of a particular product, which incentivises technology substitution; and
- sectoral benchmarks: allocation is based on the emissions intensity of a particular industry or product category, which encourages substitution between similar products. However, this would be very difficult to implement.

Having no free allocations also incentivises end-use substitution, the choice to consume different products when relative prices change. Overall, this means that the gradual movement away from free allocations to full auctioning will incentivise a greater array of domestic mitigation options.

The auctioning of units also provides revenues that can be put to other uses. Revenues may be used to reduce other taxes, to increase

Fig. 35 EU ETS grandparenting contributed to overallocation, which lowered prices. *Note* EU ETS Phase 2 and 3 prices refer to December settlement futures contracts.

Source Prices sourced from Intercontinental Exchange and Quandl (2016); emissions and allocations from the European Environment Agency (2016)

general government revenues, or be directly tied to other spending programmes. For instance, in the EU at least half of all auction revenues must be used for climate- or energy-related purposes (Fig. 36).

There are several design features that may be used to limit the amount of free allocations and increase the number of units available for auction. These include:

- limiting allocations to certain industries, in particular those industries that present a carbon leakage risk. These are usually industries that are trade-exposed, which means that they are less likely to be able to pass on costs due to overseas competition, and emissions-intensive, which means that a carbon price could have a potentially large impact on their relative competitiveness;
- differentiated assistance rates seek to reflect different levels of exposure to carbon leakage. A jurisdiction may therefore decide to provide different rates of assistance to jurisdictions that are judged to have different rates of exposure to carbon leakage risk; for instance, Australia's ETS differentiated allocation rates are based on the emissions intensity of the industry;
- envelopes cap free allocations by limiting them to a certain proportion of total units. For

instance, for Phase 3 of the EU ETS, free allocations are limited to 43% of the total cap;

- automatic cuts to assistance: for instance, Australia's ETS included a carbon productivity dividend to account for potential natural carbon efficiency improvements over time, and this reduced the rate of assistance by 1.3% per year; and
- reviews can independently assess evidence of an industry's ability to pass through costs, and therefore their exposure to carbon leakage. These reviews can be a formal mechanism with outcomes that are directly tied to rates of assistance, or they can simply provide a source of information that enables government to make informed decisions. For instance, Australia's Productivity Commission was responsible for assessing cost pass-through and leakage risks.

Figure 37 shows the rates of assistance for moderately emissions-intensive and highly emissions-intensive industries in the Australian ETS. Assistance to highly emissions-intensive industries started at 94.5% of the benchmark level, while assistance for moderately emissions-intensive industries started at 66%. Both assistance rates were subject to the carbon productivity dividend, and assistance rates then declined by 1.3% each year.

Fig. 36 The manner of assistance provided to companies determines the mitigation possibilities available. *Source* Vivid Economics

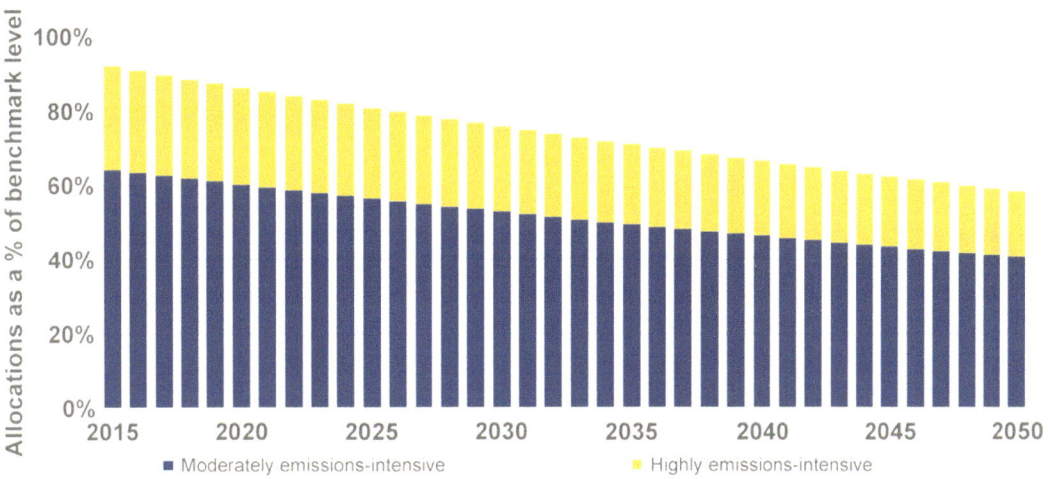

Fig. 37 Differentiated and declining assistance in the Australian ETS signalled transition to auctions. *Source* Vivid Economics

8.3.3 Governing Carbon Markets

The rules for governing a carbon market are essential to its efficient and robust functioning. The governance of a carbon market can include a wide variety of different decisions and design options. However, several design considerations common to all carbon markets are particularly important. These include decisions regarding:

- coverage of emissions;
- banking of emissions units;
- international linking and the use of offsets;

- the use of price stabilisers; and
- provisions for secondary market development.

(1) Coverage

Broad coverage and market design that encourages cost pass-through and increases the efficiency of an ETS can improve its robustness. Coverage rules determine which sectors and greenhouse gases incur a carbon price liability; cost

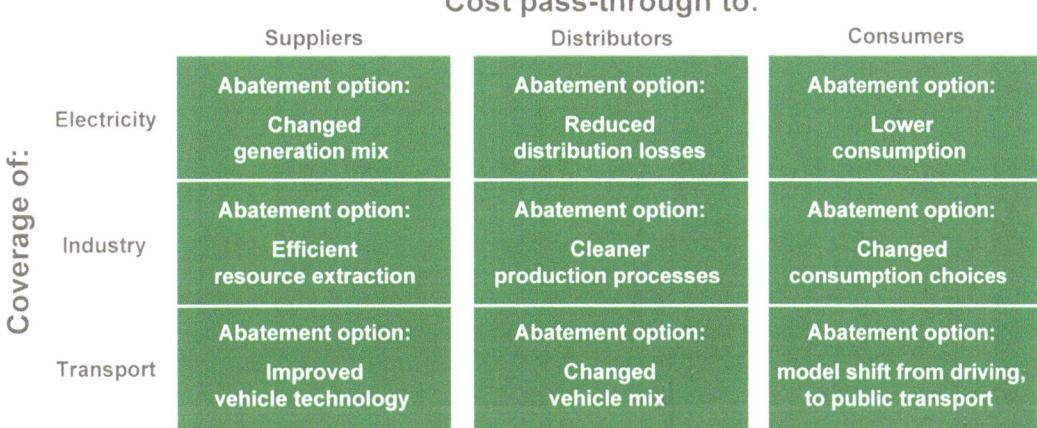

Fig. 38 Coverage decisions and the level of cost pass-through determine the mitigation options available. *Source* Vivid Economics

pass-through is determined by the broader set of rules that determine which users will face the costs and incentives following from a carbon price.

Broad coverage and cost pass-through increase market efficiency by incentivising more abatement options. Examples of the abatement options that might be incentivised through different coverage and cost pass-through arrangements are outlined in Fig. 38. When coverage is limited or cost pass-through does not occur, these options are not considered, which means that mitigation becomes more expensive.

The potential emission reductions from any given mitigation option increase with the carbon price. This means that as carbon prices rise more and more, potential mitigation options will be missed if a specific source of mitigation is not incentivised. In turn, this means that when coverage is limited, or cost pass-through is constrained, more costly abatement will be required and less abatement will be achieved at any given carbon price. This is illustrated in Fig. 39.

Narrow coverage reduces robustness by increasing the carbon market's exposure to

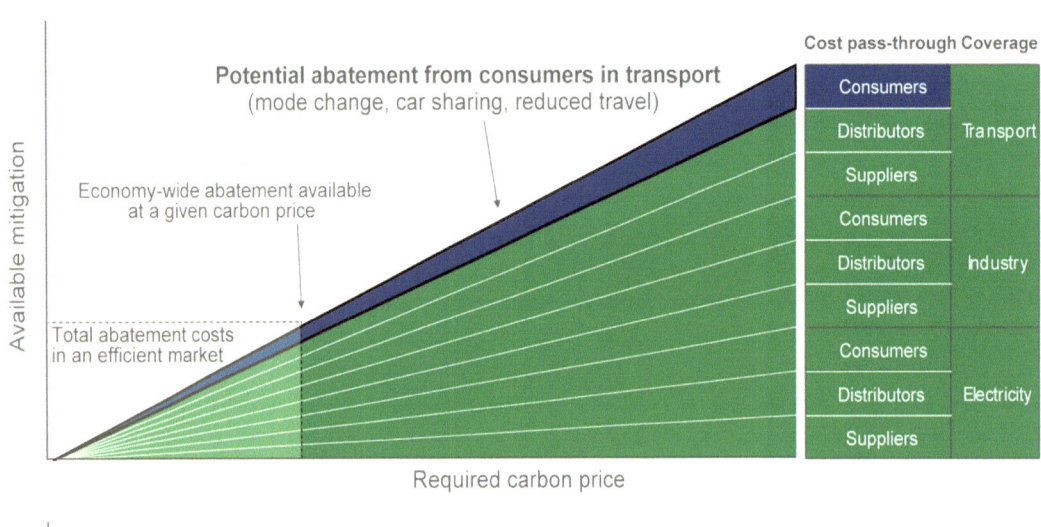

Fig. 39 When coverage or cost pass-through is limited, any given target will require more costly abatement. *Source* Vivid Economics

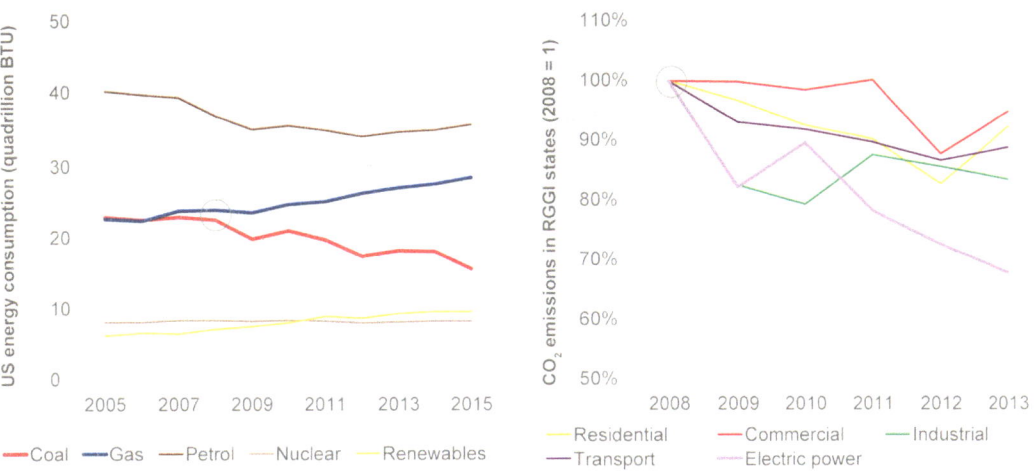

Fig. 40 The decision of RGGI member states to limit coverage to the power sector exaggerated the impacts of the unconventional gas technology shock. *Source* Calculated from U.S. Energy Information Administration (2016) data

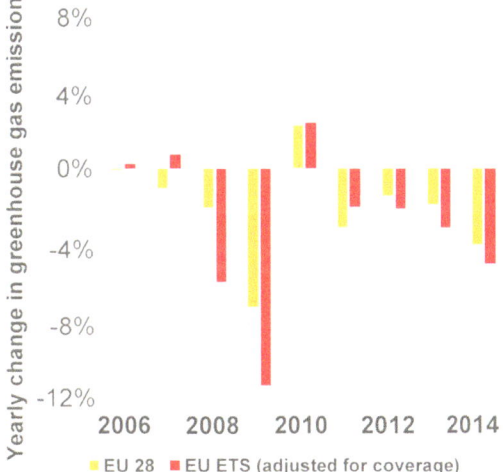

Fig. 41 Production shock concentrated in EU ETS sectors exaggerated the impact of the EU recession on demand for emission units. *Note* Production levels in EU ETS sectors are proxied by the weighted production value of NACE codes 17, 19, 20, 23, 24 and 35 using constant 2010 prices disaggregated by industry sector. This includes industries responsible for the vast majority of EU ETS emissions covered, including the manufacture of paper, coke, refined petroleum, chemicals, base metals, other mineral products and electricity and gas supply. Emissions have been adjusted to account for geographic expansion and changed coverage arrangements to ensure inter-annual consistency. *Source* OECD (2016) and Eurostat (2016)

sector-specific shocks. Shocks will often affect different parts of the economy in different ways.

Technology shocks will often drive major change in the emissions profile of a single sector, while leaving others largely unaffected. For instance, the expansion of unconventional gas extraction in the USA had a major impact on power sector emissions. This in turn was reflected in demand for units in the Regional Greenhouse Gas Initiative (RGGI), which only covered the electricity sector. In RGGI member states, emissions from electricity generation declined by more than 30% between 2008 and 2013, while emissions from other sectors declined at a far slower rate. This is shown in Fig. 40.

Economic shocks can also be concentrated in specific sectors. For instance, the EU ETS primarily covers emissions from the industrial and energy sectors. These sectors were particularly affected by the recession in 2007–08, where production levels in those sectors covered fluctuated far more than in the EU as a whole. This was reflected in emission outcomes, and therefore demand for European emission allowances. This sharp fall in demand was largely responsible for the fall in prices that occurred in 2008. This is shown in Fig. 41.

(2) Banking

The banking of carbon units can both increase the cost-effectiveness of an ETS by enabling substitution of abatement across time and improve the robustness of the market. Banking lets firms stockpile carbon units for later use. This makes caps across compliance periods function as a budget rather than a target. This in turn means that when emissions in a compliance year are lower than expected, carbon units of that vintage retain their value, as they can still be used to discharge liabilities. Banking increases efficiency as it enables emissions to be reduced at the time when mitigation is at its lowest cost; it

also increases robustness by ensuring that carbon units retain a positive value even after a shock. However, this also means that when a shock is large, banking can lead to a persistent oversupply that can continue to depress prices for many years.

The importance of banking is demonstrated by the experience of the EU ETS. Banking of units was not allowed from phase 1 to phase 2 of the ETS, which led prices to fall towards zero as the level of oversupply was revealed. Subsequently, the European recession in phase 2 of the ETS, led to large falls in industrial production, which resulted in an oversupply more than 10 times larger than in phase 1. The ability to bank units meant that prices remained low, but greater than zero. However, this has also contributed to the prolonged depression in prices in phase 3 of the ETS, as shown in Fig. 42.

(3) Linking and offsets

Linking with other emission trading systems, or with credible offset mechanisms, will increase the efficiency of carbon markets, and may strengthen its robustness. Linking to other ETSs generally stabilises demand across markets, because overall emissions become less dependent

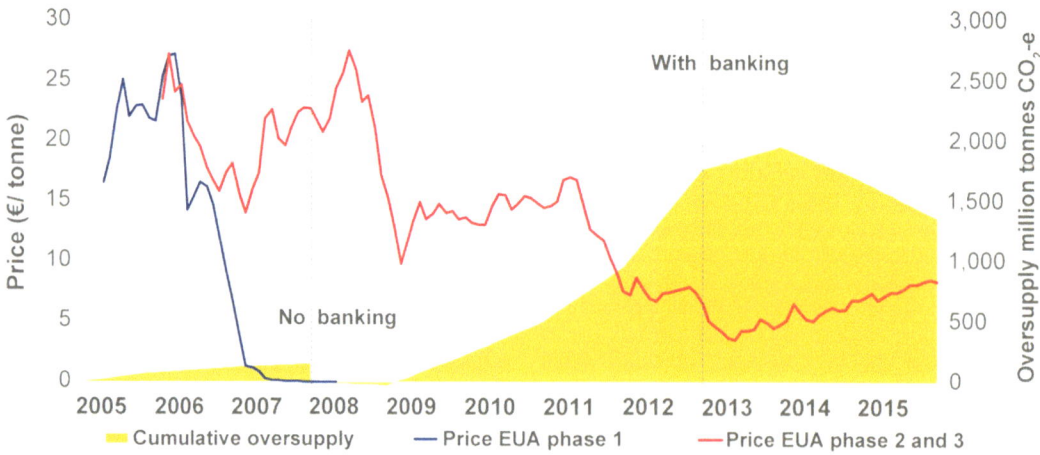

Fig. 42 Banking ensured prices remained positive despite oversupply in the EU ETS. *Note* European emission allowances (EUA) prices based on December delivery futures prices. Cumulative oversupply is calculated as total EUAs allocated (both free and auctioned) plus the number of Kyoto units surrendered, minus verified emissions. *Source* European Environment Agency (2016), Intercontinental Exchange and Quandl (2016)

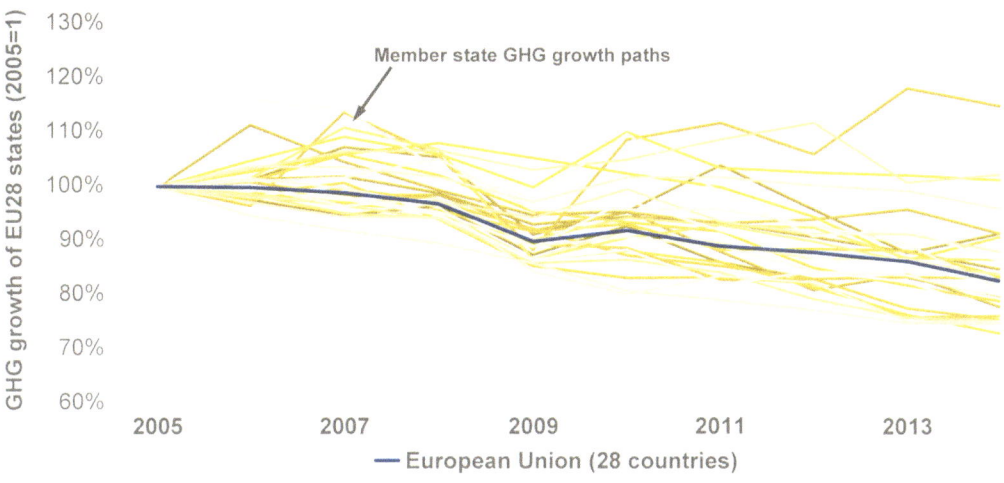

Fig. 43 Volatility in country emissions can be stabilised by linking markets. *Note* GHG = greenhouse gas emissions. *Source* Eurostat (2016)

on economic circumstances in any one market. For instance, emissions in the EU as a whole exhibit much more stability than emissions in an average EU member state, as is shown in Fig. 43. A link can affect:

- supply only, such as the use of Kyoto offsets in the EU ETS and New Zealand ETS;
- demand only, such as the proposed one-way link between the Australian ETS and the EU ETS; and
- supply and demand, such as the link between RGGI member states and California and Quebec.

A supply-only link can reduce prices within an ETS but may also be a mechanism for transmitting supply shocks through low prices. A demand-only link can increase the cost of mitigation but can also stabilise prices in the event of a demand or technology shock. Supply and demand links mean that shocks are transmitted between markets, but their effects are usually tempered. When a small market links with a larger market, however, economic circumstances in the larger market can dominate circumstances in the smaller market.

International offsets can provide substantial abatement at low cost and support the expansion of carbon markets. However, offsets often suffer from quality concerns, because determining the actual level of emission reductions from project or sectoral crediting mechanisms remains difficult. As such, offsets are often subject to a set of limits that seek to constrain their impact on the broader functioning of markets. These limits come in two main forms:

- qualitative limits allow only certain types of units to be used in an ETS (for instance, restrictions on the source country of Kyoto units in the EU ETS); and
- quantitative limits restrict the number of a certain unit class that can be used. When a limit is binding (or expected to bind), a quantitative limit will force prices to diverge.

The propensity for prices to diverge when a qualitative limit is applied or when a quantitative limit binds means that there can be a direct trade-off between efficiency and robustness. For instance, prices of New Zealand emission units closely tracked the Kyoto unit price from 2010 to 2013. However, the prospect of a ban on Kyoto units and its subsequent introduction led prices to diverge after this period. In introducing this ban, New Zealand made a trade-off between efficiency and robustness, forgoing a low-cost source of

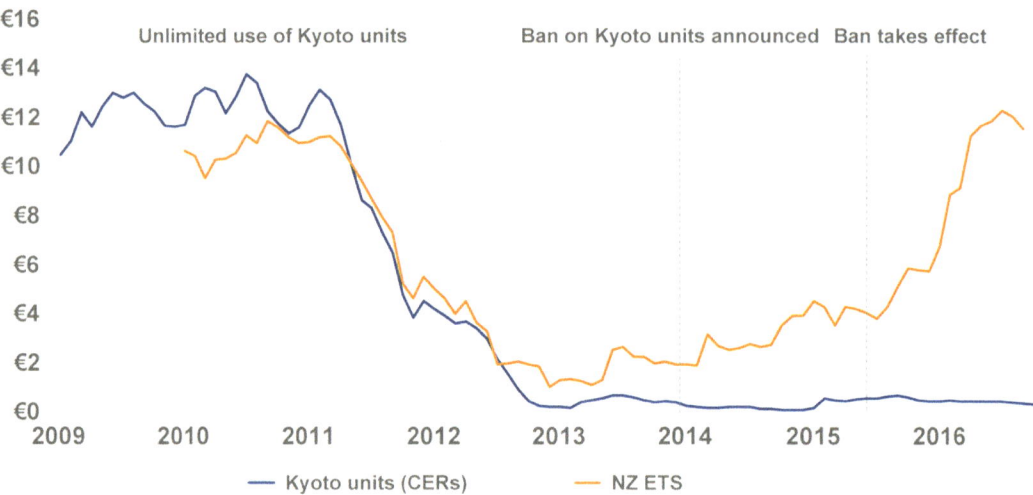

Fig. 44 Compliance rules determine relative prices. *Note* CER = certified emission reductions. *Source* Intercontinental Exchange (2016) and International Carbon Action Partnership (2016). Exchange rates from OECD (2016)

mitigation to achieve a carbon price more consistent with the long-term need to reduce emissions. This is shown in Fig. 44.

(4) **Price stabilisers**

Price stabilisers can be a useful tool to maintain a robust carbon price signal, especially in new markets. The policy compromises that are made when establishing a carbon market mean that carbon markets may be particularly fragile to economic shocks. There are two main types of price stabiliser that can be used: price-contingent stabilisers and quantity-contingent stabilisers.

Price-contingent stabilisers intervene in the market when one or more price criteria are met. These stabilisers include:

- price floors, auction reserve prices and surrender charges, which seek to maintain a minimum price or the equivalent price incentive;
- price ceilings and contingency reserves, which seek to maintain a maximum price; and
- price corridors, which seek to ensure that the price remains within a given range.

Most stabilisers target a price level by adjusting the level of supply or demand in the

market. In its purest form a price floor will provide unlimited demand, and a price ceiling will provide unlimited supply, when prices hit certain predetermined levels. More common is the use of soft-price controls like auction reserve prices, which limit supply by setting a minimum price at auction, and contingency reserves, which increase unit supply when a specific price level is reached. A price corridor refers to stabilisers that seek to maintain both a minimum and a maximum carbon price. Price corridors, in the form of auction reserve prices and contingency reserves, are used in both the Californian ETS and RGGI. The effect of these policies on prices in these markets is shown in Figs. 45 and 46.

Surrender charges do not seek to maintain a specific price but try to maintain a minimum price incentive by charging a top-up fee for each unit that is surrendered for compliance. A surrender charge is currently used in the UK to encourage greater levels of mitigation than those driven by the current low carbon price.

Quantity-contingent stabilisers are less common, the main example being the EU's market stability reserve (MSR). The MSR seeks to correct imbalances in the carbon market by adjusting supply based on the number of excess units currently in the secondary market. When the quantity of units in the secondary market is

Fig. 45 Soft price floors and ceilings may result in prices deviating outside their target range. *Source* Regional Greenhouse Gas Initiative (2016)

Fig. 46 California's auction reserve price has limited falls in the carbon price, but also reduced revenues. *Source* Climate Policy Initiative (2016)

above a certain level, units that were to be auctioned are instead placed in the MSR. When the quantity of units in the market is below a predetermined level, it reintroduces some of the previously withheld supply. As the MSR withdraws supply gradually, it may be better suited to addressing temporary imbalances rather than structural oversupply. Figure 47 demonstrates the way in which the MSR would alter supply given an illustrative emissions trajectory.

Price controls can be difficult to integrate with other policy mechanisms and may create

difficulties in linking with other markets. By making supply responsive to prices, the use of price controls may undermine the environmental integrity of a link. It is possible that if a price floor is triggered in one jurisdiction, it can increase supply to such an extent that net emissions in that jurisdiction exceed the emissions that would have occurred under business as usual. Price controls can also have difficult distributional consequences; if triggered, they may result in a net transfer of resources from one jurisdiction to another. Harmonising price

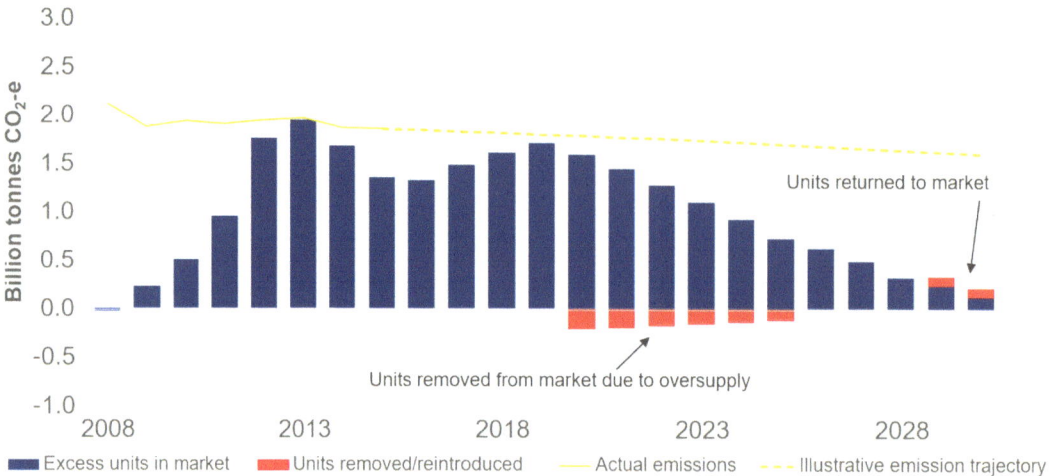

Fig. 47 The EU's market stability reserve adjusts the quantity of units in the market in response to over- or undersupply. *Note* Illustrative emissions trajectory developed for illustrative purposes only. Cumulative oversupply is calculated as the total EU emission allowances allocated (free and auctioned) plus the number of Kyoto units surrendered, minus verified emissions. *Source* European Environment Agency (2016)

controls may deal with these issues to some extent. However, this entails other complications; for instance, it may require the development of common auctioning platforms or agreeing to methods to correct for exchange rate fluctuations.

(5) **Secondary market development**

Policy design can support the development of a liquid and transparent secondary market, which increases efficiency and improves price discovery. In particular, market design should seek to support the development of institutions that lower transaction costs and reduce the counterparty risks borne by participants engaging in the market.

The market relies on central counterparties to reduce delivery risks on contracts. Active trading of emission units requires that market participants are confident that their counterparty will deliver either units or payment, as agreed. In a large market with many buyers and sellers, assessing the financial risks associated with each counterparty can be arduous. As such, markets often rely on trusted central counterparties to reduce these risks and facilitate trade. These central counterparties ensure that all parties meet minimum standards and will guarantee delivery

in the event of a default from either party. They often support both over-the-counter (OTC) trading and exchange trading platforms.

Cost reductions are sought through movement from trading in OTC markets to exchange-based trading of standardised contracts. Exchange-based trading generally develops in markets with large numbers of buyers and sellers, as the aggregation of trades provides greater liquidity. Over the course of phases 1 and 2 of the EU ETS, the market gradually moved from OTC trading to exchange-based trades. Exchange-based trading has several advantages, including providing transparent signals on pricing and reducing costs. The shift to exchange-based trading in the EU ETS is illustrated in Fig. 48.

The growth of futures markets reflects the desire to hedge risks and develop a longer-term price signal for long-lived assets. Over phases 2 and 3 of the EU ETS, futures markets expanded and have since consolidated, with liquid futures markets operating several years in advance—this is shown in Fig. 49. Nearly all trades in the EU ETS secondary market are now futures contracts. The most commonly traded contracts are within-year and December delivery futures, while large markets also exist for December delivery futures contracts running up to 2020.

Fig. 48 As markets mature and participation increases, trading tends to move to secondary exchanges. *Note* As source data were not available, the figure was created by digitalisation; as such it is indicative only

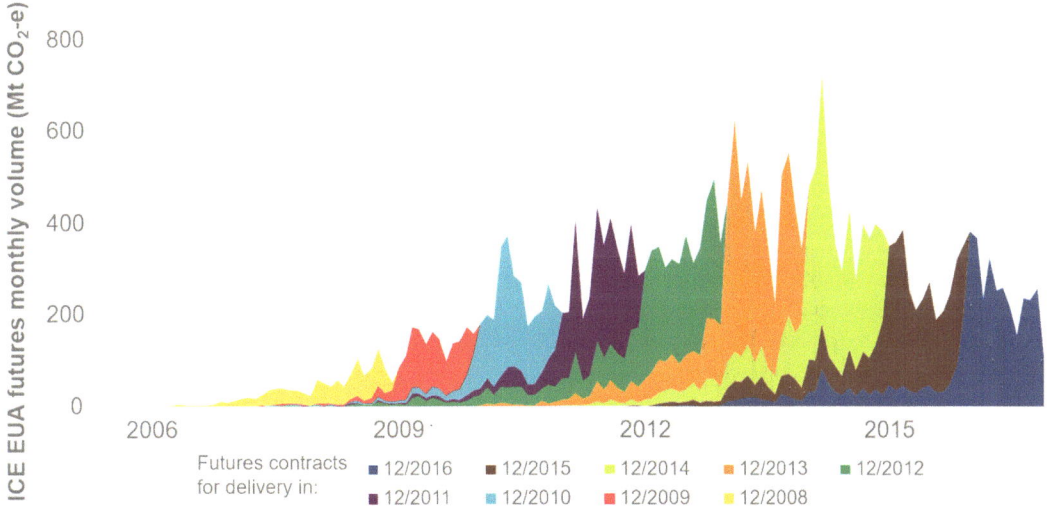

Fig. 49 The growth of futures markets provides new opportunities for risk management and improves price discovery. *Note* Volumes for December delivery futures only. ICE EUA = Intercontinental Exchange EU (emission) allowances. *Source* Intercontinental Exchange and Quandl (2016)

The EU's changing institutions and regulations have supported these market developments. In phases 1 and 2 of the EU ETS, registries were managed by member states, creating significant duplication across jurisdictions. However, trade in units was concentrated in just a few of these registries, which were linked with active exchange platforms—in particular, France, which hosted the BlueNext exchange. Trading was facilitated through open market access,

which enabled liable entities, financial institutions and other individuals and businesses to hold emission units. Instances of fraud in spot markets in 2008 and 2009 led to more stringent regulation, with spot markets regulated as financial products, whereas they were initially regulated as commodities. These problems on the spot market also accelerated the shift toward futures contracts, which were regulated as financial products from the outset.

EU ETS Phase 1 and 2

EU ETS Phase 3 and beyond

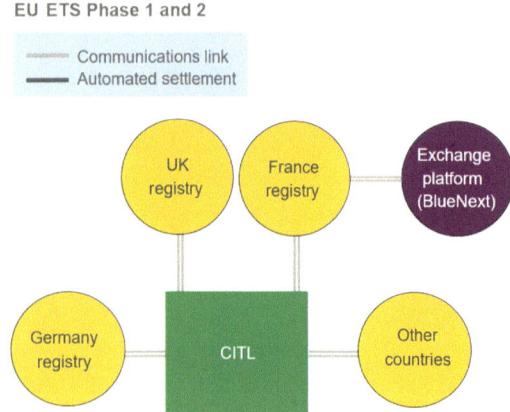

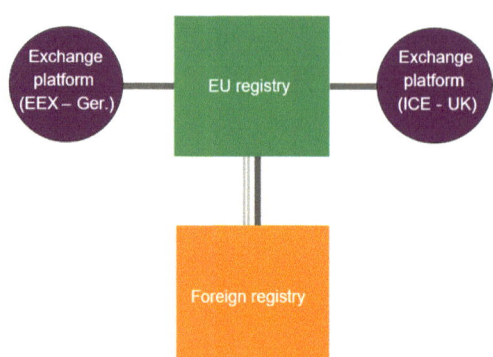

Fig. 50 The design of the EU registry system has become more centralised, and has developed to facilitate the development of exchange trading platforms. *Note* CITL = community independent transaction log; EEX = European Energy Exchange; ICE = Intercontinental Exchange. *Source* Vivid Economics

In phase 3 of the EU ETS, national registry systems were consolidated into a single EU registry, which supported the development of a more efficient and mature market. The move to a centralised market increased the ease of trade across the EU by requiring a common set of processes and approaches for accessing the registry. The move to centralised reporting increased the transparency of market information, which enabled better decision-making by market participants. The EU registry also included a new type of account for exchange platforms. This simplified the process of engaging in exchange-based trading, while ensuring that security standards were maintained. The single registry also facilitates linking with other carbon markets in the future, by enabling registry links to occur under specified circumstances. These changes in registry design are illustrated in Fig. 50.

8.3.4 Managing Policy Interactions

Carbon pricing operates within a broader policy mix. Many of these other policies will influence how a carbon market functions and its ability to reduce emissions efficiently. Policies that impact the operation of the carbon market can include those that directly target climate change mitigation, as well as those that target other policy objectives. They can be grouped into three broad categories:

- complementary policies improve the functioning of carbon markets and increase efficiency; this in turn reduces costs and the likely carbon price;
- overlapping policies duplicate some of the incentives provided by carbon markets. These may increase or decrease the overall efficiency of climate policy depending on policy design, but all overlapping policies tend to reduce the equilibrium carbon price; and
- countervailing policies provide incentives that oppose those provided by the carbon market. These will reduce the efficiency of climate policy and increase prices.

These policy groupings are outlined in Fig. 51.

(1) Complementary policies

Complementary policies are those that improve the functioning of an ETS by addressing market or regulatory failures that can impede the efficient operation of the carbon market. There is a wide variety of complementary policies that can improve market functioning in several ways. Some examples include:

- mandatory energy efficiency labelling, which reduces search costs and allows people to

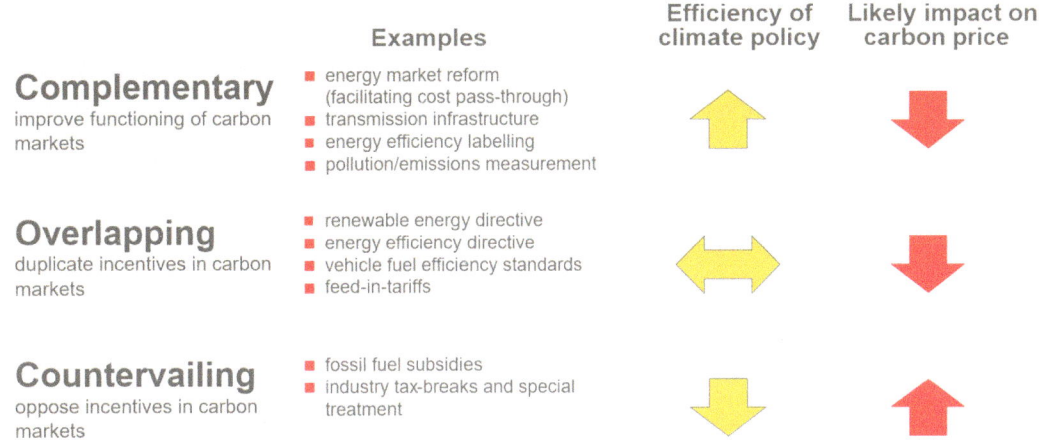

Fig. 51 The efficiency of climate policy is affected by the interaction of carbon pricing with complementary, overlapping and countervailing policies. *Source* Vivid Economics

make better decisions regarding appliance purchases. Figure 52 shows how the demand for new refrigerators shifted towards more efficient models following the introduction of energy use labels and minimum performance standards; and

- direct measurement of methane in underground coal mines for safety purposes, which allows those emissions to be priced more accurately than using a standard regional factor.

In many cases, carbon pricing can work alongside complementary policies to support

policy objectives other than emission reductions. This provides the opportunity to implement a broader package of reforms that enhance the performance of each policy and support sustainable growth.

(2) **Overlapping policies**

Overlapping policies can have a positive or negative impact on the efficiency of climate policy. Some will increase efficiency, especially when they are addressing a genuine market failure. For instance, without guidance, consumers

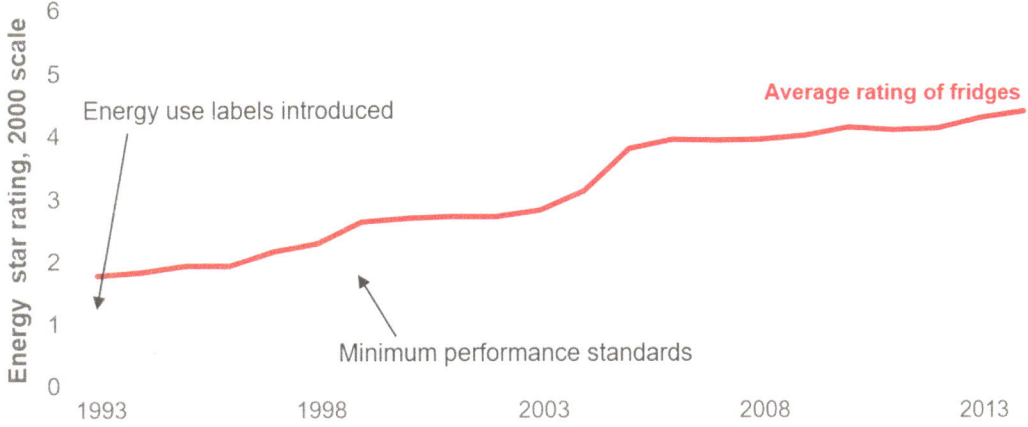

Fig. 52 Labels may have helped improve consumer choice of refrigerators in Australia. *Source* Energy Efficiency Strategies (2016)

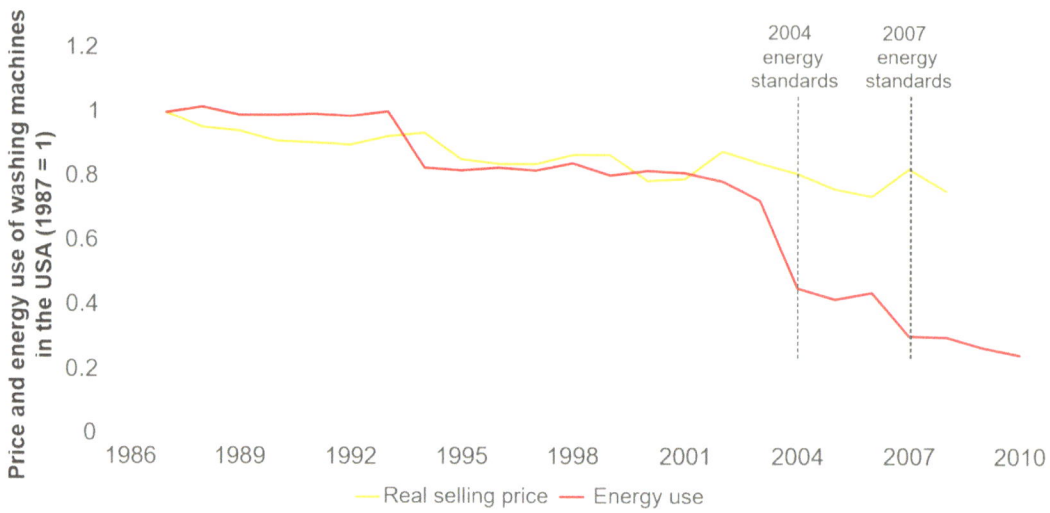

Fig. 53 Energy efficiency standards in the USA have reduced energy use at a low cost. *Source* International Energy Agency (2015)

will often purchase appliances that are inefficient and high cost to operate. This can occur for a variety of reasons. For instance, consumers may seek to reduce the time costs associated with researching the relative performance of different appliances and not consider the environmental or energy impacts of their choices. Alternatively, landlords may purchase cheap but inefficient appliances, which are higher cost because their tenants pay for high rates of energy use. Because of these and other market failures, the introduction of energy efficiency standards has reduced emissions cost-effectively in many countries. These standards improve the quality of the stock of appliances sold, while in many cases having very little impact on prices. Figure 53 shows the impact of the introduction of energy efficiency standards for washing machines in the USA, which resulted in substantial reductions in energy use while having no noticeable impact on prices.

In other cases, overlapping policies will reduce the efficiency of a carbon-pricing system, particularly when multiple policy instruments seek to address the same market failure. An example of this is the use of feed-in tariffs in Germany, where generous subsidies for the generation of solar power led to significant growth of the industry. However, this mitigation occurred at a very high cost and placed downward pressure on prices in the ETS.

California's low-carbon fuel standard is an example of an overlapping policy that probably reduces the effectiveness of the carbon price. The fuel standard is enforced through the creation and trade of certificates, expressed in tonnes per CO_2 equivalent. As such, it duplicates the incentives of the California ETS, which also requires carbon units to be surrendered for emissions from the combustion of transport fuels. This reduces the efficiency of climate policy as it means that the transport sector faces a higher effective carbon price than other sectors of the economy. This larger incentive to reduce emissions in the transport sector also puts downward pressure on the carbon price. The disparity between the effective carbon price faced by transport and other covered sectors is illustrated in Fig. 54.

(3) Countervailing policies

Countervailing policies are those that act in opposition to the incentives created by a carbon market. The most obvious example of a countervailing policy is fossil fuel subsidies. These subsidies are often large and remain in place in many jurisdictions that are also subject to an

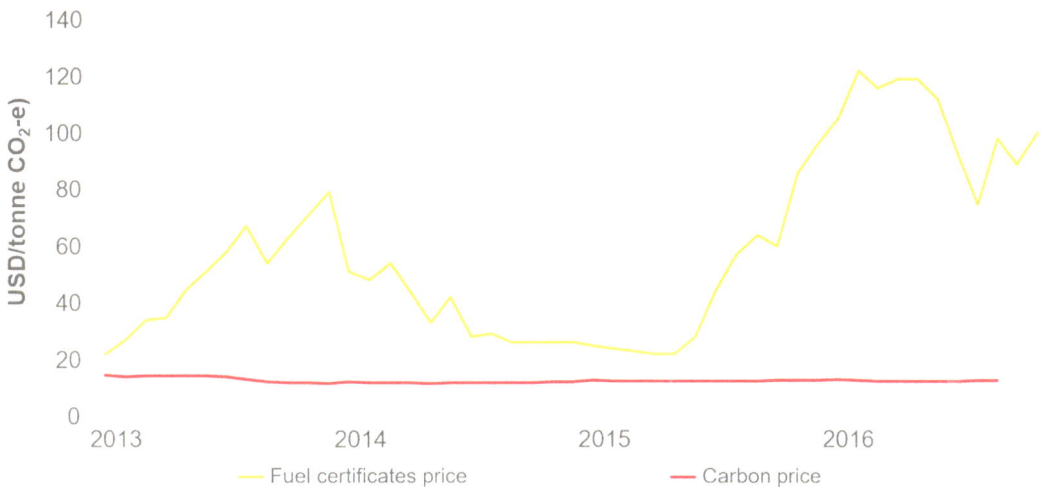

Fig. 54 California's low-carbon fuel certificates duplicate carbon price incentives, with their high price reducing both demand and price in the carbon market.

Source California Air Resources Board (2016) and Climate Policy Initiative (2016)

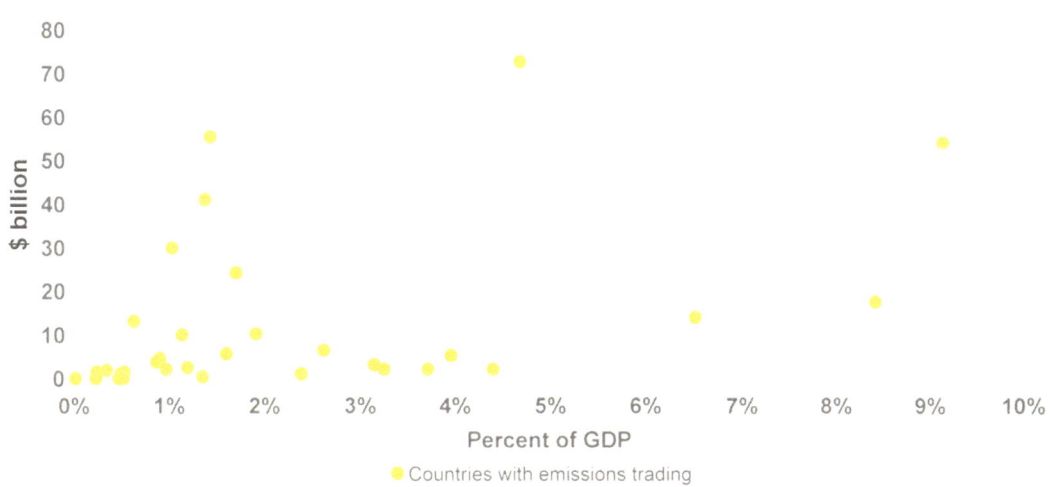

Fig. 55 Many countries retain expensive fossil fuel subsidies that counteract the goals of their carbon markets. *Source* International Monetary Fund (2015)

ETS, as shown in Fig. 55. Other countervailing policies may be less obvious. For instance, urban mobility policies counteract the impact of the carbon price, even though this was not their original intent.

In some cases, a jurisdiction may decide to retain a countervailing policy despite its impact on the functioning of the carbon market. This may occur when the jurisdiction judges that the negative impact of the policy on the efficiency of climate policy is more than outweighed by its benefits, and that alternative policies are unable to deliver the desired policy outcomes.

8.4 Evolutionary and Revolutionary Change in Carbon Markets

Changes in carbon markets can be evolutionary or revolutionary. China will shortly establish a carbon market. On December 19, 2017, the National Development and Reform Commission (NDRC) announced the official launch of China's carbon emissions trading system, starting with the power generation sector. Previously, the NDRC had issued the Plan for Building a National Carbon Emissions Permit Trading Market (for the Power Generation Sector), which outlined the design of China's carbon emissions trading system. However, the design is likely to change over time in response to shocks and changing policy priorities. Identifying the changes necessary and what they entail can help guide the future policy direction for China's carbon market. Our research shows that the cost-effectiveness of climate change mitigation will become more important over time as greater levels of higher-cost mitigation are required. This means that policymakers may wish to move from a carbon-pricing system that is less weighted towards cost-effectiveness to one where the cost-effectiveness of the policy is a more important consideration. The nature of the policy change required to give this effect depends on the nature of the problem and China's specific circumstances.

Evolutionary and revolutionary change differ in their speed, level and type of impact. In general, a change will be more revolutionary if it occurs rapidly, has a large impact and changes the nature of the system, rather than just adjusting its efficiency or the distribution of costs. For instance, policy change that results in net zero emissions before 2050 through a fundamental change in the energy system and in industrial composition would be a revolutionary change (Fig. 56).

The decision to pursue evolutionary or revolutionary change depends on the rate and level of change required, the certainty of impacts, and the political and institutional feasibility of delivery. A more revolutionary approach will be required when a faster transition or a larger change is required. Consideration must also be given to the uncertainty of impacts, with risk-averse decision makers likely to prefer evolutionary change when the level of uncertainty regarding a policy's impact is lower. The political feasibility of a change is also important. For example, evolutionary change may be more appropriate when the trade-offs required by a change are large, or where the need for change is contested. The level of institutional capacity also plays a role: if institutions are more capable, they will be more able to manage change effectively and respond to changing circumstances as they occur (Fig. 57).

Climate change has aspects that suggest a balance between evolutionary and revolutionary change. The rate and level of required emission reductions suggest the need for revolutionary change. However, the difficult politics of such a

Indicator	Revolutionary	Evolutionary
Rate	Rapid change (e.g. net zero emissions before 2050)	Gradual change (e.g. net zero emissions after 2050)
Level	Large impact (e.g. high carbon prices, radical change to industry composition)	Small impact (e.g. low carbon prices, changes competitiveness within industries)
Type	Changes the functioning of a system (e.g. liberalising energy markets)	Change to efficiency or cost burden (e.g. shift from free allocations to auctions)

Fig. 56 Evolutionary and revolutionary change differ in their rate, level and type of impact. *Source* Vivid Economics

Indicator	Revolutionary	Evolutionary
Rate	Faster transition required	Slower transition acceptable
Level	Large change is required	Change has a smaller impact
Certainty	Impacts are largely certain	Impacts are uncertain
Political feasibility	No regrets or small trade-offs with broad support	Requires significant trade-offs, widely contested
Institutional capacity	Systems and processes are sufficient to handle change	Capacity constraints make change difficult

Fig. 57 The appropriate approach to policy change should be based on the rate and level of change required, the certainty of impacts, and the political and institutional feasibility of the change. *Source* Vivid Economics

change, and limitations in institutional capacity, suggest that a rapid evolution may be more appropriate (Fig. 58).

Achieving the desired result requires careful consideration of how to sequence change, with a jurisdiction's choices heavily dependent on its specific circumstances. Some elements of climate policy may be more conducive to revolutionary change than others. The interaction of these elements is therefore a key determinant of the sequencing of policy change.

China looks likely to use a slow start to emissions trading; if it does, it should consider adopting price controls and a clear reform path to reduce the risk of becoming stuck in the transition trap. Design trade-offs in the early stages of

carbon pricing are common and can result in fragility. This means that an active approach to price management, such as a tight price corridor, can be used to ensure incentives are maintained while the market develops. This can be supported by signalling the direction of climate policy. For instance, committing to specific changes through legislation can provide clear guidance on the direction of policy to help maintain prices and public support for carbon markets in the event of a shock.

In the longer term, climate policy should develop in a manner that delivers:

- increased clarity of price signals, with the ETS playing an increasingly large role in

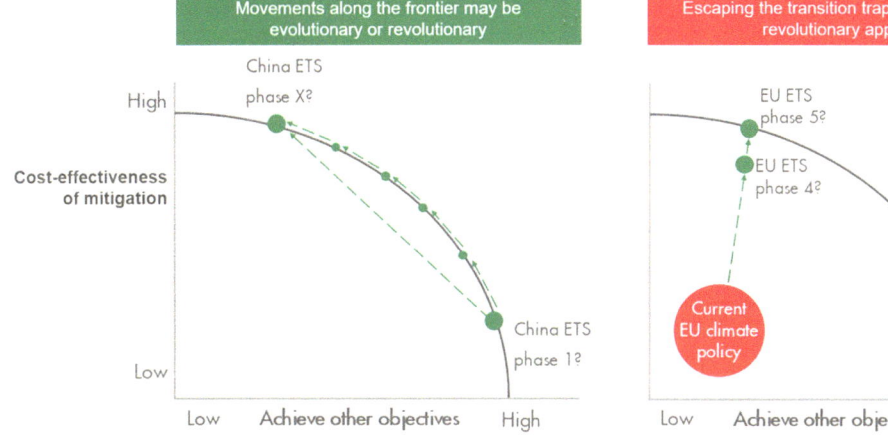

Fig. 58 The choice between evolutionary and revolutionary change is context-dependent. *Source* Vivid Economics

	Evolutionary	Revolutionary
Creation	■ Target setting based on national circumstance, gradually transitioned to global carbon goal by 2030 ■ Bottom-up cap setting to 2020 ■ Net carbon neutrality after 2050	■ Move to binding, absolute targets and top-down cap setting by 2020 ■ Targets aligned with goal of achieving global net carbon neutrality before 2050
Distribution	■ Move from grandfathering to production-based allocations by 2020 ■ Assistance to emissions intensive industries only with high assistance rates, declining gradually over time ■ Auctioning of remaining units	■ End grandfathering as soon as possible, immediate development of process level baselines ■ Assistance limited to emissions-intensive, trade-exposed industries, and quickly declines ■ 100% auctions by 2030
Governance	■ Start with low ($10-20) price floor or MSR, may also include price ceiling, both growing slowly ■ Phase in coverage starting with electricity generators and major industrial energy users ■ Slow movement to linking with international partners	■ High (>$40) price floor that increases rapidly ■ Immediate coverage of all energy, transport and industrial emissions, with full banking ■ Agriculture and land sectors covered by 2025 ■ Near-term linking with willing international partners
Policy interactions	■ ETS operates as part of a broad policy mix, with other policies playing dominant role before 2030 ■ Carbon market gradually becomes more influential in economic decisions	■ ETS plays immediate major role in investment decision-making, becomes the major driver of decarbonisation investment by 2025 ■ Other policies aligned with decarbonisation objective

Fig. 59 Evolutionary and revolutionary change can differ with regard to all aspects of scheme design. *Note* MSR = market stability reserve. *Source* Vivid Economics

economic activity and in different sectors and sources of emissions facing equivalent price signals; and

• a greater international focus, with domestic circumstances playing a progressively smaller role in determining policy relative to the broader global need to reduce greenhouse gas emissions.

Figure 59 outlines what evolutionary and revolutionary change could involve for China's climate policy. The major difference is the rate of change, with the revolutionary approach adopting more cost-effective policy approaches sooner and achieving net zero emissions earlier. In the revolutionary case, carbon pricing becomes the major driver of new energy investment in the short term, supported by a robust price corridor, broad coverage and near-term linking with international partners.

The following sections provide more detail on the differences between an evolutionary or revolutionary change in the aspects of carbon market design.

8.4.1 Creation of Carbon Units

Targets and cap setting are closely related, with changes in targets often automatically reflected in caps. Figure 60 outlines the interrelationships

and potential sequencing of policies regarding the creation of carbon units. It demonstrates the relationship between targets, caps and enabling factors, to demonstrate two potential pathways for reform.

In these scenarios targets shift over time from being heavily dependent on national circumstances to an assessment of the emission reductions required to stabilise the global climate system. Over time, emission targets move from quantified actions or targets based on emissions intensity, to targets based on absolute emissions. The shift to absolute targets is only likely to occur after robust projections of emission levels have been completed, to allow future targets to be calibrated against expected emission outcomes. Once absolute targets are established, China may choose to move from aspirational targets to binding commitments. This could include a commitment to make good any failure to achieve a targeted emissions level in a given year. At this time, China may work with international partners to establish principles for setting national targets to achieve the required global emission reductions. These principles can facilitate a shift in the factors that determine the emissions target, from a focus on national political and economic circumstances to principles-based global mitigation shares.

An evolutionary approach to creation

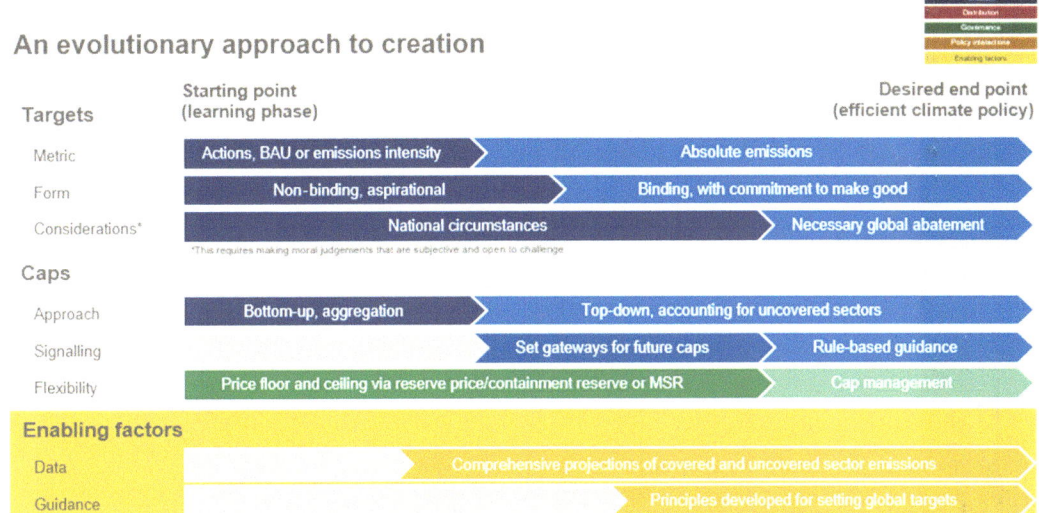

A revolutionary approach to creation

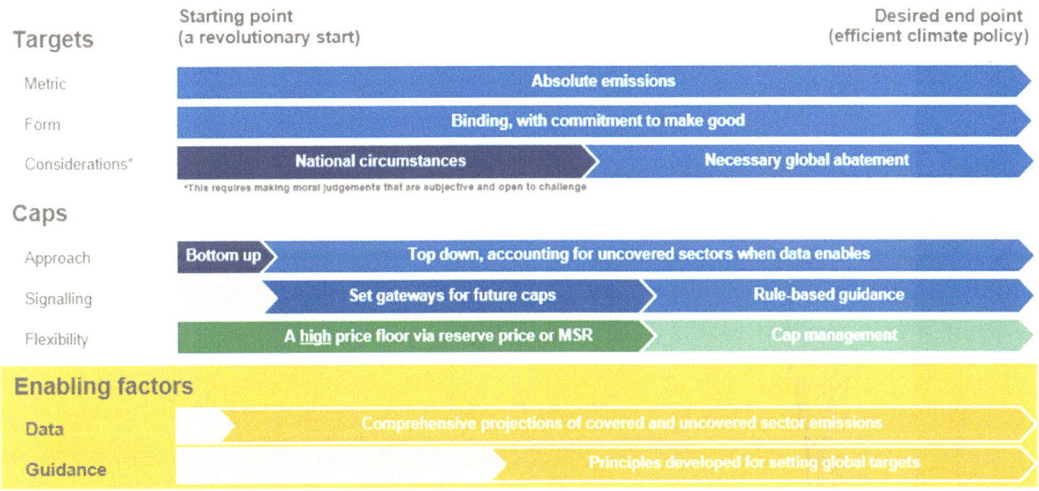

Fig. 60 Sequencing of evolutionary and revolutionary approaches to creating carbon units. *Note* BAU = business as usual. *Source* Vivid Economics

ETS caps follow a similar progression, moving from the bottom-up estimation of emissions from covered sectors to a calculation based on the required emission reductions. In the learning phase of an ETS, a jurisdiction may have insufficient information to establish a credible cap and may instead allow caps to vary, for instance, based on the level of free allocations in a given year. However, over time, the ETS cap should take on a larger role in ensuring that national targets are achieved, which would require a top-down calculation of the required cap, given targets and expected emissions in uncovered sectors. To provide guidance to markets, China may choose to establish gateways that indicate the range of future caps in the medium term. Eventually China may adopt a set of rules aligned across jurisdictions that outlines the processes for calculating caps.

The number of units released into the secondary market may be altered through the use of price flexibility mechanisms. In the early years of operation, this could include price floors, contingency reserves or market stability reserves. In the longer term, price management may be provided through flexible cap management, either independently or in concert with other jurisdictions.

A more revolutionary approach to the creation of units would see an immediate move to more stringent target conditions and a more rapid evolution of caps. China could choose to adopt absolute targets immediately and to make good any discrepancy that may be caused through uncertain uncovered sector emissions and covered sector response. Targets would be more stringent as they transition to reflect required global mitigation at an earlier stage, while changes to cap setting and price controls would also accelerate.

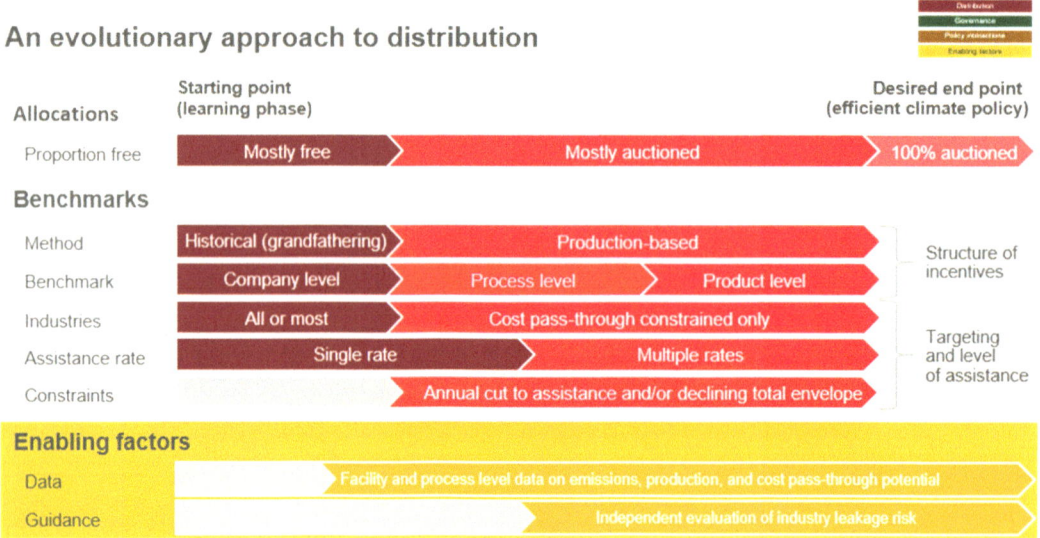

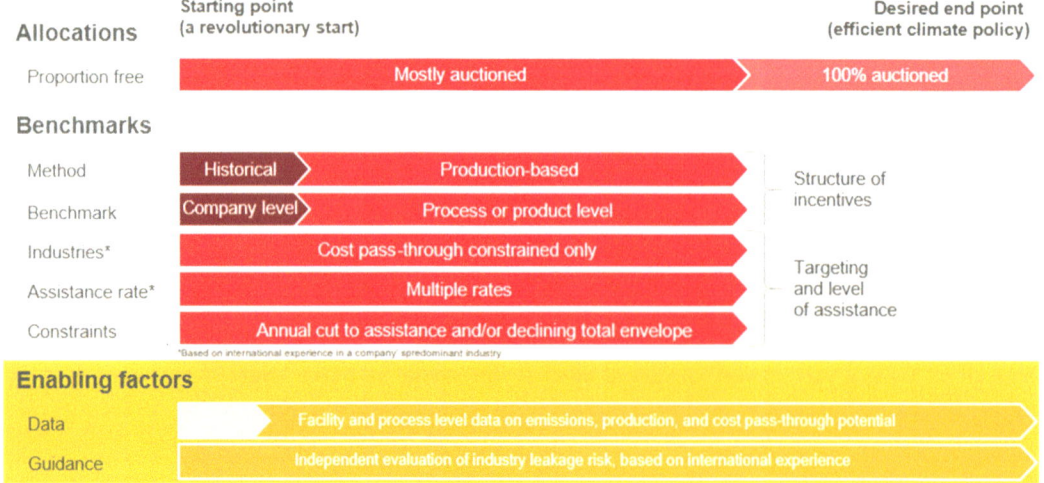

Fig. 61 Sequencing of evolutionary and revolutionary approaches to distribution. *Source* Vivid Economics

8.4.2 Distribution of Carbon Units

With the supply of carbon units determined, the next step is to decide how, and how many units should be auctioned and freely allocated. In the learning phase of the ETS, China may choose to allocate most units free of charge, as it seeks to introduce liable polluters to their obligations under the ETS and reduce the risk of carbon leakage. In this case, the method of freely allocating permits will determine which mitigation opportunities will be used. Figure 61 outlines in more detail the potential sequencing of distribution methods.

Data limitations may mean that initially units may be grandfathered or allocated according to a company's historical emissions. A quick movement away from grandfathered allocations will free up a greater proportion of units to be auctioned and reduce the risk that allocations are simply sold for profit.

Limiting allocations to those industries that are constrained by cost pass-through can be an effective means of avoiding carbon leakage while reducing costs. Determining whether an industry is constrained by cost pass-through requires facility- and process-level data to be collected to enable the assessment of emissions intensity and trade exposure. These data may also be used for benchmarking, which facilitates the movement away from grandfathered allocations. Over time, regular independent evaluations of cost pass-through can be used to more closely target assistance to those industries most at risk of carbon leakage.

The level of assistance to industries should be reduced over time to encourage greater efficiency and product substitution. This can be achieved by differentiating assistance rates by emissions intensity and/or instituting rules-based mechanisms to reduce allocations over time. In the long term, the market should move to 100% auctioning of units as the expansion and linking of carbon markets negates the risk of carbon leakage.

A revolutionary approach to allocations would prioritise the collection of data needed to assess cost pass-through and develop benchmarks. It would rapidly move to more cost-effective benchmarking methodologies and shift to the full auctioning of units at an earlier point in time.

8.4.3 Governance of Carbon Units

The rules that govern the carbon market will determine its cost-effectiveness and its robustness. As higher levels of mitigation are required, movement towards more efficient governance designs will help limit the costs associated with China's energy transition. Figure 62 provides further detail on the potential scheduling of reforms and their relationship with the allocation and distribution of carbon units.

Over time the role of an ETS can change, from first being a learning mechanism, to then being one that plays a major role within a broader policy mix, to eventually becoming the key driver of decarbonisation in an economy. This change is facilitated through changed market rules that act to broaden and deepen the impact of the carbon market on decision-making throughout the economy.

There are some design decisions, like banking, where China may choose to move to the optimal policy setting immediately or after a very brief period. This is because there are limited negative consequences of adopting banking early on, and banking can significantly improve the robustness of a carbon market.

Other decisions, like coverage, are subject to more complex trade-offs and capability constraints. For simplicity, an ETS may first cover a small number of high-emitting facilities in order to incentivise those mitigation options that have the potential to make the biggest impact. Over time this can be expanded to include all sectors of the economy with accurate measurement, reporting and verification (MRV) and, as MRV improves, can extend to all sectors.

Domestic offsets may be used in sectors that have insufficiently robust MRV to be subject to liabilities under an ETS, but where the mitigation outcomes of specific projects are well known. With a slow start to emissions trading, it may be

An evolutionary approach to governance

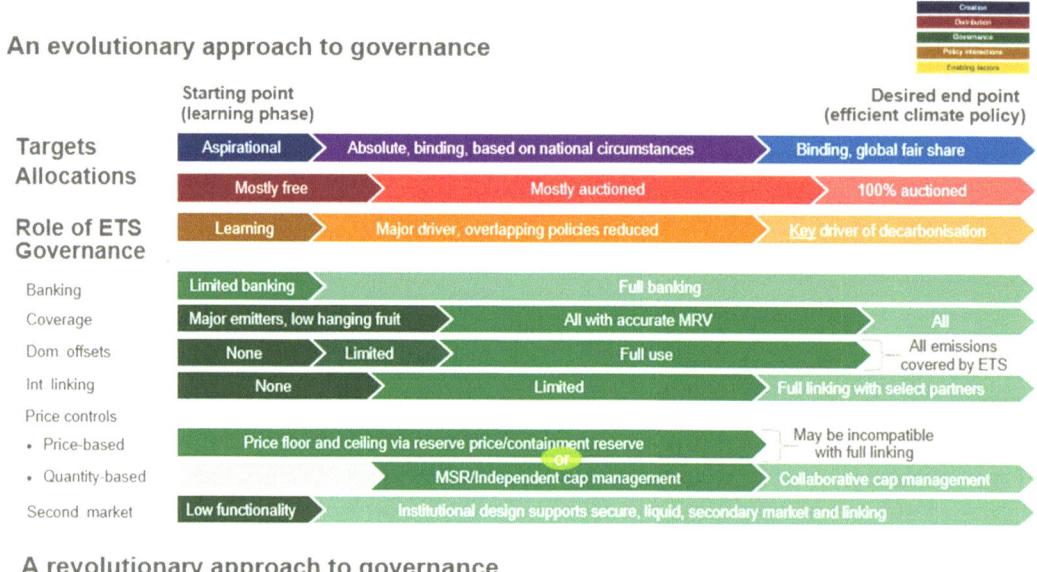

A revolutionary approach to governance

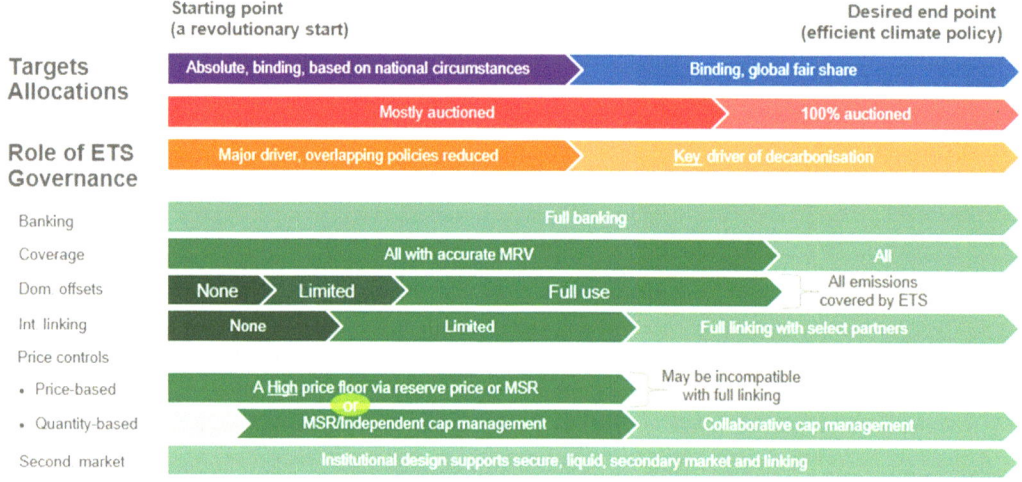

Fig. 62 Sequencing evolutionary and revolutionary approaches to governance. *Source* Vivid Economics

appropriate to apply quantitative limits to the use of domestic offset units to limit their potential impact on these potentially fragile markets. As the market matures, these limits may be removed, ensuring that domestic offsets trade at the same price as other carbon units. As MRV processes improve and more sectors are covered, the domestic offset market will gradually disappear.

Similarly, the use of international units may not be allowed in the early years of the carbon market given its fragility, but over time

movement to a fully linked global carbon market will reduce costs and increase the robustness of interlinked ETSs. Given the current state of the global carbon market, full linking may take several years to progress. In the interim, limited linking can be used, with quantitative limits used to quarantine the potential impacts of these units on markets. The choice to move to full linking will have implications across a range of policy areas. For instance, traditional price controls may be inconsistent with full linking as the divergence in prices between markets can lead to perverse effects. Fully linked jurisdictions will also need to agree to principles and rules for determining the minimum ambition of targets appropriate for each jurisdiction.

Price controls should apply from the outset to ensure that credible prices are maintained while the market is fragile. As the market develops, the price should be allowed to vary more freely, as market forces will be better able to identify the appropriate price. Direct price controls may need to be removed when jurisdictions fully link their carbon markets. At this point, linking partners may choose to empower an independent institution, like a carbon central bank, to manage caps in a manner that achieves targets while maintaining a credible price trajectory.

The development of the secondary market should proceed rapidly to increase cost-effectiveness and enable liable entities to hedge their risks. In the learning phase of an ETS, some market institutions, like the system of registry units, may not have the degree of functionality they would require as the ETS expands. Price discovery in the secondary market may be less important at this stage if a tight price corridor is in place. However, as the market expands and prices float more freely, providing the necessary institutions and market rules to support the secondary market becomes of vital importance.

A revolutionary approach to the governance of carbon markets would rapidly move to more advanced market designs and target higher prices through price controls and caps. Such an approach would have no learning phase but would instead start with robust design and rules. Banking would be allowed from the outset and coverage would be maximised. There would be more rapid movement to trade in domestic offsets and international units, with full linking occurring in the medium term. A rapid transition to a low-carbon economy would be supported initially by a high price floor or similarly robust quantity-based instrument, and by rapidly declining emission caps. The design of institutions and market rules would ensure efficient secondary market functioning, supported by exchange-based trading and liquid futures markets.

8.5 Challenges in Establishing a National Carbon Market

Establishing a carbon market is a complex and systematic task, especially the unification of China's carbon markets. Impeccable systems and institutions are needed, including determining the scope of the market, allocating allowances, choosing the allowance allocation method, building sophisticated monitoring, report and supervisory systems, and establishing a market control mechanism.[28] The design and implementation of each element has an impact on the emissions reduction outcome of the carbon market, and will have a wide and profound influence on the regional economy's development and the evolution of industry.[29]

However, due to insufficient time to prepare and gain experience, the creation of China's carbon markets is faced with some challenges: regulated tariffs, poor regulatory ability, a defective trading mechanism and poor market fluidity. Most regions are constructing carbon exchanges in accordance with administrative regulations. Total allowances in pilot regions are high, which decrease the allowance price significantly. The markets are not active enough, leading to limited trading volumes. The market supervisory and regulatory system needs further improvement. The awareness and skills of market participants need to be strengthened.[30]

Pilot regions face more difficulties than non-pilot regions as they are integrated into the national carbon market. The pilot regions have gained a lot of experience, but their existing trading systems need to be aligned with the national carbon market. They also face transitional obstacles related to the enterprises controlled, the allowance allocation methods used

---

[28]Zhang Xin, Challenges in Integrating Local Carbon Markets into the National Carbon Market, in China Economic & Trade Herald, 2015, (16), pp. 74–76.

[29]Zhang Xin, Sun Zheng, Meng Tianyu and Wang Ying, Reflections on and Suggestions for Regional Disparity in the Construction of a National Carbon Market, in China Economic & Trade Herald, 2017, (20), pp. 30–31.

[30]Zou Chunlei, The National Carbon Market is Emerging, in China Electric Power News, 2016.

and the greenhouse gas emission monitoring, report and verification system.[31]

8.5.1 The Failure to Achieve Market-Based Electricity Pricing Prevents the Carbon Market from Working Effectively

China's power industry has been dominated by coal-fired power plants for a long time. The power industry is not only a major primary energy consumer, but also a major carbon dioxide emitter. Its carbon dioxide emissions account for around 40% of the total emissions from China's fossil energy consumption. The power industry has inevitably become a major target for control, and it plays an important role in shaping the national carbon market.[32]

The Dian Gai No. 9 Order, issued by the central government in March 2015, opened a new phase in power system reform. In the previous 18 months, six key policies, including accelerating transmission and distribution tariff reform, were introduced. The pilot projects for transmission and distribution tariff deregulation were expanded to 20 regions (Shenzhen, western Inner Mongolia and 18 provinces). In addition, there are four pilot regions for comprehensive reform and two pilot regions for power sales reform. A new mechanism where "the electricity price is determined mainly by the market" will eventually be built and will be extremely beneficial to China's energy saving and emission reductions.

Overall, the 13th Five-Year Plan (2016–20) is a critical period for power reform and an initial stage for the carbon market. The challenge of how to dynamically align carbon market rules to power system reform needs to be overcome

urgently.[33] According to Li Jifeng, given that China is vigorously promoting tariff deregulation, the carbon emitted in power generation should be attributed to the power producers, and the rules of the carbon market should be designed on that basis.

Electricity tariff deregulation could improve the effectiveness of carbon markets and reduce emissions in three ways. It could facilitate technical progress in power producers and optimise the power mix; and it could encourage electricity users—both those covered and those not covered by the carbon market—to save energy. Optimisation of the power mix depends largely on energy authorities planning for power generation with renewables. The role of the carbon market should, therefore, be to reduce emissions by encouraging downstream users to save electricity.[34]

Prices in the carbon market, an important indicator of the effectiveness of the market's rules, reflect the marginal cost for market participants to achieve emissions reduction objectives. The lower the carbon price, the more effective the rules, and vice versa. Take China's carbon markets launched after 2017: calculations using a computable general equilibrium model show that the carbon cost for achieving the same emissions reduction if the power tariff is controlled is 18–32% higher than if the power tariff is deregulated. This is because electricity price control prevents the power sector from tapping the potential for low-cost emissions reduction, which intensifies the pressure on other sectors and increases the overall cost of emissions reduction. In other words, promoting the deregulation of the power tariff may not only reduce carbon emissions by the power industry, but significantly reduce emissions reduction pressure

---

[31]Li Zhuo, Obstacles to Integrate Pilot Regional Markets into National Carbon Market, in Invest Beijing 2016, (04), pp. 39–41.

[32]Zhang Lixing, Comparative Study on and Suggestions for Participation by Power Companies in Pilot Regions in Carbon Trading, in Resources Economization & Environment Protection, 2016, (12), pp. 16–17.

[33]Li Jifeng, The Design of Carbon Market's Rules should be Aligned to Price Deregulation, in China Energy News, 2016-09-12.

[34]Li Jifeng, The Design of Carbon Market Rules Should Be Aligned with Tariff Deregulation, in China Energy News, 2016-09-12.

on companies from other sectors in carbon markets.[35]

8.5.2 Carbon Emission Regulations Barely Support the Carbon Market

China has not imposed a compulsory carbon emissions permit trading system. Carbon trading in China is now mainly on a voluntary basis, i.e. businesses or individuals can voluntarily purchase carbon emission reductions. Although some provinces and municipalities, including Hubei and Jiangsu, introduced some local regulations on carbon emissions permit trading, laws have not been passed at the state level. There are no uniform settlement standards yet, nor effective supervision. In addition, it is difficult to give purchasers long-term economic incentives with voluntary emissions reduction trading.[36]

Once the basic framework of the national carbon market has been built and the market is running, each module should be managed by professional organisations. For instance, the China Securities Regulatory Commission could manage carbon trading, and the Certification and Accreditation Administration of the People's Republic of China could review and approve third parties. Establishing the carbon market will involve many departments, but China still needs one department to coordinate all other departments and mobilise forces across the country to complete the work. While China's carbon market is being established, the National Development and Reform Commission could be responsible for overall planning and coordination. The most scientific approach would be dual-level management, with the central government designing the main rules for market unification and local government enforcing them.

8.5.3 The Carbon Trading Mechanism Is not yet Mature

Compared with international carbon markets, China's trading mechanism for carbon products is not sound. Currently, there are many kinds of carbon trading market in the world, including curb exchanges, markets regulated by government, and markets that function on a voluntary basis. They differ in their method of allocating allowances and in their regulations and emissions reduction verification mechanisms. Direct trading between different markets is impossible. The international carbon finance market is highly fragmented. Although China is a developing country, any clean development mechanism projects it participates in can only be traded in as a primary market, through the International Carbon Fund. Trading and circulation between domestic sub-accounts are not allowed. Domestic products are isolated from the international carbon trading market, making China's carbon emission resources uninfluential. China is at the end of capital and resource chains. Due to the separation between China and the UN's registration mechanisms, the clean development mechanism cannot properly register carbon emission reductions.[37]

In addition, there is still a dispute over whether China's future national carbon market should choose centralised trading at a single exchange or decentralised trading at several exchanges under common rules.

The advantages of centralised trading are unified standards and centralised management, which is conducive to price discovery. It also matches China's "from top to bottom" approach to energy saving and emissions reduction. The competent authorities in central government could make flexible adjustments, creating a synergy between the carbon trading mechanism and other energy and climate policies. The planned national carbon market is expected to cover 4 billion tonnes of carbon allowances, the largest in the world. The huge size imposes very

---

[35]Lu Zhengwei and Tang Weiqi, Operation Experience from Domestic Pilot Carbon Markets and the Construction of a National Market, in Fiscal Science, 2016, pp. 81–94.

[36]Wang Chaoying and Gan Aiping, Problems in the Unification of China's Carbon Markets and Countermeasures, in Foreign Economic Relations and Trade, 2015, pp. 100–103 and 145.

[37]Feng Weiwei, China will Launch a Carbon Emissions Permit Trading Mechanism in 2017, in Energy Conservation and Environmental Protection, 2017, pp. 34–35.

strict requirements on the design and improvement of trading platforms and rules. If the trading rules of a single trading platform are not perfect at the moment the market is launched, there will be a strong impact, and the competition mechanism will not be in place and functioning correctly, which runs counter to the goal of optimising rule-related services.[38]

Decentralised trading means that trading is conducted simultaneously on many trading platforms under strictly unified calculation criteria and trading rules. Market systems are improved through free competition and the survival of the fittest. Decentralised trading is common in carbon exchanges in Europe and North America. First, unified, clear and transparent carbon emission calculation criteria should be set up to ensure that the objects traded in different markets are homogeneous and to avoid cross-market arbitrage. Second, unified registration systems should be set up to ensure the reliability of cross-market trading information. Once these conditions are met, exchanges should be allowed to carry out carbon trading at the same time. This will introduce competition among trading platforms, forcing them to improve services and motivating them to innovate, thus driving China's carbon market towards maturity and perfection.[39]

In decentralised trading, local exchanges could attempt to form a trading alliance or use other cooperative mechanisms to facilitate mutual recognition of their members. They could also work towards integration by engaging financial institutions to open accounts on behalf of clients. This would optimise the process of opening, registering and managing accounts and facilitate cross-market trading.

8.5.4 Fluidity Needs to Improve

China's pilot carbon markets have a common problem of poor liquidity, low volumes and low turnover. Currently, the daily trading volume of most pilot carbon markets is small. Although the phenomenon of zero deals on trading days during the non-performance period is becoming less and less frequent, a daily trading volume of only hundreds or dozens of tonnes is not uncommon.[40] Due to such low fluidity, it is difficult to attract financial and investment institutions to carry out stable and active carbon trading activities, which increases the likelihood that the market will be controlled by a few oligarchs.

The seven pilot carbon markets are dominated by trading through performance agreements. It is common for trading volume and value to be high when a performance period is near at hand, and low after it has ended. During the three performance years of 2013–15, the total trading volume of the seven pilot carbon markets was slightly above RMB 2 billion. This is not enough for a country like China with a huge carbon emissions reduction potential. The size of the global carbon market in 2011 was $176 billion (the EU Emissions Trading System accounted for more than 90%). Unlike the EU carbon market, which has various carbon financial products (futures, options, spot, forwards), China only has the spot carbon product, and no price discovery or risk hedging tools. China's carbon exchanges are far from mature.[41]

Moderate fluidity is the key to achieving reasonable prices and guiding companies to reduce emissions cost-effectively. Without fluidity, those companies planning to purchase allowances would not be able to buy them, and those planning to sell would not sell them. Nor would companies be able to compare the price with their own emission reduction costs and make cost-minimising emission reduction decisions. To improve fluidity, allowances must be

---

[38]Wang Junchun, A National Carbon Market is a High-Speed Train for China's Green Low-Carbon System, in Strategic and Emerging Industries of China, 2017, pp. 48–49.

[39]Chen Xiangguo and Ma Aimin, Construction of the National Carbon Market is Faced with New Challenges, in Energy Conservation and Environmental Protection 2016, pp. 24–25.

[40]Lu Zhengwei and Tang Weiqi, Operation Experience from Domestic Pilot Carbon Markets and Construction of the National Market, in Fiscal Science, 2016, pp. 81–94.

[41]Jiang Rui, Carbon Trading and Outlook on China's Carbon Market, in China Policy Review, 2017, pp. 52–56.

tight. Market players and traded products need to be diversified into futures, options and other allowance derivatives. The starting price should not be high. The policies should be progressive, which may strengthen investors' trust in the market and emissions reduction policies. More carbon asset management training should be made available for the companies controlled. Violations of regulations should be penalised.[42]

Carbon allowances should be traded in a fair environment with fluidity. To this end, risk control during trading, and supervision and management of market participants and staff, should be improved. Carbon exchanges play a unique role in building the national carbon market. Every pilot province and municipality should set up its own exchange. We know that other regions have expressed interest in building exchanges. Maintaining more than one exchange in an existing market encourages fair competition and improves services. But too many exchanges will reduce average profitability, which might lead to cut-throat competition and fragmentation of the national carbon market, resulting in local protectionism. We should pay attention to these problems when building the national carbon market.[43]

9 Strategic Transition and Structural Reform of Energy Companies

In the context of the global energy transition, international energy companies and China's state-owned companies are undergoing a series of significant strategic transitions that affect their decision-making, organisational structure, priorities, global planning, business models, finance and asset strategies, and digital platforms. In previous sections of Special report 4 we said that international oil and gas companies like BP,

Shell and Total have adapted their strategies to changing market conditions. However, international energy companies are not the only ones doing this: global electric utilities, especially grid companies, are also changing their strategies.

This section begins with the findings of research on power grids in transition by Vivid Economics, a UK-based consultancy. The section will then discuss the reform of China's state-owned energy enterprises (SOEs), since they are a key part of China's energy administration system reform. In the context of China's slowing economic growth, severe overcapacity, mounting environmental pressures, and adjustment of the energy mix, the reform of China's SOEs is an important element in the larger reform of state-owned assets. SOE reform focuses on optimisation, improving efficiency and unlocking dynamism. Both central energy enterprises and local state-owned energy companies have begun their reforms. Given its initial difficulties, the reform programme needs to be pushed continuously.

9.1 Structural Reform of Global Energy Companies— Electricity Grids in Transition[44]

Decarbonisation, decentralisation and intelligent technologies pose challenges to the global energy transition. As a result, structural reforms must be accelerated to ensure energy resources are allocated in a more efficient, safer and more sustainable way, especially renewables like wind and solar power, thus delivering the strategic goal of low-carbon energy. The structural reforms and institutional changes designed to address the challenges have significant implications for the market-oriented low-carbon reform of China's power network.

[42]Fai Li, Clear Our Mind: Make Full Preparation for the National Carbon Market, in Energy Conservation and Environmental Protection, 2014, pp. 32–33.

[43]Zhang Xin, Promoting the National Carbon Market through Effective Policies and Useful Actions, in Zhejiang Economy, 2016, pp. 24–26.

[44]The main ideas and content in this section are taken from Electricity Grids in Transition, a research report prepared for the DRC-Shell Markets Revolution Work Stream by UK-based Vivid Economics, October 2017.

9.1.1 Key Principles of Efficient Network Provision

In order to realise the economic benefits of liberalised electricity markets, international best practice suggests a set of principles for efficient network provision:

(1) **Proceed towards full liberalisation of the wider electricity system**

Effective investment in and operation of the wider electricity system is a necessary condition for electricity network efficiency. This requires liberalising sectors suitable for competition (such as fuel production, generation and retail), use of markets to procure key services (capacity, balancing) and the pricing of externalities, such as air pollution and carbon emissions.

(2) **Align incentives with public policy objectives**

Make incentives for network operators consistent with their providing a reliable and affordable supply of electricity by controlling monopolistic behaviour and ensuring prices reflect underlying costs:

- *Reform electricity network institutions*. Electricity networks are natural monopolies, with little scope for competitive markets. It is therefore critical that their incentives are aligned with public policy objectives. A monopoly faces incentives to underinvest in new infrastructure and to charge prices that are higher than its costs. A state-owned company may face incentives to prioritise short-term political objectives, rather than longer-term public policy objectives. These incentives can be mitigated through institutional reform of the electricity network. One option is to reform the network company's incentives through performance-based regulation enforced by an independent regulator. Another is to separate

network operation and ownership through the creation of an independent system operator (ISO). The UK and most European countries currently use performance-based regulation, while the USA uses the ISO model across its transmission systems, for example in Pennsylvania-New Jersey-Maryland (PJM, the transmission system in north-eastern USA), California and New York.

- *Consider the use of locational pricing*. Efficient network investment and operation make use of information on network congestion. If implemented, locational (nodal or zonal) pricing can help reveal the costs of network congestion. Nodal pricing is used in several US states, Argentina, Chile, Ireland, New Zealand, Russia and Singapore, while zonal pricing has been adopted by most European countries and Australia. However, locational pricing has disadvantages as well as advantages. Importantly, locational pricing is most effective once time-of-use pricing is fully implemented across network users.

(3) **Take further action to meet the challenges of a decarbonised system**

The electrification of energy demand and improvements in the efficiency of electrical appliances will exacerbate the uncertainty over the future volume and location of demand for transmission capacity. Flexible resources such as electricity storage and demand response can substitute for new network investment, so long as sufficient investment incentives are present:

- *Designate strategic zones for transmission-scale renewable generation to reduce planning and investment uncertainty*. Renewable energy resources may be located far from demand centres and thus require large-scale transmission investment. Uncertainty over the volume and location of generation can be mitigated through zoning.

- *Ensure there are revenues available to encourage flexible resources to offer a full range of system services.* The flexible resources needed for decarbonisation contribute several system services, such as balancing and frequency response, but there may be underinvestment if markets do not exist for these services. Several electricity markets in the West run demand curtailment markets, allowing flexible resources to generate revenues.

(4) **Prepare for the development of a decentralised electricity system and its associated digitalisation by investing in the coordination of decentralised resources—their control, balancing, security and data flows**

- *Coordinate investment in decentralised resources.* A coordination problem arises where independent developers that lack information about the plans of other developers may make similar investments, creating overinvestment or, if the developers are risk-averse, underinvestment. Either way, the result can be inefficient. Solutions include formal processes for multilateral resource planning and the publication of current and consented resources, as used by the transmission and distribution system operators of Spain and Ireland.
- *Determine how decentralised resources will be controlled.* While distribution networks today are largely passive, an active network is capable of accommodating distributed resources. As the electricity system becomes more active and complex, a single system operator may start to rely on intermediaries, such as virtual power plants and partners such as distribution system operators, to assist with system balance. New systems of control, with new computational requirements, administrative rules and institutional characteristics, may then be employed, reflecting new operational vulnerabilities.
- *Balance data transparency with security.* As information communications technology (ICT) infrastructure expands, the amount of data from the electricity system grows. The data systems need their own infrastructure, with public access to facilitate competition and optimise operations. Meanwhile, the distribution of data across resources creates new risks of cyberattack and privacy loss, solved through adequate protocols.

9.1.2 Roadmap for Efficient Network Arrangements

This section suggests a roadmap for the development of network arrangements in China, as the country simultaneously carries out large-scale investment, market reforms and decarbonisation of its electricity system. These options are based on international best practice and leading thinking on future arrangements.

The roadmap is based on the following guiding principles:

- *Strong markets need strong government.* Market-based solutions have the potential to identify and deliver cost-effective investment and operation of electricity systems. However, both markets and natural monopoly networks benefit from a strong government to take an active role in ensuring institutional incentives are aligned with public policy objectives. Network incentives can be aligned through separation of roles (unbundling) or strong regulation.
- *The institutional framework can be developed progressively.* Wholesale institutional reform is challenging and disruptive. At the outset, small changes in current practice and small-scale pilots may provide proof-of-concept sufficient to build consensus for larger-scale reforms.

The roadmap suggests the following:

(1) **Immediate actions**

- *Continue the market liberalisation programme.* Phase 1 of the DRC-Shell cooperation suggested a programme of electricity market liberalisation consistent with this. China's 13th Five-Year Plan (2016–20) aims to improve the systems by which markets play the decisive role in resource allocation. It will be important to continue with the market liberalisation programme to deliver a more advanced and efficient electricity system.

- *Rationalise investment planning.* Clearly defined metrics for reliability and economic efficiency help network planners to identify efficient investments. Meanwhile, the application of the beneficiary pays principle encourages investment that increases productivity, avoiding diverting national resources to stimulate regional output. Together, these approaches, when applied at national and regional level, facilitate greater interconnection and sharing of generation services.
- *Implement a coordinated approach to investment.* The use of a common investment framework enables coordinated planning of generation and network investment, harnessing both strategic decisions and market enterprise. For example, the framework might set out the role of generation zones for large-scale renewables alongside alternatives such as small-scale distributed generation.

Implement smart system architecture. Smart, distributed resources are crucial to affordable decarbonisation. Before building the distributed resources, the system architecture to manage it can be laid down. At a minimum, this includes: deployment of smart meters to introduce time-of-use pricing to consumers on the distribution network; upstream information and communications architecture; and R&D to develop technical solutions for the smart grid.

(2) Move towards efficient pricing

- *Deregulate prices.* Cost-reflective pricing can signal efficient investments and operations to decision-makers. The deregulation of prices can proceed sequentially, with further price reform contingent on the success of previous reforms. Price reform could commence upstream and progress downstream,

beginning with input fuels, followed by generation and network access, and ending with retail, with provision to protect retailers if wholesale prices rise above retail prices before deregulation is complete.

- *Create harmonised trading arrangements between transmission systems.* Use prices to determine interconnector flows between provincial and regional transmission systems, signalling which provinces or regions could benefit from new investment. The prices would stimulate lower-cost generators to respond to demand.
- *Implement time-of-use pricing.* Time-of-use pricing allows consumers and flexible resources to respond to variation in generation costs and demand. Again, it can proceed sequentially, starting with larger consumers, such as industrial facilities with flexible production schedules, and ending with smart household appliances.
- *Consider locational pricing.* Similarly, locational pricing can signal efficient investments and operations, but it brings disadvantages as well as advantages. China may consider locational pricing, following the implementation of time-of-use pricing, that is, signal geographical network constraints once demand peaks have been shaved. Zonal pricing is a potential intermediate step between uniform and full nodal pricing.
- *Protect end users.* The deregulation of retail electricity may result in rent-seeking by retailers, raising consumer prices. This can happen if consumers do not switch suppliers readily or for other reasons competition is not effective. Policies to protect consumers could be developed alongside any deregulation of retail electricity prices.

(3) Begin market trials

- *Create small-scale trials.* Create small-scale trials to competitively procure new transmission investments, non-network alternatives to new transmission assets, ancillary services and so on. Use competitive tenders or auctions for the trials. These may provide proof-of-concept and experience with innovative, cost-effective solutions. To be successful, procurement has to be open-access and transparent.

- *Progressively introduce market procurement.* If the competitive procurement trials are successful, they can be scaled up and wider market procurement progressively introduced, where appropriate, across each transmission system. Market procurement can help to reveal information about the relative costs and benefits of a range of technologies.

(4) **Make institutional choices**

- *Develop the transmission network institutions.* Options include the status quo, an enhanced role for market procurement and a regulated transmission system operator (TSO) or independent system operator (ISO). International experience of the regulated TSO and ISO models is yet to reveal the best performer of the two, so it matters more to adopt a good-quality institutional model early than to choose between the options.
- *Select a model of control for decentralised resources.* Initially, when the number of resources is small, the TSO may be able to control them directly. However, as the number of resources increases, and as temporal

and locational pricing become more sophisticated, the computational, commercial and contractual capacity of a single operator model may be exceeded, and new models of control may be needed.

Box 1: Balancing the electricity system

A simple example of balancing an electricity system under network constraints is shown in Fig. 63. In this example, a system operator balances a system of two cities interconnected with a capacity-constrained transmission line. The numbers in green, blue and purple are the inputs for the system operator's balancing problem. The numbers in red are the outputs.

In City A, generators can produce up to 150 MW. Generation costs $10/MWh. Consumers demand 50 MW. In City B, generators have 50 MW capacity. Generation costs $20/MWh. Consumers demand 90 MW, exceeding the local generation capacity. Consumer demand in both cities is constant and does not change with price. The interconnection between City A and City B can carry up to 80 MW.

To minimise the total cost of the system, the system operator first uses the cheap generation in City A. Generators in City A serve the local demand (50 MW) and

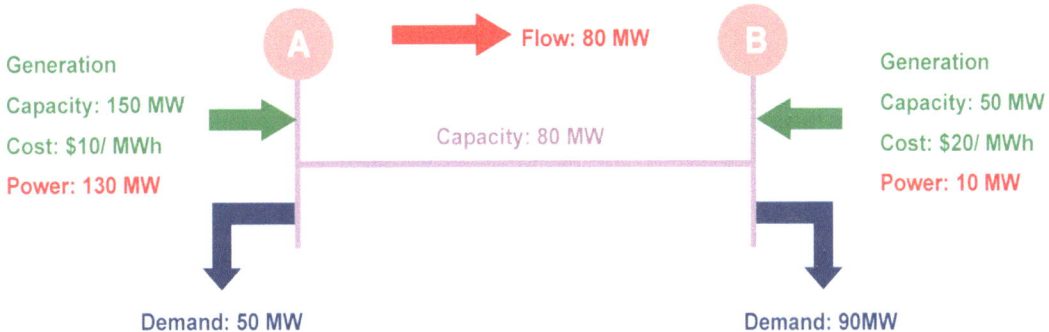

System price: $20/ MWh
Total cost: $2800

Fig. 63 Example of system balancing with network constraints. *Source* Vivid Economics

export 80 MW to City B (in total 130 MW). Because the capacity of the interconnection is fully utilised, expensive generators in City B serve the rest of the local demand (10 MW). As a result, the system price equals the cost of the generators in City B ($20/MWh), and the total cost is $2,800 (140 MW x $20/MWh).

9.1.3 Challenges in Network Provision

As electricity systems decarbonise and decentralise, the challenges of planning, delivery, operation and cost recovery are changing for network service providers.

(1) **The challenges of network provision**

The challenge of efficient provision encompasses planning, delivery, operation and cost recovery. Challenges in planning and delivery of electricity networks arise due to uncertainty over future electricity demand and the difficulty of coordinating network investment with independent generators. Challenges in operation arise from the capacity constraints of the network, the complexity of the system and the unpredictability of flows of electricity. Challenges in cost recovery arise due to the natural monopoly characteristics of electricity networks and the difficulties of mitigating monopoly behaviour with conventional regulation:

- *Planning and delivery.* The planning and delivery of electricity networks faces uncertainty over future electricity demand and difficulties in coordinating network investment with independent generators. First, the volume and location of future demand is uncertain and depends on population growth, changes in settlement patterns, changes in industrial structure, technology and economic growth. Network planners judge where network assets will be required. Second, while vertically integrated utilities plan generation and network investment simultaneously, in a liberalised electricity system, generation and network investment are carried out by separate organisations. Without coordination, generators face the risk that their revenues may be curtailed by network congestion and networks face the risk that generators will underutilise their assets.

- *Operation.* Balancing the electricity system is challenging due to system complexity and the unpredictability of electricity flows. With many sources of generation and consumption, as well as network constraints, the optimal level of production and consumption for each source is a complex calculation. Furthermore, due to the physical laws governing electricity networks, the precise flows of electricity through the network depend on the volumes of consumption and production of each generator and user and cannot be predicted in advance.

- *Cost recovery.* Electricity networks are natural monopolies. One of the roles of the system operator is to levy charges to pay the network owner. This is achieved through designing tariffs which recoup capital, operation and maintenance costs for the network owner, and passing these costs through to network users. In the case where the network operator is also the network owner, it operates as a monopoly, and faces incentives to underinvest in network infrastructure while charging high prices to consumers.

Innovative arrangements are needed to mitigate monopoly behaviour effectively. Electricity networks are characterised by high capital costs and economies of scale. For these reasons, electricity networks are natural monopolies, with a single network serving a given area. The conventional approach to mitigating monopolistic behaviour in a natural monopoly is regulation. However, regulators have imperfect information on current network costs and how these costs can be reduced over time as productivity improves. Depending on the type of regulation, network companies may face incentives to overstate their costs or to overinvest.

(2) Future changes: decarbonisation and decentralisation

Electricity systems are decarbonising and decentralising, and efficient network provision takes these changes into account. Potential changes in key characteristics of the electricity system are encompassed within two broader trends: decarbonisation of electricity and the wider energy system; and decentralisation of system resources, as summarised in Fig. 64. These changes, and their implications for the challenges of efficient network provision, are described in turn.

(1) Decarbonisation

Decarbonising an electricity system and related changes in the wider energy system involve changes to generation technologies and far-reaching changes to patterns of electricity demand. Generation technologies will shift from fossil generation to low-carbon generation, that is, a mix of carbon capture and storage, nuclear, biomass and renewables. Electricity demand will be affected by increases in demand from electrification of end-use sectors, particularly heat and transport, as well as decreases in demand from greater efficiency of electrical appliances. There will also be a shift in the profile of demand as low-carbon flexible resources (electricity storage and demand-side response) emerge to balance the relatively inflexible generation profile of nuclear and renewables.

These changes are likely to exacerbate the challenge of planning and delivery of network infrastructure. Future volumes of demand will be more difficult to forecast due to uncertainty over the level of electrification of end-use sectors and improvements in the efficiency of electrical appliances. Another element of uncertainty is the degree to which low-carbon flexible resources will reduce peak demand, and therefore the appropriate level of network capacity.

Low-carbon flexible resources can substitute for new network investments and can reduce total network costs. However, as these resources provide different system services (balancing, frequency response, network congestion mitigation) there may be underinvestment in flexible resources if markets do not exist for all the services that these resources provide.

(2) Decentralisation

Decentralisation involves a shift in electricity resources from the transmission system to the distribution system. Decentralised electricity resources comprise generation, demand response and storage. Decentralised generation includes

Fig. 64 Future electricity grids will be shaped by two broad trends: decarbonisation and decentralisation. *Source* Vivid Economics

wind and solar, which are increasingly connected to the distribution network, including at the household level. Demand response is the flexible operation of electrical equipment in response to system conditions; electric vehicles are expected to significantly increase this potential as they may be able to charge at times of high electricity generation and low demand. Decentralised storage may also increase, potentially even at the household level.

To unlock decentralised electricity resources requires a shift to a smart grid. A smart grid is characterised by dominance of controllable electricity resources (generators, storage, appliances) throughout the electricity system, the ability of users to express preferences for use of their devices, the development of operating standards to allow resources to be operated in a coordinated way, sufficient development of information and communications technology (specifically, communications bandwidth, data storage and computing power), and adequate data security and privacy protocols.

Decentralisation exacerbates both the planning and delivery, and operation challenges:

- *A coordination problem may arise between the transmission system and the various distribution systems*. In current electricity systems, the bulk of new investment occurs in the transmission system, where adequate information is available on current and planned resources to inform new investment decisions. A shift in resources to the distribution network, where adequate information is not typically available, will create a coordination problem, where investors will not have a clear understanding of system needs, nor consequently of potential returns on investment. This could result in overinvestment, underinvestment, a poor technology mix or a poor spatial distribution of resources.
- *The computational requirements of balancing a decentralised system will increase significantly*. Optimising the operation of the entire system requires determining the optimal volume of output and consumption of every resource in the system. As the number of

controllable resources increases from the limited set of large resources in a transmission system to the total set of resources across all distribution systems, the computational demands of this optimisation calculation also increase. If computing technology is not able to meet these computational demands, then intermediate levels of control are required, and only partial optimisation is possible.

Finally, decentralisation will require risks of data privacy and cybersecurity to be effectively managed. Decentralisation will be accompanied by a very significant extension of information and communications technology across all distributed resources and will create risks of data privacy and cybersecurity. Protocols to manage data use and control these risks can be developed and implemented.

9.1.4 Network Arrangements to Address Current Challenges

To address the challenges arising from decarbonisation and decentralisation, grid companies need to take effective measures to ensure efficient network arrangement. The international experience in previous electricity market reform provides some best practice arrangements that China can draw from:

- an institutional model to align incentives with public policy objectives: an institutional model is needed that mitigates monopolistic behaviour and provides networks with incentives to invest in appropriate network infrastructure and to operate infrastructure efficiently;
- strategic transmission planning: determining the appropriate profile of new transmission investment is a complex process and requires strategic planning;
- the appropriate level of locational pricing: investing in and operating networks efficiently requires an understanding of current network congestion; and
- a regime for merchant investments: entry of merchant investors has the potential to deliver

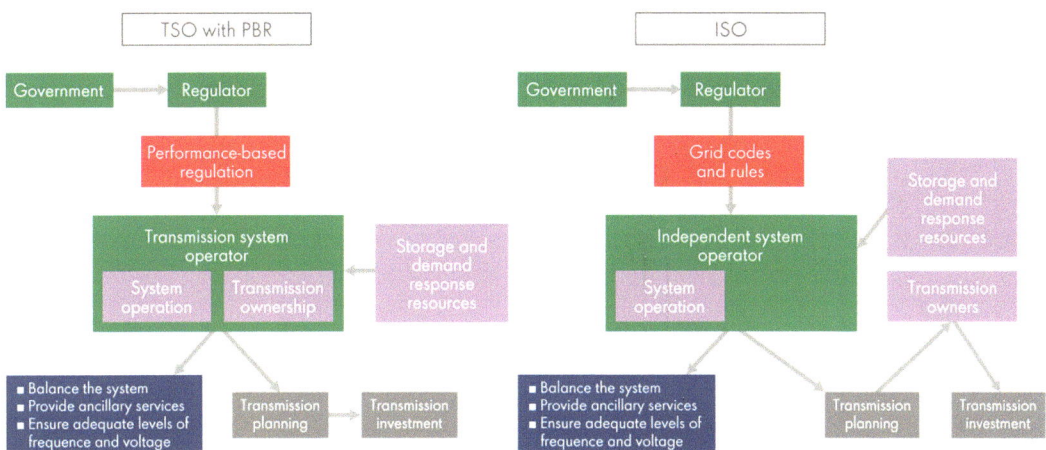

Fig. 65 The transmission system operator (TSO) and independent system operator (ISO) are the two main alternative institutional models for system ownership and operation. *Note* PBR = performance-based regulation. *Source* Vivid Economics

greater adequacy of investment than a single-owner network.

(1) An institutional model to align incentives with public policy objectives

An institutional model is needed that mitigates monopolistic behaviour and provides networks with incentives to invest in appropriate network infrastructure and to operate infrastructure efficiently. Two institutional models that can create an efficient regime are the TSO model, with performance-based regulation, and the ISO model. Figure 65 highlights the key difference between these two models: a TSO both owns and operates the transmission system, requiring strong regulation, while an ISO is a system operator that is fully separated from ownership of all network resources. A number of intermediate models also exist, for example where the system operator and transmission owner are legally separate companies but owned by the same parent company.

As shown in Fig. 64, most electricity systems have moved from vertically integrated monopoly utilities before liberalisation to one of these institutional models today. In 1985, Chile was

the first country to adopt the ISO model. The UK shifted from vertical integration to a TSO with performance-based regulation structure in 1990, with Germany following suit in 1998. Following the orders of the Federal Energy Regulatory Commission (FERC) (the US electricity regulator), the US regions of Pennsylvania-New Jersey-Maryland (PJM), and California (CAISO) transitioned from a vertically integrated structure to the ISO model in the late 1990s (Fig. 66).

(1) TSO with performance-based regulation

A transmission system operator (TSO) is an entity that both owns and operates the transmission system; it therefore has incentives for monopoly behaviour. The TSO owns all the network assets and is also responsible for planning, deployment and operation of the system. The TSO model is prevalent in most European countries.

Performance-based regulation is needed to align a TSO's incentives with public policy objectives. A TSO is difficult to regulate as it has better information than the regulator on the costs it faces. This gives rise to one of two problems. If

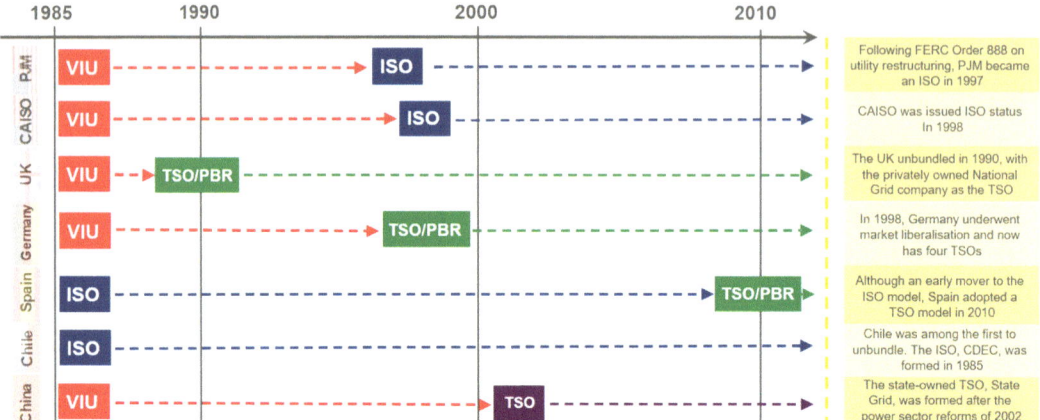

Fig. 66 Several countries shifted to the TSO and ISO models as they liberalised their electricity markets. *Note* VIU = vertically integrated utility; TSO/PBR = transmission system operator with performance-based regulation; PJM = the regional transmission organisation for the north-eastern USA; CAISO = California Independent System Operator

the regulator tries to prevent monopoly behaviour by imposing a cap on prices, known as price regulation, it estimates the price level required for the TSO to recover its costs. The TSO then has incentives to exaggerate the costs it faces, and secure price caps greater than its actual costs, a problem known as adverse selection. In this case, the TSO can underinvest and continue to set high prices. If instead the regulator tries to prevent monopoly behaviour by paying the TSO for its costs plus a regulated return (cost-of-service regulation), the TSO does not have any incentive to take necessary measures to decrease its costs, a problem known as moral hazard. In contrast to these approaches, performance-based regulation seeks simultaneously to address both monopoly profits and underinvestment. An example of performance-based regulation is the imposition of a price (or revenue) cap, which is adjusted every year by the rate of inflation and target rate of productivity growth. The TSO is constrained in exaggerating its costs by the regulator conducting benchmark analysis and detailed studies of the TSO's historical accounts. The TSO has some incentive to lower its costs because it can retain the difference

between the price cap and its actual costs as profits.

Performance-based regulation is still evolving and there is no consensus on the optimal mechanism. Several forms of performance-based regulation have been adopted and these continue to evolve. An example of their evolution is the introduction in the UK of the retail price inflation minus X (RPI-X) mechanism in 1992, and its eventual replacement by the revenue = incentives + innovation + outputs (RIIO) mechanism in 2013, which is described in Box 2. A key challenge in performance-based regulation is its high information burden. The regulator has to review the TSO's accounts and business plans and undertake benchmark analysis.

Box 2: Performance-based regulation in the UK

The UK introduced performance-based regulation in electricity networks in 1992 with the RPI-X mechanism, eventually replacing this in 2013 with the more sophisticated RIIO mechanism:

- RPI-X mechanism. Ofgem and its predecessors, the electricity market regulator in the UK, used the RPI-X mechanism until 2013. Under this mechanism, Ofgem carried out cost forecasts to determine the base revenue required by the TSO to recover its costs. Based on this base revenue Ofgem set a price cap and adjusted it each year for retail price inflation (RPI) and an assumed target rate of productivity growth calculated by statistical benchmark analysis. Ofgem reset the price cap at the end of the five-year price control period to ensure that the TSO's cost savings are passed on to end consumers. Under the RPI-X mechanism, the TSO had incentives to reduce its costs because it was able to retain the margin between the price cap and its own actual costs as profit, as shown in Fig. 67. However, the RPI-X mechanism did not adequately incentivise service quality or innovation. It allowed cost savings to be achieved through decline in quality of service, and the five-year price control periods did not provide sufficient incentives to develop new technologies (such as smart meters) with longer investment

cycles and the potential to provide cost savings over the longer term.

- RIIO mechanism. In 2013, Ofgem introduced the revenue = incentives + innovation + outputs (RIIO) mechanism to address the concerns with the RPI-X mechanism. The RIIO mechanism takes the elements of the RPI-X mechanism that work well, such as the determination of the base revenue, and adds extensive innovation and output targets to them. The RIIO mechanism determines the base revenue from output-led business plans with greater use of option analysis and scenario planning. In their business plans, companies compare the costs and benefits of options for delivering long-term outputs under various scenarios and assess the value of keeping these options open. Moreover, financial and reputational incentives strengthen the incentive structure. Ofgem rewards or penalises companies when they achieve or miss their output targets respectively. Reputational incentives do not have a financial element, but they affect Ofgem's evaluation of base revenue in the next review periods. A longer control period, of eight years, encourages

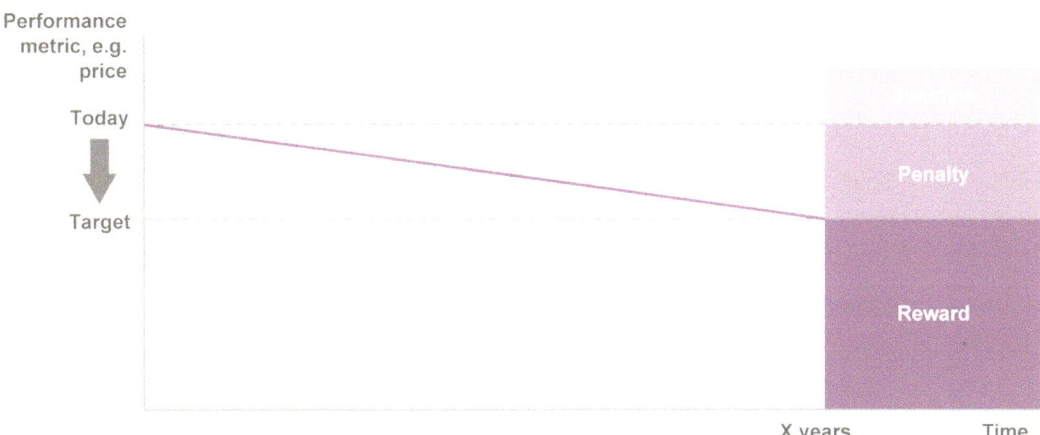

Fig. 67 Under the RPI-X mechanism, the TSO takes measures to decrease its costs because it retains revenues from cost savings. *Source* Vivid Economics

the TSO to focus on longer-term investments, such as smart meters. RIIO provides a wider set of performance incentives than did RPI-X, but at a cost of greater complexity and reduced transparency.

(2) Independent system operator

An independent system operator (ISO) is a system operator that is fully separated from ownership of all network resources. The ISO performs all system operation functions, including allocating network capacity among generators and consumers to respect the physical characteristics of the network, carrying out residual balancing to ensure electricity is delivered to where it is most valued, and maintaining the stability of the electricity system. In an ISO model, the transmission network resources are owned by one or more transmission owners (TOs). The ISO levies charges from generators and/or consumers for use of the transmission system and pays these charges to the TOs. Typically, an ISO is also responsible for planning new transmission investments, either mandating TOs to carry out these investments or organising competitive tenders to facilitate delivery of investment. An ISO is usually a non-profit entity.

Separation of operation from ownership removes the system operator's incentives to charge monopoly prices. An ISO may be a non-profit organisation or may earn profits on revenues for system operation. As an ISO does not earn profits determined by revenues from use of the transmission system or by the costs of transmission investments, it has no incentive to charge tariffs that are higher than needed to recover the investment costs of the transmission network, or underinvest in transmission assets. Unlike a TSO, therefore, an ISO does not need performance-based regulation.

The ISO's functions can be specified by its mandate and its behaviour can be governed by a

set of rules. The mandate could be to minimise the total cost of meeting a given reliability standard. Rules could govern processes for transmission planning and investment; providing grid connections to new system resources; administering competitive tenders for new network assets; levying charges for network use; and monitoring market power in the electricity system.

Nevertheless, it is desirable to have in place a mechanism to incentivise the ISO. It is unlikely to be possible to specify a set of rules that perfectly incentivise the management to meet the ISO's mandate. That is, to encourage the ISO to carry out the planning, investment and operation of the transmission system to strike the best possible balance between reliability of electricity supply and economic efficiency of the network. Imposition of financial penalties on the ISO is likely to be a poor incentive mechanism as ISO revenues are likely to be small relative to the welfare losses arising from poor performance in operating the system. Instead, well-designed management incentives may be needed.

The ISO model is prevalent in North and South America. Chile, Argentina and Peru were early adopters of the ISO model. There are a large number of ISOs in the USA, each covering a transmission network.

(2) Strategic transmission planning

The design of new transmission investment is a complex process and requires strategic planning. Specific challenges are the uncertainty in volume and location of future demand; the coordination problem facing generation and network investors; and the large number of possible transmission investments. These challenges can be mitigated by strategic planning. Key characteristics of strategic planning include:

- *Define the objectives of new transmission investment.* Transmission planning is more effective when the objectives of new investment are clearly defined. These include reliability (security and adequacy of supply) and economic efficiency (reduction in the total cost per unit output). Defining these criteria allows the benefits of new transmission infrastructure to be measured.

Estimate and compare the benefits of each proposed investment. Cost-benefit analysis is a key planning tool to identify proposals that meet investment objectives. Reliability and economic efficiency can be assessed with electricity system modelling, and modelling of multiple scenarios can help identify least regrets infrastructure investment under uncertainty. If available, locational pricing provides a clear signal of congestion costs, and can substantiate the economic benefits of new network infrastructure.

Consult all relevant stakeholders. As the costs of investments are borne by network users, they have incentives to ensure that only the most valuable network infrastructure is developed. Stakeholder consultation can elicit views from generators, consumers (municipalities, consumer interest groups), and connected transmission and distribution systems to inform the cost-benefit analysis.

(3) **Appropriate level of locational pricing**

Investing in networks efficiently requires an understanding of current network congestion. New network investments that relieve significant network congestion are particularly valuable.

Many electricity systems operate a system of uniform transmission pricing, which does not signal network congestion. Under a system of uniform pricing, the price of electricity is the same at every network connection, regardless of the degree of congestion. It does not signal the need to invest in solutions to relieve network congestion, such as new network investment, generation, or non-network alternatives to new transmission assets, such as electricity storage and demand-side response. While, in principle, the system operator can signal the need to invest through location-specific transmission charges, the rate of these charges is difficult to determine if the electricity price does not signal network congestion. Furthermore, uniform pricing results

in re-dispatch costs, where some plant that is scheduled to generate is compensated for curtailment if the network is revealed to be congested.

Nodal pricing can help signal network congestion. Nodal pricing, or locational marginal pricing, is a price mechanism that reflects the cost of supplying additional electricity at a given network connection (node), given the demand for electricity, transmission constraints and options for local generation at that node. When there is no network congestion, overall demand is met at least cost and all nodal prices are the same. When network congestion occurs, demand is met by costlier local generation rather than cheap generation from another node, raising prices at congested nodes. Nodal pricing therefore signals the need to invest in solutions to relieve network congestion, such as new network investment, new local supply and non-network alternatives to new transmission assets, such as electricity storage and demand-side response. Nodal pricing is briefly illustrated in Box 3.

Box 3: Nodal pricing

Figure 68 provides a simple example of balancing an electricity system under nodal pricing. In this example, a system operator balances a system of two cities interconnected with a capacity-constrained transmission line under nodal pricing. The numbers in green, blue and purple are the inputs for the system operator's balancing problem. The numbers in red are the outputs.

In City A, generators can produce up to 150 MW. Generation costs $10/MWh. Consumers demand 50 MW. In City B, generators have 50 MW capacity. Generation costs $20/MWh. Consumers demand 90 MW, exceeding the local generation capacity. Consumer demand in both cities is price-inelastic. In other words, consumer demand is constant and does not change with price. The interconnection between City A and City B can carry up to 80 MW.

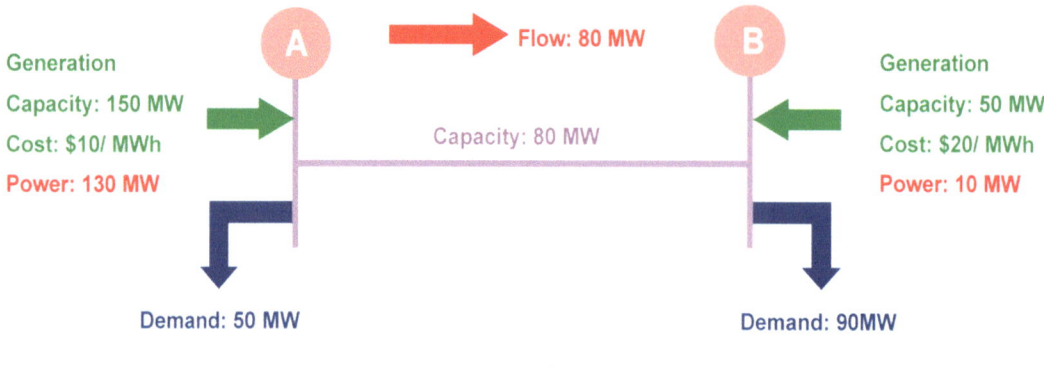

Fig. 68 Nodal pricing reflects the cost of supplying additional electricity at a given node. *Source* Vivid Economics

To minimise the total cost of the system, the system operator first uses the cheap generation in City A. Generators in City A serve the local demand (50 MW). That is why the nodal price in City A equals the local generation cost ($10/MWh). Generators in City A also export 80 MW to City B. Since the capacity of the interconnection is fully utilised, expensive generators in City B serve the rest of the local demand (10 MW). As a result, the nodal price in City B equals the local generation cost ($20/MWh). The total cost is $1,500 (= 140 MW * $10/MWh + 10 MW * $20/MWh).

However, nodal pricing has disadvantages as well as advantages. First, vulnerability to market power may arise because the segmentation of the electricity market into smaller locational markets increases the concentration of generators at each node with a supply deficit. Some authors challenge this view and argue that the network architecture is the main driver of market power, instead of the pricing mechanism. Nodal pricing can also reduce liquidity in long-term contracting, such as financial transmission rights and contracts for differences. This problem is addressed in the USA by averaging nodal prices into trading hub prices to provide liquidity to market participants.

Zonal pricing, another form of locational pricing, may provide a useful compromise between uniform and nodal pricing. Zonal pricing addresses the complexity introduced by large numbers of nodes by aggregating nodes into zones. Similar to dispatch under nodal pricing, the system operator first dispatches generation, given transmission constraints between zones. If transmission lines in a given zone are congested, the system operator has to re-dispatch generation in that zone to alleviate congestion. As a result, zonal pricing provides some of the benefits of nodal pricing in terms of signalling network congestion but does not fully eliminate the re-dispatch costs associated with uniform pricing.

Locational pricing may be considered once time-of-use pricing is fully implemented. Wider changes to improve the flexibility of the electricity system through electricity storage and demand response are expected to reduce demand and generation peaks, which would automatically reduce network congestion relative to an inflexible system. These changes require time-of-use pricing to be fully implemented across all system resources (including end users) to be fully effective.

(4) A regime for merchant investments

The entry of merchant investors has the potential to deliver greater adequacy of investment than a single-owner network. Merchant transmission investors are third-party developers of transmission projects. If locational pricing is implemented, a merchant transmission investor has the incentive to invest in a new transmission link when the revenues from use of that link are greater than the investment cost. Therefore, in principle, merchant competition offers the potential to increase the adequacy of the transmission infrastructure by investing where an incumbent is not willing to do so. This might be the case if the incumbent is an unregulated monopoly, or a poorly regulated TSO.

Other attractive properties of the merchant model are the ability to include non-network alternatives to new transmission assets in planning processes, reduce the risk for consumers and minimise investment costs. In liberalised power markets, potential merchant investors could invest in new transmission capacity, or enter the generation market to supply local generation to a node that is served by a congested transmission link. Investment risk is transferred from regulated transmission owners and consumers to the merchant. As the merchant is the beneficiary of any cost saving, construction costs may also be minimised.

However, merchant investment alone is not sufficient to ensure overall adequacy of the network, underscoring the importance of a well-designed institutional model. Transmission investments exhibit economies of scale, where large-capacity investments carry only a small cost premium relative to small investments. As large-capacity investments offer significant additional benefits at little additional cost, they are socially desirable; however, as these additional benefits are reflected in lower locational prices (due to lower congestion), they are less desirable for private investors. In this setting, merchants will tend to underinvest in new network capacity. Alternatively, with an ISO institutional model, the ISO can also ensure overall network adequacy by planning new network

capacity and delivering new investment at minimal cost by running competitive tendering processes.

Merchant transmission investments have been implemented in the USA, Australia and Argentina. In the USA, merchant investment is promoted by the FERC Order 1,000, and a number of projects are in progress or have been completed in recent years. Nearly all merchant-led investments have been on interconnectors, that is, links between distinct networks. Here merchants alleviate coordination and cost allocation issues between different system operators.

9.1.5 Network Arrangements to Address Future Challenges

This section discusses new arrangements to address the future changes of decarbonisation and decentralisation. They comprise:

- Strategic generation zones: Strategic generation zones coordinate investment in transmission and generation assets and connect remote renewable energy resources to large population centres.
- Markets for flexibility services: Flexible resources, such as electricity storage and demand response, can substitute for new network investment, as well as providing a range of different system services. A simple set of markets for each system service can reward flexible resources and avoid underinvestment.
- System for controlling decentralised resources: While distribution networks today are largely passive, they will need to become active to accommodate distributed resources. Distributed resources increase the complexity of the electricity system. If the system is too complex for a single system operator to balance, a hierarchy of resource control will be needed, with intermediaries such as virtual power plants and distribution system operators interacting with the transmission system operator. The hierarchy of resource control may reflect computational requirements, institutional characteristics and operational vulnerability.

- Coordinated investment in decentralised resources: Investment in generation and storage resources should meet the needs of the whole electricity system. A decentralised electricity system may not provide adequate information to investors on system needs, raising the risk of inefficient investment. A coordinated approach to investment can mitigate this risk.
- Open-access, public data on system conditions and resources: Market participants need a degree of access to market information to facilitate a level playing field for competition. A data exchange could be part of an efficiently functioning energy system, but would need to be both secure and accessible.
- Accommodating future innovations: Innovations in electricity networks include new network structures and peer-to-peer electricity trading, which may offer significant benefits. These innovations can be facilitated through pilots and early-stage funding.

(1) Strategic generation zones

The decarbonisation of electricity generation and the wider energy system heightens the challenge of planning and delivering network infrastructure. The capacity of new transmission investment will be more difficult to determine due to greater uncertainty over the level of total electricity demand (due to electrification of end-use sectors and improvements in the efficiency of electrical appliances) and peak demand, as flexible resources contribute to smoother generation and consumption profiles. The degree to which generation will be centralised—that is, connected to the transmission network—will be difficult to forecast.

In many countries, renewable energy resources are located in areas that are distant from large population centres, and thus require large-scale transmission investment. For example, in the UK, the majority of electricity demand is located in the

south of England, while a large proportion of onshore wind resources are located in Scotland and offshore wind resources in the North Sea.

In a liberalised electricity system, investors in generation and network investment face a coordination problem. While vertically integrated utilities are able to plan generation and network investment simultaneously, in a liberalised electricity system, generation and network investment are carried out by different institutions. This gives rise to a coordination problem, where generation investors face the risk that their revenues may be reduced due to inadequate network investment, and network investors face the risk that generators will underutilise their new investments. This can lead to underinvestment.

This coordination problem can be mitigated with strategic generation zones. If a strategic decision is made to exploit a large renewable resource that is distant from large population centres, generation investors may be given incentives to invest there. This may require an overarching strategic plan to be developed by an institution, such as a government agency, with sufficient authority to determine the location of both transmission and generation investment. It may also require a credible, long-term regime for network connection to be developed to reduce stranded asset risks for generators. For example, in the UK, nine offshore wind farm zones of varying size were identified within UK waters to deliver 33 GW of potential offshore wind capacity. The Crown Estate, the statutory owner of seabed rights, asked renewable energy developers to bid for exclusive rights to develop offshore wind farms within the zones. The Electricity Networks Strategy Group, a high-level forum bringing together key stakeholders in electricity networks, including the Crown Estate, then identified key transmission investments needed to meet future demand, given the expected location of future generation. The areas identified by this exercise are shown in Fig. 69.

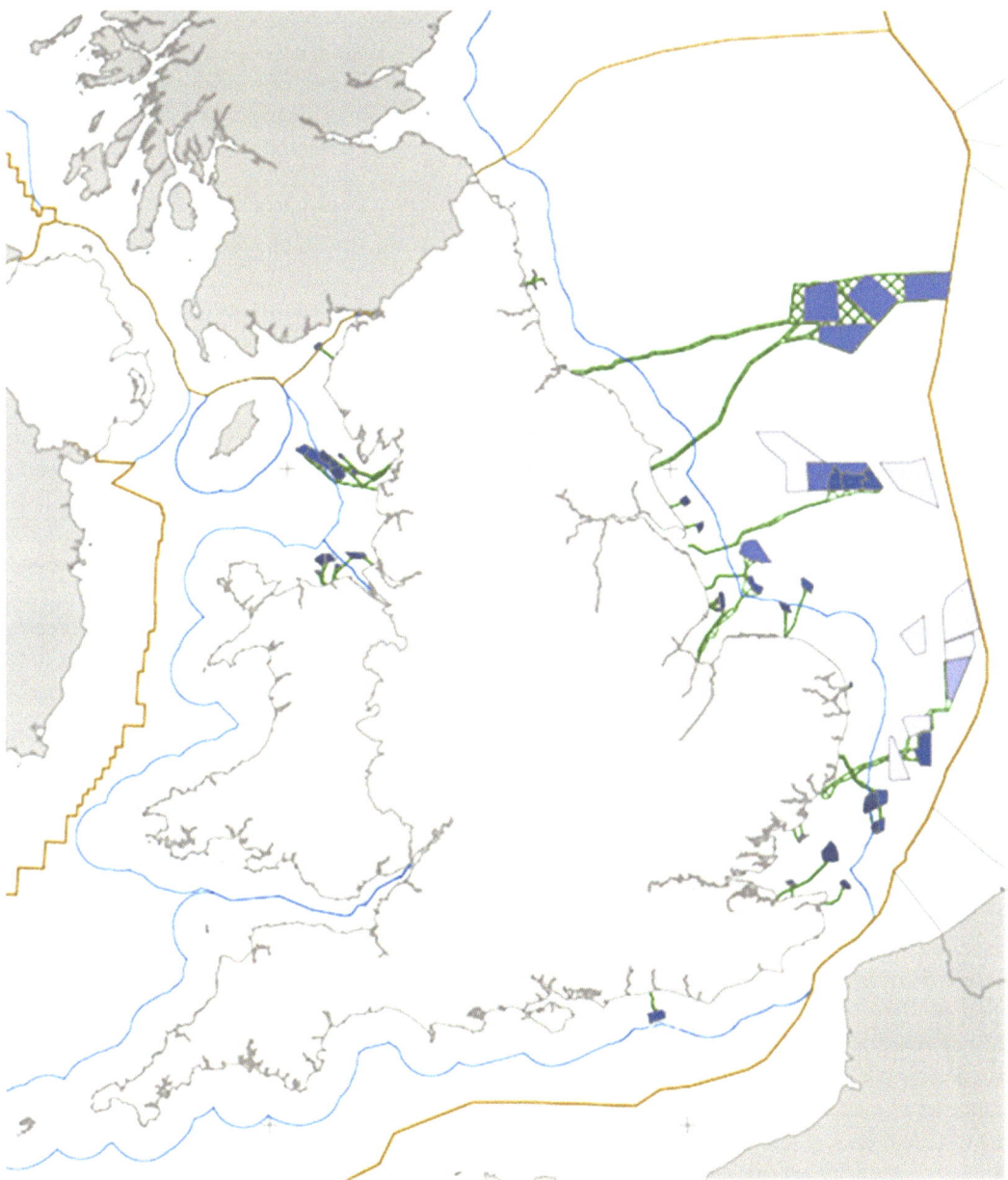

Fig. 69 UK offshore wind farm zones identified for development of renewable energy capacity. *Source* The Crown Estate (2017)

(2) **Markets for flexibility services**

Decarbonisation will require flexible resources, electricity storage and demand response, which offer non-network alternatives to new transmission assets. Electricity networks are costly, long-lived assets and investments are made in the presence of uncertainty over the future spatial and temporal profile of generation and demand. It will be increasingly valuable to enable flexible resources, such as electricity storage and demand response, to substitute for new network investments where possible, or defer new network investments until the future profiles of generation and demand are better understood. As discussed above, nodal pricing provides a locational signal for non-network alternatives.

Flexible resources provide a range of system services, which may each be rewarded. As well as substituting for or deferring new network investments, services provided by flexible resources include balancing and system stability. To provide balancing services, electricity storage can hold surplus energy until it is needed, and demand response can shift demand to when electricity is being generated. To provide system stability, electricity storage and demand response can adjust system voltage and frequency. In order for flexible resources to deploy at volumes that bring the greatest benefits, mechanisms exist to reward them for the system services they provide.

A simple set of markets for each system service allocates existing flexible resources and signals investment need. Storage and demand response are already actively engaged in balancing supply and demand in several wholesale markets. However, markets for system stability are typically not developed enough to allow flexible resources to participate to their fullest extent. Procurement mechanisms for system services typically specify these services in terms of the properties of the thermal generators that historically provided them, which often closes these mechanisms to new, low-carbon flexible resources. Recent attempts to create procurement mechanisms for flexible resources have resulted in a patchwork of complex and mutually inconsistent set of mechanisms. For example, in Great Britain, electricity storage facilities providing a short-term ancillary service called enhanced frequency response are not allowed to participate in the capacity mechanism. A set of simple markets to reward all system services provided by flexible resources is needed to ensure adequate investment to provide non-network alternatives.

(3) Model for control of decentralised resources

While distribution networks today are largely passive, they will need to become active to accommodate distributed resources. As demand on the distribution network is typically inflexible, generators on the transmission system operate flexibly to meet demand and maintain system security. As decentralised electricity resources (distributed generation, storage and demand response) are deployed in the distribution network, these resources will also need to operate flexibly.

Distributed resources increase the complexity of the electricity system. The optimisation of electricity system operation involves finding the optimal volume of output and consumption of every resource in the power system. A conventional, centralised electricity system typically includes a limited number of large transmission-connected generators, suppliers and large industrial consumers with flexible demand. However, a decentralised electricity system will include a very large number of small decentralised resources. The number of resources to be optimised might increase by a factor of several million.

A similar increase in computational demand would occur if the temporal resolution at which resources are controlled increases. For example, shifting from hourly to real-time (second-by-second) settlement of all dispatch and consumption actions would increase the computational demands of optimal system balancing by a factor of 3,600.

If the system is too complex for a single system operator to balance, a hierarchy of resource control will be needed. A full optimisation of a decentralised electricity system would have very significant computational requirements. If computing technology is not able to meet these requirements, then control of the system will be distributed.

Virtual power plants (VPPs) and distribution system operators have a role to play in a hierarchy of resource coordination. VPPs, also known as aggregators, could coordinate (aggregate) decentralised resources and coordinate them individually to present the system operator at the transmission level with a level of net generation (or consumption). If the system is complex, more than two levels of resource coordination might be needed, with some VPPs coordinating the activity of smaller VPPs further down the hierarchy. Distribution system operators are VPPs that coordinate all resources in a given distribution system, either directly or via intermediate VPPs.

Hierarchies of resource control provide only partial optimisation of the electricity system. In order to optimise the whole electricity system, a single optimising agent must know the demand and supply curves for each system resource. Where no single agent has this information, only partial optimisation is possible, as groups of resources for which information is available must be optimised separately. Markets between groups of resources will be required to coordinate their operation to balance the whole system. However, supply and demand curves in electricity systems change in real time, while the process of price discovery in decentralised markets takes place over time through multiple iterative trades. Therefore, markets between multiple levels of resource coordination can provide only a partial optimisation of the electricity system.

A decision has to be made about how distributed resources will be coordinated, with several possibilities available. Four possible models for operating distributed system

resources are currently being discussed in the literature and are illustrated in Fig. 70. These models are:

- *Whole-system operator.* This model involves only one level of control. The TSO carries out constrained dispatch of the whole electricity system. In other words, it carries out least-cost dispatch across all system resources at the transmission and distribution system levels, taking capacity constraints at both network levels into account. In this model, distribution network operators retain their current, minimal functions of basic planning and operation of the distribution network.

- *Whole-system operator with distribution system operator.* This model has two levels of control. First, the TSO carries out economic dispatch of the whole electricity system, including constrained dispatch of the transmission system. In other words, it carries out least-cost dispatch across all system resources at the transmission and distribution system levels, but only takes capacity constraints at the transmission network level into account. Second, distribution system operators (DSOs) modify the operation of distributed resources

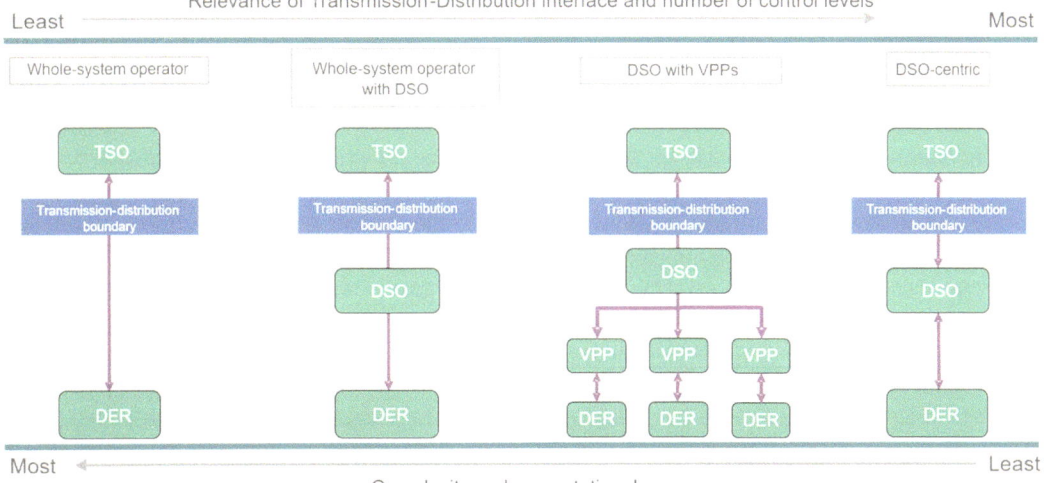

Fig. 70 There are several possible models for coordinating distributed energy resources. *Note* TSO = transmission system operator; DSO = distribution system operator; DER = distributed energy resources; VPP = virtual power plant.. *Source* Vivid Economics

to take into account capacity constraints at the distribution network level.

- *Virtual power plants with distribution system operator.* This model involves three levels of control. First, VPPs provide the transmission-level system operator with offer curves for generation or demand reduction for the resources they operate. Second, the transmission-level system operator carries out constrained dispatch of the transmission system. Third, DSOs modify the operation of distributed resources to take into account capacity constraints at the distribution network level.

- *Distribution system operator.* This model involves two levels of control. First, DSOs act as sole distribution-level VPPs, controlling all system resources at the distribution level, and carry out constrained dispatch of the whole distribution system. DSOs provide the transmission-level system operator with bid or offer curves for the generation and demand resources they operate. Second, the transmission-level system operator carries out constrained dispatch of the transmission system.

Key criteria in this decision are the computational requirements, institutional characteristics and operational vulnerability of each model. As explained above, systems with high computational requirements, or small improvements in computing technology, will require more levels of resource coordination, while systems with low computational requirements, or large improvements in computing technology, will require fewer levels of resource coordination, and—with sufficiently developed computing technology— potentially a whole-system operator. Preferences for particular institutional characteristics are also relevant. For example, the whole-system operator and distribution system operator models involve coordination of resources by a single operator. Consumers may or may not have concerns over price, quality of service and privacy. If they have concerns, they might prefer a model involving control of resources by VPPs, which compete to meet customer requirements. Finally, the models

may have different degrees of vulnerability to failures in information and communications technologies (ICT), such as those caused by cyberattacks.

Hierarchies of resource control may be needed until ICT capabilities are sufficiently developed. While in the near term a single operator may be able to optimise the electricity system with relatively low volumes of distributed resources, once sufficient volumes of distributed resources are deployed, the task of optimisation may be too great for a single operator. Hierarchies of control may be established early to ensure that increasing volumes of distributed resources can be accommodated if improvements in computing technology fail to keep pace with increases in computational requirements. Even if computing technology improves to the point where a fully distributed system can be optimised by a single operator, an increase in the temporal resolution of system control (for example, from half-hourly settlement towards real time) would result in significant further increases in computational requirements. A shift from hierarchies of control to a whole-system operator model would only be viable if improvements in computing technology are sufficient to meet the requirements of optimising a fully distributed system in real time.

(4) Coordinated investment in decentralised resources

Electricity systems with largely centralised resources provide adequate information to investors on system needs. Electricity systems periodically require investment in new resources, such as new generation plants. In theory, developers invest in new resources in response to a price signal in the wholesale market. These new investments are typically large. In principle, this leads to a coordination problem, whereby either several investors might plan to develop a similar resource (overinvestment) or investors might not invest in required resources due to the risk that other investors might do so (underinvestment). In practice, these risks are minimised because the transmission system operator knows which resources are under development and awaiting grid connection and is able to make this information public.

However, a decentralised electricity system may not provide adequate information, risking inefficient investment. As resources shift to the distribution system level, the same coordination problem may arise. This is because, unless adequate procedures are introduced, no single market participant knows which resources are under development and awaiting grid connection across all electricity networks in the system. The consequence is, again, inefficient investment: over-investment, underinvestment, a poor technology mix or a poor spatial distribution of resources.

The risk of inefficient investment can be mitigated through a coordinated approach to planning new assets. As set out in Fig. 71, this could be achieved either through formal processes for resource planning and decision-making, or through provision of information.

Formal processes for resource planning and decision-making by system operators at the transmission and distribution system levels have been introduced in Spain and Ireland. Processes include formal collaboration, with the TSO and

DSOs planning infrastructure and generation investment together. For example, in Spain, some regional administrations founded evaluation boards. In these boards the administration, TSO, DSO and developers coordinate investment plans and grid connection requests. The TSO and DSO analyse and approve investment plans together. Thereby, the TSO and DSO minimise the network development and project costs as well as project risks. In Ireland, under the group processing approach, investment plans of developers are collected in batches and then submitted to the TSO and DSO for consideration. Then, the TSO or DSO process the plans that are most suited to their system. This approach coordinates the development of the transmission and distribution systems and efficiently allocates scarce capacity.

An alternative approach is to provide adequate information. For example, compulsory registration in a publicly available database on application for planning or grid connection consent would provide investors with an understanding of the pipeline of future resources, and allow them to evaluate potential investments against the expected system requirements.

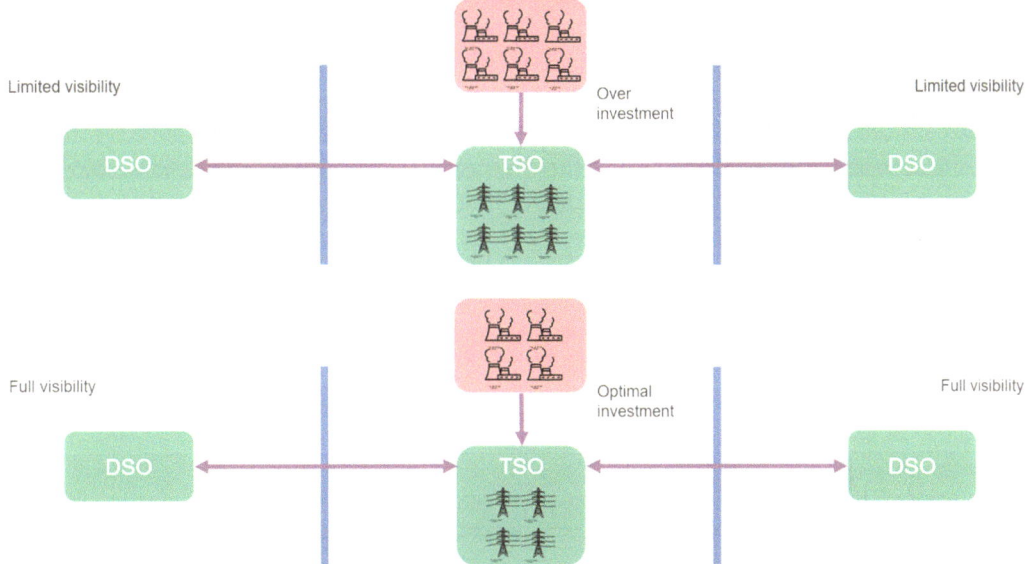

Fig. 71 A coordinated approach can ensure efficient investment in electricity system resources. *Note* TSO = transmission system operator; DSO = distribution system operator. *Source* Vivid Economics

(5) **Open-access, public data on system conditions and resources**

The development of ICT infrastructure is expanding the amount of data and information available. Smart meters in buildings track the profile of energy consumption every second, offering a new source of information on energy consumption and user behaviour. On the electricity grid, sensors and wide area networks monitor the reliability of electricity grids, providing real-time information on network conditions.

Market participants need a degree of access to market information to facilitate a level playing field for competition. The Council of European Energy Regulators, for example, has identified limited access to information as a key barrier to entry for new market participants. Access to information about distributed resources and network conditions may also grow in importance in the future. This information may also provide a foundation for new opportunities for system balancing, as new participants could enter the market and find more efficient balancing solutions.

A data exchange could be part of an efficiently functioning set of future energy networks but would need to be managed carefully to mitigate risks while being accessible. A data exchange is a secure store of data, for example, on customer use patterns, available distributed resources, local prices and network conditions. The availability of this data raises privacy concerns and a balance would therefore need to be struck between accessibility and protection. Use of data exchanges will require adequate institutional arrangements. For example, if DSOs, who participate in markets for electricity system services, ran data services then gave themselves preferential access. In the long term, ICT developments may make operation of data exchanges by centralised authorities unnecessary, particularly if the trading of electricity shifts towards a peer-to-peer level.

(6) **Accommodating future innovations**

Innovations in electricity networks include new network structures and peer-to-peer electricity trading, which may offer significant benefits. New network structures, including microgrids and fractal grids, have the potential to make electricity systems more resilient to failure, as discussed above. Peer-to-peer trading through a distributed data management platform, such as blockchain, offers the potential to lower transaction costs and reduce the role of intermediaries in electricity markets.

These innovations can be facilitated with arrangements required to deliver efficient networks, today and in the future. Most fundamentally, liberalised electricity markets provide a supportive environment for the development, demonstration and adoption of innovations. More specifically, an institutional model that aligns system operator incentives to public policy objectives will be needed to mitigate any incentive for incumbents to block the spread of innovations. A model for control of decentralised resources will also be needed to allow innovations such as new network structures and peer-to-peer electricity trading the opportunity to participate in electricity markets.

It is possible that over time these and other innovations will drive or enable larger changes that have the potential to restructure the electricity system more significantly. It will be worthwhile monitoring new technologies and business models so that policy and regulation can respond appropriately, to realise value and address risks.

(1) **New network architectures**

Microgrids and fractal grids are innovative network architectures. Both of these new network architectures provide greater resilience than the radial links of conventional distribution networks. Microgrids achieve resilience through redundancy in generation, while fractal grids achieve resilience through redundancy in network infrastructure.

A microgrid is a small-scale, partially self-sufficient network incorporating both generation and demand sources. Microgrids may be connected to the local distribution network, importing or exporting electricity according to system conditions, but may also disconnect from the distribution network and operate as an island. As a microgrid can meet a degree of its own demand, it is more resilient to wider system failures, caused by a fault or cyberattack, than a radial network, and is well suited for critical functions such as healthcare, military installations or data centres. As microgrids are self-sufficient, they may require more on-site generation than conventional networks. The deployment of on-site generation in microgrids may result in a larger total volume of generation assets in the wider electricity system, implying a degree of asset redundancy and an increase in costs. In principle, the redundancy can be mitigated if sufficient generation is deployed to serve only essential loads when islanded. A microgrid can aggregate its resources like a virtual power plant to coordinate electricity trading with the wider power system. This aggregation can be accomplished by a central controller, or potentially by peer-to-peer communication of individual microgrid resources without a central controller. Figure 72 shows the structure of a microgrid with some of these features.

A fractal grid is a new network structure currently at the concept stage. A fractal grid combines the economy of radial networks, where redundancy is minimal, with the resilience of meshed networks, where nodes are connected by multiple links. The fractal grid achieves these properties through use of a fractal, or recursive pattern, where multiple sets of links with the same structure are connected together in a parent-child relationship. Failure of a single link does not prevent power flows between nodes, and the fractal architecture can accommodate microgrids that are able to island themselves in case of wider system failure. Proponents of fractal grids note that most urban spatial areas already have a fractal structure, so a fractal grid system could be easy to develop in cities. There are several fractal grid demonstration projects. For example, CleanSpark's Fractal Grid has a federated structure that connects microgrids in a

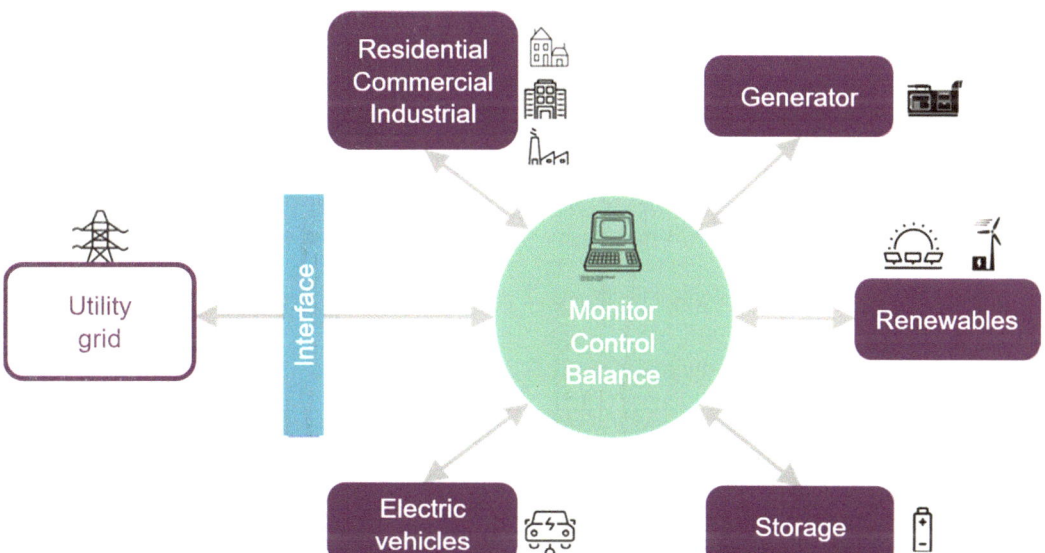

Fig. 72 A microgrid contains generation and flexible resources, as well as sources of electricity demand. *Source* Vivid Economics

parent-child relationship. Microgrids in the fractal grid can share their generation and services with other microgrids to shave peak demand and thereby increase the reliability of the whole system, or they can island themselves to manage generation and load independently. The Fractal Grid Demonstration at Camp Pendleton military camp in California and NRECA's Agile Fractal Grid are other examples of microgrids in a parent-child relationship.

(2) **Peer-to-peer electricity trading**

Peer-to-peer electricity trading could allow individual owners of small-scale generation, storage and demand resources to participate in electricity markets. Currently, electricity is traded between generators and large electricity suppliers. As discussed above, virtual power plants are also likely to enter the electricity market. In addition, developments in peer-to-peer electricity trading could facilitate individual owners of small-scale generation, storage and demand resources to trade electricity. Peer-to-peer energy trading is being trialled at pilot scale in several small microgrid networks. For example, the Brooklyn Microgrid in New York connects generators, distribution lines, batteries and load sources, with trades and electricity flows tracked though the blockchain distributed ledger.

If peer-to-peer trading is widespread, there may be a reduced role for intermediaries in operating the electricity system. Some commentators suggest that blockchain could automate the active participation of large numbers of distributed resources, such that an intermediary, for example, a virtual power plant, is not required. In future, peer-to-peer electricity trading might take place not only within a microgrid, but between resources at the level of the distribution network, and potentially the transmission network. It is theoretically possible that sufficient automation might reduce the role of system operators at the distribution or even transmission system levels, though it is likely that their core roles, managing network constraints and maintaining system security, will remain.

9.2 Analysis of Policy on the Reform of State-Owned Energy Enterprises in China

9.2.1 The Reform of State-Owned Enterprises Since the 12th Five-Year Plan (2011–15)

In November 2013, the Third Plenary Session of the 18th CPC Central Committee deliberated and adopted the Decision of the CPC Central Committee on Several Major Issues Concerning Comprehensively Deepening Reform, marking the start of this round of reform of state-owned enterprises. General Secretary Xi clearly explained the status of state-owned enterprises and the background of the reform in his statement on the Decision: "State-owned enterprises are an important force that promotes the modernisation of the country and safeguards the common interests of the people. After years of reform, state-owned enterprises have, on the whole, been integrated with the market economy. Meanwhile, they need to be further reformed due to the presence of some problems and drawbacks". The key points for further deepening reform are identified in the Decision: "China will perfect the state-owned asset management system, strengthen the supervision of state-owned assets primarily by means of capital management, reform the system of authorised operation of state-owned capital, establish several state-owned capital operation companies, and support the reorganisation of state-owned enterprises into state-owned capital investment companies if conditions permit".

As for important measures for deepening the reform of state-owned enterprises, the Decision sets out that, China will, first: "actively develop a mixed ownership economy", "allow more state-owned sectors, and sectors with other types of ownership, to develop into a mixed ownership economy", "allow non-state capital to participate in state-owned capital investment projects" and "encourage non-public ownership enterprises to participate in the reform of state-owned enterprises, and encourage the development of mixed ownership enterprises held by non-public capital"; and second, "impel state-owned enterprises to perfect the modern enterprise system", "accurately define the functions of different state-owned enterprises", "increase state-owned capital investment in non-profit enterprises", "separate national railway network infrastructure from railway passenger and freight transport, and liberalise competitive businesses in natural monopoly industries owned by state-owned capital" and "improve the corporate governance structure to achieve coordinated operation and effective balance".

Since then, General Secretary Xi has talked about the reform of state-owned enterprises on several occasions.

In March 2014, he said in his speech during the deliberation of the Anhui delegation at the Second Session of the 12th National People's Congress that: "the basic policy for the development of a mixed ownership economy has been identified. Detailed rules and regulations are the key to, and decisive factor for, success. The experience and lessons gained from the past reform of state-owned enterprises will be drawn on. The reform of state-owned assets will not be an opportunity for profiteering".

In August 2014, he said at the Fourth Meeting of the Leading Group of the CPC Central Committee for Comprehensively Deepening the Reform that: "state-owned enterprises must be well developed, especially those administered by the central government that dominate the major industries and key sectors and that impact national security and the national economy. They are an important pillar of CPC administration and the economic foundation of socialist state power".

In June 2015, he said at the 13th meeting of the Leading Group of the CPC Central Committee for Comprehensively Deepening the Reform that: "upholding the leadership of the CPC is a unique advantage of the state-owned enterprises of our country" and that China will "make state-owned enterprises bigger and stronger and able to continuously enhance the dynamism, dominance, influence and resilience of the state-owned economy", "prevent loss of state assets" and "accelerate the formation of a state-owned asset supervision system characterised by comprehensive coverage, clear division of labour, coordination and effective restrictions".

In July 2015, he said in a survey in Jilin that: "state-owned enterprises are an important force that promotes the modernisation and safeguards the common interests of the people. China will unswervingly uphold the important role of state-owned enterprises in the development of the country and resolutely develop state-owned enterprises and make them bigger and stronger". "China will deepen the reform of state-owned enterprises, improve the corporate governance model and operating mechanism, truly establish enterprise status as market players and enhance their internal dynamism, market competitiveness and leadership." "The reform of state-owned enterprises will preserve and appreciate state-owned capital, enhance the competitiveness of the state-owned economy and amplify the functions of state-owned capital."

In May 2016, he said at the 13th Meeting of the Financial Leading Group of the CPC Central Committee that: "China will promote the reform of state-owned enterprises". "It should be emphasised that the disposal of 'zombie state-owned enterprises' means the reform of state-owned enterprises and the strategic adjustment of the state-owned economy."

In July 2016, he said in his speech at the National Forum on the Reform of State-Owned Enterprises that: "state-owned enterprises, which are an important force that strengthens the country and safeguards the common interests of the people, must grow bigger and stronger rationally and emphatically and continuously enhance their dynamism, influence and resilience to preserve and appreciate state-owned assets. China will unswervingly deepen the reform of state-owned enterprises, make efforts to innovate systems and mechanisms, speed up the formation of a modern enterprise system, and activate and energise the enthusiasm, initiative and creativity of all kinds of talent in state-owned enterprises".

In October 2016, he said in his speech at the National Working Conference on Party-Building of State-Owned Enterprises that: "state-owned enterprises should not just be developed but well developed. China will require all relevant departments and state-owned enterprises in all regions to adjust to the developments and changes in the economy at home and abroad and in accordance with the decisions and arrangements of the CPC Central Committee on the reform and development of state-owned enterprises. They should follow the policies that preserve and appreciate state-owned assets, enhance the competitiveness of the state-owned economy and amplify the functions of state-owned capital, deepen their reform and improve their operational management, strengthen the supervision of state-owned assets and unswervingly make state-owned enterprises bigger and stronger".

In October 2017, he made a concise and systematic exposition of issues concerning the reform of state-owned enterprises in his report to the 19th National Congress of the Communist Party of China (NCCPC): "China will perfect the management of state-owned assets; reform the state-owned capital authorisation system; speed up the optimisation, structural adjustment and strategic restructuring of the state-owned economy; preserve and appreciate state-owned assets; impel state-owned capital to grow bigger and stronger and prevent the loss of state-owned assets; and deepen the reform of state-owned enterprises, develop a mixed ownership economy and foster world-class enterprises with global competitiveness".

Premier Li Keqiang has also spoken many times about the reform of state-owned enterprises.

In March 2016, he said in his government work report that: "China will promote development through reform and tackle firmly the most difficult problems affecting the quality and efficiency of state-owned enterprises; restructure state-owned enterprises, especially those that are managed by the central government, through innovation and mergers; promote equity diversification reform and carry out pilot projects to introduce boards of directors, a professional management system, mixed ownership and employee stock ownership; deepen the reform of enterprise employment systems and explore and

establish a compensation system for high-level talent and managers through market-oriented recruitment methods; accelerate the restructuring and formation of state-owned capital investment and operational companies; transform the functions of state-owned asset supervision and administrative institutions by means of capital management, and preserve and appreciate state-owned assets; give greater autonomy to local government with regard to the reform of state-owned enterprises; and accelerate the divestment of those state-owned enterprise functions unrelated to operations, solve problems left over by history, downsize state-owned enterprises and enhance their core competitiveness".

In May 2016, he said at a State Council Executive Meeting that: "China will promote development through reform and will tackle the hardest problems to downsize central government-administered state-owned enterprises and improve their quality and efficiency, and with the courage and determination to make prompt and resolute decisions".

In December 2017, he said at a State Council Executive Meeting that: "China will implement the arrangements identified at the 19th NCCPC, focus on deepening the reform of state-owned enterprises and make efforts to preserve and appreciate state-owned assets".

9.2.2 Use the 1+N Policy System to Reform State-Owned Enterprises

In November 2013, the Decision of the CPC Central Committee on Several Major Issues Concerning Comprehensively Deepening Reform was released. On December 30 of that year, the Leading Group of the CPC Central Committee for Comprehensively Deepening the Reform, under the leadership of General Secretary Xi, was formed. In August 2015, the CPC Central Committee and the State Council jointly issued the Guiding Opinions on Deepening Reform of State-Owned Enterprises. Policies on the reform of state-owned enterprises were introduced within the framework of the Guiding Opinions.

A complete 1+N policy system has been created for the new round of state-owned enterprise reform. "1" refers to the Guidance, which is the overall idea behind the policy system. "N" refers to policies that cover such key areas as leadership of the CPC, the state-owned asset management system, mixed ownership, classification of state-owned enterprises, supervision and solving problems left over by history. "N" includes:

- the Opinions on Rationally Identifying and Strictly Standardising Remuneration and the Business Expenses of Heads of Central Government-Administered State-Owned Enterprises, deliberated and adopted in August 2014;
- the Plan of Reform of the Compensation System for Heads of Central Government-Administered State-Owned Enterprises, implemented in January 2015;
- the Several Opinions on Further Deepening Electricity System Reform, introduced in March 2015;
- the Opinions on State-Owned Enterprises' Development of the Mixed Ownership Economy, introduced in September 2015;
- the Several Opinions on Upholding the Party's Leadership and Strengthening Party Building in Deepening Reform of State-Owned Enterprises, introduced in September 2015;
- the Guiding Opinions on Encouraging and Standardising the Introduction of Non-State-Owned Capital into Projects Invested in by State-Owned Enterprises, deliberated and adopted in September 2015;
- the Several Opinions on Reforming and Perfecting the State-Owned Asset Management System, introduced in October 2015;
- the Opinions on Strengthening and Improving the Supervision of State-Owned Assets of Enterprises and Preventing the Loss of State-Owned Assets, introduced in October 2015;
- the Circular on Printing and Distributing Support Documents for Electricity System Reform, implemented in November 2015

(related support documents are: the Opinions on Promoting the Reform of Power Transmission and Distribution Prices, the Opinions on Implementing the formation of Electricity Markets, the Opinions on Establishing and Operating Power Trading Institutions, the Opinions on the Orderly Liberalisation of Power Generation and Consumption Planning, the Opinions on Implementing Electricity Retail Reform, and the Guiding Opinions on Strengthening and Regulating the Supervision and Administration of Private Coal-Fired Power Plants);

- the Guiding Opinions on the Definition and Classification of Functions of State-Owned Enterprises, introduced in December 2015;
- the Guiding Opinions on Further Regulating and Strengthening the Management of State-Owned Assets of Administrative Institutions, introduced in December 2015;
- the Interim Procedures for the Administration of Stock Ownership and Dividend Incentives of State-Owned Scientific and Technological Enterprises, implemented in March 2016;
- the Circular of the State Council on Printing and Distributing the Work Plan for the Divestment of Functions Unrelated to the Operations of State-Owned Enterprises and for Solving Problems Left Over by History, introduced in March 2016;
- the Regulations on the Supervision and Administration of Transactions of State-Owned Assets of Enterprises, implemented in July 2016;
- the Guiding Opinions on Restructuring and Reorganising Central Government-Administered State-Owned Enterprises, implemented in July 2016;
- the Opinions on Establishing an Accountability System for Illegal Operations and Investments by State-Owned Enterprises, implemented in August 2016;
- the Opinions on the Employee Stock Ownership Pilot Project of State-Held Mixed Ownership Enterprises, introduced in August 2016;
- the Implementation Plan for Improving the Functional Classification Assessment of

Central Government-Administered State-Owned Enterprises, implemented in August 2016;

- the Review of Policies and Suggestions on Supporting Reform of State-Owned Enterprises, introduced in August 2016;
- the Circular on Doing a Good Job of Stock Ownership and Dividend Incentives of Central Government-Administered Scientific and Technological Enterprises, implemented in October 2016;
- the Opinions on the Pilot Implementation of the Functions and Powers of Boards of Directors for Central Government-Administered State-Owned Enterprises, deliberated and adopted in December 2016;
- the Guiding Opinions on Innovating Government's Resource Allocation Methods, introduced in January 2017;
- the Regulations on the Supervision and Administration of Investment by Central Government-Administered State-Owned Enterprises, introduced in January 2017;
- the Plan of the State-Owned Assets Supervision and Administration Commission of the State Council (SASAC) for Transforming Functions by Means of Capital Management, introduced in April 2017;
- the Guiding Opinions of the General Office of the State Council on Further Improving the Corporate Governance Structure of State-Owned Enterprises, introduced in April 2017;
- the Circular on Doing a Good Job of Declaring and Liquidating State-Owned Capital Operation Budgets for the Divestment of the Central Government-Administered State-Owned Enterprise Functions of Water, Power and Heat Supply and Property Management for Employee Residential Areas in 2017, introduced in April 2017;
- the Several Opinions on Deepening Oil and Natural Gas System Reform, introduced in May 2017; and
- the Implementation Plan for the Transformation of Central Government-Administered State-Owned Enterprises into a Corporate System, introduced in July 2017.

9.2.3 Steady Progress in Priorities

First, the Ten Reform Pilots have been implemented since 2016. The 10 pilots were carried out at central government-administered state-owned enterprises to create experience that can be extended into other companies. The 10 pilots had the following objectives:

(1) **Reform the state-owned asset management system**

State-owned capital investment and operational company reform pilots: China National Cereals, Oils and Foodstuffs Corporation (COFCO), State Development & Investment Corporation (SDIC), Shenhua Group, Baosteel, Wuhan Iron and Steel Corporation (WISCO), China Minmetals Corporation (CMC), China Merchants Group (CMG), China Communications Construction Company (CCCC), China Poly Group (state-owned capital investment company reform pilot) and China Chengtong and China Reform Holdings Corporation (CRHC, state-owned capital operation company reform pilot).

Central government-administered state-owned enterprise merger pilots: China State Construction Engineering Corporation (CSCEC) and China National Materials (Sinoma), China Ocean Shipping Company (COSCO) and China Shipping Container Lines (CSCL), and China Power Investment Corporation (CPI) and State Nuclear Power Technology Corporation (SNPTC).

(2) **Develop mixed ownership**

Pilot mixed ownership reform in several key sectors: power, oil, natural gas, rail, civil aviation, telecommunications and the military.

The first batch comprises nine pilots, the second batch consists of 10 pilots and the third batch comprises nine. The 28 pilots include China Unicom, China Telecom, China Railway Group (CREC), China Railway Construction Corporation (CRCC), China Railway Corporation (CRC), China National Aviation Holding Company (CNAH), China Eastern Airlines, China Southern Airlines, State Grid Corporation of China (SGCC), China Southern Power Grid

(CSG), POWERCHINA, China Nuclear Engineering & Construction Corporation (CNEC), China Energy Engineering Corporation (CEEC), State Power Investment Corporation (SPIC), Harbin Electric Corporation (HE), China National Petroleum Corporation (CNPC), China National Offshore Oil Corporation (CNOOC), China Petroleum & Chemical Corporation (Sinopec), China State Shipbuilding Corporation (CSSC), China North Industries Group Corporation (CNGC), China South Industries Group Corporation (CSGC), China Shipbuilding Industry Corporation (CSIC), Aero Engine Corporation of China (AECC), China Aerospace Science & Industry Corporation (CASIC), China Aerospace Science and Technology Corporation (CASC), Aviation Industry Corporation of China (AVIC), and China National Building Material (CNBM).

Pilots of employee stock ownership of mixed ownership enterprises comprise third-level subsidiaries of Shenhua Group, China National Machinery Industry Corporation (Sinomach), China Baowu, COSCO, COFCO, CMG, China Energy Conservation and Environmental Protection (CECEP), CNBM, China Academy of Building Research (CABR), and CREC.

(3) Improve the modern enterprise system

Pilot to implement the functions and powers of boards of directors: CECEP, CNBM, Sinopharm, Xinxing Cathay International Group, China Baowu, SDIC and China General Nuclear Power Group (CGN).

Pilot for the market-based selection and employment of managers: Baosteel, Xinxing Cathay International Group, Sinopharm, CECEP, SDIC and China Railway Signal & Communication Corporation (CRSC).

Pilot to recruit professional managers: China Baowu Group, Xinxing Cathay International Group and CNBM.

Pilot to reform emolument differentiation in enterprises: Scheduled for launch in 2018 by the Ministry of Human Resources and Social Security (MOHRSS).

(4) Prevent the loss of state-owned assets

Pilot for state-owned enterprise information disclosure: COFCO, CSCEC, SPIC and China Southern Airlines.

(5) Solve the problems left over by history

Pilot the divestment of state-owned enterprise functions that are unrelated to operations and solve the problems left over by history: This will be carried out in selected provinces and cities.

Second, the merger and reorganisation of central government-administered state-owned enterprises is an important measure to optimise the distribution of state-owned capital and adjust the industrial sector. It is just one of the Ten Reform Pilots mentioned above. Six pilot enterprises have completed their merger and reorganisation. They include CSCEC-Sinoma (China National Building Material, CNBM), COSCO-CSCL (China COSCO Shipping Group) and CPI-SNPTC (State Power Investment Corporation). In recent years there have been other mergers and reorganisations of state-owned enterprises, such as CSR-CNR (CRRC), MCC-CMC, Sinotrans and CSC-CMG, HKCTS-CITS (China National Travel Service Group Corporation), Sinolight-CNACGC (Poly Group), Baosteel-WISCO (China Baowu) and Shenhua Group-China Guodian (National Energy Investment Group).

Third, excess capacity in the coal industry is being resolved. Capacity cutting by state-owned coal enterprises is an important part of supply-side structural reform. In July 2016, the State-Owned Assets Supervision and Administration Commission of the State Council (SASAC) identified the tools for solving excess coal capacity in central government-administered state-owned enterprises. SASAC proposed that: (i) central government-administered state-owned enterprises should cut capacity by 15% within five years and 10% within two years; and (ii) central government-administered state-owned enterprises involved in the coal industry on a non-core basis should withdraw from the industry—only professional coal enterprises and coal power integration enterprises should remain. All the central government-administered coal enterprises achieved their capacity-cutting targets for 2016 and 2017. SDIC and Poly Group have transferred their coal businesses to China Coal Group. In addition, local government-administered state-owned coal enterprises are actively cutting capacity and have achieved their capacity-cutting targets for 2016 and 2017.

9.3 The Reform of State-Owned Energy Enterprises

9.3.1 The Idea Behind the Reform

Based on a review of the speeches and policies on the new round of state-owned enterprise reform, the basic idea behind the reform can be summed up in the following six points:

First, ensuring national energy security wlll continue to be an important goal. China is a major energy producer and consumer, with a balanced structure of supply and demand. Total demand is large and has been relatively stable following the economic downturn, with low growth expected in the years to come. There are objective constraints, such as local resource endowment and long-distance transport. Seasonal and partial shortages are, therefore, difficult to avoid. Ensuring national energy security remains an important goal of this round of reform of state-owned energy enterprises. In the future, China's energy supply system will be continuously optimised on the basis of local conditions to ensure that the foundation of national energy security will not change.

Second, state-owned enterprises will become bigger and stronger. General Secretary Xi has continually emphasised that China will make state-owned enterprises and state-owned capital grow bigger and stronger. Scale is one of the characteristics of state-owned energy enterprises and will remain so in the future. According to data from the National Bureau of Statistics, in 2015 state-owned enterprises in the coal, petroleum,

chemical and power sectors accounted for 47.2% of state-owned industrial enterprise assets and 40.2% of their employees. The aggregate assets of state-owned energy enterprises amounted to 18.03% of the total assets of all industrial enterprises. Central government-administered energy enterprises include China National Petroleum Corporation (CNPC), China Petroleum & Chemical Corporation (Sinopec), China National Offshore Oil Corporation (CNOOC), National Energy Investment Group (renamed after the Shenhua Group-China Guodian merger), China Coal Group, SPIC (after the CPI-SNPTC merger), China Huaneng Group (CHNG), China Huadian Corporation (CHD) and China Datang Corporation (CDT), as well as many local government-administered state-owned energy enterprises that are a pillar of their local economy. These enterprises have good industrial foundations and advantages of scale, technology and market, reflecting the strategic pattern of state-owned capital in the energy industry. In the future, the issues of how to make central government-administered energy enterprises stronger and bigger will be the focus of their reform and development. The smooth progress of the pilots to improve central government-administered energy enterprises is an example for the reform and improvement of local government-administered state-owned energy enterprises.

Third, the roadmap for improving state-owned capital in state-owned energy enterprises will be created. The Guiding Opinions on Restructuring and Reorganising Central Government-Administered State-Owned Enterprises, which was implemented in July 2016, stipulates that these enterprises "will significantly improve their support capabilities in the sectors that impact national security, such as national defence, energy, transport, food, information and ecology", and that they will "enhance their dominance in the key industries that affect the national interest, people's livelihoods and the national economy, such as major infrastructure, key resources and public services", and that they will "drive industries like new energy, new materials, aerospace and intelligent manufacturing". Specifically, the following efforts will be made:

(1) China will implement sole proprietorship or shareholding in fields like: water conservation, hydropower and navigation-hydropower integration hubs in important river basins; natural monopolies like oil and natural gas trunk pipeline networks, power grids and nuclear power; and reduce excess capacity in the coal industry;

(2) China will encourage central government-administered state-owned enterprises in power, new energy and oil and gas pipelines to fund specialist joint-stock platforms; and

(3) China will reorganise central government-administered state-owned enterprises in the upstream and downstream coal, power and metals industries, which will provide a model for state-owned capital in the energy sector.

Fourth, state-owned energy enterprises will be reorganised into state-owned capital investment companies. The function of these companies will be to optimise the distribution of state-owned capital through the entry or withdrawal or flow of state-owned capital in the energy industry, and to pool the advantages of state-owned energy enterprises and enhance their operational efficiency and international competitiveness. The state-owned capital investment company reform pilot is at SDIC and the Shenhua Group. The goal of the Guiding Opinions on Restructuring and Reorganising Central Government-Administered State-Owned Enterprises is to "create a mechanism for the entry or withdrawal or flow of state-owned capital" and to "reorganise central government-administered energy enterprises into state-owned capital investment and operational companies". Some local governments have identified the reform pilot of reorganising provincial government-administered energy enterprises, such as those engaged in the coal business, into local state-owned capital investment companies.

Fifth, state-owned energy enterprises will be encouraged to develop mixed ownership. The

development of mixed ownership is an important measure to reform the state-owned capital management system and the internal mechanisms of state-owned enterprises. The 1+N policy system includes special policy arrangements, such as the Opinions on the Development of the Mixed Ownership of State-Owned Enterprises, which was introduced in September 2015, and the Opinions on the Employee Stock Ownership Pilot of State-Held Mixed Ownership Enterprises, which was launched in August 2016. The mixed ownership reform of state-owned enterprises is described in other documents as well. The Guiding Opinions on Restructuring and Reorganising Central Government-Administered State-Owned Enterprises emphasises that China will maintain state-owned capital's control of key industries and sectors, while supporting the participation of non-state-owned capital. The Opinions on Deepening Electricity System Reform, introduced in March 2015, states that China will "steadily reform, liberalise and open electricity retail to private capital in an orderly manner", "encourage private capital to invest in power distribution" and "gradually open power distribution investment to market participants that meet requirements, and encourage the development of distribution businesses based on mixed ownership". The Opinions on Deepening Oil and Natural Gas System Reform, introduced in May 2017, states that China will "improve the corporate governance structure of state-owned oil and gas enterprises and encourage them to diversify their ownership structure and allow mixed ownership, if conditions permit". In the coal sector, Shenhua Group has been identified as one of the pilots for mixed ownership reform of central government-administered state-owned enterprises.

Sixth, the supply-side structural reform of state-owned energy enterprises will be deepened. Downsizing state-owned energy enterprises is a key element of supply-side structural reform. At present, it primarily involves three aspects.

(1) *Tackling the hardest problems in the reform of state-owned energy enterprises*. This primarily refers to deepening electricity and oil

and natural gas system reform. Relevant government departments have produced support documents like Opinions on Further Deepening Electricity System Reform in March 2015 and Opinions on the Reform of Power Transmission and Distribution Prices, Opinions on Creating Electricity Markets, Opinions on Implementing and Operating Power Trading Institutions, Opinions on the Orderly Liberalisation of Power Generation and Consumption Planning, Opinions on Implementing Electricity Retail Reform, Guiding Opinions on Strengthening and Regulating the Supervision and Administration of Private Coal-Fired Power Plants, and Opinions on Deepening Oil and Natural Gas System Reform.

(2) *Steadily implementing capacity cutting in the coal and coal-fired power industries.* China has intensified capacity cutting in the coal industry since 2016. Resolving excess coal production has become one of the key tasks in the reform of state-owned energy enterprises. In February 2016, the State Council issued Opinions on Resolving Excess Capacity to Stop Losses and Develop the Coal Industry. Since then, support documents on human resources, finance, safety supervision, land, quality inspection, taxation and environmental protection have been published, and major coal-producing provinces, municipalities and autonomous regions have identified their capacity-cutting targets, plans and measures. The Guiding Opinions on Restructuring and Reorganising Central Government-Administered State-Owned Enterprises of July 2016 stipulates that China will "reduce excess capacity and speed up the elimination of outdated capacity" in industries like coal. In July 2017, 16 ministries and commissions jointly issued Opinions on Promoting Supply-Side Structural Reform to Prevent and Resolve the Risk of Excess Coal-Fired Power Capacity, which started capacity cutting in the coal-fired power industry. The document states that China will ensure power security while "eliminating outdated capacity",

"shutting down illegal projects" and "strictly controlling the scale of new capacity"; and that China will "encourage large power generation groups to reorganise and integrate upstream and downstream enterprises in coal and power to unlock synergies". The coal-fired power industry has identified its capacity-cutting targets for the 13th Five-Year Plan (2016–20). These include stopping or suspending the construction of 150 GW of coal-fired power generation and eliminating outdated capacity of more than 20 GW.

(3) *Solving problems left over by history and divesting state-owned enterprises of functions unrelated to their operations.* In March 2016, China introduced the Circular of the State Council on Printing and Distributing the Work Plan for Accelerating the Divestment of State-Owned Enterprise Functions Unrelated to Their Operations and Solving Problems Left Over by History. The circular also identified these two issues as one of the Ten Reform Pilots.

9.3.2   Major Progress

Progress has been made in the reform of state-owned energy enterprises that have been conducted in an orderly manner and in accordance with the six aspects of the basic idea, as summarised in Table 18.

9.4   Three Conclusions on the Reform of State-Owned Energy Enterprises

9.4.1   The Key to the New Round of Reform of State-Owned Energy Enterprises Is to Improve the Efficiency of State-Owned Capital Investment in the Energy Industry

This round of reform of state-owned enterprises focuses on the state-owned capital management system. Specifically, for state-owned energy enterprises, the goal is to increase the efficiency

Table 18 Summary of progress in the reform of state-owned energy enterprises

Aspect	Progress
Ensuring national energy security	The responsibility of ensuring energy supply and energy and technological reserves has been fulfilled
Making state-owned energy enterprises grow bigger and stronger	(1) The state-owned energy enterprises that have reorganised are operating stably (2) Reorganisation of central and local government-administered state-owned energy enterprises will continue (3) Integration across the value chain has taken place, which represents a transformation of the industry (4) State-owned energy enterprises have been downsized through measures like streamlining the business, eliminating outdated and inefficient production capacity and divesting state-owned enterprises of unrelated operations (5) The modern enterprise system has been improved; the role of the Party has been activated and loss of state-owned assets prevented
Optimising the distribution of state-owned capital	(1) The idea of state-owned capital investment in different energy fields has been identified (2) Horizontal and vertical integration and the reorganisation of state-owned energy enterprises, such as the merger of Shenhua Group and China Guodian, have started (3) SDIC and Poly Group have transferred their coal operations to China Coal Group—a result of the policy of withdrawing central government-administered state-owned enterprises from coal if it is not their core business
Reorganising state-owned energy enterprises into state-owned capital investment companies	(1) Two central government-administered state-owned enterprises, SDIC and Shenhua Group, are in the state-owned capital investment company reform pilot (2) Some local government-administered state-owned energy enterprises are also involved in local state-owned capital investment company reform pilots
Carrying out mixed ownership reform	Mixed ownership reform has been carried out to varying degrees at second- and third-level subsidiaries of state-owned energy enterprises
Deepening supply-side structural reform	(1) The documents on deepening electricity system reform and on oil and natural gas system reform have been introduced and a roadmap for marketisation has been released (2) The annual capacity-cutting plan for state-owned energy enterprises in the coal industry has been oversubscribed for two years in a row (3) Capacity cutting has been promoted in the coal industry (4) Solving problems left over by history and the divestment of state-owned enterprise functions that are unrelated to operations have been accelerated. Some state-owned energy enterprises have already completed this work. For example, the initiative for CNPC and Sinopec to divest unrelated operations affects more than 2 million households near the oil fields. In terms of the workload needed to divest water, power and heat supply and property management for employee residential areas, CNPC and Sinopec account for a third of the total of all the central government-administered state-owned enterprises. The provincial government-administered state-owned energy enterprises have produced a schedule for this work. Some state-owned energy enterprises in Shandong, Henan, Hebei and Shaanxi have achieved their targets

of state-owned capital investment in the energy industry. Under the new management system, the border between the government and enterprises is clearer. As a contributor of state-owned capital, the government is not engaged in managing asset operations, which is the business of the enterprises, but in managing capital and continuously optimising the distribution of state-owned capital by leveraging liquidity and improving the return on state-owned capital. The reform addresses three levels:

- the responsibility of government departments for state-owned enterprise regulations, and the relationship between the government and state-owned enterprises, should be optimised;
- the corresponding internal reform of state-owned enterprises should be deepened; and
- government should delegate some management and investment powers to state-owned enterprises, which in effect means decentralising some power. Specifically, for state-owned energy enterprises, the government should optimise the distribution of state-owned capital in the energy industry through guarantees of state-owned capital for key and strong sectors and withdrawal from inefficient areas to improve the return on capital.

As part of the reform of the state-owned capital management system, state-owned energy enterprises need to carry out internal reforms, such as reorganising themselves into state owned capital investment companies, establishing a modern enterprise system with internal mechanisms and opening up to mixed ownership. The necessity of reorganising state-owned energy enterprises into state-owned capital investment companies should proceed on a case-by-case basis. In the reform pilot of turning central government-administered state-owned enterprises into state-owned capital investment companies, only two energy enterprises (SDIC and Shenhua Group) are involved. For those

state-owned energy enterprises that transition into state-owned capital investment companies, their internal systems and mechanisms should be reformed in accordance with the new state-owned asset management system. The reason for this is that the goal of the reform of the state-owned capital management system is to improve the return on investment of state-owned capital. The mixed ownership system reform pilot involves CNPC, Sinopec and central government-administered state-owned enterprises in the power industry.

9.4.2 State-Owned Energy Enterprises Should Become Bigger and Stronger

Making state-owned energy enterprises bigger and stronger is a realistic requirement for national energy security and social stability. China is a major energy consumer, working in close cooperation with the global supply market in the three major fossil energy areas: oil, natural gas and coal. It should be emphasised that energy supply is not only an economic issue but also a social responsibility. First, state-owned energy enterprises shoulder the social responsibility of ensuring energy supply. Second, since state-owned energy enterprises have large workforces and the areas where they are located are often endowed with energy resources, they shoulder the social responsibility of stabilising employment and promoting regional industrial restructuring. Bearing this dual (economic and social) responsibility, state-owned energy enterprises need to become bigger and stronger. Through years of development, they have laid a good foundation for growing bigger and stronger.

Growing stronger comes before growing bigger, which suggests that the former is harder. State-owned energy enterprises have already been big, but this does not mean they are strong. Generally, they are subject to heavy historical burdens, high interest rates on debt, poor use of resources and low profit margins. Improving return on investment of state-owned capital in the energy industry means that the operational effi-ciency of state-owned energy enterprises must be improved.

Therefore, the key to this round of reform of state-owned energy enterprises is to optimise their internal structure in business, staffing, assets and capital, and strengthen their capabilities in management, operation and investment to help them become bigger. Through optimisation and strengthening, state-owned energy enterprises can use their scale and industry advantages to develop gradually into internationally competitive energy enterprises.

9.4.3 Many Difficulties Remain in the Reform of State-Owned Energy Enterprises

This round of reform of state-owned enterprises, which focuses on improving the state-owned capital management system, is steadily making progress. Despite some achievements, considerable difficulties remain.

First, there are institutional barriers to withdrawing state-owned capital from inefficient energy assets. Characterised by their large size and diversified operations, state-owned energy enterprises have varying degrees of inefficient assets. The losses made by state-owned energy enterprises in these inefficient assets can be stopped by the state withdrawing its capital. Relevant support policies for system reform (by means of capital management) and supply-side structural reform (through capacity cutting) are in place. However, withdrawing state-owned capital from inefficient state-owned energy assets is difficult. The difficulties include resettling employees, repaying debt and preserving assets. Once inefficient assets are removed, the repositioning and resettlement of employees and debt repayment should be handled in a timely manner. Thanks to the central government, local government and state-owned enterprises, the repositioning and resettlement of employees is being addressed through various channels. However, the problem of debt repayment remains a difficulty. A debt may be shelved or borne by the parent company. But this also involves the

debt-owning banks, which makes institutional progress difficult. Besides, a market-based valuation of assets is hard to make. The value of inefficient assets is low and they often create huge losses for state-owned energy enterprises. However, the government requirement to preserve and appreciate the value of state-owned assets means that their estimated value rarely equals their market value, and their estimated value is unacceptable to private capital.

Second, electricity system reform and oil and natural gas system reform are slow-paced. Although both are industrial management system reforms, they are closely related to the reform of state-owned enterprises because both industries are dominated by state-owned capital. Both involve the liberalisation of some business areas. The mixed ownership reform pilot involves central government-administered state-owned enterprises in these two industries. The slow pace of the two industry reforms is driven primarily by three factors: (i) market liberalisation has an impact on energy security; (ii) the influence of the existing benefits structure on traditional enterprises has been broken; and (iii) the Chinese market system still requires continuous improvement. These three factors are difficult to change quickly and the implementation of the reforms will take time and government determination.

Third, state-owned energy enterprises lack the ability to manage investment. State-owned energy enterprises were historically industrial groups focused on production. Under the new state-owned capital management system, they are required to maximise the investment function of state-owned capital. However, they lack not only human resources for and experience in capital investment, but also the requisite internal management system. An organisational structure can be gradually established and optimised, but reforming investment mechanisms is difficult.

Releasing the dynamism of state-owned energy enterprises is an important means to improve operational efficiency. Internal mechanism reform and large system reform complement each other. Since the reform of the state-owned capital management system has yet to be fully implemented, the reform of state-owned enterprises is difficult to implement. Reform involves defining the responsibility of state-owned asset administrative departments, the division of labour between government departments, and the government's authorisation to delegate power to state-owned enterprises.

Fourth, mixed ownership reform is difficult to implement. The reform pilot for mixed ownership of state-owned energy enterprises involves second- and third-tier non-core subsidiaries and is difficult to roll out at scale. The mixed ownership reform aims primarily to make state-owned capital open to private capital, creating a mix of state-owned and private capital. Such a mix is not simply adding or subtracting capital but integrating different governance mechanisms. Private enterprise might reject the capital management system and mechanisms of state-owned enterprises, while state-owned enterprises might not accept the management system of private enterprise. As things stand, institutional integration is the biggest obstacle. Questions such as "Which businesses should be open to private capital?" and "How should state-owned assets be valued and priced?" are still unanswered. The businesses favoured by private enterprise are not necessarily open to them, while those open to them are not necessarily attractive to them. The asset prices, which are based on the government policy of preserving and appreciating state-owned assets, are likely to be higher than the expectations of private capital. These are the real problems underlying mixed ownership reform.

Special Report 5: International Energy Cooperation and Governance

Wei Jigang and Peter Webb

1 Executive Summary

Currently, the global energy landscape is undergoing major shifts in terms of demand, supply, technology, structure, market and investment. Countries have general requirements for improved air quality and reduced emissions and pollution. International energy cooperation and governance must adapt to these new changes; promote the transition to clean, low-carbon, efficient and secure global energy; and drive high-quality development of global energy.

DRC Team Lead of Special Report 5:
Wei Jigang from the Research Department of Industrial Economy, Development Research Center (DRC) of the State Council of China.

Shell Team Lead of Special Report 5:
Peter Webb, Government Relations Advisor, Shell International B.V.

Contributors:
Rob Bailey, Daniel Quiggin, Felix Preston and Sian Bradley from the Department of Energy, Resources and Environment, the Royal Institute of International Affairs; Hong Tao from DRC; and Chen Jinxiao from the Chinese Academy of Social Sciences.

W. Jigang (✉)
Research Department of Industrial Economy,
Development Research Center (DRC) of the State
Council of China, Beijing, China

P. Webb
Government Relations Advisor, Shell
International B.V., The Hague, the Netherlands

1.1 Making Global Energy Governance Fit for the Future

Energy security has traditionally focused on oil and gas supply security and price stability and created a set of international energy governance regimes. Today, three key trends are reshaping the nature and focus of international energy cooperation and governance. First, most governments and companies recognise that energy security must be delivered alongside rapid action to reduce greenhouse gas emissions. Second, there are increasing concerns over air pollution at the national level. Third, policy encouragement for lower carbon, cleaner options has helped spark technological advances and declining prices. International energy governance needs to adapt to new market realities and shifts in energy mix in order to remain relevant and effective.

However, the current global energy governance regime suffers from two principal deficiencies in its ability to facilitate a low-carbon, secure energy transition. First, the principal energy security regime of the International Energy Agency (IEA) does not include emerging economies, which account for a major, and growing, share of consumption. Second, even though the energy sector is responsible for around two-thirds of global greenhouse gas emissions, climate objectives have not been fully integrated into energy governance arrangements.

© The Author(s) 2020
Shell International B.V. and the Development Research Center (DRC) of the State Council of the People's Republic of China (eds.), *China's Energy Revolution in the Context of the Global Energy Transition*, Advances in Oil and Gas Exploration & Production, https://doi.org/10.1007/978-3-030-40154-2_6

Harmonisation of global energy and climate governance regimes could be achieved by a long-term agenda under the auspices of the G20, beginning with the alignment of international organisation mission statements. This requires steps to build the G20's leadership and oversight function by establishing a long-term series of G20 energy ministerial meetings and building an energy secretariat function.

1.2 Investment Regimes for a Low-Carbon, Energy-Secure Transition

Allocating investment during a swift energy transition, while maintaining oil and gas supply security and reliability of electricity networks, requires clear market signals and the incorporation of transition risks into the cost of capital. However, the global coverage and level of carbon pricing are insufficient, fuel subsidies distort investment decisions, and transition risks are poorly understood by capital markets.

As significant heavy industries embark on domestic carbon pricing, it is in all countries' interests to ensure a level international playing field through the widespread adoption of carbon pricing elsewhere. And it is in all countries' interests to ensure sufficient global investment in hydrocarbon exploration and development through the transition, in particular to avoid supply crunches.

International momentum to propagate carbon pricing and improve the incorporation of transition risks into financial decision making is growing. G20 states collaborate with each other to support the development, adoption and implementation of the recommendations of the Financial Stability Board's Taskforce on Climate-related Financial Disclosures. G20 states continue to support the phase-out of fossil fuel subsidies and work through multiple channels to promote carbon pricing through lesson sharing, clarify regimes for linking markets under the United Nations Framework Convention on Climate Change, and explore the use of border adjustment measures to encourage wider adoption and reduce leakage.

1.3 Security of Supply of Metals and Minerals for Future Energy Systems

Concerns are growing about the security of supply of metals and minerals critical to the manufacture of technologies and infrastructure for future energy systems. Compared to markets for oil and gas or agricultural commodities, markets for metals and minerals have poor information and weak governance.

Secure supply of affordable raw materials will be crucial to the economic security of all countries.

Supply chain shortages and trade disputes could be avoided by countries working together to improve information, data and price transparency across the supply chain, while also supporting dialogue and developing win-win technology and investment packages with existing, new and emerging producers to encourage trade and prevent export controls.

Circular economy strategy could also help reduce the long-term import needs of these critical materials. There should be greater focus on research, development and demonstration (RD&D) of enhanced recovery and recycling of metals and minerals from in-use stocks, enabling the greater use of secondary materials.

Reducing reliance on critical metals and minerals could be furthered by forming unconventional alliances with regional neighbours to support RD&D of substitute technologies for those low-carbon energy technologies most dependent on critical metals and minerals.

1.4 Electricity Reliability During the Shift from Molecules to Electrons

Steep reductions in the cost of clean energy technology are accelerating the shift from molecules to electrons. The nature of energy security is therefore also changing, not least because an electricity-focused energy system requires real-time management for system reliability. Interconnection of national electricity grids can help balancing, reduce electricity costs and provide opportunities for trade.

1.5 Strategic Path for International Energy Cooperation

International cooperation is critical to achieving global energy objectives. The global challenge of enhancing energy security over the course of the next half century, while decarbonising the energy sector, will not be achieved through governments acting unilaterally in isolation from one another.

In Fig. 1, the proposed international cooperation options are shown on a matrix. The horizontal axis represents a spectrum from bilateralism to multilateralism, and the vertical axis represents a continuum from market-led approaches to more direct policy interventions.

As noted above, these proposals should be viewed as part of a broad package, since many could be delivered together, and implementing some depends on progress in others. However, what is obvious is that, while the options span a range of market- and policy-led approaches, few are based on a transactional approach. Instead, the proposals cluster among regionalism and multilateralism in the form of efforts to establish, strengthen and propagate common rules and norms.

Patience will be required, as the success of many of these proposals rests on wider political dynamics beyond the energy sector, not least China's relations with the USA and other countries. Moreover, realising the win-wins that are available will often require compromise. In addition, the Belt and Road Initiative can provide a platform for regional and plurilateral governance by establishing rules of the road; alternatively, it could provide an arena for country-by-country bilateral deals.

G20 plays a central role in many of the proposals. It provides a potential avenue through which China and partner countries can pursue reforms and has a track record of reforming international organisations and establishing new ones. Its membership has critical mass in key areas such as emissions, energy consumption and

investment, and R&D. And it has existing workstreams in relevant areas such as infrastructure, green finance, investment and fossil fuel subsidies.

As such, the G20 could evolve into the apex organisation for global energy governance. But this will not happen overnight, and considerable effort would be needed to build and sustain its capacity to do so.

1.6 China Should Develop Future-Oriented Win-Win International Energy Cooperation and Shoulder Greater Responsibilities in Global Energy Governance

As a large country with high levels of energy production, energy consumption and trade volumes, as well as increasing dependence on foreign energy, China's development has had significant and far-reaching impacts on the global energy landscape. Moreover, China is beginning to be a global leader in the deployment of energy technologies. As a major manufacturer and exporter of low-carbon technologies, China stands to benefit economically from rapid global decarbonisation; and as a country with high levels of exposure to climate risks, China also stands to benefit from accelerated global emissions reductions. Finally, as the world's largest energy consumer, China is exposed to market disruptions and it benefits from resilient international energy markets.

To establish a stable, effective and sustainable energy supply system, China must implement a more inclusive international energy cooperation strategy and promote win-win collaborations. Strategically, China should deepen international energy cooperation in five aspects: multi-level international energy cooperation partners, multi-channel international energy cooperation methods, diversified international energy cooperation forms, multi-area international energy cooperation content, and multi-task international energy cooperation processes.

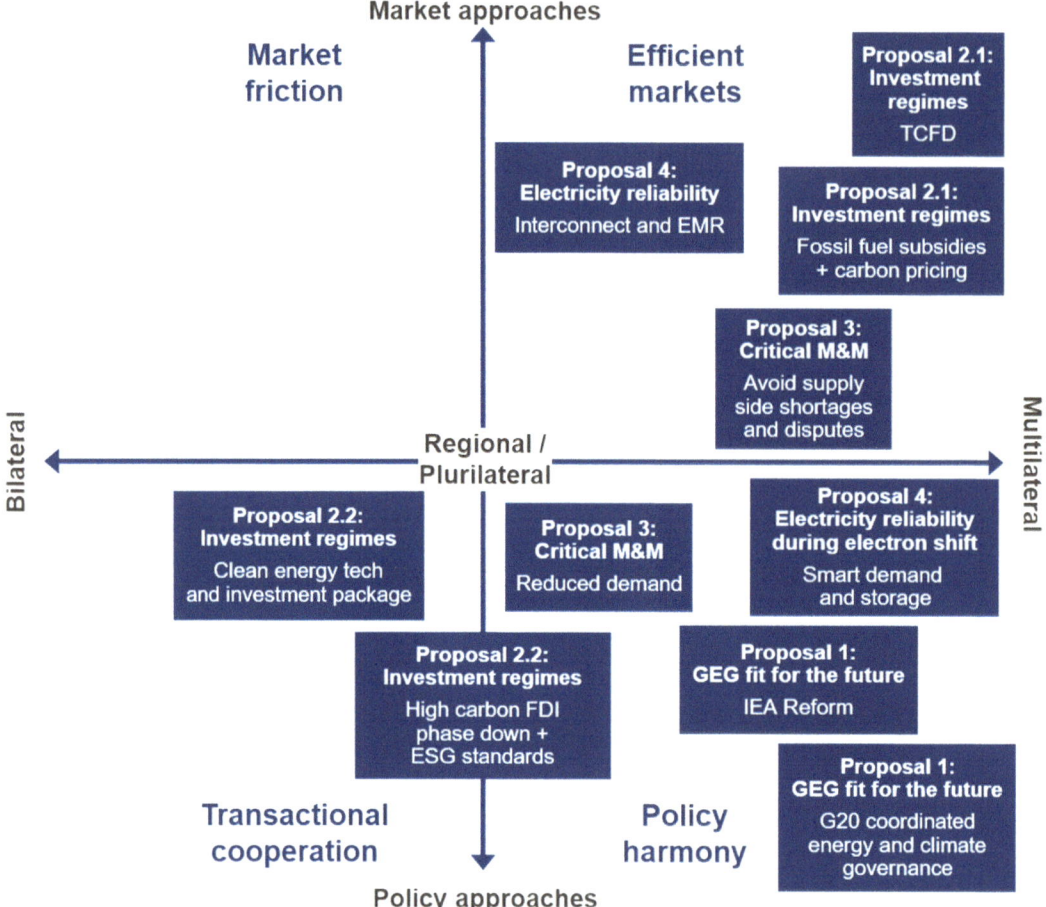

Fig. 1 Strategic path for international energy coopera-tion. *Note* TCFD = Taskforce on Climate-related Finan-cial Disclosures; FDI = foreign direct investment; ESG = environmental, social and governance; GEG = global economic governance; IEA = International Energy Agency. *Source* IEA (2017); BP Energy Outlook 2018

Due to its leadership role among emerging economies, and its position as the world's largest energy consumer and greenhouse gas emitter, China is uniquely placed to drive global energy governance reforms.

China should undertake a major power's responsibilities in global energy governance. It should drive change of the international energy governance landscape dominated by developed economies to make the global energy governance regime more representative and better reflect the interests of developing and emerging economies. And it should propose building a shared future in global energy to deliver win-win results with the world. China has a responsibility to reduce emissions, promote the energy revolution and lead in the energy transition, and be open to the outside world.

First, China should participate deeply in the current cooperation between, and reform of, existing international energy governance institu-tions. Specifically, China should actively advo-cate or lead the amendment of international mechanisms and the establishment of new ones. It should integrate with the international energy market and participate in global energy gover-nance in depth, thus becoming an active, responsible, constructive and reliable contributor.

In particular, China should: (i) have compre-hensive strategies and a roadmap for its participation in global energy governance; it should assess the cost and benefits of participation and identify the methods and support systems needed for participation; (ii) meet the expectations of the international community, such as joining the IEA's emergency response mechanism and improving data quality; (iii) establish an internal conference and consultation mechanism for international energy issues and allow the authorised departments to attend international conferences to make China's voice heard in a more powerful way; (iv) collaborate with G20 emerging economies to promote IEA reform; and (v) identify the international energy conferences and activities to participate in and the department and officials who should take part.

Second, China should strengthen regional energy security and align it with socioeconomic development. Specifically, China should: (i) form a community of common interests and bolster energy security through regional and multilateral cooperation; (ii) collaborate with global or regional organisations to establish stable and cooperative relations between energy suppliers and users; (iii) participate in regional multilateral governance to promote regional energy security and the joint development of all parties' energy economies; (iv) better manage its different governance roles at regional and multilateral levels; (v) insist on mutually beneficial bilateral cooperation and enhanced multilateral cooperation; and (vi) improve energy security while hedging against the possible failure of IEA reform.

Third, China should strengthen its participation in global energy governance and its ability to: (i) set international energy agendas, especially those representing emerging and developing economies; (ii) expertly apply international energy rules, especially the laws and regulations governing international energy trade and financial investment; (iii) optimise domestic energy governance and undertake effective energy diplomacy to reform international energy cooperation and make it more inclusive of developing countries; (iv) have sufficient human resources with the right training for international energy governance; (v) enable its energy companies to participate in the international energy market; and (vi) build discussion platforms with the help of non-governmental and international organisations, and enhance research on, and capacity building for, participation in global energy governance.

2 Introduction

As energy systems around the world undergo major shifts, global energy governance[1] (institutions, mechanisms and norms) and international energy cooperation must evolve to remain relevant. Given the high transaction costs associated with the creation or the reform of multilateral institutions, changes in the past have often been slow and piecemeal. Yet abrupt economic and energy security crises have, at times, led to more dramatic changes—most notably after the 1973 oil shock.

Most governments and companies recognise that energy security must be delivered alongside rapid action to reduce greenhouse gas emissions. Air quality is another driver of change, especially in developing and emerging economies. Meanwhile, policy encouragement for lower carbon, cleaner options has played a role in sparking technological advances and declining prices.

Achieving the Paris Agreement's goals of limiting global warming to well below 2°C, while ensuring energy security through the transition, presents a major challenge to both national and collective energy policymaking. Without concerted domestic action and cooperation at unprecedented scales, the world is unlikely to see the tectonic shifts in technology deployment and capital allocation that are required.

The transformation of energy markets and investments is changing the terms of energy

[1]See page 3 of the Chatham House – DRC report, Navigating the New Normal: China and Global Resource Governance, for a comprehensive definition of global resource governance, including energy governance. Available at https://www.chathamhouse.org/sites/files/chathamhouse/publications/research/2016-01-27-china-global-resource-governance-preston-bailey-bradley-wei-zhao-final.pdf.

choices. As transport and heating become increasingly electrified, the focus of policy is beginning to shift from the governance and security of molecules to that of electrons. The expansion of electricity networks, rapid deployment of electric vehicles and growth in storage capacity could become a self-reinforcing dynamic, creating new trade patterns and interdependencies.

It is in all countries' national interests to ensure energy security. Much has been made of China and the EU's growing dependence on oil imports, in contrast to North America's emerging age of energy independence, following the shale gas revolution. However, as the nature of energy choices shift, the emphasis of energy policy and international cooperation is expanding to include the governance of grids and cross-border trade in electricity, as well as the supply security of critical metals and minerals for the low-carbon economy, such as lithium, cobalt and nickel.

Governments also have broader strategic interests that are linked to energy governance and cooperation, from economic performance to quality of life, including clean air and water. These policy areas are inextricably linked to domestic energy choices—regardless of a country's level of dependence on overseas imports or its role as an energy exporter—and are facing increasing public scrutiny. Table 1 describes some of the key strategic interests of major economies and how they might be affected by a revised perspective on energy security and cooperation that encompasses energy transition.

Governments will need to use a combination of unilateral, bilateral and multilateral approaches to manage these challenges effectively. While effective engagement and joint leadership in a multilateral context require significant political capital in the short term, in the longer term such an approach can help mitigate the substantial transaction costs associated with managing multiple bilateral relationships. It can also help reinforce emerging international norms and provide a consistent structure for international cooperation.

With the exception of the Paris Agreement, progress in multilateral forums has been slow and complicated. Governments need to begin laying the foundations for multilateral approaches early and be willing to play the long game, while at the same time leveraging unilateral, bilateral and thematic approaches to expedite progress at the multilateral level and plug the gaps in the existing energy governance architecture. These approaches will need to be resilient and able to adapt to changing political and economic contexts, climate shocks and new information.

Pragmatic approaches will be needed—in many cases working within existing institutions and groupings, and in others working to reform them. The G20, for example, accounts for more than 80% of global GDP, global primary energy demand, and global emissions respectively, and has the collective capacity to drive change on various fronts including investment, infrastructure, access to energy and fossil fuel subsidies. The International Energy Agency (IEA), by contrast, is declining in relevance as hydrocarbon demand growth shifts away from its member OECD countries towards emerging markets, and as increasingly complex energy systems break down and make the distinction between producers and consumers less distinct.

Box 1: Definitions of international cooperation and global energy governance

While no official definition of global governance or international cooperation exists, here we define the two in the following terms. Global governance can be undertaken by a coalition of national governments, market actors, regulators or lawmakers, within a formal or informal multilateral organisation. Decisions within these organisations, structures and mechanisms influence hard rules such as national laws, regulations and standards, but also soft rules such as norms and cultural practices. Rules can arise from the bottom up, for instance Internet governance, as well as from the top down, in the case of the United Nations.

Table 1 The key strategic interests of major economies and the potential impacts of a revised perspective on energy security and cooperation

	Trade dependencies	Access to critical metals for low-carbon energy	Cross-border electricity and interconnections	Leadership in clean energy technologies	Energy—a key component of international cooperation
China	Oil security is a top priority. China is the world's largest net oil importer. Concern over instability in the Middle East (which supplies 50% of imports) as well as strategic maritime choke points (straits of Hormuz and Malacca). As an associate country of the IEA, China does not formally participate in the IEA's emergency mechanism. Gas is rising in strategic importance. The Russia-China gas pipeline is due to start by 2020. Shale gas development is slower than expected. China is now the third largest importer of LNG, mainly from Australia and Qatar	China depends on access to global markets for a range of commodities. It could depend on imports for as many as 39 out of 45 key minerals by 2020. A major player in metals and minerals markets, China is the key producer of rare earth elements and a major importer and processor of most metals. Circular-economy policies could dramatically reduce import needs and energy consumption, especially if saturation of metals and minerals in the economy is reached sooner than expected	Increasing interconnectedness among Belt and Road Initiative countries. China plans to increase long-distance exports (e.g. to Central Asia and Europe) from western China through investment in ultra-high voltage transmission lines. China's domestic electricity market reforms would need to advance alongside or ahead of ambitious plans for transmission links with neighbours	Ambitious but relative renewable targets. China aims to increase the share of non-fossil energy in energy consumption to 15% by 2020 and to 20% by 2030, and to increase installed renewable power capacity to 680 GW by 2020. China is capturing the low-carbon market: it is the world's largest manufacturer of solar photovoltaic cells and BYD is the world's largest electric vehicle manufacturer	Advancing energy governance is a priority for the G20 and was put on the agenda by China at the Hangzhou summit in 2016. The Belt and Road Initiative will play an important role, from electricity interconnections to opportunities to mine critical materials. China is now among the leaders on climate change and has a strong strategic interest in rapid decarbonisation globally. Bilateral relations will remain key –with Russia, the Middle East and Australia
USA	Towards energy independence? The shale revolution has been a game-changer. US net imports of petroleum are equal to about 25% of US petroleum consumption— the lowest level	US dependence on foreign sources of rare earth elements (REE) is "a very real concern". It is overwhelmingly reliant on China for REEs (more than 70% of imports, 89% including indirect imports). It has not been economic to	Power links with regional neighbours are growing. The USA is importing more electricity from Canada and Mexico. Recent and proposed transmission projects have the potential to increase	Decentralised leadership. 29 states require electric utilities to deliver a proportion of electricity from renewables by a given date. 118 mayors have endorsed a goal to power their communities with 100% renewable	US energy policy has been transformed by the shale revolution—oil imports are under control; gas is a growing export sector—primarily constrained by export restrictions —and coal is feeling the brunt

(continued)

Table 1 (continued)

	Trade dependencies	Access to critical metals for low-carbon energy	Cross-border electricity and interconnections	Leadership in clean energy technologies	Energy—a key component of international cooperation
	since 1970. Resuscitating a collapsing coal sector is a priority for the Trump administration, but this will prove challenging due to the competitiveness of gas, and increasingly of renewables	mine these minerals in the USA in recent years	bidirectional trade. US electricity networks are not well connected. The three main interconnections operate largely independently from each other with limited transmission between them	energy. There has been strong reaction at city and state level to the decision to leave the Paris Agreement	of cheap alternatives. Nevertheless, the USA is more energy-secure than it has been for decades. The greatest risk is that it loses industrial competitiveness to countries at the leading edge of clean energy
EU	EU security concerns are focused on gas—around 90% of oil demand is met by imports (including from Norway). The EU spends five times more on oil imports than gas. But gas supplies are far more concentrated: almost half of imports come from Russia. Member states vary in their national interests due to, for example, energy mix and supply routes	A rising focus on raw material supplies. The EU has identified 27 critical raw materials used in its high-tech manufacturing sectors. In addition to maintaining competitiveness, these materials are needed for the EU's push on renewable energy. The strategy is designed to support a range of actions across production, efficiency and trade	The EU is leading globally in electricity market integration—both interconnections and market coupling. EU-wide rules enable rising electricity flows and enhance efficiency. The EU is encouraging links with neighbouring countries under its Energy Union. Member states are learning to integrate renewable energy in large volumes, although the impacts on utilities have not been resolved	Clean energy tech is critical for climate goals and competitiveness. The renewable energy directive sets a binding target of 20% final energy consumption from renewable sources by 2020. The UK, France and Germany have all announced dates for ending internal combustion engine sales, pushing for EVs	EU concerns: Russian gas imports, rising cybersecurity risks and the impact of renewables on existing utility business models. But renewables, EVs and efficiency measures are also seen as a huge opportunity. Individual member state interests vary according to their energy and industrial mix
Japan	Japan imports almost all its fossil fuels, which amount to around 90% of energy consumption, following the Fukushima nuclear shutdown. It is also the largest oil consumer and net importer in the	Dependence on a single source for rare earth elements has led to a push for diversification. Japan aims to secure 60% of supply from outside China and has been importing small amounts from India. As for other major		Renewable energy is the key to Japan diversifying away from fossil fuel imports as well as nuclear. The large-scale introduction of feed-in tariffs started in 2012, following the Fukushima disaster	Energy security priorities is one factor in Russia-Japan rapprochement over disputed territories

(continued)

Table 1 (continued)

	Trade dependencies	Access to critical metals for low-carbon energy	Cross-border electricity and interconnections	Leadership in clean energy technologies	Energy—a key component of international cooperation
	world. Japan is reliant on seaborne imports from the Middle East through maritime choke points. Two-thirds of fossil fuel imports pass through the South China Sea	economies, the concern is a mix of securing industrial competitiveness in high-tech sectors, ambitions to generate more renewable energy, and military applications			
India	Despite significant domestic resources, India is heavily dependent on overseas fuels, importing 80% of its oil. With limited strategic storage, it is exposed to supply disruptions. It is the fifth largest LNG importer— 14 mmtpa, mainly used in power generation and fertiliser production. It is the largest coal importer (from Australia and Indonesia), despite having large coal reserves	India could become highly dependent on China for critical minerals. One assessment identifies 13 critical materials needed to feed India's growing economy. Unless these are produced domestically, India will become import-dependent, especially from China. These materials are needed for emerging tech sectors and to meet ambitious goals for renewables, LEDs and EVs	Increasing access to energy and grid stability are national priorities. 240 million people have no access to electricity, and there is a huge push to bring access to villages. The country has a power surplus, but power cuts are the norm. Grid integration between states is required. India has pioneered LEDs to improve efficiency and reduce grid pressure	India is fast becoming a leader in low-carbon technology. Renewable power targets are for 175 gigawatts (GW) by 2022, with 100 GW of solar, 60 GW of wind, 10 GW of bioenergy and 5 GW of small hydro. In 2017 a new record in solar auction bids was set at 2.62 rupees/kWh	India is set to become a global energy player, driving demand growth in the coming years. Energy security concerns have focused on oil imports from the Middle East. Expanding access to electricity and investing in better infrastructure are top domestic priorities. But India is switching its focus to affordable clean energy and the associated industrial sectors
Brazil	Push to secure export market share. Brazil's oil exports are rising, and in early 2017 were 65% higher than the previous year, with a record high of more than 1.46 million bpd. Petrobras recently cut petrol prices below import parity to regain	An $8.4-billion rare earth elements deposit was discovered in 2012. In 2016 production reached 1,100 Mt, up from 880 Mt in 2015	Part of Brazil's power system is connected to Argentina, Uruguay and Paraguay. These interconnections are used in case there is excess energy generation in one country and a lack of energy supply in another, or in emergency cases	Massive renewables generation. Hydro accounts for up to 70% of power generation, making Brazil the second-largest producer globally, though droughts caused by climate change could pose a challenge in the long run. Brazil's energy efficiency programme	There is significant potential for cooperation with China—Brazil has expertise in generation and China in smart meters and electronics. Chinese companies are influential in Brazil's power sector; along with

(continued)

Table 1 (continued)

	Trade dependencies	Access to critical metals for low-carbon energy	Cross-border electricity and interconnections	Leadership in clean energy technologies	Energy—a key component of international cooperation
	market share. Even so, refined petroleum remains one of Brazil's largest imports due to the scale of domestic consumption			mandates utilities to invest $250 million annually	foreign investment funds, they are seen as likely bidders in asset sales— Eletrobras and Cemig are planning to divest, including some hydropower assets
Saudi Arabia	Attempt to shift away from critical dependency on export revenue. Traditionally, Saudi Arabia is critically dependent on fossil fuel exports, and is economically damaged by the current low oil prices. Vision 2030 is intended to diversify and shift away from oil dependence The IPO of Saudi Aramco is the first step in a longer process		The state-owned Saudi Electricity Company (SEC) plans to privatise all its power stations by 2020	Target of generating 9.5 GW per year from renewable sources by 2023 through investments of $30–50 billion. On-track to exceed the target in 2017	In 2017 Saudi Arabia and China deepened their energy relationship with more than 20 agreements on oil investments and renewable energy. China has discussed taking a stake in Saudi Aramco
South Korea	Lacking domestic energy reserves, South Korea relies on imports for about 98% of its fossil fuel consumption. With no international oil or natural gas pipelines the country relies exclusively on tanker shipments of LNG and crude oil. South Korea is highly dependent on the Middle East for its oil supply	South Korea has traditionally been heavily reliant on China for its rare earth imports and has actively attempted to diversify its import base		World leader in energy storage technology and is developing the world's largest energy storage system. South Korea is heavily reliant on fossil fuels, but has announced plans to phase out coal plants and increase the share of renewables	Since the lifting of sanctions, South Korea has increased shipments of oil from Iran, and crude oil imports have risen 26.5% per year since 2016 as Iran has sought to boost its market share in the context of OPEC production limits. The two countries recently signed memoranda of understanding on energy cooperation

International cooperation is a broader set of collaborative actions. Agreements are informal and compliance less enforceable. International cooperation tends to be negotiated through diplomatic channels, whereas global governance tends to occur via multilaterals. In practice there is considerable overlap between global governance and international cooperation, for instance, informal dialogue or cooperation between China and the USA in the run-up to COP21 facilitated the Paris Agreement.

It is widely accepted that energy systems are undergoing revolutionary change. But it remains unclear how global energy governance (institutions, mechanisms and norms) and international energy cooperation should respond. As a consequence of the high transaction costs associated with the reform of multilateral institutions, or indeed from the creation of new organisations, changes in the past have been slow and evolutionary. Yet abrupt economic and energy security crises have, at times, led to a more dramatic change—most notably after the 1973 oil shock.

However, building long-term coalitions and the structures to support them tends to be incremental, as the alignment of nation states' objectives requires multiple rounds of negotiations. A key question is whether the main trends and drivers reshaping energy technologies and producer-consumer dynamics will be reflected in a renewal of global energy governance. Another is what role China will play in this process.

2.1   Strategic Direction of Travel

China will face several key strategic decisions to achieve energy security and grid reliability, while accelerating the energy transition to improve air quality and lower emissions. These strategies will sit along a spectrum of unilateral to multilateral approaches, enabling policy-orientated or market-led approaches, as illustrated in Fig. 2.

All actions will require leadership from government, especially in the initial design and implementation, be they market or policy approaches. Market approaches are generally aimed at enabling rules-based commerce and trade, and ensuring a level playing field for energy markets. Policy approaches imply a more direct role for governments, such as in strategic oil storage or bilateral investment frameworks.

Overall, China's increased engagement on energy through cooperative mechanisms or multilateral arrangements is likely to lead to sustained, mutually beneficial outcomes from a wider number of countries. China's leadership could provide the foundations of trust and willingness to act.

The calculation of costs, benefits and risks by China's partners will change, depending on the engagement of great powers with multilateral organisations. If China focuses primarily on unilateral action and multiple bilateral agreements, confidence in wider collective approaches is likely to erode, resulting in fragile energy security, higher transaction costs, a breaking down of trade, reduced overseas investment and slower progress to reduce emissions. This risk would be amplified if the USA steps back from investing in rules-based trade and wider multilateral arrangements.

China will, of course, weigh the price of entry to any multilateral or plurilateral arrangements—such as steps towards greater transparency—just as existing members will weigh the impact of Chinese membership or enhanced engagement (such as reduced voting shares), relative to the benefits. A potential leadership role for China in international energy cooperation is of particular relevance for the Belt and Road Initiative, where the trust and engagement of many partners is needed to underpin a win-win approach.

Given the open-market focus of many multilaterals, China will also need to consider making government interventions, not just domestically, but at the international level. While a command and control approach to international cooperation

Market approaches

Market friction

China pursues market reforms in isolation; domestic market functioning is improved but opportunities for deeper market integration and accompanying benefits from trade, investment and technology are not realised.

Efficient international markets

China maximises the potential of enterprise to deliver low-cost energy security and emission reductions through alignment of domestic and international markets.

Transactional cooperation

China focuses on supporting SOEs through domestic policy levers and case-by-case bilateral deals with partner countries. This allows for pragmatism at the cost of complexity and inefficiency.

Policy harmony

China collaborates proactively on new policy initiatives and reforms, and to harmonise interventions e.g. alignment on standards and regulations, subsidy reforms, targets etc.

Unilateral

Multilateral

Policy approaches

Fig. 2 The two key variables within which China's future strategic international actions should be considered. *Note* SOEs = state-owned enterprises

and governance is possible, multilaterals tend to be designed to facilitate common rules and mechanisms for the efficient functioning of markets. If China continues its policy of moving from quantity- to quality-led growth, and liberalising markets domestically, then enhanced international cooperation and engagement with multilaterals in global economic governance is a natural course of action that will benefit all countries.

Taking this context as its starting point, this report sets out four priority areas for cooperation on global energy governance:

(1) making global energy governance fit for the future explores the potential governance and cooperative actions China could take with its international G20 partners to reform the IEA and make the G20 the hub of global energy governance;

(2) investment regimes for low-carbon, energy-secure transition explore investment and market mechanisms that China could expedite, enabling optimal allocation of capital across an increasingly complex set of energy sector objectives, while maintaining traditional energy security;

(3) security of supply of metals and minerals for future energy systems explores supply and demand measures that China could take to avoid supply shortages and trade disputes, while minimising the need for additional extraction capacity—both of which could prevent future limitations to clean energy deployment; and

(4) the shift from molecules to electrons explores a regional supply-side strategy that if adopted by China could accelerate electricity market harmonisation and trade,

alongside an international infrastructure standardisation and investment strategy that could enable demand-side balancing—both of which could increase the reliability of future electricity networks as electrification and renewable generation accelerates.

3 New Trends in International Energy Cooperation

3.1 Energy Security Has Traditionally Focused on Oil and Gas Supply Security and Price Stability

Energy security has historically been regarded as a function of global geopolitics, with securing supply at affordable prices the key policy goal for import-dependent countries. This put the spotlight on avoiding a significant interruption of fuel—especially oil—produced in a few key countries and traded in global markets. At the national level, this conception of energy security was reflected in aspirations for energy independence, not least in the USA.

Given the concentration of conventional energy supplies in a handful of regions, much attention has been given to the risk of a supply disruption through critical maritime choke points like the Strait of Hormuz and the Strait of Malacca. Dependence on key gas pipeline routes, especially those between Russia and Europe, have also been a source of concern, although these are typically seen as important bilateral or regional issues rather than risks that can be managed by the international community.

This traditional concept of energy security has created a set of dynamics in past decades that consolidated into two distinct groups of producers and consumers, which joined forces to strengthen their respective bargaining power. The 1970s fuel price crisis still looms large over the institutions that emerged from that time to address the risk of a further disruption, not least the International Energy Agency (IEA). The International Energy Forum (IEF), and more recently the G20, have provided a valuable space for dialogue between producers and consumers, but it has rarely led to concrete policy action.

In recent years, there have been efforts to expand partnerships between OECD countries and emerging economies (especially China), since the latter are not members of the IEA but account for a rapidly increasing share of global demand. The OECD association agreements with China and India have made significant progress, especially on technical cooperation and knowledge sharing, but it remains challenging to ensure that existing international organisations reflect the evolving geographical distribution of energy consumption—as well as adapting to wider shifts in energy security.

3.2 New Trends Are Changing the Nature of Energy Security

Today, three key trends are reshaping the nature and focus of international energy cooperation and governance.

First, with increasing awareness of global climate threats, combined with rising concerns over air pollution at the national level, it has become clear to most governments and companies that energy security must be delivered alongside action to rapidly reduce greenhouse gas emissions internationally. The Paris Agreement of December 2015, adopted by all 196 parties of the United Nations Framework Convention on Climate Change (UNFCCC), was followed by reports of global temperature records in 2015 and 2016—underscoring the scale of the challenge to keep global warming to well below 2°C above pre-industrial levels and pursue efforts to limit the temperature increase to 1.5°C.

This means that most national, regional and city governments are factoring in climate-friendlier or lower-carbon options in their national energy planning, especially for the power sector. Since the aggregation of all current nationally defined pledges to the Paris Agreement fall short of the 2°C target, 2018 will witness the beginning of the "facilitative dialogue" to revisit those pledges before the agreement enters into force in 2020. The decision of the USA to withdraw from the Paris Agreement (see Box 2) is a blow to this process, but it could also be a catalyst for higher ambition among other parties and at the city and state level in the USA itself. The EU, China, India and other key countries have restated their ongoing commitment since the US announcement.

Taking a longer view, policy activity on climate change has undoubtedly broadened and deepened in recent decades. In fact, the number of climate change related laws and policies adopted globally has doubled every 4–5 years since 1997 to more than 1,250. The annual peak in new legislation and executive action occurred in 2010, with 127 laws passed, but fell to 45 in 2016. This slowing in recent years suggests a shift towards implementation and consolidation (Fig. 3).

Domestic concerns over air pollution are another important driver of change, along with other localised environmental impacts and water scarcity. The World Health Organization (WHO) estimates that 92% of the global population lives in regions where air quality is below the minimum recommended WHO guidelines, resulting in an estimated 3 million premature deaths per year. Communities living in the Western Pacific and South East Asia experience the greatest proportion of premature deaths, 88% as a direct result of poor air quality due to the

burning of coal and biomass. The health benefits of switching from coal are clear: in the USA between 1945 and 1960, winter all-age mortality rates dropped by 1% as a direct result of switching from coal gas, and by 3% in winter infant mortality.

Box 2: The USA and the Paris Agreement

In June 2017, President Trump announced that the USA will withdraw from the Paris Agreement. Losing the world's second largest emitter was a heavy blow to the agreement, which is less than two years old and has yet to be implemented. Yet Paris was designed to withstand upheavals of this sort. It has a flexible structure based on voluntary pledges to reduce emissions. Governments set these individually based on national circumstances, rather than through quid pro quo negotiations as happens during trade rounds at the World Trade Organization. Consequently, US withdrawal doesn't destabilise the agreement by short-changing other countries.

It also seems unlikely to trigger a domino effect. Many countries, cities and companies have confirmed their strong commitment to implementing the Paris Agreement, including 12 US states (accounting for a third of US GDP) and more

Fig. 3 Cumulative number of climate laws passed in 159 countries

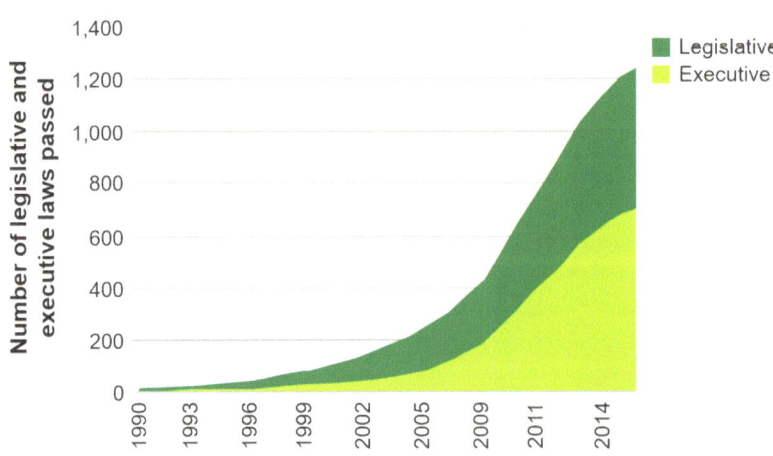

than 200 mayors. A "G6" statement omitting the USA in the recent G7 leaders' communiqué reaffirmed their commitment. China and the EU, the first and third largest emitters in the world, are set to announce a new partnership on climate action. India, the fourth largest emitter, has also reconfirmed its commitment to Paris.

The decision has obvious implications for global emissions, but the impact might not be as large as expected for three reasons. First, even if the USA had decided to remain in the agreement, it would probably not meet its emissions pledge, which President Trump has criticised. It is, after all, non-binding. Second, much American policy and regulatory action on climate change happens at the state level, and many states remain committed to emissions reduction. Third, emission reductions are increasingly the result of market forces rather than national targets. Republican Texas produces a quarter of all US wind power because it makes economic sense. Shale gas has squeezed coal out of the US power mix because it is cheaper.

The second trend reshaping the nature of international energy cooperation can be seen in the policy encouragement for lower-carbon options. This is occurring alongside dramatic technological advances and declining prices, which in turn are driving the transformation of energy markets and investments, changing the terms of energy choices. The most notable transformations are the shale revolution in the USA, the falling cost of renewable energy, and digitalisation of the electricity sector. For instance, the technology-driven shale revolution in the USA has enabled oil production to more than double, from a low of 4.3 mb/d in September 2008 to 9.6 mb/d in July 2015. Output dipped as the oil price crashed in late 2014, but at the start of May 2017 production stood at 9.3 mb/d. The well-known decline in solar photovoltaic prices continues, with prices in India down almost 40% between 2016 and 2017.

This transformation of energy markets and investments is driving the third key shift, from molecules to electrons. With emerging electricity network expansion, electric vehicle (EV) and storage capacity growth are likely to complete a self-reinforcing circle, and drive new patterns of trade, new vulnerabilities and new interdependencies. Further, EV lithium-ion battery manufacturing capacity is set to increase more than sixfold between 2016 and 2020, with manufacturers selling new lithium-ion batteries into the stationary storage market at low prices.

These three structural shifts or drivers are challenging traditional conceptions of energy security and shifting the focus of energy policy-makers. These can be separated into four areas, although in practice they are highly interconnected:

- from fuels to technologies: in light of the technology revolution in shale, renewables and, on the demand side, the focus on competitiveness, are as much about technology access and leadership as about securing fuel supplies;
- from molecules to electrons: a growing focus on efficient and resilient electricity systems that can manage variable production from renewables, integrate battery storage and EVs, and deliver investment in grid infrastructure;
- access to other minerals and metals: the shift in materials needed for low-carbon energy will raise new questions for energy and resource governance. Examples of key materials include lithium, cobalt, rare earth elements and precious metals; and
- digital disruption: as in other sectors, digital technology promises to deliver new opportunities for efficiency, from demand-side management to autonomous vehicles. In parallel, it has also increased concerns over cybersecurity in the energy sector.

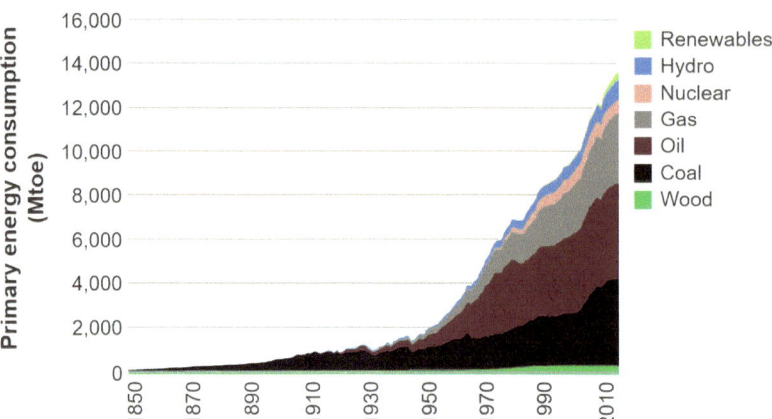

Fig. 4 Global primary energy consumption by fuel type, 1850–2015. *Source* Chatham House analysis

3.3 Energy Demand Growth and Access to Energy

Global primary energy consumption growth has proven remarkably stable since 1850, growing at 2.4 ± 0.08% per year, and it doesn't appear to be slowing. Current consumption is around 77 gigajoules (GJ) per capita; in the developed world it is 177 GJ per capita. If the projected 11 billion global population in 2100 consumes the equivalent of the developed world today, energy consumption will increase fivefold. Such a five-fold increase is equivalent to a growth rate of 1.9%, below but remarkably similar to the historic growth rate since 1850. Relative to 2014, the IEA's 2040 2°C scenario implies an increase of 9% in primary energy consumption, a significant departure from the 85% increase implied by historical trends. Meeting this target will require redoubling policy support for the deployment of new technologies, alongside significant investment in energy efficiency. At the same time, we may be on the cusp of a demand revolution, which makes it possible to do much more with much less (Fig. 4).

Providing power for the 1.2 billion individuals who in 2016 lacked access to electricity is no easy task. While this figure is 15 million fewer than in 2015, 244 million people in India and 632 million in sub-Saharan Africa still lack access. There have already been significant shifts in geographical patterns of energy consumption—caused primarily by the growth of China, India

and other emerging economies—which have driven nearly all growth since the turn of the 21st century. Between 2000 and 2015 for example, China was responsible for more than 40% of global oil demand growth, while consumption in Japan, the USA and EU fell by 20% (Fig. 5).

Several recent studies have found that increasing per capita wealth or income by 1% decreases per capita energy intensity by around 0.3%; in other words, wealthier countries tend to have a lower energy intensity. Further, those technological advances which have reduced energy intensity correlate with those that have increased per capita wealth. In other words, decreases in energy intensity are due to countries becoming wealthier, not by producing the same level of wealth with less energy. This technologically enabled wealth creation and reduction

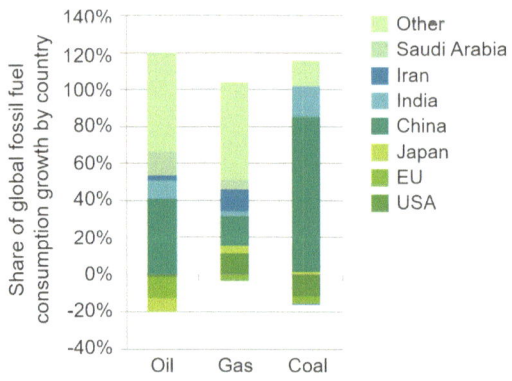

Fig. 5 Share of global fossil fuel consumption growth by country, 2000–15. *Source* Chatham House analysis

in energy intensity is synonymous to the move from industrial to service sector driven growth, occurring in countries across South East Asia. This trend of quantity to quality is exemplified by electrification—the greater use of electrons rather than molecules—alongside digitalisation and the growing Internet of things.

Box 3: Critical areas for a 2°C world and the energy transition

The aggregate of all nationally determined contributions pledges far from meets the CO_2 reductions required under the Inter-governmental Panel on Climate Change mitigation pathways. Increased political ambition and expansion of climate policies and legislation are, therefore, required to ensure the world doesn't exceed 2°C, which could result in runaway global warming.

Turning to the International Energy Agency's New Policies Scenario (NPS), which represents 2016 global climate policies, by 2040 a 2°C world will require 16.7% less primary energy consumption relative to the NPS scenario. Renewable generation, as a share of the energy mix,

will need to increase by 70% (excluding biomass and hydropower). Coal, oil and gas consumption will need to decline simultaneously by 52%, 30% and 23% respectively (Fig. 6). Given that the past 25 years has witnessed coal and oil growth of 71.2% and 42.5% respectively, these reductions require a reversal of investment trends, alongside a reversal in consumption and production of fossil fuels (Fig. 7).

Ensuring the appropriate level of investment in energy infrastructure will be critical. Figure 8 shows the difference in capex accumulation by fuel type, alongside additional investment in the power and energy efficiency sectors, under the NPS and 2°C scenarios. This climate policy driven imperative is compounding a strong and growing argument that the oil, and potentially the gas sector, is in limbo.

Investment in conventional exploration and production (E&P) projects is at historic lows. Many projects are high cost and high risk, which combined with low oil prices, is limiting companies' ability to self-finance new E&P projects.

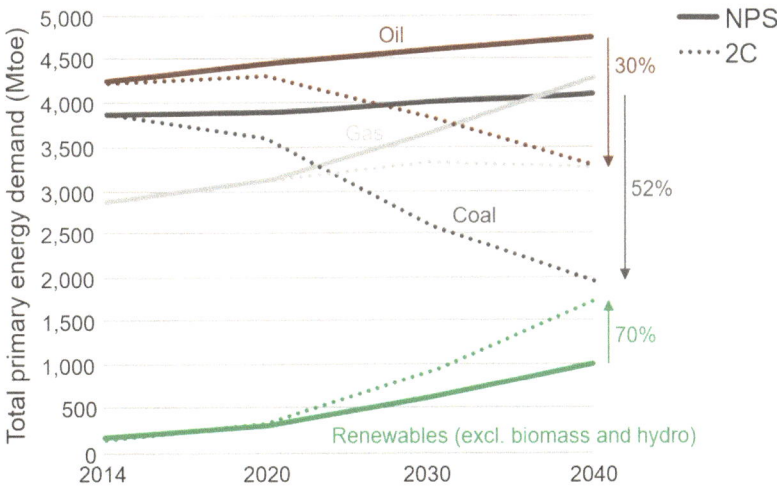

Fig. 6 Global energy mix. *Source* Chatham House analysis

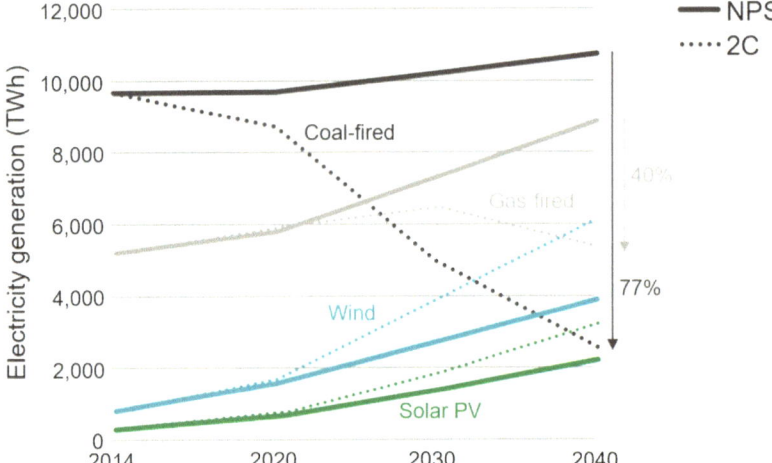

Fig. 7 Global electricity generation. *Source* Chatham House analysis

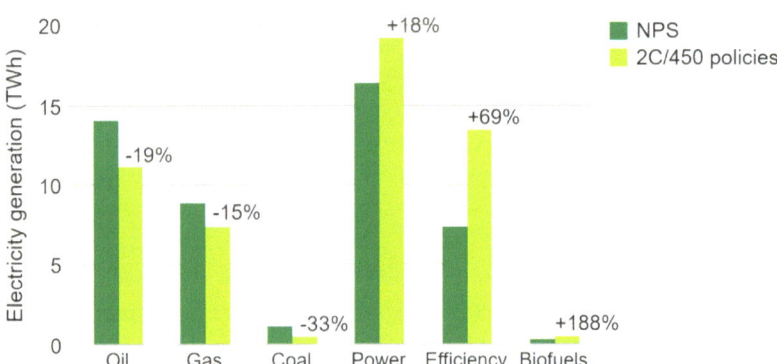

Fig. 8 Global energy accumulated capex in the energy industry, 2012–35

3.4 New Concepts of Energy Security and Energy Cooperation

Traditional energy security dialogues and mechanisms, including those for oil and gas security, will continue to play a key role in delivering energy security, perhaps for several decades. Today, these remain the focus of policymakers, especially in countries where demand and exports are rising. The challenge for policymakers will be how to adapt these instruments to new market realities and shifts in the energy mix in order to remain relevant and effective.

Meanwhile, the complex nature of these shifting trends points to a need for mainstream energy governance within a range of international processes, especially those related to shifting finance at scale. Pressure to phase out coal and reduce carbon liabilities in the global financial structure, for instance, will have profound impacts on global energy systems.

Moreover, gaps in the existing structure could be increasingly apparent given the shifts in energy markets, technology and politics. Technical and policy support for energy efficiency and clean technologies has become a core competency of institutions such as the IEA, and new institutions focused solely on renewable energy, such as the International Renewable Energy Agency have emerged, as well as initiatives such as Mission Innovation and Breakthrough Energy. But until now, there has been a division between traditional oil market security questions and the wider conception of energy security, which is arguably now essential.

It remains challenging to enhance the legitimacy and representativeness of current energy governance structures through reform processes.

In part due to difficulties around potential treaty reform, voting rights and other areas, some of the fastest-growing energy markets—including China—have been left with relatively little influence over international energy cooperation at the IEA. This is not only a question for energy policy—it reflects, for example, how China and other rising powers have struggled to make their voices heard at the International Monetary Fund and World Bank, and initial US reservations about the Asian Infrastructure Investment Bank.

Following the financial crisis, the G20 has increased its role in energy, bringing in both established and emerging energy powers. The G20 has delved into traditional energy governance—codified in its energy principles—but also provides a platform for innovative governance reforms with cross-cutting impacts for energy systems. These range from the G20 Task Force on Climate-related Financial Disclosures to the introduction of the Green Finance Study Group during China's presidency of the G20 in 2016. The German government also made carbon prices a priority during its presidency of the G20 in 2017.

4 Major Factors Driving Global Energy Governance Transformation

4.1 Shifts in Energy Consumption Patterns and the Role of Key Institutions and Groupings

The global energy landscape is being redrawn as the weight of new consumers diminishes the market power of the old. Non-OECD oil demand overtook that of the OECD in 2011, and China is now the world's largest renewable energy investor, around double that of the USA. The growing market power of emerging economies, as well as the changing fossil fuel balance in global and national energy mixes, has led to questions about the representativeness and agency of existing mechanisms and groupings.

Two prominent examples are the International Energy Agency (IEA), where voting shares still reflect consumption levels of the 1970s and emerging economies are not members (although they have recently explored partnership arrangements); and the Asian Infrastructure Investment Bank (AIIB), established by China in 2016 in response to a perceived gap in the existing international arrangements.

As Fig. 9 shows, the IEA member countries' share of primary energy consumption fell from 61 to 42% between 1980 and 2013. IEA member countries' share of production also dropped from 41 to 30% over the same period. This declining representation undermines the IEA's ability to fulfil its core functions. In 2009 one of the IEA's founding fathers, Henry Kissinger, put the dilemma to the governing board:

> [The fact that] nations outside of the OECD (and the IEA) now account for the majority of global energy consumption is a change of great significance...the fact that this bloc of rising major consumers resides outside of the cooperative framework of the IEA compromises the organization's ability to effectively address global energy security and climate concerns and concurrently deprives these nations of the full benefits of IEA membership.

Today, the G20—which includes many of the IEA member countries but also emerging economies—represents a much greater proportion of energy markets and greenhouse gas emissions. Its members accounted for 82% of consumption and 73% of production in 2013, as well as around 80% of greenhouse gas emissions. The G20 collectively represents almost 40% more primary energy consumption and production relative to the IEA or OECD. Even if China was to join the IEA, the G20 would still represent around 18% more primary energy consumption and around 24% more production—one reason IEA Association Agreements have also been developed with Indonesia, Thailand, Singapore, Morocco and India.

Of course, these organisations vary considerably in nature and in their expectations of membership. The IEA is a treaty-based organisation and members agree to rules and

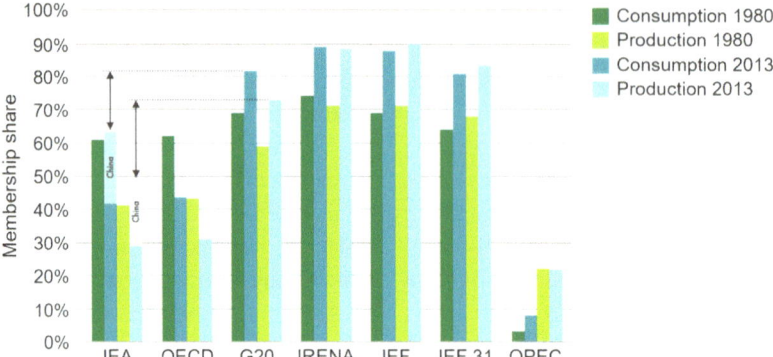

Fig. 9 Selected international energy and governance membership organisations and their respective primary energy consumption and production share of global totals.

Note Membership in 1980 represented by current membership. *Source* Chatham House analysis of data from the U.S. Energy Information Administration (EIA)

mechanisms that in theory bring binding obligations in key areas (such as the commitment to hold emergency oil stocks equivalent to at least 90 days of net oil imports). In contrast, the International Energy Forum is a forum for informal dialogue between the major petroleum producer and consumer countries. It has led to formal cooperation in one major area—the Joint Organisations Data Initiative.

The G20 has agreed to action in multiple areas of energy policy since it took on a more important role in international governance following the financial crisis, partly in response to concerns about high and volatile energy prices. Almost every G20 presidency country has introduced or expanded the remit in the energy agenda, so that it now has initiatives ranging from fossil fuel subsidies, green finance and access to energy to encouraging data sharing. Its success in moving these agendas forward has been mixed, though it remains the primary forum for discussing subsidies (Fig 10).

In 2014 the G20 adopted the G20 Principles of Energy Collaboration. Of the nine principles, the second calls for international energy institutions to become more representative and inclusive of emerging and developing economies. Regular meetings are now held by the G20 energy ministers, but as the G20 is a forum and not an agency, effective global energy governance is likely to need support institutions to engage with this working group. Another area of

action within the G20 is the consolidation of the G20's Energy Sustainability Working Group role (Figs. 11 and 12).

4.2 The Challenge of Phasing Down Coal

There is wide agreement that phasing down coal within the global energy mix is critical to achieving the Paris Agreement, as well as safeguarding public health and the environment. Analysis from University College London, UK, for example, suggests that more than 80% of global coal reserves must remain unburned in order to retain a 50% chance of keeping the climate within 2°C of warming, even with application of carbon capture and storage.

Norms have emerged in multilaterals such as the World Bank and across governments including the UK, which have agreed not to finance coal-fired power except in the most extreme circumstances. At the same time, a growing fossil fuel divestment movement has emerged in OECD countries, helping to push institutional investors and others away from coal and the most polluting fossil fuels; this now includes more than 700 institutions and around $5.46 trillion assets under management.

Real-world trends are underpinning these emerging norms around the future of coal. First is the decline of the coal industry: in the USA, coal

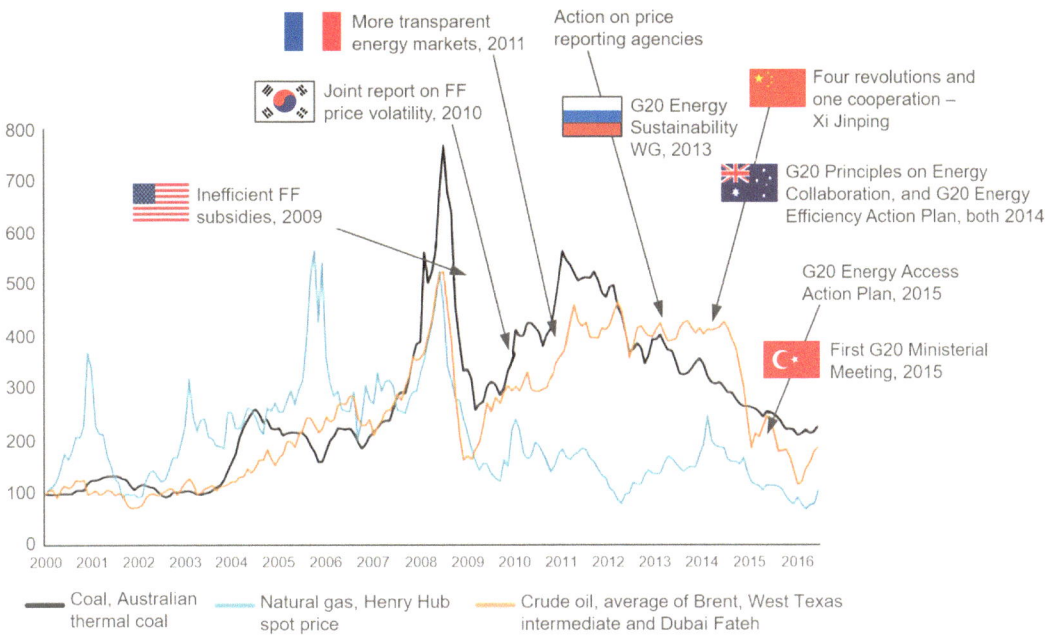

Fig. 10 G20 initiatives on energy. *Note* FF = fossil fuels; WG = Working Group

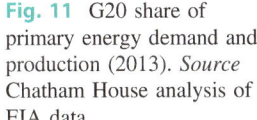

Fig. 11 G20 share of primary energy demand and production (2013). *Source* Chatham House analysis of EIA data

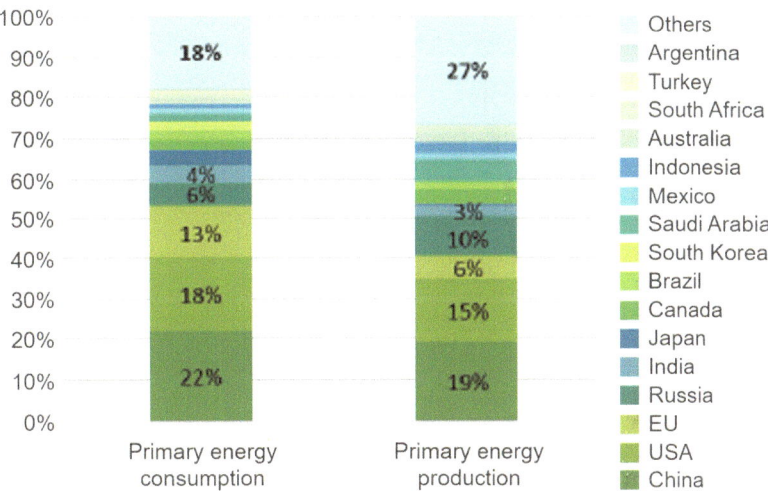

has been displaced by cheap gas, which contributed to the country's largest coal producer, Peabody, filing for bankruptcy protection in 2016. Elsewhere major miners have divested their steam coal assets into bad banks, such as BHP's spin-off South32; or to emerging market state-owned enterprises, such as Rio Tinto's divestment of its Australian coal assets to Chinese SEO Yancoal in early 2017. Second, is the impact of the rapidly declining price of renewables—even in coal strongholds like South Africa and India—where renewable auctions and wind power in particular continue to set record low prices.

Major coal producer-consumers have made similar commitments at home, not least because

Fig. 12 Reserves and current production (2015). *Source* Chatham House analysis of BP Statistical Review

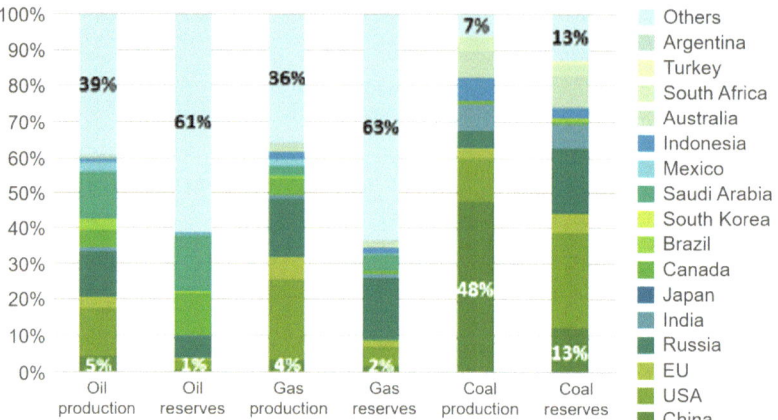

they typically align with broader strategies to restructure domestic industries and drive prosperity through innovation. China's coal consumption may have peaked as early as 2013 according to the IEA, and it is expected to plateau over the coming 5-10 years. China's 13th Five-Year Plan (2016–20) and nationally determined contributions (NDC) to the Paris Agreement have set it on course to reduce coal production and consumption—China may even be able to meet its NDC goals before the 2030 target.

Meanwhile, India's coal demand is currently growing slower than expected—this is important since the IEA expects India to account for 50% of global growth in coal-fired power to 2040. Coal stockpiles have been growing, and domestic producers have been asked to slow production. India has introduced coal taxes, 40% of which are earmarked to subsidise renewables. South East Asia markets—where coal demand is expected to triple to 2040, according to the IEA's New Policies Scenario—have also set ambitious renewable energy targets in their NDC submissions.

Box 4: The role of gas

Current global gas-fired power station capacity factors are just below 40%, compared to 55% for coal-fired. As more renewables are integrated into the grid, the ability of gas-fired power stations to respond quickly to variations in the supply-demand balance will become essential. In conditions of consistently varying output, gas generators perform better relative to coal, both in terms of economic and emission performance (Fig. 13).

Developments in China reflect those globally, with regulators in the process of reforming gas markets and the government setting ambitious targets within the 13th Five-Year Plan (2016–20) to almost double consumption to 350–380 billion cubic metres (bcm) by 2020. This would require a 13–15% compound annual growth rate which, given the 12% growth in the first 10 months of 2016, appears likely. In 2014, the power sector accounted for around 14% of the 187 bcm of gas demand in China, under the 13th FYP; this is set to rise to 26–29% by 2020. Globally, gas generation capacity has grown from 0.84 terrawatts (TW) in 2000 to 1.47 TW in 2015, an increase of 75%, but still below the coal capacity of 1.85 TW.

The shift from coal to gas in the power sector is likely to be accompanied by a shift in the residential space heating sector. In the 1980s, coal provided less than 1% of space heating energy demand in the USA. China is somewhat behind. In 2012, in the urban regions of northern China, district

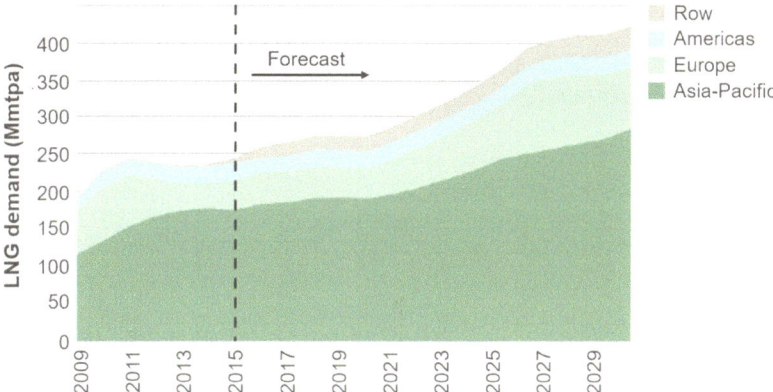

Fig. 13 Liquefied natural gas (LNG) demand by region. *Source* Chatham House analysis

heating systems warmed 45% of the urban floor area (the world's largest; 90% of this heat was fuelled by coal). With rising pressure on the Chinese government to reduce local air pollution, it is no surprise that the 13th FYP targets increased gas consumption in the residential sector. In the first quarter of 2016, a larger than expected switch from coal to gas in the heating sector pushed Chinese gas consumption to a record high.

This shift to gas should be viewed within context—the IEA 2°C scenario holds global primary consumption of gas to a modest 14.1% increase between 2015 and 2040, equivalent to 0.5% annual growth.

4.3 Transformation of the Electricity Sector

Policy encouragement for lower-carbon, climate friendly options has led to the rapid deployment of renewable energy, especially wind and solar. This has resulted in significant challenges for existing utility business models and the need for electricity market reform in countries that are further down the curve of renewables deployment. Although electricity markets are primarily

the concern of national or subnational authorities, there is growing appetite for technical cooperation and experience sharing in this area of energy policy. In particular, how to learn from the successes and failures of countries in the vanguard, while recognising the very different market contexts and designing appropriate policy responses accordingly.

In electricity markets where renewables have significantly increased their share of production, electricity prices have declined significantly. For instance, in California, where solar photovoltaic provides 13.2% of electricity, the production-weighted average price of utility-scale solar fell by 38.1% between 2014 and 2015, causing a 25.5% decline in gas power prices. Falling power prices are the result of the merit-order effect, where relatively more expensive fossil fuel generators are displaced by renewable generators. These structural declines in prices in power networks with a high penetration of renewables is resulting in fossil fuel generators extracting value during periods of low renewable generation, when they can rapidly ramp up output.

Renewable energy prices keep falling, but even in 2014 onshore wind was delivering electricity in Europe at $50 per megawatt-hour (MWh) without subsidies, compared to $45–140/MWh for fossil fuel power plants. In August 2016, Chile witnessed power auctions in which solar won contracts at half the price of coal power, resulting in power prices collapsing

Fig. 14 Levelised cost of electricity (LCOE) of various generators—global, historical, Chinese and US forecasts. *Source* Chatham House Analysis

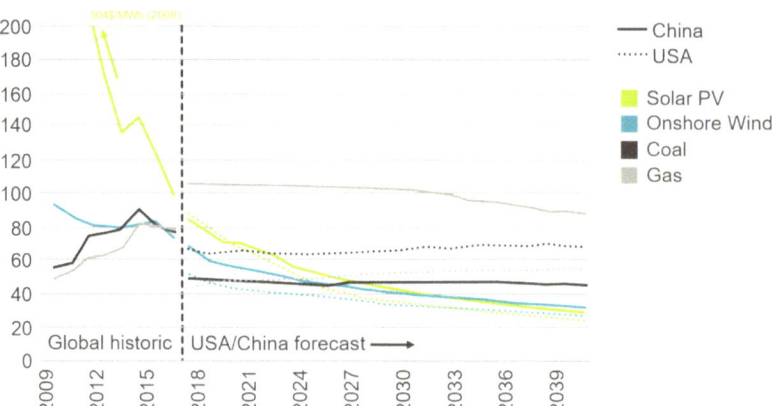

by 40%. In December 2016 the average capex of solar fell below that of onshore wind in 58 non-OECD countries, a significant milestone as solar prices have fallen far faster than most analysts had forecast.

The historical trend of costs rising for fossil fuel generators while those for renewables fall is clear. As Fig. 14 illustrates, between 2009 and 2016 the global levelised cost of electricity (LCOE) for coal- and gas-fired power stations increased by 63% and 39% respectively, while solar and onshore wind declined by 67% and 20% respectively. In China, both solar and onshore wind LCOEs are likely to be cheaper than coal and gas by 2028; both are already cheaper than gas power stations. In the USA, the point at which solar and wind become cheaper than coal and gas is likely to occur as early as 2024. In Germany solar and onshore wind LCOEs are already cheaper than both coal and gas generators, while in the UK onshore wind LCOE is already lower than coal generators.

Currently, 42% of primary energy is used to produce electricity, which in turn makes up 18% of final energy consumed by end users. With electricity increasingly supplied by renewables, the governance of low-carbon energy will play an important role in ensuring secure and affordable energy, improving air quality and mitigating climate change. As Fig. 15 shows, renewables' share of the energy mix increased from a negligible base in 1990 to 2.8% of primary energy consumption in 2015.

Globally, electricity generation increased by 29.2% between 2006 and 2015, while renewables' share of this generation increased from 19.7 to 24.2% over the same period. This trend of renewables increasing its share of electricity supply is particularly prevalent in the EU, where renewables almost doubled its share from 15.8 to 30% in 2006–15. China, by comparison, increased its share from 16.1% in 2006 to 25% in 2016 (Fig. 16).

This shift in the energy mix towards renewables is likely to accelerate, given that renewables increased its share of new capacity additions from 8.5% in 2002 to 42.6% in 2015 (Fig. 17). Globally, China looks set to continue to be the single greatest contributor of new renewable capacity additions. The shifting capital allocation of some independent oil companies is acknowledgement of this shift. For example, Equinor invested $2.3 billion in offshore wind, and by 2030 aims to invest 15–20% of capital expenditure in renewables.

Electrification of transport and heating (the shift from molecules to electrons) is likely to accelerate over the next 15 years, increasing the share of global energy supplied by the electricity sector. By 2040, 25% of the global car fleet is expected to be electric, which could increase global electricity demand by 11%, relative to 2015. This would add 2,700 terrawatt-hours to global electricity demand, more than double the global generation from wind and solar in 2015 (Figs. 18 and 19).

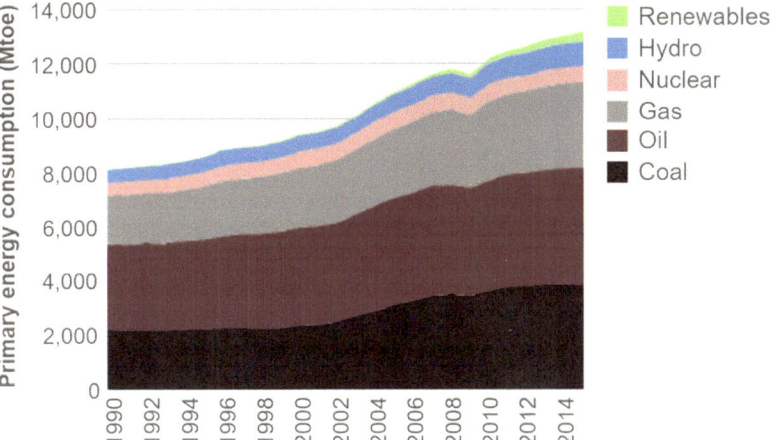

Fig. 15 Global primary energy consumption. *Source* Chatham House analysis

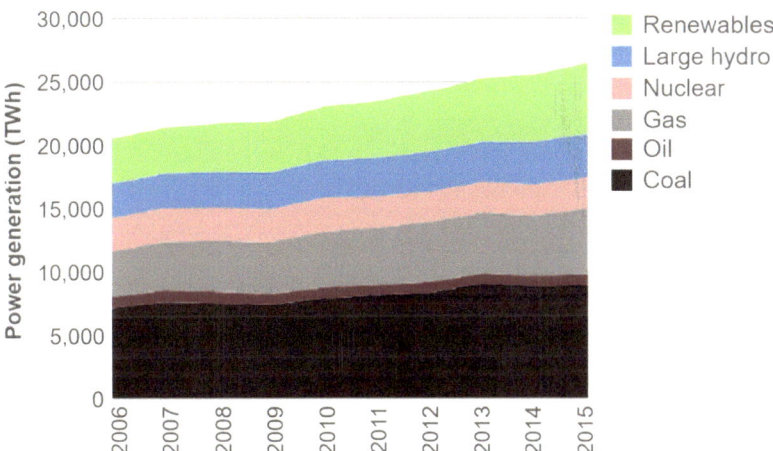

Fig. 16 Global power generation by source. *Source* Chatham House analysis

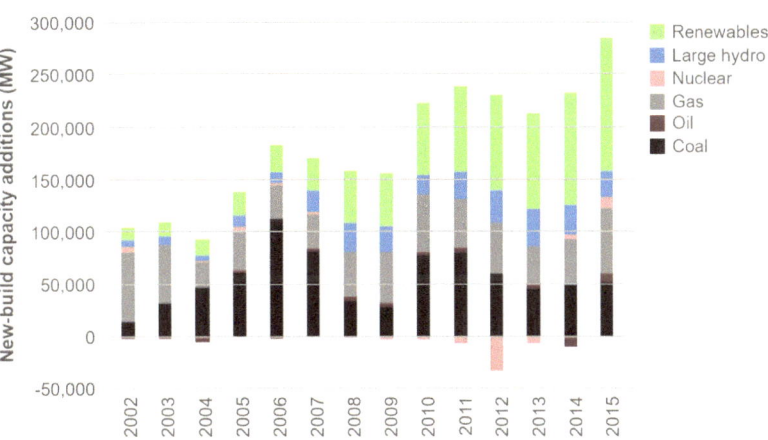

Fig. 17 New-build capacity additions, 2001–15. *Source* Chatham House analysis

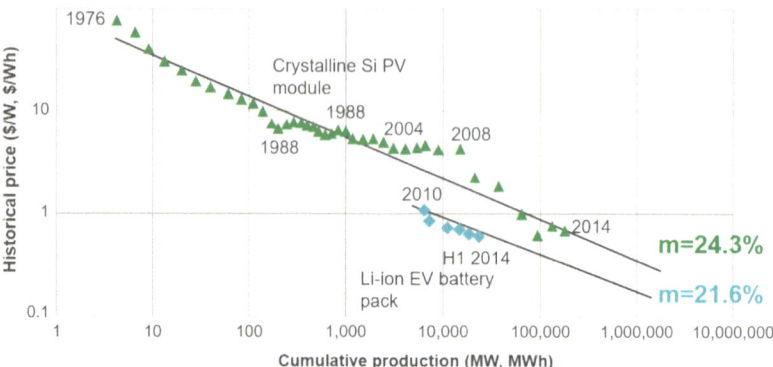

Fig. 18 Swanson's Law—lithium-ion EV battery experience curve compared with solar PV experience curve

Box 5: From quantity to quality: Digitalisation and the Internet of things

Electrification, digitalisation and the Internet of things are increasingly central to high-added value economic activity. In many countries, the growing importance of electricity as an energy carrier has occurred alongside a shift from industrial- and infrastructure-driven growth to an economy driven by services.

Decoupling electricity-based services from emissions will only be possible with enhanced system flexibility, enabling a range of intermittent and dispatchable low-carbon power to replace traditional fossil fuel base load generation, without compromising security of supply. Battery storage—whether stationary or in vehicles—will clearly be an important component in enabling such flexibility.

But there is hope that the digitalisation of the energy sector and the Internet of things will unlock huge efficiency opportunities, while simultaneously enhancing the quality of energy services. DeepMind has already proven the capacity for machine learning to find unforeseen efficiency gains in a Google data centre and is now working with National Grid in the UK on a pilot project.

Within the smart energy household the idea is that devices and appliances connect to the Internet via smart meters. These appliances and devices can provide more services to the consumer while also responding to signals to reduce demand, providing balancing and flexibility on the demand side. An example is the Nest self-learning thermostat, which optimises heating and cooling of homes and businesses to conserve energy.

In the UK the Electricity Networks Strategy Group foresees the smart, digital grid becoming the "enabler for a radical departure from the operation of the current power system, with extensive balancing on the demand side". By 2020, nearly 800 million smart meters are expected to be installed globally. In China in 2015, 447 million smart meters had been installed.

Commercial application of demand management is already bearing fruit in some countries. But in all likelihood the scale of the opportunity will only be revealed with advances in machine learning and innovations in the devices and the systems that integrate them in the home or office.

Cheaper electricity storage will facilitate new energy choices and reshape energy governance. The declining cost of EV batteries is enabling cheaper stationary battery storage, creating a virtuous circle of electrified transport supplied by renewables and storage. Excess EV battery production capacity is resulting in manufacturers

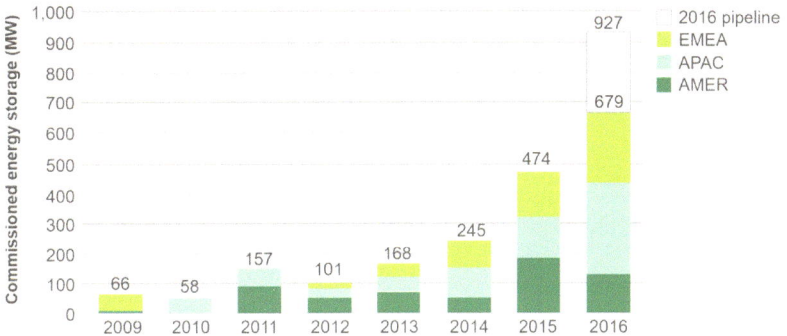

Fig. 19 Global commissioned energy storage by region, 2009–16 (MW). *Note* EMEA = Europe, the Middle East and Africa; APAC = Asia-Pacific; AMER = the Americas

selling new lithium-ion batteries into the stationary storage market at low prices. In 2016, lithium-ion EV batteries provided 90% of utility-scale solar photovoltaic stationary storage, with almost 1 GWh of lithium-ion storage commissioned or in pipeline. Energy company Total invested $2.5 billion in the battery manufacturer Saft in 2016 and the solar company SunPower. In late 2016, Bloomberg New Energy Finance forecast that EV battery manufacturing learning rates of around 19% can be expected in the future, with prices falling to $109/kWh by 2025 and $73/kWh by 2030.

Box 6: Cooperation in grid infrastructure planning and capacity allocation

Meeting the forecast growth in global electricity demand is expected to require a vast expansion of the use of renewable energy, much of it from variable sources, such as wind and solar.

While decentralised renewably generated electricity and local use is expected to be a key feature of decarbonised energy in most countries, seasonal variations in many parts of the world will lead to the need for huge storage capabilities, as well as large-scale transfer of electricity across continental distances. Greater transmission of electricity between regions is likely, therefore, to be critical.

There are currently around 12 proposals for regional grids, with interconnection capacity likely to double by 2025 (Fig. 20). Electricity network expansion, interconnection and cooperation will be a key consideration of the Belt and Road Initiative. Regional expansion of electricity grids raises important questions of investment security, infrastructural lock-in and the dangers of stranded assets.

Trade agreements and regionally coordinated planning of generation and storage capacity will be required to minimise investment risk within these large-scale infrastructure projects. Further, regionally coordinated dispatch of storage and generation, in response to varying demand patterns, will require new agencies to govern efficient operation of power markets, alongside national power market reforms and harmonisation between regional neighbours. One such example is the EU's Capacity Allocation and Congestion Management regulations to ensure effective cooperation between grid operators, power exchanges and regulators between the 28 EU member states (Fig. 21).

The transformation and expansion of the electricity sector raises an important dilemma—the institutional home for renewable energy and grid governance remains unclear. The International Renewable Energy Agency (IRENA) was established as an intergovernmental organisation in 2009, in order to increase

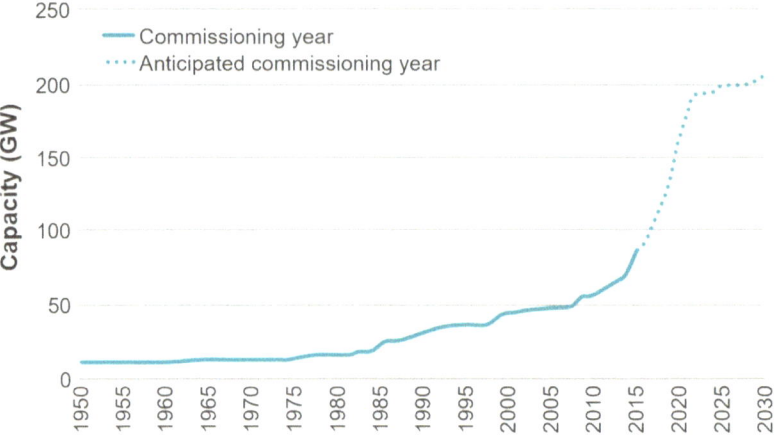

Fig. 20 Growth in high-voltage interconnection capacity

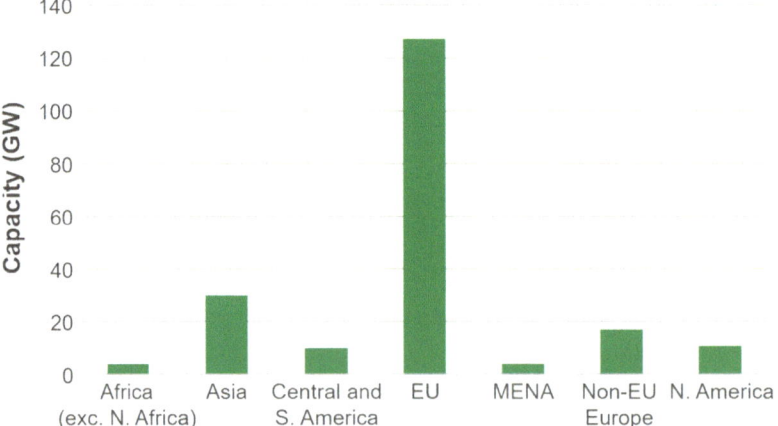

Fig. 21 Regional interconnector capacity planned and commissioned in 2030. *Note* MENA = Middle East and North Africa

the uptake of renewable energy worldwide and address some of the challenges outlined in this section. The German government led a coalition of the willing in designing IRENA's mission. Recognising that existing energy institutions had placed more emphasis on fossil fuels, IRENA seeks to be a global voice and knowledge source for renewables, providing policy advice and serving as a network hub for member states. It is open to all UN members, lending it a high degree of legitimacy, and members have equal weight in decision-making. However, its authority and technical capacity still lags someway behind that of traditional energy governance mechanisms, such as the IEA and policy processes like those of the G20.

4.4 New Critical Commodities for Low-Carbon Energy Transition

With sales of battery electric vehicles (BEVs) and plug-in hybrids expected to top 2 million per year by 2020, there are growing concerns as to how markets and supply chains will respond to increased demand for EV critical metals such as lithium, cobalt, nickel and manganese. As these materials are difficult to replace with substitutes, governments are keen to safeguard against supply shortages. While no physical scarcity currently exists, supply risks are associated with the concentration of resources within a handful of countries and regions. Figure 22 illustrates the distribution of these materials.

Imminent low-carbon critical metal supply constraints are unlikely. This is supported by several studies showing that lithium supply may be less problematic than previously thought. However, before 2030 and the commercialisation of substitute technologies, such as lithium sulphur, which is cobalt-free and has significantly reduced lithium content, price volatility and supply chain bottlenecks could appear after 2025. Most investment in the battery sector in the past five years has poured into the major lithium-ion manufacturers. Around 100 companies are currently competing for investment in the advanced battery space, each receiving $5 million per year on average. Larger scale investment is required to bring forward the point at which clear winners emerge and commercialisation of next-generation battery technology is achieved. Increased financial support for the U.S.-China Clean Energy Research Center and Mission Innovation could, for example, help achieve this goal.

Although supply constraints are unlikely, for EVs to achieve price parity with internal combustion engine vehicles, the aggregate cost of all materials in lithium-ion batteries needs to decline. This is especially true given price parity is estimated at $100/kWh and material input costs are currently around $100/kWh. Action to alleviate potential supply constraints will have the joint effect of reducing material input costs. To this end, international cooperation to promote open data on critical metals and minerals reserves

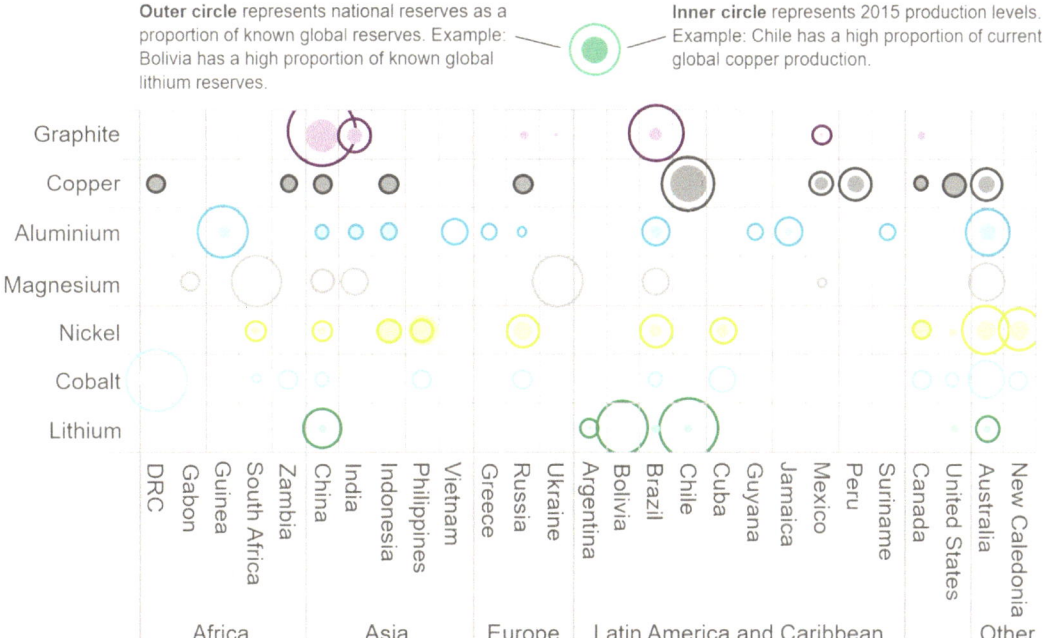

Fig. 22 Production and reserves of critical materials for lithium-ion batteries in selected countries, 2015

and trade, alongside the mapping of in-use stocks of materials to enhance recycling collection rates, is surely a win-win.

Box 7: The circular economy and resource efficiency

China is, and will continue to be, a resource-reliant economy: as the pace of economic growth slows, demand for materials is set to continue growing. Globally, resource consumption is expected to more than double by 2050, while resource efficiency could save the global economy $2.9 trillion per year by 2030.

It is possible to determine the per capita income level at which per capita demand for a particular material saturates. Across the UK, USA, Germany and Japan, apparent domestic consumption of steel saturates around 0.5–0.8 tonnes per capita, once an income threshold of $1,200 GDP/capita is reached.

As saturation effects kick in and economies establish a stock of primary materials in physical infrastructure and consumer goods, the circular economy can provide a means to further decouple materials demand from GDP. A circular economy is one in which the stock of materials within the economy is reused, remanufactured or repurposed as secondary materials, reducing the demand for primary material production and extraction. Globally, the circular economy is central to the next wave of energy and resource productivity enhancement. This is especially true for China, which produces 46% of global aluminium, 50% of steel and 60% of the world's cement.

Globally, iron and steel production accounts for 10% of total energy demand. Primary production of steel requires vast amounts of coking coal. Production of secondary steel, in electric arc furnaces from in-use stocks that have come to the end of their life, accounts for around 29% of steel production. Secondary steel saves 740 kg of coal, 1,400 kg of iron ore and 120 kg of limestone for every tonne produced. Ensuring there is an optimal mix between primary and secondary production will be essential as China and other countries decouple energy and resource consumption from economic growth.

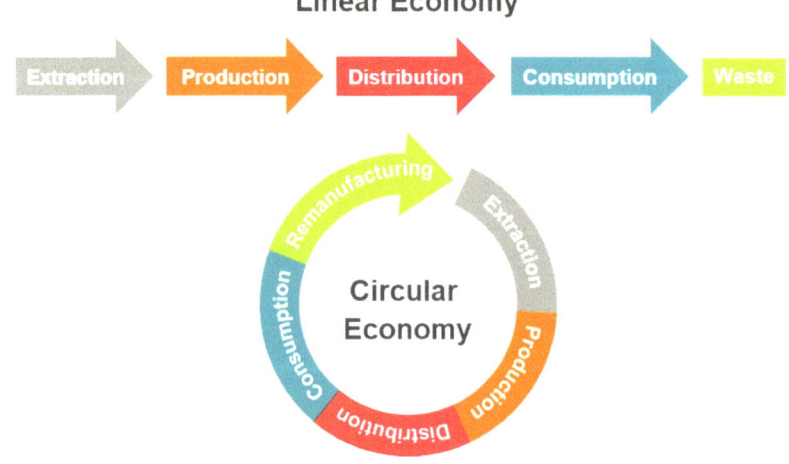

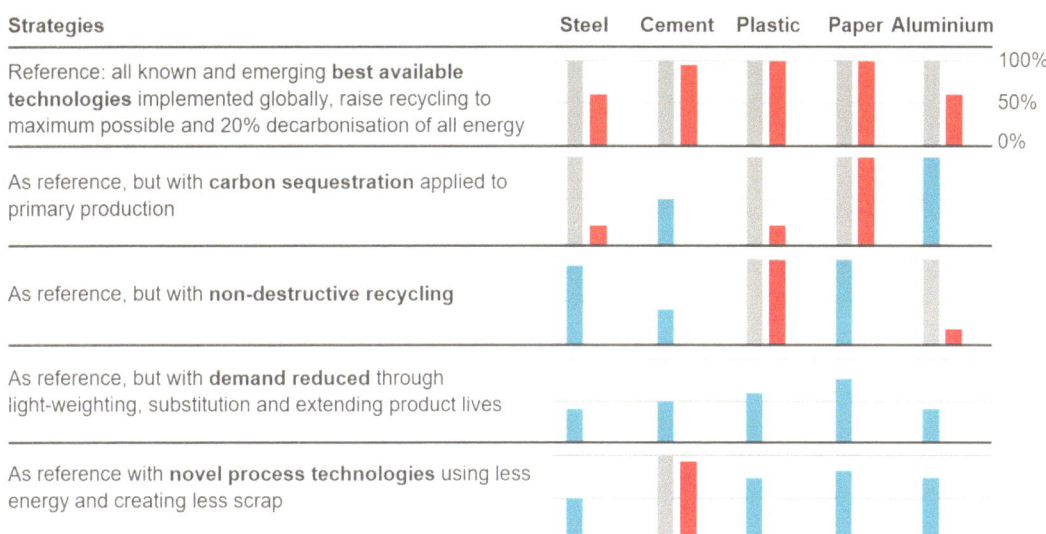

Strategies	Steel	Cement	Plastic	Paper	Aluminium

Fig. 23 Predicted 2050 emissions for the five key materials under various future strategies. *Note* The blue bar shows how extensively the strategy must be implemented to reach the target set by the Intergovernmental Panel on Climate Change. If 100% implementation is insufficient, the red bar shows the excess emissions relative to the target

Capturing the benefits of the circular economy at the global scale is no easy task. While many of the technologies already exist to use previously wasted materials, standards and regulatory barriers often prevent progress. In 2008, China passed the Circular Economy Promotion Law, while in 2015 the European Commission adopted a Circular Economy Action Plan. At the international level, focus could be directed at cooperation between jurisdictions on reducing non-tariff barriers and growing the markets for circular products and services, investment and export opportunities. This could be achieved through greater engagement with the G7 Alliance on Resource Efficiency, a knowledge-sharing forum formed in 2015 (Fig. 23).

4.5 Changing Political Dynamics for Petroleum Producers

The emergence of new technologies like renewables, battery storage and EVs is not confined to the electricity sector. As Fig. 24 illustrates, the emergence of horizontal drilling techniques and the expansion of the number of rigs used to exploit shale oil has reversed US oil production. The US shale industry can now respond quickly to changes in the oil price. Project execution times have fallen and shale companies have cut costs, enabling them to survive in low oil price environments. Advances in horizontal drilling technology and fracturing processes, as well as huge cost reductions in extraction, have brought about a structural change in oil markets. This has undermined the ability of conventional producers to use production targets to influence oil prices (Fig. 25).

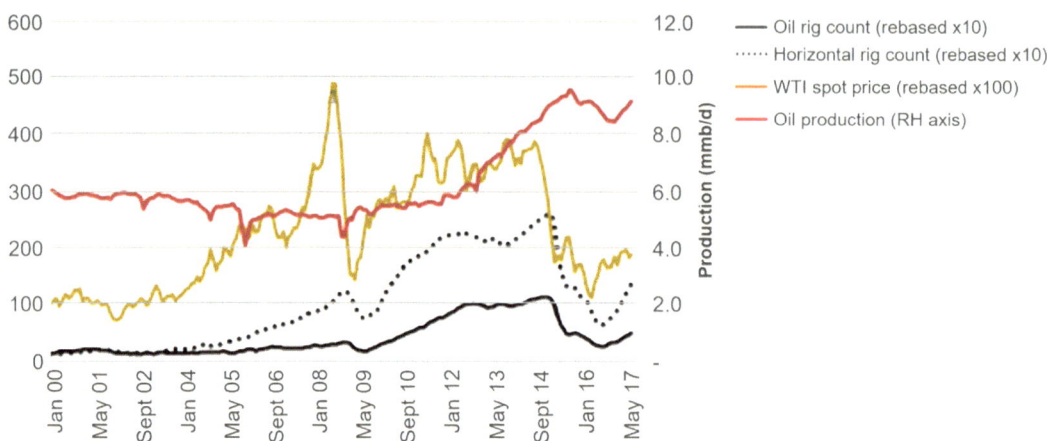

Fig. 24 Oil production in the USA in response to shale oil *Source* Chatham House analysis of EIA and Baker Hughes data

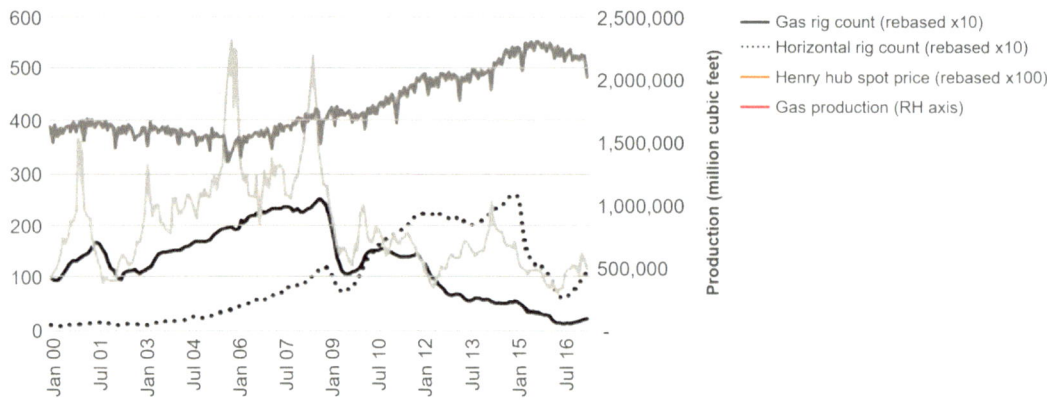

Fig. 25 Gas production in the USA in response to shale gas. *Source* Chatham House analysis of EIA and Baker Hughes data

Combined with the fast-growing EV sector, this shift presents major challenges to traditionally oil-led energy cooperation and governance. As Fig. 26 illustrates, transition to EVs is likely to displace at least 2% of current oil demand within 10 years, and more than 14% by 2040. EV transition and the associated displacement of oil demand may occur faster than currently anticipated, amplifying the shift of energy governance away from oil-led cooperation. Further technological innovation would be required to displace oil from aviation and maritime, as well as petrochemicals. As global oil demand peaks (potentially as soon as 2020), plateaus and

declines, the focus on oil-led energy cooperation and governance may start to shift elsewhere.

At the same time, with control over accessible conventional oil resources largely in the hands of national oil companies (NOCs) and their governments, independent oil companies (IOCs) are struggling to find viable business models or public acceptance for exploration and production projects in more geologically or politically challenging areas, such as deep water or the Arctic. This is reflected in 2017 planned expenditure, with falls of 7% and 15% for European and US IOCs respectively, while NOCs plan to increase spending by 9%. Further, after falling

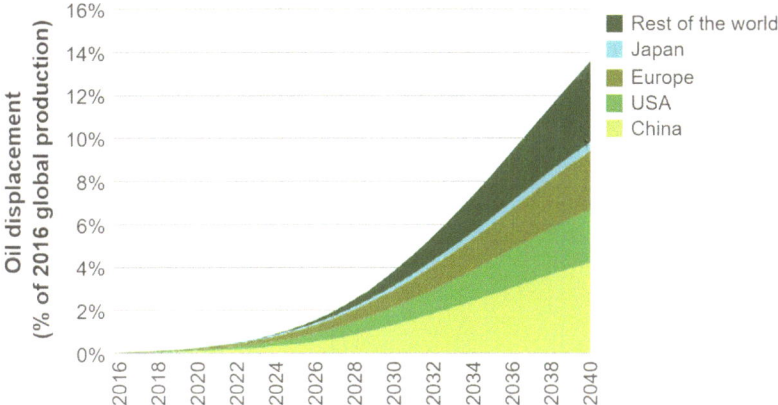

Fig. 26 Crude oil displacement from electric vehicle sales. *Source* Chatham House analysis

34% in 2016, offshore spending is expected to decline by an additional 20–25% in 2017 and floating rigs to fall from 133 in 2016 to 120 in 2017.

In line with the trend of switching from coal to gas (see Box 7) and the electrification of transport, IOCs are increasingly switching their focus to natural gas. Gas production in the second quarter of 2016 by Shell, BP and Exxon-Mobil, as a share of total production, stood at 50%, 47% and 41% respectively (Fig. 27). Natural gas could deliver lower emissions relative to coal; however, the climate impact will depend on managing methane leakage. Methane (CH4) accelerates climate change and has a greater global warming potential than CO_2 over a short time horizon. Leakage rates of 1.5% from the production of natural gas increase the climate impact of natural gas by 50%. One recent study shows leakage rates at gas power stations and refineries are much higher than previously thought, up to 120 and 90 times greater respectively.

For producer countries, lower oil prices have led to a reassessment of the role of hydrocarbons in their economies, as budgetary pressures increase in line with indebtedness and macroeconomic volatility. As the 2016 analysis by the International Monetary Fund shows in Fig. 28, government revenues of the countries in the Cooperation Council for the Arab States of the Gulf have rapidly declined with the oil price. Saudi Arabia's revenues fell by more than $100

billion and it borrowed $17.5 billion to finance its budget. Saudi Arabia has ramped up its efforts to diversify its domestic energy mix, issuing tenders for solar and wind energy investment worth between $30 and 50 billion.

At the same time, lower oil prices are giving fiscal breathing space for countries that provide consumption subsidies for their citizens or industries. This in turn may affect the willingness of these countries to subsidise fossil fuel consumption, should higher prices return. In the OECD, fossil fuel subsidies stood at $51 billion, or $784 per person, in 2013. In 2014, global fossil fuel consumption subsidies stood at $493 billion, a reduction of $39 billion year-on-year, with oil subsidies representing more than half of all consumption subsidies. While consumption subsidies have fallen in recent years, fossil fuel consumption has been supported over the long term. Between 2003 and 2015 the consumption weighted mean net tax on petrol fell by 13.3% across 157 countries. Of these 157 countries, 84 either reduced consumption subsidies or increased petrol taxes. But as consumption has shifted towards nations that have maintained subsidies or lowered petrol taxes, subsidy reforms have been undermined at a global level. Following the agreement by the G20 in September 2009 to phase out inefficient fossil fuel subsidies, as of mid-2015 the unweighted mean net tax was unchanged across the 20 states. However, this commitment has not been reaffirmed at the latest G20 meetings (Fig. 29).

Fig. 27 IOC natural gas production and as a share of total production

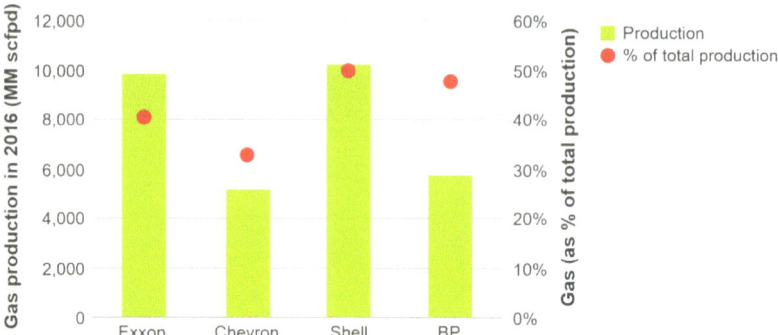

Fig. 28 GCC government revenue and expenditure (per cent of non-oil GDP, weighted average)

Fig. 29 Estimates for global fossil fuel consumption subsidies and subsidies for renewables

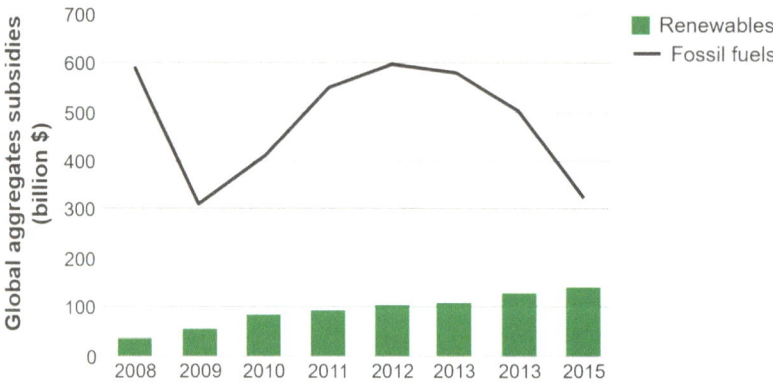

Box 8: Imperative areas for cooperative action

Figure 30 shows a set of imperatives for a low-carbon, energy-secure transition that responds to the main trends reshaping the global energy system and driving the energy transition. We also highlight examples of specific actions the international community could pursue to help deliver on these imperatives.

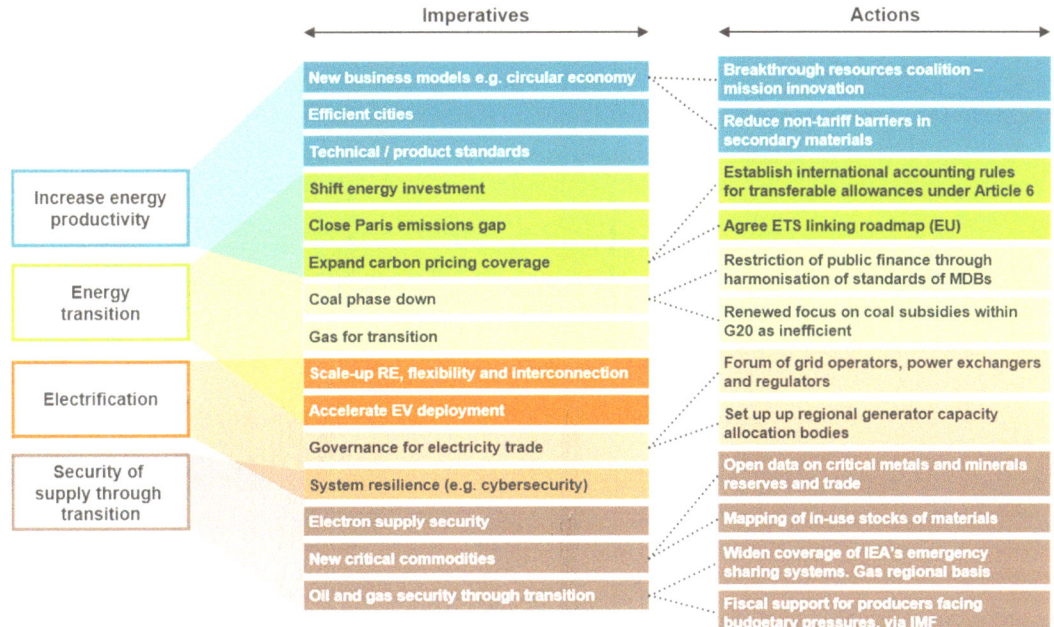

Fig. 30 Imperatives for a low-carbon, energy-secure transition, alongside examples of actions that could enable those imperatives to be met. *Note* Article 6 = Article 6 of the Paris Agreement allows countries to cooperate with each other when implementing their nationally determined contributions by using market-based mechanisms; ETS = EU Emissions Trading System; IEA = International Energy Agency; IMF = International Monetary Fund

5 Five Pillars of International Cooperation on Energy

There are five pillars of global governance and international cooperation within the energy sector: trade, investment, innovation, security of supply, and cross-border externality. These pillars are critical when analysing the effectiveness of agreements and actions taken at the international level.

5.1 Trade

Despite the nascent transformation of the energy sector, energy trade remains dominated by fossil fuels, with Chinese demand a key market driver. In the past 15 years, traded volumes of oil have increased 14%, coal has more than doubled, and natural gas and liquefied natural gas (LNG) have increased by almost 60%.

Reflecting the broad trend of declining demand among OECD countries and strong growth among emerging economies, China became the world's second largest crude oil importer in 2009 and is set to surpass the USA to become the largest. India is now the third biggest importer, although with imports 42% lower than China. In the context of the US tight oil revolution, the centre of gravity of international oil trade has shifted east, with non-IEA consumer countries accounting for a growing share of demand, and non-OPEC countries providing a growing share of supply. Net importing countries, including China, still view high dependency on oil from the Gulf states of the Middle East as an energy security vulnerability.

In 2016, global trade in LNG increased at its fastest rate in five years as new facilities in the USA and Australia began to challenge Qatar's market dominance. Between 2009 and 2015 imports of LNG grew 30.5% globally, with the Asia-Pacific region and China representing

71.1% and 8.1% respectively of the 245 million tonnes of demand in 2015. Global demand growth of 72.3% by 2030, relative to 2015, is expected. The world's biggest LNG plant is in Australia, having cost $200 billion. US exports have been driven by the Sabine Pass terminal in Louisiana, owned and operated by Cheniere Energy, which ships LNG to markets such as Japan and Chile.

These new LNG trade flows are beginning to draw new import–export interdependencies, resulting in new bilateral trade agreements such as the May 2017 agreement between the USA and China. The bilateral agreement aims to consolidate the export of US LNG to China into long-term trade contracts, which accounted for 7% of China's imports in March 2017.

A clearer understanding of coal markets and the links between coal producing and consuming countries is needed to avoid potential roadblocks to implementation of the Paris Agreement, and to address potential cross-border financial exposure to coal investments.

The market for traded coal has grown rapidly since the turn of the century, doubling in weight to around 1.4 billion tonnes and increasing fourfold in value to almost $100 billion. Coal trade has also increased as a share of global coal consumption, rising from 18.5% in 2000 to 23.5% in 2014. Today, six coal superpowers dominate trade: Australia and Indonesia account for around 60% of exports while China, India, Japan and South Korea provide 60% of imports. This trade is underpinned by foreign direct investment flows into coal mining, infrastructure and generation capacity within the Asia-Pacific region.

Trade is a common area for international cooperation, because countries can mutually benefit from deeper trade, which facilitates access to cheaper goods and new markets. However, the political economy of trade integration is not straightforward because although benefits may occur in the economy, trade can create winners and losers at the sectoral level; governments may also wish to protect certain sectors of the economy for industrial strategic reasons. Governance arrangements are therefore

needed to set the rules for how countries trade with one another—in particular to mitigate tit-for-tat barriers and provide mechanisms for negotiation and dispute resolution. At the multi-lateral level, these rules are set by the World Trade Organization (WTO), although many regional and bilateral trade agreements exist between countries.

Trade in energy (as a natural resource) falls largely outside the WTO's scope due to sovereignty considerations and other factors. Some regional agreements have tried to bridge this gap for energy, notably the Energy Charter Treaty and the energy chapter of the North American Free Trade Agreement. For processed products and manufactures, however, normal WTO rules apply—these energy-related trades have become a political battleground.

While provisions in the General Agreement on Tariffs and Trade (GATT) do not include natural resources, a service relating to natural resources (exploration, exploitation, technical testing and transport) is subject to GATT regulations unless provided by government authorities. This had led to calls for a more coherent framework, specifying which rules apply to which resource type, which qualifications are needed for a resource to be considered a good or a service, and including important issues for oil and gas, such as investment protection.

With the rising importance of renewable energy, both for energy supply and as an important area for manufacturing competitiveness, there has been rising scrutiny over tariffs on low-carbon and environmental goods, which raise their costs and slow down diffusion rates. Proposals to eliminate these tariffs have been made at the WTO by countries accounting for 86% of global trade in these goods. Yet at the same time, some of the same countries have become embroiled in serious trade disputes over specific low-carbon products in which there is particularly fierce competition. It is estimated that roughly 14% of WTO disputes since 2010 relate at least in part to renewable energy. Many of these disputes concern renewable energy subsidies and local content requirements, which countries and states have used to support domestic industrial sectors; there are also several disputes over the pricing of low-carbon exports such as solar panels, which have led to increases in import duties. These disputes have raised prices, damaging the deployment of renewable energy sources.

The trade in electricity is a special case. Electricity is not traded globally, unlike most other products and services that are subject to international competition. Furthermore, in its classificaton by the WTO, electricity is treated as both a good and a service and is therefore subject to different tariffs and rules. In addition to the interconnectors needed for the flow of electricity across borders, there is the issue of market coupling, i.e. arrangements that allow efficient trading of energy (in this case electricity) between markets. These tend to be managed under bilateral and regional arrangements, the most extensive being those in the EU and with its neighbours.

Increasing trade in electricity is driving a growing number of agreements on electricity trade via interconnectors. As well as governing operation, agreements can also mitigate the investment risks associated with large infrastructure projects. For example, in 2016, the North Sea countries of Belgium, Denmark, France, Germany, Ireland, Luxembourg, the Netherlands, Norway and Sweden signed an energy cooperation agreement to build missing interconnectors and allow greater trading of energy, while further integrating their respective national energy markets. In July 2015, the EU adopted trading regulations on Capacity Allocation and Congestion Management. The regulations were designed to integrate more renewables into the grid and enable effective cooperation regionally between grid operators, power exchanges and regulators of the 28 EU member states, saving EU consumers €2.5–4 billion.

Growing deployment of new technologies—especially batteries—is increasing trade in critical raw materials such as lithium, cobalt, polysilicon and rare earth elements. Goldman Sachs estimates that for every 1% increase in the penetration of plug-in battery electric vehicles, lithium carbonate equivalent (LCE) demand will increase by 70,000 tonnes per year, equating to around half the current annual demand for LCE.

However, compared to energy commodities, international metals and minerals markets are relatively ungoverned: transparency is poor and when cooperation occurs, it has often taken the form of collusion; markets are also regularly disrupted by unilateral export controls.

5.2 Investment

Finance and investment regimes are fast becoming new frontiers of energy governance, driven in part by efforts to shift capital from high-carbon to low-carbon energy sources, as well as concerns over disruption to the business models of utilities. More than 60% of global emissions are caused by investments in, and operation of, long-life infrastructure, which have the potential to lock in emissions far into the future. The IEA 2°C pathway requires a tripling of annual investment in low-carbon power infrastructure by 2035, and an eightfold increase in energy efficiency investments. Today, shadow carbon prices are already used as investment screening tools by many oil majors, and attention is growing among financial regulators and policymakers.

Expansion of shadow carbon prices and greater investment screening within the fossil fuel sectors have become commonplace. There is growing scrutiny from institutional investors, particularly regarding so-called transition risk to assets and business models arising from decarbonisation of the economy. A major game-changer is the G20 Financial Stability Board's Task Force on Climate-related Financial Disclosures (TCFD), which has published recommendations on climate disclosure. Recently, investors voted for ExxonMobil to publish annual assessments of the impacts of a low oil demand scenario under a 2°C target. The first step under the TCFDs implementation path is illustrated in Fig. 31. Coordination among financial policymakers and regulatory authorities will be necessary if climate-related financial disclosures are to encourage portfolio rebalancing.

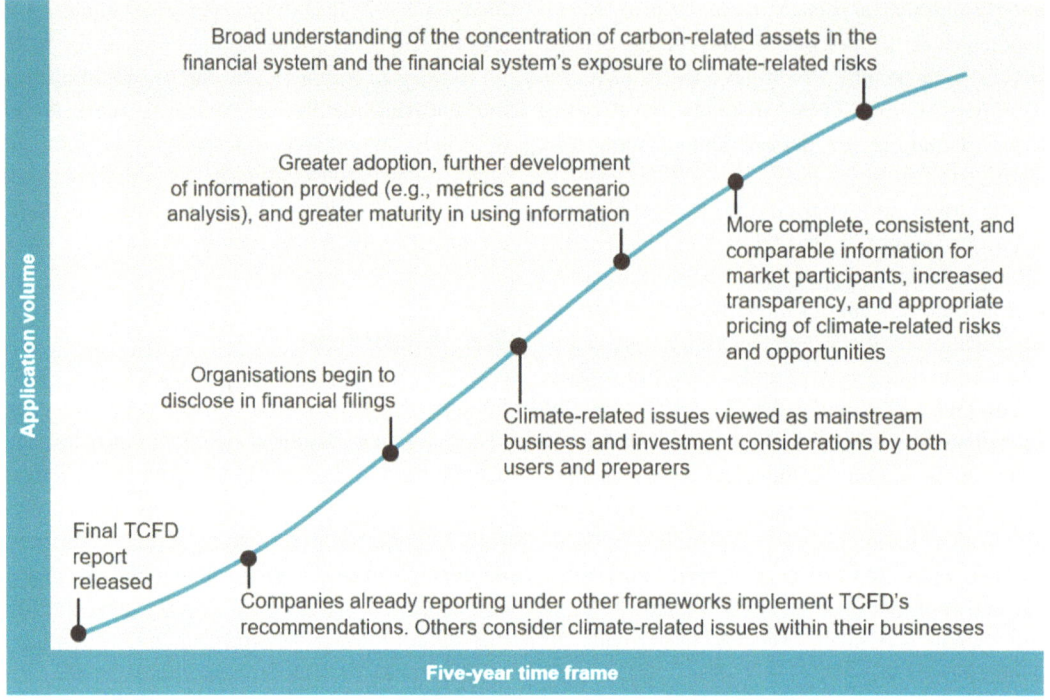

Fig. 31 TCFD implementation path of recommendations on climate-related financial disclosures

Attention is also shifting to the role of development finance institutions and multilateral development banks, including the Asian Infrastructure Investment Bank (AIIB) and New Development Bank, as capital providers. Although most institutions are doing something to mainstream climate change into their portfolios and operations, progress has been patchy and uneven. There is a lack of common frameworks to evaluate the social, economic and sustainability of projects across multilateral development banks.

One of the clearest governance challenges revolves around financial flows to coal mining and coal-fired power generation. The implementation of multilateral guarantees continues to come under scrutiny from civil society; the International Finance Corporation, for instance, has allegedly financed 41 new coal projects since the World Bank commitment in 2013. The AIIB has stated that it would not invest in coal but is at risk of relenting on this commitment following heavy lobbying by coal exporting countries such as Australia. Meanwhile Japan, China and South Korea account for most of the G20's $24 billion in export-import bank and development finance packages that support coal.

The IEA 2°C pathway will also require the stranding of high-carbon assets, which cannot be operated fully over their lifetime to respect the remaining carbon budget. Although there is much bilateral cooperation on energy infrastructure investment, there is a lack of rules or standards for aligning investments with low-carbon priorities. There may be opportunities for the G20 to address this in the first instance, building both on its global infrastructure initiative and the TCFD.

Hydrocarbon investments commonly face significant political risks, by nature of the countries targeted and the complexities of resource curse dynamics and the long timescales over which assets operate and recoup costs. Cooperative approaches have emerged to help manage these risks, through pooling, dispute resolution and norm setting, in particular the World Bank's Multilateral Investment Guarantee Agency and the Energy Charter Treaty.

5.3 Innovation

Despite rapid progress on some technologies such as solar photovoltaic, wind power, LED lightbulbs and batteries, the pace of deployment is too slow to meet climate goals, according to the IEA. Chatham House research has found that innovations within the energy sector take 20-30 years to penetrate the mass market. Accelerating the rate at which future innovations reach the mass market could be achieved by a publicly backed energy patent pool, enabling market disrupters to access patents.

Mission Innovation and the Breakthrough Energy Coalition are new, significant global attempts to scale up R&D funding and focus on the next generation of technological breakthroughs needed to achieve the Paris Agreement, such as advanced battery chemistries like metal-air. However, many key technologies are held up not by lack of research and development but by lack of policy support and investment. Reaching critical levels of deployment will require wider alignment of supply, demand and market factors. Ensuring rapid penetration of electric vehicles, autonomous driving and wireless charging, for example, requires cooperation and coherence of standards and regulations across jurisdictions.

Most energy technologies are part of complex global technology systems. Their development does not often follow a linear logic or evolve within the boundaries of individual economic sectors. Many breakthrough innovations occur when different fields interact. For example, innovation in solar photovoltaic technologies has benefited from developments in consumer and industrial electronics, and advances in concentrated solar power derive from aerospace and satellite technologies.

The IEA's Tracking Clean Energy Progress 2017 report shows that many necessary technologies are held up not by lack of innovation, but by lack of policy support and investment. It also reveals where some of the most urgent action may be needed to accelerate technology deployment and adoption.

Most energy models anticipate large-scale application of negative emissions technologies, such as bio-energy with carbon capture and storage (BECCS) or carbon capture and use. However, there are significant uncertainties and potential trade-offs. Electrification also creates new challenges for innovation at the system level. For instance, strategies to bring down the costs of electricity storage, expansion of grid infrastructure and coordinated reforms of power markets may need to be aligned.

There is increasing interest in the potential for new business models to unlock energy and resource savings through substitution, digitalisation, sharing and reuse, among others. China and the EU both have circular economy strategies that could be used as a platform for cooperation (Fig. 32).

5.4 Supply Security

With instability in the Middle East unlikely to recede in the foreseeable future, and the possibility of a supply crunch from conventional resources on the horizon, security of supply remains a political priority for many importing governments. In theory, international cooperation can help manage the risks of major supply disruption by setting rules and modalities for how governments will coordinate in the event of a shock, so avoiding an all-out scramble for supply that could heighten a crisis. However, IEA members' share of energy trade is declining, leading to growing questions over the effectiveness of the IEA emergency response mechanism in the event of a serious supply shock. Key issues therefore include whether and how emerging economies can be integrated into the IEA regime, or whether regional approaches may offer more practical solutions and if this is the case, how coordination can be managed among them. Other related concerns include maritime governance, particularly around critical choke points.

Meanwhile, the growing adoption of renewable energy sources and transformation of electricity grids create new challenges of supply security. Key issues are likely to include:

Fig. 32 Analysis of clean technology progress. *Source* International Energy Agency, Tracking Clean Energy Progress 2017

1. cooperation to improve battery technology and seasonal storage investment to bring down the costs of electricity storage;

2. expansion of grid infrastructure regionally and continentally to smooth supply–demand imbalances and exploit opportunities to export electricity from regions with high renewable capacities; and

3. coordinating reforms of power markets in line with grid expansion to enable capacity markets, or equivalent market mechanisms, to provide

adequate and affordable fast-reacting generators to balance intermittent renewable supply.

Due to the increasing demand for certain metal resources for low-carbon technology manufacture, the geographic concentration of resources such as lithium and cobalt in fragile regions, and the difficulty of substitution, governments and businesses are increasingly keen to safeguard against supply shortages in growing markets.

The World Trade Organization (WTO) has been of little use in managing supply security as it is focused on avoiding import restrictions rather than avoiding export controls. Article XI of GATT 1994 requires that exports should not be subject to quantitative restrictions "other than duties, taxes or other charges"—but it does not fix a maximum level for border taxes, except for those countries which accepted them within their accession agreements (including China). Article XI has been interpreted as not prohibiting export taxes, which explains why "more than 1/3 of notified export restrictions are in resource sectors, where export taxes on natural resources appear to be twice as likely as export taxes in other sectors". In recent years, several key raw material suppliers have used export controls. Indonesia has introduced a ban on exports of unprocessed ores. Vietnam has also imposed restrictions on iron ore and copper, and export taxes have recently been debated in Brazil and India. In 2012, the USA, the EU and Japan lodged a complaint with the WTO over Chinese export quotas of rare earth minerals, and the quota system was withdrawn in late 2014. There have been various proposals for voluntary avoidance of export restrictions in times of crisis or extreme price volatility, which such restrictions tend to exacerbate.

Data is also an important area for cooperation. Markets that are less transparent tend to be more prone to instability, as market participants have less confidence in fundamentals and so are more prone to erratic behaviour such as panic buying or hoarding. Governance arrangements to improve transparency and accessibility of data, in particular on stocks, in oil markets have emerged from the International Energy Forum in the form of the Joint Organisations Data Initiative—an approach that has been mirrored in soft commodity markets with the Agricultural Market Information System. As the shift from molecules to electrons continues, and energy interdependencies become increasingly linked to data-intensive electricity grids, data will become increasingly fundamental to energy cooperation and governance.

5.5 Cross-Border Externality

By its very nature, a cross-border externality cannot be resolved without cooperation. The text-book example of cooperation and governance to resolve a cross-border externality is the Montreal Protocol to deal with ozone-depleting gases (see Appendix). Climate change has proven more challenging, essentially because of the initial lack of readily available low-cost substitutable technologies and the greater entanglement of high-carbon (as compared to high chlorofluorocarbons, CFC) activities within economies. The politico-economic challenges of action on climate change have been eased significantly by rapid falls in low-carbon technology costs and growing awareness of the co-benefits of low-carbon policies. This facilitates an approach that allows governments to unilaterally determine national ambition based on domestic circumstances within a multilateral framework for transparency and reporting (see Appendix).

Post-Paris, there is growing momentum around cooperation on carbon pricing as a market-based mechanism to internalise the externality. Linking carbon pricing mechanisms provides a means to broaden coverage, reduce decarbonisation costs and minimise carbon leakage.

Box 9: Carbon pricing

The Energy Transitions Commission, along with many other commentators, view carbon pricing as an essential component to drive the energy transition. Around 13% of global CO_2 emissions are covered by carbon pricing in one form or another. Typical prices are around \$10/tonne. Coverage and prices are likely to rise in the coming years. By the end of 2017 it is anticipated that 30% of global emissions will fall under carbon pricing systems, with countries such as Canada ratcheting floor prices up to \$25/tonne or more within the next five years.

As many of the emission trading systems (ETS) have experienced low and unstable prices, the desired effect of stimulating changes within the energy system has been weaker than expected. However, many companies have now adopted shadow carbon prices, under the anticipation that carbon pricing would impact returns on their investments. But the weaker than expected development of ETS has led to concerns over companies dropping these screening measures. This highlights the need to ensure greater coverage, with stable, predictable and significant prices.

The lull in momentum to expand carbon pricing, either as a trading system or as a tax, has recently reversed. The Carbon Pricing Leadership Coalition (CPLC) represents 21 countries, five Canadian and US states, and almost 140 global corporations. In January 2016 the CPLC challenged global governments to double carbon pricing coverage by 2020 and double again by 2030. Further, the Climate Leadership Council recently proposed a $40/tonne carbon tax and dividend scheme in the USA, including a carbon border adjustment tax.

The pathway to the national Chinese ETS is under way. However, the uncertainty over allowance holders within China's regional ETS pilots being able to transfer to the national ETS is hampering the 2017 intended launch of the national ETS. Lack of transparent market information within the pilots has also reduced the volumes traded relative to the cap, compared to other ETS such as California.

While a global carbon market is unlikely in the short to medium term, linking existing carbon markets is generally accepted to increase the effectiveness of national markets. Considering the national ETS in China is due to commence in 2017, one significant policy consideration is the future linking to the EU or South Korean market. If carbon clubs or bilateral links to other ETS are to emerge, greater transparency within the Chinese ETS will be critical. Given the potential size of the Chinese ETS, if it fails to deliver a stable and effective carbon price, carbon trading is likely to face a considerable challenge in convincing international policymakers that continued development of carbon markets is an effective climate policy pathway.

6 Making Global Energy Governance Fit for the Future

Much of the current global energy governance architecture has evolved from arrangements to address oil security concerns stemming from the oil shocks of the 1970s. However, the effectiveness of these mechanisms—specifically the emergency response mechanism of the International Energy Agency (IEA)—has been gradually eroded by the shift in hydrocarbon demand from IEA member countries of the OECD to emerging economies outside of the regime.

More recently, with around two-thirds of global greenhouse gas emissions coming from the energy sector, decarbonisation has moved to the centre of the energy policy agenda. An increasingly analytical focus on low-carbon transition is evident within organisations like the IEA and with the establishment of new multilateral institutions, such as the German-led International Renewable Energy Association. Meanwhile, multilateral development banks, such as the World Bank, have scaled up their technical assistance and financial commitments for cleaner energy.

Although outside the sphere of traditional energy governance, the establishment of the United Nations Framework Convention on Climate Change in 1990 and the Paris Agreement in 2015 have had profound implications. The latter provides the pillars of a new global climate regime: a transparency framework for the measurement and reporting of emission reductions and actions; a long-term goal to reduce global greenhouse gas emissions to net zero and limit temperature rises to well below 2°C; and a process of pledge and review through which governments progressively update the ambition of their nationally determined contributions every five years in order to incrementally increase collective ambition.

The strong overlap between the global climate regime objective of limiting global warming to well below 2°C and the challenges of managing a low-carbon, secure transition means there are opportunities to harmonise climate and energy governance. At the same time, representation in traditional energy governance institutions and mechanisms must be broadened, even as they focus on energy transition; membership beyond OECD countries is critical if the IEA is to remain a relevant institution.

Box 10: IEA expansion

One of the principal barriers to expanding IEA membership to include emerging economies has been that the International Energy Programme (IEP) treaty, which provides the legal framework for member countries, requires that signatory countries be members of the OECD. This means that expanding membership of the IEA would require treaty change. It would also require reforms to the voting rights of countries within the IEA, which currently reflect their levels of consumption in 1973; this would inevitably create winners and losers and would give an incoming producer such as China significant voting power.

One way to effectively become a member of the IEA without signing the IEP treaty would be to follow Norway's path. Norway has a separate partnership agreement with the IEA under which it acts as a full member country for all intents and purposes, although it has not signed the IEP. Such a path could potentially be available to emerging economies. This would, of course, require the emerging economies concerned to follow the same rules—including those on data sharing and

emergency stock coordination—as existing member countries.

A roadmap to expansion could conceivably be agreed between the IEA and emerging economies setting out key milestones that both sides would need to reach in order for the partnership to become politically and technically feasible. These might include: (i) the IEA opening a new secretariat office in Asia to deepen its links with officials in Asian consumer countries; (ii) the IEA cooperating with China on electricity market reform pilots, providing technical support to strengthen ties at the municipal level and demonstrate its capabilities beyond the oil market; and iii) emerging economies developing the necessary technical and institutional infrastructure to meet IEA requirements in terms of information and data sharing, stock management, demand restraint and agreeing to peer review.

6.1 Cooperation Opportunity

Due to its leadership role among emerging economies, and its position as the world's largest energy consumer and greenhouse gas emitter, China is uniquely placed to drive global energy governance reforms.

Targeted outcome: Representative global energy governance institutions providing harmonised energy and climate public goods.

To achieve such an outcome, China could adopt two distinct strategic cooperation pathways:

(1) reform the IEA to make it more representative, effective and relevant; and
(2) harmonise global energy and climate governance to facilitate the transition to a low-carbon, energy-secure future.

6.2   The G20 as a Forum for Reform

The G20 could provide a suitable forum to promote global energy governance reform. It has a track record of international organisation reform (emerging economies used it to reform the International Monetary Fund following the global financial crisis of 2008) and experience of establishing new international mechanisms, such as the Agricultural Market Information System. The G20's membership provides a balance of OECD and emerging economies, and of producers and consumers of fossil-fuel based energy. It therefore has the requisite critical mass on key transition issues such as energy consumption and production (23% and 19% respectively), emissions (30% of global total, double that of the USA) and energy investment (21%).

Moreover, many aspects of the current G20 agenda are highly relevant, including the Principles for Energy Collaboration it adopted in 2014 (which included its long-running commitments to fossil fuel subsidy reform), the emergence of green finance under China's presidency in 2016, and progress on climate-related financial disclosures through the Financial Stability Board.

Sustaining a reform process through the G20 would not be without challenges. First, the G20 is an informal club of governments, without a secretariat to provide support and ensure delivery. Second, and relatedly, the G20 struggles to sustain focus as its agenda tends to shift with the priorities of successive presidencies. Consequently, any ambitious energy governance reform agenda would need to include a strategy to sustain G20 engagement.

Box 11: Strengthening oil security in Asia

Asia consumes a third of the world's oil. About two-thirds of that are imports of crude oil.

Only two ASEAN Plus Three countries (Japan and South Korea) are party to the OECD-based IEA with its emergency response mechanism and obligations to hold at least 90 days of oil imports as compulsory stocks.

Asia and ASEAN lack mechanisms to respond to supply disruptions collectively rather than competitively: major disruptions would lead to a scramble for supplies and large spikes in international oil prices. The ASEAN Petroleum Supply Agreement (APSA) supports, on a best endeavours basis, member countries experiencing a supply disruption equivalent to 10% of consumption over 30 days. These efforts may include a coordinated stock release, relying on cooperation between the state companies of the ASEAN countries. A major disruption would likely generate considerable uncertainty over how Asian governments would respond. Given APSA is currently set to expire in 2023, and global oil prices are relatively low, now is a good time to push for expansion of regional strategic oil which could calm markets if there was a disruption to supply. This expanded APSA agreement could include;

- increasing   government–controlled stockholding in non-IEA, ASEAN Plus Three countries to IEA-equivalent levels: IEA members Japan and South Korea hold more than double the IEA target of 90 days of net imports; ASEAN countries less than half; estimates for China suggest around 50 days;
- a commitment to coordinate the release of stocks, to limit free-riders on stocks released by other countries;
- a commitment to maintain, on a pro rata basis, exports of products refined during a disruption of crude oil supplies. Middle East crude is refined in South Korea, Japan, India and Singapore for export as product. Without such a commitment, the product trade would transmit disruption, resulting in a scramble for supplies; and

- developing an understanding with the Middle East and other crude oil exporters on how they would allocate remaining supplies in the event of disruption to their own refineries in the region or pro rata to all their customers.

Given the current engagement of the ASEAN Council on Petroleum with the ASEAN Plus Three Oil Stockpiling Roadmap forum and Energy Security Forum, the above improvements to APSA could be achieved by enhanced engagement with these forums.

6.3 Rationale for Action

Expanding the membership of the IEA to include emerging economies is crucial if the IEA's emergency response mechanism—the only international coordination mechanism for oil stock release and sharing—is to remain relevant. The alternative is a patchwork of national resilience measures and bilateral producer-consumer arrangements which risk a damaging scramble in the event of a major market dislocation. A more inclusive emergency response mechanism is therefore in the interests of all oil-consuming countries. However, as the world's largest oil importer, China has a particular interest in ensuring international market resilience. China will also benefit indirectly as a result of reduced risk to vulnerable neighbouring oil-importing countries.

Harmonising climate and energy governance regimes with globally agreed decarbonisation objectives could help rationalise global governance, achieving closer alignment between international organisations. In particular, mechanisms and political processes to increase ambition on decarbonisation of the energy sector provide a significant market opportunity for China as a major exporter of low-carbon goods, such as solar photovoltaic panels and batteries. It is also consistent with President Xi's stated wish

to protect the Paris Agreement in response to the US withdrawal.

Box 12: Energy Charter Treaty

The Energy Charter Treaty (ECT) is legally binding and establishes a multilateral framework for cooperation between countries in the energy industry, covering trade, transit, investments and energy efficiency. The treaty contains dispute resolution procedures.

The ECT originated in Russia-EU relations. However, in 2009 Russia announced it did not intend to become a contracting party. In 2015, 72 countries (to date 83 countries), including China, signed the International Energy Charter (IEC); expanding and modernising the ECT. While the IEC is a political declaration and is non-binding, it is a first step towards the international equivalent of the legally binding ECT.

China's accession to the ECT was inhibited by concerns over international arbitration cases. However, as the role and remit of the Belt and Road Initiative (BRI) expands, a common set of rules and mechanisms under the ECT/IEC could be beneficial to China, allowing a common basis for negotiation and agreements, and binding non-WTO BRI countries. Further, a legally binding IEC could provide a low-transaction-cost means to mitigate overseas energy infrastructure investment risks. The benefits to China outweigh the risks of joining, given the balance between China's outward and inward investments, which will accelerate under BRI. The short-term benefits apply primarily to investments in central and eastern European countries (which are already signatures to the ECT), especially for non-WTO members. Given the ECT's origins in the EU and the lack of signatories from ASEAN countries, China could work with the EU and ASEAN Plus Three to develop the IEC towards a legally binding framework.

6.4 Strategic Pathways

China could work through the G20 to reform the
IEA and formalise a long-term agenda under the
auspices of an enhanced G20 to harmonise global
energy and climate governance.

(1) IEA reform

China could work with emerging economies
(including other association countries—India and
Indonesia) within the G20 to initiate a process for
IEA reform at a summit chaired by an emerging
economy (to ensure its priority on the agenda).
This could include a five-year roadmap to partnership for the emerging economies based on the
precedent of Norway, which is a partner country
rather than a full member having never signed
the International Energy Programme treaty.

The push for IEA reform could be strengthened by China first cooperating to enhance
regional energy security, which has certain
structural weaknesses—for example, through
ASEAN Plus Three (see Box 11), the International Energy Charter (see Box) and regional
cooperation on gas (see Box 13). This would still
allow China to improve its energy security and
hedge against the possible failure of IEA reform,
for example were it to be blocked by member
countries. It would also signal to IEA member
countries that China is serious about strengthening energy security governance.

Box 13: Regional gas security

Asia is the world's largest consumer of
liquefied natural gas (LNG), and the most
vital incremental demand market. As such,
the region has known extremely high spot
prices in periods of limited supply. While
new supplies seem set to outstrip demand
in the coming years, buyers should not
become complacent. China should view
the backdrop of well-supplied LNG markets as an opportune time to promote
regional frameworks to enhance natural

gas security and promote greater flexibility
of supplies.

These could include a concerted push
for less rigid contracts with volumes
delivered through open destinations or
more flexible redelivery clauses. A regional pricing hub, that would be indexed
to gas rather than oil prices, would also
provide more accurate pricing signals to
shift volumes from one destination to
another. Finally, storage capacity will
support a more robust and flexible gas
market. China has already issued targets to
increase its gas storage capacity and could
work with other countries to develop similar infrastructure, which could then be
coordinated more effectively in times of
shortage. North-east Asia would be the
most natural starting point given the concentration of large buyers there, but
ASEAN Plus Three would also offer a
convenient framework, bringing together
regional   consumers   and   producers
(Fig. 33).

Box 14: Mission statements of key international organisations

The Energy Charter Treaty provides a
multilateral framework for energy cooperation that is unique under international law.
It is designed to promote energy security
through the operation of more open and
competitive   energy   markets,   while
respecting the principles of sustainable
development and sovereignty over energy
resources.

The International Energy Agency works
to ensure reliable, affordable and clean
energy for its 29 member countries and
beyond. Its mission is guided by four main
areas of focus: energy security, economic
development, environmental   awareness
and engagement worldwide.

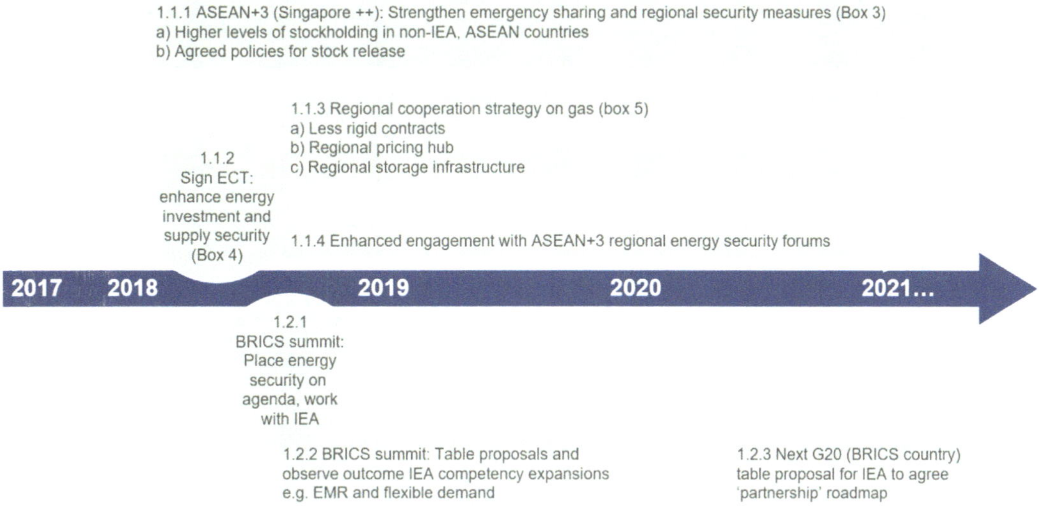

1.1.1 ASEAN+3 (Singapore ++): Strengthen emergency sharing and regional security measures (Box 3)
a) Higher levels of stockholding in non-IEA, ASEAN countries
b) Agreed policies for stock release

1.1.3 Regional cooperation strategy on gas (box 5)
a) Less rigid contracts
b) Regional pricing hub
c) Regional storage infrastructure

1.1.2
Sign ECT:
enhance energy
investment and
supply security
(Box 4)

1.1.4 Enhanced engagement with ASEAN+3 regional energy security forums

| 2017 | 2018 | 2019 | 2020 | 2021... |

1.2.1
BRICS summit:
Place energy
security on
agenda, work
with IEA

1.2.2 BRICS summit: Table proposals and
observe outcome IEA competency expansions
e.g. EMR and flexible demand

1.2.3 Next G20 (BRICS country)
table proposal for IEA to agree
'partnership' roadmap

Fig. 33 Timeline of action plans under strategic pathway: reform the IEA to make it more representative, effective and relevant. *Note* ASEAN Plus Three = the Association of Southeast Asia Nations and the three East Asia nations of China, South Korea and Japan; ECT = Energy Charter Treaty; BRICS = Brazil, Russia, India, China and South Africa; EMR = electricity market reforms

The International Energy Forum, as the neutral facilitator of open dialogue on energy with key global oil and gas actors, helps ensure energy security and transparency.

The International Monetary Fund's fundamental mission is to help ensure stability in the international system. It does so in three ways: keeping track of the global economy and the economies of member countries; lending to countries with balance of payments difficulties; and giving practical help to members.

The International Partnership for Energy Efficiency Cooperation seeks to accelerate the adoption of energy efficiency policies and practices. It assists member countries to identify and share proven, innovative practices and data on energy efficiency to better inform decision makers.

This also serves to foster bilateral and multilateral initiatives between countries.

The International Renewable Energy Agency is an intergovernmental organisation that supports countries in their transition to a sustainable energy future. It serves as the principal platform for international cooperation, a centre of excellence, and a repository of policy, technology, resource and financial knowledge on renewable energy.

The United Nations Framework Convention on Climate Change (UNFCCC) provides the foundation for multilateral action to combat climate change and its impacts on humanity and ecosystems.

The World Bank has set two ambitious goals to push extreme poverty to no more than 3% by 2030, and to promote shared prosperity and greater equity in the developing world.

(2) **Harmonisation of global energy and climate governance**

China could seek to initiate a long-term agenda, under the auspices of the G20, to help align global energy governance and climate governance regimes to accelerate the transition to a low-carbon, energy-secure future. First, groundwork steps could seek to position the G20 as a forum for political leadership and consensus on the transition to a low-carbon, energy-secure future through a series of statements and commitments on secure, long-term decarbonisation and collective efforts to raise the ambition of nationally determined contributions (NDCs) (see Box).

This could be followed up by the G20 launching a global energy governance reform agenda to align international organisation mission statements with the common goal of a secure, low-carbon energy transition as enshrined in the Paris Agreement and UN Sustainable Development Goals (see Box). This would need to be complemented with steps to build the G20's leadership and oversight function by: (i) establishing a long-term series of G20 energy ministerial meetings; and (ii) building its institutional capacity through an energy secretariat function distributed among the relevant international organisations.

Box 15: Increasing NDC ambition

Closing the gap between the emission reductions pledged under existing NDCs, and those needed to achieve the Paris goals, will be addressed through a process every five years at the UNFCCC, where governments progressively increase the ambition of their NDCs.

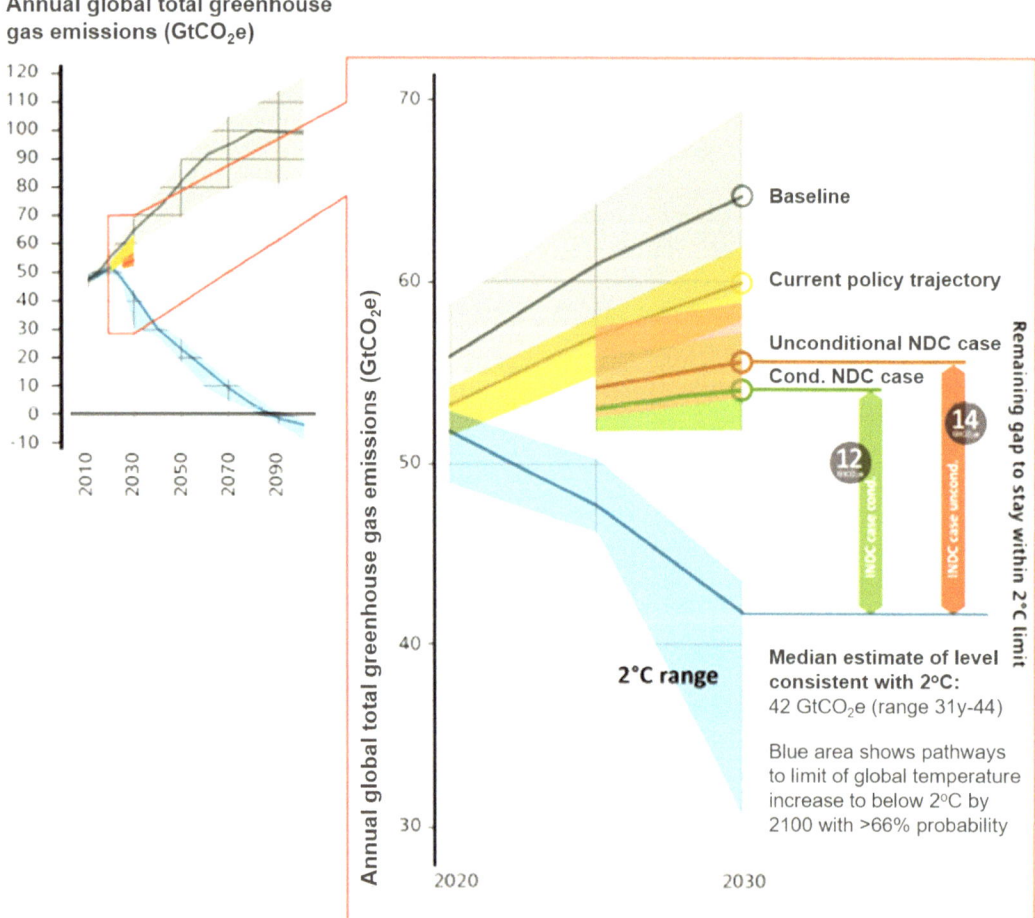

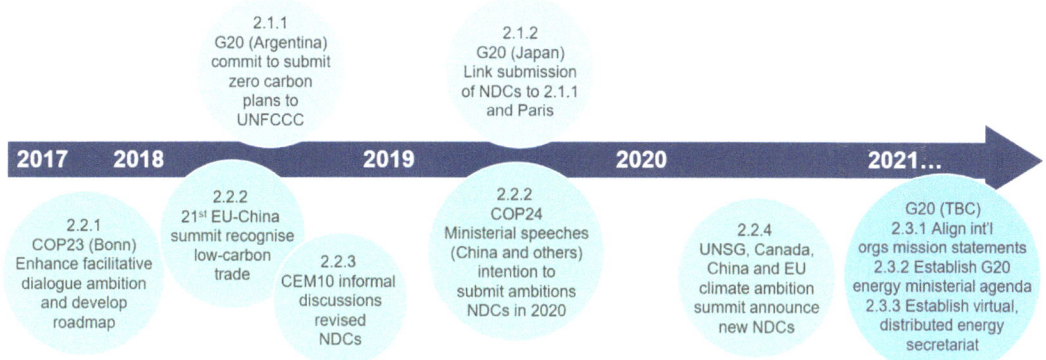

Fig. 34 Timeline of action plans under the strategic pathway: harmonise global energy and climate governance to facilitate the transition to a low-carbon, energy-secure future. *Note* UNSG = United Nations Secretary-General; COP23 = 23rd Conference of the Parties to the United Nations Framework Convention on Climate Change (UNFCCC); CEM10 = 10th Clean Energy Ministerial

As the largest emitter in the world, China will be critical to any credible global effort to increase NDC ambition. Chinese ambition (alongside that of the USA) was similarly crucial to setting expectations in the run-up to Paris in 2015. Although it would be unrealistic to suppose that a similar US-China dynamic can be recreated with the current US administration, China could seek other developed country partners.

Importantly for China, leveraging its own NDCs to increase the ambition of other developed country NDCs makes sound economic sense. The resultant demand for low-carbon goods in developed economies could provide large export opportunities for Chinese manufacturers.

This could begin with diplomatic efforts to achieve progressively more ambitious language on emissions at the G20, culminating in a statement of intent on NDC ambition in 2020 that is repeated at five-yearly intervals in increasingly determined terms. Supportive activities could include using EU-China summits and the Clean Energy Ministerial as forums to signal intent and discuss new NDCs and, more ambitiously, a summit with the UN Secretary-General, EU and

other progressive governments to announce revised NDCs in early 2020 (Fig. 34).

6.5 Risks and Obstacles

The principal risks to these strategies are political: US antipathy towards climate policy would make progress through the G20 more difficult and may strengthen the positions of less progressive governments such as Saudi Arabia and Russia (although it should be noted that following the US withdrawal from the Paris Agreement, both restated their support). The Germany-chaired 2017 summit ultimately saw a G19 statement on climate change that excluded the USA, establishing the precedent for circumventing its obstructiveness. For more ambitious reforms, it may be necessary to wait until a more receptive US administration is in place.

Other political risks relate to the expansion of the IEA—in particular, some member countries may be resistant for fear of having their voting shares diluted. By seeking expansion of the IEA through partnership arrangements—such as that with Norway—rather than full participation (which would entail reforming the International Energy Programme treaty as well), this issue could potentially be avoided. IEA members would benefit from expansion and an empowered IEA when the next energy crisis emerges.

The G20's lack of a supporting institutional infrastructure and its fluctuating agenda present an obstacle to the forum acting as a leadership and oversight mechanism for the energy transition. This could be addressed directly through reforms to build a distributed secretariat function and initiate an energy ministerial track to facilitate such a function. However, formalising the G20 to this extent is likely to be met with caution from governments reluctant to accept further international obligations.

7 Investment Regimes for Low-Carbon, Energy-Secure Transition

Attention regarding the risks of the rapidly accelerating energy transition is growing among policymakers, central banks, regulators, institutional investors and financiers. Allocating investment during a swift energy transition, while maintaining oil and gas supply security and reliability of electricity networks, requires clear market signals and a level playing field.

Several investment mechanisms, finance sector rules and market structures could provide this clarity, given the correct reforms. These mechanisms, rules and structures fall roughly into two categories: first, international governance arrangements to help optimise global capital

allocation; and second, engagement within countries to improve national governance of the energy sector and access to finance and technologies.

Box 16: The challenge of optimal capital investment across the energy system

Ensuring the appropriate level of investment in the oil and gas sectors is achieved —in parallel to investment in the growing renewable, electric vehicle and battery storage markets—is a fine balancing act. The Energy Transitions Commission, and New Climate Economy, anticipate that a 2°C world requires $21.3 trillion invested across all energy sectors—including exploration, production and distribution— between 2015 and 2030. As Fig. 35 illustrates, this implies an average of $566 billion per year in hydrocarbons and $500 billion per year in renewables, which equates to increases of 31% and 68% respectively on 2016 expenditure.

Upstream oil and gas expenditure will be cut by an estimated $1 trillion between 2015 and 2020, as a result of continued low prices and slack global demand. In 2016, upstream oil and gas investment

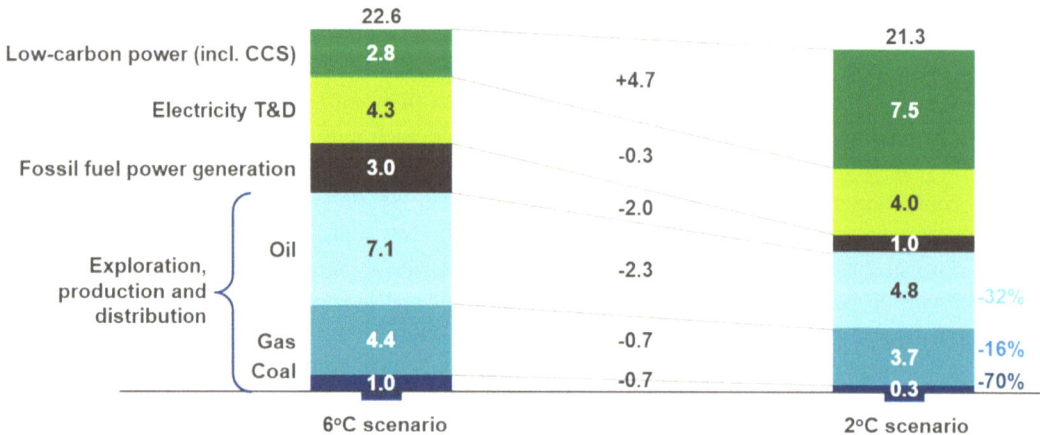

Fig. 35 Investment requirement in energy, 2015-30 ($ trillion, constant 2010 dollars). *Note* CCS = carbon capture and storage; T&D = power transmission and distribution

fell 26% due to reduced drilling, which comprises two-thirds of the sector's total investment. Offshore spending is expected to decline by 20–25% in 2017 alone. Yet these aggregate figures mask very different approaches to commercial risk in upstream investments between international and national oil companies (IOCs and NOCs): while European and US IOCs are expected to scale back investment in 2017 by a further 7% and 15% respectively, NOCs plan to increase spending by 9%. Meanwhile, investment in renewables fell by 23% in value in 2016; although more capacity was added to the global grid than ever before, thanks to declining costs. Both these trends, across the renewables and oil and gas sectors, are contrary to the investment required under a 2°C energy-secure world. This is partly due to the market not receiving correct signals regarding risks and benefits and lack of clarity over how the energy transition will progress.

The construction of market structures and investment mechanisms should remain technology-agnostic, allowing the market to select the optimal technologies. On the basis of the market providing a level playing field, with climate risks evaluated in costs and benefits, this agnostic stance should be applied to clean energy as well as fossil fuel technologies. That said, there are certain technologies where high capital costs and network expansion costs require greater support. Examples of such technologies, in relation to China's 13th Five-Year-Plan (2016–20), are offshore wind and high-voltage

power transmission. In such instances, multilateral development banks and public-private partnerships could be used to direct additional investment, alongside partnering with countries that have expertise with a given technology (Fig. 36).

7.1 Global Governance of Investment Regimes

Uncoordinated or unaligned investment policies will be insufficient to achieve energy security and the long-term goal of the Paris Agreement.

The Financial Stability Board Taskforce on Climate Disclosure is a potential game-changer. Its recommendations for companies on voluntary, consistent climate-related financial disclosures are a clear first step to providing better information to investors and other market players on the risks to business models and balance sheets presented by the low-carbon transition—so-called transition risks. These recommendations may require further development, for instance on their application by state-owned enterprises and in less liberalised markets. The metrics by which companies disclose in different sectors will also need to be tested and refined, and the scenarios under which these metrics sit require methodological standardisation across geographies and industries. While recommendations can be voluntarily adopted by companies, state backing by China and its G20 partners

Fig. 36 Global investment in energy supply by fuel

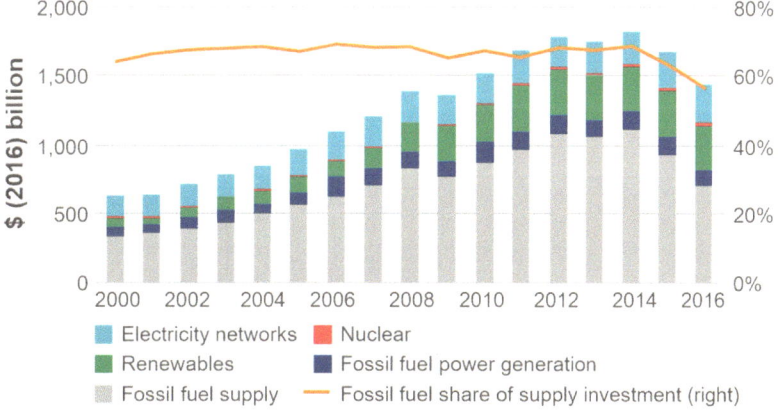

will send a strong signal—stimulating implementation.

Ultimately, incorporating transition risks into the cost of investment capital will help allocate investments to where they are needed to provide energy security through the transition, though further development of the recommendations across diverse jurisdictions and sectors is required to mitigate unintended negative consequences of disclosures.

Carbon pricing can also help, by providing a market signal through demand for energy products and services; it is the most efficient economic mechanism to deliver decarbonisation. The failure to implement carbon pricing regimes and increase their global coverage will significantly hinder the achievement of national climate policy and international commitments under the Paris Agreement—to the detriment of all countries. The slowdown in momentum to expand carbon pricing, either as a trading system or as a tax, has recently taken off again. The Carbon Pricing Leadership Coalition (CPLC) represents 21 countries, five Canadian and US states, and almost 140 global corporations. In January 2016 the CPLC challenged global governments to double carbon pricing coverage by 2020 and double again by 2030. The Republican-led Climate Leadership Council recently proposed establishing a $40/tonne carbon tax and dividend scheme within the USA.

Early emissions trading system (ETS) adopters have lessons to offer on the impacts of conflicting policies and poor adoption strategies on establishing stable and effective pricing. For instance, the cost-effectiveness of an ETS and achieving an effective price can be undermined by policymakers inadequately attempting to balance the trade-off of reduced industrial competitiveness by introducing competing policies or reducing sectorial coverage. If these lessons can be leveraged through lesson sharing—along with designing for compatibility and robust monitoring, reporting and verification procedures—the likelihood of success of China's ETS and others will increase.

Carbon leakage may undermine the effectiveness of carbon pricing (carbon leakage is where the restructuring of high-carbon industries, for example, encourages an outflow of capital and technologies beyond national borders). Regional and international coordination to link carbon markets, through mechanisms such as Article 6 of the Paris Agreement on transferable allowances, could minimise carbon leakage. Studies have shown that in a non-linked EU ETS, energy emission leakage peaks at 16%. If however, the EU and China linked, once the Chinese ETS is established, energy emission leakage could peak at 8.5% (see Figs. 37 and 38). However, linking carbon pricing schemes is not straightforward and is a risky endeavour. For example, linking carbon trading schemes between Europe, California and Quebec led to price volatility, not stability. This is likely to be due to differences in the regulations governing each jurisdiction, and the interactions of

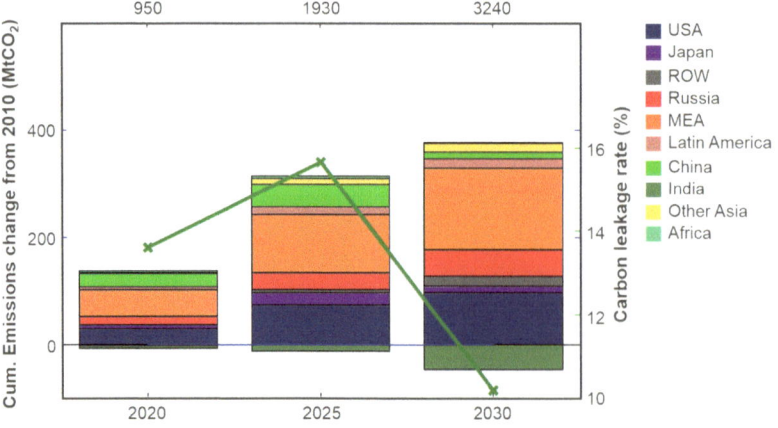

Fig. 37 EU ETS non-linked: energy emission leakage peaks at 16%

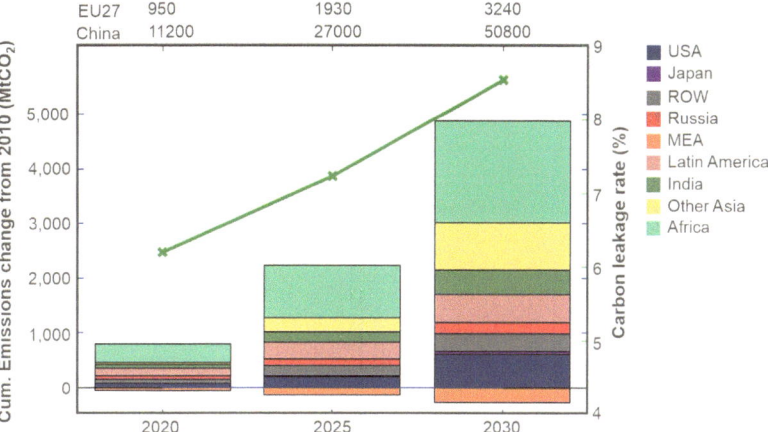

Fig. 38 EU-China ETS linked: energy emission leakage peaks at 8.5%

conflicting parallel climate policies between jurisdictions.

Another way to promote carbon pricing and mitigate leakage could be through the use of border adjustment measures (BAMs). Countries without (a sufficient) carbon price would experience a compensatory tax on their exports to countries within the scheme, so would be incentivised to apply a price themselves in order to improve market access. The recent proposal by the Climate Leadership Council on a US carbon tax included a proposed border adjustment tax. Legal opinion indicates that WTO rules do not preclude the use of BAMs if carefully designed. However, this does not mean that affected countries would not launch dispute proceedings.

Fossil fuel subsidies undermine carbon pricing. Continued multilateral efforts to phase out fossil fuel subsidies can help avoid the creation or continuation of havens for high carbon capital.

Lower oil prices have given political and fiscal breathing space for countries such as Indonesia, which successfully phased down consumption subsidies for citizens or industries as fossil fuel prices collapsed. Of 157 countries, 84 either reduced consumption subsidies or increased petrol taxes. But as consumption has shifted towards nations that have maintained subsidies or lowered petrol taxes, subsidy reforms have been undermined at the global level (Fig. 39).

7.1.1 Cooperation Opportunity

As China strives to open up its domestic markets to foreign investment, commercial transparency and the disclosure of climate-related risk will be key factors in building the confidence of international investors. At the same time, China is establishing the world's largest carbon market; the success of which will be partly contingent on minimising policy conflicts, particularly where

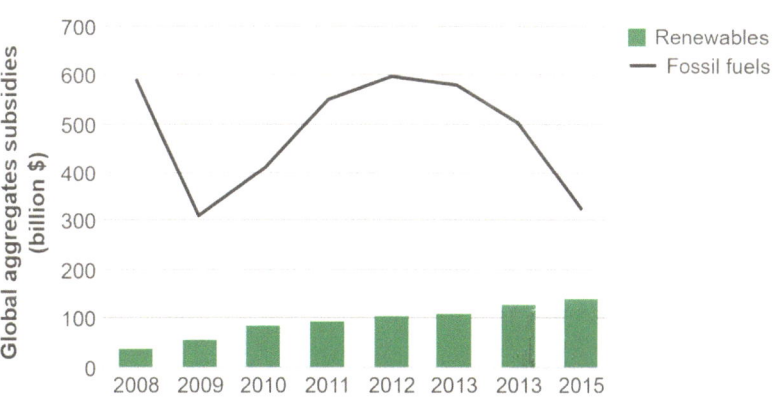

Fig. 39 Estimates for global fossil-fuel consumption subsidies and subsidies for renewables

the distortions associated with fossil fuel subsidies and the impacts associated with carbon leakage are concerned.

Given the political importance of managing the transition to a market-led economy and slower but higher quality economic growth at home, policy support for measures to disclose climate-related financial risks and develop effective carbon pricing mechanisms are well aligned with China's strategic priorities.

7.1.2 Outcome

The targeted outcome is a global market risk disclosure structure and carbon pricing mechanisms that optimally allocate capital across all aspects of the global energy system, maintaining energy security while achieving the goals of the Paris Agreement.

To achieve such an outcome China could adopt two distinct strategic governance pathways:

(1) collaborate with G20 states to support the development, adoption and implementation of FSB TCFD energy sector recommendations, and;

(2) support the phase-out of fossil fuel subsidies and establish carbon pricing mechanisms that are fit for a low-carbon, energy-secure system.

7.1.3 Using the G20 and Other Multilateral Platforms to Drive Reform

Given that the TCFD was established by the Financial Stability Board, whose board members include the central bankers of all G20 major economies, the G20 is well placed to drive further development and adoption of TCFD recommendations. Further, while the commitment to fossil fuel subsidy phase-down wasn't reaffirmed at the G20 2017 Hamburg meeting, the G20 has been pivotal in global subsidy reform since 2009. China could therefore use the experience of G20 to drive reforms in partnership with other leading G20 economies.

There is currently no single institutional or organisational home for the governance and coordination of carbon pricing. However, China could work with several multilateral organisations to expand the effectiveness of carbon pricing, such as at the UNFCCC on transferable allowances, the EU and the Carbon Pricing Leadership Coalition on linking and carbon clubs, or with the WTO on border adjustment measures.

7.1.4 Rationale for Action

In the case of China, greater transparency of capital markets may help foster confidence in foreign investors. More broadly, as a major energy consumer, China will benefit from optimal capital allocation in the energy sector through the transition, which would help avoid possible supply crunches that may result from investors failing to appreciate transition risks and investment requirements. Finally, as a carbon-pricer, China has an interest in international carbon pricing to avoid damaging the competitiveness of its heavy industries.

From the perspective of China's partners, Chinese-state backing of TCFD recommendations, fossil fuel subsidy reform, and the establishment of effective carbon price mechanisms will all help stimulate development and implementation. Further, China's partners will benefit from an equalisation between increasingly penalised OECD carbon-intensive sectors and the competitive advantage of Chinese carbon-intensive exports.

7.1.5 Strategic Pathways

(1) **G20 states support development, adoption and implementation of FSB TCFD recommendations**

Developing and adopting the TCFD framework could be achieved by China working with G20 member states and future Chairs to establish a trilateral working group under the G20 process. This could bring together the Financial Stability Board (FSB), the G20 Energy Ministerial and energy sector participants in the Business 20 dialogue, with a view to unlocking some of the challenges of TCFD development and implementation. In the

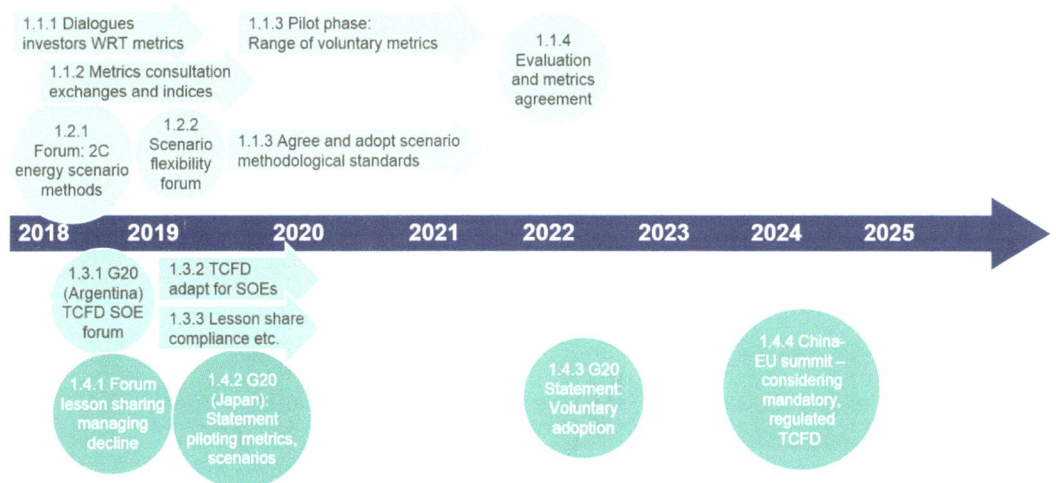

Fig. 40 Timeline of action plans under the strategic pathway: collaborating with G20 states to support the development and adoption of FSB TCFD energy sector recommendations

first instance, this could include: (i) definition and testing of disclosure metrics; and (ii) methodological standardisation of energy scenarios across geographies and industries under which disclosures are made. The recommendations also require: (iii) adaptation for public finance, including state-owned enterprises and policy banks; and (iv) building momentum of G20 governments committed to FSB TCFD recommendations and facilitating implementation (Fig. 40).

(2) **Fossil fuel subsidies and carbon pricing mechanisms fit for purpose**

Global governance of an internationally coordinated phase-down of fossil fuel subsidies could be achieved by China working with G20 partners to foster a renewed G20 focus on fossil fuel subsidies. Establishing effective carbon pricing mechanisms could be enabled by China working with a variety of international forums to encourage international lesson sharing of best practice in establishing domestic carbon prices, alongside international cooperation to minimise carbon leakage (Fig. 41).

7.1.6 Risks and Obstacles

Where the development and adoption of FSB TCFD energy sector recommendations is concerned, confidence building measures will be critical. High-level commitment to the development of FSB TCFD metrics and scenarios would help send a clear message to the market and have the potential to encourage widespread voluntary adoption. Mandatory adoption will be politically difficult, but with strong state backing is unlikely to be necessary, as was learnt from the Financial Stability Forum recommendations on Enhancing Market and Institutional Resilience after the 2008 financial crisis.

Renewed G20 focus on fossil fuel subsidy phase-out will face similar obstacles to previous subsidy reform efforts, where progress has been undermined by the inconsistencies in reporting. The process was further undermined by the US position in 2017, which ultimately resulted in the G20 Hamburg Communique's failure to reaffirm fossil fuel subsidy commitments. The USA is a key player in ensuring subsidy reform and has worked with China to peer-review progress on subsidy reform in the past.

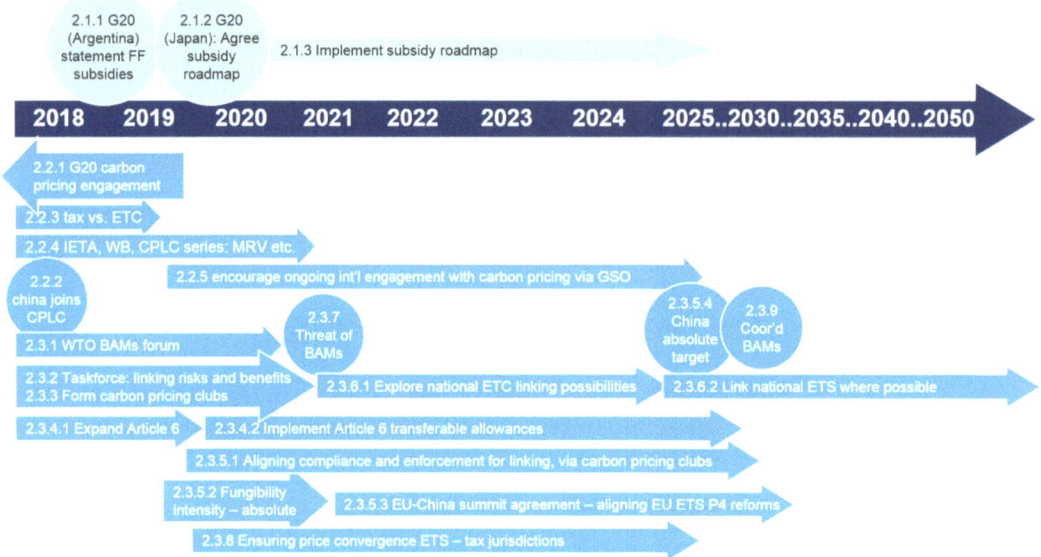

Fig. 41 Timeline of action plans under the strategic pathway: global phase-out of fossil fuel subsidies and establishing carbon pricing mechanisms fit for a low-carbon, energy-secure system

There is a significant risk from linking the China emissions trading system with other jurisdictions. Should this result in failure—for example problems with monitoring, reporting and verification (MRV), price instability or a prolonged price collapse—it could represent an insurmountable setback to the internationalisation of carbon trading. Cooperation with other carbon-pricing jurisdictions to establish clubs of free-trading economies with border adjustment measures on non-member countries may therefore be a less risky way to encourage propagation.

infrastructure. By some estimates, the scale of these investments could be up to $900 billion.

China has an interest in ensuring good energy sector governance in partner countries where it invests. This will help mitigate the risks of instability and underinvestment in these countries, enhancing China's energy security as a result (see Box 17). China can also cooperate with partner countries to support them through the transition, especially by providing access to low-carbon technologies produced by Chinese companies (see Box 18).

The way that China approaches investments along the BRI will be a measure of China's commitment to green and sustainable growth, both at home and abroad for many of its international partners.[2] Learning from past experience and minimising the risk of investment disputes with host countries and underperforming assets —by adhering to high standards of social and environmental governance and developing

7.2 International Cooperation on National Energy Sector Investment

Over the past decade, China has rapidly increased its foreign direct investment (FDI) in overseas energy sectors as part of its Go Out strategy. Over the coming decades, the Belt and Road Initiative (BRI) looks set to channel unprecedented outflows of Chinese capital into energy resources and infrastructure, and higher up the value chain in new technologies and

[2]See Navigating the New Normal for an in-depth exploration of China's evolving approach to FDI and the BRI: https://www.chathamhouse.org/publication/navigating-new-normal-china-and-global-resource-governance.

appropriate risk management tools—is one key aspect of this. At the same time, Chinese oil companies have paid around 20% above the industry average for oil and gas assets. Together, these factors have affected the profitability of Chinese FDI in the hydrocarbon and mining sectors, leading some Chinese officials to question the performance of such investments.

Avoiding carbon leakage is another factor, as China's domestic economy undergoes structural transition and as high-carbon capital and technologies seek an escape valve in the BRI. President Xi Jinping proposed an "international coalition for green development" at the first Belt and Road Forum for International Cooperation in Beijing in 2017; and greening the BRI has risen up the policy agenda.

Box 17: Minimising investment risks through enhanced governance

The risks to fossil fuel and clean energy infrastructure investments in host countries, especially in developing countries whose regulatory and governance structures are weak, can be minimised by the development and implementation of strong international standards. Targeted governance mechanisms have evolved to help minimise the risk of some of the negative governance and societal impacts associated with resource development and trade. These range from the transparency of resource revenues under the Extractive Industries Transparency Initiative (EITI) to certifying that raw mineral supply chains are conflict-free through the OECD and the due diligence guidelines of the China Chamber of Commerce of Metals Minerals & Chemicals Importers & Exporters (CCCMC).

The EITI aims to enhance transparency around resource-related payments to governments. The EITI transparency standards require disclosure along the value chain as to how revenue makes its way through the government and contributes to the economy. At the same time, environmental, social and governance (ESG) risk frameworks that originated in multilateral banks, such as the International Finance Corporation's (IFC) Environmental and Social Performance Standards, are increasingly referenced in the market.

China is not an EITI member country, but its companies are active in implementing EITI standards in countries like Iraq and Mongolia. While there is the potential for tension between the initiative's mechanism and China's core foreign policy principle of non-interference, China has approved EITI's principles and strongly supported the EITI in international forums. The EITI is perhaps the most high-profile of a suite of governance mechanisms and technical competencies that resource investors have developed, in order to safeguard relationships and investments with host governments and communities, often in collaboration with multilateral banks. Figure illustrates some of the areas of governance support and capacity building around a hydrocarbon development (Fig. 42).

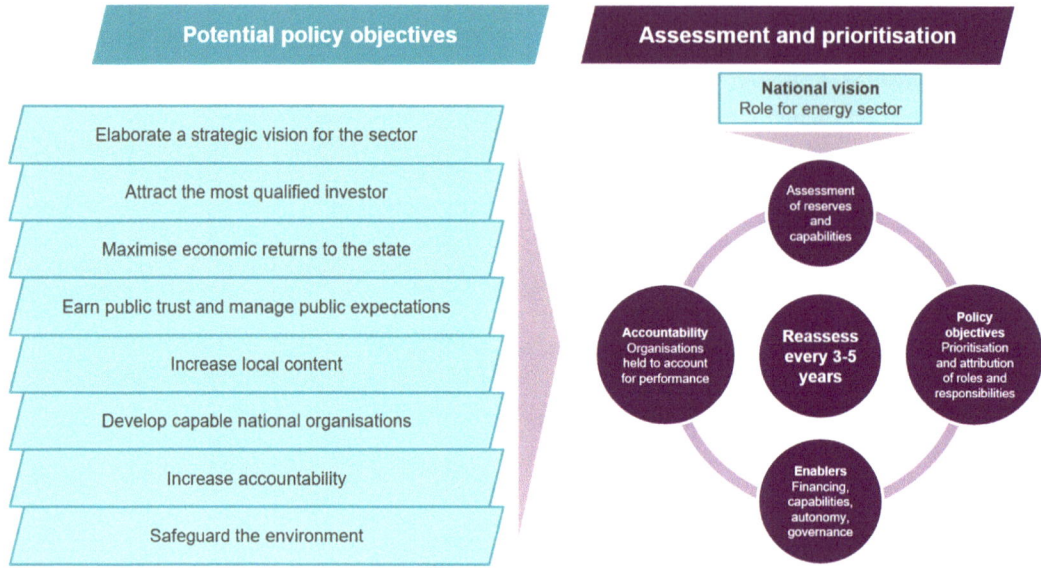

Fig. 42 Recommendations for incremental resource governance improvements by the Chatham House New Petroleum Producers Discussion Group

As clean energy FDI expands along the BRI and across developing countries, the environmental, social and governance (ESG) risk frameworks that underpin fossil fuel investments are equally relevant. ESG standards that originated in multilateral banks as minimum investment standards—such as the International Finance Corporation's (IFC) Environmental and Social Performance Standards—are increasingly referenced in capital markets. The Equator Principles, for example, apply to all commercial bank lending and require investments to meet IFC performance standards. These de-risking tools are equally applicable to low-carbon investments, because they are scaled and as public-private partnerships increase in importance. They are also crucial to the development of the Chinese-led Asian Infrastructure Investment Bank, which has committed to meet or exceed the multilateral investment standards of its peers.

Box 18: Candidate countries for investment and technology packages

Many developing countries, some of which are in the BRI, could greatly benefit from clean energy technology and investment packages. The Climate Vulnerable Forum (CVF) represents those countries most at risk from the impacts of climate change, and therefore with most to gain from such packages. All 48 members of the CVF have pledged to ensure 100% of energy production is from renewable sources "as rapidly as possible", and by 2030–50 at the latest.

The CVF applied more pressure at the UNFCCC Bonn Climate Change Conference in May 2017. The forum's formal statement by Ethiopia, read as follows:

> In accordance with… the Paris Agreement, we are of course calling for a rapid scaling up of predictable, adequate and sustainable financing that should be balanced and readily accessible, together [with] capacity and technological assistance from developed countries to address climate risks.

Of the 48 members, 13 are BRI countries, expanding to 27 with the inclusion of developing countries in Africa. Given China's interest in exporting clean energy technologies to BRI and African countries, combined with the need to ensure that Chinese companies retain their societal licence to operate, providing packages for these countries has the potential to be a win-win approach.

7.2.1 Cooperation Opportunity

China is in a unique position to work with BRI partner countries on a low-carbon, energy-secure transition by creating the frameworks for FDI and energy infrastructure development in clean energy technologies. At the same time China could work with others to enhance the governance and transparency of energy sector investments in line with best practice standards, such as the EITI.

Targeted outcome: stable energy transitions in partner countries, facilitated by:

- harmonised portfolio standards and strategies of multilateral development banks (MDBs), policy banks and export credit agencies (ECAs);
- investment and technology packages for developing country partners; and
- support for energy sector good governance.

To achieve such an outcome China could adopt two intrinsically linked strategic cooperation pathways:

(1) multilateral coordination on phasing down FDI and development finance for high-carbon projects, while implementing international best practice on environmental, social and governance standards; and

(2) cooperation strategies with developing countries, MDBs and climate finance providers to increase access to clean energy technologies and alternative development pathways.

7.2.2 The BRI as a Focal Point for Cooperation with Financial Institutions and Regional Forums

BRI countries provide an obvious focus for cooperation in this regard. Other developing country groupings with high demand for renewable technologies that could be met under (2) above include the Africa Renewable Energy Initiative (AREI) through the Forum on China-Africa Cooperation (FOCAC) and the Climate Vulnerable Forum. Supporting investment packages could be developed in collaboration with developed country partners via MDBs and international coordination and support mechanisms such as the OECD Development Centre and the International Renewable Energy Agency (IRENA).

7.2.3 Rationale for Action

Benefits for China include:

- better governance in oil and gas producing developing countries, improving overall energy security;
- improved societal licence to operate for Chinese companies by joining the EITI, for example, and engaging in technology transfer projects; and
- contributing to sustainable investments in BRI countries.

Benefits for partner countries include access to climate finance and low-carbon technologies, and support with national governance reforms where needed.

7.2.4 Strategic Pathways

(1) **Phasing down high-carbon FDI and development finance, and enhancing best practice ESG standards**

The phase-down of FDI into high-carbon sectors could be achieved by China working to: (i) establish BRI and G20 working groups to align the standards and strategies of MDBs and

Fig. 43 Timeline of action plans under the strategic pathway: multilateral coordination of phasing down high-carbon FDI and development support, and enhancing best practice

other development actors towards both fossil fuels and clean energy technologies, with a clear and common roadmap for transition (the Asian Infrastructure Investment Bank could play a leading role in this). Committing to eliminate support for unabated coal projects could be a first step. At the same time: (ii) enhanced transparency and governance of energy infrastructure FDI could be achieved by implementing EITI and by expanding best practice in transparency and environmental and social good governance of investments in clean energy sectors (Fig. 43).

(2) **Increasing access to clean energy technologies in developing countries**

Cooperation strategies with developing countries, MDBs and climate finance providers to increase access to clean energy technologies, through the provision of technology and investment packages, could be achieved by China and the BRI working with the EU, OECD Development Centre and IRENA. This cooperation strategy would work to: (i) support research on clean energy deployment requirements and innovative financing packages in developing countries. The BRI could also work via FOCAC and other regional forums to: (ii) establish regionally focused clean energy technology and investment packages, and deployment commitments. And: (iii) establish an international taskforce on financing Sustainable Development Goal 7 (access to affordable, reliable, sustainable

and modern energy for all) including deployment commitments (perhaps linked to the G20 energy principles)—this would further strengthen the support of clean energy technology packages in non-BRI countries (Fig. 44).

7.2.5   Risks and Obstacles
While the alignment of FDI standards and strategies will entail significant political capital and time, working initially with bilateral or small groupings of MDBs, such as the Asian Infrastructure Investment Bank and major policy banks, could represent an initial achievable step. There is also great political and commercial sensitivity around payments between and to governments, complicating efforts to increase the transparency of fossil fuel and linked infrastructure investments.

8   Security of Supply of Metals and Minerals for Future Energy Systems

The accelerating deployment of low-carbon energy technologies—in particular, electric vehicle (EV) batteries—is resulting in rapidly increasing demand for metals and minerals such as lithium, cobalt, polysilicon and rare earth elements. Goldman Sachs estimates that for every 1% increase in the market penetration of EVs, lithium carbonate equivalent demand will increase by around half the current annual demand.

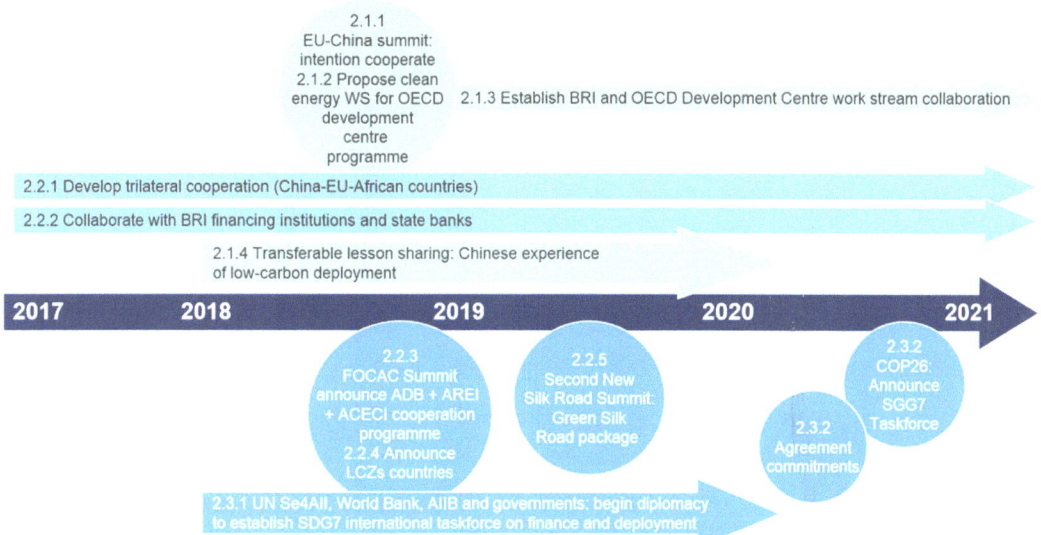

Fig. 44 Timeline of action plans under the strategic pathway: cooperation strategies with developing countries, MDBs and climate finance providers to increase access to clean energy technologies

Supply disruptions in metals and minerals markets have traditionally had less visible impacts on consumers than those in fossil fuel and food markets, for instance. As a result, they have received less attention from policymakers, and the regulation and data associated with these markets are generally considered national competencies. Compared to energy commodities, there is relatively little in the way of international governance and cooperation for metals and minerals markets. The market distortions of recent years—from anticompetitive practices such as export cartels to the imposition of unilateral export restrictions—have exposed many gaps in metals and minerals governance.

Box 19: WTO limitations in resolving disputes over metals and minerals and energy resources

While the WTO provides the main global institutional framework for coordinating and governing international trade, the trade in natural resources falls largely outside its remit. Separate chapters were included in the WTO texts to address issues related to agriculture and textiles, but not natural

resources. Some regional agreements have tried to bridge this gap for energy, notably the Energy Charter Treaty and the North American Free Trade Agreement's energy chapter. Several reasons for the special treatment of natural resources in trade agreements have been cited:

1. the geographical distribution of natural resources is highly uneven and production is largely immovable;
2. the sovereign control of resource-endowed countries over their natural resources and their right to exploit them for economic and social development is recognised in international law;
3. trade in natural resources is often perceived as a key national security issue;
4. the extraction and consumption of resources produce severe negative externalities compared with most other traded products; and
5. a large proportion of resource trade takes place under long-term contracts, often through dedicated infrastructure

such as pipelines or liquefied natural gas tankers.

The exhaustibility of natural resources, and the potential for infringements on national security, provide two potential justifications for trade restrictions under WTO rules, although these must be implemented in a non-discriminatory way.

In recent years, several key raw material suppliers have used export controls. Indonesia has introduced a ban on exports of unprocessed ores. Vietnam has also imposed restrictions on iron ore and copper, and export taxes have recently been debated in Brazil and India. In 2012, the USA, EU and Japan lodged complaints with the WTO over Chinese export quotas of rare earth minerals, and the quota system was withdrawn in late 2014. There have been various proposals for voluntary avoidance of export restrictions in times of crisis or extreme price volatility, which such restrictions tend to exacerbate.

With the shift from molecules to electrons, and the large-scale deployment of low-carbon energy technologies, new energy security concerns are emerging. First is the security of critical metals and minerals supply chains, and second is the availability of and access to new technologies in metals and minerals production, processing and recycling.

Many metals and minerals are important inputs to low-carbon energy technologies, but the extent to which they are critical depends on a variety of factors. Criticality is dependent on a commodity's importance to economic growth, combined with risks to supply shortages, and will vary between different consumer countries. Take the example of electric vehicles (EV); increasing EV uptake is of growing importance to China in terms of tackling urban air quality, but it is also critical to China's structural transition and its role as a leading country in high-tech industries, including EV manufacturing.

With EV sales likely to surpass 2 million per year by 2020, there are growing concerns as to how markets and supply chains will respond to increased demand for EV-critical metals such as lithium, cobalt, nickel and manganese. As these materials are difficult to substitute, governments are keen to safeguard against supply shortages. While no physical scarcity currently exists, supply risks are associated with the concentration of resources within a handful of countries and regions.

8.1 Cooperation Opportunity

Ensuring the security of metals and minerals for scaling up clean energy technologies will mean engaging with a range of new supply chains and producers. As the largest producer, consumer and trader of metals and minerals, and the largest manufacturer of renewable energy components, China is uniquely placed to lead on the establishment of multilateral metals and minerals institutions and policy processes. It also has the most to gain from stable markets; for example, a 1% shift in the price of iron ore could have cost China $800 million in 2013.

Targeted outcome: Greater global coordination and cooperation on the supply and demand of critical metals and minerals help mitigate the risks of disruption to the deployment of clean energy technologies.

To achieve such an outcome, China could adopt two distinct strategic cooperation pathways:

(1) avoid supply-side shortages and trade disputes in the global supply chain of critical metals and minerals required for clean energy technology deployment; and
(2) minimise the need for additional extraction capacity by enabling countries to lower their dependency on the global supply of critical metals and minerals

8.2 Working with Key Partners to Achieve Cooperation

Enhancing the governance of metals and minerals supply chains will mean forging partnerships with key players, some of whom are established partners and others newer partners. On the supply chain side, the China Chamber of Commerce of Metals Minerals & Chemicals Importers & Exporters could work with the European Commission and United States Geological Survey, both of which already maintain data and information on reserves and production.

Dialogues between producers and consumers will also be vital, China could encourage G20 states to engage constructively with regional forums, such as the Community of Latin American and Caribbean States (CELAC) and the Forum on China-Africa Cooperation (FOCAC). On the demand side, circular economy-based partnerships and breakthrough resource coalitions on secondary materials could reduce primary demand. China could work with the Clean Energy Ministerial and Mission Innovation to forge such partnerships.

8.3 Rationale for Action

By ensuring that the development and deployment of low-carbon technologies are not affected by supply chain disruption and shortages of critical inputs, both strategic cooperation pathways help facilitate improved urban air quality, as well as supporting China's progress up the manufacturing value chain. Ultimately, if a large part of China's development strategy is concerned with positioning itself as an exporter of batteries and EVs, its economic security will be closely linked to the security of the necessary raw materials.

Enhanced market cooperation and transparency could reduce international concerns over China's stockpiling of metals. At the production level, adherence to high standards of transparency and environmental and societal risk management can help reduce the risk of underperforming investments, particularly where the

societal licence to operate is compromised by lack of investment in the host country, or labour and environmental violations.

From the perspective of China's partners, China's dominance in the production of rare earth elements makes it a critical partner in any efforts to reform the governance of metals and minerals markets. For China and its partners, fostering innovation and R&D into enhanced recovery, recycling and substitution helps diversify supply and increase substitution. Developing substitute technologies is aligned with China's domestic ambitions to enhance R&D capacity and develop new business models.

8.4 Strategic Pathways

(1) **Avoiding supply shortages and trade disputes**

Supply chain shortages and trade disputes could be avoided by China working with partners to develop: (i) improved information, data and price transparency across the supply chain, while also: (ii) supporting dialogues and developing win-win technology and investment packages with existing, new and emerging producers to support trade and prevent export controls. Resolution of trade disputes that do arise would require China to work with major economies to reaffirm their commitment to WTO rules and dispute resolution mechanisms, while working long term to address gaps in rules around the trade of raw metals and minerals (Fig. 45).

(2) **Reducing the demand for increased production**

Minimising the need for additional extraction capacity could be achieved by: (i) accelerated R&D, innovation and deployment of enhanced recovery and recycling of metals and minerals from in-use stocks, enabling the greater use of secondary materials. This could be led by a Clean Energy Ministerial breakthrough resources coalition, including industrial partners and national bodies. There should be greater focus on future plans for

Fig. 45 Timeline of action plans under the strategic pathway: the avoidance of supply-side shortages and trade disputes in the global supply chain of critical metals and minerals required for low-carbon technology deployment

accelerating such R&D. China's circular economy strategy could also help reduce its long-term import needs for these critical materials. Reducing reliance on critical metals and minerals could be furthered by forming unconventional alliances with regional neighbours, including the EU and Japan, to support: (ii) R&D and piloting of substitution technologies for those low-carbon energy technologies most dependent on critical metals and minerals (Fig. 46).

8.5 Strategic Direction of Travel

For many critical metals and minerals, the concentration of production within specific regions and countries, facilitates the possibility of this strategy being achieved with multiple bilateral dialogues. However, the growing number of countries deploying clean energy technologies at scale naturally expands the number of nations with an interest in the stable supply of critical metals and minerals. So, while the centre of gravity in these dialogues and market reform

Fig. 46 Timeline of action plans under the strategic pathway: minimising the need for additional extraction capacity by lowering the dependency of countries on critical metals and minerals

mechanisms sits with China, should China choose to pursue this strategy of utilising multi-lateral platforms and regional forums, it is likely to deliver greater win-win and optimal outcomes for all countries.

8.6 Risks and Obstacles

Data sharing on metals and minerals markets is politically challenging, not least because of its links to both national security and financial markets (where stockpiles and metal warehousing are concerned). However, if China chose to take a lead in this area, this could increase trust and demonstrate a commitment to greater transparency. Progress has been possible in other, arguably more sensitive, areas such as energy and food.

At the same time, relatively little governance architecture exists at present, so there is minimal risk of disrupting or undermining international norms. Relatively low prices take some of the heat out of metals and minerals markets and provide political space for governance reforms. China already has strong trade and investment links with Australia, Argentina, the Democratic Republic of the Congo and Chile—although these will require sensitive management as the geopolitics of individual players on the supply side shifts, and as international competition for access to resource contracts increases.

9 Electricity Reliability During the Shift from Molecules to Electrons

The pace at which solar photovoltaic, wind and electric vehicle battery pack prices keep falling is accelerating the fundamental shift of the energy transition—from molecules to electrons.

As early as 2014, onshore wind delivered electricity in Europe at $50 per megawatt-hour (MWh) without subsidies, compared to $45–140/MWh for fossil fuel power plants. The trend of renewables becoming increasingly cheaper than fossil fuel generation across jurisdictions is clear. In China, both solar and onshore wind's levelised cost of energy is likely to be cheaper than coal and gas by 2028 (both are already cheaper than gas power stations). In Germany, solar and onshore wind are already cheaper than coal and gas; in the USA they are likely to become cheaper as early as 2024. It is therefore no surprise that renewables have increased their share of global new capacity additions from 8.5% in 2002 to 42.6% in 2015.

One of the principal solutions for achieving security of supply or balanced electricity systems with a high penetration of renewables, while reducing curtailment, is the increased trade of electricity through power interconnectors. While high voltage cross-border interconnector capacity is expanding rapidly—growing by 81% in the 10 years up to 2015—and is likely to double by 2025, China, and Asia as a whole, has limited interconnection to its neighbours, relative to its renewable power generation. In Europe, around 23 MW of interconnection capacity exists per GWh of renewable generation, whereas China's interconnection to Russia is equivalent to 3 MW per GWh.[3] Based on announced and planned projects, this will rise to 9 MW per GWh. More interconnections in China would reduce wind (17%) and solar (10%) curtailment and stem the rising amount of hydropower dumped in Sichuan and Yunnan provinces—142 and 314 TWh respectively in 2014—a fivefold and sixfold increase on the previous year.

EV penetration rates are also set to increase dramatically. In 2016, EV sales grew by 55%, 20 times greater than internal combustion engine (ICE) vehicles. EV lithium-ion battery costs have collapsed by three-quarters over the past six years, with a further 30% reduction likely by the end of the decade as global manufacturing capacity looks set to increase sixfold by 2020.[4] The point at which the total cost of running an EV reaches parity with ICE vehicles is widely expected in the next decade or so. By 2040, Bloomberg New Energy Finance (BNEF) expects EVs to account for more than half of all new vehicle sales and a

---

[3]Chatham House calculations.

[4]Relative to 2016, based on company announcements.

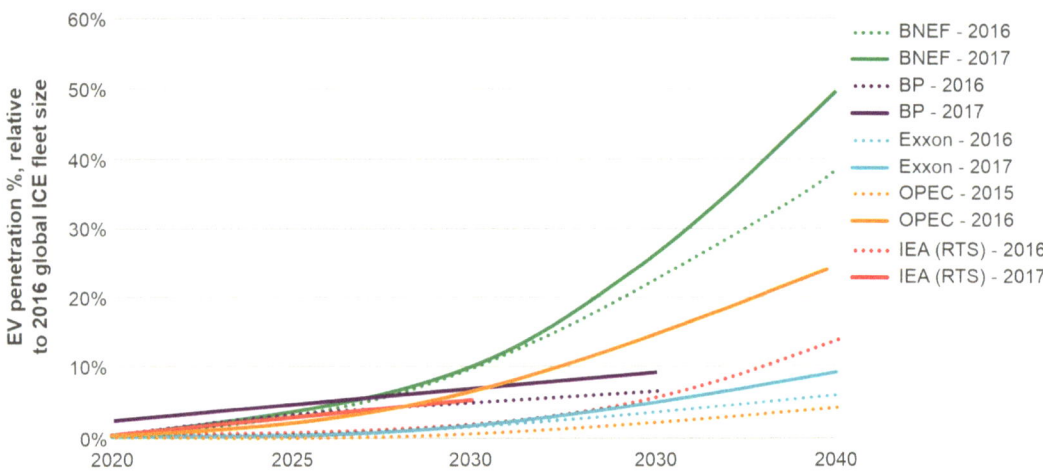

Fig. 47 Various forecasts of global EV fleet size as a percentage of 2016 ICE fleet size, and how those forecasts have increased over the past few years

third of all vehicles on the road. Goldman Sachs forecasts a global fleet of 83 million EVs by 2030. Even OPEC more than quadrupled its forecast between 2015 and 2016 to 266 million EVs globally by 2040 (Fig. 47).

Unless the expansion of EV charging infrastructure is staggered, the market penetration of EVs will reach limitations as inadequate grid reinforcement and increased peak demand will compound the difficulties of maintaining reliability of electricity supply. In the UK, the National Grid forecast of 9 million EVs by 2030 would increase electricity demand by around 8%. However, peak demand would increase by

around 13% if charging is not staggered. Smart staggered charging requires two-way communication, which isn't currently built into standard EV charging points (Fig. 48).

9.1 Cooperation Opportunity

As the largest renewables and EV battery manufacturer and generator of renewable power in the region, China could take the lead in cooperation and governance reforms to facilitate cross-border electricity integration. The BRI could provide a suitable forum for cooperation.

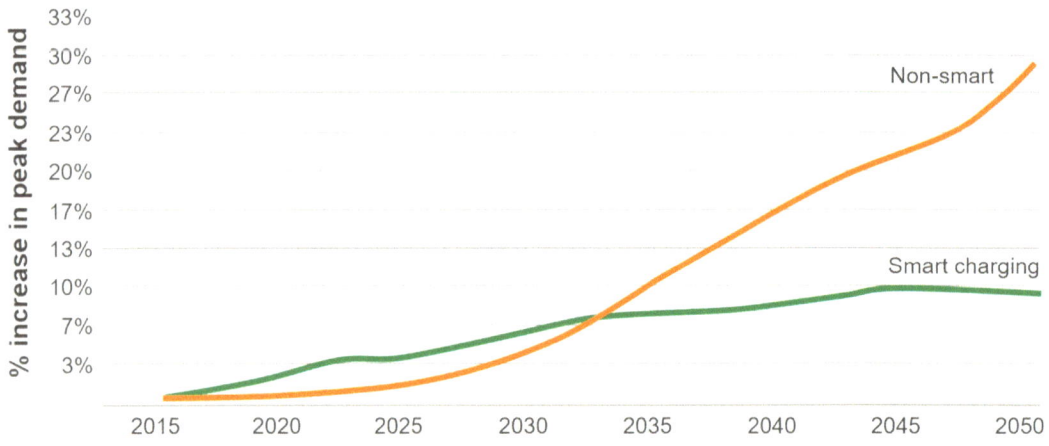

Fig. 48 eV peak electrical demand in the UK, depending on adoption of smart charging

Targeted outcome: Regional electricity systems capable of integrating high penetration of renewables, and the use of surplus power, while connecting millions of electric vehicles.

In order to achieve such an outcome, the complexity of connecting electricity systems across borders via interconnectors requires governments to take the lead in several areas. China and its regional partners require harmonised national electricity markets to trade electricity effectively through interconnectors. The avoidance of stranded assets and ensuring the future connection of high renewable supply with high demand regions requires regional planning to avoid lock-in and lock-out. Further, as electricity markets become increasingly interconnected the risks of cyberattack are best mitigated by strengthening regulations across jurisdictions.

Staggered smart EV charging, such that demand is spread over the evening when drivers return home to flatten and reduce peak demand, requires international standardisation of charging infrastructure. Scaling up the deployment of lithium-ion stationary battery storage requires alignment of investment strategies towards lithium-ion.

To achieve such an outcome China could adopt two distinct strategic cooperation pathways:

(1) regionally connected, coordinated and secure electricity markets enabling cross-border trade of electricity through interconnectors; and
(2) international cooperation to accelerate smart management of EV electricity demand and vast deployment of cost-effective electricity storage.

9.2 Strategic Regional Forums to Drive Reforms

The BRI could provide a platform for the regional planning of interconnectors, as well as working with international partners to learn and implement crucial lessons in the electricity market reforms required to connect electricity markets effectively. Working in partnership with

ASEAN Plus Three could enable greater participation and buy-in from regional partners. On EV charging and deployment of affordable storage, IRENA is well placed to host a China-led initiative to drive the agenda internationally.

9.3 Rationale for Action

While both strategic pathways facilitate the shift from molecules to electrons—improving air quality and maintaining reliability of electricity supply—further benefits, or win-wins, for China and its international partners can be identified.

China can benefit from opportunities to export surplus electricity and reduce electricity costs through grid interconnection and balancing: regional electricity trade saves consumers money and decreases capacity margin requirements for China and its partners. Cooperation on energy interdependence can build trust among partners, creating wider benefits.

Internationally standardised smart EV charging stations reduce all countries' EV deployment limits by lowering peak demand, while removing limits to interoperability between various EV and charging infrastructure manufacturers—enabling China to export more EVs. Further, China and its partners benefit mutually from the reduced need to strengthen grid capacity and the increased potential this allows for more renewables. Lower EV integration limits improve air quality in cities, which is especially important for China, while reduced EV import costs benefit China's partners.

As the global leader in lithium-ion, solar photovoltaic and wind turbine manufacturing, China will benefit from increased exports.

9.4 Strategic Pathways

(1) **Regional cross-border trade of electricity**
China could use the BRI to establish two multilateral clubs that would work in parallel to enable interconnector-facilitated cross-border trade of electricity. This may be more effective working in partnership with the ASEAN Plus Three forum. The first club

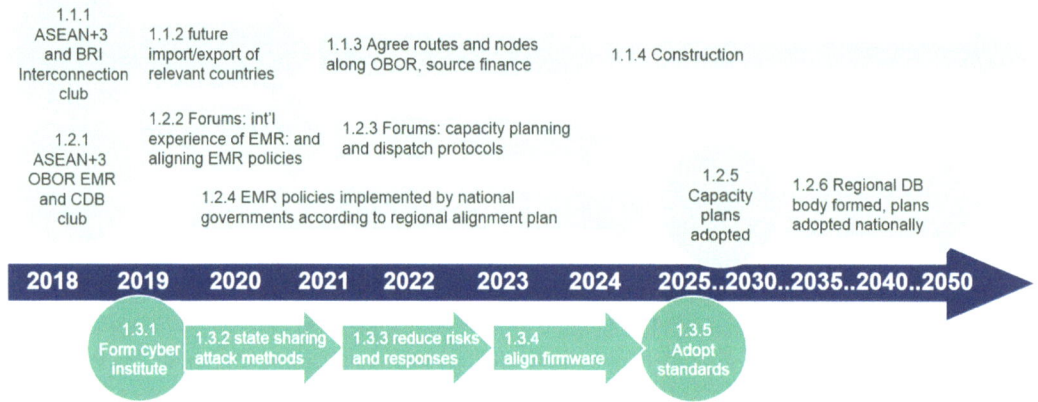

Fig. 49 Timeline of action plans under the strategic pathway: regional cross-border trade of electricity

would focus on the planning and construction of mutually beneficial interconnectors. The second club would focus on the alignment of regional national electricity market reforms (EMR), leading to the coordination of regional capacity, dispatch and balancing (CDB). China could simultaneously work with G20 states to form a grid cybersecurity multilateral to ensure the security of increasingly interconnected electricity systems (Fig. 49).

(2) **Smart EV charging and cost-effective electricity storage**

China could work in partnership with the International Renewable Energy Agency (IRENA) to achieve international coordination and standardisation of smart EV charging stations and the global alignment of electricity storage investment towards lithium-ion batteries (Fig. 50).

9.5 Risks and Obstacles

China's partners could be unwilling to lock themselves into dependency on Chinese electricity exports, particularly given local opposition to interconnector routes. Further, there may be opposition to the greater remit and power of ASEAN Plus Three and BRI in planning interconnector routes, compounding the fact that both currently lack the technical expertise. As such, the most politically feasible option would be to connect high demand and low supply regions within China.

Regarding electricity market reforms (EMR) and capacity, dispatch and balancing (CDB), regional differences in phases of EMR and political willingness to align reforms may prove difficult to overcome, especially given that national regulators currently lack enforcement capability and capacity. Given this, a stepped

Fig. 50 Timeline of action plans under the strategic pathway: smart EV charging and deployment of cost-effective electricity storage

approach to EMR could be adopted, such that the benefits of incremental reform are tangible to each partner country.

Cooperation on internationally coordinated grid cybersecurity requires a high degree of trust between states, as well as between states and corporations, and has proven elusive so far, though this is not a precondition for cooperation on grid interconnection.

Conclusions

This paper has explored in detail four cooperation options that China could pursue to help facilitate the shift to a low-carbon, secure energy system—a system based on quality, both on the supply and the demand side, rather than quantity. From this follows several observations:

1. International Cooperation is Critical to Achieving Chinese and Global Energy Objectives

The global challenge of enhancing energy security over the course of the next half century while decarbonising the energy sector will not be achieved through governments acting unilaterally in isolation to one another. Energy security—whether of oil and gas supply or the reliability of (increasingly interconnected) electricity grids—will be greatest when governments have shared rules for how markets are governed during normal operation and, critically, disruption. Such frameworks build trust, reduce system costs and mitigate the risk of market failures should disruptions occur. Similarly, the costs of decarbonisation can be minimised through cooperative actions and arrangements to align standards, avoid carbon leakage and extend carbon pricing, and maximise aggregate demand for low-carbon technologies. Nor can the two objectives—energy security and decarbonisation—be considered in isolation: each has implications for the other.

2. There Is No Silver Bullet Governance Reform or Cooperation Strategy

This paper has expanded on four opportunities for international cooperation. There are several others that could be explored. It is clear that intergovernmental cooperation and governance reforms will be needed in many different areas. A country can pursue international cooperation along a spectrum of interventions, ranging from simple pragmatic bilateral approaches to bolstering regional mechanisms and ambitious multilateral reform agendas with higher risks of failure but potentially larger returns. Different approaches will often be mutually reinforcing and form distinct parts of an overall strategy: the opportunities for cooperation considered in this paper include different mixes of bilateral, regional and multilateral elements.

3. The Prize for All Countries Is Substantial

A country's dominance in specific areas of low-carbon manufacturing, and its ambition to extend this to new green technologies, means it can benefit from emissions reduction efforts in other countries, which will increase demand for its exports. Establishing strong governance of critical metals for low-carbon manufacturing, on which China's export base will increasingly depend, is important for China's long-term economic security. As the world's largest energy consumer and importer, China has a strong interest in establishing effective governance for fuel security and grid interconnection. China's energy security will depend on successful global efforts to limit climate change, so supporting other countries' transition is likely to be a priority.

4. Win-Wins Are Available, but Will Require Compromise

The international community has a shared interest in collective energy security and low-cost decarbonisation, creating opportunities for win-win cooperation. However, political compromises will often be needed to move forward. For example, expanding IEA membership will require concessions from existing member countries, but also from China and other emerging economies, to improve data transparency and agree to emergency sharing rules.

5. G20: Challenge and Opportunity

The G20 offers a potential avenue through which China and partner countries can pursue reforms. Its membership has critical mass in key areas such as emissions, energy consumption and investment and R&D, as well as workstreams in relevant areas such as infrastructure, green finance, investment and fossil fuel subsidies. It also has a track record of reforming international organisations and establishing new ones. However, the G20 also presents significant challenges, not least a lack of institutional capacity and focus. In the short term, the largest challenges may be political, with the current US administration unlikely to support many of the reforms explored in this paper.

6. Belt and Road Initiative: A Platform for Cooperation

China's Belt and Road Initiative (BRI) provides various opportunities for cooperation with partner countries: on energy security, possibly through the Energy Charter Treaty (ECT); on low-carbon energy and infrastructure, in partnership with multilateral development banks; and on electricity network integration. China is capable of building formal governance structures associated with the BRI. But it will need to consider the opportunity cost of creating new governance arrangements in the BRI (instead of at the global level) and the appetite of BRI countries to agree new governance arrangements beyond the existing ones they participate in (e.g. ECT, IEA, ASEAN Plus Three).

7. China's Strategic Path

Ultimately, how China seeks to pursue enhanced cooperation is a strategic question for China. At one extreme, China could adopt a pragmatic, transactional approach based on bilateral deals with one country after another. At the other extreme, China might seek to pursue its objectives through multilateral institutions. There are pros and cons to each—for example, while individual bilateral deals may be easier to make, they might not provide global goods and could become too numerous to manage effectively. On the other hand, multilateral approaches may be better suited to providing global goods but can take a long time to negotiate or reform.

Plotting the different cooperation options considered above on the matrix presented in the introduction is revealing (see below) (Fig. 51).

While the options span a range of market- and policy-led approaches, very few are based on a transactional approach. Instead, the proposals cluster among regionalism and multilateralism, in the form of efforts to establish, strengthen and promote common rules and norms.

---

10   China's International Energy Cooperation and Strategies

China has become the world's second-largest economy, with sustained and rapid economic development and continuously growing national strength, since its reform and opening up began in 1978. With its growing economy, China's demand for energy is increasing, turning it into a country with high levels of energy production, consumption and trade, as well as an increasing dependence on foreign energy.

China is also becoming a leader in global energy technology deployment. China's development has had a significant and far-reaching impact on the global energy landscape. Influenced by world politics, the global economy and technological revolution, the international energy landscape displays new features and changes and brings new opportunities and challenges to different countries. It is difficult for any one country to address critical energy issues on its own. The question of how to reform global energy governance and guarantee every country's energy interests through international cooperation has become a major issue for China and other

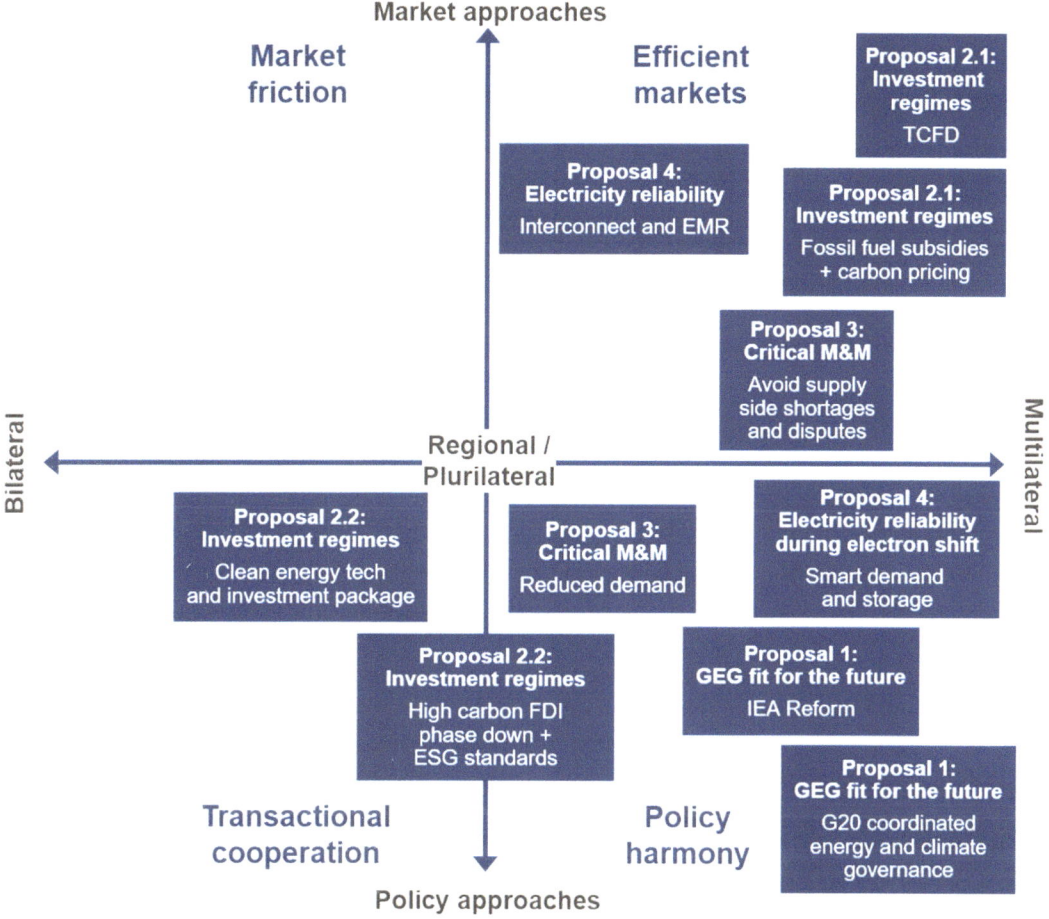

Fig. 51 Each international energy cooperation and governance opportunity placed within the spectrum of unilateral to multilateral and policy-led to market-led approaches

countries in this new era. How China builds international cooperation strategies and better participates in global energy governance is not only related to its own development but to that of the world as well.

10.1 China's Relations with Major Energy Trade Partners

As China's economic output increases, its demand for energy grows as well. International energy trade—an important way to meet energy needs—plays an increasingly important role globally. As China's dependence on foreign energy will not change in the short term, the importance of its energy imports far outweighs that of its exports. Energy trade will play a significant role for some time to come and be an important driver for domestic economic growth. Rising energy dependence means that the international energy landscape has more and more impact on China's economic growth and energy security. China's relations with its major energy trading partners, including Russia, the Middle East, Central Asia, the USA, ASEAN and Latin America, are therefore crucial.

10.1.1 Russia

Energy exports are an important industry that shore up the Russian economy. Before 2011, Russia's oil and gas exports to Asia-Pacific countries accounted for only 3% and 5% respectively of its total hydrocarbon exports. The US and European economic sanctions against Russia accelerated its energy cooperation with Asia-Pacific. With their agreement on joint energy cooperation, China and Russia established extensive cooperation in oil, natural gas, nuclear power, electric power, coal and other areas. Russian oil is mainly transported by sea, rail and pipeline to China. As China borders on Russia, road and pipeline transport are the most cost-effective and safest ways of trading oil and natural gas. The China-Russia Energy Cooperation Committee (formerly the Energy Negotiation Mechanism at Vice Premier Level) established a new high-level platform for China-Russia energy cooperation.

Energy cooperation between China and Russia focuses on oil trade. In 2015, China became the largest buyer of Russian oil. According to China's General Administration of Customs, in 2016 Russia replaced Saudi Arabia for the first time as China's largest supplier of crude oil. In 2016, China imported 381 million tonnes of crude oil, an increase of 13.57% over the previous year, and its imports from Russia rose by almost 25% to 1.05 million barrels per day. China-Russia energy investment and technology cooperation has been upgraded from direct trade to joint exploration, joint development and benefit-sharing, cooperation across the value chain, and from upstream to downstream areas. By way of new joint ventures and share acquisitions, China has been gradually participating in Russia's crude oil refining, refined oil product processing, liquefied natural gas processing, oil and gas sales and other areas. In the future, China will expand into fields like technology research and development, equipment manufacturing and engineering services.

10.1.2 Middle East

The Middle East comprises 17 countries, including Saudi Arabia, the United Arab Emirates, Iran, Iraq and Israel. It is an important hub between the west and the east and a point where the Silk Road Economic Belt and the 21st Century Maritime Silk Road meet.

China-Middle East energy cooperation is one of the keys to China's energy security. As China's economy and energy demand grow, energy cooperation will be one of the most strategic areas in China-Middle East relations.

China started to import petroleum and related products from the Middle East in the 1990s. Gradually, the Middle East became China's largest source of oil imports. From 1996 to 2009, China's crude oil imports from the Middle East accounted for 45-50% of China's total oil imports. In 2011, China imported about 130 million tonnes of crude oil from the Middle East, accounting for more than 50% of total imports that year. In 2013, imports reached 146.54 million tonnes or 52% of total imports. In 2014, crude oil imports from the Middle East increased to 171.7 million tonnes, which was 46% of the total.

Cooperation on oil and gas between China and the Middle East countries has made remarkable progress. CNPC, Sinopec and CNOOC have developed many large oil field projects in Iran and Iraq. In particular, China-Iran oil and gas cooperation has great potential in dozens of large oil and gas fields, and in auxiliary project construction, engineering and technical services. In addition, China also cooperates with Middle East countries on training in clean energy.

The Belt and Road Initiative offers valuable opportunities for cooperation between China and the Middle East. Responding to the initiative and strengthening the cooperation with China will help Middle East countries develop their domestic economies. In addition, the region's petroleum exporting countries depend more on the Chinese market during low oil price cycles, which makes cooperation even more attractive.

10.1.3 Central Asia

The five countries of central Asia (Kazakhstan, Turkmenistan, Uzbekistan, Kyrgyzstan and Tajikistan) are a hub for the Silk Road Economic

Belt. They have been an important economic corridor linking China with Western Asia and Europe since ancient times.

Central Asia has abundant oil and gas reserves, ranking third in the world after the Middle East and Russia. More than 70% of its crude oil output is exported. In 2015, Kazakhstan produced 85.6 million tonnes of crude oil, of which more than 70 million tonnes went to other countries; Turkmenistan produced 74.16 billion cubic meters of natural gas and exported more than 50 billion cubic meters. Since 2000, China and Central Asia have been deepening their cooperation in oil and gas trading. Now, China is Kazakhstan's third-largest oil importer, and Uzbekistan's and Turkmenistan's largest gas importer. At the end of the 12th Five-Year Plan (2011–15), China's oil and gas imports from Central Asia accounted for 1.85% and 48% respectively of its total oil and gas imports. Thanks to Central Asia, China's overseas energy supply structure has been optimised. China's main imports from the Central Asian countries are primary energy products. In 2014, for example, China's oil and gas imports from Central Asia accounted for 34% and 36% respectively of its total imports from this region. At the same time, China exports to Central Asia processed energy products and manufactured goods, such as oil country tubular goods, petroleum machinery and chemicals, which made up 4.91%, 5.15% and 3.52% respectively of China's total exports to Central Asia. China has built power transmission links and substations, thermal power plants and other projects for Kyrgyzstan, Uzbekistan and other Central Asian countries.

In addition to energy trade, China's cooperation with Central Asia's energy industry is also developing from upstream energy exploration and production to cooperation along the entire value chain. Future cooperation will prioritise downstream industries, such as petroleum refining and petrochemicals. In the future, it will be further extended to energy-related industries like petroleum equipment manufacturing and production services.

The long-term energy cooperation between China and the Central Asian countries is based on mutual trust. The Belt and Road Initiative provides a new platform and opportunities for diversification. The Central Asian countries can expand into and build energy export markets in North-east Asia, South Asia and the Pacific Rim countries.

10.1.4 USA

The USA and China are the world's largest energy producers and consumers. The two countries regarded energy as one of the important cooperation areas in 1979 when they established diplomatic relations. Over the past 40 years, despite differences and obstacles, China and the USA have been expanding and deepening energy cooperation and made great progress. Especially in the 21st century, the USA's energy independence strategy offers an unprecedented opportunity for China-US energy cooperation against the background of energy and economic globalisation. The two countries have enormous potential for cooperation in areas like natural gas, nuclear power and renewable energy.

Bilateral and multilateral energy cooperation mechanisms—like the APEC Energy Working Group, the Asia-Pacific Partnership on Clean Development and Climate, and the Carbon Sequestration Leadership Forum—strengthen China-US cooperation in the energy sector. From 2004, China-US energy policy dialogue became an important communication channel for the two governments and a platform for the two countries' energy sectors to discuss topics like energy trends, energy efficiency, resource and environmental protection. The U.S.-China Oil and Gas Industry Forum—co-hosted by China's National Energy Administration, the U.S. Department of Energy and the U.S. Department of Commerce—has been actively promoting the two countries' cooperation in oil and gas.

In 2008, the U.S.-China Ten-Year Framework for Cooperation on Energy and Environment was signed during the fourth China-U.S. Strategic Economic Dialogue, making energy cooperation the focus of the conference for the first time.

10.1.5 South Asia and ASEAN

South Asia covers eight countries, including India, Pakistan, Bangladesh and Sri Lanka. Five of the eight countries share a border with China.

Bangladesh, Sri Lanka and the Maldives are on the Maritime Silk Road. South East Asia, which is part of the BRI's China–Indochina Peninsula Corridor, is a strategic channel between China and the Indian Ocean.

China has obvious advantages in the field of energy development and equipment manufacturing, as well as strong technical strength in terms of coal exploitation and renewable energy, including hydropower. Energy cooperation between China and ASEAN countries has a good foundation. At present, China Southern Power Grid (Yunnan and Guangxi) is interconnected with the power grids in Laos, Myanmar and Vietnam, and industries like power transmission, power exchange and transnational meter reading have been developed. Power trading is beginning to take shape. China Southern Power Grid started to transmit power to Laos in 2009, and Thailand purchases electricity from Yunnan via Laos. In October 2008, six 100 MW power generating units at Shweli River Hydropower Station, Myanmar's largest hydropower build-operate-transfer (BOT) project, were integrated with China Southern Power Grid and started to transfer power to China. Electricity has gradually become a new growth point in China's trade with ASEAN countries.

Cooperation in power investment between China and ASEAN countries continues to evolve. In recent years, companies like State Power Investment Corporation, China Huaneng Group, China Guodian Corporation, China Huadian Corporation, China Datang Corporation and China Three Gorges Corporation work with ASEAN countries to develop power resources. For example, the companies have built hydropower stations and coal-fired power plants on a BOT basis in Laos and Vietnam; and State Grid Corporation of China jointly operates the Philippines' power transmission network.

In 2015, China's power companies invested $317 million in the 10 ASEAN countries in projects of more than $30 million, and $4.63 billion in engineering, procurement and construction (EPC) projects of more than $100 million. Overall, China and the ASEAN countries have a large trade volume of energy products and energy equipment. Energy products mainly include crude oil, natural gas, coal, coke and refined petroleum products; energy equipment is both imported and exported, including complete equipment packages for thermal power plants and hydropower stations, equipment for nuclear power plants, and new energy generation equipment for wind and solar power. The China-Myanmar oil and gas pipelines interconnect China with ASEAN. They act as an energy channel for China to the Indian Ocean and play an important role in diversifying China's oil and gas imports. In addition, China cooperates with Thailand on marsh gas, solar photovoltaic and biomass power.

10.1.6 Latin America

Oil and natural resources in Latin America are abundant but underexplored and underdeveloped. This brings opportunities for energy cooperation and labour export between China and Latin America.

Latin America is one of the first destinations of China's Go Out strategy. However, resource nationalisation in Latin America and geopolitical implications from the USA have hindered energy cooperation between China and Latin America. Rising international oil prices make Latin America more eager to shake off US control and strengthen cooperation with China in finance, technology, infrastructure and other sectors.

At present, China's cooperation with Venezuela, Brazil, Peru, Ecuador and other Latin American countries focuses mainly on energy trade, oil and gas, engineering and technical services, and labour export. It covers fields like exploration, exploitation and refining of oil and gas resources, and pipeline construction. Latin American countries need a great deal of investment in oil and gas exploration and development and engineering and technical services. In the future, China and Latin America will deepen their cooperation in those fields and in new energy as well.

10.2 Problems and Challenges Facing China's Future International Energy Cooperation

International energy cooperation is an important means to improve energy security and promote sustainable energy development in all countries. Energy cooperation enables a single country to reap benefits that it cannot acquire alone. It is therefore important that China and other countries cooperate to safeguard their energy interests through multilateral partnerships and multi-level international cooperation mechanisms.

China's international energy cooperation is faced with a series of real and potential problems and challenges.

10.2.1 Problems

First, most of China's international energy cooperation is through dialogue or general cooperation. China mainly joins organisations that are established for coordination or dialogue, but rarely joins alliances or organisations subject to laws or regulations. Due to its membership in only a few international energy cooperation organisations, China has a low level of international cooperation. At the regional level, China participated in more cooperation organisations with legally binding agreements. In the Asia-Pacific region, although China plays an important role in international organisations (such as APEC, Shanghai Cooperation Organisation and ASEAN), energy cooperation in this region, especially in East Asia (including South East Asia and North East Asia), lacks the legal and institutional framework and relevant energy cooperation organisations.

Second, China's overseas energy investment is still at an early stage. In oil, China is gradually developing from a small- to a large-scale investor and is broadening the scope of its investments. These include oil field production management and technical services, oil field development, engineering and construction, risk exploration, oil trade, equipment export, and oil and gas asset acquisition.

Third, China does not have pricing power in the international energy market. Lacking control and influence, China has been disadvantaged in international energy competition for a long time, so it can only passively accept the implications of rising international oil prices. In terms of pricing, China has no perfect oil futures market; it has yet to establish a pricing mechanism for imported oil based on procurement risk and, therefore, does not have the muscle to influence international oil prices. If the price of oil rises, China has to pay a premium that western countries do not. As a result of the high cost of importing crude oil, China's oil products are less competitive and its oil product companies less profitable, but household consumption expenditure keeps rising.

Fourth, China faces a complex political environment in energy diplomacy (the political risks of energy security are even greater than commercial risks). China's huge oil demand and heavy dependence on imported oil put great pressure on the international oil market. China's development of oil resources in the South China Sea and the East China Sea has led to regional conflicts over resources. China's energy diplomacy, which is directed at strategic interests and traditional geopolitics, will also impact the international order led by the USA.

In addition, China needs to address several issues related to international energy cooperation, including long-term energy cooperation strategies, establishing high-level coordination mechanisms and risk guarantee mechanisms for international cooperation, supportive financial and fiscal taxation policies, and technical support and basic research work.

10.2.2 Challenges

The challenges that China faces are either internal or external.

Internally, China's main challenge is insufficient international energy cooperation. First, Chinese energy companies are usually good at managing domestic projects but lack international management experience. Unlike the domestic market, international projects have a high level of political, economic, legal and social uncertainty. A major challenge faced by Chinese energy companies is how to integrate, optimise and use their international assets efficiently to obtain a good return, even when they lack knowledge and experience. In terms of design and profitability, Chinese energy companies have a large gap to bridge compared to their large international competitors.

Moreover, the energy acquired through direct economic cooperation is on a small scale. Direct investment in infrastructure and manufacturing is massive but insufficient in other economic fields of cooperation. Due to its lack of trading experience in international energy commodities and the absence of a fully established energy reserve system, China is incapable of effectively regulating international oil prices, and is therefore continuously exposed to fluctuations in international oil prices.

China also suffers from a lack of internationally experienced talent. The ideas, knowledge, experience and capability that come with international experience are needed to support international business operations and sharpen competitiveness. China should develop a corps of internationally experienced talent, equipped with the requisite global outlook and business capabilities to enable its energy companies to implement the Go Out strategy.

The main external challenge is conflicting approaches to cooperation. Countries differ in how they resolve conflicts and facilitate dialogue and cooperation in energy. Binding, authoritative and relevant international regimes could lower countries' transaction costs and facilitate the formation of international cooperation, according to new liberal institutionalists. As for energy security, energy shortage will promote international cooperation on energy governance, rather than lead to energy conflicts, thereby helping to eliminate such conflicts. In reality, the approach of most countries to energy cooperation combines the two ideas above or falls between one or the other. The landscape of multilateral energy cooperation is based on different answers to different problems. However, the perception that the West is drifting between the two ideas will hinder China's participation in energy governance to a large extent.

The second external challenge is the different conceptions of energy security held by different countries. Energy experts can be characterised as arguing that energy security can be ensured by markets or geostrategy. Markets convert energy from a strategic commodity into ordinary goods. Geostrategy views access to energy as a strategic issue. Experts who hold the former view consider China's dependence on imported energy will help it fit in with the global energy market system. Those who take the latter view, argue that China's efforts to reduce its dependence on energy markets will lead to conflict and threaten the stability of the international energy market. One expert pointed out that the USA classifies China's energy security strategies as part of a "strategic" paradigm and defines its own energy security policies as "market-oriented". Other experts indicated that US energy security policies are based on "hegemonic stability theory", while China's energy security strategies are based on "peaceful development theory". These different conceptions of energy security are the root cause of distrust and, as a result, poor energy cooperation between nations.

The third external challenge is the conflict of interest between participants in global energy governance, which has a negative impact. For example, long-term China-US structural conflict makes China doubt the reliability of the International Energy Agency, which is dominated by the USA. Because of differences between China and Russia on energy cooperation in Central Asia, the Shanghai Cooperation Organisation Energy Club faced a bumpy start. Historical hostility and issues between China and Japan impede energy cooperation in North-east Asia. Several countries politicise energy issues or dominate other countries' diplomacy in international dialogues on energy.

The fourth external challenge is the pressure of international competition. Although progress has been made in overseas business, Chinese energy companies are still far behind the oil majors in terms of capital operations, human resources, technology and equipment, bidding capability for international projects, and business experience. Large countries rich in oil and gas resources usually have a mature international energy development market. Developed economies, with strong international business capabilities and immense power, dominate the market. They enjoy a relatively monopolistic position and form strategic partnerships, which make it difficult for Chinese companies to access these markets.

The fifth is the harm caused by the China threat theory. Hidden behind energy resources are complex political relationships between countries. As international energy issues became politicised, the China energy threat theory started to spread among western countries in the early 21st century. These countries argued that China's growing energy demand would have serious impacts on geopolitics and international energy security. Even worse, this distortion is combined with the theory that China is an economic and neo-colonialist threat in many energy-producing countries in Central Asia, West Asia, Africa and Latin America. Because western countries maliciously hype up the theory, and some countries blindly believe them, the theory seriously impacts China's overseas energy strategy and its cooperation with other countries.

The sixth challenge are the risks to China's cooperation with resource countries. These are typically political, security or legal risks, or weak infrastructure. For instance, the cooperation between China and resource countries in transnational oil and gas development faces many risks and challenges. Countries with abundant oil and gas resources in Asia, Africa and Latin America are important energy cooperation partners for China. But many of them, especially those in the Middle East and North Africa, wrestle with security issues, long-standing and more recent disputes, and sociopolitical instability. This makes international cooperation very demanding. China's economic and trade cooperation in energy with these countries is confronted by enormous risk.

The seventh external challenge is that China's energy consumption and import structures are becoming similar to those of neighbouring countries, leading to fierce competition with Japan, India, South Korea and other countries in the international energy market. Historical disputes and political obstacles impede energy security cooperation among the East Asia countries, which makes the current regional cooperation mechanism difficult to improve.

10.3 China's National Interest in Energy

Energy: The 13th Five-Year Development Plan for Energy (2016-20) states that China will seize the great opportunities of the Belt and Road Initiative (BRI) to promote the interconnection of energy infrastructure across BRI countries and regions. China will implement the strategy of opening up energy markets, increase international cooperation on production capacity and actively participate in global energy governance.

Electricity: As stated in the 13th Five-Year Development Plan for the Power Industry, China will push forward the construction of cross-border power transmission and interconnected power grids, encourage businesses to take part in overseas power projects, and promote international cooperation on electric power technologies, equipment, standards, grid upgrades and engineering services.

Coal: According to the 13th Five-Year Development Plan for the Coal Industry, China will drive international coal trade and expand overseas engineering, procurement and construction projects and technical services to strengthen the international competitiveness of China's coal industry. China will also enhance coal production cooperation within the international community to promote the extraction and use of overseas coal resources. China will invest in upstream and downstream assets and build associated infrastructure to bring win-win results.

Oil: As stated in the 13th Five-Year Development Plan for the Oil Industry, China will improve international oil cooperation by exploring different ways of collaboration, increasing overseas investment and optimising the investment structure. China will also interconnect oil infrastructure in countries within the BRI.

Natural gas: As stated in the 13th Five-Year Development Plan for Natural Gas, China will deepen cooperation with gas producers by diversifying gas supply, safeguarding cross-border pipelines to ensure supply security, enhancing cooperation with gas consumers, accelerating the formation of regional gas markets and building more pricing power in those markets.

Renewable energy: According to the 13th Five-Year Development Plan for Renewable Energy, China will actively promote global energy transition and drive the formation of a global renewable energy value chain to sharpen the international competitiveness of the renewable energy industry. China will also promote high-profile cooperation projects in the BRI countries to help consulting, contracting, equipment manufacturing and other businesses in the sector to expand their overseas markets.

Box 20: Climate change risk assessment by the Chinese government

A risk assessment of climate change includes: (i) future global greenhouse gas emission pathways; (ii) direct risks to the climate from global greenhouse gas emissions, and; (iii) risks caused by the interaction between climate and human beings.

1. Future global greenhouse gas emission pathways

Current policies of the major countries and regions follow the medium to high emissions pathway, i.e. greenhouse gas emissions will continue to rise in the coming decades, then stabilise before decreasing gradually. The low emissions pathway calls for intensive technical innovation in wind power, nuclear energy, solar energy storage, biofuels, carbon storage, large-scale energy conservation and other energy technologies; otherwise, it will be hard to achieve near-zero emissions by the end of this century.

Exploitation of coal reserves, oil shale and methane hydrate will increase the likelihood of countries taking the high emissions pathway. Emissions impact the climate, so all pathways except near-zero emissions will increase risk over time.

2. Direct risks to the climate from global greenhouse gas emissions

The risk of climate change is nonlinear and may intensify at any time. Many factors may lead to sudden or irreversible change, causing high uncertainty.

Any emissions pathway can cause a global rise in temperature. The ability of human beings to endure heat stress is limited. Current weather conditions exceed the work safety threshold in some tropical countries. Future weather conditions could go beyond the fatal heat threshold. Crop tolerance of high temperatures is also limited. Crop yields will decrease substantially in future if weather conditions exceed the critical threshold. Therefore, climate change will bring enormous risks to global food security.

Water resource pressure intensifies with population increases. In some regions, climate change could increase the risk of acute water shortage. The probability of extreme drought on farmland will increase significantly. Although climate change mitigates water shortage in South and East Asia, it increases the risk of flooding, especially in the high emissions pathway. Frequent flooding will result in greater loss of life, crops and assets, and the need for extensive flood control at higher cost.

Climate models show that due to the inertia of the climate system, when the global temperature rises by 2°C, global sea level will rise by 10–15 meters with the melting of the ice cap. But when this will happen is highly uncertain.

3. Risks caused by the interaction between climate and human beings

The risks of climate change are systemic. The interaction between climate and complex human systems can lead to high levels of risk. The frequent occurrence of extreme weather events has a huge impact on global food security which, combined with policies and market reaction, will result in unprecedented hikes in the global price of grain.

Climate change also increases the probability of extreme events. For example, during the 2007–11 drought in Syria, the interaction between climate change and grain export restrictions, resource pressure and poor national governance led to turmoil.

Extreme water and land shortages can lead to regional conflicts and large-scale migration, which increases the risk of state failure and can even threaten the stability of developed countries.

10.4 China's Strategic Vision for International Energy Cooperation

To establish a stable, effective and sustainable energy supply system, China must strengthen international energy cooperation to deliver a win-win situation. Strategically, China should deepen international energy cooperation in five areas: multi-level international energy cooperation partners, multi-channel international energy cooperation methods, diversified forms of international energy cooperation, multi-area international energy cooperation content, and multi-task international energy cooperation processes (Fig. 52).

10.4.1 Multi-level International Energy Cooperation Partners

Participants in international energy activities include not only energy exporters, importers and transit countries, but also international energy organisations and international energy companies. Therefore, in international energy cooperation, China needs to have multi-level partners.

In addition to strengthening cooperation with governments, China should also work with supranational, intergovernmental and international non-governmental organisations, as well as with international energy companies. Currently, Russia, the Arab states of the Persian Gulf, and countries in Latin America and North Africa are the major energy exporters. Because energy exporters are usually dominant in international energy cooperation, China needs to strengthen its cooperation with them and develop collaborative energy projects through mergers and acquisitions, equity participation and bidding. Currently, the USA, Western Europe and Japan are the major energy importers. There is potential for both competition and cooperation between China and energy importers, so it is important to strengthen communication and cooperation with large oil consumers (such as the USA, Japan, India and others) and reduce or avoid friction. Besides, China needs to enhance its cooperation with international oil exporting and consumption organisations, in particular the International Energy Agency (IEA) and the Organization of the Petroleum Exporting Countries (OPEC), and actively participate in the Energy Charter Conference, the International Energy Forum and United Nations energy conferences.

China also needs to develop international energy cooperation in crude oil with large international energy companies. It should do this by integrating oil resources and related channels in international energy markets and by establishing strategic alliances with these companies and national oil companies to gain more access to resources and projects.

Fig. 52 China's strategic
vision of future international
energy cooperation

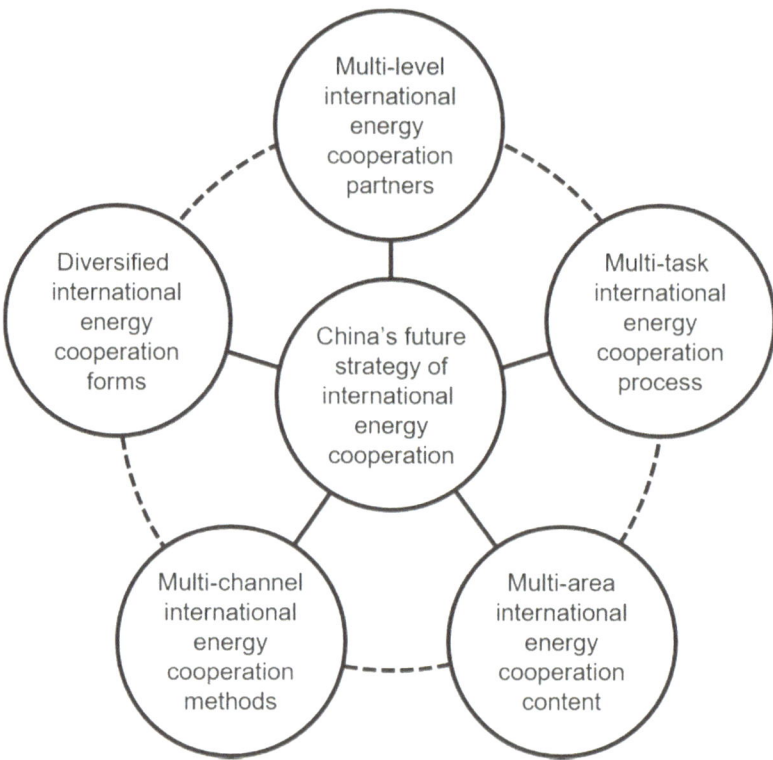

10.4.2 Multi-channel International Energy Cooperation Methods

China should adopt multi-channel methods in international energy cooperation, including multilateral, regional and bilateral approaches.

Multilateral energy cooperation is the most important method in international energy collaboration. To raise its influence and make its voice heard, China must participate extensively in international energy organisations, strengthen multilateral energy cooperation and security, bolster the legal and regulatory framework on multilateral energy cooperation, protect its own energy security and maximise its benefits. It is also important that China promotes extensive global dialogue with energy exporters, consumers and transit countries, and establishes and improves the international multilateral energy diplomacy mechanisms, thus achieving a balance in global energy interests and a stable energy market.

China should strengthen regional energy cooperation and promote regional integration of energy markets. Asia is an important energy supplier and a fast-growing market in energy consumption, so it plays an extremely important role in international energy supply and demand. China, as a member of Asia-Pacific Economic Cooperation (APEC), can actively promote energy cooperation in Asia. China should help form organisations for energy cooperation in East Asia, actively coordinate energy policies with East Asia countries, include oil and gas development in the Russian Far East in the East Asia energy cooperation framework to maintain energy security in the region, and support common interests. China should promote energy cooperation under the framework of the Shanghai Cooperation Organisation and work for the energy security and sustainable economic development of its members.

China should develop bilateral energy cooperation with other countries and establish good bilateral energy relationships to lay a strategic foundation for international energy cooperation. China should develop a long-term energy cooperation programme to strengthen efforts in energy diplomacy and improve energy cooperation mechanisms and foreign economic policies. Diversified external supply of oil and gas is a requirement of China's energy strategy, so China needs to develop bilateral energy relationships with the main oil producers in the Persian Gulf, North Africa, the Black Sea and Caspian Sea region, and the Mediterranean. China should enhance energy dialogue and participate in energy development with these regions, make them more interdependent with China, improve economic and trade ties, and secure oil delivery to the Chinese market with free trade agreements.

10.4.3 Diversified Forms of International Energy Cooperation

China's international energy cooperation should be implemented through different forms of international cooperation, including energy trade, agreements and investment.

In international energy trade, China should continue to actively import energy; compete extensively in the oil-rich regions; develop energy cooperation relations in Latin America, West Africa and South East Asia to diversify oil import supply; open multiple import channels that are stable, of high quality and low prices; and enhance its energy security with more international supply sources. While improving energy transport by rail, China needs to expand into new modes of energy delivery (especially pipelines), open up new seaways, enhance international cooperation to maintain maritime security, and actively explore new routes or types of transport to avoid choke points. In this way, China will diversify its delivery routes for overseas oil and maintain a leading role in maritime transport. China needs to actively participate in the international oil futures market and use its position as a big importer to curb price fluctuations and impacts, strengthen the hand of Chinese companies in external negotiations, reduce the cost of imports, strive for more control over international oil prices, and raise China's status in the international oil market.

International energy cooperation agreements involve technology and services, exploration and exploitation, environmental protection and new energy development. China should actively participate in international energy cooperation agreements to drive wide-ranging and deep collaborative relationships with other countries through mining taxation agreements, production sharing agreements, risk services, joint ventures, and so on. China should explore specific energy cooperation projects and sign agreements on oil and gas exploration with other governments. These agreements should include trade and investment, joint projects and labour export services, and upstream and downstream activities. Chinese companies can use their human and technological advantages to explore and develop oil resources in oil-producing countries and provide services in the form of engineering, procurement and construction, and earn a share of oil in proportion to their (the companies') input.

In terms of cooperation in international energy investment, China should encourage domestic companies to open overseas energy markets and increase China's share of international energy through direct investment. China should protect domestic companies in overseas energy markets by signing bilateral investment agreements with energy producers and by establishing overseas risk-taking exploration funds. China should accelerate reform of the overseas investment approval system, improve overseas investment laws and regulations, and effectively guarantee companies' rights in overseas investment and operations.

10.4.4 Multi-area International Energy Cooperation Content

International energy cooperation covers exploration, development, transport, processing and reserves of oil and gas, as well as environmental protection.

China can develop extensive collaborative relations with other countries in energy use, environmental protection, energy saving, management systems, laws and regulations. China's strategy of sustainable energy requires it to become more efficient in energy development and use, secure energy supply and protect the environment. China's dependence and impact on the international energy market opens great opportunities for its international energy and environmental industries. China's enormous demand for energy requires it to adopt comprehensive and diversified energy supply strategies, enhance international cooperation in the development and deployment of alternative energy technologies, and scale up the deployment of renewable and new energy. China needs to strengthen international energy information exchange and data sharing and increase the transparency and stability of international energy markets. It also needs to improve international energy cooperation by establishing a multilevel international dialogue mechanism covering the government, business community, academia and non-governmental organisations.

10.4.5 Multi-task International Energy Cooperation Processes

First, build a new international energy governance platform. Within the framework of the Belt and Road Initiative (BRI), China will establish new regional energy governance organisations, expand cooperation with existing global energy governance organisations, improve the dialogue mechanisms, promote the reform of multilateral energy laws and regulations, create energy trade rules and standards, and establish new regional energy financial systems.

Second, enable the interconnection of infrastructure networks. China will strengthen the construction of Asia-Europe infrastructure, build oil and gas pipelines in Asia, develop new combined sea and road transport routes, and construct and improve cross-border power transmission systems to create all-round, multilevel, sophisticated and interconnected energy networks.

Third, enhance international oil and gas cooperation. China will deepen industry-wide cooperation and seek the joint development of oil and gas resources, promote the diversification of

overseas oil and gas resources and channels, establish world-class transnational oil and gas companies and international petroleum trading centres, and deeply involve itself in the global oil and gas pricing system.

Fourth, strengthen cooperation in clean energy and energy efficiency technologies, and lead the regional energy transition. China will increase the export of energy-efficient technologies and make more effort to strengthen cooperation in energy technology innovation and cooperation mechanisms to promote regional renewable energy development.

Fifth, drive equipment manufacturers and service companies to adopt China's Go Out policy. China will enhance international cooperation in energy equipment manufacturing, services and engineering; actively participate in equipment bidding activities for exploration and exploitation projects of shale gas, coalbed methane and offshore oil and gas in BRI countries; and encourage renewable energy equipment manufacturers of solar photovoltaic panels and wind turbines to Go Out to overseas markets. China will also encourage domestic energy think tanks to participate in the activities, research, consulting and assessment services of the International Energy Agency, and provide intellectual support to deepen international energy cooperation.

In summary, China should become an active participant in international energy cooperation and help strike a balance in international energy relations through cooperation. The Chinese government should regard the advantages of energy cooperation as an important prop to economic development and its geostrategic interests. The government should proactively carry out energy diplomacy and adopt flexible policies based on China's national interests. It should develop extensive relations with all types of international energy participants to ensure China's energy security and maintain a stable international energy market, thus facilitating sustainable economic development in China and globally.

11 China's Role and Responsibilities in Global Energy Governance

11.1 China's Role in Global Governance Reform Is Increasingly Important

From an outsider to an insider, from a follower to an influencer: could China promote and lead the reform of global energy governance?

11.1.1 China Has Become a Country with the Strongest Global Economic and Energy Influence—This Influence Will Last

(1) **Economy: China has become the world's second largest economy and will remain open to maintain fast growth. Its global influence is far-reaching**

Despite China undergoing structural adjustment and entering a new economic phase of slower growth, the scale, speed and quality of its economic rise remain the highest in the world. Eventually, China may overtake the USA in terms of the size of its economy, which means the relations between China and India may need to be taken into long-term consideration.

New changes are occurring in China's foreign economic relations: the first is the change in scale from being a small trading country to a large emerging economy; and the second is the structural change resulting from China's comparative advantages.

China is going to implement new strategies of openness, namely those of strengthening its emerging power, improving its international competitiveness and building mutually beneficial win-win relations with the international community. New benefits in international competition and new patterns in China's openness need effective global governance to safeguard financial security, resources and energy.

(2) **Energy: China has become the world's most influential country in energy and its influence is increasing**

In terms of energy supply and demand, China is the centre of the global energy industry, as it is the world's largest energy consumer. China will continue to be the largest growing energy market and is already transitioning to low-carbon energy.

(3) **Trends: Forecasting global economic and energy trends is important for China, enabling it to expand its influence in international energy**

The world economy is recovering, but uncertainty remains. In contrast, China's economic growth and energy development are largely ensured. The reconstruction of the global energy governance system is accelerating and the global energy landscape is adjusting. As the interests of energy producers and consumers differ, the focus of competition shifts to winning control over energy pricing, currency settlement, and transition and reform.

(4) **Inclusiveness: China's participation in global energy governance is essential to improve inclusiveness and universality**

The existing global energy governance system does not meet the requirements of emerging economies, including China. These countries want to take part in the governance system and address global energy supply and demand. The main defects of the current governance framework are: participants are not well represented; an effective dialogue mechanism between energy suppliers and consumers is not in place; the system does not address the diversity of energy supply, and; it does not meet multiple objectives, such as combating climate change or promoting technology transfer.

Global energy governance is evolving into a new system that is inclusive, multi-polar and diversified, and China's participation is essential to the universality of this system. Governance reform needs to match future requirements. The centre of energy trade is moving to the Asia-Pacific region. Climate change should be taken into consideration and electricity should be a long-term focus area.

China's participation is beneficial to the reform of global energy governance, especially to improving the system's inclusiveness and universality. Deepening the cooperation between traditional governance institutions and developing countries is in the interest of all parties. China is an indispensable partner with the largest economic volume among developing and emerging economies.

Global energy governance institutions represented by the International Energy Agency (IEA) are inviting emerging economies to participate. This benefits the emerging economies and highlights the IEA's position in global energy governance. China has also been actively involved in energy governance activities and international cooperation organisations. China's participation is essential to the universality of the future global energy governance system.

11.1.2 China Has a Deeper Involvement in Global Energy Governance and a More Open and Flexible Attitude

(1) **Changes: China exerts more and more influence on global energy governance**

China has become an active participant in, and contributor to, global energy governance. Will China become the initiator, promoter and leader of global energy governance reform?

In the future, when China markedly increases its economic strength, it will continue to open up and deepen its involvement in the international energy market. China has goals in common with other countries in energy security—its energy security is largely at one with the world's. Global energy governance plays an increasingly important role in ensuring common energy security and

China will become more and more important in global energy governance.

(2) **Method: China's foreign energy relations are more open and flexible. The aim is to shift from self-sufficiency to win-win cooperation**

As China moves from self-sufficiency to win-win cooperation, its foreign energy relations become more open and connected to the international energy market. China is increasingly opening up to the outside world, but it will never compromise its interests, especially in energy security. Although it is debatable whether the market can ensure energy security, China's market-oriented reform of its energy sector is an irreversible trend, especially in oil, gas and electricity.

(3) **Participation: China participates more proactively in global energy governance, including global mainstream and leading regional governance organisations**

China is proactively and constructively participating in global energy governance, including the International Energy Agency, G20, United Nations Framework Convention on Climate Change and others. China is also proactively leading regional and multilateral governance, exploring new governance channels and seeking win-win cooperation.

As China and the rest of the world open up to each other, China has rapidly improved its ability to participate in global governance, although gaps remain.

(4) **Attitude: An open, market-oriented energy security concept is conducive to China's participation in global energy governance**

Against the backdrop of energy globalisation, China needs to be prepared to work alongside and with market forces and ensure energy security within the context of an open, market-oriented energy security framework.

China cannot step away from the international energy market and international political and economic environment. The question is: how can China compete in the global market and achieve energy security through market-oriented means.

(5) **Focus: When developing strategies and plans, China needs to pay more attention to international energy cooperation and participate actively in global energy governance**

According to the 13th Five-Year Development Plan for Energy (2016-20), China will assess the domestic and international energy sector to develop programmes, make full use of domestic and international markets and resources, comprehensively implement its strategy of opening up energy markets and international cooperation, seize the great opportunities brought by the Belt and Road Initiative, promote the interconnection of energy infrastructure, enhance cooperation on international production capacity, and actively participate in global energy governance. The plan defines several tasks: participating in major energy events, defining rules and regulations for international platforms and organisations, strengthening cooperation with regional organisations, and collaborating to ensure regional energy security.

11.1.3 China Will Impact Future Global Energy Governance

(1) **Reforms: Pragmatic reform on both sides and increased cooperation are prerequisites for China's substantive participation in the mainstream governance framework**

At present, China relies mainly on mainstream mechanisms in global energy governance. Substantive participation calls for reform and compromise on both sides. China's participation in the mainstream framework, led by international organisations like the IEA, requires effort from both sides. Apart from China's willingness, capacity, and strategic appeal, global energy governance needs a more inclusive framework,

so reform of these international organisations has become an irreversible trend.

(2) **Multilateral common ground: China is currently participating in global energy governance under the existing framework but has shown initiative by creating more flexible practices**

The governance mechanisms led by China, or involving China, are concerned mainly with regional or multilateral common ground. The Belt and Road Initiative, especially with regard to the construction of energy infrastructure, will test the effectiveness of the emerging governance framework and China's influence on it.

(3) **Taking the initiative: China will play an active role in the global energy governance system, but improvements to the system are necessary**

China's willingness and capability are the main factors that influence China's involvement in the global energy governance system, both of which depend on China's attitude and strategic choices. China's participation in global energy governance requires it to create the right conditions for its involvement. In terms of its own interests, the costs, benefits and risks are the key to China's further involvement in the global energy governance system. With respect to its approach, should China work to improve the existing mechanisms or make a radical restart?

(4) **Implications: China's participation in global energy governance will have deep and extensive impacts**

China's participation in global energy governance will affect the existing governance framework and its participants in profound and different ways. First, how will China's participation or actions affect the objectives and operation of the existing governance system? Second, how will China's participation affect the interests of other participants in the existing system? As can be seen above, win-win governance thanks to China's participation will become possible.

Box 21: The concept of global energy governance

Governance is an important part of global energy cooperation. It consists of international narratives, norms, rules and formal and informal organisations that have direct or indirect impact on energy production, trade and consumption.

1. Global energy governance is a series of arrangements centred on the governments of sovereign countries. It includes multilateral conventions, international and regional organisations and bilateral partnerships.

2. It also includes the arrangements by which important roles are played by non-state organisations, such as global or regional energy exchanges, arbitration mechanisms, supply chain initiatives, and so on. Energy governance organisations usually involve alliances with different stakeholders, including governments, companies, international organisations, cities, and non-governmental organisations. These stakeholders all participate in an orderly, predictable international framework that is related to energy production, trade and consumption.

Box 22: The advantages and disadvantages of conventional governance systems

1. International Energy Agency (IEA): An experienced, professional organisation with strong executive power, but lacking inclusiveness and the ability to develop. How distant is China to it? What is China's relation to it? How many conditions (such as OECD membership) are there and what will be the cost of reform?

2. Group 20 (G20): A highly authoritative and representative organisation covering most of the major energy suppliers and consumers but lacking a dedicated secretariat and executive power. Will the G20 Energy Sustainability Working Group become a platform for leaders of large countries to demonstrate their energy and political power?

3. Belt and Road initiative (BRI): An initiative promoted and led by China and covering most economies, except those in North and South America. It has difficulty balancing supply and demand and in providing services other than the construction of energy infrastructure. The impact of BRI on the mainstream energy governance framework, and its relationship with the IEA and the G20, are not clear.

Box 23: China's participation in global energy governance

China's participation in global energy governance, which has been continuously expanded and deepened, is closely related to the global political, economic and energy landscape.

China was an outsider in all international energy mechanisms before it opened its economy, after which it gradually integrated with the world and began to participate in global energy governance. In 1983, China became a member of the World Energy Council. During this initial period, China was cautious and focused on exploring and studying international rules. Since the 1990s, with the development of its economy, China began to actively participate in global energy governance. In 1991, China joined the APEC Energy Working Group. It then participated in regional energy cooperation mechanisms. However, China's involvement was not deep at that time, so the symbolism of its participation outweighed the effects. Entering the 21st century, and with increasing globalisation, China started to make more effective contributions to global energy issues. Its role shifted from active participant to active influencer. It has developed cooperation in various forms with the International Energy Agency (IEA), the International Energy Forum, the Organization of Petroleum Exporting Countries (OPEC), the Energy Charter, the International Renewable Energy Agency and the International Atomic Energy Agency as a member, ally, dialogue partner or observer, and played an important role in several major energy issues. In particular, China became the promoter of some major energy issues through the G20, BRICS countries (Brazil, Russia, India, China, and South Africa), Asia-Pacific Economic Cooperation (APEC) and Shanghai Cooperation Organisation.

China is also a founding member of many international energy organisations, such as the International Energy Forum, Joint Organisations Data Initiative, International Partnership for Energy Efficiency Cooperation, and the Clean Energy Ministerial. China has also actively participated in and hosted many international energy conferences. In November 2015, China became one of first IEA associate members. In 2016, China hosted the G20 Summit in Hangzhou, where global energy governance was an important topic.

Source IEA, China's Engagement in Global Energy Governance (2016)

Box 24: The relationship between the Belt and Road Initiative and international cooperation on energy and global energy governance

Energy cooperation is an important part of the Belt and Road initiative (BRI). Strategically, it is necessary to clarify the position of energy cooperation and its scope and role, as well as the impacts of energy cooperation on BRI and global energy governance.

1. BRI can bring dividends to regional energy cooperation. First, energy cooperation can lead, influence and drive other actions during the implementation of BRI. Second, energy cooperation is the foundation for building a community of common political, economic and energy security.

2. Energy cooperation in BRI requires certain conditions to thrive. These include defining mutual benefits, win-win cooperation and incremental progress, a common understanding of cooperation and open economies, prioritising the role of the market and making companies the main actors, implementing a strategy of effective checks and balances, and bilateral cooperation within a multilateral environment.

3. Key energy collaboration areas in BRI are strengthening cooperation in the construction of energy infrastructure to improve regional energy trade and energy security; interconnecting oil, gas and electricity facilities; cooperating in energy production, processing and use to guarantee regional energy security; establishing regional energy

cooperation platforms and mechanisms to enhance energy cooperation and security.

4. There are differences between energy cooperation within BRI and mainstream global energy governance mechanisms. First, BRI has new requirements for global energy governance, such as: a top-level inclusive platform for negotiating win-win energy cooperation, a relevant and professional platform for promoting the interconnection of energy infrastructure, and a platform for flexible competition and coordination that ensures long-term investment in energy. Second, energy cooperation in BRI makes the inadequacies of global energy governance more obvious, such as the low level of inclusiveness, lack of multilateral negotiation mechanisms to promote the interconnection of energy, lack of rules and regulations facilitating regional energy trade and investment, lack of financial mechanisms to ensure long-term energy investment, and lack of a dialogue mechanism that covers political, economic and energy security.

5. The international community is cautious about the goals and energy cooperation proposed by BRI, especially with regard to benefits that can be achieved in energy supply and energy demand. Some countries are concerned about the goals of BRI, its ability to implement projects sustainably and the risk of using a market-oriented model.

Box 25: China should have a long-term plan for global energy governance

1. China needs to clearly recognise how future trends in global energy governance will positively or negatively influence China and other emerging economies, as this will affect China's willingness to participate in international governance and its strategies of participation. This is related to the compromises that can be made with other participants and the conditions, costs, benefits and risks of China's participation.

2. China should also recognise its unique status as a large developing economy that has an increasing need for energy, while undergoing an energy transition. China needs to understand correctly how it differs from the core interests of other participants (mainly developed countries) in the global energy governance framework. And, China needs to appreciate the complexity of, and potential changes in, its relations with other participants, as these are closely tied to economic development and the pressures of controlling climate change.

3. China must clearly recognise that energy security is the main goal of its participation in global energy governance and the driver of its external energy relations. It is impossible to convince China in the short term that external markets alone can ensure

energy security; domestic market reform is also required to create the necessary conditions for energy security. China's energy security continues to be goal-oriented, i.e. the continuous, sufficient and economical supply of energy is the primary goal; efficiency and fairness are secondary. China's energy security cannot be achieved simply by joining the mainstream governance system or international energy organisations.

11.2 China's Responsibilities in Global Energy Governance

11.2.1 China Should Shoulder the Responsibilities of a Major Power in Global Energy Governance

(1) In terms of global energy governance, if China takes the responsibilities of a major power, China and the world will benefit

China should actively participate in and improve global energy governance by ensuring energy security is the key factor and by dealing with the connection between climate governance and trade governance.

(2) With regard to climate change, China has signed the Paris Agreement and will assume its emission reduction responsibilities in a fair way

China will assume its emission reduction responsibilities at its own pace. The energy transition is an internal driver for China, more so than an external one.

(3) The transformation of the domestic energy system is already in progress, driven by the goals of China's energy revolution

China's transition to a clean, low-carbon energy system is under way. This benefits both China and the world. The import and use of natural gas and other low-carbon energy is part of this energy transition.

(4) China will continue to open up and insist on proactive and win-win cooperation with others in international energy

China will actively participate in mainstream global energy governance and cooperate regionally with neighbouring countries. Promoting energy infrastructure construction under the Belt and Road Initiative helps developing countries escape from energy poverty quickly and enables China to develop its energy equipment industry.

At the strategic level, China has proposed goals and responsibilities for its participation in global energy governance and international energy cooperation. The strategy clearly requires China to actively participate in global energy governance and shoulder its international responsibilities and obligations. China will seek to reform global energy governance, consolidate and improve its bilateral and multilateral energy cooperation mechanisms, and actively participate in international institutional reform.

11.2.2 Regional and Multilateral Cooperation Will Coordinate Energy Security with Socioeconomic Development

(1) Multilateral international cooperation is important, both for China and globalisation

As the global energy market is now highly integrated, "in an interdependent situation, rational and self-interest-driven actors will take the international mechanism as a way to increase their ability to reach a mutually beneficial agreement".

(2) **China has the responsibility to form a community of common interests and promote common energy security through regional and multilateral cooperation**

China has the responsibility to develop deep cooperation with global and regional organisations and establish stable cooperative relations with others in energy supply and demand. As China participates in regional multilateral governance, it has the responsibility to promote common regional energy security and the joint development of all parties' energy economies.

(3) **In the future, China will have the responsibility to assume different governance roles at regional and multilateral levels**

Experience shows that China can promote and lead multilateral energy governance. China's role varies in different governance frameworks. In the future, China will have the responsibility to shoulder different governance roles.

(4) **China has the responsibility to promote equal and mutually beneficial bilateral and multilateral cooperation**

Bilateral, multilateral and global energy cooperation can be described as: "bilateral cooperation lays the foundation, multilateral cooperation improves quality, and global governance seeks common ground". Bilateral cooperation has fewer requirements and is easier to achieve. It serves as the foundation for more extensive and deeper international energy cooperation and global energy governance. At present, China has the responsibility to manage its relationship with those countries with which it has close energy cooperation.

In the medium and long terms, China will have the responsibility to evolve bilateral forms of energy cooperation into a more extensive and deeper multilateral cooperation.

11.2.3 China's Responsibilities Are to Improve Its Energy Security and Make Global Energy Security More Inclusive and Collaborative

(1) **China's energy security is critical to global energy security**

China ensures its energy security by participating in global energy governance. This is beneficial to energy suppliers and to countries dependent on energy imports. Global energy governance with multilateral participation, including China's, is more inclusive and more likely to achieve common energy security.

(2) **Better international energy cooperation and global energy governance would strengthen China's energy security, energy transition and long-term development**

The Chinese government has a new security strategy based on mutual trust, mutual benefits, equality and coordination. The government emphasises the need to address differences, enhance mutual trust through dialogue, resolve disputes through negotiation, and strengthen security through cooperation. Globally, China should take as its starting point the need to improve the basic rules of the international system, actively advocate or lead the improvement of international mechanisms or establish new ones, improve its ability to set agendas and become a major policymaker in international regulations, and become an active contributor to global energy governance—one perceived as responsible and predictable.

11.3 Predicting the Future of Global Energy Governance

Global energy governance reform is about inclusiveness and looking forward to the future. Does an ideal form of governance exist or not?

11.3.1 In the Short and Medium Terms, It Will Become Increasingly Urgent to Reform Mainstream Governance Organisations

(1) **New mechanisms and new institutions are needed, but it will take time**

There is a need to create new mechanisms and institutions to accomplish new missions in global energy governance. New mechanisms and new institutions will be more beneficial to the future of global energy governance, but this cannot be accomplished at one stroke. New mechanisms and new institutions can solve many problems that existing mechanisms and institutions cannot solve, but it is not an easy task to create them.

(2) **Reform should be pragmatic**

There is a consensus that global energy governance requires new mechanisms and new institutions, but this does not mean that new institutions need to be formed within a short time. Establishing new institutions costs time and money and may be impeded by political factors. It may therefore be more realistic to reform the existing mechanisms and institutions. To do so, it is necessary to be fully aware of the difficulties of reform and devise pragmatic plans.

A mainstream governance institution like the International Energy Agency should deepen its reform of inclusiveness, reformulate strategic objectives and transform itself into a global energy institution. More emerging economies should be admitted, and the economic transition of OPEC countries facilitated. The financial derivatives market and global energy pricing should be examined for reform and a more universal alternative explored. Consensus and trust should be built to avoid differences of interest after expansion and inclusion.

11.3.2 In the Long Term, Global Energy Governance Will Extend Beyond Energy Supply and Demand

(1) **Future global energy governance should ensure global energy security**

A more inclusive and balanced global energy governance system should look to the future and strengthen global energy security.

(2) **New missions and tasks in global energy governance will be more diverse, inclusive and balanced**

New missions and tasks to achieve a more inclusive and balanced global energy governance system should: improve the governance framework and strengthen international cooperation and policy alignment; mitigate climate change and regional pollution; cover financial investment, bulk commodity trading, information and data sharing; and improve technical progress and technology transfer in energy.

This can be achieved in the following ways: (i) consolidate the common security concept of global energy; (ii) address environmental challenges like climate change and local pollution; (iii) strengthen multilateral international cooperation mechanisms and share best practice in energy development and use; (iv) use existing energy governance mechanisms; (v) build jointly new international cooperation frameworks, including those for energy and finance; (vi) extend the existing financial analysis of commodity markets into the energy sector; (vii) establish a dialogue on policies that improve stability in key energy-producing regions; (viii) improve energy market transparency and the quality of data; (ix) establish effective strategic reserve mechanisms and cooperation methods; (x) improve fairness in energy trading and energy investment; and (xi) improve jointly energy efficiency, energy innovation and new energy technologies.

11.3.3 Will the Ideal Energy Governance Framework Be G20 + IEA + ...? And Will It Be Truly Global?

(1) **All organisations, governance and frameworks have the potential to lead global energy**

The G20 focuses on general issues, while the IEA concentrates on specific ones. The two organisations do not operate at the same level, nor do they have substantive competition or cooperation. However, both have the potential for global energy governance and display different types of leadership in global social development and energy security.

First, both organisations comprise member countries that are independent economies (except for the EU). They are comprehensive, inclusive and responsive to the needs of emerging and developing countries. They have the ability to build more mature and stable relations between energy producers (including OPEC countries) and consumers.

Second, they complement each other in terms of professionalism, influence and executive power, but both need to compromise. The G20 should significantly improve its Energy Working Group, while the IEA needs to reform to become a global institution. At the same time, the two organisations should form a closer working relationship by, for example, enhancing international laws and regulations for investment protection, and by supporting emerging economies and countries to develop their own low-carbon strategy.

Third, it is essential that they coordinate the relationship between oil and gas exporting and importing countries and balance the interests of countries like Saudi Arabia and Russia that rely heavily on oil and gas exports.

Fourth, the path of combining two organisations into one and the position of the successor organisation is uncertain, so questions like expanding the number of members still need to be considered. In addition, it is important to manage the relationship with other international institutions, such as the United Nations Framework Convention on Climate Change (UNFCCC), the Organization of the Petroleum Exporting Countries (OPEC), the International Energy Forum, the Clean Energy Ministerial, the International Renewable Energy Agency, the International Partnership for Energy Efficiency Cooperation, the Energy Charter Treaty, and others in the energy sector. Institutions outside energy that have a much broader coverage include the Organisation for Economic Co-operation and Development (OECD), the World Trade Organization (WTO) and the World Bank.

(2) **The successor governance organisation will still need all parties to eliminate differences of interest, political risk and other uncertainties**

The biggest advantage of the successor governance organisation is that it will be inclusive and representative, which plugs the gap in international cooperation and international market supervision. The disadvantage is that it requires long-term and unremitting efforts from multiple parties.

China's Energy Research Institute believes that the G7 and the BRICS countries (Brazil, Russia, India, China and South Africa) can build a new energy governance organisation based on equal participation. This mechanism could provide guidance for existing energy governance institutions like the IEA, the World Bank and the WTO, and broaden cooperation with them.

Another option is the G7 + BRIC (Brazil, Russia, India, and China) + six international organisations (IEA, OPEC, UNFCCC, WTO, the International Monetary Fund and the World Bank).

The G7 (or G20) + BRICS + IEA + another option has a broad base of participants that cover much of the world economy and global energy. In terms of energy consumption, it covers the main developed countries and emerging economies; and in terms of energy security, it covers issues like strategic oil reserves and crisis management, producer and consumer organisations, climate change, energy markets, price and financial regulations, and energy poverty and energy equality. However, as there are many participants and their interests vary greatly, the political risks of different parties failing to compromise needs to be considered.

11.3.4 Plan Ahead, Face the Challenges and Find Solutions in Global Energy Governance

(1) The connection between climate governance and energy governance will become increasingly prominent in the next two decades

As we approach 2030, the link between energy security and carbon reduction will intensify. The question of energy security could be replaced by climate change, which could seriously alter the status of global energy governance.

First, energy security will not be replaced by climate change in the short term. Fossil fuels will remain the world's main energy carriers for the next 20 years at least, though the possibility cannot be ruled out that local trade flows of oil and gas will change.

Second, emerging economies are undergoing both an economic and an energy transition. Increased electrification and the lower cost of decarbonised energy will accelerate the substitution of power for conventional fossil fuels. This will change the focus from energy security to power security and adjust the action plan to ensure energy security converges with combating climate change.

Third, for now, global energy governance is related to global trade governance. Differences in carbon taxation are likely to imperil the current WTO free trade principles and may also affect bilateral trade interests. Different countries

develop differently, so the carbon tax and its impacts should be evaluated accordingly. It should be noted that the convergence of energy security with combating climate change is a long process.

(2) Energy governance will impact the energy security of developing countries

The need for policies and actions in the energy sector to address climate change are becoming incontestable. As a result, the prevailing global energy governance values are in transition. This could threaten the energy security of emerging economies, including China's. Developed countries are happy to see emerging economies pledge huge nationally determined contributions but neglect the cost. If carbon cost is added to oil and gas products in global trade too early, the cost to oil and gas importing countries will increase. This will impact their energy security and may be detrimental to exporters as well.

11.4 Policy Proposals

All participants should actively contribute to the reform of global energy governance.

11.4.1 For the Chinese Government, This Means Strategy Formulation, Internal Development, Active Participation and Achieving Breakthroughs at Home

(1) Develop strategies for cooperation with, and the reform of, international energy governance organisations

It is in China's interests to firmly adhere to multilateral norms and international rules, seize opportunities and address challenges together with the international community, and actively and inclusively participate in global energy governance. This requires China to research and develop national strategies for participation in international energy cooperation and global energy governance.

This study proposes that China should: (i), draw up strategies and a roadmap for its participation in global energy governance and evaluate the cost and benefits of participating in global energy governance in different scenarios. This would include identifying the various methods of engagement and providing the necessary support systems; (ii), actively respond to the expectations of the international community for China's participation in global energy governance, such as the IEA's oil emergency response mechanism and the improvement of data quality; (iii), establish an internal conference and consultation mechanism to reach agreement on major international issues, using authorised delegates, to make China's voice heard at international conferences; (iv), prepare a list of international energy conferences and activities to attend and identify the department and officials who should participate and the goals they should achieve; (v), release transparent national energy policies on a regular basis, including white papers on China's domestic energy policies and external energy relations, and; (vi), set up an emergency system (like strategic oil reserves) for potential energy security problems (like armed conflict).

(2) Develop domestic capabilities to strengthen China's engagement in global energy governance

Boosting China's capabilities to modernise energy governance is a prerequisite for its participation in global energy governance.

First, China should develop the capability to set international energy agenda, especially those representing emerging economies and developing countries (soft power).

Second, China should develop the capability to expertly apply international energy regulations, especially those on international energy trade and financial investment.

Third, China should optimise domestic energy governance and develop a modern governmental structure for energy diplomacy, focusing on reforming the domestic energy governance

system and international energy cooperation mechanism.

Fourth, China should improve the development of human resources for international energy governance.

Fifth, China should improve the capability of energy companies to participate in international energy market activities and fully serve the international energy market.

Sixth, China should build discussion platforms with the help of unofficial and international organisations and enhance research and capacity building for participation in global energy governance.

(3) Opening up to one another and fostering faith in the international energy market can support energy security

Future global energy governance will be dependent on China and the world opening up to each other. China should be aware that the international energy market, when it functions well, can ensure energy security through strong and liquid trading to match supply with demand.

There is global consensus that the international energy market ensures energy security under normal economic and geopolitical circumstances. Dependence on imported energy (mainly oil and gas) is no longer a measure of energy security. In the future, energy shortage may not be as critical as before and energy commodities procured from the international market may be cheaper than those produced in China—this is thanks to globalisation, countries opening up to each other, the energy transition and China's energy revolution.

Second, opening up China's energy resources in certain fields not only involves energy development but also investment and trade. Win-win energy development can be achieved by defining areas for international cooperation, learning from international energy development experience and advanced technologies, and using foreign investment to make domestic energy markets more competitive. China's further integration into the global energy market will contribute to the energy security of the whole world.

(4) **Making breakthroughs, integrating different governance organisations and sharing China's successes**

The costs, benefits and risks of China participating in energy governance mechanisms differ from one mechanism to another. China is more likely to make a breakthrough in a governance organisation in which it has a leadership role (such as energy cooperation in the Belt and Road Initiative). Even if the aims and jurisdictions of international energy cooperation and global energy governance differ, the benefits of cooperation and China's successful experience help to improve global energy governance.

In 2016, China made breakthroughs in its energy cooperation with the G20 and the IEA. China became an associate member of the IEA, the current global energy governance organisation, indicating China's progress in mainstream global energy governance. The G20 is a global economic governance mechanism, of which China is a founding member. The G20 provides an arena for China to enhance its leadership in global energy governance.

Promoting energy cooperation in the Belt and Road Initiative and reforming global energy governance are equally important to China and mainstream global energy governance organisations like the IEA. This study proposes that China:

(i)   enhance transparency of all policies and strengthen mutual trust to create a win-win situation and an energy cooperation mechanism that works under normal economic and geopolitical circumstances;

(ii) actively reform global energy governance. Reform includes comprehensive cooperation with existing mainstream international organisations and the promotion of China's policies and proposals, active participation in the formulation and reform of multilateral laws and rules of energy governance, and advancing the view that all parties share equally the benefits and risks of energy governance and compete in an orderly way; and

(iii) actively innovate governance platforms and promote dialogue between them, such as that between the Belt and Road Initiative and global organisations like the IEA, OPEC, SCO, EU, OECD and WTO.

11.4.2 The International Community Should Make the Mainstream Governance System More Inclusive

(1) **The existing global energy governance framework does not meet the needs of a globalised world, so reform to make it more inclusive is urgent**

The current governance mechanism is unable to solve risks effectively—political, legal, and the risk of corruption during energy investment. Some energy governance mechanisms are regionally focused and have not been applied globally; others are becoming less effective with time.

Despite twists and turns, globalisation is irreversible. So, global governance is essential and global energy governance is an important part of global economic governance. Policies that integrate energy suppliers with consumers are becoming increasingly tighter. In particular, as the USA becomes energy independent and the oil and gas consumption of developed countries peaks (developed countries are unlikely to increase consumption), only China and India will have the potential to increase oil and gas consumption. Emerging and developing economies are, therefore, the cooperation partners that the IEA and OPEC strive for.

Thanks to globalisation, energy security is not as critical as before. Only by improving inclusiveness can the challenges of global energy governance be addressed. Inclusiveness means not only that the IEA framework should cover more oil and gas consumers, but also that a higher-level framework should be in place to manage relations between the IEA and OPEC and form a new community of energy interests.

(2) **The G20 and IEA are the main global governance institutions—their integration requires political compromise**

At present, the global energy governance framework is divided into three levels. The first level is the United Nations and G20. They are not energy governance organisations, but they can deliver strategic policy orientations that have a broad consensus and influence energy governance. The second level is represented by professional energy governance organisations like the IEA. They are energy clubs whose members are developed economies. Non-OECD members that are not part of OPEC are often partners of the IEA. OPEC, as a producer organisation, is the IEA's counterpart, but its executive power is now declining. Organisations at this level tend to be well run, strongly executive and professional. The third level is represented by emerging professional energy governance organisations like the International Renewable Energy Agency (IRENA), the International Energy Forum (IEF) and the Energy Charter. They exert vertical governance in most cases and often cooperate with the G20 and other organisations at higher levels.

Integrating the first and second levels (the G20 with the IEA and OECD) requires political compromise, which means it will be a long-term process. The willingness of OECD countries to compromise in the non-energy field will affect the IEA. Mainstream energy governance organisations like the IEA need to reform and become more inclusive. Integrating the interests of the IEA and OPEC is harder than integrating the

interests of energy-consuming countries like China and India. It can only be achieved through global governance at the first level, which is a long-term process.

Integrating global energy governance with social and economic development is important. The inclusive Belt and Road Initiative framework complements the professionalism of IEA energy governance. They could learn from one another. Integrating global energy governance into social and economic development helps broaden consensus on global energy governance.

(3) **High-level G20-IEA energy governance would help coordinate climate control and lower-level energy governance**

Global energy governance within the G20-IEA framework would be at a higher level and more inclusive than the current IEA and would be more professional and executive in terms of energy governance than the current G20. Global energy governance under the integrated G20 and IEA would be similar to global climate control in terms of level, effectiveness and inclusiveness.

In the short term, the fate of climate control lies in the energy transition and energy governance, so the G20-IEA framework would help create a win-win situation. The benefits for oil and gas importers are obvious, as it subjects OPEC oil and gas exporters to a system of checks and balances and encourages them to transition to cleaner energy.

In the long term, consensus on climate control helps suppliers and consumers reach agreement on energy. This agreement goes beyond any of the previous energy security approaches (sufficient, sustainable and affordable) centred on oil and gas security in the narrow sense. Instead, it focuses on the sustainability of energy security (green, effective, fair, etc.) and helps push supply and demand towards clean, low-carbon energy.

11.4.3 Stakeholders Are Looking to the Future and Planning for Global Collaborative Energy Security

(1) International non-energy organisations should keep pace with global energy governance reform

Organisations like the WTO should consider introducing energy trade rules to address the changing modes of global trade. Current WTO trade rules need to be reformed to meet the need for inclusiveness, digital intelligence, efficiency and convenience in the context of globalisation. In energy, these requirements are: first, there may be changes in the flow and types of oil trade, and a globally unified liquefied natural gas (LNG) market could come into existence; second, despite the strict controls of oil and gas producers, the trade rules in upstream oil and gas and other services are global and homogeneous; third, the vision of transnational or transcontinental power trading can be realised.

The World Bank, International Monetary Fund and other organisations should plan early for investment in the low-carbon age. In the future, especially after 2030, global oil and gas consumption could peak or stay at a high level, with growth no longer possible. In a low-carbon power system, electricity generated with fossil energy will decline. It is likely that the financial return from investments in fossil energy, especially power, will be lower than previously. It will be important that these organisations adapt and invest in sustainable energy in the low-carbon age.

(2) The future global energy governance framework and the Belt and Road Initiative should provide oil and gas exporting countries with the opportunity to transition

Supplies of oil and gas are expected to increase in the future. China and India are located in the geographical mid-point between Russia, Saudi Arabia and other OPEC countries and the USA. Russia and Saudi Arabia should speed up their transition to reduce their over-dependence on oil and gas exports. The Belt and Road Initiative provides them with opportunities for social development and economic transition, including the transition of energy. China is strengthening its cooperation with the Gulf states to interconnect energy infrastructure. Multilateral consensus is expected to be formed and converted into investment. The future global energy governance framework and leading organisations should also make efforts to broaden consensus beyond energy.

(3) Only with the comprehensive and collaborative transition of energy consumers and energy exporters can common global energy security be achieved

The world is a whole, so the interests of energy producers, consumers and exporters should be taken into consideration to achieve balance. This requires long-term concerted effort by international organisations and the countries concerned. Global energy governance will become more significant in this long-term process, so it needs to reform itself continuously to keep pace with the times.

Appendix

Annex A.1: Inventory of Multilateral Global Governance Mechanisms, Institutions and Forums

Mechanism/process/organisation	Relevant function	Trade	Investment	Innovation	Security of supply	Externals
Major institutions with energy/climate gov. remit						
World Trade Organization (WTO)	Trade rules incl. Environmental Goods Agreement, and dispute resolution					
G20	Energy principles and activities on subsidies removal, climate risk disclosure, and energy for all					
OECD	Information and policy guidance, trade and investment conventions					
Multilateral development banks e.g. World Bank, AIIB	Policy and tech. assistance; investment guarantees (MIGA), dispute resolution (ICSID) and investment standards (IFC); gas flaring					
United Nations Development Programme (UNDP)	Climate adaptation, energy access, resilience					
Bilateral investment treaties (BIT)	Trade rules, recourse to arbitration					
Shanghai Cooperation Organisation (SCO)	Economic, security and resource cooperation					
Arctic Council	Intergovernmental forum (Arctic environment, navigation)					
UN Convention on the Law of the Sea (UNCLOS)	International navigation, boundary disputes, marine environment					

(continued)

Mechanism/process/organisation	Relevant function	Trade	Investment	Innovation	Security of supply	Externals
BRICs	Cooperation on energy security and efficiency					

Major energy and climate organisations

UNFCCC	Emissions reduction under INDCs, investment					
UN Sustainable Development Goals (SDGs)	Access to Energy					
International Energy Agency (IEA)	Emergency Response Mechanism, technology roadmaps					
Org. of the Petroleum Exporting Countries (OPEC)	Production quotas					
International Energy Forum (IEF)	Ministerial dialogue, data (Joint Organisations Database Initiative)					
International Renewable Energy Agency (IRENA)	Policy and technical assistance, technology					
Energy Charter Treaty	Trade rules, fuel in transit, dispute resolution, energy efficiency					
Extractive Industries Transparency Initiative (EITI)	Transparency of revenues					

Other significant agencies/processes

Global Green Growth Initiative (GGGI)	Policy and technical assistance					
Mission Innovation	Finance energy R&D					
Clean Energy Ministerial	Scaling up clean energy					
International Emissions Trading Association (IETA)	International framework for trading in greenhouse gas reductions					
Cities 40 (C40)	Transport, green buildings					
Climate and Clean Air Coalition (CCAC)	Improving air quality					
Climate Invest Funds (CIF)	Scaling up climate, clean energy investment					

(continued)

Mechanism/process/organisation	Relevant function	Trade	Investment	Innovation	Security of supply	Externals
Other						
Kimberley Process, Better Gold, Dodd-Frank	Supply chain transparency					
Equator Principles	Investment standards					
ISO (oil), Euro (Auto),	Product standards					
ASEAN and ASEAN Plus Three	Petroleum Security Agreement					
Comb. Maritime Taskforces	Security of sea lanes					
International Maritime Organization (IMO)	Fuel efficiency, fuel standards, environment					

Annex A.2: Relevant Actors and International Experience

See Tables 2, 3, 4, 5, 6.

Table 2 Relevant actors and international experience in the strategic pathway of reforming the IEA to make it more representative, effective and relevant

	Regional oil and gas security	IEA reform process
International experience	Energy Charter Treaty, G20 Energy Ministerial Communiqué 2016, G20 Principles on Energy Collaboration	IMF precedent, Norway precedent, Russian membership of Nuclear Energy Agency
State actors	National Energy Administration (NEA) of China, Ministry of Economy, Trade and Industry (METI) of Japan, and Korea Energy Economics Institute (KEEI), Indonesia, Laos, Malaysia, Myanmar, Philippines, Thailand and Vietnam	China, India and Indonesia (existing G20 association members) plus (e.g. Russia, Brazil, South Africa—partner countries) to make proposal for IEA reform at G20
Non-state actors	ASEAN Centre for Energy	
Governmental expert group (GEG)/multilaterals	IEA, ASEAN Plus Three, ECT, BRICS, OECD, ASEAN Council on Petroleum	G20, IEA, OECD, IEP
GEG reform	Development of ASEAN Plus Three oil emergency sharing mechanisms	IEA treaty reform on OECD membership and voting shares

Table 3 Relevant actors and international experience in the strategic pathway of developing and adopting the TCFD framework

	TCFD: metric determination	TCFD: scenario methodologies	TCFD: adapting to state-owned enterprises (SOEs)	TCFD building momentum
International experience	Financial Stability Forum: Enhancing Market and Institutional Resilience, Paris Agreement, Montreal Protocol			
State actors	G20 member states and their finance ministries	G20 member states and their finance ministries	G20 member states and respective SOEs	G20 member states and their finance ministries
Non-state actors	Exchange operators, multinational energy and extraction companies	Multinationals that maintain energy scenarios	SOEs	Energy and extractive industry companies
GEGs/multilaterals	WB, C40, FSB	IEA, ETC, FSB, UNFCCC, IPCC	FSB	CEM, C40, G20, FSB

Note TCFD = Task Force on Climate-related Financial Disclosures

Table 4 Relevant actors and international experience in the strategic pathway of rationalising public energy finance

	Renewed G20 focus on phasing out fossil fuel subsidies	Multilateral coordination on phasing down high carbon FDI and development support	International sharing on establishing effective carbon pricing	International cooperation on minimising carbon leakage
International experience	G20 2009 and process	OECD MDB reforms	ETS—UK, EU, Australia, Canada, California, South Korea, Brazil Tax—Mexico, Japan, South Africa	Switzerland-EU, UK-EU, California-Quebec, Climate Leadership Council (BAMs)
State actors	G20 member states and their finance ministries	NDRC, MOFCOM or SASC	Common: MOFCOM, NDRC, National Energy Commission	
Non-state actors	IOCs and extractives	SOEs	Certifiers/verifiers (PwC, KPMG); fossil fuel companies, industrial sectors and utilities	International corporations that have already adopted an internal shadow carbon price (e.g. IOCs)
GEGs/multilaterals	G20	OECD, WB, AIIB, IFC, ADB, WTO, IETA Article 6, GGGI, UNFCCC, CPLC and the Carbon Pricing Panel, IMF, VCS		
GEG reform	Re-establish G20 FF subsidy reform programmes	MDB investment standards and portfolios	Dedicated G20 and/or World Bank working group; expanded CPLC/Carbon Pricing Panel membership	WTO rules on border adjustment measures, expanded CPLC/Carbon Pricing Panel membership

Table 5 Relevant actors and international experience in the strategic pathway of regionally connected, coordinated and secure electricity markets enabling cross-border trade of electricity through interconnectors

	Regional planning of interconnector routes	Regional alignment of national EMR leading to coordination of CDB	Cooperation on internationally coordinated grid cybersecurity
International experience	EU interconnection, US interstate	EU market integration, EU Capacity Allocation and Congestion Management regulations	Ukrainian power grid cyber attack DoE: Cybersecurity for Critical Energy Infrastructure
State actors	State Grid Corporation of China, China Southern Power Grid, NDRC, National Energy Administration		
	MOFCOM, NDRC	Power sector regulators (NEC)	
Non-state actors	Siemens, GE	DNO; utilities, generator operators	Symantec, Honeywell, Verizon
GEGs/multilaterals	WTO, ECT, MIGA, ICSID, IFC, BIT	IEA, ECT, IRENA, IEF, GIZ, CEM, 21st Century Power Partnership, ISGAN	IEA, IMPACT, ITU
GEG reform	Regional assistance from exiting GEGs to China and Asia		Reform of IEA to create initiative on cybersecurity?

Table 6 Relevant actors and international experience in the strategic pathway of international cooperation to accelerate smart management of EV electricity demand and deployment of cost-effective electricity storage

	Internationally standardised smart EV charging stations	International alignment of electricity storage investment towards lithium-ion batteries
International experience	Dutch charging service providers: Open Smart Charge Protocol (OSCP), GreenFlux and Enerxis	Tesla deployment in California, UK govt. investment in battery deployment, MDB reforms
State actors	State Grid Corporation of China, China Southern Power Grid, NDRC, Standardization Administration of the PRC	
Non-state actors	DNO, EV manufacturers	Battery and EV manufacturers (BYD, CATL), DNO
GEGs/multilaterals	IRENA, IEA, IEC, IOS, IEEE	World Bank, AIIB, ADB, MIGA, ICSID, IFC, CCCMC, CEM, Mission Innovation, GGGI, CIF, IFC, BIT
GEG reform	Current international standard setting effective, widespread international engagement and adoption required	Coordinated reform and alignment of investment standards and strategies of MDBs

Annex A.3: The Development of Emergency Response Measures for Oil

The 1967 and 1973 oil crises raised the spectre of potentially large long-term politically induced supply disruptions. As the political situation in the Middle East became more fragile there was growing awareness and concern about the implications of the various choke points involved in the international oil trade, such as the Strait of Hormuz and Bab el Mandab between the Horn of Arica and Yemen.

In this context there was a growing view that security of supply would best be served by coordinated responses revolving around increased storage. The first response came from an initiative launched by Henry Kissinger as US Secretary of State and involved the creation of the International Energy Agency (IEA) in November 1974. Central to this was the IEA's Emergency Sharing Mechanism. This required the members, who were basically members of the OECD (France initially excluded itself but later joined), to maintain 90 days of oil consumption (crude and products) in storage. A series of rules were then laid down to govern the terms under which the stocks would be released. The European Commission, feeling rather outdone by the IEA, which was an American initiative, then introduced its own emergency scheme along similar lines, although the relationship between the two schemes was never clarified. In 1978, the USA began its Strategic Petroleum Reserve (SPR) that involved building very large crude storage capacity in salt caverns in Louisiana. The creation of the SPR led to much discussion about the role of freeriding on storage, since any release of stocks by the SPR would dampen price spikes internationally in what was a global market. More recently other countries, notably China, have been building up their strategic reserves, helped by the oil price collapse since June 2014.

In the 1980s, further supply security measures were effectively introduced as the result of the development of paper and futures markets. These initially emerged as informal forward markets but developed into formal markets in the late 1980s—New York Mercantile Exchange (NYMEX) and the International Petroleum Exchange (IPE), which later became the Intercontinental Exchange (ICE) in London. This allowed consumers (and producers) to hedge against price volatility.

These mechanisms were tested on only a few occasions. The first was the loss of crude supplies associated with the Iranian Revolution in the late 1970s and the Iran-Iraq war in the 1980s. This was not a real test because the SPR, while having been filled, had not yet built physical pumping capacity to recover the crude. In a very disrupted market where the IEA decided not to invoke the Emergency Sharing Mechanism, the result was everyone for themselves with very intense competition between US and Japanese companies. The result was effectively the second oil price shock. The next event was the loss of Iraqi and Kuwaiti crude supplies following Iraq's invasion of Kuwait in 1990. The price effects of this were initially dampened because Saudi Arabia was carrying significant spare capacity that it pushed into the market. At the start of the campaign to liberate Kuwait in 1991, the IEA released stocks, but all this did was to aggravate price volatility. In 2005, the IEA again announced a stock release in response to Hurricane Katrina and its impact on oil logistics in the Gulf of Mexico. There were few takers of crude on offer.

There is a lively debate in the academic literature on the effectiveness of such mechanisms and their optimal design. However, the consensus among industry insiders is that in the event of a really serious disruption—for example closing the Strait of Hormuz—the IEA's scheme would fall to pieces as governments looked to their own national interests.

Annex A.4: The International Renewable Energy Agency (IRENA)

IRENA was established as an intergovernmental organisation in 2009, in order to increase the uptake of renewable energy worldwide. Having recognised that existing energy institutions put renewable energy at a disadvantage compared to other energy sources, the German government led a coalition of the willing in designing IRENA's mission. Its focus is on being a global voice and knowledge source for renewables (including renewable technologies and patents), providing policy advice on renewables and serving as a network hub for member states. It is open to all UN members, lending it a high degree of legitimacy, and all members have equal weight in the decision-making process. Given its soft mandate and focus on information, stakeholder engagement and agency are relatively high, with member states generally sending officials from their energy rather than climate or environment ministries to IRENA council meetings.

Nonetheless, there was initial resistance to the creation of IRENA. Reasons for abstaining varied, from reservations about renewables and fears of action against fossil fuel and nuclear interests, to scepticism about the creation of a new international organisation (given inefficiencies and deadlock in existing ones) and the financial burden imposed on member states. Several IEA member states, including France, the UK, USA, Canada, Italy, Japan and Australia did not want to create a rival to the IEA. Some emerging economies raised sovereignty concerns: Brazil was concerned that IRENA would try to limit its use of biofuels and large hydropower, for instance. While it took many years to gain political traction, most of the original abstainers ultimately joined. Membership now stands at 75 countries, including China, and 26 of 28 IEA member states.

Many of these stakeholders are unlikely supporters at first glance. This includes the United Arab Emirates (UAE) and Nigeria, both OPEC countries. The UAE has taken a significant role in funding several IRENA initiatives and largely kept the organisation afloat in its initial phases with the support of Germany. The decision to locate the headquarters in Abu Dhabi sent a strong political message about IRENA and renewables in that even oil exporting countries

recognise their economic potential and are willing to contribute to IRENA. This message has become stronger following the collapse in oil prices in recent years, which has forced exporters to reconsider the role of fossil fuels in supporting economic growth and in turn, political stability.

Annex A.5: The G20's Growing Role in Global Energy Cooperation

The G20 is the most prominent of several global governance initiatives adopting a coalition of the willing approach. The global financial crisis in 2008 gave it new impetus and prompted it to broaden its scope from economic cooperation to energy and climate change. Like shifts in energy governance before it, this expansion was partly driven by exposure to high prices and volatility in oil markets; removing distortions and inefficiencies in energy markets was identified as a key line of defence. Moreover, at a time when non-OECD oil demand was about to overtake that of the OECD, and the representativeness and agency of the IEA were increasingly questioned, the G20 offered a forum that included established and emerging powers, as well as accounting for more than 80% of total primary energy demand and at least 50% of global fossil fuel trade.

The commitment to phase out inefficient and wasteful consumption subsidies was made at the G20 Pittsburgh Summit in 2009. Designing a robust reporting, but politically acceptable, mechanism proved challenging, given that price reform remains primarily an area of national competency. Gaps in the conceptual framing of the commitment—from lack of consensus regarding what constitutes a subsidy to varying interpretations of terms like "inefficient", "wasteful consumption", "rational circumstances" and "market stability"—led to discrepancies in the reporting of existing subsidies and those required to be phased out. While some have questioned the effectiveness of the commitment for this reason, it has successfully placed the issue of subsidies on the international agenda and helped create the longer-term conditions for reform. When oil prices collapsed, many governments took advantage of the political space to reform consumption subsidies.

The G20's role in energy cooperation has grown since, with a commitment to improve the transparency of energy data made at the St Petersburg Summit in 2013, and adoption of the G20 energy principles at the Brisbane Summit in 2014. Perhaps most importantly, while emerging economies have struggled to effect fundamental reforms at the multilateral level, they have been able to make a growing contribution to energy dialogue via the G20; for instance, China's use of its presidency in 2016 to champion green finance. Moreover, the Hangzhou Consensus reaffirmed "the importance of energy collaboration towards a cleaner energy future and sustainable energy security with a view to fostering economic growth", drawing perhaps the clearest links yet between global economic and energy and climate governance.

Annex A.6: Case Studies, Going it Alone Versus Cooperative Action

Case study 1: IEA

Pooled economic and political capital is more powerful than a single country acting alone. In response to the 1973 oil crisis, the USA took unilateral and multilateral action simultaneously. Unilaterally, the USA banned oil exports, which helped to reduce and stabilise oil prices in the short term. At the same time the USA was a founding member of the IEA in 1974, central to which was the emergency response mechanism (ERM), a multilateral organisation and cooperative mechanism to enable the group of consumers to exert greater market power over oil producers. The unilateral response of the oil export ban was arguably a short-term measure designed to address an acute short-term problem. The establishment of the IEA and ERM took longer to

structure and implement than the USA's unilateral export ban but has arguably been more effective in achieving the desired objective of stabilising prices.

Before the ban was lifted in 2015, many critics argued that lifting the ban would increase production, as producers would begin to sell to the global market. The Council on Foreign Relations estimated production would rise by 1.2 mb/d, translating into a 1% increase in US GDP. Disentangling the impact of decreasing global oil prices on the US oil industry after lifting the ban and how US producers would otherwise be impacted, is challenging. After lifting the ban at the end of 2015, year-on-year production declined by 0.5 mb/d, but this could have been much greater if the unilateral approach hadn't been reversed.

These two distinct approaches exhibit simultaneous unilateral and multilateral responses to the same issue. Dichotomising between collaboration and going it alone is somewhat of a fallacy, as countries have used a mixture of approaches to achieve the same objective.

Reforming the IEA by making it more representative of production and consumption to enable the IEA to fully achieve its remit presents significant challenges. Principally, countries such as China and India would need to move from associates to full members. The joint political will to overcome these hurdles, doesn't appear to exist. This is mainly due to the IEA's archaic voting system. Simply put, voting shares are determined on the basis of the oil consumption data from 1973, leaving China and India with little voting influence. If updated to more recent oil consumption data, the existing memberships' voting influence would significantly diminish. Modifications to the IEA's formal treaty would require ratification by the IEA's member states at the domestic level, which would likely run into significant domestic political challenges. These hurdles could be overcome by mutual effort, if the pooled political will and determination existed within the current IEA membership. Indeed, a coalition between the associate members would enable greater aggregate political capital, which could be leveraged to bring about modifications to the IEA's voting system. This stalemate is the context in which the questions of new institutional creation or formalised partnerships have arisen.

Case study 2: IRENA

The literature is unclear on the process by which states form new international institutions. This is made more complex as institutional formation often runs in parallel to diverging from existing institutions. What is clear is that new organisations are difficult to establish, especially as new institutions sometimes perform overlapping functions with those in existence, leading to fragmentation. Further, new institutions have high start-up costs, which can be avoided by simply working within, and reforming, existing organisations.

The formation of IRENA in 2009 can help probe these questions. IRENA was instigated by Germany, Spain and Denmark to balance the IEA's support for fossil fuels and nuclear power, in favour of greater support for renewable technologies. By forming a coalition of the willing (or coalition of dissatisfied states), rather than a bland consensus, the Germany-led coalition gradually grew in membership to become IRENA. In a broad sense, states chose to join during the formation stage to influence the institutional structure and objectives. Many governments had specific and unique drivers: the Obama administration, for example, viewed IRENA as a timely way to demonstrate a different approach to carbon options from that of the Bush administration.

It is too early to determine if IRENA has the teeth to achieve its goals. However, its formation represents a relatively unique departure from the crisis catalyst that drove the formation of the UN and the Bretton Woods institutions, one which may be difficult to repeat. The formation of IRENA as an entity with its own governing structures differs to the formation of the

International Partnership for Energy Efficiency Cooperation (IPEEC). Formed around the same time as IRENA, the eventual nesting of IPEEC within the IEA again illustrates that the formation of IRENA is relatively unique. It could be that the costs associated with institutional reform of the IEA outweighed those of institutional creation, leading to the formation of IRENA as an entity within its own right.

Case study 3: Carbon trading

Unilateral action can play a catalytic role in the development of an international or multilateral response. This interplay between unilateral and multilateral responses is clearly demonstrated in the parallel systems of the UK and EU emissions trading systems (ETS). The UK was able to implement carbon trading faster by going it alone, or unilateral action, but then extended that knowledge and expertise to the collaborative and multilateral approach of the EU ETS. This catalytic effect can also be observed in the recent proposal by the Climate Leadership Council's carbon tax proposal, which included a proposed border adjustment tax:

> Border adjustments for the carbon content of both imports and exports would protect American competitiveness and punish freeriding by other nations, encouraging them to adopt carbon pricing of their own.

The merits of this strategy are debatable. However, the intention is clear: by taking unilateral action a future interlinked carbon pricing system could develop, achieved in the first instance without multilateral cooperation. Such a unilateral decision by the USA would have far-reaching implications on trade and hence on global energy governance.

While the border adjustment tax could, in theory, be implemented more swiftly, the downside would likely be a trade war. While the border adjustment tax is not technically an import tariff, it is unlikely to comply with WTO rules. Even if it did, other countries would likely start to impose similar border adjustment taxes to protect their exporting industries and hence a trade war would likely ensue, damaging domestic growth across nations. Hence a multilateral, more complex approach, would likely lead to a more optimal outcome. The multilateral response would likely involve the linking of carbon markets through international accounting rules for transferable carbon allowances or credits under Article 6 of the Paris Agreement, or through the Emissions Mitigation Mechanism (EMM). Establishing the EMM would require extensive negotiations and leadership from the big emitters, such as China. Negotiations would likely take considerable effort and time, longer than the unilateral proposal of a border adjustment tax.

Case study 4: Oil spills

Oil spills and the environmental damage caused by them is an international challenge, as oil tankers cross borders and sea lanes and navigate international waters. Oil spills are an example of where an international response is stimulated by a crisis or shock. In 1967, the UK took the controversial decision to set fire to the ship-wrecked Torrey Canyon supertanker to prevent leaking oil from spreading. Although contained within the borders of the UK, the oil spill also contaminated 80 km of the French coastline; hence the domestic action of the UK had a unilateral dimension to it. This led to several international conventions, including the recognition of the right of coastal states to take unilateral measures "to prevent grave and imminent danger to their coastline from oil pollution". In this example, unilateral action catalysed multilateral agreements and standards, which in turn protected the rights of states to act unilaterally. In the 1970s, the USA unilaterally forced oil tankers entering US ports to comply with double-hull standards, catalysing the international adoption of the 1973 MARPOL Convention and 1978 Protocol on maritime pollution, which extended US standards internationally.

Case study 5: Chlorofluorocarbons and the Montreal Protocol

A multilateral response stimulated by the perceived threat of unilateral trade sanctions can be observed in the banning of chlorofluorocarbons (CFCs) to prevent the depletion of stratospheric ozone. As the hole in the ozone layer grew over Antarctica in the early 1970s, the public became concerned over increased skin cancer rates. By the mid-1970s it had become clear that CFCs were the cause. Between the mid-1970s and 1980s, led by the USA, several countries took domestic action and imposed bans and reductions in the use of CFC-based aerosols. International negotiations had been underway for some time but came to a head in the mid-1980s.

A key motivation of Europe and Japan to reach an agreement was the fear that the USA would take unilateral action and impose trade sanctions, if talks failed. In 1987, the Montreal Protocol was signed, with a clear tapering schedule for the production and use of CFCs globally. It should be noted, that another key component on reaching an agreement was the availability of substitutes for CFCs, which limits the replicability of this approach in other areas, and in particular climate change.

Nevertheless, the Montreal Protocol is widely acknowledged as the world's most effective environmental treaty. 98% of the production and consumption of ozone-depleting substances (ODS) has now ended; and the ozone layer is projected to recover to its pre-Antarctic ozone hole state over the next 50 or so years, and to its pre-industrial state in about 500 years. At the same time, the Protocol has made a major contribution to slowing the rate of global warming: almost all ODS, including CFCs and hydrochlorofluorocarbons (HCFCs), are themselves powerful greenhouse gases, typically far more powerful even than hydrofluorocarbons (HFCs).

Another breakthrough was reached in October 2016, when nearly 200 countries agreed to phase down global emissions of highly climate-warming HFCs. The deal amends the Montreal Protocol, and requires cuts in HFC use starting in 2019 for wealthier countries, with all other nations significantly reigning in consumption by the end of the 2040s. The move could prevent up to 0.5°C in global warming above pre-industrial levels by the end of the century.

Much of the success of the Montreal Protocol has been attributed to its institutional design. The supreme decision-making body of the ozone regime is the Meeting of the Parties (MOP), with preparatory discussions taking place in the Open-ended Working Group (OEWG). These bodies have proved, in contrast to those in many other multilateral environmental agreements (MEAs), generally harmonious forums for resolving the key political, technical and financial issues faced by the parties. A sense of community has traditionally prevailed, which has been helpful in resolving disputes (although unfortunately, as discussed below, this positive atmosphere has not extended to discussions on HFCs, which have been comparatively acrimonious).

Nearly 30 years after its inception, the ozone regime has now reached a mature stage, where its institutions are well established and its main objective—the phase-out of the production and consumption of ODS—has been largely met. The Montreal Protocol, however, is still faced with several challenges, and not just that of rising HFC use: illegal trade in ODS, and the disposal of banks of stored ODS, for example, remain important concerns. Nonetheless, it is true to say that the Montreal Protocol is widely accepted as a highly effective regime. The result of this, however, is that the issue of ozone protection has largely dropped down the international agenda, receiving only limited political attention.[5]

---

[5]The text above draws heavily on a Chatham House Research Paper by Stephen O. Andersen, Duncan Brack and Joanna Depledge (2014).

Annex A.7: The World Trade Organization and Energy Resources

While the WTO, with its 164 member countries, provides the main global institutional framework for coordinating and governing international trade—trade in natural resources, including fossil fuels, falls largely outside its remit (though some rules have implications for resource reserves and related services). Separate chapters were included in the WTO texts to address issues related to agriculture and textiles, but not natural resources. Some regional agreements have tried to bridge this gap for energy, notably the Energy Charter Treaty and the energy chapter in the North American Free Trade Agreement (NAFTA), and there were efforts to address energy within the Transatlantic Trade and Investment Partnership negotiations.

Several reasons for the special treatment of natural resources in trade agreements have been cited. First, the geographical distribution of natural resources is highly uneven and location of production is largely immovable. Nearly 90% of oil reserves are located within just 15 countries. Second, the sovereign control of resource-endowed countries over their natural resources and their right to exploit them for economic and social development is recognised in international law (see UN General Assembly Resolution 1803 of 1962). Third, trade in natural resources is often perceived as a key national security issue by exporters and importers. Fourth, the extraction and consumption of resources produce severe negative externalities compared with most other traded products. Finally, a large proportion of resource trade takes place under long-term contracts, often through dedicated infrastructure such as pipelines or liquefied natural gas tankers. The exhaustibility of natural resources, and the potential for infringements on national security grounds, provide two potential justifications for trade restrictions under WTO rules, although these must be implemented in a non-discriminatory way.

While these reasons may be broadly accepted, in practice there are often blurred lines or different interpretations between countries. Most importantly, the point at which a natural resource becomes a saleable commodity is often hard to distinguish and remains controversial. A further complexity is that while the General Agreement on Trade in Services (GATS) provisions do not include natural resources, a service relating to natural resources (exploration, exploitation, technical testing, transport) is subject to the GATS disciplines unless provided by government authorities. This had led to calls for a more coherent framework, specifying the rules that apply to each resource type, the qualifications needed for a resource to be considered a good or a service, and important issues for oil and gas, such as investment protection.

Another key issue is that while supply security concerns in consumer countries focus on the ability to import goods, the WTO is focused on avoiding import restrictions. Article XI of the General Agreement on Tariffs and Trade (GATT) 1994 requires that exports should not be subject to quantitative restrictions "other than duties, taxes or other charges", but it does not fix a maximum level for border taxes, except for those countries which accepted them within their accession agreements (including China). Article XI has been interpreted as not prohibiting export taxes, which explains why "more than a third of notified export restrictions are in resource sectors, where export taxes on natural resources appear to be twice as likely as export taxes in other sectors". In recent years, several key raw material suppliers have used export controls. Indonesia has introduced a ban on exports of unprocessed ores. Vietnam has also imposed restrictions on iron ore and copper, and export taxes have recently been debated in Brazil and India. In 2012, the USA, EU and Japan lodged complaints with the WTO over Chinese export quotas of rare earth minerals, and the quota system was withdrawn in late 2014. There have been various proposals for voluntary avoidance of export restrictions in times of crisis or extreme price volatility, which such restrictions tend to exacerbate.

Although there has been growing concern over the functioning of resource markets, the much delayed Doha round is unlikely to introduce major changes to the treatment of extractive industries at the WTO, partly because a significant constituency within the WTO negotiations argue that "access to, and use of, natural resources as well as the right to regulate, should remain outside the scope of negotiations". Past negotiations and instruments, such as the Generalized System of Preferences (GSP), concerning natural resources in GATT and WTO, were dominated by concerns about developing country dependency on raw material exports and structural inequalities in the international economic system as a potential cause (dependency theory). Discussions during the resource boom reflected concerns around high resource prices, extreme price volatility, increasing restrictions on trade and the diminishing ability of the WTO dispute settlement regime to solve resource trade disputes.

Annex A.8: Nuclear Power: Decline or Resurgence?

Nuclear power generation, as a share of global electricity supply, has decreased from a peak in 2006, when it provided 15%, to 10.8% in 2015. Beyond public and political concerns over safety, financing and the economics of nuclear power have hampered deployment. Long construction times and high capital expenditure result in unfavourable risk-reward profiles, from the perspective of potential investors. Until 2013 the global average levelised cost of energy (LCOE) was $93–94/MWh, but this jumped to $140/MWh in 2014. Estimates for the first half of 2015 put the high point valuation at $258/MWh, rising to $290/MWh in the second half. This is the direct result of infrastructure costs increasing to $3.72–7.5/MWh. Gas-fired generator build costs at the time were between $0.46–1.65/MWh. The near 50% increase in nuclear LCOE is due to underestimating the cost of building third-generation reactors. These third-generation reactors are designed to be safer, with passive safety systems built into their design, rather than safety systems iteratively added to old designs. Due to design complications, delays and cost overruns the costs have risen.

As China gains experience in new nuclear deployment, initial capital costs will likely fall. If nuclear is to play an increasing role in power provision, international cooperation around technology sharing will be required, with China likely to play a leading role.